T0174391

Proceedings *of the* Sixth International Conference *on* Difference Equations

Augsburg, Germany 2001

New Progress *in* Difference Equations

Proceedings *of the* Sixth International Conference *on* Difference Equations

Augsburg, Germany 2001

New Progress *in* Difference Equations

Edited by

Bernd Aulbach
Saber Elaydi
Gerasimos Ladas

CRC PRESS

Boca Raton London New York Washington, D.C.

Library of Congress Cataloging-in-Publication Data

International Conference on Difference Equations (6th : 2001 : Augsburg, Germany)
   Proceedings of the Sixth International Conference on Difference Equations, Augsburg, Germany, 2001 : new progress in difference equations / edited by Bernd Aulbach, Saber Elaydi, Gerasimos Ladas.
     p. cm.
   Includes bibliographical references and index.
   ISBN 0-415-31675-8 (alk. paper)
   1.Difference equations—Congresses. I. Title: Proceedings of the 6th International Conference on Difference Equations, Augsburg, Germany, 2001. II. Title: New progress in difference equations. III. Aulbach, Bernd, 1947-IV. Elaydi, Saber, 1943- V. Ladas, G. E. VI. Title.

QA431.I15145 2004
515'.625--dc22   2003070008

Visit the CRC Press Web site at www.crcpress.com

© 2004 by Chapman & Hall/CRC

No claim to original U.S. Government works
International Standard Book Number 0-415-31675-8
Library of Congress Card Number 2003070008
Printed in the United Kingdom by Biddles/IBT Global 1 2 3 4 5 6 7 8 9 0
Printed on acid-free paper

Contents

Preface   xi

List of Contributors   xiii

Opening Lecture   1

Difference Equations and Boundary Value Problems 3
A. N. Sharkovsky

Of General Interest   23

"Real" Analysis Is a Degenerate Case of Discrete Analysis 25
D. Zeilberger

On the Discrete Nature of Physical Laws 35
F. Iavernaro, F. Mazzia and D. Trigiante

Discrete Dynamical Systems   49

Linear Self-Assemblies: Equilibria, Entropy and Convergence Rates 51
L. Adleman, Q. Cheng, A. Goel, M.-D. Huang and H. Wasserman

Synchronization in a Discrete Circular Network 61
S. S. Cheng, C.-J. Tian and M. Gil'

Bifurcation of Periodic Points in Reversible Diffeomorphisms 75
M.-C. Ciocci and A. Vanderbauwhede

Evolution of the Global Behavior of a Class of Difference Equations 95
A. S. Clark and E. S. Thomas

A Survey of Exponential Dynamics 105
R. L. Devaney

Farey's Rule for Bifurcations of Periodic Trajectories
in a Class of Multi-Valued Interval Maps 123
V. V. Fedorenko

Combinatorics of Angle-Doubling: Translation Principles
for the Mandelbrot Set 131
K. Keller

The Inflation and Perturbation of Nonautonomous Difference
Equations and Their Pullback Attractors 139
P. Kloeden and V. Kozyakin

A Short Introduction to Asynchronous Systems 153
V. Kozyakin

A Local-Global Stability Principle for Discrete Systems and
Difference Equations 167
U. Krause

Stability Implications of Bendixson Conditions for
Difference Equations 181
C. C. McCluskey and J. S. Muldowney

Optimal Topological Chaos in Dynamic Economies 189
K. Nishimura, T. Shigoka and M. Yano

Dynamics of the Tangent Map 199
H. Oliveira and J. S. Ramos

Thresholds, Mode Switching and Complex Dynamics 207
H. Sedaghat

Computation of Nonautonomous Invariant Manifolds 215
S. Siegmund

Dynamic Equations on Time Scales   229

Exponential Functions and Laplace Transforms
for Alpha Derivatives 231
E. Akin-Bohner and M. Bohner

Integration on Measure Chains 239
B. Aulbach and L. Neidhart

Asymptotic Formulae for Dynamic Equations on Time Scales
with a Functional Perturbation 253
S. Castillo and M. Pinto

Oscillation of a Matrix Dynamic Equation on a Time Scale 267
L. Erbe

Continuous Dependence in Time Scale Dynamics 279
B. M. Garay, S. Hilger and P. E. Kloeden

On the Riemann Integration on Time Scales 289
G. Sh. Guseinov and B. Kaymakçalan

Cauchy Functions and Taylor's Formula for Time Scales $\mathbb{T}$ 299
R. J. Higgins and A. Peterson

Embedding a Class of Time Scale Dynamics into
O.D.E. Dynamics with Impulse Effect 309
J. López Fenner

An Oscillation Criterion for a Dynamic Sturm-Liouville Equation 317
Z. Opluštil and Z. Pospíšil

Two Perturbation Results for Semi-Linear Dynamic
Equations on Measure Chains 325
C. Pötzsche

Miscellaneous on Difference Equations   **335**

Conjugate Singular and Nonsingular Discrete
Boundary Value Problems 337
R. P. Agarwal and D. O'Regan

Asymptotic Solutions of a Discrete Schrödinger Equation
Arising from a Dirac Equation with Random Mass 349
B. Aulbach, S. Elaydi and K. Ziegler

Existence of Bounded Solutions of Discrete Delayed Equations 359
J. Baštinec, J. Diblík and B. Zhang

Difference ϕ-Laplacian Periodic Boundary Value Problems:
Existence and Localization of Solutions 367
A. Cabada and V. Otero-Espinar

Asymptotic Behavior of Solutions of $x_{n+1} = p + \frac{x_{n-1}}{x_n}$ 375
E. Camouzis and R. DeVault

Limit Behavior for Quasilinear Difference Equations......................383
M. Cecchi, Z. Došlá and M. Marini

Difference Equations in the Qualitative Theory of
Delay Differential Equations 391
J. Čermák

Properties of a Class of Numbers Related to the Fibonacci,
Lucas and Pell Numbers 399
F. M. Dannan

Oscillation Theory of a Class of Higher Order Sturm-Liouville
Difference Equations 407
O. Došlý

A Transformation for the Riccati Difference Operator 417
J. Elyseeva

On the Dynamics of $y_{n+1} = \frac{p+y_{n-2}}{qy_{n-1}+y_{n-2}}$ 425
E. A. Grove, G. Ladas and L. C. McGrath

On the Difference Equation $y_{n+1} = \frac{y_{n-(2k+1)}+p}{y_{n-(2k+1)}+qy_{n-2l}}$ 433
E. A. Grove, G. Ladas, L. C. McGrath and H. A. El-Metwally

Almost Periodic Solutions in a Difference Equation 453
Y. Hamaya

Discrete Quadratic Functionals with Jointly Varying Endpoints
via Separable Endpoints 461
R. Hilscher and V. Zeidan

On Finite Difference Potentials 471
A. Hommel

Moment Equations for Stochastic Difference Equations 479
K. Janglajew

Convergence of Solutions in a Nonhyperbolic Case
with Positive Equilibrium 485
C. M. Kent

Strongly Decaying Solutions of Nonlinear Forced Discrete Systems 493
M. Marini, S. Matucci and P. Řehák

Multidimensional Volterra Difference Equations 501
R. Medina and M. Gil'

Constructing Operator-Difference Schemes for Problems
with Matching Boundaries 507
R. V. N. Melnik

On Difference Matrix Equations 515
E. Pereira and J. Vitória

On Some Difference Equations in the Context of q-Fourier Analysis 523
Λ. Ruffing and M. Simon

Nonoscillation and Oscillation Properties of Fourth Order
Nonlinear Difference Equations 531
E. Schmeidel

A Computational Procedure to Generate Difference Equations
from Differential Equations 539
P. G. Vaidya and S. Angadi

Difference Equations for Multiple Charlier and Meixner Polynomials ... 549
W. Van Assche

Author Index   559

Preface

The series of International Conferences on Difference Equations and Applications (ICDEA) has established a tradition of conferences moving around the world. Following the first five meetings which were held, respectively, in San Antonio/USA (1994), Veszprém/Hungary (1995), Taipei/Taiwan (1997), Poznań/Poland (1998), Temuco/Chile (2000), the sixth meeting took place in Augsburg/Germany from July 30 to August 3, 2001. It brought together more than 100 scientists from 24 countries including a substantial number of world renowned mathematicians working in the field of difference equations.

The conference in Augsburg was organized by Bernd Aulbach and sponsored by the University of Augsburg and the following non-profit organizations whose support is gratefully acknowledged: Deutsche Forschungsgemeinschaft, Bayerisches Staatsministerium für Wissenschaft, Forschung und Kunst, Gesellschaft der Freunde der Universität Augsburg, Albert-Leimer-Stiftung, Mathematischer Verein der Universität Augsburg.

This volume contains a selection of papers on difference equations which have been presented at the 6th ICDEA and accepted for publication after peer review. It covers the latest progress in a wide range of topics from the theory of difference equations and its applications, and it is organized as follows:

- **Opening Lecture**, by Alexander N. Sharkovsky

- **Of General Interest**, with contributions by the invited speakers Doron Zeilberger and Donato Trigiante

- **Discrete Dynamical Systems**, including contributions by the invited speakers Sui Sun Cheng, Robert L. Devaney, Vladimir V. Fedorenko, Peter Kloeden, Victor Kozyakin, Ulrich Krause, James S. Muldowney, Kazou Nishimura, Stefan Siegmund, André Vanderbauwhede

- **Dynamic Equations of Time Scales**, including contributions by the invited speakers Martin Bohner, Lynn Erbe, Stefan Hilger, Peter Kloeden, Allan Peterson, Manuel Pinto

- **Miscellaneous on Difference Equations**, including contributions by the invited speakers Ravi P. Agarwal, Saber Elaydi, Gerry Ladas, Walter Van Assche

Apart from some outstanding lectures, the 6th ICDEA had another highlight, the foundation of the **International Society of Difference Equations** (ISDE). During a particular meeting the conference participants discussed the possibility of establishing a society with the aim of coordinating the various activities in the growing field of difference equations. As a result of this meeting, it was decided to found such a society as a medium where its members can communicate and interact (free of charge) to the benefit and further promotion of the field of difference equations and its applications.

The following have been elected as representatives of the new "International Society of Difference Equations" (ISDE): Bernd Aulbach as President, Sui Sun Cheng, Ondřej Došlý, Saber Elaydi, István Györi, Johnny Henderson, Billur Kaymakçalan, Peter Kloeden, Gerry Ladas, Allan Peterson, George Sell, Alexander Sharkovsky, Donato Trigiante and Jian She Yu as members of the Advisory Board, and Hassan Sedaghat as the editor of the associated (already existing and independent) electronic newsletter on "Difference Equations and Discrete Dynamical Systems".

Information on the ISDE can be found on `http://web.umr.edu/~isde/`. The newsletter is on `http://mywebpages.comcast.net/dedds/`.

Contributors

L. Adleman

Department of Computer Science
University of Southern California
Los Angeles, California 90089, USA
adleman@usc.edu

R. P. Agarwal

Department of Mathematical Sciences
Florida Institute of Technology
Melbourne, Florida 32901–6975, USA
agarwal@fit.edu

E. Akin–Bohner

Department of Mathematics
Florida Institute of Technology
Melbourne, Florida 32901, USA
eakin@math.unl.edu

S. Angadi

Mathematical Modelling Unit
National Institute of Advanced Studies
Bangalore 560012, India
savita@nias.iisc.ernet.in

B. Aulbach

Institut für Mathematik
Universität Augsburg
86135 Augsburg, Germany
aulbach@math.uni-augsburg.de

J. Baštinec

Department of Mathematics
Brno University of Technology
616 00 Brno, Czech Republic
bastinec@feec.vutbr.cz

M. Bohner

Department of Mathematics
Florida Institute of Technology
Melbourne, Florida 32901, USA
bohner@umr.edu

A. Cabada

Departamento de Análise Matemática
Universidade de Santiago de Compostela
15782, Santiago de Compostela, Spain
cabada@usc.es

E. Camouzis

The American College of Greece
6 Gravias Street, Aghia Paraskevi
15342 Athens, Greece
e_camouzis@yahoo.com

S. Castillo

Departamento de Matemática
Universidad del Bío-Bío
Casilla 5-C, Concepción, Chile
scastill@ubiobio.cl

M. Cecchi

Department of Electronics and Telecommunication
University of Florence
50139 Florence, Italy
cecchi@det.unifi.it

J. Čermák

Institute of Mathematics
Brno University of Technology
616 69 Brno, Czech Republic
cermakh@um.fme.vutbr.cz

Q. Cheng

Department of Computer Science
University of Southern California
Los Angeles, California 90089, USA
qcheng@cs.usc.edu

S. S. Cheng

Department of Mathematics
Tsing Hua University
Hsinchu, Taiwan 30043, R. O. C.
sscheng@math2.math.nthu.edu.tw

M.-C. Ciocci

Department of Pure Mathematics and Computer Algeb
University Gent, Krijgslaan 281
9000 Gent, Belgium
mcc@cage.rug.ac.be

A. S. Clark

Department of Mathematics
University at Albany, SUNY
Albany, NY 12222, USA
aron@krikkit.math.albany.edu

F. M. Dannan

Department of Mathematics
Faculty of Science, Qatar University
Doha, Qatar
fmdannan@qu.edu.qa

R. L. Devaney

Department of Mathematics
Boston University
Boston, MA 02215 USA
bob@bu.edu

R. DeVault

Department of Mathematics
Northwestern State University
Natchitoches, LA 71497, USA
richarddevault@yahoo.com

J. Diblík

Department of Mathematics
Brno University of Technology
616 00 Brno, Czech Republic
diblik@feec.vutbr.cz

Z. Došlá

Department of Mathematics
Masaryk University
662 95 Brno, Czech Republic
dosla@math.muni.cz

O. Došlý

Department of Mathematics
Masaryk University
662 95 Brno, Czech Republic
dosly@math.muni.cz

S. Elaydi

Department of Mathematics
Trinity University
San Antonio, Texas 78212, USA
selaydi@trinity.edu

H. A. El-Metwally

Mathematics Department
Faculty of Science
Mansoura University
Mansoura, Egypt

J. Elyseeva

Department of Mathematics
Moscow State University of Technology
Moscow, Russia
elyseeva@mtu-net.ru

L. Erbe

Department of Mathematics and Statistics
University of Nebraska-Lincoln
Lincoln, NE 68588-0323, USA
lerbe@math.unl.edu

V. V. Fedorenko   Institute of Mathematics
                          National Academy of Sciences of Ukraine
                          01601 Kiev, Ukraine
                          vfedor@imath.kiev.ua

B. M. Garay   Mathematics Institute
                          Technical University
                          1521 Budapest, Hungary
                          garay@math.bme.hu

M. Gil'   Department of Mathematics
                          Ben Gurion University
                          Beer Sheva 84105, Israel
                          gilmi@cs.bgu.ac.il

A. Goel   Department of Computer Science
                          University of Southern California
                          Los Angeles, California 90089, USA
                          agoel@cs.usc.edu

E. A. Grove   Department of Mathematics
                          University of Rhode Island
                          Kingston, RI 02881-0816 USA
                          grove@math.uri.edu

G. Sh. Guseinov   Department of Mathematics
                          Atılım University
                          06836 Incek, Ankara, Turkey
                          guseinov@sci.ege.edu.tr

Y. Hamaya   Department of Information Science
                          Okayama University of Science
                          Okayama 700-0005, Japan
                          hamaya@mis.ous.ac.jp

R. J. Higgins   Department of Mathematics and Statistics
                          University of Nebraska-Lincoln
                          Lincoln, NE 68588-0323, USA
                          rhiggins@math.unl.edu

S. Hilger   Didaktik der Physik und Mathematik
                          Katholische Universität Eichstätt
                          85071 Eichstätt, Germany
                          stefan.hilger@ku-eichstaett.de

R. Hilscher

Department of Mathematics
Michigan State University
East Lansing, MI 48824-1027, USA
hilscher@math.msu.edu

A. Hommel

Institute of Mathematics and Physics
Bauhaus–University of Weimar
99421 Weimar, Germany
angela.hommel@bauing.uni-weimar.de

M.-D. Huang

Department of Computer Science
University of Southern California
Los Angeles, California 90089, USA
huang@usc.edu

F. Iavernaro

Dipartimento di Matematica
Università di Bari
70125 Bari, Italy
felix@dm.uniba.it

K. Janglajew

Institute of Mathematics
University of Bialystok
15-267 Bialystok, Poland
jang@math.uwb.edu.pl

B. Kaymakçalan

Department of Mathematics
Georgia Southern University
Statesboro, GA 30460, USA
billur@gasou.edu

K. Keller

Department of Mathematics and Computer Science
University of Greifswald
17487 Greifswald, Germany
keller@mail.uni-greifswald.de

C. M. Kent

Department of Mathematics
Virginia Commonwealth University
Richmond, Virginia 23284-2014, USA
cmkent@mail1.vcu.edu

P. E. Kloeden

Fachbereich Mathematik
Johann Wolfgang Goethe Universität
60054 Frankfurt am Main, Germany
kloeden@math.uni-frankfurt.de

V. Kozyakin

Institute for Information Transmission Problems
Russian Academy of Sciences
101447 Moscow, Russia
kozyakin@iitp.ru

U. Krause

Fachbereich Mathematik und Informatik
Universität Bremen
28343 Bremen, Germany
krause@math.uni-bremen.de

G. Ladas

Department of Mathematics
University of Rhode Island
Kingston, RI 02881-0816 USA
gladas@math.uri.edu

J. López Fenner

Departamento de Ingeniería Matemática
Universidad de La Frontera
Temuco, Chile
jlopez@ufro.cl

M. Marini

Department of Electronics and Telecommunication
University of Florence
50139 Florence, Italy
marini@ing.unifi.it

S. Matucci

Department of Electronics and Telecommunications
University of Florence
50139 Florence, Italy
matucci@det.unifi.it

F. Mazzia

Dipartimento di Matematica
Università di Bari
70125 Bari, Italy
mazzia@dm.uniba.it

C. C. McCluskey

Department of Mathematics and Statistics
University of Victoria
P. O. Box 3045 STN CSC
Victoria BC, Canada, V8W 3P4

L. C. McGrath

Department of Mathematics
University of Rhode Island
Kingston, RI, 02881-0816 USA

R. Medina

Departmento de Ciencias Exactas
Universidad de Los Lagos
Casilla 933, Osorno, Chile
rmedina@ulagos.cl

R. V. N. Melnik

University of Southern Denmark
Mads Clausen Institute
6400, Denmark
rmelnik@mci.sdu.dk

J. S. Muldowney

Department of Mathematical and Statistical Sciences
University of Alberta
Edmonton AB, Canada, T6G 2G1
jim.muldowney@ualberta.ca

L. Neidhart

Institut für Mathematik
Universität Augsburg
86135 Augsburg, Germany
neidhart@math.uni-augsburg.de

K. Nishimura

Institute of Economic Research
Kyoto University, Sakyoku
Kyoto 606-8501, Japan
nishimura@kier.kyoto-u.ac.jp

H. Oliveira

Departamento de Matemática
Instituto Superior Técnico
1049-001 Lisboa, Portugal
holiv@math.ist.utl.pt

Z. Opluštil

Department of Mathematics
Masaryk University
662 95 Brno, Czech Republic
oplustil@math.muni.cz

D. O'Regan

Department of Mathematics
National University of Ireland
Galway, Ireland
donal.oregan@nuigalway.ie

V. Otero-Espinar

Departamento de Análise Matemática
Universidade de Santiago de Compostela
15782, Santiago de Compostela, Spain
vivioe@usc.es

E. Pereira

Department of Informatics
Universidade da Beira Interior
6200-Covilhã, Portugal
edgar@noe.ubi.pt

A. Peterson

Department of Mathematics and Statistics
University of Nebraska-Lincoln
Lincoln, NE 68588-0323, USA
apeterso@math.unl.edu

M. Pinto

Departamento de Matemática
Universidad de Chile
Casilla 653, Santiago, Chile
pintoj@uchile.cl

Z. Pospíšil

Department of Mathematics
Masaryk University
662 95 Brno, Czech Republic
pospisil@math.muni.cz

C. Pötzsche

Institut für Mathematik
Universität Augsburg
86135 Augsburg, Germany
poetzsche@math.uni-augsburg.de

J. S. Ramos

Departamento de Matemática
Instituto Superior Técnico
1049-001 Lisboa, Portugal
sramos@math.ist.utl.pt

P. Řehák

Department of Mathematics
Masaryk University Brno
603 00 Brno, Czech Republic
rehak@math.muni.cz

A. Ruffing

Department of Mathematics
Munich University of Technology
80333 München, Germany
ruffing@ma.tum.de

E. Schmeidel

Institute of Mathematics
Poznań University of Technology
60-695 Poznań, Poland
eschmeid@math.put.poznan.pl

H. Sedaghat

Department of Mathematics
Virginia Commonwealth University
Richmond, VA 23284-2014, USA
hsedagha@vcu.edu

A. N. Sharkovsky

Institute of Mathematics
National Academy of Sciences of Ukraine
01601 Kiev, Ukraine
asharkov@imath.kiev.ua

T. Shigoka

Institute of Economic Research
Kyoto University, Sakyoku
Kyoto 606-8501, Japan
shigoka@pop.nifty.com

S. Siegmund

Department of Mathematics
University of California
Berkeley, CA 94720, USA
siegmund@math.berkeley.edu

M. Simon

Department of Mathematics
Munich University of Technology
Arcisstrasse 21
80333 München, Germany

E. S. Thomas

Department of Mathematics
University at Albany, SUNY
Albany, NY 12222, USA
et392@math.albany.edu

C.-J. Tian

College of Information Engineering
Shenzhen University
Shenzhen 518060, P. R. China
tiancj@szu.edu.cn

D. Trigiante

Dipartimento di Energetica
Università di Firenze
50134 Firenze, Italy
trigiant@cesit1.unifi.it

P. G. Vaidya

Mathematical Modelling Unit
National Institute of Advanced Studies
Bangalore 560012, India
pgvaidya@nias.iisc.ernet.in

W. Van Assche

Department of Mathematics
Katholieke Universiteit Leuven
3001 Leuven, Belgium
walter@wis.kuleuven.ac.be

A. Vanderbauwhede

Department of Pure Mathematics and Computer Algebra
University Gent, Krijgslaan 281
9000 Gent, Belgium
avdb@cage.rug.ac.be

J. Vitória

Department of Mathematics
Universidade de Coimbra
3000-Coimbra, Portugal
jvitoria@mat.uc.pt

H. Wasserman

Department of Computer Science
University of Southern California
Los Angeles, California 90089, USA
Halwass@aol.com

M. Yano

Department of Economics
Keio University
Tokyo 108-8345, Japan
myano@tkd.att.ne.jp

V. Zeidan

Department of Mathematics
Michigan State University
East Lansing, MI 48824-1027, USA
zeidan@math.msu.edu

D. Zeilberger

Department of Mathematics
Rutgers University
New Brunswick, NJ 08903, USA
zeilberg@math.rutgers.edu

B. Zhang

Department of Applied Mathematics
Ocean University of Quingdao
266033 Quingdao, China
bgzhang@public.qd.sd.cn

K. Ziegler

Department of Physics
University of Augsburg
86135 Augsburg, Germany
ziegler@physik.uni-augsburg.de

Opening Lecture

Difference Equations and Boundary Value Problems 3
A. N. Sharkovsky

Difference Equations and Boundary Value Problems

A. N. SHARKOVSKY [1]

Institute of Mathematics
National Academy of Sciences of Ukraine
01601 Kiev, Ukraine
E-mail: asharkov@imath.kiev.ua

Abstract For the research on nonlinear boundary value problems of mathematical physics, in particular, to a simulation of self-arising chaotic evolutions in deterministic systems, the application of low-dimensional dynamics is very promising. As it is known, there are boundary value problems which directly lead to the study of difference equations with continuous argument, and thus allows to effectively use the theory of dynamical systems, especially given by one-dimensional maps. This paper presents an overall picture of the study of some simple boundary value problems and their associated difference equations with continuous time.

Keywords Difference equation with continuous time, Boundary value problem, Reduction method, Difference equation with discrete time

AMS Subject Classification 37, 39A11

1 Introduction

Among boundary value problems (BVP) for partial differential equations (PDE) there are certain classes of problems that are reducible to difference, differential-difference and other relevant equations. These classes consist mainly of problems for which the representation of the general solution for the PDE is known.

An effective study of such problems by the "reduction method" has become possible only in the last 20-30 years due to appreciable advances in the theory of difference equations with continuous time, which in turn has gained momentum through the progress in the theory of difference equations with discrete time or, more exactly, in the theory of dynamical systems, specifically given by one-dimensional maps. It is just the fact that even trajectories of dynamical systems on the real line can easily exhibit very complicated behavior

---

[1]Supported by the Ministry of Education and Science of Ukraine, State Fund of Fundamental Investigations, Project No 01.07/00081.

up to a quasi-random one, which led to the understanding that deterministic evolutionary processes described by quite simple but nonlinear BVP can be found to be actually indistinguishable in long-time behavior from random processes.

Here we will try to show how the reduction method may be used, and how much this method might be profitable, on the example of very simple nonlinear BVP. A new direction, that such a reduction opens up, has been considered in the book [19] and in a series of papers by the author and his colleagues (see References [1-26] in this paper). At present time, investigations on the theory of difference equations with continuous time and the opportunities of their applications, unfortunately, are still not widespread enough.

2   Reduction of BVP to difference equations: some examples

The wave equation

$$w_{tt} - w_{xx} = 0, \qquad x \in [0,1], \quad t \in R^+, \tag{1}$$

with the local boundary conditions

$$H_0(w, w_t, w_x)\,|_{x=0} = 0, \qquad H_1(w, w_t, w_x)\,|_{x=1} = 0, \tag{2}$$

or with the nonlocal boundary conditions

$$H_0(w, w_t, w_x)\,|_{x=0} = 0, \qquad H_{10}(w, w_t, w_x)\,|_{x=0} = H_{11}(w, w_t, w_x)\,|_{x=1}, \tag{3}$$

is a classic example of a BVP which is reducible to difference equations with continuous time (DE) or differential-difference equations (DDE).

The general solution of Eq.(1) can be written as follows

$$w(x,t) = u(t+x) + v(t-x), \tag{4}$$

where u and v are arbitrary, generally speaking, C^2-smooth functions. This allows, substituting (4) either into (2), or into (3), to obtain corresponding DDE or, in more simple cases, DE. For example, substituting (4) into (2), we get relationships between u, v and their derivatives:

$$G_0(u(\tau), v(\tau), u^{'}(\tau), v^{'}(\tau)) = 0, \qquad \tau \geq 0, \tag{5}$$
$$G_1(u(\tau+2), v(\tau), u^{'}(\tau+2), v^{'}(\tau)) = 0, \qquad \tau \geq -1,$$

where

$$G_i(z_1, z_2, z_3, z_4) = H_i(z_1 + z_2, z_3 + z_4, z_3 - z_4), \qquad i = 0, 1.$$

The relationships (5) are a system of differential-difference equations of the first order for u and v. In the case that conditions (3) are boundary conditions for BVP then substituting (4) into (3), we get for u and v also a system of DDE but already one of the second order.

Of course, such systems of DDE at specific boundary conditions are found to be a system of pure difference equations, and two such examples for the wave equation are considered in Examples 5 and 6 below. The simplest general situation for this to be so is provided when both functions $H_0(w_1, w_2, w_3)$ and $H_1(w_1, w_2, w_3)$ from the boundary conditions (2) either depend on w_1 only (i.e., these boundary conditions do not involve derivatives) or, in contrast, are independent of w_1. Another more interesting example of reducibility to DE can be obtained when one of the functions H_0 and H_1 is independent of w_2 and w_3 and the other depends linearly on these; for instance, H_0 does not depend on w_2 and w_3 and H_1 is of the form

$$H_1(w_1, w_2, w_3) = p(w_1)w_2 + q(w_1)w_3.$$

In this case, Syst.(5) can be rearranged to a so-called completely integrable DDE, which can in turn be reduced to a one-parameter family of DE (for a discussion of completely integrable DDE, see [19], Part III, Ch.1, and the references given there).

It makes sense to note also that any initial conditions for BVP can be transformed (by standard means) to corresponding initial conditions for DE or DDE.

2.1   Simplest examples

Here we give several examples of such problems generating typical difference equations. In fact, we consider problems where a BVP is reduced to

- autonomous difference equation (example 1);

- non-autonomous difference equation (example 2);

- a family of autonomous difference equations (example 3).

Example 1.   We start with the simplest nonlinear BVP consisting of the simplest PDE

$$\partial w/\partial t \; = \; \partial w/\partial x, \quad x \in [0,1], \;\; t \in R^+, \tag{6}$$

and the (nonlocal) boundary condition

$$w \mid_{x=1} \; = \; f(w) \mid_{x=0}, \tag{7}$$

where f is a C^1-smooth function. On substituting the general solution of (6)

$$w(x,t) \; = \; u(x+t), \tag{8}$$

where u is an arbitrary C^1-smooth function, into the boundary condition (7) we obtain the *autonomous difference equation*

$$u(\tau + 1) = f(u(\tau)), \quad \tau \in R^+. \tag{9}$$

Every initial condition for this BVP

$$w(x, 0) = \varphi(x), \quad x \in [0, 1], \tag{10}$$

with φ being a C^1-smooth function, gives the initial condition

$$u(\tau) = \varphi(\tau), \quad \tau \in [0, 1), \tag{11}$$

for the DE (9). Here it should be noted that both a solution of BVP generated by the initial data φ and the corresponding solution of DE (9) will be C^1-smooth if and only if $\varphi(1) = f(\varphi(0))$ and $\varphi'(1) = f'(\varphi(0))\varphi'(0)$, which is a usual assumption in BVP theory, the so-called C^1-smooth *consistency conditions*.

The solution $u_\varphi(x, t)$ of DE (9) for the initial condition (11) can be written in the form

$$u_\varphi(\tau) = f^n(\varphi(\{\tau\})), \quad n \le \tau < n + 1, \quad n = 0, 1, 2, ..., \tag{12}$$

with f^n standing for the n-th iteration of f (i. e., $f^n = f \circ f^{n-1}$ and $f^0(x) \equiv x$), and hence the corresponding solution $w_\varphi(x, t)$ of BVP can be written in the form

$$w_\varphi(x, t) = f^{[t+x]}(\varphi(\{t + x\})), \quad t \in R^+, \ x \in [0, 1], \tag{13}$$

with $[\cdot]$ and $\{\cdot\}$ standing for the integral and fractional parts of a number.

Example 2. The second example deals with the PDE

$$\partial w / \partial t = \partial w / \partial x + d(t), \quad x \in [0, 1], \ t \in R^+. \tag{14}$$

Here we insert into the PDE (6) a nonautonomous term in its simplest form. With the general solution of (14)

$$w(x, t) = u(x + t) + D(t), \quad \text{where } D(t) \text{ is an antiderivative of } d(t), \tag{15}$$

and the boundary condition (7), for the function

$$v(\tau) = u(\tau) + D(\tau) \tag{16}$$

we obtain the *nonautonomous difference equation*

$$v(\tau + 1) = f(v(\tau)) - D(\tau + 1), \quad \tau \ge 0. \tag{17}$$

Example 3. Let us return to equation (6)

$$\partial w / \partial t = \partial w / \partial x, \quad x \in [0, 1],$$

but replace the boundary condition (7) with that involving derivatives

$$w_t \mid_{x=1} = g(w) w_t \mid_{x=0}, \quad g \text{ is a } C^0\text{-smooth function.} \tag{18}$$

In this case, the BVP can be reduced to a *family of difference equations*. Indeed, integrating the boundary condition (18) gives

$$w \mid_{x=1} = f(w) \mid_{x=0} + \gamma, \tag{19}$$

where f is an antiderivative of g and γ is an arbitrary constant. On substituting the general solution of (6) $w(x,t) = u(x+t)$ into this new boundary condition, we obtain the one-parameter family of difference equations

$$u(\tau + 1) = f(u(\tau)) + \gamma, \quad \tau \in R^+. \tag{20}$$

Since $w(x,t)$, a solution of BVP, has to be continuous, the function u corresponding to the initial condition $w(x,0) = \varphi(x)$ is a solution of a solitary one of the equations of the family, namely, the equation with γ as follows

$$\gamma = \varphi(1) - f(\varphi(0)). \tag{21}$$

2.2 Continuity via discreteness: "continuous lattice" of uncoupled oscillators

When the term "difference equation" is used, the argument of the equation is, as a rule, meant to vary discretely. With the reduction of a BVP to a DE (whenever possible) we always arrive at an equation whose argument varies continuously, and in the simplest cases this equation is of the form (9).

Certainly, the properties of solutions of DE (9) should be closely connected to the properties of the analogous *difference equation with discrete time*

$$u_{n+1} = f(u_n), \quad n \in Z^+, \tag{22}$$

or, what amounts to the same thing, to the properties of the dynamical system given by the map

$$u \quad \mapsto \quad f(u). \tag{23}$$

What are the distinctions in the behavior of the solutions of the difference equations (9) and (22)? Every solution u_n, $n \in Z^+$, of Eq. (22) is determined uniquely by the value $u_0 \in R : u_n = f^n(u_0)$. Every solution $u(\tau)$, $\tau \in R^+$, of DE (9) is determined by its values on the interval $[0,1)$: $u_\varphi(\tau) = f^{[\tau]}(\varphi(\{\tau\}))$, $\tau \geq 0$. Thus, every solution of DE (9) consists of a continual family of solutions of Eq. (22). The dynamics of solution of DE (9) can be treated as dynamics of continuum of uncoupled oscillators: at every point $\tau \in [0,1)$, there is disposed the same oscillator $u \mapsto f(u)$; its oscillations are independent of oscillations at other points from the interval $[0,1)$, and therefore, if the map f possesses the so-called sensitive dependence on initial data, the states of oscillators that were very close in an initial moment can be very different as time goes on. And as a result, just the independence of oscillators in such a continual family causes the cascade process of structure to emergence until arbitrary small scales and the appearance in the time limit of "random" structures.

Due to the possibility of reducing a BVP to a DE, it is possible to interpret each of the three BVP presented in Examples 1–3 as a problem on dynamics of a continuum of uncoupled oscillators: at every point $x \in [0, 1)$ we have "a pendulum", oscillations of which are independent of the oscillations of "pendulums" at other points from $[0, 1)$. In the problem (6), (7), "a pendulum" is the same for all the initial states φ and for all the points $x \in [0, 1)$, namely, $w \mapsto f(w)$. In the problem (6), (18), "its own pendulum" corresponds to every initial state φ, but it is the same at all the "suspension" points x (namely, $w \mapsto f(w) + \varphi(1) - f(\varphi(0))$). Finally, in the problem (14), (7) if the "forcing" $d(t)$ is a 1-periodic function, we have the family of "pendulums" $w \mapsto f(w) - D(x)$, for every point $x \in [0, 1)$ there is "its own pendulum" (i.e. the law of pendulum oscillation depends on the suspension point).

2.3   More examples

It is very easy to modify each of the previous BVP in order to obtain a BVP which will be reduced to a differential-difference equation only (and cannot be reduced to a DE) and, hence, which already can be interpreted as a lattice of coupled oscillators (and for which dynamics will be essentially different from that described above, in particular, the self-stochasticity phenomenon will be impossible).

Example 4.   Let us modify the BVP of the example 3, namely, replace the boundary condition (18) by the following:

$$\nu w_t \mid_{x=1} \; = \; g(w) \mid_{x=0} . \tag{24}$$

In this case, substituting the general solution of PDE $w(x, t) = u(x + t)$ into (24), we obtain the DDE

$$\nu u'(\tau + 1) = g(u(\tau)), \tag{25}$$

which cannot be reduced exactly to a DE.

In the examples 1–3 we had one-dimensional BVP and obtained, after reduction, difference equations of first order. There are also many-dimensional BVP which can be reduced to a difference equation whose order is only one (see, for instance, [3, 6, 7]), and, to the contrary, there are one-dimensional BVP reducible to a system of difference equations.

Let us consider two typical examples for the wave equation (1).

Example 5. At first we consider an example of a BVP with a local boundary conditions, namely, with the conditions

$$w\,|_{x=0}\ =\ 0, \qquad\qquad w_x\,|_{x=1}\ =\ h(w_t)\,|_{x=1}\,. \tag{26}$$

Taking into account the first condition and also the representation of the general solution of equation (1)

$$w(x,t)\ =\ u(t+x)+v(t-x),$$

we conclude that $v=-u$. Using the second boundary condition we obtain for the function

$$z(\tau)\ =\ u'(\tau-1),\quad \tau\geq 0,$$

the difference equation (of first order)

$$z(\tau+2)+z(\tau)\ =\ h(z(\tau+2)-z(\tau)). \tag{27}$$

Thus, the problem (1), (26) is reduced also to one difference equation, in which the dependence of $z(\tau+2)$ on $z(\tau)$ is given implicitly and can be many-valued (see, for example, [13, 18]).

Therefore, each initial condition for the BVP

$$w(x,0)=\varphi(x),\qquad w_x(x,0)=\psi(x),\qquad x\in[0,1], \tag{28}$$

is transformed into the initial condition for the difference equation (27)

$$z(\tau)\ =\ \begin{cases} (-\varphi(1-\tau)+\psi(1-\tau))\,/\,2 & \text{for } 0\leq\tau\leq 1,\\ (\varphi(\tau-1)+\psi(\tau-1))\,/\,2 & \text{for } 1<\tau\leq 2. \end{cases}$$

Example 6. Now we consider an example of the BVP with the nonlocal boundary conditions

$$w_x\,|_{x=0}\ =\ aw_t\,|_{x=0},\ a>0, \qquad\qquad w\,|_{x=1}\ =\ h(w\,|_{x=0}). \tag{29}$$

With the general solution of the equation (1)

$$w(x,t)\ =\ u(t+x)+v(t-x),$$

we obtain

$$u(\tau)\ =\ \frac{1+a}{2}z(\tau+1)\ -\ A,\qquad v(\tau)\ =\ \frac{1-a}{2}z(\tau+1)\ +\ A, \tag{30}$$

where A is an arbitrary constant, z is the general solution of the difference equation of second order

$$z(\tau+2)\ =\ \frac{2}{1+a}h(z(\tau+1))\ +\ bz(\tau),\qquad b=-\frac{1-a}{1+a}. \tag{31}$$

Putting

$$y(\tau) = \frac{1}{b} z(\tau + 1), \quad f(y) = \frac{2}{a - 1} h(b y), \tag{32}$$

we conclude that for $a \neq 1$, this BVP is reduced to the *system of difference equations*

$$\begin{aligned} y(\tau + 1) &= f(y(\tau)) + z(\tau) \\ z(\tau + 1) &= b\, y(\tau). \end{aligned} \tag{33}$$

To this system there corresponds a two-dimensional map, namely,

$$\begin{aligned} y &\mapsto f(y) + z, \\ z &\mapsto b\, y, \end{aligned} \tag{34}$$

which is Henon's map if $f(y) = 1 - My^2$ and is Lozi's map if $f(y) = 1 - M|y|$.

2.4   Which properties of maps specify the behavior of solutions of DE and BVP?

We start with several sufficiently obvious properties of solutions for DE and BVP. Difference equations, as such, do not necessitate any smoothness (or even continuity) of their solutions. However, just the C^1-smooth solutions should be used when investigating BVP, and we need to assume the non-linearity f and the initial data φ to be C^1-smooth and also the C^1-smooth consistency conditions to hold.

1. *DE (9), and hence BVP (6),(7), has bounded solutions different from constant ones if and only if the map f possesses at least one nondegenerate bounded invariant interval.*

2. *If the values of $\varphi(t)$, $t \in [0, 1]$, fall into one of bounded invariant intervals of the map f, then the solution u_φ of DE (9) (and correspondingly the solution w_φ of BVP (6),(7)) is bounded and, moreover, takes on values from just this invariant interval for all $t \geq 0$.*

3. *If for some $t' \in [0, 1]$, the value $\varphi(t')$ does not belong to any invariant bounded interval, then the solution u_φ (and correspondingly, w_φ) is unbounded.*

4. *If for a function φ, among its values are those fallen into two disjoint bounded invariant intervals of the map f, for instance, $[a, b]$ and $[c, d]$, $b < c$, then the solution u_φ (and correspondingly w_φ) oscillates without undamping and the amplitude of its oscillations on any interval $[T, T+1]$ is no smaller than $|b - c|$.*

Two more simple but important properties of the long-time behavior of solutions are given next.

1. *If γ_+ is an attracting fixed point of the map f and a function φ takes on values only from the basin of the point γ_+, then the solution u_φ (and correspondingly w_φ) tends uniformly to the constant solution $u(t) \equiv \gamma_+$ (and correspondingly to the constant solution $w(x,t) \equiv \gamma_+$), as $t \to \infty$.*

2. *If γ_- is a repelling fixed point of the map f and a function $\varphi(t) \not\equiv \gamma_-$ is such that $\varphi(t_*) = \gamma_-$ with a certain $t_* \in (0,1)$, then for any $\varepsilon > 0$ and $\delta > 0$ there exists $N > 0$ such that whatever $i > N$, the ε-neighborhood of the set $u_\varphi(V_\delta(t_* + i))$, with $V_\delta(\cdot)$ being the δ-neighborhood of a point, contains the interval*

$$Q_f(\gamma_-) := \cap_{\delta > 0} \cap_{j > 0} \overline{\cup_{i > j} f^i(V_\delta(\gamma_-))},$$

which is called the domain of influence of the point $z = \gamma_-$, or, in other words,

$$\mathrm{Lt}_{i \to \infty} \, x_\varphi(V_\delta(t_* + i)) \;=\; Q_f(\gamma_-), \tag{35}$$

where $\mathrm{Lt}_{i \to \infty} \cdot$ stands for the topological limit of a sequence of sets. Then, in particular,

$$\sup_{t \in V_\delta(t_* + i)} |x'_\varphi(t)| \;\longrightarrow\; \infty \quad as \quad i \to \infty.$$

An analogous situation will be for the solution $w_\varphi(x,t)$ but for values (x,t) such that $(x + t) \in V_\delta(t_* + i)$, i.e. $t_* + i - \delta < x + t < t_* + i + \delta$, as $i \to \infty$.

The theory of one-dimensional maps provides insight into why and how chaotic (but smooth) solutions of DE and BVP develop, in particular, to clarify *self-structuring and self-stochasticity phenomena* inherent in such solutions. Here we will point out once more the very important properties of one-dimensional maps, which result in the "chaotization" of solutions for BVP and their attendant DE.

- For self-structuring phenomena those are:

  - "complex" topological organization of a basin (i.e. a domain of attraction) for attracting cycles or cycles of intervals;
  - self-similarity of graphs for map iterations.

- For self-stochasticity phenomena those are:

  - existence of an absolutely continuous invariant measure (a.c.i.m.);
  - the availability of a.c.i.m. is not an extraordinary point.

Self-structuring in solutions of DE or BVP is due to the complex structure of the basin of an attracting cycle (or so-called cycle of intervals) of its associated one-dimensional map. Such a basin, as a rule, is a union of countably many intervals, more precisely, the basin of a cycle (or a cycle of intervals) for a map f is of the form $\bigcup_{i=0}^{\infty} f^{-i}(B)$ with B being the *domain of immediate attraction* (or *sink of basin*) of the cycle (that is, B being the set that consists of m intervals, each attracted by the cycle and, besides, containing one of the cycle points (or intervals); here m is for the period of the cycle). This explains why time-evolving cascade processes of appearance of structures arise in solutions of DE and BVP and how they carry on. These structures are coherent and, moreover, may be found to be self-similar. Such properties have their origin in the geometry of the graphs of iterations of one-dimensional maps. In fact, when n is large enough, the graph of f^n is self-similar at Misiurewicz's points — that is, at repelling periodic points and their preimages. Usually a map f has many Misiurewicz's points. For instance, there is a countable number of such points if f possesses a cycle of period > 2.

Self-stochasticity occurs in solutions of DE and BVP when its associated map has an invariant measure absolutely continuous with respect to the Lebesgue measure. If for a map $u \mapsto f(u)$ there exists a.c.i.m., then one can find through this measure the probability of a trajectory $f^n(u)$ lying in one or another of the regions of the phase space when n is large enough. This *temporal stochasticity* of trajectories $u_{n+1} = f(u_n), n = 0, 1, 2, \ldots$, transforms into a *spatial-temporal stochasticity* of solutions of DE or BVP with associated map f. The reason is that at every fixed $\tau_* \in [0,1]$ there is an "oscillator" (of its own), namely, $u(\tau_*) \mapsto f(u(\tau_*))$, and for all τ these "oscillators" act independently of each other (although, maybe, following the same law). In case f has a.c.i.m., the deterministic function $u_{n+1}(\tau) = f(u_n(\tau)), \tau \in [0,1]$, behaves like a random function of τ for n large enough, and, moreover, tends (in special metric) as $n \to \infty$ to a certain random process, whose distributions can be described in terms of the invariant measure of the map f.

3   Difference equations and boundary value problems as dynamical systems

If it is necessary to investigate the asymptotic behavior of the solutions of evolutionary problems, it is usually convenient to transfer to the corresponding dynamical system (DS) (if it is possible) and to apply the achievements of DS theory. We often follow such an approach.

Evolutionary BVP for PDE generally induce infinite-dimensional dynamical systems on their spaces of initial states. For parabolic PDE (such as Navier-Stokes equation), the attractor of the associated DS is in many cases a finite-dimensional subset of the phase space. Another situation occurs in problems for hyperbolic equations which we Fconsider here: the phase space of the associated DS (which consists of smooth functions), as a rule, does not contain its attractor. Therefore, when analyzing evolutions given by our BVP, we have to consider classes of functions, which are wider than classes of smooth functions, and to use special metrics in order to complete the phase space of the original DS and to describe by this means the asymptotic behavior of trajectories and, in particular, to construct the attractor of the DS.

For a description of the asymptotic behavior of trajectories it is expedient to use (as is accepted in the theory of dynamical systems) the concept of the *ω-limit set of a trajectory*. We use also a standard notion of attractor but here we have to consider not only the original space C^k but also a new space C^* obtained via the completion of the original space, for example, by the use of some proper metric ρ^*.

By the *attractor* $\mathcal{A}$ we mean the smallest closed set in the phase space C^* which has the property that $\omega[\varphi] \subset \mathcal{A}$ for all $\varphi \in C^k$ outside of a set of first Baire category (with respect to C^k-topology).

The attractor $\mathcal{A}$ can easily be shown to be well-defined. If the dynamical system is uniformly continuous on C^k with respect to ρ^*, it induces (by continuity) the dynamical system C^* and, in particular, determines a motion on the attractor $\mathcal{A}$.

For many DE and BVP, the attractor will consist of discontinuous multivalent functions, which could result in a very complicated long-time behavior of (smooth) solutions. In this case, we typically observe a self-structuring phenomenon whose description calls for the notions of self-excited structures, cascade process of appearance of structures, self-similarity, fractal structures, etc. Moreover, for these problems, there may occur a self-stochasticity phenomenon which lies in the fact that the attractor of a deterministic DS contains random functions.

3.1 Long-time behavior of solutions

As was noted above, both DE (9) and BVP (6), (7) generate infinite-dimensional dynamical systems on the space of smooth functions $[0, 1] \to R^1$. DE (9) generates a dynamical system with discrete time $\varphi_0 \mapsto \varphi_n$, $n \in Z^+$, where $\varphi_n = \varphi_n(\tau) = u(\tau + n)$, $\tau \in [0, 1]$, and $u(t)$ is a solution of DE (9) with initial condition $\varphi_0(t)$. The boundary value problem (6), (7) generates the dynamical system with continuous time $w_{(0)} \mapsto w_{(t)}$, $t \in R^+$, where $w_{(t)} = w_{(t)}(x) = w(x, t)$, and $w(x, t)$ is the solution of the BVP with initial condition $w(x, 0) = w_{(0)}(x)$.

We restrict our consideration to the class of unimodal functions with negative Schwarzian derivative which is well investigated and has the properties we would like to consider.

As is known, for the difference equation (22), the periodicity or asymptotic periodicity of solutions are not typical properties: if, for example, we take $f(x) = \lambda x(1-x)$, where λ is a parameter and $x \in I = [0, 1]$, then the Lebesgue measure of the set of parameters such that the map f has an a.c.i.m. is positive (Jakobson, 1981); when the map f has an a.c.i.m., the map possesses periodic and asymptotically periodic trajectories, however, the trajectory of almost every (in measure) point from I is not an asymptotically periodic one.

Therefore, infinite-dimensional DS induced by DE (9) and BVP (6), (7) are in a certain sense even simpler than one-dimensional DS given by f: almost all trajectories on the attractor are periodic almost always (that is, for almost all λ) and, hence, almost all trajectories in phase space (and solutions of generating problems) are also asymptotically periodic almost always [19].

It makes sense to give some explanations and more precise formulations.

If the map f has a Lyapunov unstable point and this point falls into "the initial set" $\varphi([0,1))$, then the trajectory is not compact in $C^r([0,1),R)$ whatever $r \geq 0$, and the ω-limit set can be empty. Hence, to characterize the asymptotic behavior of trajectories (or appropriate solutions), it is necessary to take a space with a large stock of functions and with a metric other than the metric of uniform convergence.

One possibility is to use the Hausdorff metric for the graphs of functions, as in [19, 20], or the equivalent metric

$$\rho^\Delta(w_1, w_2) = \sup_{\varepsilon>0} \min\{\varepsilon, \sup_x \Delta(w_1^\varepsilon(x), w_2^\varepsilon(x))\},$$

where $w^\varepsilon(x) = w(V_\varepsilon(x))$, $V_\varepsilon(x)$ is the ε-neighborhood of a point x,

$$\Delta(A_1, A_2) = \max\{\sup_{y\in A_2} \sigma(y, A_1), \sup_{y\in A_1} \sigma(y, A_2)\},$$

and σ is the Euclidean distance.

If we complete the space of C^1-functions $[0,1] \to R$ via this metric, we obtain a compact space, denoted further by $C^\Delta([0,1],R)$. The dynamical system on C^1 will induce a dynamical system on C^Δ. As a consequence (see, for example, [19]), for "almost every" function f (from families of functions which we consider), almost every "initial point" from C^1 determines a trajectory, the ω-limit set of which is a finite set in C^Δ, and points of this set form a periodic trajectory of the dynamical system on C^Δ.

Proposition 1 *Let the attractor of a map f be an attracting cycle (or a cycle of transitive intervals) of period m. Then for every initial function from $C^1([0,1],I)$ which are not constant on any subinterval of $[0,1]$, the trajectory of DS on C^1, corresponding to this initial function, has an ω-limit set which is a cycle with period m for the DS on C^Δ, and every "point" of this ω-limit set is generally a proper upper semicontinuous function.*

Thus, the attractor of the dynamical system induced by DE (9) or BVP (6), (7) is in C^Δ a set consisting of periodic trajectories with period m.

In the case where the map f has not only periodic and asymptotically periodic trajectories (for example, if f has a cycle of period $\neq 2^i, i \geq 0$), there exist trajectories (and hence, solutions of both DE (9) and BVP (6), (7)) which are not asymptotically periodic. For example, such one is a trajectory (in C^Δ), corresponding to initial functions $\varphi \in C([0,1],I)$ which have intervals $L_\varphi \subset [0,1)$ with $\varphi(t)|_{L_\varphi} \equiv const$, if, in addition, the trajectory of the map

f passing through the point $x_0 = \varphi(t)|_{L_\varphi}$ is not periodic or asymptotically periodic.

A similar statement holds also for other DE and BVP.

3.2 Self-similarity of solutions

Thus, dynamical systems induced by either a BVP or its associated DE usually have in $C^\Delta([0,1], I)$ an attractor $\mathcal{A}^\Delta$ containing discontinuous functions. For this reason, a "point" $\zeta \in \mathcal{A}^\Delta$ as function $\zeta : [0,1] \to R$, can be of a very complicated organization from the geometric viewpoint, which is of a great importance in simulating actual turbulent processes. In particular, the graph of $\zeta(x)$ can be a locally self-similar set (in the plane (x, ζ)) and can have fractal (box-counting) dimension > 1.

For the one-dimensional BVP (6), (7) it is possible to use the concept of "self-similarity" in its simplest form. By the *self-similarity* of the graph of a function $\zeta(x)$ at a point x_* we mean that there exists a neighborhood V of x_* and a contracting diffeomorphism $\sigma : V \to V$ with x_* being its fixed point such that the graph of $\zeta(x)|_V$, as set in the plane (x, ζ), is invariant under the transformation $(x, \zeta) \mapsto (\sigma(x), \zeta)$; $1/\sigma(x_*)$ gives the self-similarity scaling.

The self-similarity of graphs of solutions and the self-similarity and fractality of "points" of ω-limit sets are evidently a consequence of the self-similarity of iterations of the associated map f and, accordingly, of the self-similarity and fractality of the topological limits of the graphs of certain subsequences of iterations of the map f. For DE (9) and BVP (6), (7) this can be seen well from the representation of solutions (12) and (13), respectively. It is known, that the self-similarity of iterations of the associated map f takes place in the limit for so-called Misiurewicz's points of f, that are repelling periodic points of f and all their preimages.

In the case where an ω limit set is a cycle, it is possible to propose an exact and clear formulation on the self-similarity property. Every initial function φ gives a linear ordering of the points of the ω-limit set $\omega^\Delta[\varphi]$, which just conforms to moving on this set; for example, for the solutions of BVP, we put ζ_τ, $0 \le \tau < m$, for functions from $\omega^\Delta[\varphi]$ to which the sequences $w_\varphi(x, mi + \tau)$ converge in metric ρ^Δ, as $i \to \infty$.

Proposition 2. *If for the BVP (6), (7) with an initial function φ, the ω-limit set $\omega^\Delta[\varphi]$ is a cycle, then*

- *the graph of each (upper semicontinuous) function $\zeta_\tau \in \omega^\Delta[\varphi]$ is self-similar at each point x such that the fractional part of $x + \tau$ equals τ^* and $\varphi(\tau^*)$ is a Misiurewicz's point of the map f with*

$$\frac{d}{d\tau} f^k(\varphi(\tau)) \ne 0 \quad at \quad \tau = \tau^*,$$

where $k \geq 0$ is a number of steps in which $\varphi(\tau^*)$ falls in a cycle; in addition, the self-similarity scaling is equal to the multiplicator of the cycle in which the point $\varphi(\tau^*)$ falls;

- *whatever function $\zeta \in \omega^{\Delta}[\varphi]$, the fractal dimension of the graph of $\zeta(x)$ is greater than 1 if the map f has a cycle of period $\neq 2^k, k \geq 0$, on the smallest invariant interval of the map f which contains the set $\varphi([0,1])$.*

3.3 Self-stochasticity of solutions

If for some φ, the value of a function $\zeta_t \in \omega[\varphi]$ is a nondegenerate interval at every point x from the interval $[0,1]$ or, at least, from some subinterval of $[0,1]$, then the corresponding solution w_φ behaves extremely irregularly and *there is no way of telling certainly its values when t is large enough.*

In this case, it is often possible to use a more "exact" metric, enabling one to measure the distance between deterministic and random functions:

$$\rho_0^{\#}(w_1, w_2) = \sup_{\varepsilon > 0} \min\{\varepsilon, \sup_x \int_R |F_{w_1}^{\varepsilon}(x, z) - F_{w_2}^{\varepsilon}(x, z)| \, dz\},$$

where

$$F_w^{\varepsilon}(x, z) = \frac{1}{2\varepsilon} \int_{x-\varepsilon}^{x+\varepsilon} F_w(y, z) \, dy,$$

$F_w(x, z)$ is the distribution function of $w(x)$ (for each pair (x, z), $F_w(x, z)$ is the probability of $w(x) < z$). More precisely, it is necessary to use a similar metric $\rho^{\#}$ that takes into account all finite-dimensional distributions (see [22, 23]). If we complete the space of C^1-functions $[0,1] \to R$ via this metric, we obtain a (non-compact!) space, denoted by $C^{\#}([0,1], R)$. Then "points" of the ω-limit set can also be random functions.

Proposition 3 *If the map f has an ergodic a.c.i.m. μ, and the support of the measure μ consists of m intervals, then for the DS generated by the BVP (6), (7) the ω-limit set of each trajectory with a nonsingular initial function is a cycle in $C^{\#}$ with period m; moreover, each "point" of this ω-limit set is a random function whose distribution is given in terms of the measure μ.*

3.4 "Universalities" for infinite-dimensional dynamical systems

One might hope that infinite-dimensional dynamical systems, generated by BVP and DE considered above, have to inherit certain properties of one-dimensional dynamical systems given by their associated map f. Bifurcations of periodic trajectories and so-called universal properties of these bifurcations are just such properties. First of all, this is the universal order of bifurcations determined by the following ordering of the natural numbers:

$$1 \prec 2 \prec 2^2 \prec \ldots \prec 5 \cdot 2^2 \prec 3 \cdot 2^2 \prec \ldots \prec 5 \cdot 2 \prec 3 \cdot 2 \prec \ldots \prec 9 \prec 7 \prec 5 \prec 3,$$

and also the universal rate for such bifurcations and the universal change of amplitude of arising oscillations, given by the known Feigenbaum-Coullet-Tresser constants $\delta = 4.6992\ldots$ and $\alpha = 2.5029\ldots$, respectively.

It is convenient to demonstrate these facts for the BVP (6), (18) of Example 3. This BVP is reduced to the one-parameter family of difference equations (20)

$$u(\tau + 1) = f(u(\tau)) + \gamma, \quad \tau \in R^+,$$

where f is an antiderivative of g and $\gamma = \gamma[\varphi] = \varphi(1) - f(\varphi(0))$. Thus, for the investigation of this problem we have the family of maps

$$f_\gamma : u \mapsto f(u) + \gamma,$$

in which different bifurcations can take place if the parameter γ is varied.

For the sake of simplicity, we assume here that the function g (in the boundary condition (18)) satisfies the conditions

- g is a C^2-smooth function with $g'(u) < 0$ for $u \in R$ and $g(\bar{u}) = 0$ for some $\bar{u} \in R$,

- $(u - \bar{u})\, g''(u - \bar{u}) \geq 0$.

Examples of such a function are linear functions of the form $g(u) = -k(u - \bar{u})$ with $k > 0$.

Under these conditions on the function g, f_γ is a convex function with negative Schwarzian. There always exists an interval $\Gamma(g)$ such that for $\gamma \in \Gamma(g)$ the map f_γ has bounded nondegenerate invariant intervals, the largest of which, denoted by I_γ, contains all these intervals. (Namely, $I_\gamma = [u_0, f_\gamma^{-1}(u_0)]$ and $\Gamma(g) = [\gamma_{min}, \gamma_{max}]$, where $u_0 = u_0(\gamma)$ is the left fixed point of f_γ (i.e., the least of the two roots of the equation $f_\gamma(u) = u$), $f_\gamma^{-1}(u_0)$ is the preimage of u_0, γ_{min} is from the solution (u, γ) of the system $f_\gamma(u) = u$, $f_\gamma'(u) = 1$, and γ_{max} is from the solution (u, γ) of the system $f_\gamma(u) = f_\gamma^{-1}(u_0)$, $f_\gamma'(u) = 0$.) Therefore, the bounded nonconstant solutions of the problem (6), (18) arise if and only if $\varphi \in B(g)$, where

$$B(g) = \left\{\varphi \in C^1([0, 1], I) : \gamma[\varphi] \in \Gamma(h) \text{ and } \varphi(x) \in I_{\gamma[\varphi]} \text{ for } x \in [0, 1]\right\}.$$

Let $\gamma(n)$ denote the lower bound of those γ such that f_γ has a cycle of period n, and let $\gamma^*(2^i) = \inf_{s>0} \gamma(2^i(2s + 1))$.

As known, the following relation holds: $\gamma(n) < \gamma(n')$ for $n \prec n'$.

If, for example, φ_ξ, $\xi \in (\xi_1, \xi_2)$, is a family of functions from $B(g)$ depending continuously on the parameter ξ, and $\gamma_j := \gamma[\varphi_{\xi_j}]$, $j = 1, 2$, then the following useful criteria for solutions of BVP (6), (18) and the associated DE via the corresponding infinite-dimensional DS in C^Δ and $C^\#$ can be given:

If the map f_{γ_1} has no cycles of period n_1 and f_{γ_2} has a cycle of period n_2, then for any n, $n_1 \prec n \prec n_2$,

- *there exists an interval $\Xi_n \subset (\xi_1, \xi_2)$ such that for any $\xi \in \Xi_n$ the ω-limit set $\omega^\Delta[\varphi_\xi]$ is a cycle of period n;*

- *if $n_1 \neq 2^i$, there exists a subset $\Xi' \subset \Xi_n$ of positive Lebesgue measure such that for any $\xi \in \Xi'$ the ω-limit set $\omega^\#[\varphi_\xi]$ is a cycle of period n whose points are random functions.*

We will say that a solution w_φ is asymptotically periodic in t if the corresponding ω-limit set $\omega^\Delta[\varphi]$ is a cycle. Then we can characterize the long-time behavior of solutions of the problem (6), (18) in the following way:

Proposition 4

- *For any $\varphi \in \mathcal{P}_i := \{\varphi \in B(g) : \gamma(2^i) < \gamma[\varphi] < \gamma(2^{i+1})\}$ the solution w_φ is asymptotically 2^i-periodic in t, and for each ζ from the ω-limit set $\omega^\Delta[\varphi]$ the fractal dimension of the graph of ζ equals 1, i.e., it coincides with its topological dimension (the reason is that the function ζ is many-valued on a finite or countable set only).*

- *For almost any $\varphi \in \mathcal{Q}_i := \{\varphi \in B(g) : \gamma^*(2^{i+1}) < \gamma[\varphi] < \gamma^*(2^i)\}$ the solution w_φ is asymptotically periodic in t with the period divisible by 2^i; there exists a constant $C > 0$ such that for any $\varepsilon > 0$ one can find a 2^i-periodic step function $P_{\varphi,\varepsilon}$ and a positive number $\theta(\varphi, \varepsilon) > 0$ such that*

$$\mathrm{mes}\left\{(x,t) : |w_\varphi(x,t) - P_{\varphi,\varepsilon}(x,t)| > C\alpha^{-i} \ \text{for} \ \theta < t < \theta+1\right\} < \varepsilon$$

$$\text{for} \ \theta > \theta(\varphi, \varepsilon).$$

  For any $\varphi \in \mathcal{Q}_i$ (excluding $\varphi \equiv const$) the fractal dimension of the graph of $\zeta \in \omega^\Delta[\varphi]$ is greater than 1.

- *For any $\varphi \in \mathcal{R}_i := \{\varphi \in \mathcal{Q}_i : \text{the map } f_{\gamma[\varphi]} \text{ has a.c.i.m.}\}$ the solution w_φ is asymptotically periodic in t with the period divisible by 2^i and, moreover, the ω-limit set $\omega^\#[\varphi]$ is a cycle whose points are random functions.*

- *The following universal relations hold:*

$$\lim_{i \to \infty} \frac{\nu(\mathcal{P}_i)}{\nu(\mathcal{P}_{i+1})} = \lim_{i \to \infty} \frac{\nu(\mathcal{Q}_i)}{\nu(\mathcal{Q}_{i+1})} = \lim_{i \to \infty} \frac{\nu(\mathcal{R}_i)}{\nu(\mathcal{R}_{i+1})} = \delta,$$

  where ν is the measure on $B(g)$, given by the formula $\nu(\Phi) = \mathrm{mes}\{\gamma : \gamma = \gamma[\varphi], \varphi \in \Phi\}$ for any open (in C^1-topology) set $\Phi \subset B(g)$.

4 Conclusion: ideal turbulence

The term *turbulence*, having first appeared in hydrodynamics, is now used actively in a broader sense for chaotic fluctuations of different characteristics of continuous media.

Many effects that characterize the phenomenon of turbulence, such as the cascade process of birth of coherent structures of decreasing scales and self-stochasticity, can be observed in simple (deterministic but infinite-dimensional) dynamical systems generated by one- and two-dimensional BVP, examples of which were considered above. Such BVP bear no direct relation to hydrodynamics but they occur frequently, for example, in the studies of electromagnetic fields.

In the series of our works, where such BVP are investigated, we use the term *ideal turbulence* (i.e. turbulence in ideal media without any internal resistance) or, which is the same, the term *dry turbulence* (i.e. turbulence in "dry" water from von Neumann's terminology). This term was suggested by the author in 1983 [11].

The excitation of turbulence in BVP is connected with the non-compactness of the trajectories of the corresponding dynamical system. When describing the asymptotic behavior of such a system, we used two metrics allowing us to complete the original space of smooth functions with upper semi-continuous or random functions. With these metrics, a definition of turbulence (mathematically well defined) can be given. For example [3, 7], *we say that an initial state generates strong turbulence if in the phase space completed with upper semi-continuous functions, the trajectory corresponding to this state has an ω-limit set containing at least one upper semi-continuous function whose graph has fractal dimension greater than its topological dimension. And, we say that an initial state generates stochastic turbulence if in the phase space completed with random functions the corresponding ω-limit set contains at least one random function.*

With these definitions, we can give criteria of turbulence for each of the BVP considered above. For instance, for the problem (6), (7), strong turbulence will arise if the map f has a cycle of period $\neq 2^k$, $k \geq 0$, and, moreover, the stochastic turbulence will be generated if this map has a smooth invariant measure. The turbulence arising in this problem is almost independent of an initial state, in contrast with the problem (6), (18) (Example 3). As one can see from Statement 4, for the latter problem the turbulence is determined just by initial data, because the one-dimensional map associated with the problem (and determining the dynamics of solutions) depends on the parameter given by initial data.

* * *

In the BVP considered, the diameters of structures can decrease to zero. This does not agree with the discrete nature of time and space. It is natural to ask: can mathematical models using such BVP describe adequately the behavior of real physical objects (without any additional conditions or assumptions) when time is sufficiently large?

Is it possible to realize a computer investigation of such a BVP and to obtain results which will be coincident with reality? Maybe it is possible to draw some parallel between the dynamics of real objects, if the dynamics is simulated by such kind of BVP (at least, on certain space-time scales), and the dynamics observed in computer experiments.

Questions of this kind are considered in [15, 16, 17].

In particular, we consider situations where the dynamics obtained in numerical investigations via discretization of space and time in these models are probably closer (in a certain sense) to reality than exact solutions of the originals.

* * *

We believe that the reduction of a BVP to a DE (of course, if this is a possibility) has considerable value for the investigation of this BVP; such a reduction allows immediate invocation of the results of the theory of difference equations with continuous argument, which has a number of outstanding distinctions from the theory of difference equations with discrete argument.

References

[1] Ivanov, A.F. and Sharkovsky, A.N., Oscillations in singularly perturbed delay equations, in *Dynamics Reported, New Series* **1**, Springer Verlag, 1991, 165-224.

[2] Romanenko, E.Yu., On chaos in continuous difference equations, in *World Scientific Ser. in Applicable Analysis* **4**, World Scientific, 1995, 617-630.

[3] Romanenko, E.Yu. and Sharkovsky, A.N., From one-dimensional to infinite-dimensional dynamical systems: Ideal turbulence, *Ukrain. Math. J.* **48** (1996), 1604-1627 (in Ukrainian), *Ukrain. Math. J.* **48** (1996), 1817-1842.

[4] Romanenko, E.Yu. and Sharkovsky, A.N., Dynamics of solutions for simplest nonlinear boundary value problems, *Ukrain. Math. J.* **51** (1999), 810–826 (in Russian), *Ukrain. Math. J.* **51** (1999), 907-925.

[5] Romanenko, E.Yu. and Sharkovsky, A.N., From boundary value problems to difference equations: A method of investigation of chaotic vibrations, *Int. J. Bifurcation and Chaos* **9** (1999), 1285-1306.

[6] Romanenko, E.Yu., Sharkovsky, A.N. and Vereikina, M.B., Self-structuring and self-similarity in boundary value problems, *Int. J. Bifurcation and Chaos* **5** (1995), 1407-1418, and in *Thirty years after Sharkovskii's theorem: New perspectives* (Proc. Conf.), World Scientific, **8**, 1995, 145-156.

[7] Romanenko, E.Yu., Sharkovsky, A.N. and Vereikina, M.B., Structural turbulence in boundary value problems, in *Proc. 1st Intern. Conf. Control of Oscillations and Chaos*, St.Petersburg, v.3 , 1997, 492-497.

[8] Romanenko, E.Yu., Sharkovsky, A.N. and Vereikina, M.B., Self-stochasticity in deterministic boundary value problems, *Nonlinear Boundary Value Problems* 9, Inst. Appl. Math.and Mech., NAS of Ukraine, Donetsk, 1999, 174-184.

[9] Sharkovsky, A.N., Coexistence of cycles of continuous transformation of the straight line into itself, *Ukrain. Math. J.* 16(1964), 61-70 (in Russian), and *Int. J. Bifurcation and Chaos* 5 (1995), 1263-1273, and in *"Thirty years after Sharkovskii's theorem: New perspectives"* (Proc. Conference), World Scientific, 1995, 1-11.

[10] Sharkovsky, A.N., On oscillations whose description is given by autonomous difference and differential-difference equations, in *Proc. of VIII ICNO (1978)*, *Academia, Prague* 2, 1979, 1073-1078.

[11] Sharkovsky, A.N., "Dry" turbulence, in *Short Communications. Intern. Congress Mathematicians, 1983*, Warszawa, 1983, 10 (12), p.4.

[12] Sharkovsky, A.N., "Dry" turbulence, in *Nonlinear and Turbulent Processes in Physics* (Proc. Conf., ed. by Sagdeev, R.Z.), Gordon and Breach, 3, 1984, 1621-1626.

[13] Sharkovsky, A.N., Ideal turbulence in an idealized time-delayed Chua's circuit, *Int. J. Bifurcation and Chaos* 4 (1994), 303-309.

[14] Sharkovsky, A.N., Universal phenomena in some infinite-dimensional dynamical systems, *Int. J. Bifurcation and Chaos* 5 (1995), 1419-1425, and in *"Thirty Years after Sharkovskii's Theorem: New Perspectives"* (Proc. Conf.), World Scientific, 1995, 157-164.

[15] Sharkovsky, A.N., Berezovsky, S.A., Computer turbulence, in *Proc. Intern. Conf. Self-Similar Systems (Dubna, 1998)*, Joint Inst. Nuclear Reseach., Dubna, Russia, 1999, 251-259.

[16] Sharkovsky, A.N., Berezovsky, S.A., Transitions of "correct-incorrect" numerical calculations for solutions of some problems. "Phase transitions" in computer turbulence, in *Proc. 2nd Intern. Conf. Control of Oscillations and Chaos (COC'2000)*, St.Petersburg, Russia, 2000, Vol.1, 6-9.

[17] Sharkovsky, A.N., Berezovsky, S.A., Phase transitions in correct-incorrect calculations for some evolution problems, in *Proc. of the European Conf. Iteration Theory (ECIT'2000)* (to appear).

[18] Sharkovsky, A.N., Deregel, Ph. and Chua, L.O., Dry turbulence and period-adding phenomena from a 1-D map with time delay, *Int. J. Bifurcation and Chaos* **5** (1995), 1283-1302.

[19] Sharkovsky, A. N., Maistrenko, Yu. L. and Romanenko, E. Yu., *Difference Equations and Their Applications*, Naukova Dumka, Kiev, 1986 (in Russian), and Kluwer Academic Publisher, Dordrecht, 1993.

[20] Sharkovsky, A.N., Maistrenko, Yu.L. and Romanenko, E.Yu., Asymptotical periodicity of difference equations with continuous time, *Ukrain. Math. J.* **39** (1987), 123-129 (in Russian).

[21] Sharkovsky, A.N., Maistrenko, Yu.L. and Romanenko, E.Yu., Attractors of difference equations and turbulence, in *Plasma theory and nonlinear and turbulent processes in physics (Kiev, 1987)*, World Scientific, Singapore, 1988, 520-536.

[22] Sharkovsky, A.N. and Romanenko, E.Yu., Ideal turbulence: Attractors of deterministic systems may lie in the space of random fields, *Int. J. Bifurcation and Chaos* **2** (1992), 31-36.

[23] Sharkovsky, A.N. and Romanenko, E.Yu., Problems of turbulence theory and iteration theory, in *Proc. of the European Conf. Iteration Theory (ECIT-91)*, World Scientific, Singapore, 1992, 242-252.

[24] Sharkovsky, A.N. and Romanenko, E.Yu., Difference equations and dynamical systems generated by some classes of boundary value problems, in *Proc. Steklov Math.Inst.* (to appear).

[25] Sharkovsky, A.N., Romanenko, E.Yu. and Vereikina, M.B., Self-structuring and self-stochasticity in difference equations and some boundary value problems, in *Proc. Intern. Conf. "Self-similar systems" (Dubna, 1998)*, Joint Inst. Nuclear Reseach., Dubna, Russia, 1999, 237-250.

[26] Sharkovsky, A.N. and Sivak, A.G., Universal phenomena in solution bifurcations of some boundary value problems, *Nonlinear Math. Physics* **1** (1994), 147-157.

Of General Interest

"Real" Analysis Is a Degenerate Case of Discrete Analysis 25
D. Zeilberger

On the Discrete Nature of Physical Laws 35
F. Iavernaro, F. Mazzia and D. Trigiante

"Real" Analysis Is a Degenerate Case of Discrete Analysis

DORON ZEILBERGER [1]

Department of Mathematics, Rutgers University
New Brunswick, NJ 08903, USA
E-mail: zeilberg@math.rutgers.edu

The ICDEA Conferences: An Asymptotically Stable Recurrence

In one of yesterday's invited talks, Gerry Ladas outlined briefly the history of the previous conferences, and how successful they were. I totally agree. But no recursive sequence can exist without *initial conditions*. Hence, special credit and thanks should go to Saber Elaydi, whose initial idea it was, in 1994. Well done, Saber!

The current term in this sequence, $ICDEA_6$, is a huge success, thanks to the efficient and *friendly* organization of Bernd Aulbach and his gang of young assistants.

Discrete Analysis: Yet Another Cinderella Story

There are many ways to divide mathematics into *two-culture* dichotomies. An important one is the Discrete vs. the Continuous. Until almost the end of the 20th century, the continuous culture was dominant, as can be witnessed by notation. An important family of Banach spaces of *continuous* functions is denoted by L^p, with a *capital L*, while their discrete analogs are denoted by the lower-case counterpart l^p. A function of a *continuous* variable is denoted by $f(x)$, where the continuous output, f, is written at the same level as the continuous input x, but if the input is discrete, then the function is given the derogatory name *sequence*, and written a_n, where the continuous output, a, looks down on the discrete input, n.

Indeed, the conventional wisdom, fooled by our misleading "physical intuition", is that the real world is *continuous*, and that discrete models are necessary evils for approximating the "real" world, due to the innate discreteness of the digital computer.

Ironically, the opposite is true. The

REAL REAL WORLDS (PHYSICAL AND MATHEMATICAL) ARE DISCRETE.

---
[1] Supported in part by the NSF

Continuous analysis and geometry are just degenerate approximations to the discrete world, made necessary by the very limited resources of the human intellect. While discrete analysis is conceptually simpler (and truer) than continuous analysis, technically it is (usually) much more difficult. Granted, real geometry and analysis were necessary simplifications to enable humans to make progress in science and mathematics, but now that the *digital* Messiah has arrived, we can start to study discrete math in greater depth, and do *real*, i.e. discrete, analysis.

When we watch a movie we have the *appearance* of continuity, but in fact it consists of a discrete sequence of frames. When we look at a photograph, we have the *semblance* of a continuous image, but it is really a collection of *discrete* pixels. On a more fundamental level, we now know that energy and matter, and probably time and space too, are discrete, as described so charmingly in Professor Trigiante's invited talk given in this conference two days ago.

Don't Worry, the Continuous Heritage is not a Total Waste

I will show later that while the efforts of Cauchy, Weierstrass, Dedekind and many others for a 'rigorous' foundation of analysis were misguided, a lot (and perhaps most) of continuous analysis can be salvaged as a special degenerate case of "discrete symbolic analysis".

My Perhaps Not So Foolish 'April Fool's Jokes'

As Dr. Peter Menacher, the eloquent and erudite *Oberbürgermeister* of Augsburg, said in yesterday's lovely reception at the magnificent (and mathematically tiled!) City Hall, Augsburg has seen many royalties, starting with its namesake, Emperor Augustus. Now each self-respecting king or duke had a *court jester*, also known as the *fool*. Of course, that 'fool' was usually the least foolish person in the whole kingdom, but his position enabled him to get away with much more freedom of speech than any other subject, since it was all 'in jest'.

Analogously, my own best ideas, far surpassing anything in my 'serious' papers, are contained in my annual *April Fool's* jokes, sent to my E-correspondents and posted on my website. This way I can express my 'off the wall' ideas without being considered a crackpot.

For example (2001), the idea of computerizing Tim Gowers's plan for studying the asymptotics of the Ramsey numbers $R(n, n)$, published in **Ekhad and Zeilberger's personal journal**, `http://math.rutgers.edu/~zeilberg/pj.html`, or my idea (1995) for proving the Riemann Hypothesis, also published there. But the most promising idea is in my 1999 'joke', entitled: 'Mathematical Genitalysis: A Powerful New Combinatorial Theory that Obviates Mathematical Analysis', that was also published in the 'Personal Journal'.

The main thrust of that article was the concept of 'symbolic discretization', akin to, but much more powerful than, 'numeric discretization'. I believe that this *crazy* idea has a great potential. But, even more important, it suggests a truly *rigorous* and *honest* foundation for the whole of mathematics.

Towards a FINITE (and hence RIGOROUS) Foundation of Mathematics

(i) The mathematical (and physical) universe is a huge (but FINITE) **DIGITAL** computer.

(ii) The traditional real line is a meaningless concept. Instead the *real* **REAL** 'line', is neither real, nor a line. It is a *discrete* necklace! In other words $R = hZ_p$, where p is a huge and unknowable (but fixed!) prime number, and h is a tiny, but *not* infinitesimal, 'mesh size'.

Hence even the *potential infinity* is a meaningless concept.

Since h is so tiny, and p is so large, and both are unknowable, they should be denoted by symbols, like h and c in physics, and π and e in math. This also explains why traditional real analysis did so well in modeling nature, the same way that Newtonian physics approximated nature so well, as long as you didn't travel too fast or penetrated with too high energy.

It is probably possible to *deconstruct* the whole of traditional mathematics along finitism, but I doubt whether it is worth the effort. Let's just redo a few basic definitions.

The True Derivative

Leibnitz and Newton defined the derivative by

$$Df(x) = \frac{f(x+h) - f(x)}{h} \quad ,$$

where h is *infinitesimal*, whatever that means. Then Cauchy and Weierstrass found a 'rigorous' definition:

$$Df(x) = \lim_{h \to 0} \frac{f(x+h) - f(x)}{h} \quad ,$$

using the notion of *limit*, whatever it means. But the only TRUE definition is

$$Df(x) := \frac{f(x+h) - f(x)}{h} \quad ,$$

where h is the Fundamental mesh size, a Mathematical Universal constant, that unlike Planck's constant, we will never know, but it is very tiny. Since it is so tiny, we keep it as a *symbol*, but remember that it *signifies* a fixed constant.

When Einstein discovered General Relativity he already had the mathematical framework for it, Riemannian Geometry. Luckily, discrete calculus also already exists, but there $D = \Delta_h$. (Speaking of Δ, I love the logo of this conference that is a graphic pun featuring the finite difference operator Δ turned into the Pascal triangle mod 2 fractal.)

Let's recall the

Product Rule:

$$D(fg) = f(Dg) + (Df)g + h(Df)(Dg) \quad,$$

which implies Leibnitz's rule:

$$D^n(fg) = \sum_{i+j+r=n} \frac{n!}{i!j!r!}(D^{i+r}f)(D^{j+r}g)h^r \quad.$$

Integration is *not* a 'limit' of Riemann sums, but rather *is* a Riemann sum:

$$\int_a^b f(x)dx := h \cdot \sum_{i=0}^{(b-a)/h} f(a+ih) \quad.$$

REAL (i.e. discrete) analysis is conceptually simpler than traditional 'real' (continuous) analysis, and of course is much truer. But it is, on the whole, technically more difficult. Hence 'Naked Brain' humans had no choice but to pursue the latter kind.

My First Love: DISCRETE Analytic Functions

These are functions defined on the lattice $hZ + ihZ$, satisfying:

$$\frac{f(m+(1+i)h) - f(m)}{(1+i)h - 0} = \frac{f(m+ih) - f(m+h)}{ih - h} \quad. \qquad (Duffin)$$

In other words, the two "derivatives" along the two diagonals of any unit square $\{m, m+h, m+(1+i)h, m+ih\}$ are the same. Now $(Duffin)$ can be rewritten as

$$\frac{f(m) + f(m+h)}{2} \cdot h + \frac{f(m+h) + f(m+h+ih)}{2} \cdot ih +$$

$$\frac{f(m+h+ih) + f(m+ih)}{2} \cdot (-h) + \frac{f(m+ih) + f(m)}{2} \cdot (-ih) = 0 \quad,$$

which means that the "integral" of an analytic function around any fundamental square is zero, and since the "integral" over any closed simple (discrete) 'curve' is a (finite!) sum of integrals over fundamental squares, we have immediately both Cauchy's and Morera's theorems!

The theory of discrete analytic functions was initiated by Jacqueline Ferrand-Lelong[2]. Dick Duffin made it into a full-fledged theory, and it was further developed by myself (in my Ph.D. thesis), and several others.

The main stumbling block in the further development of the theory of discrete analytic functions is the fact that the property of being discrete-analytic is not preserved under multiplication. But using the discrete Leibnitz rule one can express the derivative of a product, and then the product is "almost analytic". So I am sure that the full arsenal of *continuous* complex analysis can be discretized, but the details might be too complicated for humans.

Continuous Analysis is a DEGENERATE (not LIMITING) case of Discrete Symbolic Analysis

So one should be able to develop a full theory for discrete analysis, with an *arbitrary* mesh-size h. But *now* we declare that h does not represent a specific quantity (the mesh-size) but rather represents itself, i.e., stays as a symbol. Then continuous analysis is just the *degenerate* case $h = 0$ of the full h-theory.

So the following is the only valid definition of the classical derivative

$$f'(x) := \frac{f(x+h) - f(x)}{h}\Big|_{h=0} \quad ,$$

which in Maple® would read: `subs(h=0, simplify((f(x+h)-f(x))/h))`.

For example,

$$(x^2)' = \frac{(x+h)^2 - x^2}{h}\Big|_{h=0} = \frac{2xh + h^2}{h}\Big|_{h=0} = 2x + h\big|_{h=0} = 2x \quad .$$

Another example is

$$(a^x)' = \frac{a^{x+h} - a^x}{h}\Big|_{h=0} = a^x \cdot \frac{a^h - 1}{h}\Big|_{h=0} = a^x \ln a$$

where, *by definition*

$$\ln a := \frac{a^h - 1}{h}\Big|_{h=0} \quad .$$

Using this definition, we can recover all the properties of ln, for example:

$$\ln(ab) = \frac{(ab)^h - 1}{h}\Big|_{h=0} = \frac{((ab)^h - b^h) + (b^h - 1)}{h}\Big|_{h=0} =$$

$$= \frac{b^h(a^h - 1) + (b^h - 1)}{h}\Big|_{h=0} = b^0 \cdot \frac{a^h - 1}{h}\Big|_{h=0} + \frac{b^h - 1}{h}\Big|_{h=0} = \ln a + \ln b \quad .$$

---

[2] A brilliant mathematician. She was the classmate of Roger Apéry and tied with him for first-place, but unlike Apéry, Ferrand was a **tala** (one of those that **von(t a la) messe**).

Neo-Pythagoreanism or: Anaxogoras deserved to be drowned

It is utter nonsense to say that $\sqrt{2}$ is *irrational*, because this presupposes that it exists, as a *number* or *distance*. The truth is that there is no such number or distance. What does exist is the *symbol*, which is just shorthand for an *ideal* object x that satisfies $x^2 = 2$. In Maple® notation: `sqrt(2)=RootOf(x**2=2)`.

The fundamental metric in plane geometry is not *length* but *area*. In the discrete plane $(hZ)^2$, the area of a region is simply the number of lattice points in the interior of that region. So the *true* Pythagorean theorem is not $c^2 = a^2 + b^2$, but rather $c^2 = a^2 + b^2 + O(h)$. The notion of *distance* is usually not defined. What *does* make sense is *distance squared* between point A and point B, which, *by definition*, is the area (i.e. the number of lattice points in the interior) of the square, one of whose sides is AB.

Interface with Numerics: Interval Arithmetics

Whenever one wants to do fully rigorous analytical calculations on the computer, one uses *interval arithmetics*, where one represents a 'real' number by a closed (or open) interval it is known to belong to. While this is done for computational reasons, we can also do it philosophically, and only talk about intervals $[a, b]$, where a and b are *rational* (and hence meaningful) numbers.

The statements $e = 2.718281828...$ and $\pi = 3.14159...$ are *meaningless*, while $e = 2.718281828$, $\pi = 3$, $\pi^2 = 10$, and $\pi = 355/113$, while wrong, are at least *meaningful*. On the other hand $\pi \in [31415/10000, 31416/10000]$ is correct and meaningful.

One has the obvious rules: $[a, b] + [c, d] = [a + c, b + d]$ etc.

Blessed Are The Δifference Equations for They Shall Inherit Math

Project 1: Use difference equations to prove the Riemann Hypothesis.

I believe that the fundamental equations, both theoretically and practically, are *difference* equations rather than *differential* equations. And indeed they are all over mathematics and will become more and more prominent with the advent of computers, both as substitutes to differential equations and for their own sake.

Recall that the prime number theorem $\pi(x) - x/\log(x) = o(x/\log(x))$ is equivalent to $R(x) := \psi(x) - x = o(x)$, where

$$\psi(x) := \sum_{p^m \le x} \log p \quad .$$

Tchebychev proved that, for large x, $A_1 x \le \psi(x) \le A_2 x$, by using the extremely simple *recurrence*:

$$(\log 2) \cdot x + O(log(x)) = \psi(x) - \psi(x/2) + \psi(x/3) - \psi(x/4) + \cdots \quad ,$$

that yields $A_1 = \log 2$ and $A_2 = 2 \log 2$.

This was considerably improved by Tchebychev himself and James Joseph Sylvester, who found other, more complicated, but still linear, recurrences, that brought A_1 up and A_2 down.

The next step was realized by Erdös and Selberg, who combined two simple recursive inequalities for $R(x) := \psi(x) - x$. The first one is linear (in $|R(x)|$):

$$|R(x)| \leq \frac{1}{\log x} \sum_{n \leq x} |R(x/n)| + O(\frac{x \log \log x}{\log x}) \quad,$$

while the second one is quadratic:

$$\sum_{n \leq x} \frac{\log n}{n} R(n) = - \sum_{n \leq x} \frac{1}{n} R(n) R(x/n) + O(x) \quad,$$

from which it follows that $|R(x)| \leq \sigma x$ for $x > x_\sigma$, for every $\sigma > 0$. Hence $R(x) = o(x)$.

Exercise: Find more powerful recurrences (alias *difference equations*) that would imply the stronger statement $R(x) = O(x^{1/2+\epsilon})$, for every $\epsilon > 0$. Collect \$1,000,000.

Project 2: Find a Rigorous Proof of Fermat's Last Theorem

Andrew Wiles's alleged 'proof' of FLT, while a crowning *human* achievement, is not rigorous, since it uses continuous analysis, which is meaningless. I do believe that it is possible to convert it into a rigorous proof in an analogous way to converting a proof in combinatorics that uses *convergent* (and hence meaningless) power series into a proof that uses *formal* (and hence fully kosher) power series. But the end result would be very artificial.

I hope that one of you would be able to find a completely elementary (and hence fully rigorous) proof of FLT using recurrences, possibly along the following lines.

Let's define

$$W(n; a, b, c) := (a^n + b^n - c^n)^2 \quad.$$

W satisfies lots of recurrences, e.g.,

$$\Delta_a^{2n+1} W = 0 \quad.$$

I am almost sure that there exists a *polynomial*, discoverable by computer, with *positive coefficients* such that

$$W(n; a, b, c) = P(W(n; a - 1, b, c), W(n; a, b - 1, c), \ldots, W(n - 1; a, b, c), \ldots)$$

for $n > 3$. Since $W > 0$ for $n = 3$ and $abc > 0$, FLT would follow. Of course it suffices that P is a rational function whose numerator and denominator have positive coefficients.

Analogy:

Theorem (Askey-Gasper, 1977)

Let $A(m, n, k)$ be the Maclaurin coefficients of $R := (1 - x - y - z + 4xyz)^{-1}$, i.e.:

$$\frac{1}{1 - x - y - z + 4xyz} = \sum_{m,n,k \geq 0} A(m, n, k) x^m y^n z^k \quad ,$$

then $A(m, n, k) > 0$ for all $m, n, k \geq 0$.

The obvious recurrence

$$A(m, n, k) = A(m - 1, n, k) + A(m, n - 1, k) + A(m, n, k - 1)$$
$$-4A(m - 1, n - 1, k - 1) \quad ,$$

is *useless* because of the *minus* sign on the right side, but Gillis and Kleeman (1979) came up with another recurrence:

$$mA(m, n, k) = (m+n-k)A(m-1, n, k) + 2(m-n+k-1)A(m-1, n, k-1) \quad ,$$

which immediately implies positivity, by induction, in $m \geq n \geq k \geq 0$, and by symmetry, for all $m, n, k \geq 0$. This recurrence is ingenious, but once found, is completely routine. Just check (or let Maple® do it if you are too lazy) that

$$\frac{\partial}{\partial x}R = (1 + 2z)(x\frac{\partial}{\partial x} - y\frac{\partial}{\partial y} + z\frac{\partial}{\partial z} + 1)R + 2(y\frac{\partial}{\partial y} - z\frac{\partial}{\partial z})R \quad .$$

Now this differential equation satisfied by R is not only routine to *verify*, it is also routine to *discover*, once you tell the computer what to look for. Let's hope that a similar recurrence would be found for FLT.

Philosophical Conclusion

I am not a professional philosopher of mathematics, nor an expert logician or foundationalist, but I think that the philosophy that I am advocating here is called *ultrafinitism*. If I understand it correctly, the ultrafinitists deny the existence of *any* infinite, not even the potential infinity, but their motivation is 'naturalistic'; i.e., they believe in a 'fade-out' phenomenon when you keep counting.

Myself, I don't care so much about the natural world. I am a platonist, and I believe that *finite* integers, *finite* sets of *finite* integers, and all *finite* combinatorial structures have an existence of their own, regardless of humans (or computers). I also believe that *symbols* have an independent existence. What is completely meaningless is any kind of *infinite*, actual or potential.

So I deny even the existence of the Peano axiom that every integer has a successor. Eventually we would get an overflow error in the big computer in the sky, and the sum and product of any two integers is well-defined only if the result is less than p, or, if one wishes, one can compute them modulo p. Since p is so large, this is not a practical problem, since the overflow in our earthly computers comes so much sooner than the overflow errors in the big computer in the sky.

However, one can still have 'general' theorems, provided that they are interpreted correctly. The phrase 'for *all* positive integers' is meaningless. One should replace it by: 'for finite or symbolic integers'. For example, the statement: "$(n + 1)^2 = n^2 + 2n + 1$ holds for all integers" should be replaced by: "$(n + 1)^2 = n^2 + 2n + 1$ holds for finite or symbolic integers n". Similarly, Euclid's statement: 'There are infinitely many primes' is meaningless. What is true is: if $p_1 < p_2 < \ldots < p_r < p$ are the first r finite primes, and if $p_1 p_2 \ldots p_r + 1 < p$, then there exists a prime number q such that $p_r + 1 \leq q \leq p_1 p_2 \ldots p_r + 1$. Also true is: if p_r is the 'symbolic r^{th} prime', then there is a symbolic prime q in the discrete symbolic interval $[p_r + 1, p_1 p_2 \ldots p_r + 1]$.

By hindsight, it is not surprising that there exist undecidable propositions, as meta-proved by Kurt Gödel. Why should they be decidable, being meaningless to begin with? The tiny fraction of first-order statements that are decidable are exactly those for which either the statement itself or its negation happens to be true for *symbolic* integers. A priori, every statement that starts "for every integer n" is completely meaningless.

I hope to expand this line of thought that may be called 'ultrafinite computerism' or 'ansatz-centric formalism' (as opposed to Hilbert's *logocentric* formalism) in the future.

On the Discrete Nature
of Physical Laws

FELICE IAVERNARO, FRANCESCA MAZZIA

Dipartimento di Matematica, Università di Bari
Via Orabona 4, I-70125 Bari, Italy
E-mail: felix@dm.uniba.it, mazzia@dm.uniba.it

and

DONATO TRIGIANTE

Dipartimento di Energetica, Università di Firenze
via C. Lombroso 6/17, I-50134 Firenze, Italy
E-mail: trigiant@cesit1.unifi.it

Abstract We present a few examples of problems where the struggle between the discrete and continuous conception of the physical world is more evident. The matter gives rise to several considerations of different nature, even extraneous both to mathematics and physics. Leaving aside a philosophic discussion, which is appealing but formidable, we state the most obvious, i.e., that the continuous world is too smooth for the needs of applications.

Keywords Motion of the electron, Difference equations, Complex dynamics, Wave equation, Solitons

AMS Subject Classification 65L05, 65L20, 37F45, 35L05, 74J35

1 Introduction

Our starting point is the following remark. It is obvious that a discrete set is strictly contained in a continuous set and then it is an approximation (in the sense of a part) of it. Nevertheless, things completely change if one considers discrete functions, i.e., functions defined on a discrete set of points. The richness of possible behaviors drastically grows with respect to that of continuous functions.

Starting from the last decades of the nineteenth century, many mathematicians (Weierstrass, Dini, Peano, to name a few) defined curves (later on called fractals) which were considered like monsters because their shapes were far from the traditional shapes inherited from Euclidean geometry. But those curves, even with a low degree of smoothness, were still continuous. Discrete functions are not granted by such property. Consequently, the richness of behaviors further increases as the recent flourishing of strange curves (Julia and Mandelbrot sets, etc) shows. By all means, the use of the computer has enormously encouraged such flourishing.

What about the impact of such curves on other scientists? Those whose fields were closely related to mathematics with a long tradition, such as physicists, had a reaction similar to that of mathematicians: the monsters should be kept apart, although, as we shall see, a few individuals (for example Poincaré) realized the need of more general models describing the physical world, for example by considering a discrete time and then discrete equations.

Other sciences, for example biology, whose approach to mathematics is more recent, had a more favorable attitude, certainly because the object of their studies has a more evident discrete nature.

Although firmly rooted in nineteenth century mathematics, numerical analysis has had a great development after the advent of computers. Computers work on discrete quantities and numerical analysis played a central role in transforming continuous objects (almost universally used by mathematicians and physicists) into discrete ones as needed by computers. There is, however, a constraint in this job: the solutions of the discrete equations should have a behavior similar to those of the continuous equations. In other words, numerical analysis is as a train on a track. Of course, the constraint defined by the track is the one said above, i.e., the necessity that continuous and discrete solutions look alike. There is no blame on this: even on the track there was so much to do and the enormous development of scientific calculus gives a clear evidence of the work done.

It is also clear, however, that outside the track, there is a large land whose exploration has started only recently. Numerical analysts were aware of this, but since the land was outside their scopes, they did not pursue analyzing what they considered the land of errors, instabilities, etc. For example, strange solutions of chaotic nature, called *ghost solutions*, are generated by the simple Euler or mid-point methods in ranges of the step-number h different from those usually used by numerical analysts [12, 13].

Is our *forma mentis* which strongly influences us to consider such non-conventional solutions a by-product of mathematics or, on the contrary, does it describe physical reality?

In this paper we shall analyze three examples, taken from physics and other disciplines which seem to show that the land outside the track has its right to exist in the physical world. In other words, it may happen that the continuous description of the real world is only an approximation of it, which indeed could be of discrete nature.

2   The dynamics of the electron

The problem of stating the equations describing the motion of the electron, started just after its discovery (Thomson 1896). It continues to be a problem, as shown by papers which still appear on it. In the review paper [2] Caldirola gave a historical overview of this classical model. A very rich bibliography can be found in the more recent paper [4]. The names of the greatest physicists that have written on it (Poincaré, Lorentz, Abraham, Fermi, Dirac etc.) clearly show that it is a formidable one, whose solution entails the basic ideas of physics.

At the beginning (Abraham, 1903) the electron was considered as a rigid sphere, but very soon the model revealed to be inconsistent with other physical requirements. It was corrected by Lorentz (1904): the spherical shape is maintained for the electron at rest while it contracts with ratio $(1 - \beta^2)^{1/2}$ (where $\beta = v/c$, v, c are the velocity of the particle and of the light, respectively) in the direction of the motion. This model was more successful, although, few years later, it was considered not in agreement with the requests of the newly born special relativity theory. In the simplest form (when $\beta << 1$), the equation predicted by the model for the dynamics of the electron, having radius R and charge e, under the action of an external force F, is

$$m_0 \frac{dv}{dt} = F + \Gamma_0, \tag{1}$$

where m_0 is the rest mass of the electron, given by

$$m_0 = \frac{2}{3} \frac{e^2}{Rc}.$$

An extra force Γ_0 appears here for the first time. Under many forms it was destined to be a source of troubles for this and all the subsequent models. The form that it assumes in the Lorentz model is

$$\Gamma_0 = \frac{2}{3} \frac{e^2}{c} \frac{d^2v}{dt^2} + \sigma_1 R + \sigma_2 R^2 + \dots$$

where the coefficients σ_i contain higher order derivatives of the velocity. Few comments are needed.

- The fact that it depends on higher order derivatives of the velocity makes such an equation extraneous to the tradition of physics, either Newtonian or relativistic and even quantistic, where the equations are always of second order. As a matter of fact, it implies that the motion of the electron, to be completely determined, would need not only the traditional quantities such as the initial position and the initial velocity, but also, for example, the initial acceleration.

- If one tried to avoid this by letting $R \to 0$ (point electron), then the rest mass m_0 would diverge.

Many efforts were made to give a meaning to Γ_0. It was said that it is the effect of the radiation of the electron due to the self-generated electromagnetic field.

Few years later, the Lorentz equation (1) was abandoned because it is not relativistic invariant.

An improved model, still non relativistic was provided later by Schott (1912) and Page (1918). We report their equation since it is interesting for our reasoning. In short notation, i.e., after summing the series contained in their equations, it can be written as

$$\frac{m_0 c}{2R} \left(1 - e^{-\frac{2R}{c}\frac{d}{dt}} \right) v(t) = F. \tag{2}$$

The above equation still contains derivatives of any order. It is interesting, however, to observe that the exponential is nothing but the reverse shift operator, well known in discrete mathematics. In fact, by posing

$$h = \frac{2R}{c}, \tag{3}$$

the reverse shift operator is defined by

$$E_h^{-1} v(t) = v(t - h)$$

and satisfies

$$E_h = e^{\frac{2R}{c}\frac{d}{dt}}.$$

With this notation, the above equation (2) could be written as

$$\frac{m_0 c}{2R} \left(v(t) - v(t - h) \right) = F, \tag{4}$$

showing the essential discrete nature of the equation. More importantly, the terms containing higher order derivatives of the velocity have disappeared. But accepting the discrete equation would imply the introduction of a new entity in the physical world, i.e., a sort of atom of time. This was eventually done many decades later. Before that, for a long time, physicists were concerned about the meaning to give to terms in the expansion of $e^{-\frac{2R}{c}\frac{d}{dt}}$. Especially the third term, the one containing the second derivative of the velocity, which survived in the following model, due to Dirac, was object of intensive studies in the three successive decades, starting from the forties. As already said, it was considered the effect of the radiation of the moving electron.

The model provided by Dirac, in the fundamental work of 1938, in fact, while it solves the problem of diverging of the rest mass, still contains the radiating term. The equation, when $\beta << 1$, is

$$m\frac{dv}{dt} = F + \frac{2}{3}\frac{e^2}{c^3}\frac{d^2 v}{dt^2}.$$

This equation is similar to the one obtained by Lorentz, except that this time m remains finite even though it is the difference between two diverging terms.

This equation has been extensively studied, although its solutions are often in contrast with both common sense and experimental physical evidence.

We report here two of the most paradoxical behaviors of the solutions in two simple cases.

a) If the external force is due to a positive electric charge, for example a proton, then the solution requires that the electron approaches it until a non zero minimum distance, then stops and comes back with increasing velocity and acceleration. [1]

b) If the external force is a pulse, then the solution prescribes that the electron starts to accelerate before the pulse is active. It is obvious that this contradicts the causality principle.

2.1   The chronon

In the fifties, Caldirola began to consider the discrete nature of the equation. He gave to the quantity h defined by (3) and having the dimension of time, the name *chronon*. Moreover, he assumed that

1. the chronon is a universal elementary time;

2. a force, acting upon the electron at a given instant τ of the proper time, induces a sudden transition from the state at time $\tau - 2h$ to the state at time τ;

3. for $h \to 0$ the new equations must recover those of classical dynamics;

4. the laws must be relativistic invariant.

He obtained a difference equation which reduces to the Dirac one when $h \to 0$. The space component of the obtained four dimensional equation in the non relativistic approximation, i.e., $\beta << 1$, is

$$\frac{m_0}{2h} \left(v(t) - v(t - 2h) \right) = F, \tag{5}$$

essentially similar to (4).

Suppose now we are in a position of having accepted the discrete solutions described by the previous equation and try to approximate it by a continuous one. The first thing to do is to expand the term $v(t - 2h)$ in Taylor series. The second term would be the one containing the second derivative of the velocity, i.e., the radiating term.

Of course the introduction of the elementary time cannot be taken thoughtlessly, although it has been invoked by many important physicists. Starting from Mach and Poincaré (and even before), the idea has shown up several times in the last century. A more recent physicist who has used the discrete time is the Nobel laureate G. 't Hooft [5, 6].

---

[1]The beautiful discussion of this case is due to G. Zin. One of us (D.T.) was introduced to this problem by him; his unique personality, both as a mathematician and in social life, was outstanding.

2.2   Overlapping with numerical analysis

Returning to the discrete equation of the electron, the most interesting part for us, numerical analysts, is the sequel of the story.

Once the discrete equation for the velocity has been accepted, one has to define the equation relating the position $r(t)$ to $v(t)$ (Caldirola called such equation *transmission law*). The continuous relation is, obviously, $v = dr/dt$. On the contrary, the discrete equations are not univocal, as we know very well. In our terminology, one may use explicit Euler, implicit Euler, the trapezoidal rule, the mid-point rule, etc.

The only difference between us and physicists, when facing the problem of choosing among such discretizations, is that we have in mind the continuous problem and we try to mimic its solutions. They must be guided by the physical implications that a choice would imply. The final results do not seem to be different. Caldirola chose the trapezoidal rule, i.e.,

$$r(t) - r(t - 2h) = h \left(v(t) + v(t - 2h) \right),$$

saying that this is the most stable choice. In fact we know that the trapezoidal rule is perfectly stable for fixed h; i.e., his *boundary locus* is exactly the imaginary axis (see, for example, [1]).

In other cases, for example in the case of a radiating electron, Caldirola proposed the use of what we call explicit Euler while in the case of absorbing electron the implicit Euler is used instead. He called such equations the retarded and the advanced forms, respectively. It is clear why. The explicit Euler method uses information from the last point (past), while the implicit one uses information from the new point (future).

It is also clear why they describe radiating and absorbing particles. A conservative system should have the Jacobian of the Hamiltonian function with imaginary eigenvalues. When, for example, the explicit Euler is used, the imaginary axis is outside the absolute stability region of the method and then the electron seems to lose energy (radiation). The contrary would happen when using the implicit Euler.

At this point the question becomes of physical nature: is the radiation (or absorption) a physical necessity? If yes, then no objection can be raised in using such non symmetrical methods. If not, then one more condition, for example of conservative type, should be added in deriving the equations.

For the non radiating, non absorbing electron, Caldirola also used the midpoint rule, i.e.,

$$r((n + 1)2h) - r((n - 1)2h) = 4hv(2nh).$$

No objection until the external force is not rapidly varying. On the contrary, when this is the case (in our terminology the *stiff* case), the nature of the problem may significantly change. We know, in fact, that the absolute stability region of such method is only a small segment on the imaginary axis. If the eigenvalues of the Jacobian multiplied by h went outside such segment, closed orbits could open and the electron would run away, just as in the case of the field generated by the proton discussed for the Dirac equation. A reminiscence of this could be the fact, as asserted in [4], that the discrete approach implies the existence of an upper limit for the eigenvalues of the Hamiltonian. Again, is this a real physical requirement, or it is only due to the chosen discretization?

There are other questions which look familiar to us, for example the question of layer solutions when the force strongly varies within an interval of time of order h, which reminds us of the stiffness nature of the problem. But we do not pursue this matter further.

Apart from the above consideration, which would suggest a more strict collaboration among physicists and discrete mathematicians, the new formulation does not suffer from the troubles of the continuous equations. In particular, the motion of the electron is uniquely defined by its initial position and velocity. In a certain sense this is more comfortable than believing to be obliged to know also its initial acceleration or, even worse, all the initial derivatives.

3 The Verhulst equation

The continuous Verhulst equation

$$p'(t) = rp(t)\left(1 - p(t)\right), \qquad r > 0, \tag{6}$$

describes the evolution of a biological species in an environment with limited resources. It is just a generalization of the Malthus law, for which the resources were considered unlimited. Its solutions, although clearly showing the experimental phenomenon of saturation, are nothing at all special, from the mathematical point of view.

On the contrary, its discrete twin

$$p_{n+1} = p_n + rp_n(1 - p_n), \tag{7}$$

has literally created a revolution in many sciences. It introduced, in the seventies, the concept of chaotic behavior. It is known that such behavior is defined by the existence, in the solution, of infinitely many periodic motions.

Few years before, most probably, such kind of solutions would have been placed in the lumber-room of other unconventional solutions. But the demand for applications had made the time ready to accept such solutions. In fact, already in 1964 (see [8]), Lorenz was studying such a discrete equation and its periodic solution in relation to climate evolution. The same equation arrived a few years later when May was studying biological problems (see [9, 10]).

This equation has become so popular that it can be considered as one of the superstar equations and hence its fame has gone beyond the limits of specialized literature. Here we use it to reaffirm once again the *leitmotif* of the present discussion; that is, approximating the continuum by means of the discrete may significantly change the nature of a problem if one is well-disposed to move his mind beyond its track.

Suppose that (6) is not analytically solvable and we (numerical analysts) are asked to simulate the continuous dynamical system by means of a discrete one. This is the analogue of what happens when one numerically solves the linear test problem $y' = \lambda y$, $\lambda < 0$ and finds conditions (A-stability) to retain, inside the discrete solution, the qualitative asymptotic behavior of the continuous one. Indeed, the analytic solution of the continuous Verhulst equation clearly shows the global asymptotic stability nature of its equilibrium $\bar{p} = 1$, no matter how the initial condition p_0 or the growth parameter $r > 0$ are chosen; we would like to reproduce this behavior after the application of a suitable method. With this in mind, we discretize our nonlinear problem (6) using explicit and implicit Euler obtaining, respectively,

$$p_{n+1} = p_n + hrp_n(1 - p_n), \tag{8}$$

and

$$p_{n+1} = p_n + hrp_{n+1}(1 - p_{n+1}), \tag{9}$$

where $h > 0$ is the step-length. Setting $\lambda = hr$ we see that (8) is nothing but the discrete logistic model (7) (in passing observe that in this case the equation is not always explicitly solvable), while (9) defines an implicit map. We observe that $\bar{p} = 1$ remains an equilibrium point of both (8) and (9) and we are interested in studying its stability character. By linearization, one immediately realizes that while no constraint on λ must be imposed in (9) in order to get asymptotic stability, the condition $\lambda \in [0, 2]$ is to be fulfilled in case we decide to adopt equation (8). Hence, as it happens for the linear test problem, we are inclined to suggest the use of implicit Euler because of its independent qualitative behavior with respect to changes in the parameter λ. The solutions of the simpler model represented by equation (7) have a great significance in biology also when $\lambda > 2$ and in particular under chaotic regime.

The differences between the behaviors of the solutions of (8) and (9) are better emphasized in the frame of complex dynamics. The parameter r (and hence λ) and p_0 are now allowed to assume complex values and, consequently, the difference equations are viewed as maps $\mathbb{C} \to \mathbb{C}$. Again, if we put aside for a while our final goal to mimic the solutions of the continuous model (6) (that is, we avoid the reasoning of a numerical analyst and look outside the track), the most interesting and beautiful results concerning the dynamics of discrete solutions are obtained when those maps come from explicit methods. For example, it is well known that (8) is analytically conjugate to the map $z_{n+1} = z_n^2 + c$ and therefore its stability region

$$\mathcal{D} = \{\lambda \in \mathbb{C} \mid p_n \text{ is bounded}\}$$

is the renowned Mandelbrot set (p_0 is chosen as the critical point of the map). More generally it is often the case that $\mathcal{D}$ turns out to be a fractal when explicit methods are used to discretize the equation (6).

Example 3.1 *The two-steps Adams-Bashforth method applied to (6) takes the form*

$$p_{n+1} = (1 - \frac{\lambda}{2})p_{n-1} + \frac{3}{2}\lambda p_n + \frac{\lambda}{2}p_{n-1}^2 - \frac{3}{2}\lambda p_n^2. \tag{10}$$

Its solution now depends on the two initial conditions p_0 and p_1; this means that (10) defines a map $\mathbb{C}^2 \to \mathbb{C}^2$. Figure 1 reports its stability region ($[p_0, p_1]$ is again the critical point of the map). Some details of the fractal shape of $\mathcal{D}$ are visible in Figure 2: contrary to what happens for the quadratic map (9), the set $\mathcal{D}$ appears to be not connected.

Figure 1: The stability region $\mathcal{D}$ for the two-steps Adams-Bashforth method applied to the Verhulst equation

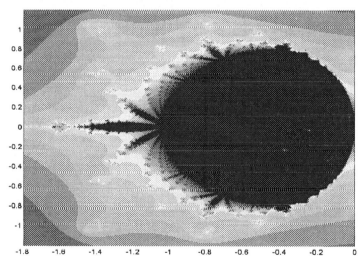

Once again we realize that the discrete world is far reacher than the continuous one: it is the physical interpretation of the problem that shall indicate which mathematical model is preferable. As an example, the study of discrete dynamical systems possessing chaotic solutions has become usual in many application fields (biology, engineering, medicine, etc.)

4   Solitons and the discrete wave equation

The ghost we are dealing with in this section did not appear as outcome of a computer or as a solution of a mathematical equation. It first appeared in 1834

Figure 2: Magnification of a region close the boundary of $\mathcal{D}$

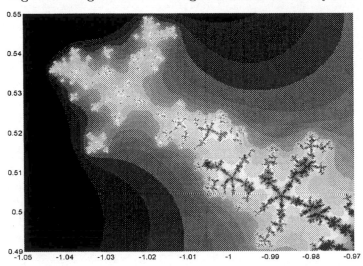

on a river and it had the privilege to be literally chased. The description of this event is unique in histories of both mathematics and physics and deserves to be reported. The writer is J. Scott Russel:

> I was observing the motion of a boat which was rapidly drawn along a narrow channel by a pair of horses, when the boat suddenly stopped, not so the mass of water in the channel which it had put in motion, it accumulated round the prow of the vessel in a state of violent agitation, then suddenly leaving it behind, rolled forward with great velocity, assuming the form of a large solitary elevation, a rounded, smooth and well-defined heap of water, which continued its course along the channel apparently without change of form or diminution of speed. I followed it on horseback...

Needless to say, Russel's witness faced great hostility and skepticism from the scientific community (see [11]).

As outcome of a computer it appeared only one century later when three eminent physicists, E. Fermi, J. Pasta and S. Ulam, decided to explain heat diffusion by means of elementary interactions in a solid body. They discretized the space variable of the wave equation, leaving the time continuous. By the way, we note that such a technique became popular later on and assumed the name of *method of lines*. More precisely they introduced a nonlinear term involving the space variables, to give an account, by means of a dynamical system from mechanics, for an irreversible macroscopic process described by the Fourier law. Then they simulated, with the help of a computer, the obtained equations. We quote from Newell the description of the result.

Now a great surprise was encountered (...). The energy did not thermalize! In fact, after being initially contained in the lowest mode and then flowing back and forth among several low-order modes, the energy eventually recollected into the lowest modes (...) and then the process repeated itself. (...) The system seems to behave like a system of linearly coupled harmonic oscillators whose motion on a torus is quasiperiodic. But how could this be? Why didn't the nonlinearity excite all the Fourier modes?

The following comment by Newell is also interesting:

The FPU experiment had failed to produce the expected result and indeed the results it did produce challenged (...) the basic thinking of physicists of the day. *Nevertheless, since it was not connected with what was regarded at that time as frontier physics,(...) it could easily have been dismissed, as it was by many, as an anomalous curiosity. (...) Fortunately the curious result of the FPU experiment was not ignored by all.*

The sequel of the story is also interesting. The solitary waves, called *solitons*, have been accepted as an important stone of the modern physical world. But still, the discrete nature of the equations which had shown them for the first time were, in a certain sense, rejected. In fact Kruskal and Zabusky did, in the reverse way, what Caldirola had done for the Dirac equation (or what numerical analysts usually do): they found a continuous equation, just expanding in Taylor series the discrete quantities of the FPU model. They imposed that the continuous solutions maintained similar properties as the discrete ones.

As for the Dirac equation, such continuous newborn has order higher than two (actually it is fourth order).

4.1 The complete discrete wave equation

In ([7]) the existence of soliton-like solutions was stated for the linear wave equation discretized with respect both the space and the time. The starting point was the study of the behavior of the discrete harmonic oscillator

$$y_{n+1} - 2\gamma y_n + y_{n-1} = 0 \tag{11}$$

where $\gamma = 1 - \omega^2 \tau^2/2$, with ω^2 the coefficient of the linear spring and τ the elementary discrete time. The obtained solutions are very different from the well-known continuous ones. For example it is easily deduced that the existence of bounded solutions requires that $\gamma \in [-1, 1]$. In such a case the solution of (11) is

$$y_n = A\sin(\alpha n\tau + \vartheta); \qquad \cos\alpha\tau = \gamma.$$

Without going into details, we summarize the main differences.

a) Contrary to what happens in the continuous model, the solutions of (11) are in general non periodic except for suitable (in most cases irrational) values of the constant γ which form a dense subset of $[-1, 1]$. This implies that periodic solutions are generated only when γ is known with infinite precision. If the precision is finite, then the motion will be non periodic and will mimic a combination of periodic solutions with many different periods (chaotic regime).

b) If one defines a probability density function both for the discrete and the continuous solutions, then such density functions turn out to be the same. In other words, even if the solutions are very different in the two cases, they have the same distribution.

c) In the continuous case there is a one-to-one correspondence between the minimum period and the frequency. In the discrete case, to a fixed period, say $M\tau$ (we suppose M even for simplicity), there correspond $M/2-1$ values of the frequency, namely $\alpha = \frac{2s\pi}{M\tau}$, $s = 1, \ldots, M/2 - 1$.

Point *c)* is crucial for an in-depth analysis of the discrete wave equation

$$u_{n,j+1} - 2u_{n,j} + u_{n,j-1} = q^2(u_{n+1,j} - 2u_{n,j} + u_{n-1,j}), \qquad (12)$$

because it states the possibility of combining periodic solutions with the same period but different frequencies to generate a localized pulse that travels without dispersion in a generally dispersive field. The positive constant q depends on the properties of the medium and contains information of the discrete time unit τ and the space elementary quantity, say σ. As for the constant γ, solutions that have a physical meaning are obtained when $q \in [0, 1]$.

Of course, as in the continuous case, the solutions of (12) can be studied by expanding in linear waves and are therefore linear combinations of plane waves of the form $Ae^{i(\pm kn\sigma \pm \omega j\tau)}$, where n, j are integers while k is the wavelength and ω the frequency. The phase and group velocities are, respectively,

$$\begin{cases} u_{ph} = \dfrac{\sigma}{\tau} \left| \dfrac{\sin^{-1}(q\sin(k\sigma/2))}{k\sigma/2} \right|, \\ u_{gr} = \dfrac{\sigma}{\tau} q \cos k\sigma/2 \left(1 - q^2 \sin^2(k\sigma/2)\right)^{-1/2}, \end{cases}$$

and one gets $u_{ph} = u_{gr} = \sigma/\tau$, for $q = 1$. This limit case is interesting in many respects. As $q \to 1$, all the waves assume the same velocity σ/τ, which turns out to be the maximum allowed in that medium (the existence of a finite limit speed is a direct consequence of the discrete nature of the model).

By continuity when $q \simeq 1$ the dispersion phenomenon is almost absent and the solutions possess the typical features of solitons. For example, periodic (when $q = 1$) or quasi-periodic (when $q \simeq 1$) solutions may be linearly combined to generate a localized pulse that travels, maintaining its shape unchanged (or almost unchanged, see Figure 3) even after a collision with another pulse (Figure 4).

Figure 3: A pulse (soliton-like solution) traveling in a field with a small dispersion factor ($q = .99$).

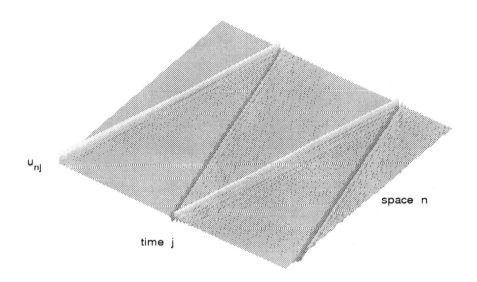

References

[1] Brugnano, L. and Trigiante, D., *Solving ODEs by Linear Multistep Initial and Boundary Value Methods*, Gordon & Breach, Amsterdam, (1998).

[2] Caldirola, P., A relativistic theory of the classical electron, *La Rivista del Nuovo Cimento* **13** (1979).

[3] Dirac, P. M., Classical theory of radiating electron, *Proc. Roy. Soc.* 167A **148** (1938).

[4] Farias, R. H. A. and Recami, E., Introduction of a quantum of time (chronon), and its consequences for quantum mechanics, available at URL *http://www.arXiv.org/abs/quant-ph/9706059*.

[5] 't Hooft, G., Quantization in space and time in 3 and in 4 space-time dimensions, lectures given at NATO Advanced Study Institute. In Cargese 1996, Quantum Fields and Quantum Space Time, available at URL *http://www.arXiv.org/abs/gr-qc/9608037*.

[6] 't Hooft, G., Quantization of point particles in 2+1 dimensional gravity and space-time discreteness, available at URL *http://www.arXiv.org/abs/gr-qc/9601014*.

Figure 4: Collision between two pulses in absence of dispersion ($q = 1$).

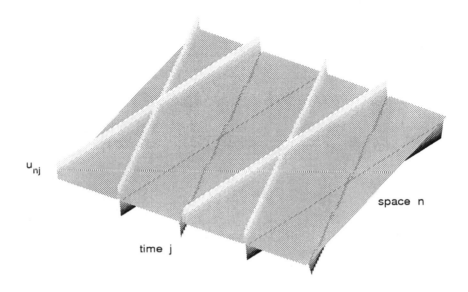

[7] Iavernaro, F. and Trigiante, D., Soliton-like solutions of the discrete wave equation, in *Proc. First World Congr. Nonlinear Analysts*, (Eds. Lakshmikantham, V.), Walter de Gruyter-Berlin-New York 1996, 1193-1202.

[8] Lorenz, E. N., The problem of deducing the climate from the governing equations, *Tellus* **XVI** (1964), 1-11.

[9] May, R. M., Biological populations with non overlapping generations. Stable points, Stable Cycles and Chaos, *Science* **186** (1974), 546-647.

[10] May, R. M., Simple mathematical models with complicated dynamics, *Nature* **261** (1976), 469-467.

[11] Newell, A. C., Solitons in Mathematics and Physics, *CBMS 48*, SIAM, Philadelphia, (1985).

[12] Yamaguti, M. and Matano, H., Euler's finite difference scheme and chaos, *Proc. Japan Acad.* **55** Ser. A (1979), 78-80.

[13] Yamaguti, M. and Ushiki, S., Chaos in numerical analysis of ordinary differential equations, *Physics 3D* (1981), 618-626.

[14] Zin, G., Su alcune questioni di elettrodinamica classica relative al moto di un elettrone, *Nuovo Cim. VI* **3** (1949).

Discrete Dynamical Systems

Linear Self-Assemblies: Equilibria, Entropy and Convergence Rates 51
L. Adleman, Q. Cheng, A. Goel, M.-D. Huang and H. Wasserman

Synchronization in a Discrete Circular Network 61
S. S. Cheng, C.-J. Tian and M. Gil'

Bifurcation of Periodic Points in Reversible Diffeomorphisms 75
M.-C. Ciocci and A. Vanderbauwhede

Evolution of the Global Behavior of a Class of Difference Equations 95
A. S. Clark and E. S. Thomas

A Survey of Exponential Dynamics 105
R. L. Devaney

Farey's Rule for Bifurcations of Periodic Trajectories
in a Class of Multi-Valued Interval Maps 123
V. V. Fedorenko

Combinatorics of Angle-Doubling: Translation Principles
for the Mandelbrot Set 131
K. Keller

The Inflation and Perturbation of Nonautonomous Difference
Equations and Their Pullback Attractors 139
P. Kloeden and V. Kozyakin

A Short Introduction to Asynchronous Systems 153
V. Kozyakin

A Local-Global Stability Principle for Discrete Systems and
Difference Equations 167
U. Krause

Stability Implications of Bendixson Conditions for
Difference Equations 181
C. C. McCluskey and J. S. Muldowney

Optimal Topological Chaos in Dynamic Economies 189
K. Nishimura, T. Shigoka and M. Yano

Dynamics of the Tangent Map 199
H. Oliveira and J. S. Ramos

Thresholds, Mode Switching and Complex Dynamics 207
H. Sedaghat

Computation of Nonautonomous Invariant Manifolds 215
S. Siegmund

Linear Self-Assemblies: Equilibria, Entropy and Convergence Rates

LEONARD ADLEMAN [1]

Departments of Computer Science and Molecular Biology
University of Southern California (USC)
Los Angeles, California 90089, USA
Email: adleman@usc.edu

Q. CHENG [2] and ASHISH GOEL

Department of Computer Science, USC
Emails: qcheng@cs.usc.edu, agoel@cs.usc.edu

and

MING-DEH HUANG [3] and HAL WASSERMAN

Department of Computer Science, USC
E-mails: huang@usc.edu, halwass@aol.com

Abstract Self-assembly is a ubiquitous process by which objects autonomous-ly assemble into complexes. In the context of computation, self-assembly is important to both DNA computing and amorphous computing. Thus a well developed mathematical theory of self-assembly will be useful in these and other domains. As a first problem in self-assembly, we will explore the creation of linear polymers. Polymers are chains of molecular units. For examples, a molecule of DNA is a polymer made from the bases adenine, guanine, cytosine and thymine; and proteins are polymers formed from the twenty amino acids. In our model the dynamics of a linear polymerization system is determined by difference and differential equations with initial conditions, and we investigate the equilibrium behavior of such a system. A good characterization of the equilibria of these systems will allow us to predict which computations can be carried out with adequate yields and what quantities of initial substrates these computations require. The rates at which such systems approach equilibria

---

[1] Research supported by grants from NASA/JPL, NSF, ONR, and DARPA.
[2] Research supported by NSF Grant CCR-9820778.
[3] Research supported by NSF Grant CCR-9820778.

are equally important since convergence rates are an obvious measure of the time complexity of computation through self-assembly.

Keywords Self-assembly, Linear polymerization, Difference equations, Differential equations

AMS Subject Classification 39A11, 68Q05, 92B05

1   Introduction

Self-assembly is a ubiquitous process by which objects autonomously assemble into complexes. Atoms react to form molecules; molecules react to form crystals; cells coalesce to form organisms. In the context of computation, self-assembly is important to both DNA computing [1, 7, 8, 10, 11] and amorphous computing [3]. Thus we believe that a well developed mathematical theory of self-assembly will be useful in a variety of domains.

In this paper we study reversible tile systems and ask the following basic questions: Do reversible self-assembling systems achieve equilibria? Do there exist good characterizations of these equilibria? What is the number of steps required for these self-assemblies to converge to equilibria? These questions were first asked by Adleman [2] and partially answered in the context of what he called the irreversible n-linear polymerization problem.

Our motivations for studying the equilibria and convergence rates of reversible systems are manifold. Since self-assembly is universal, a good characterization of the equilibria of these systems will allow us to predict which computations can be carried out with adequate yields and what quantities of initial substrates these computations require. Convergence rates are an obvious measure of the time complexity of computation through self-assembly.

As a first problem in self-assembly, we will explore the creation of linear polymers. We model linear polymerization systems as linear tile systems, in which the tiles are assumed to lie on a line. In our model the dynamics of a linear polymerization system is determined by difference and differential equations with initial conditions, and we investigate the equilibrium behavior of such a system. Motivated by thermodynamics and physical chemistry [5], we define the *profit* of the system as the sum of a term related to information-theoretic entropy and a term related to bond stability. The latter term is motivated by the definition of free energy in chemical reactions. We show that profit is maximized at equilibrium and prove a weak equivalent of the second law of thermodynamics. We conclude with a study of an important class of linear self-assembling systems: the reversible and irreversible cases of n-linear polymerization. For these systems, we give tight bounds on the number of steps required to reach distance ε from equilibrium, showing in most cases that the distance from equilibrium decays exponentially over time.

2   Linear self-assemblies

We assume that there are n types of tiles T_1, ..., T_n, arranged on the doubly infinite line, with α_i being the relative fraction of tiles of type i, $\alpha_i > 0$, $\sum_i \alpha_i =$. Some of these tiles may be bonded to their neighbors. By a slight abuse of notation, we will use T_i as well as i to refer to the type of a tile. We are also given (sticking, breaking) probabilities (σ_{ij}, τ_{ij}), $1 \le i \le n, 1 \le j \le n$, $0 \le \sigma_{ij} < 1, 0 < \tau_{ij} \le 1$.

Let r_{ij} denote the fraction of T_i tiles that are bonded to a T_j tile on the right. Let l_{ij} be the fraction of T_j tiles that are bonded to a T_i tile on the left. Thus, r_{ij}, l_{ij} change with time and are initially 0. Define r_i to be the fraction of tiles of type T_i that are unbonded on the right, and similarly define l_j to be the fraction of tiles of type T_j that are unbonded on the left. Clearly[4], $r_i + \sum_j r_{ij} = 1$, and $l_j + \sum_i l_{ij} - 1$.

Define $R_{ij} = \alpha_i r_{ij}$, $R_i = \alpha_i r_i$, $L_{ij} = \alpha_j l_{ij}$, and $L_j = \alpha_j l_j$. Clearly, $\sum_j R_{ij} + R_i = \alpha_i$, and $\sum_i L_{ij} + L_j = \alpha_j$. R_{ij} can be interpreted as the fraction of neighboring pairs of tiles such that the left tile is of type T_i, the right tile is of type T_j, and there is a bond between the two. R_i is then the fraction of tiles which are free to the right and are of type T_i. It follows that $R_{ij} = L_{ij}$ and that $\sum_i R_i = \sum_j L_j$. We define $F = \sum_i R_i$; F denotes the fraction of tiles that are unbonded to their right (or equivalently, unbonded to their left).

As a result of a toss, an existing bond between tile T_x and T_y breaks with probability τ_{xy}. Thus a fraction τ_{xy} of bonds of type $T_x T_y$ break as a result of a toss.

Now consider the formation of new bonds. Concentrate on a specific tile of type T_x that is unbonded to its right before the toss. After the toss, the right surface of such a tile comes in contact with a previously unbonded left surface of a tile of type T_y with probability L_y/F; these two surfaces then form a bond with probability σ_{xy}. Thus, a fraction $(L_y/F) \cdot \sigma_{xy}$ of free right surfaces of tiles of type T_x form a bond with a tile of type T_y as a result of a toss.

Therefore the update rule is

$$\Delta R_{xy} = -R_{xy} \cdot \tau_{xy} + R_x \cdot (L_y/F) \cdot \sigma_{xy}.$$

If we assume continuous-time model, then

$$\dot{R}_{xy} = -R_{xy} \cdot \tau_{xy} + R_x \cdot (L_y/F) \cdot \sigma_{xy}. \tag{1}$$

In either case, at (bond) equilibrium, $R_{xy} \cdot \tau_{xy} = R_x \cdot (L_y/F) \cdot \sigma_{xy}$, for all x, y.

Note that F, R_x, and L_y all are (linear) functions in R_{xy}. Hence the set of equlibrium conditions is a set of polynomial conditions on R_{xy}. The set

---

[4]All summation indices range from 1 to n, unless stated to the contrary.

of R_{xy} which satisfies all the equilibrium conditions, where $0 \leq R_{xy} \leq 1$, is called the *equilibrium set*.

In the continuous model, the set of differential equations in (1) defines a dynamical system on the convex region defined by $0 \leq R_{xy} \leq 1$ for all x, y, and $0 \leq \sum_y R_{xy} \leq \alpha_x$ for all x.

Note also that in formulating the update rule, we implicitly assume that what happens to the right of a tile is independent of what happens to its left. We refer to this as the *independence assumption*.

A supertile is a maximal contiguous set of tiles that are bonded together. For a supertile Γ, $\{\Gamma\}$ denotes the fraction of supertiles of type Γ. Consider a supertile $\Gamma = T_{i_1} T_{i_2} T_{i_3} \ldots T_{i_k}$. Given independence, $\{\Gamma\} = (L_{i_1}/F) \cdot r_{i_1,i_2} \cdot r_{i_2,i_3} \ldots r_{i_{k-1},i_k} \cdot r_k$. Let T_i denote the last tile in Γ_1 and T_j denote the first tile in Γ_2. Given independence and assuming bond equilibrium, we have

$$\frac{\{\Gamma_1 \Gamma_2\}}{\{\Gamma_1\}\{\Gamma_2\}} = \frac{r_{ij} F}{r_i L_j} = \frac{R_{ij} F}{R_i L_j} = \frac{\sigma_{ij}}{\tau_{ij}}.$$

We say that the system is at *strong equilibrium* if for all supertiles Γ_1 and Γ_2, $\frac{\{\Gamma_1 \Gamma_2\}}{\{\Gamma_1\}\{\Gamma_2\}} = \frac{\sigma_{ij}}{\tau_{ij}}$, where T_i is the last tile in Γ_1 and T_j is the first tile in Γ_2. Therefore

Theorem 2.1 *Independence and bond equilibrium imply the system is at strong equilibrium.*

2.1   Total entropy density

Let $\Gamma(t)$ be the tile-type that occupies position t on the line. Let $B(t) = 1$ if tiles at positions $t - 1$ and t are bonded together. The information-theoretic entropy of a discrete random variable X that draws its values from a countable universe U is given by

$$\mathbf{H}[X] = -\sum_{x \in U} \mathbf{Pr}[X = x] \ln \mathbf{Pr}[X = x].$$

We define the internal entropy density S_I of a linear self-assembly as the information-theoretic entropy of the tuple $\langle \Gamma(t), B(t) \rangle$ given the values of $\langle \Gamma(t'), B(t') \rangle$ for all $t' < t$. Intuitively, this is the additional amount of information stored in a tile-position, given a description of all the tile-positions to its left. Given the independence assumption,

$$S_I = \mathbf{H}[\langle \Gamma(t), B(t) \rangle | \Gamma(t - 1)] = \sum_i \alpha_i \mathbf{H}[\langle \Gamma(t), B(t) \rangle | \Gamma(t - 1) = T_i].$$

If the tile at position $t - 1$ is T_i then the probability of the tile at position t being of type T_j with a bond between the two tiles is r_{ij}, and the probability of the tile at position t being of type T_j without a bond between the two tiles is $r_i \cdot (L_j/F)$. Consequently,

$$\mathbf{H}\left[\langle\Gamma(t), B(t)\rangle | \Gamma(t-1) = T_i\right] = -\sum_j r_{ij} \ln r_{ij} - \sum_j \frac{r_i L_j}{F} \ln \frac{r_i L_j}{F}$$

and therefore,

$$S_I = \sum_i \alpha_i \sum_j \left(r_{ij} \ln \frac{1}{r_{ij}} + \frac{r_i L_j}{F} \ln \frac{F}{r_i L_j}\right).$$

With some algebraic manipulation one can show that

$$S_I = \sum_i \alpha_i \ln \alpha_i + \sum_i R_i \ln \frac{1}{R_i} + \sum_j L_j \ln \frac{1}{L_j} + F \ln F + \sum_{ij} R_{ij} \ln \frac{1}{R_{ij}}$$

Define the external entropy density S_X of the linear assembly as

$$S_X = \sum_{ij} R_{ij} \ln \frac{\sigma_{ij}}{\tau_{ij}}.$$

Define the total entropy density S of a linear self-assembly as the sum of its internal and external entropy densities.

Intuitively, the internal entropy density exactly captures the amount of information-per-tile-position stored in the linear assembly. The definition of external entropy density is motivated by a corresponding definition in physical chemistry. If r_f and r_r are the equilibrium forward and reverse reaction rates of a chemical process, then the process releases energy proportional to $\ln r_f/r_r$, and this energy is then available to increase the entropy of the external environment. As a consequence, the total entropy of the universe is maximized when the sum of the internal entropy of a system and the energy released by the system (appropriately normalized) is maximized.

As it turns out, $-S$ is a Lyapunov function for the dynamical system defined by the system of differential equations in Eq 1. More precisely, the following theorem can be proven.

Theorem 2.2 *1.* $\dfrac{\partial S}{\partial R_{xy}} = 0$ *iff* $\dfrac{R_{xy} \cdot F}{R_x \cdot L_y} = \dfrac{\sigma_{xy}}{\tau_{xy}}$. *Thus, the bond equilibrium conditions derived from the update rules for R_{xy} are exactly the same as those that characterize the set of stationary points of S, as a function in R_{xy} in the unit hypercube $0 \le R_{xy} \le 1$.*

2. S is a concave function; that is,

$$\sum_{v_{ij}, v_{xy}} \frac{\partial^2 S}{\partial R_{ij} \partial R_{xy}} v_{ij} v_{xy} \le 0$$

for any vector $(v_{11}, v_{12}, \cdots, v_{nn})$ and the equality holds only when $(v_{11}, v_{12}, \cdots, v_{nn})$ is zero vector. Thus, every point in the equilibrium set, which is also stationary point of S, is a local maximum for S.

3. Local maximum does not occur on the boundary.

4. $\dot{S} = \sum \frac{\partial S}{\partial R_{xy}} \dot{R}_{xy} > 0$ as long as $\dot{R}_{xy} \neq 0$ for some bond x, y, and $\dot{S} = 0$ when $\dot{R}_{xy} = 0$ for all x, y.

Note that 2 and 3 of the Theorem imply that there is a unique maximum and the maximum is an interior point, and 4 implies that the total entropy density of a linear self-assembly is increasing until (bond) equilibrium is reached, at which point, the total entropy density is maximized, and henceforth the system remains at equilibrium. But how fast does a linear self-assembly system approach this equilibrium? This question seems to be quite complicated in general. In what follows we will focus on an interesting subcase called the *n-linear polymerization* and analyze the convergence rates for such systems.

3   *n*-linear polymerization

In the cell, DNA and proteins are created by using highly specialized enzymes (polymerases) or enzyme complexes (ribosomes) to catalyze the addition of successive units to one end of a growing chain. In the case of some synthetic polymers and in DNA computation, the product is created by the recursive concatenation of smaller polymers. Here we wish to model the latter form of polymerization. This is our motivation for *n-linear polymerization*, which we define as follows:

Start with equal quantities of tiles $T_1, \ldots, T_n$, where $n \geq 2$. The gluing rule is that the right side of each T_i, for $1 \leq i \leq n - 1$, can glue to T_{i+1} only. Hence the allowable combinations are all subsequences of $T_1 T_2 \ldots T_n$. Here we consider the *uniform* case, in which all allowable bonds have the same on-rate $\sigma \in [0, 1]$ and off-rate $\tau \in [0, 1]$.

Let time-dependent variables $v_1, \ldots, v_{n-1}$ be defined as follows: v_i is the portion of the T_i tiles which are currently free to the right. We assume below that $v_1 = v_2 = \cdots = v_{n-1}$. This assumption is justified for several reasons. First, if it is ever true, then it is preserved over time. Hence it is justified if the initial state has this symmetry (for example, if we employ the natural initial state $v_1 = \cdots = v_{n-1} = 1$). Second, we shall prove in Section 3.3 that, even if this symmetry does not initially hold, the system rapidly evolves toward it.

Hence let v denote the common value of $v_1, \ldots, v_{n-1}$. How does v evolve over one time-step? Of the $1 - v$ portion of T_i tiles currently bonded to the right, each will become free with probability τ. Meanwhile, of the v portion of

T_i tiles currently free to the right, each will become bonded with probability σ if the tile to its right is a T_{i+1}. What then is the probability that the tile to the right of a free T_i is a T_{i+1}? Since all T_1 tiles are free to the left, while a v portion of $T_2, \ldots, T_n$ tiles are free to the left, this probability is $v/[v(n-1)+1]$. Hence the update rule for v (where v' denotes the new value) is given by:

$$v' - v = \frac{-v^2 \sigma}{v(n-1)+1} + (1-v)\tau. \tag{2}$$

Now assume $\tau \neq 0$ and let $\rho := \sigma/\tau$. Then it is easily seen that, for v on legal domain $[0, 1]$, this transformation has a unique fixed point $v = \beta$, where β is defined as follows:

$$\beta = \frac{(n-2) + \sqrt{n^2 + 4\rho}}{2(\rho + n - 1)}. \tag{3}$$

Hence, when $v = \beta$, the system is at bond equilibrium. But then we can specify a strong equilibrium for the system, namely when each T_i (for $1 \leq i \leq n - 1$) is free to the right with *independent* probability β. Indeed, since this state has bond equilibrium and independence, it follows from Theorem 2.1 that it is a strong equilibrium.

3.1 Dynamics of irreversible polymerization

Here we consider n-linear polymerization in the irreversible case ($\tau = 0$, $\sigma \neq 0$). For this case, the equilibrium is evidently for all tiles to form complete $T_1 T_2 \ldots T_n$ strands. Equivalently, the v variable, representing the fraction of gluable surfaces free, goes to zero. We will next determine the rate at which v goes to this equilibrium.

Theorem 3.1 *For irreversible n-linear polymerization with on-rate σ, and for $\varepsilon > 0$, let t be the smallest time such that the v variable, starting from any initial value, is guaranteed to be at most ε if time elapsed is at least t. Then t is at least*

$$\frac{1}{2\sigma(e-1)}[1/(e\varepsilon) - 1] + \frac{n-1}{2\sigma}\lfloor \ln(1/\varepsilon) \rfloor$$

and at most

$$\frac{e}{\sigma(e-1)}[e/\varepsilon - 1] + \frac{n-1}{\sigma}\lceil \ln(1/\varepsilon) \rceil.$$

The lower bound in the theorem is based on the observation from (2) that v is multiplied at each step by a factor of

$$1 - \frac{v\sigma}{v(n-1)+1} = 1 - \frac{\sigma}{n-1}\left[1 - \frac{1}{v(n-1)+1}\right]; \tag{4}$$

hence, if the current value of v is v_0, then v is multiplied at each step by a factor which is at least $1 - 1/Q$ where

$$Q = \left[\frac{\sigma}{n-1} \left[1 - \frac{1}{v_0(n-1)+1} \right] \right]^{-1} = \frac{n-1}{\sigma} + \frac{1}{\sigma v_0}.$$

The matching upper-bound can be proven based on the observation that as long as $v \geq v_0/e$, v is multiplied at each step by a factor which is at most

$$1 - \frac{\sigma}{n-1} \left[1 - \frac{1}{v_0(n-1)/e+1} \right].$$

Therefore the time to reduce v to ε is $\Theta((1/\varepsilon + n\log(1/\varepsilon))/\sigma)$. We can also show that the time for half of the tiles to form complete $T_1 T_2 \ldots T_n$ strands is $\Theta((n\log n)/\sigma)$. It is perhaps surprising that the latter bound, as a function of n, is as large as $\Theta(n\log n)$. A priori, it is conceivable that a large volume of complete $T_1 T_2 \ldots T_n$ strands could be assembled in just $\Theta(\log n)$ steps. This would be achieved if at step one many polymers of size 2 were created, at step two polymers of size 4, and so on. But the above bound shows that this is in fact not what occurs, and indeed that, roughly speaking, complete $T_1 T_2 \ldots T_n$ strands are created by the addition of, on the average, just one tile every $\log n$ steps.

3.2   Dynamics of reversible polymerization

We now turn to reversible n-linear polymerization. Here we will prove that v in fact converges exponentially fast to equilibrium value β.

Theorem 3.2 *For reversible n-linear polymerization, the v variable, starting from any initial value, is guaranteed to be within distance ε from equilibrium value β if time elapsed is at least $\max\{3, 1/\tau\} \cdot \ln(1/\varepsilon)$.*

The theorem is proved by considering the current distance Δ from equilibrium, and showing that $|\Delta|$ is multiplied at each step by a factor which is at most $\max\{1 - \tau, 0.71\}$.

Observe that, for the irreversible case, the time for v to reach distance ε from equilibrium, expressed as a function of ε, is $\Theta(1/\varepsilon)$. For the reversible case, on the other hand, this time is $O(\log(1/\varepsilon))$. This may seem paradoxical but is in fact easily explained. At irreversible equilibrium, the quantities of tile strands which need to combine are brought down to zero, so that it becomes increasingly difficult for the remaining fragments to find each other.

3.3 Uniqueness of the n-linear equilibrium

Irreversible n-linear polymerization evidently goes to a unique equilibrium – i.e., a strong equilibrium corresponding to all tiles forming complete $T_1 T_2 \ldots T_n$ strands. We now turn to the reversible case, for which, in Section 3, we specified a particular equilibrium – a strong equilibrium in which each gluable surface is free with independent probability β. The following theorem states that any system must in fact go to this unique equilibrium.

Theorem 3.3 *Starting from any initial state, a reversible n-linear polymerization system will go to a unique equilibrium, namely the state in which each gluable surface is free with independent probability β, where β is as given in (3).*

For $1 \leq i \leq n - 1$, let v_i denote the portion of the T_i tiles which are currently free to the right. The idea of the proof is to show that $|v_i - v_j|$ is multiplied at each time-step by a factor which is at most $\max\{1 - \tau, 2/3\}$. Hence all of $v_1, \ldots, v_{n-1}$ are rapidly squeezed together toward a common value.

References

[1] Leonard M. Adleman. Molecular computation of solutions to combinatorial problems. *Science*, 266:1021–1024, November 11, 1994.

[2] Leonard M. Adleman. Towards a mathematical theory of self-assembly. Technical Report 00-722, Department of Computer Science, University of Southern California, 2000.

[3] H. Abelson, D. Allen, D. Coore, C. Hanson, G. Homsy, T. Knight, R. Nagpal, E. Rauch, G. Sussman and R. Weiss. Amorphous computing. *AI Memo 1665*, August 1999.

[4] G.E. Bredon. *Topology and Geometry*, Springer-Verlag, 1993.

[5] D. Eisenberg and D. Crothers. *Physical Chemistry with Applications to the Life Sciences*, Benjamin Cummings, 1979.

[6] Centre for Quantum Computation web page, *http://www.qubit.org*.

[7] Paul Rothemund. Programmed self-assembly using lateral capillary forces. *Manuscript*.

[8] Paul Rothemund and Erik Winfree. The program-size complexity of self-assembled squares. *STOC 2000*.

[9] D. Welsh. Percolation and the random cluster model: Combinatorial and algorithmic approaches, in *Probabilistic Methods for Algorithmic Discrete Mathematics*, M. Habib, C. McDiarmid, J. Ramirez-Alfonsin, B. Reed (Eds.), Springer, 1998.

[10] Erik Winfree, Furong Liu, Lisa A. Wenzler, and Nadrian C. Seeman. Design and self-assembly of two-dimensional DNA crystals. *Nature*, 394:539-544, 1998.

[11] Erik Winfree. PhD thesis. Cal Tech, 1998.

Synchronization in a Discrete Circular Network

SUI SUN CHENG

Department of Mathematics, Tsing Hua University,
Hsinchu, Taiwan 30043, R. O. C.
E-mail: sscheng@math2.math.nthu.edu.tw

CHUAN-JUN TIAN

College of Information Engineering
Shenzhen University, Shenzhen 518060, P. R. China
E-mail:tiancj@szu.edu.cn

and

MICHAEL GIL'

Department of Mathematics, Ben Gurion University
P. O. Box 653, Beer Sheva 84105, Israel
E-mail: gilmi@cs.bgu.ac.il

Abstract Synchronization is a common feature in complex dynamical systems. In this paper, we consider a temporal-spatio network system which is discrete in time and also in space. We find sharp conditions for all its solutions to be partially or fully synchronized. Our results will likely find applications in digital neural networks theory.

Keywords Synchronization, Discrete heat equation, Spectral radius

AMS Subject Classification 93C55, 93C10, 39A11

1 Introduction

The phenomenon of synchronization is widespread in nature and in technology. The fact that various objects operate in harmony seems to be a manifestation of self-organization in nature. For an introduction, reference [1] can be consulted. Synchronization can also occur in network models where time and space are both assumed to be discrete. An example is the following information distribution problem of a ring. Suppose we are interested in the state values at equally spaced abstract neuron units $z_1, ..., z_n$ of a "cycle" (so that z_1 is communicating with only z_n and z_2, z_2 only with z_1 and z_3, etc.) during different time periods $t = 0, 1, 2, \ldots$. We may let $u_i^{(t)}$ be the state values of the point z_i in the time period t. During the time period t, if the state value $u_1^{(t)}$ is larger than $u_2^{(t)}$, information will "flow" from the point z_1 to the point z_2. The subsequent change of state value at the point z_2 is $u_2^{(t+1)} - u_2^{(t)}$, and, under a first order approximation assumption, it is reasonable to postulate that it is proportional to the difference $u_1^{(t)} - u_2^{(t)}$, say, $\gamma \left(u_1^{(t)} - u_2^{(t)} \right)$, where γ is a positive rate constant. Similarly, information will flow from the point z_3 to the point z_2 if $u_3^{(t)} > u_2^{(t)}$. Thus, it is reasonable that the total effect is

$$u_2^{(t+1)} - u_2^{(t)} = \gamma \left(u_1^{(t)} - u_2^{(t)} \right) + \gamma \left(u_3^{(t)} - u_2^{(t)} \right) = \gamma \left(u_3^{(t)} - 2u_2^{(t)} + u_1^{(t)} \right).$$

By similar considerations, we may then obtain the system of equations

$$
\begin{aligned}
u_1^{(t+1)} - u_1^{(t)} &= \gamma \left(u_n^{(t)} - u_1^{(t)} \right) + \gamma \left(u_2^{(t)} - u_1^{(t)} \right), \\
u_2^{(t+1)} - u_2^{(t)} &= \gamma \left(u_1^{(t)} - u_2^{(t)} \right) + \gamma \left(u_3^{(t)} - u_2^{(t)} \right), \\
\ldots &= \ldots \\
u_n^{(t+1)} - u_n^{(t)} &= \gamma \left(u_{n-1}^{(t)} - u_n^{(t)} \right) + \gamma \left(u_1^{(t)} - u_n^{(t)} \right),
\end{aligned}
$$

which describes the evolution process of the state values as t tends to infinity.

There are many questions of interests which we can raise regarding the evolutionary system. An interesting one is whether the state values $u_1^{(t)}, \ldots, u_n^{(t)}$ are ultimately in synchronization so that

$$\lim_{t \to \infty} \left| u_i^{(t)} - u_j^{(t)} \right| = 0, \ 1 \le i, j \le n.$$

In this note, we will consider such a problem for a slightly more general system of equations of the form

$$
\begin{cases}
u_1^{(t+1)} = f \left(u_1^{(t)} \right) + \gamma_t \left(f \left(u_n^{(t)} \right) - 2f \left(u_1^{(t)} \right) + f \left(u_2^{(t)} \right) \right), \\
u_2^{(t+1)} = f \left(u_2^{(t)} \right) + \gamma_t \left(f \left(u_1^{(t)} \right) - 2f \left(u_2^{(t)} \right) + f \left(u_3^{(t)} \right) \right), \\
\ldots = \ldots \\
u_n^{(t+1)} = f \left(u_n^{(t)} \right) + \gamma_t \left(f \left(u_{n-1}^{(t)} \right) - 2f \left(u_n^{(t)} \right) + f \left(u_1^{(t)} \right) \right),
\end{cases}
\tag{1}
$$

for $t = 0, 1, 2, \ldots$, where $f : R \to R$ is a Lipschitz function which satisfies

$$|f(u) - f(v)| \le \Gamma |u - v|, \ u, v \in R,$$

for some fixed positive constant Γ. There are many Lipschitz functions. For example, the tent map g defined by

$$
g(x) = \begin{cases}
0 & x < 0 \\
2x & 0 \le x \le 1/2 \\
2(1 - x) & 1/2 \le x \le 1 \\
0 & x > 1
\end{cases},
$$

is a Lipschitz function with Lipschitz constant 2. Note that the diffusion constant γ is now replaced by a time-dependent sequence $\{\gamma_t\}_{t=0}^{\infty}$.

Note that if we denote the vector col $\left(u_1^{(t)}, ..., u_n^{(t)}\right)$ by $u^{(t)}$, then our system can be written in the form $u^{(t+1)} = F\left(u^{(t)}\right)$. Thus, given an initial distribution $u^{(0)} = \phi$, it is easily seen that we can calculate $u^{(1)}, u^{(2)}, ...$ successively and in a unique manner from (1). Such a sequence $\left\{u^{(0)}, u^{(1)}, ...\right\}$ is said to be a solution of (1).

We say that a solution $\left\{u^{(t)}\right\}_{t=0}^{\infty}$ is synchronized if $\lim_{t\to\infty}\left|u_i^{(t)} - u_j^{(t)}\right| = 0$ for $1 \le i, j \le n$. More generally, let Ω be a subset of $\{1, 2, ..., n\}$, we say that a solution $\left\{u^{(t)}\right\}_{t=0}^{\infty}$ is Ω-partially synchronized if $\lim_{t\to\infty}\left|u_i^{(t)} - u_j^{(t)}\right| = 0$ for all i and j in Ω.

We remark that when (1) is asymptotically stable, i.e., when every solution of (1) tends to zero as t tends to infinity, then every solution of (1) is synchronized. Thus, any asymptotic stability criterion is also a synchronization criterion for (1). However, it is easy to find examples in which a solution is synchronized, yet it does not tend to zero as t tends to infinity (see last section).

For the sake of convenience, we will use $\rho(A)$ to denote the spectral radius of a square matrix A. We will also assume that $n \ge 2$ to avoid the trivial case $n = 1$.

2   Synchronization criteria for $2 \le n \le 4$

We will derive several sufficient conditions for the solutions of (1) to be ultimately synchronized. As we will see shortly, it is better to distinguish the cases where n is small and where n is sufficiently large. We first consider the case where $n = 2$. Then our system is

$$\begin{cases} u_1^{(t+1)} = (1 - 2\gamma_t) f\left(u_1^{(t)}\right) + 2\gamma_t f\left(u_2^{(t)}\right), \\ u_2^{(t+1)} = 2\gamma_t f\left(u_1^{(t)}\right) + (1 - 2\gamma_t) f\left(u_2^{(t)}\right). \end{cases} \tag{2}$$

Suppose $n = 2$. We assert that if

$$\limsup_{t\to\infty} |1 - 4\gamma_t| < \frac{1}{\Gamma}, \tag{3}$$

then every solution of (1) is ultimately synchronized.

Indeed, note that

$$\left|u_1^{(t+1)} - u_2^{(t+1)}\right| = |1 - 4\gamma_t|\left|f\left(u_1^{(t)}\right) - f\left(u_2^{(t)}\right)\right| \le \Gamma|1 - 4\gamma_t|\left|u_1^{(t)} - u_2^{(t)}\right|. \tag{4}$$

If (3) holds, then there are constant $d \in (0, 1)$ and positive integer T such that

$$\Gamma|1 - 4\gamma_t| < d < 1, \ t > T,$$

so that

$$\left|u_1^{(T+n)} - u_2^{(T+n)}\right| \le d \left|u_1^{(T+n-1)} - u_2^{(T+n-1)}\right| \le \ldots \le d^n \left|u_1^{(T)} - u_2^{(T)}\right|$$

for $n = 1, 2, \ldots$. If we now let $n \to \infty$, we see that

$$\lim_{t \to \infty} \left|u_1^{(t)} - u_2^{(t)}\right| = 0.$$

Next, we consider the case where $n = 3$. Our system is now

$$\begin{cases} u_1^{(t+1)} = f\left(u_1^{(t)}\right) + \gamma_t \left(f\left(u_3^{(t)}\right) - 2f\left(u_1^{(t)}\right) + f\left(u_2^{(t)}\right)\right), \\ u_2^{(t+1)} = f\left(u_2^{(t)}\right) + \gamma_t \left(f\left(u_1^{(t)}\right) - 2f\left(u_2^{(t)}\right) + f\left(u_3^{(t)}\right)\right), \\ u_3^{(t+1)} = f\left(u_3^{(t)}\right) + \gamma_t \left(f\left(u_2^{(t)}\right) - 2f\left(u_3^{(t)}\right) + f\left(u_1^{(t)}\right)\right). \end{cases} \tag{5}$$

We assert that if

$$\limsup_{t \to \infty} |1 - 3\gamma_t| < \frac{1}{\Gamma}, \tag{6}$$

then every solution of (1) is synchronized.

Indeed, note that

$$\left|u_1^{(t+1)} - u_2^{(t+1)}\right| = |1 - 3\gamma_t| \left|f\left(u_1^{(t)}\right) - f\left(u_2^{(t)}\right)\right| \le \Gamma |1 - 3\gamma_t| \left|u_1^{(t)} - u_2^{(t)}\right|,$$

which is similar to (4). Thus by similar reasonings, $\lim_{t \to \infty} \left|u_1^{(t)} - u_2^{(t)}\right| = 0$. The case where $\lim_{t \to \infty} \left|u_1^{(t)} - u_3^{(t)}\right| = 0$ is similarly proved.

Suppose $n = 4$. We assert that if

$$\limsup_{t \to \infty} \rho \begin{pmatrix} 1 - 3\gamma_t & \gamma_t \\ \gamma_t & 1 - 3\gamma_t \end{pmatrix} < \frac{1}{\Gamma}, \tag{7}$$

then every solution of (1) is $\{1,2\}$-, $\{3,4\}$-, $\{1,4\}$- and $\{2,3\}$-partially synchronized, and if

$$\limsup_{t \to \infty} |1 - 2\gamma_t| < \frac{1}{\Gamma}, \tag{8}$$

then every solution of (1) is $\{2,4\}$- and $\{1,3\}$-partially synchronized.

Indeed, we may infer from (1) that

$$\begin{pmatrix} u_1^{(t+1)} - u_2^{(t+1)} \\ u_4^{(t+1)} - u_3^{(t+1)} \end{pmatrix} = \begin{pmatrix} 1 - 3\gamma_t & \gamma_t \\ \gamma_t & 1 - 3\gamma_t \end{pmatrix} \begin{pmatrix} f\left(u_1^{(t)}\right) - f\left(u_2^{(t)}\right) \\ f\left(u_4^{(t)}\right) - f\left(u_3^{(t)}\right) \end{pmatrix} \tag{9}$$

and

$$u_2^{(t+1)} - u_4^{(t+1)} = (1 - 2\gamma_t)\left(f\left(u_2^{(t)}\right) - f\left(u_4^{(t)}\right)\right).$$

If (7) holds, then there is some $d \in (0,1)$ and positive integer T such that

$$\left\| \begin{pmatrix} u_1^{(T+n)} - u_2^{(T+n)} \\ u_4^{(T+n)} - u_3^{(T+n)} \end{pmatrix} \right\|$$

$$\leq \Gamma \left\| \begin{pmatrix} 1 - 3\gamma_t & \gamma_t \\ \gamma_t & 1 - 3\gamma_t \end{pmatrix} \right\| \left\| \begin{pmatrix} u_1^{(t)} - u_2^{(t)} \\ u_4^{(t)} - u_3^{(t)} \end{pmatrix} \right\|$$

$$\leq d^n \left\| \begin{pmatrix} u_1^{(T)} - u_2^{(T)} \\ u_4^{(T)} - u_3^{(T)} \end{pmatrix} \right\|$$

for $n = 1, 2, 3, ...$, where $\|\cdot\|$ denotes any naturally induced norm. As $n \to \infty$, we see that

$$\lim_{t \to \infty} \begin{pmatrix} u_1^{(t)} - u_2^{(t)} \\ u_4^{(t)} - u_3^{(t)} \end{pmatrix} = 0.$$

Similarly, we may show that

$$\lim_{t \to \infty} \begin{pmatrix} u_1^{(t)} - u_4^{(t)} \\ u_2^{(t)} - u_3^{(t)} \end{pmatrix} = 0.$$

When (8) holds, the assertion that

$$\lim_{t \to \infty} \left| u_2^{(t)} - u_4^{(t)} \right| = \lim_{t \to \infty} \left| u_1^{(t)} - u_3^{(t)} \right| = 0$$

is similarly proved.

We remark that when $\gamma_t = \gamma$ for all l, then condition (7) holds if, and only if, $0 < \gamma < 1/2$, and (8) holds if, and only if, $0 < \gamma < 1/2$; thus every solution of (1), where $n = 4$, is synchronized, when $0 < \gamma < 1/2$.

3 Synchronization criteria for $n \geq 5$

It is necessary to distinguish the case where n is even and the case where n is odd. To motivate our main result for the case where n is even, let us first consider the case where $n = 6$. Then from (1), we see that

$$\begin{pmatrix} u_1^{(t+1)} - u_5^{(t+1)} \\ u_2^{(t+1)} - u_4^{(t+1)} \end{pmatrix} = \begin{pmatrix} 1 - 2\gamma_t & \gamma_t \\ \gamma_t & 1 - 2\gamma_t \end{pmatrix} \begin{pmatrix} f\left(u_1^{(t)}\right) - f\left(u_5^{(t)}\right) \\ f\left(u_2^{(t)}\right) - f\left(u_4^{(t)}\right) \end{pmatrix},$$
$$\tag{10}$$

$$\begin{pmatrix} u_2^{(t+1)} - u_6^{(t+1)} \\ u_3^{(t+1)} - u_5^{(t+1)} \end{pmatrix} = \begin{pmatrix} 1 - 2\gamma_t & \gamma_t \\ \gamma_t & 1 - 2\gamma_t \end{pmatrix} \begin{pmatrix} f\left(u_2^{(t)}\right) - f\left(u_6^{(t)}\right) \\ f\left(u_3^{(t)}\right) - f\left(u_5^{(t)}\right) \end{pmatrix},$$
$$\tag{11}$$

$$
\begin{pmatrix} u_3^{(t+1)} - u_1^{(t+1)} \\ u_4^{(t+1)} - u_6^{(t+1)} \end{pmatrix} = \begin{pmatrix} 1 - 2\gamma_t & \gamma_t \\ \gamma_t & 1 - 2\gamma_t \end{pmatrix} \begin{pmatrix} f\left(u_3^{(t)}\right) - f\left(u_1^{(t)}\right) \\ f\left(u_4^{(t)}\right) - f\left(u_6^{(t)}\right) \end{pmatrix},
$$

$$(12)$$

and

$$
\begin{pmatrix} u_1^{(t+1)} - u_2^{(t+1)} \\ u_6^{(t+1)} - u_3^{(t+1)} \\ u_5^{(t+1)} - u_4^{(t+1)} \end{pmatrix} = \begin{pmatrix} 1 - 3\gamma_t & \gamma_t & 0 \\ \gamma_t & 1 - 2\gamma_t & \gamma_t \\ 0 & \gamma_t & 1 - 3\gamma_t \end{pmatrix} \begin{pmatrix} f\left(u_1^{(t)}\right) - f\left(u_2^{(t)}\right) \\ f\left(u_6^{(t)}\right) - f\left(u_3^{(t)}\right) \\ f\left(u_5^{(t)}\right) - f\left(u_4^{(t)}\right) \end{pmatrix}.
$$

If

$$
\limsup_{t \to \infty} \rho \begin{pmatrix} 1 - 2\gamma_t & \gamma_t \\ \gamma_t & 1 - 2\gamma_t \end{pmatrix} < \frac{1}{\Gamma}, \tag{13}
$$

then by reasonings similar to those above, we see that every solution is $\{1,5\}$-, $\{2,4\}$-, $\{2,6\}$-, $\{3,5\}$-, $\{3,1\}$-, and $\{4,6\}$-partially synchronized. Note that if we let G be the graph with vertices $1, 2, ..., 6$ and edges $\{1,5\}$, $\{2,4\}$, $\{2,6\}$, $\{3,5\}$, $\{3,1\}$, and $\{4,6\}$, then G has exactly two connected components, one of them has the vertices $1, 3, 5$, while the other has the vertices $2, 4, 6$. In other words, every solution of (1) is $\{1,3,5\}$- and $\{2,4,6\}$-partially synchronized. Similarly, if

$$
\limsup_{t \to \infty} \rho \begin{pmatrix} 1 - 3\gamma_t & \gamma_t & 0 \\ \gamma_t & 1 - 2\gamma_t & \gamma_t \\ 0 & \gamma & 1 - 3\gamma_t \end{pmatrix} < \frac{1}{\Gamma}, \tag{14}
$$

then every solution is $\{1,2\}$-, $\{3,6\}$- and $\{4,5\}$-partially synchronized. As a direct consequence, if (13) and (14) hold simultaneously, then every solution of (1) is synchronized.

In view of the similarities between (10), (11) and (12), it is reasonable to suspect that there is a reason for it. Indeed, this is due to the fact that system (1) is invariant with respect to cyclic labeling of the points $z_1, z_2, ..., z_n$ of the ring. More precisely, let us say that σ is a rotation mapping of the vector $(z_1, z_2, ..., z_n)$ if $\sigma(z_1) = z_n, \sigma(z_2) = z_1, ...,$ and $\sigma(z_n) = z_{n-1}$. For example, the vector $(1, 2, ..., n)$ after one rotation becomes $(n, 1, 2, ..., n-1)$. After one more rotation, it becomes $(n-1, n, 1, 2, ..., n-2)$, etc. We say that an equation involving a vector of the form $(z_1, z_2, ..., z_n)$ is rotated if each z_i is replaced with $\sigma(z_i)$. For instance, if we regard the first equation in (1) as an equation involving the vector $(u_1, ..., u_n)$, then after one rotation, it becomes the second equation in (1). Clearly, if each equation in (1) is rotated, the resulting system is identical to the original one, that is, (1) is invariant with respect to rotation. In view of this observation, it is then easy to see that (10) implies (11) and (12), since (11) is obtained by from (10) after one rotation, while (12) is obtained after two rotations.

The above observations motivate the following results.

LEMMA 1. Suppose $n = 2m$ where $m \geq 3$. Let

$$
A_t = \begin{pmatrix}
1-2\gamma_t & \gamma_t & 0 & 0 & \cdots & 0 \\
\gamma_t & 1-2\gamma_t & \gamma_t & 0 & \cdots & 0 \\
0 & \gamma_t & 1-2\gamma_t & \gamma_t & \cdots & 0 \\
\cdot & \cdot & \cdot & \cdot & \cdots & \cdot \\
0 & \cdot & \cdot & \cdot & \cdots & 1-2\gamma_t
\end{pmatrix}_{(m-1)\times(m-1)} , \quad t \geq 0.
$$

(15)

If $\limsup_{t\to\infty} \rho(A_t) < 1/\Gamma$, then every solution of (1) is $\{1,3\}$-, $\{2m,4\}$-, $\{2m-1,5\}$-, ... $\{m+3, m+1\}$–partially synchronized.

Indeed, it is easily verified that

$$
\begin{pmatrix}
u_1^{(t+1)} - u_3^{(t+1)} \\
u_{2m}^{(t+1)} - u_4^{(t+1)} \\
u_{2m-1}^{(t+1)} - u_5^{(t+1)} \\
\cdots \\
u_{m+3}^{(t+1)} - u_{m+1}^{(t+1)}
\end{pmatrix}
= A_t \cdot
\begin{pmatrix}
f\left(u_1^{(t)}\right) - f\left(u_3^{(t)}\right) \\
f\left(u_{2m}^{(t)}\right) - f\left(u_4^{(t)}\right) \\
f\left(u_{2m-1}^{(t)}\right) - f\left(u_5^{(t)}\right) \\
\cdots \\
f\left(u_{m+3}^{(t)}\right) - f\left(u_{m+1}^{(t)}\right)
\end{pmatrix}.
$$

Then by reasonings we have explained before, we see that

$$
\lim_{t\to\infty} \left(\left| u_1^{(t+1)} - u_3^{(t+1)} \right|, ..., \left| u_{m+3}^{(t+1)} - u_{m+1}^{(t+1)} \right| \right) = 0.
$$

We remark that in view of the rotation invariance of (1) explained above, it is easily seen that the conditions in Lemma 1 imply that every solution of (1) is $\{1,3,...,2m-1\}$- and $\{2,4,...,2m\}$-partially synchronized.

LEMMA 2. Suppose $n = 2m$ where $m \geq 3$. Let

$$
B_t = \begin{pmatrix}
1-3\gamma_t & \gamma_t & 0 & 0 & \cdots & 0 \\
\gamma_t & 1-2\gamma_t & \gamma_t & 0 & \cdots & 0 \\
0 & \gamma_t & 1-2\gamma_t & \gamma_t & \cdots & 0 \\
\cdot & \cdot & \cdot & & \cdots & \cdot \\
\cdot & \cdot & \cdot & \gamma_t & 1-2\gamma_t & \gamma_t \\
\cdot & \cdot & \cdot & 0 & \gamma_t & 1-3\gamma_t
\end{pmatrix}_{m\times m} , \quad t \geq 0.
$$

(16)

If $\limsup_{t\to\infty} \rho(B_t) < 1/\Gamma$, then every solution of (1) is $\{1,2\}$-, $\{2m,3\}$-, $\{2m-1,4\}$-, ..., $\{m+3,m\}$- and $\{m+2, m+1\}$-partially synchronized.

This follows easily from the fact that

$$
\begin{pmatrix}
u_1^{(t+1)} - u_2^{(t+1)} \\
u_{2m}^{(t+1)} - u_3^{(t+1)} \\
u_{2m-1}^{(t+1)} - u_4^{(t+1)} \\
\cdots \\
u_{m+2}^{(t+1)} - u_{m+1}^{(t+1)}
\end{pmatrix}
= B_t \cdot
\begin{pmatrix}
f\left(u_1^{(t)}\right) - f\left(u_2^{(t)}\right) \\
f\left(u_{2m}^{(t)}\right) - f\left(u_3^{(t)}\right) \\
f\left(u_{2m-1}^{(t)}\right) - f\left(u_4^{(t)}\right) \\
\cdots \\
f\left(u_{m+2}^{(t)}\right) - f\left(u_{m+1}^{(t)}\right)
\end{pmatrix}.
$$

We remark that in view of the rotation invariance of (1) explained above, it is easily seen that the conditions in Lemma 2 imply that every solution of (1) is $\{1, 2m\}$-, $\{2, 2m - 1\}$-, ..., $\{m, m + 1\}$-partially synchronized. Note that the graph with vertices $1, 2, 3, ..., 2m$ and edges $\{1, 2\}$, $\{2m, 3\}$, ..., $\{m + 2, m + 1\}$, $\{1, 2m\}$, $\{2, 2m - 1\}$, ..., $\{m, m + 1\}$ is connected. Thus the conditions of Lemma 2 imply every solution is synchronized.

THEOREM 1. Suppose $n = 2m$ where $m \geq 3$. If $\limsup_{t \to \infty} \rho(A_t) < 1/\Gamma$, then every solution of (1) is $\{1, 3, ..., 2m - 1\}$- and $\{2, 4, ..., 2m\}$-partially synchronized, and if $\limsup_{t \to \infty} \rho(B_t) < 1/\Gamma$, then every solution of (1) is synchronized, where A_t and B_t are defined by (15) and (16) respectively.

Next, we deal with the case where $n = 2m + 1$ for $m \geq 3$. Let us first consider the special case $n = 5$. Note that

$$
\begin{pmatrix} u_1^{(t+1)} - u_2^{(t+1)} \\ u_5^{(t+1)} - u_3^{(t+1)} \end{pmatrix} = \begin{pmatrix} 1 - 3\gamma_t & \gamma_t \\ \gamma_t & 1 - 2\gamma_t \end{pmatrix} \begin{pmatrix} f\left(u_1^{(t)}\right) - f\left(u_2^{(t)}\right) \\ f\left(u_5^{(t)}\right) - f\left(u_3^{(t)}\right) \end{pmatrix},
$$
(17)

and

$$
\begin{pmatrix} u_1^{(t+1)} - u_3^{(t+1)} \\ u_5^{(t+1)} - u_4^{(t+1)} \end{pmatrix} = \begin{pmatrix} 1 - 2\gamma_t & \gamma_t \\ \gamma_t & 1 - 3\gamma_t \end{pmatrix} \begin{pmatrix} f\left(u_1^{(t)}\right) - f\left(u_3^{(t)}\right) \\ f\left(u_5^{(t)}\right) - f\left(u_4^{(t)}\right) \end{pmatrix}.
$$
(18)

In view of the similarities between (9) and (17) and (18), it is easily seen if

$$
\limsup_{t \to \infty} \rho \begin{pmatrix} 1 - 3\gamma_t & \gamma_t \\ \gamma_t & 1 - 2\gamma_t \end{pmatrix} < \frac{1}{\Gamma},
$$
(19)

then every solution of (1) is $\{1, 2\}$-, $\{3, 5\}$-, $\{1, 3\}$-, $\{5, 4\}$-partially synchronized. Note again that (18) holds in view of the rotational invariance of (1). This motivates the following result.

THEOREM 2. Suppose $n = 2m + 1$ where $m \geq 2$. Let

$$
C_t = \begin{pmatrix} 1 - 2\gamma_t & \gamma_t & 0 & 0 & \cdots & 0 \\ \gamma_t & 1 - 2\gamma_t & \gamma_t & 0 & \cdots & 0 \\ 0 & \gamma_t & 1 - 2\gamma_t & \gamma_t & \cdots & 0 \\ \cdot & \cdot & \cdot & \cdot & \cdots & \cdot \\ 0 & 0 & 0 & 0 & \cdots & 1 - 3\gamma_t \end{pmatrix}_{m \times m}, \quad t \geq 0.
$$

If $\limsup_{t \to \infty} \rho(C_t) < 1/\Gamma$, then every solution of (1) is synchronized.

Indeed, note that

$$
\begin{pmatrix}
u_1^{(t+1)} - u_3^{(t+1)} \\
u_{2m+1}^{(t+1)} - u_4^{(t+1)} \\
u_{2m}^{(t+1)} - u_5^{(t+1)} \\
\cdots \\
u_{m+3}^{(t+1)} - u_{m+2}^{(t+1)}
\end{pmatrix}
= C_t \cdot
\begin{pmatrix}
f\left(u_1^{(t)}\right) - f\left(u_3^{(t)}\right) \\
f\left(u_{2m+1}^{(t)}\right) - f\left(u_4^{(t)}\right) \\
f\left(u_{2m}^{(t)}\right) - f\left(u_5^{(t)}\right) \\
\cdots \\
f\left(u_{m+3}^{(t)}\right) - f\left(u_{m+2}^{(t)}\right)
\end{pmatrix}.
$$

Hence $\limsup_{t\to\infty}\rho\left(C_t\right) < 1/\Gamma$ implies

$$
\lim_{t\to\infty}\left(\left|u_1^{(t+1)} - u_3^{(t+1)}\right|, \ldots, \left|u_{m+3}^{(t+1)} - u_{m+2}^{(t+1)}\right|\right) = 0.
$$

Next, in view of the rotational invariance of system (1), we see that every solution is also $\{1,2\}$, $\{2m+1,3\}$-, $\{2m,4\}$-, , ..., and $\{m+3, m+1\}$-partially synchronized. Note that the graph with vertices $1,2,3,...,2m+1$ and edges $\{1,3\}$, $\{2m+1,4\}$, ..., $\{m+3, m+2\}$, $\{1,2\}$, $\{2m+1,3\}$, ..., $\{m+3, m+1\}$ is connected. This shows that every solution of (1) is synchronized.

4   A special case

Let us take f to be the identity function and take $\{\gamma_t\}$ to be the constant sequence $\{\gamma\}$. In this case, (2) reduces to

$$
\begin{pmatrix} u_1^{(t+1)} \\ u_2^{(t+1)} \end{pmatrix} = \begin{pmatrix} 1 - 2\gamma & 2\gamma \\ 2\gamma & 1 - 2\gamma \end{pmatrix} \begin{pmatrix} u_1^{(t)} \\ u_2^{(t)} \end{pmatrix},
$$

and condition (3) becomes $0 < \gamma < 1/2$. Note that this condition is sharp since when $\gamma = 1/2$, then

$$
\begin{pmatrix} 1 - 2\gamma & 2\gamma \\ 2\gamma & 1 - 2\gamma \end{pmatrix} = \begin{pmatrix} 0 & 1 \\ 1 & 0 \end{pmatrix}.
$$

If we take $u_1^{(0)} = -1$ and $u_2^{(0)} = 1$, then since

$$
\begin{pmatrix} 0 & 1 \\ 1 & 0 \end{pmatrix}\begin{pmatrix} -1 \\ 1 \end{pmatrix} = -1\begin{pmatrix} -1 \\ 1 \end{pmatrix},
$$

we see that the corresponding solution of (2) is given by

$$
\left\{ \begin{pmatrix} -1 \\ 1 \end{pmatrix}, \begin{pmatrix} 1 \\ -1 \end{pmatrix}, \begin{pmatrix} -1 \\ 1 \end{pmatrix}, \ldots \right\},
$$

which is not a synchronized solution. Note further that if we take $u_1^{(0)} = 1$ and $u_2^{(0)} = 1$, then the corresponding solution is

$$\left\{ \begin{pmatrix} 1 \\ 1 \end{pmatrix}, \begin{pmatrix} 1 \\ 1 \end{pmatrix}, \begin{pmatrix} 1 \\ 1 \end{pmatrix}, \dots \right\},$$

which is synchronized but does not tend to zero as $t \to \infty$.

Similarly, (6) becomes $0 < \gamma < 2/3$. This condition is sharp since when $\gamma = 2/3$, (5) becomes

$$\begin{pmatrix} u_1^{(t+1)} \\ u_2^{(t+1)} \\ u_3^{(t+1)} \end{pmatrix} = \frac{1}{3} \begin{pmatrix} -1 & 2 & 2 \\ 2 & -1 & 2 \\ 2 & 2 & -1 \end{pmatrix} \begin{pmatrix} u_1^{(t)} \\ u_2^{(t)} \\ u_3^{(t)} \end{pmatrix}.$$

Under the initial condition

$$\begin{pmatrix} u_1^{(0)} \\ u_2^{(0)} \\ u_3^{(0)} \end{pmatrix} = \begin{pmatrix} -1 \\ 1 \\ 0 \end{pmatrix},$$

it is easily verified that the corresponding solution is

$$\left\{ \begin{pmatrix} -1 \\ 1 \\ 0 \end{pmatrix}, \begin{pmatrix} 1 \\ -1 \\ 0 \end{pmatrix}, \begin{pmatrix} -1 \\ 1 \\ 0 \end{pmatrix}, \dots \right\},$$

which is not a synchronized solution of (5).

Next, note that the eigenvalues of

$$\begin{pmatrix} 1 - 3\gamma & \gamma \\ \gamma & 1 - 3\gamma \end{pmatrix}$$

are $1 - 2\gamma$ and $1 - 4\gamma$. Thus condition (7) becomes

$$\max \{ |1 - 2\gamma|, |1 - 4\gamma| \} < 1,$$

or $0 < \gamma < 1/2$. This condition is sharp. Indeed, if $\gamma = 1/2$, then (9) becomes

$$\begin{pmatrix} u_1^{(t+1)} - u_2^{(t+1)} \\ u_4^{(t+1)} - u_3^{(t+1)} \end{pmatrix} = \frac{1}{2} \begin{pmatrix} -1 & 1 \\ 1 & -1 \end{pmatrix} \begin{pmatrix} u_1^{(t)} - u_2^{(t)} \\ u_4^{(t)} - u_3^{(t)} \end{pmatrix}.$$

Since -1 is an eigenvalue of

$$\frac{1}{2} \begin{pmatrix} -1 & 1 \\ 1 & -1 \end{pmatrix}$$

and the corresponding eigenvector is $\mathrm{col}(-1, 1)$. If we let

$$u_1^{(0)} = -1, \ u_4^{(0)} = 1, \ u_2^{(0)} = u_3^{(0)} = 0,$$

then since

$$\left(\frac{1}{2}\right)^t \begin{pmatrix} -1 & 1 \\ 1 & -1 \end{pmatrix}^t \begin{pmatrix} -1 \\ 1 \end{pmatrix} = (-1)^t \begin{pmatrix} -1 \\ 1 \end{pmatrix},$$

we see that neither $\left\{\left|u_1^{(t)} - u_2^{(t)}\right|\right\}_{t=0}^{\infty}$ nor $\left\{\left|u_4^{(t)} - u_3^{(t)}\right|\right\}_{t=0}^{\infty}$ converge to zero. When $n = 5$, the matrix C_t becomes

$$F(\gamma) = \begin{pmatrix} 1 - 3\gamma & \gamma \\ \gamma & 1 - 2\gamma \end{pmatrix}$$

which has eigenvalues: $-\frac{5}{2}\gamma + 1 + \frac{1}{2}\sqrt{5}\gamma$, and $-\frac{5}{2}\gamma + 1 - \frac{1}{2}\sqrt{5}\gamma$. Thus the condition $\limsup_{t \to \infty} \rho(C_t) < 1/\Gamma$ becomes

$$\max\left\{\left|-\frac{5}{2}\gamma + 1 + \frac{1}{2}\sqrt{5}\gamma\right|, \left|-\frac{5}{2}\gamma + 1 - \frac{1}{2}\sqrt{5}\gamma\right|\right\} < 1,$$

which is equivalent to

$$0 < \gamma < \frac{4}{5 + \sqrt{5}}.$$

When $\gamma = 4/(5 + \sqrt{5})$, the matrix C_t becomes

$$\begin{pmatrix} \frac{-7+\sqrt{5}}{5+\sqrt{5}} & \frac{4}{5+\sqrt{5}} \\ \frac{4}{5+\sqrt{5}} & \frac{-3+\sqrt{5}}{5+\sqrt{5}} \end{pmatrix}.$$

The number -1 is an eigenvalue of it and the corresponding eigenvector vector is $\mathrm{col}\left(1, -\frac{1}{2}\sqrt{5} + \frac{1}{2}\right)$. If we let

$$u_1^{(0)} = 1, \ u_5^{(0)} = -\frac{1}{2}\sqrt{5} + \frac{1}{2}, \ u_2^{(0)} = u_3^{(0)} = 0,$$

then since

$$\begin{pmatrix} \frac{-7+\sqrt{5}}{5+\sqrt{5}} & \frac{4}{5+\sqrt{5}} \\ \frac{4}{5+\sqrt{5}} & \frac{-3+\sqrt{5}}{5+\sqrt{5}} \end{pmatrix}^t \begin{pmatrix} 1 \\ -\frac{1}{2}\sqrt{5} + \frac{1}{2} \end{pmatrix} = (-1)^t \begin{pmatrix} 1 \\ -\frac{1}{2}\sqrt{5} + \frac{1}{2} \end{pmatrix},$$

we see from (17) that neither $\left\{\left|u_1^{(t)} - u_2^{(t)}\right|\right\}_{t=0}^{\infty}$ nor $\left\{\left|u_5^{(t)} - u_3^{(t)}\right|\right\}_{t=0}^{\infty}$ converge to zero. This shows that the condition $\limsup_{t \to \infty} \rho(C_t) < 1/\Gamma$ is sharp.

Next, we look at the case $n = 6$ and the matrix

$$C_t = \begin{pmatrix} 1 - 3\gamma & \gamma & 0 \\ \gamma & 1 - 2\gamma & \gamma \\ 0 & \gamma & 1 - 3\gamma \end{pmatrix},$$

which has eigenvalues $1 - 3\gamma, 1 - \gamma$, and $1 - 4\gamma$.

The condition $\limsup_{t\to\infty} \rho(B_t) < 1/\Gamma$ becomes

$$\max\{|1-3\gamma|, |1-\gamma|, |1-4\gamma|\} < 1,$$

or, $0 < \gamma < 1/2$. When $\gamma = 1/2$, the matrix C_t becomes

$$\frac{1}{2}\begin{pmatrix} -1 & 1 & 0 \\ 1 & 0 & 1 \\ 0 & 1 & -1 \end{pmatrix},$$

which has the eigenvalue -1 and corresponding eigenvector $\mathrm{col}(1, -1, 1)$. By reasonings similar to those described above, we see that there is a solution of (1) which is not $\{1, 2\}$-, $\{3, 6\}$-, nor $\{4, 5\}$-partially synchronized. This shows that the condition $\limsup_{t\to\infty}\rho(B_t) < 1/\Gamma$ in Lemma 2 is sharp.

Finally, the matrix

$$\begin{pmatrix} 1-2\gamma & \gamma \\ \gamma & 1-2\gamma \end{pmatrix}$$

has the eigenvalues $1 - \gamma$, and $1 - 3\gamma$. By similar reasonings, we see that the condition $\limsup_{t\to\infty}\rho(A_t) < 1/\Gamma$ in Lemma 1 is equivalent to $0 < \gamma < 2/3$ which is also sharp.

We remark that when $\gamma_t = \gamma$, the matrices A_t, B_t and C_t can be written as $A_t = I_{m-1} + \gamma\tilde{A}_{m-1}$, $B_t = I_m + \gamma\tilde{B}_m$, and $C_t = I_m + \gamma\tilde{C}_m$ respectively, where

$$\tilde{A}_{m-1} = \begin{pmatrix} -2 & 1 & 0 & 0 & \cdots & 0 \\ 1 & -2 & 1 & 0 & \cdots & 0 \\ 0 & 1 & -2 & 1 & \cdots & 0 \\ \cdot & \cdot & \cdot & \cdot & \cdots & \cdot \\ 0 & \cdot & \cdot & \cdot & \cdots & -2 \end{pmatrix}_{(m-1)\times(m-1)}$$

$$\tilde{B}_m = \begin{pmatrix} -3 & 1 & 0 & 0 & \cdots & 0 \\ 1 & -2 & 1 & 0 & \cdots & 0 \\ 0 & 1 & -2 & 1 & \cdots & 0 \\ \cdot & \cdot & \cdot & & \cdots & \cdot \\ \cdot & \cdot & \cdot & 1 & -2 & 1 \\ \cdot & \cdot & \cdot & 0 & 1 & -3 \end{pmatrix}_{m\times m}$$

and

$$\tilde{C}_m = \begin{pmatrix} -3 & 1 & 0 & 0 & \cdots & 0 \\ 1 & -2 & 1 & 0 & \cdots & 0 \\ 0 & 1 & -2 & 1 & \cdots & 0 \\ \cdot & \cdot & \cdot & \cdot & \cdots & \cdot \\ 0 & \cdot & \cdot & \cdot & \cdots & -2 \end{pmatrix}_{m\times m}.$$

The eigenvalues of the matrix $\tilde{A}_{m-1}$ are given by [2]

$$-2 + 2\cos\frac{k\pi}{2m-1}, \quad k = 1, 2, ..., m-1.$$

The eigenvalues of the matrix $\tilde{B}_m$ can then be calculated as

$$-4, \ -2 + 2\cos\frac{k\pi}{2m-1}, \ k = 1, 2, ..., m-1.$$

The eigenvalues of $\tilde{C}_m$ are given by [2]

$$-2 + 2\cos\frac{2k\pi}{2m+1}, \ k = 1, 2, ..., m.$$

From these observations, and by means of reasonings similar to those described above, it is not difficult to show that conditions $\limsup_{t\to\infty}\rho(A_t) < 1/\Gamma$ and $\limsup_{t\to\infty}\rho(B_t) < 1/\Gamma$ in Theorem 1 are now replaced by

$$0 < \gamma < \frac{2}{\max_{1\le k\le m-1}\left(2 - 2\cos\frac{k\pi}{2m-1}\right)} = \frac{1}{1 - \cos\frac{(m-1)\tau}{2m-1}},$$

and

$$0 < \gamma < 1/2$$

respectively, while the condition $\limsup_{t\to\infty}\rho(C_t) < 1/\Gamma$ in Theorem 2 is replaced by

$$0 < \gamma < \frac{2}{\max_{1\le k\le m}\left(2 - 2\cos\frac{2k\pi}{2m+1}\right)} = \frac{1}{1 - \cos\frac{2m\pi}{2m+1}}.$$

Furthermore, these conditions are sharp.

References

[1] V. S. Afraimovich, V. I. Nekorkin, G. V. Osipov and V. D. Shalfeev, *Stability, Structures and Chaos in Nonlinear Synchronization Networks*, World Scientific, 1994.

[2] R. T. Gregory and D. Karney, *A collection of Matrices for Testing Computational Algorithms*, Wiley-Interscience, 1969.

Bifurcation of Periodic Points in Reversible Diffeomorphisms

MARIA-CRISTINA CIOCCI

and

ANDRÉ VANDERBAUWHEDE[1]

Department of Pure Mathematics and Computer Algebra
University Gent, Krijgslaan 281, B-9000 Gent, Belgium
E-mail: mcc@cage.rug.ac.be, avdb@cage.rug.ac.be

Abstract In this paper we survey a general Liapunov-Schmidt type of reduction for the study of the bifurcation of periodic points from a symmetric fixed point in families of reversible diffeomorphisms. The approach is strongly interwoven with normal form theory for reversible mappings and also addresses the stability problem for the bifurcating periodic points. The paper concludes with an application to subharmonic bifurcation in reversible vectorfields.

Keywords Reversible diffeomorphisms, Periodic points, Liapunov-Schmidt reduction, Normal forms, Subharmonic bifurcation

AMS Subject Classification 37G15, 37G40, 34C25

1   Introduction

For the study of subharmonic bifurcations from a given periodic orbit in a parametrized family of autonomous systems it is a standard approach to introduce the Poincaré map associated to the periodic orbit and to study the bifurcation of periodic points in this map. This can be done either by a normal form approach or by an appropriate Lyapunov-Schmidt type of reduction (see e.g. [5]). When the system is reversible (in a sense to be defined further on) and the periodic orbit is symmetric then also the Poincaré map will be reversible, and this additional structure should be taken into account when performing the reduction and analyzing the bifurcations. The aim of this note is to give a brief survey on how this can be done; for more details we refer to the Ph. D. thesis [1] of the first author and to the papers [2] and [3]. In the same thesis [1] and in the paper [4] one can find similar results on symplectic diffeomorphisms.

---

[1]Supported by The Fund for Scientific Research – Flanders (Belgium).

2   Preliminaries

We start with the definition of a reversible diffeomorphism. Let V be a (finite-dimensional) state space and $\Gamma \subset \mathcal{L}(V)$ a compact group of linear operators acting on V. Also, let $\chi : \Gamma \longrightarrow \mathbb{Z}_2 := \{1, -1\}$ be a non-trivial group-homomorphism (such group-homomorphism is usually called a *character* of Γ). A diffeomorphism $\Phi \in C^\infty(V)$ is then called Γ-*reversible* if

$$\Phi(\gamma \cdot x) = \gamma \cdot \Phi^{\chi(\gamma)}(x), \qquad \forall x \in V, \forall \gamma \in \Gamma.$$

The basic case appears when $\Gamma = \{I_V, R\}$, with $R \in \mathcal{L}(V)$ a linear involution (i.e. $R^2 = I_V$), and when χ is given by $\chi(I_V) = 1$ and $\chi(R) = -1$; a diffeomorhism $\Phi \in C^\infty(V)$ is then R-*reversible* if

$$R \circ \Phi \circ R = \Phi^{-1}. \tag{1}$$

For the sake of simplicity we will in this paper restrict to this simple case.

Such R-reversible diffeomorphisms arise for example as stroboscopic maps for periodic non-autonomous time-reversible systems: Φ is then the time-T-map associated to a T-periodic system of the form

$$\dot{x} = f(t, x), \tag{2}$$

with $f : \mathbb{R} \times V \to V$ such that $f(t+T, x) = f(t, x)$ and $f(-t, Rx) = -Rf(t, x)$. The flow $\varphi(t, x)$ of this system will be such that $\varphi(t, \varphi(T, x)) = \varphi(t + T, x)$ and $\varphi(t, Rx) = R\varphi(-t, x)$, and hence $\Phi := \varphi(T, \cdot)$ will satisfy (1). In a similar way one can consider an autonomous reversible system on a finite-dimensional space X, of the form

$$\dot{x} = f(x), \tag{3}$$

with $f : X \to X$ smooth and such that $f(R_0 x) = -R_0 f(x)$ for some linear involution $R_0 \in \mathcal{L}(X)$. A periodic orbit κ of (3) is *symmetric* if $R_0(\kappa) = \kappa$; one can show that these symmetric periodic orbits are precisely those orbits which have two intersection points p_1 and p_2 with $\text{Fix}(R_0) := \{x \in X \mid R_0 x = x\}$. To construct a Poincaré-map associated with κ we take two R_0-invariant transversal sections Σ_1 and Σ_2 to κ, at respectively p_1 and p_2. The Poincaré-map $\Phi : \Sigma_1 \to \Sigma_1$ can then be written as

$$\Phi = \Phi_{2 \to 1} \circ \Phi_{1 \to 2},$$

where $\Phi_{1 \to 2}$ is the "halfway" Poincaré-map from Σ_1 to Σ_2, and $\Phi_{2 \to 1}$ the second halfway Poincaré-map starting at Σ_2 and arriving at Σ_1. The reversibility of (3) together with $R_0(\Sigma_1) = \Sigma_1$ and $R_0(\Sigma_2) = \Sigma_2$ imply that $\Phi_{2 \to 1} = R_0 \circ \Phi_{1 \to 2}^{-1} \circ R_0$ and

$$\Phi = R_0 \circ \Phi_{1 \to 2}^{-1} \circ R_0 \circ \Phi_{1 \to 2}.$$

Now identify Σ_1 with a vectorspace V in such a way that $p_1 \in \Sigma_1$ corresponds to $0 \in V$, and denote the restriction of R_0 to V by R; the mapping $\Phi : V \to V$

is then an R-reversible diffeomorphism, with $\Phi(0) = 0$. Periodic points of Φ bifurcating from the fixed point $x = 0$ correspond to subharmonic periodic orbits of (3) bifurcating from the symmetric periodic orbit κ. We will return to this problem of subharmonic bifurcation in Section 6.

Consider now a parametrized family of R-reversible diffeomorphisms with a symmetric fixed point; more precisely, let $\Phi : V \times \mathbb{R}^m \to \mathbb{R}$ (with $m \geq 0$) be C^∞-smooth and such that

(H1) (i) $\Phi(0, \lambda) = 0$ for all $\lambda \in \mathbb{R}^m$; let $A_\lambda := D_x\Phi(0, \lambda)$;

  (ii) $A_0 \in \mathcal{L}(V)$ is invertible; hence $\Phi_\lambda = \Phi(\cdot, \lambda)$ is a local diffeomorphism for small λ;

  (iii) $R \circ \Phi_\lambda \circ R = \Phi_\lambda^{-1}$, with $R \in \mathcal{L}(V)$ a linear involution.

Fix some integer $q \geq 1$, and consider the following problem:

($\mathbf{P}_q$) Find, for all small $\lambda \in \mathbb{R}^m$, all q-periodic points of Φ_λ close to the fixpoint $x = 0$.

Solving ($\mathbf{P}_q$) means to find all solutions $(x, \lambda) \in V \times \mathbb{R}^m$ near $(0, 0)$ of the fixpoint equation

$$\Phi_\lambda^{(q)}(x) = x, \qquad \text{with } \Phi_\lambda^{(q)} := \Phi \circ \cdots \circ \Phi \quad (q \text{ times}). \tag{4}$$

The problem ($\mathbf{P}_q$) has what we call an *implicit D_q-symmetry*. To explain what we mean by that let us denote by $\mathcal{S}_\lambda$ the solution set:

$$\mathcal{S}_\lambda := \left\{ x \in V \mid \Phi_\lambda^{(q)}(x) = x \right\}.$$

Clearly $x \in \mathcal{S}_\lambda$ implies $\Phi_\lambda(x) \in \mathcal{S}_\lambda$, and also $Rx \in \mathcal{S}_\lambda$, as can be seen from $\Phi_\lambda^{(q)}(Rx) = R\Phi_\lambda^{(-q)}(x)$. Now observe that $\Phi_\lambda^{(q)}$ acts as the identity on $\mathcal{S}_\lambda$, and hence Φ_λ generates a $\mathbb{Z}_q = \mathbb{Z}/q\mathbb{Z}$-action on $\mathcal{S}_\lambda$. The operator R on the other hand generates a $\mathbb{Z}_2$-action, as follows from $R^2 = 1$. Finally, since $R \circ \Phi_\lambda = \Phi_\lambda^{-1} \circ R$ we conclude that Φ_λ and R generate a D_q-action on $\mathcal{S}_\lambda$. (As a reminder: D_q is the symmetry group of a regular q-gone; it contains $2q$ elements and can be generated by the rotation over $2\pi/q$ and a reflection, both in the plane). We call this D_q-symmetry implicit because it appears only on the (yet to determine!) solution set $\mathcal{S}_\lambda$.

The approach to the problem ($\mathbf{P}_q$) which we describe in the next sections is based on a Liapunov-Schmidt type of reduction which lowers the dimension of the problem and leads to algebraic bifurcation equations; this will be done in such a way that the D_q-symmetry becomes *explicit* and can be used to bring the bifurcation equations in a kind of canonical form (at least in the simplest cases). Moreover we explain the relation with normal form theory for reversible diffeomorphisms, and show how these normal forms can be used to obtain some stability properties for the bifurcating periodic points. In the final section we apply these results to the problem of subharmonic bifurcation in reversible vectorfields.

3   Orbit space formulation and reduction

The basic idea behind our approach is to replace the equation (4) for the q-periodic points of Φ_λ by an equivalent equation in an appropriate orbit space, and to perform the Liapunov-Schmidt reduction on this equivalent equation. The starting point is the observation that a q-periodic point $x \in V$ of Φ_λ generates a q-periodic orbit

$$\left(\Phi_\lambda^i(x)\right)_{i\in\mathbb{Z}} \in V^{\mathbb{Z}}, \qquad \text{with } \Phi_\lambda^{i+q}(x) = \Phi_\lambda^i(x), \ \forall i \in \mathbb{Z}.$$

This motivates us to define an *orbit space* Y_q by

$$Y_q := \left\{y = (y_i)_{i\in\mathbb{Z}} \in V^{\mathbb{Z}} \mid y_{i+q} = y_i, \ \forall i \in \mathbb{Z}\right\} \cong V^q. \tag{5}$$

The mapping Φ_λ can be lifted to this orbit space by defining $\widehat{\Phi}_\lambda : Y_q \to Y_q$ as

$$\widehat{\Phi}_\lambda(y) := \left(\Phi_\lambda(y_i)\right)_{i\in\mathbb{Z}}, \qquad \forall y \in Y_q. \tag{6}$$

We also need the *shift operator* $\sigma \in \mathcal{L}(Y_q)$ given by

$$(\sigma \cdot y)_i := y_{i+1}, \qquad \forall i \in \mathbb{Z}, \forall y \in Y_q. \tag{7}$$

With these definitions it is easy to prove the following result.

Lemma 3.1 *Let $\lambda \in \mathbb{R}^m$, and let $x \in V$ be a solution of (4). Define $y \in Y_q$ by $y_i := \Phi_\lambda^{(i)}(x)$ for all $i \in \mathbb{Z}$. Then*

$$\widehat{\Phi}_\lambda(y) = \sigma \cdot y. \tag{8}$$

Conversely, if $y \in Y_q$ solves (8) (for some $\lambda \in \mathbb{R}^m$) then $x := y_0 \in V$ is a q-periodic point of Φ_λ, i.e. x satisfies (4).

Lemma 3.1 gives a one-to-one relation between the q-periodic points of Φ_λ and the solutions of (8); using this relation the problem $(\mathbf{P}_q)$ amounts to finding all solutions $(y, \lambda) \in Y_q \times \mathbb{R}^m$ of (8) in a neighborhood of $(0,0)$.

An important property of (8) is that this equation is $\mathbb{Z}_q$-*equivariant*: it follows from $\sigma^q = I$ that σ generates a $\mathbb{Z}_q := \mathbb{Z}/q\mathbb{Z}$-action on Y_q, and the definitions immediately imply that

$$\widehat{\Phi}_\lambda \circ \sigma = \sigma \circ \widehat{\Phi}_\lambda, \qquad \forall \lambda \in \mathbb{R}^m. \tag{9}$$

Also, the reversibility of Φ_λ is inherited by $\widehat{\Phi}_\lambda$. Indeed, we have

$$\widehat{\Phi}_\lambda \circ \rho = \rho \circ \widehat{\Phi}_\lambda^{-1}, \tag{10}$$

with the linear operator $\rho \in \mathcal{L}(Y_q)$ defined by

$$(\rho \cdot y)_i := R y_{-i}, \qquad \forall i \in \mathbb{Z}, \forall y \in Y_q. \tag{11}$$

Since $\rho^2 = I$ it follows from (10) that $\widehat{\Phi}_\lambda$ is ρ-reversible. Moreover, the relation $\rho \circ \sigma = \sigma^{-1} \circ \rho$ implies that σ and ρ generate a D_q-action on the orbit space Y_q.

As a first step towards solving (8) we consider the linearized equation

$$\widehat{A}_0 \cdot y = \sigma \cdot y, \tag{12}$$

with $\widehat{A}_0 := D\widehat{\Phi}_{\lambda=0}(0) \in \mathcal{L}(Y_q)$ the lift of $A_0 := D\Phi_{\lambda=0}(0) \in \mathcal{L}(V)$. The reversibility of $\Phi_{\lambda=0}$ implies that of A_0 and $\widehat{A}_0$: $RA_0R = A_0^{-1}$ and $\rho\widehat{A}_0\rho = \widehat{A}_0^{-1}$. A result similar to that of Lemma 3.1 shows that $\mathrm{Ker}\,(\widehat{A}_0 - \sigma)$ (the solution space of (12)) and $\mathrm{Ker}\,(A_0^q - I)$ are isomorphic. For a straightforward application of the Liapunov-Schmidt reduction to (8) we have to determine complementary subspaces of respectively $\mathrm{Ker}\,(\widehat{A}_0 - \sigma)$ and $\mathrm{Im}\,(\widehat{A}_0 - \sigma)$ in Y_q; since at this point in the analysis we do not want to impose any spectral conditions on A_0 (except the fact that A_0 must be invertible) this can not be done in general. In particular, we do not want to exclude the possibility that A_0 and $\widehat{A}_0$ are not semisimple, and therefore we can not assume that we have the splittings

$$V = \mathrm{Ker}\,(A_0^q - I) \oplus \mathrm{Im}\,(A_0^q - I) \quad \text{and} \quad Y_q = \mathrm{Ker}\,(\widehat{A}_0 - \sigma) \oplus \mathrm{Im}\,(\widehat{A}_0 - \sigma).$$

which are frequently taken as a starting point for a Liapunov-Schmidt reduction. The remedy for this difficulty consists in using the semisimple part S_0 of A_0 to obtain appropriate splittings of the spaces V and Y_q. We now describe how this can be done.

We know from elementary algebra that A_0 has a unique *semisimple-nilpotent decomposition* (S-N-decomposition for short), which means that there exist unique linear operators $S_0 \in \mathcal{L}(V)$ and $N_0 \in \mathcal{L}(V)$ such that $A_0 = S_0 + N_0$, S_0 is semisimple (i.e., complex diagonalizable), N_0 is nilpotent and $S_0N_0 = N_0S_0$. Then RS_0R is the semisimple part of RA_0R, and S_0^{-1} that of A_0^{-1}; combined with the reversibilty of A_0 and the uniqueness of the S-N-decomposition this implies that also S_0 is R-reversible. Some further straightforward algebra and the fact that σ is semisimple (since $\sigma^q = I$) gives the following.

Lemma 3.2 *Let $A_0 = S_0 + N_0$ be the S-N-decomposition of A_0. Then*

- S_0 *is R-reversible: $RS_0R = S_0^{-1}$;*
- S_0^q *is the semisimple part of A_0^q;*
- $Ker(A_0^q - I) \subset Ker(S_0^q - I)$;
- $\widehat{A}_0 = \widehat{S}_0 + \widehat{N}_0$ *is the S-N-decomposition of $\widehat{A}_0$;*
- $\rho \circ \widehat{S}_0 \circ \rho = \widehat{S}_0^{-1}$;
- $(\widehat{S}_0 - \sigma)$ *is the semisimple part of $(\widehat{A}_0 - \sigma)$;*
- $Ker(\widehat{A}_0 - \sigma) \subset Ker(\widehat{S}_0 - \sigma) \cong Ker(S_0^q - I)$;
- $Y_q = Ker(\widehat{S}_0 - \sigma) \oplus Im(\widehat{S}_0 - \sigma)$;
- $(\widehat{A}_0 - \sigma)$ *is invertible on $Im(\widehat{S}_0 - \sigma)$.*

Next we introduce the *reduced phase space* for the problem $(\mathbf{P}_q)$; this is the subspace U of V given by

$$U := \operatorname{Ker}(S_0^q - I), \tag{13}$$

i.e., U is the *generalized eigenspace* of A_0 corresponding to those eigenvalues which are q-th roots of unity. These eigenvalues are the *resonant eigenvalues* for the problem $(\mathbf{P}_q)$. Observe that in a standard Liapunov-Schmidt reduction one would work with the *eigenspace* corresponding to the resonant eigenvalues, while here we will use the generalized eigenspace. The following lemma summarizes the main properties of U.

Lemma 3.3 *Let U be the reduced phase space defined by (13), and let $\zeta : U \to Y_q$ be the linear operator given by*

$$\zeta(u) := \left(S_0^i u\right)_{i \in \mathbb{Z}}, \qquad \forall u \in U. \tag{14}$$

Then the following holds:

- S_0 *generates a $\mathbb{Z}_q$-action on U;*
- U *is invariant under R;*
- ζ *is an isomorphism of U onto $\operatorname{Ker}(\widehat{S}_0 - \sigma)$;*
- $\zeta(S_0 u) = \sigma \cdot \zeta(u)$ *for all $u \in U$;*
- $\zeta(Ru) = \rho \cdot \zeta(u)$ *for all $u \in U$.*

Now we have all ingredients needed to perform our reduction of the equation (8); this reduction will be based on the splitting

$$Y_q = \zeta(U) \oplus \operatorname{Im}(\widehat{S}_0 - \sigma) \tag{15}$$

of the orbit space Y_q. Since $(\widehat{S}_0 - \sigma) \in \mathcal{L}(Y_q)$ is semisimple we have $Y_q = \operatorname{Ker}(\widehat{S}_0 - \sigma) \oplus \operatorname{Im}(\widehat{S}_0 - \sigma)$, which combined with $\operatorname{Ker}(\widehat{S}_0 - \sigma) = \zeta(U)$ gives (15). It follows that each $y \in Y_q$ can be written in a unique way as $y = \zeta(u) + w$, with $u \in U$ and $w \in W := \operatorname{Im}(\widehat{S}_0 - \sigma)$. Then $\sigma \cdot y = \zeta(S_0 u) + \sigma \cdot w$, and the equation (8) splits as

$$\begin{aligned}(a) \quad & \Psi_\lambda(u, w) = S_0 u, \\ (b) \quad & \Theta_\lambda(u, w) = \sigma \cdot w;\end{aligned} \tag{16}$$

the mappings $\Psi_\lambda : U \times W \to U$ and $\Theta_\lambda : U \times W \to W$ are uniquely determinated from the relation

$$\Phi_\lambda(\zeta(u) + w) = \zeta(\Psi_\lambda(u, w)) + \Theta_\lambda(u, w), \qquad \forall(u, w, \lambda) \in U \times W \times \mathbb{R}^m.$$

Equation (16b) can (locally near $(u, w, \lambda) = (0, 0, 0)$) be solved for $w = w_\lambda^*(u)$ by the implicit function theorem; bringing this solution into (16a) gives the *determining equation*

$$\Phi_{r,\lambda}(u) = S_0 u, \tag{17}$$

with $\Phi_{r,\lambda} : U \to U$ defined by

$$\Phi_{r,\lambda}(u) := \Psi_\lambda(u, w_\lambda^*(u)), \qquad \forall (u, \lambda) \in U \times \mathbb{R}^m. \tag{18}$$

We call $\Phi_{r,\lambda}$ the *reduced mapping* for the problem $(\mathbf{P}_q)$. We refrain from calling (17) the "bifurcation equation" since in case some of the resonant eigenvalues are non-semisimple it is possible to make a further reduction to the subspace $\mathrm{Ker}\,(A_0^q - I)$ (which is then a proper subspace of U).

The next lemma gives some basic properties of $\Phi_{r,\lambda}$; they follow easily from the definitions.

Lemma 3.4 *The reduced mapping $\Phi_{r,\lambda}$ has the following properties:*

- $\Phi_{r,\lambda}(0) = 0$ *for all* λ;
- $D\Phi_{r,\lambda=0}(0) = A_0|_U$;
- $\Phi_{r,\lambda}$ *is* $\mathbb{Z}_q$-*equivariant:* $\Phi_{r,\lambda} \circ S_0 = S_0 \circ \Phi_{r,\lambda}$;
- $\Phi_{r,\lambda}$ *is* R-*reversible:* $R \circ \Phi_{r,\lambda} \circ R = \Phi_{r,\lambda}^{-1}$.

The main result is of course the relation between the solutions of $(\mathbf{P}_q)$ and the solutions of the determing equation.

Theorem 3.5 *Under the foregoing conditions there exists a smooth mapping* $x^* : U \times \mathbb{R}^m \to V$, *with* $x^*(0, \lambda) = 0$, $D_u x^*(0, 0) \cdot u = u$ *and* $x^*(Ru, \lambda) = Rx^*(u, \lambda)$ *for all* (u, λ), *and such that the following holds: for all sufficiently small* $(x, \lambda) \in V \times \mathbb{R}^m$ *we have* $\Phi_\lambda^{(q)}(x) = x$ *if and only if* $x = x^*(u, \lambda)$ *for some small* $u \in U$ *for which the determining equation* $\Phi_{r,\lambda}(u) = S_0 u$ *is satisfied; moreover, we have for such* (u, λ) *that* $x^*(S_0 u, \lambda) = \Phi_\lambda(x^*(u, \lambda))$.

Observe that the $\mathbb{Z}_q$-equivariance of $\Phi_{r,\lambda}$ implies that for each solution $u \in U$ of (17) also the other points $S_0 u, S_0^2 u, \ldots, S_0^{q-1}$ on the $\mathbb{Z}_q$-orbit of u solve the same equation; the last statement of Theorem 3.5 says that the mapping $x^*(\cdot, \lambda)$ lifts this solution orbit of (17) to a full q-periodic orbit of Φ_λ.

It is interesting to consider for a moment the special case where we assume that the mappings Φ_λ are such that

$$\Phi_\lambda \circ S_0 = S_0 \circ \Phi_\lambda. \tag{19}$$

One can then verify that

$$x^*(u, \lambda) = u \qquad \text{and} \qquad \Phi_{r,\lambda} = \Phi_\lambda|_U,$$

and the reduction result of Theorem 3.5 takes the following form: *for sufficiently small* $(x, \lambda) \in V \times \mathbb{R}^m$ *we have that* x *is a* q-*periodic point of* Φ_λ *if and only if* $x = u \in U$ *and* $\Phi_\lambda(u) = S_0 u$.

Returning to the general case we can look for the q-periodic points of the reduced mapping $\Phi_{r,\lambda}$, and apply Theorem 3.5 to this new problem (i.e., we perform our reduction on the equation $\Phi_{r,\lambda}^{(q)}(u) = u$). It follows from Lemma 3.4 that $\Phi_{r,\lambda}$ satisfies the condition (19), and one can easily see that the reduced phase space for $\Phi_{r,\lambda}$ is the space U itself (there is no further reduction). The reduction result for the special case then leads to the following conclusion: *for each sufficiently small $(u, \lambda) \in U \times \mathbb{R}^m$ we have that u is a q-periodic point of $\Phi_{r,\lambda}$ if and only if u satisfies the determining equation $\Phi_{r,\lambda}(u) = S_0 u$. The q-periodic orbits of $\Phi_{r,\lambda}$ are therefore necessarily also orbits under the natural $\mathbb{Z}_q$-action on U.* This conclusion allows us to reformulate the reduction theorem as follows.

Theorem 3.6 (Main Reduction Theorem) *Let Φ_λ ($\lambda \in \mathbb{R}^m$) be a parameterized family of R-reversible diffeomorphisms, satisfying the hypotheses (H1). Let $q \geq 1$, and define the reduced phase space U by (13). Then there exists a parameterized family of reduced R-reversible diffeomorphisms $\Phi_{r,\lambda}$: $U \to U$ such that for each sufficiently small $\lambda \in \mathbb{R}^m$ there is a 1-to-1 relation between the small q-periodic orbits of Φ_λ and the small q-periodic orbits of $\Phi_{r,\lambda}$. Moreover, $\Phi_{r,\lambda}$ is equivariant with respect to the natural $\mathbb{Z}_q$-action on U, and all small q-periodic orbits of $\Phi_{r,\lambda}$ are necessarily $\mathbb{Z}_q$-orbits; they are given by the solutions of the determining equation (17).*

We have observed before that the problem $(\mathbf{P}_q)$ has an implicit D_q-symmetry, in the sense that there is a natural D_q-action on the solution set. With the reduction the $\mathbb{Z}_q$-part of this implicit symmetry has become *explicit*: indeed, the determining equation (17) is equivariant with respect to the Z_q-action generated by S_0 on U. To get a full D_q-equivariance one has to go yet one step further, by showing that for small $(u, \lambda) \in U \times \mathbb{R}^m$ the equation (17) is equivalent to the equation

$$\mathcal{G}(u, \lambda) := S_0^{-1} \Phi_{r,\lambda}(u) - S_0 \Phi_{r,\lambda}^{-1}(u) = 0. \tag{20}$$

Using the $\mathbb{Z}_q$-equivariance and the R-reversibility of $\Phi_{r,\lambda}$ and S_0 it is easily seen that the mapping $\mathcal{G} : U \times \mathbb{R}^m \to U$ is D_q-equivariant:

$$\mathcal{G}(S_0 u, \lambda) = S_0 \mathcal{G}(u, \lambda) \qquad \text{and} \qquad \mathcal{G}(Ru, \lambda) = -R\mathcal{G}(u, \lambda).$$

Replacing the determining equation (17) by the equivalent D_q-equivariant equation (20) makes the full D_q-symmetry explicit. As will be illustrated by the examples of Section 6 this D_q-equivariance is important when solving (20) explicitly.

4   Normal form of reversible diffeomorphisms

In order to apply the reduction result of Theorem 3.6 on concrete examples one needs some method to calculate or approximate the reduced diffeomorphism $\Phi_{r,\lambda}$. One possible approach is just to follow the reduction procedure as outlined in the foregoing section; since the reduction is based on splitting the original equations and applying the implicit function theorem it is in principle possible to obtain the Taylor expansion of $\Phi_{r,\lambda}(u)$ at $(u, \lambda) = (0, 0)$ up to any desired order. A different approach which we briefly describe in this section consists in first bringing the original mapping Φ_λ in an appropriate normal form; it appears that it is then very easy to obtain (a good approximation of) the reduced mapping $\Phi_{r,\lambda}$ from this normal form.

We first consider normal forms for general diffeomorphisms. Let us assume that $\Phi : V \times \mathbb{R}^m \to V$ is a smooth mapping satisfying the following properties:

(H2) (i) $\Phi(0, \lambda) = 0$ for all $\lambda \in \mathbb{R}^m$;

   (ii) $A_0 := D_x\Phi(0,0) \in \mathcal{L}(V)$ is invertible; hence $\Phi_\lambda = \Phi(\cdot, \lambda)$ is a local diffeomorphism (near $x = 0$) for small λ.

Next to the S-N-decomposition $A_0 = S_0 + N_0$ of A_0 there is also the so-called *semisimple-unipotent decomposition*

$$A_0 = S_0 e^{\mathcal{N}_0}, \tag{21}$$

with $S_0 \in \mathcal{L}(V)$ semisimple, $\mathcal{N}_0 \in \mathcal{L}(V)$ nilpotent, and $S_0\mathcal{N}_0 = \mathcal{N}_0 S_0$. This decomposition is unique, with S_0 the same as in the S-N-decomposition and with $e^{\mathcal{N}_0} = I + S_0^{-1}N_0$. Fix some $k \geq 1$; starting from (21) and using Taylor expansions at $x = 0$ one can then determine order by order a parameter-dependent polynomial vectorfield $Z_\lambda : V \to V$ with $Z_\lambda(0) = 0$ and $DZ_{\lambda=0}(0) = 0$, and such that

$$\Phi_\lambda = S_0 e^{\mathcal{N}_0 + Z_\lambda} + \mathcal{O}(\| \cdot \|^{k+1}); \tag{22}$$

here the exponential of $\mathcal{N}_0 + Z_\lambda$ stands for the time-one-map corresponding to the vectorfield $\mathcal{N}_0 + Z_\lambda$. The normal form reduction then consists in using near-identity transformations to bring the vectorfield Z_λ in a form which satisfies certain additional conditions. In order to do so we need the following technical result.

Lemma 4.1 *Given a semisimple $S_0 \in \mathcal{L}(V)$ there exists a scalar product $\langle \cdot, \cdot \rangle$ on V such that for each $A \in \mathcal{L}(V)$ we have $AS_0 = S_0 A$ if and only if $A^T S_0 = S_0 A^T$ (with the transpose taken w.r.t. $\langle \cdot, \cdot \rangle$).*

Using this scalar product one can prove the following.

Theorem 4.2 (Normal Form Theorem) *Assume that Φ_λ satisfies (H2). Then there exists for each $k \geq 1$ a parameter-dependent near-identity transformation $\Psi_{k,\lambda} : V \to V$ such that*

$$\Psi_{k,\lambda}^{-1} \circ \Phi_\lambda \circ \Psi_{k,\lambda} = S_0 e^{\mathcal{N}_0 + Z_\lambda} + \mathcal{O}(\| \cdot \|^{k+1}), \tag{23}$$

with $Z_\lambda(0) = 0$, $DZ_{\lambda=0}(0) = 0$ and such that

$$S_0 \circ Z_\lambda = Z_\lambda \circ S_0 \qquad and \qquad e^{t\mathcal{N}_0^T} \circ Z_\lambda = Z_\lambda \circ e^{t\mathcal{N}_0^T}, \quad \forall t \in \mathbb{R}. \tag{24}$$

Under the foregoing conditions one calls

$$\Phi_\lambda^{NF} := S_0 e^{\mathcal{N}_0} + Z_\lambda$$

the *normal form* of Φ_λ up to order k. The first condition in (24) implies that Φ_λ^{NF} commutes with S_0, while the second condition is equivalent with

$$DZ_\lambda(x) \cdot \mathcal{N}_0^T x = \mathcal{N}_0^T Z_\lambda(x), \qquad \forall (x, \lambda) \in V \times \mathbb{R}^m. \tag{25}$$

In case Φ_λ is R-reversible we find that $\mathcal{N}_0$ and Z_λ in (22) will be R-reversible *vectorfields*:

$$R\mathcal{N}_0 = -\mathcal{N}_0 R \qquad \text{and} \qquad R \circ Z_\lambda = -Z_\lambda \circ R. \tag{26}$$

Moreover, there exists a scalar product on V which next to the property given by Lemma 4.1 is also such that R is orthogonal: $RR^T = I$. The near-identity transformation $\Psi_{k,\lambda}$ given by the Normal Form Theorem can then be chosen such that it commutes with R, and as a consequence the reversibility of Φ_λ is maintained after the normal form transformation. It follows that modulo a near-identity transformation we can assume that

$$\Phi_\lambda = \Phi_\lambda^{NF} + \mathcal{O}(\| \cdot \|^{k+1}) \tag{27}$$

with (in particular)

$$S_0 \circ \Phi_\lambda^{NF} = \Phi_\lambda^{NF} \circ S_0 \qquad \text{and} \qquad R \circ \Phi_\lambda^{NF} \circ R = (\Phi_\lambda^{NF})^{-1}. \tag{28}$$

One can also arrange to have (25) satisfied, but this will be of no particular help in determinig the reduced diffeomorphism $\Phi_{r,\lambda}$; it may however be very helpful in solving the determining equation (see Section 6 for an example).

Theorem 4.3 *Assume (H1), (27) and (28). Then the mappings x^* and $\Phi_{r,\lambda}$ given by Theorem 3.5 are of the form*

$$x^*(u, \lambda) = u + \mathcal{O}(\|u\|^{k+1}) \qquad and \qquad \Phi_{r,\lambda}(u) = \Phi_\lambda^{NF}(u) + \mathcal{O}(\|u\|^{k+1}). \tag{29}$$

This means that we obtain an approximation of $\Phi_{r,\lambda}$ by first bringing Φ_λ in normal form up to the desired order and then restricting the normal form to the reduced phase space U. Setting $\Psi_\lambda^{NF} := S_0^{-1} \Phi_\lambda^{NF} = e^{\mathcal{N}_0} + Z_\lambda$ one has

$$\Phi_\lambda = S_0 \Psi_\lambda^{NF} + \mathcal{O}(\| \cdot \|^{k+1}) \qquad \text{and} \qquad \Phi_{r,\lambda}(u) = S_0 \Psi_\lambda^{NF}(u) + \mathcal{O}(\|u\|^{k+1}). \tag{30}$$

Up to terms of order k the determining equation (17) takes the form

$$\Psi_\lambda^{NF}(u) = u, \tag{31}$$

which for sufficiently small (u, λ) is equivalent to

$$\mathcal{N}_0(u) + Z_\lambda(u) = 0. \tag{32}$$

This means that the solutions of the determining equation (i.e., the q-periodic points of $\Phi_{r,\lambda}$) can be approximated by the equilibria $u \in U$ of the normal form vectorfield $\mathcal{N}_0 + Z_\lambda(\cdot)$ (both this vectorfield and Ψ_λ^{NF} leave the subspace U invariant). Observe also that due to (24) and (26) the equation (32) is D_q-equivariant.

5 Stability of bifurcating periodic points

In this section we describe how the foregoing reduction and normal form results can also be used to determine the stability properties of bifurcating periodic orbits. Let $x \in V$ be a q-periodic point of Φ_λ, i.e. $\Phi_\lambda^{(q)}(x) = x$; the (linear) stability of the corresponding periodic orbit $y = (y_i)_{i \in \mathbb{Z}} = (\Phi_\lambda^{(i)}(x))_{i \in \mathbb{Z}} \in Y_q$ is then determined by the eigenvalues of $D\Phi_\lambda^{(q)}(x) \in \mathcal{L}(V)$. Observe that by the chain rule

$$D\Phi_\lambda^{(q)}(y_i) = D\Phi_\lambda(y_{i+q-1}) \cdot D\Phi_\lambda(y_{i+q-2}) \cdots D\Phi_\lambda(y_{i+1}) \cdot D\Phi_\lambda(y_i), \quad (33)$$

from which it follows that the spectrum of $D\Phi_\lambda^{(q)}(x_i)$ is independent of $i \in \mathbb{Z}$; therefore we can just take $i = 0$ and study $D\Phi_\lambda^{(q)}(x)$. For bifurcating periodic orbits we have $x = x^*(u, \lambda)$ and $y_i = x^*(S_0^i u, \lambda)$, with $(u, \lambda) \in U \times \mathbb{R}^m$ small and such that $\Phi_{r,\lambda}(u) = S_0 u$; substituting this into (33) (with $i = 0$) shows that the stability of such bifurcating periodic orbit is determined by the eigenvalues of $D\Phi_\lambda^{(q)}(x^*(u, \lambda) = \mathcal{D}(u, \lambda) \in \mathcal{L}(V)$, with $\mathcal{D} : U \times \mathbb{R}^m \to \mathcal{L}(V)$ defined by

$$\mathcal{D}(u, \lambda) := D\Phi_\lambda(x^*(S_0^{q-1}u, \lambda)) \cdots D\Phi_\lambda(x^*(S_0 u, \lambda)) \cdot D\Phi_\lambda(x^*(u, \lambda)). \quad (34)$$

Moreover, as we will see in Section 6, periodic orbits bifurcating from a symmetric fixed point (here taken to be $x = 0$) will typically themselves also be symmetric; this means that $Ru = S_0^j u$ and $Rx^*(u, \lambda) = \Phi_\lambda^{(j)}(x^*(u, \lambda))$ for some $j \in \mathbb{Z}$. For convenience of formulation we will say that $u \in U$ is *symmetric* if the $\mathbb{Z}_q$-orbit through u is invariant under R, i.e. if $Ru = S_0^j u$ for some $j \in \mathbb{Z}$. If (u, λ) solves the determining equation (17) (or the equivalent equation (20)) and if u is symmetric then the corresponding q-periodic orbit of Φ_λ is also symmetric. A straightforward calculation shows the following result for such symmetric orbits.

Lemma 5.1 *Let $(u, \lambda) \in U \times \mathbb{R}^m$ be sufficiently small and such that u is symmetric. Then there exists a linear involution $T \in \mathcal{L}(V)$ (i.e. $T^2 = I$) such that*

$$T\mathcal{D}(u, \lambda)T = \mathcal{D}(u, \lambda)^{-1}.$$

This implies that if $\mu \in \mathbb{C}$ is an eigenvalue of $\mathcal{D}(u, \lambda)$ with u symmetric then also μ^{-1} is an eigenvalue. As a consequence we can only have a weak form of stability for symmetric periodic orbits, namely when all eigenvalues of $\mathcal{D}(u, \lambda)$ are on the unit circle; the only alternative is to have eigenvalues both inside and outside the unit circle, in which case the periodic orbit is unstable. In what follows we restrict the discussion to such symmetric periodic orbits.

   For $(u, \lambda) = (0, 0)$ we find $\mathcal{D}(0, 0) = A_0^q$, which implies that $\mu = 1$ is an eigenvalue of $\mathcal{D}(0, 0)$ with algebraic multiplicity equal to dim U, and with geometric multiplicity equal to the sum of the geometric multiplicities of the resonant eigenvalues of A_0. Now we make an additional hypothesis:

(H3) All non-resonant eigenvalues μ of A_0 (i.e., $\mu^q \neq 1$) are simple and lie on the unit circle (i.e. $|\mu| = 1$).

For small (u, λ) the eigenvalues of $\mathcal{D}(u, \lambda)$ are close to those of $\mathcal{D}(0,0)$; in particular, if (H3) holds then the eigenvalues of $\mathcal{D}(u, \lambda)$ which are not close to $\mu = 1$ will be simple. If moreover u is symmetric this can be combined with Lemma 5.1 to conclude that the eigenvalues of $\mathcal{D}(u, \lambda)$ which are not close to $\mu = 1$ will be simple and on the unit circle. Consequently, under the hypothesis (H3) the stability of bifurcating symmetric periodic orbits will be determined by the *critical eigenvalues* of $\mathcal{D}(u, \lambda)$, that is by those eigenvalues which are close to $\mu = 1$. The total multiplicity of these critical eigenvalues is equal to $\dim U$.

To calculate the critical eigenvalues of $\mathcal{D}(u, \lambda)$ one can put $\mathcal{D}(u, \lambda)$ in block form using the splitting $V = \mathrm{Ker}\,(S_0^q - I) \oplus \mathrm{Im}\,(S_0^q - I) = U \oplus \mathrm{Im}\,(S_0^q - I)$, and prove that there exists a similarity transformation which makes this block form diagonal; the critical eigenvalues are then the eigenvalues of the block corresponding to the subspace U. This diagonalization procedure is relatively easy to work out when Φ_λ is in normal form up to a sufficiently high order, namely when $\Phi_\lambda = S_0 \Psi_\lambda^{NF} + \mathcal{O}(\|\cdot\|^{k+1})$ with $\Psi_\lambda^{NF} = S_0^{-1}\Phi_\lambda^{NF} = e^{\mathcal{N}_0 + Z_\lambda}$ such that

$$S_0 \circ \Psi_\lambda^{NF} = \Psi_\lambda^{NF} \circ S_0 \qquad \text{and} \qquad R \circ \Psi_\lambda^{NF} \circ R = (\Psi_\lambda^{NF})^{-1}. \tag{35}$$

In this case the procedure outlined above gives the following result.

Theorem 5.2 *Assume that Φ_λ satisfies (H1) and is in normal form up to order $k \geq 2$, i.e., we have (30) and (35). Then there exists a smooth mapping $\widehat{D} : U \times \mathbb{R}^m \to \mathcal{L}(U)$, with*

$$\widehat{D}(u, \lambda) = D\Psi_\lambda^{NF}(u)\big|_U + \mathcal{O}(\|u\|^k),$$

such that for all sufficiently small solutions (u, λ) of the determining equation (17) the critical eigenvalues of $\mathcal{D}(u, \lambda)$ are given by the q-th powers of the eigenvalues of $\widehat{D}(u, \lambda)$.

At the end of Section 4 we noticed that up to terms of order k the solutions of (17) are given by the fixed points of $\Psi_\lambda^{NF}\big|_U$, or equivalently, by the zeros of $(\mathcal{N}_0 + Z_\lambda)\big|_U$. According to the foregoing result and under the hypothesis (H3) the stability of the corresponding periodic orbit of Φ_λ is determined up to terms of order $(k - 1)$ by the eigenvalues of

$$D\Psi_\lambda^{NF}(u)\big|_U = e^{(\mathcal{N}_0 + DZ_\lambda(u))}\big|_U ;$$

this means that up to higher order terms the stability properties of the bifurcating periodic solutions of Φ_λ are the same as those of the corresponding equilibria of the normal vectorfield $\mathcal{N}_0 + Z_\lambda$ restricted to the reduced phase space U. If u is symmetric (i.e. $RS_0^j u = u$ for some $j \in \mathbb{Z}$) then

$$(RS_0^j) \circ D\Psi_\lambda^{NF}(u) \circ (RS_0^j) = (D\Psi_\lambda^{NF}(u))^{-1}$$

and

$$(RS_0^j) \circ DZ_\lambda(u) \circ (RS_0^j) = -DZ_\lambda(u),$$

which again leads to either a weak form of stability or instability. In applications the challenge will usually be to take the approximation order k sufficiently large such that the higher order terms do not disturb the picture obtained from the study of the normal form.

6   Application

In this last section we show how the foregoing methods and techniques can be used to study the problem of subharmonic bifurcation from a symmetric periodic orbit κ in the reversible system (3). As described in Section 2 this problem can be put into the form (4) by introducing a Poincaré-map P associated to κ. It is possible to construct P such that it inherits the reversibility of the vectorfield f; the fixed point of P corresponding to κ will then also be symmetric. Periodic points bifurcating from this fixed point correspond to bifurcating subharmonic solutions. We first describe the set-up and hypotheses in detail.

Let X be an even-dimensional state space $(\dim X = 2n)$, and $R_0 \in \mathcal{L}(X)$ a linear involution with $\dim(\text{Fix } R_0) = n$. Consider a smooth vectorfield $f : X \to X$ which is R_0-reversible ($f(R_0x) = -R_0F(x)$ for all $x \in X$), and denote by $\phi_t(x)$ the flow of the system

$$\dot{x} = f(x). \tag{36}$$

The reversibility of f implies that $\phi_t(R_0x) = R_0\phi_{-t}(x)$ for all $(t,x) \in \mathbb{R} \times X$. Let $p_1 \in \text{Fix } R_0$ and $T_0 > 0$ be such that $\phi_{T_0}(p_1) = p_1$ and $\phi_t(p_1) \neq p_1$ for all t such that $0 < t < T_0$; then p_1 generates a symmetric T_0-periodic orbit $\kappa := \{\phi_t(p_1) \mid t \in \mathbb{R}\}$ which has a second intersection point with Fix R_0, namely $p_2 := \phi_{T_0/2}(p_1)$. Now $R_0f(p_1) = -f(R_0p_1) = -f(p_1)$, and since R_0 is semisimple it follows that $X = \mathbb{R}f(p_1) \oplus V$ for some R_0-invariant subspace V which must then necessarily contain Fix R_0. This implies $p_1 \in V$, and we can use V as a transversal section to κ and construct a Poincaré-map $P : V \to V$ which is well-defined near p_1 and such that $P(p_1) = p_1$. Moreover, from $R_0(V) = V$ and $\phi_t(R_0x) = R_0\phi_{-t}(x)$ we get $R_0 \circ P \circ R_0 = P^{-1}$. Let $R := R_0|_V$, and define $\Phi : V \to V$ by $\Phi(x) := P(p_1 + x) - p_1$ (for all $x \in V$); then $\dim V = 2n - 1$, $\dim(\text{Fix } R) = n$, Φ is R-reversible and $\Phi(0) = 0$.

Next we have to consider those eigenvalues of $A_0 = D\Phi(0) = DP(p_1)$ which are roots of unity. It is a standard result that the eigenvalues of A_0 can be obtained from those of the monodromy matrix $M = D\phi_{T_0}(p_1)$; indeed, we have

$$A_0x = Mx + \gamma(x)f(p_1), \qquad \forall x \in V, \tag{37}$$

where $\gamma \in \mathcal{L}(V; \mathbb{R})$ is such that the right hand side of (37) belongs to V for all $x \in V$. Clearly $R_0MR_0 = M^{-1}$, and therefore if $\mu \in \mathbb{C}$ is an eigenvalue of

M then so are $\bar{\mu}$, μ^{-1} and $\bar{\mu}^{-1}$. A further consequence is that $(\det M)^2 = 1$, and since M belongs to the connected component of the identity in $\mathrm{GL}(X;\mathbb{R})$ we conclude that $\det M = 1$. Taking all this into account it follows that if $\mu = -1$ is an eigenvalue of M (i.e., a Floquet multiplier) then its multiplicity must be even. Also, $\mu = 1$ is always a multiplier, since $Mf(p_1) = f(p_1)$; the (algebraic) multiplicity of $\mu = 1$ must necessarily be even. Generically, $\mu = 1$ will be a non-semisimple multiplier with multiplicity equal to 2; in such case both $\mathrm{Ker}\,(M - I)$ and $\mathrm{Ker}\,((M - I)^2)$ are R_0-invariant, and since R_0 is semisimple we can find an R_0-invariant and one-dimensional complement U_0 of $\mathrm{Ker}\,(M - I)$ in $\mathrm{Ker}\,((M - I)^2)$. Then there exists a unique vector $u_0 \in U_0$ such that $Mu_0 = u_0 + f(p_1)$. Using the R_0-invariance of U_0, the R_0-reversibility of M and the fact that $R_0 f(p_1) = -f(R_0 p_1) = -f(p_1)$, it follows that $R_0 u_0 \in U_0$ and $M R_0 u_0 = R_0 u_0 + f(p_1)$; the uniqueness of u_0 then implies $R_0 u_0 = u_0$. Using (37) this gives the following possibilities for the eigenvalues of A_0 which are q-th roots of unity:

- $q = 1$: the eigenvalue $\mu = 1$ has always odd multiplicity, and there is at least one eigenvector u_0 which belongs to $\mathrm{Fix}\,R$; typically $\mu = 1$ will be a simple eigenvalue, with $\mathrm{Ker}\,(A_0 - I) = \mathbb{R}u_0$;
- $q = 2$: if $\mu = -1$ is an eigenvalue then it must have even multiplicity; generically this multiplicity will be equal to 2, and the eigenvalue will be non-semisimple;
- $q \geq 3$: eigenvalues of the form $\mu = \exp(\pm 2\pi i p/q)$, with $q \geq 3$, $0 < p < q$ and $\gcd(p,q) = 1$ can have any multiplicity but will typically be simple eigenvalues.

We next consider the bifurcation of q-periodic points from the fixed point $x = 0$ for different choices of q, each time assuming the simplest possible hypotheses for the resonant eigenvalues. To obtain the bifurcating solution branches we will concentrate on the normal form part of the equations, neglecting the higher order terms; a more careful analysis shows that the results obtained in this way persist when the neglected terms are taken into account.

6.1   The primary branch

We start with the bifurcation of fixed points, i.e., we take $q = 1$ in our reduction scheme. According to the foregoing discussion we can generically assume that $\mu = 1$ is a simple eigenvalue of A_0, with an eigenvector $u_0 \in V$ such that $Su_0 = u_0$, $N_0 u_0 = 0$ and $Ru_0 = u_0$. Then $U = \{\alpha u_0 \mid \alpha \in \mathbb{R}\}$, and the normal form vectorfield Z restricted to U takes the form $Z(\alpha u_0) = g(\alpha)u_0$; the reversibility of Z implies that $g(\alpha) = -g(\alpha)$, i.e. $g(\alpha) \equiv 0$. So the equation (32) is satisfied for all $u = \alpha u_0 \in U$, meaning that the fixed point $x = 0$ of Φ belongs to a one-parameter family of symmetric fixed points, given by $\{x^*(\alpha u_0) \mid \alpha \in \mathbb{R}\}$. For the original reversible system (36) the conclusion is that under the generic assumption for the multiplier $\mu = 1$ the symmetric periodic orbit κ belongs to a one-parameter family of such symmetric periodic orbits. In view of what follows we will call this the *primary branch* of symmetric periodic orbits.

6.2 Period-doubling bifurcations

Next we consider period doubling, corresponding to $q = 2$. We assume again that $\mu = 1$ is a simple eigenvalue of A_0 (with eigenvector u_0 such that $Ru_0 = u_0$); we also assume that $\mu = -1$ is a non-semisimple eigenvalue with algebraic multiplicity equal to 2 and with eigenvector v_0. Then $S_0 v_0 = -v_0$, $N_0 v_0 = 0$, and since $\mathrm{Ker}\,(A_0 + I)$ is invariant under R (by the reversibility) we must have either $Rv_0 = v_0$ or $Rv_0 = -v_0$; we can without loss of generality assume that $Rv_0 = v_0$, since in the other case ($Rv_0 = -v_0$) we have $RS_0 v_0 = v_0$, and replacing R by RS_0 in what follows the same analysis goes through. Both $\mathrm{Ker}\,(A_0 + I) = \mathbb{R}v_0$ and $\mathrm{Ker}\,((A_0 + I)^2)$ are R-invariant, and we can find a one-dimensional and R-invariant complement W_0 of $\mathbb{R}v_0$ in $\mathrm{Ker}\,((A_0 + I)^2)$. There exists a unique element $w_0 \in W_0$ such that $A_0 w_0 = -w_0 - v_0$; from this one deduces that $A_0 R w_0 = -R w_0 + v_0$, and hence $R w_0 = -w_0$. The reduced phase space U is then given by

$$U = \{\alpha u_0 + \xi v_0 + \eta w_0 \mid \alpha, \xi, \eta \in \mathbb{R}\},$$

and the restrictions of S_0, N_0 and R to U by

$$\left\{ \begin{array}{rcl} S_0(\alpha u_0 + \xi v_0 + \eta w_0) & = & \alpha u_0 - \xi v_0 - \eta w_0, \\ N_0(\alpha u_0 + \xi v_0 + \eta w_0) & = & \eta v_0, \\ R(\alpha u_0 + \xi v_0 + \eta w_0) & = & \alpha u_0 + \xi v_0 - \eta w_0. \end{array} \right.$$

The normal form vectorfield Z (restricted to U) can be written as

$$Z(\alpha u_0 + \xi v_0 + \eta w_0) = g(\alpha, \xi, \eta) u_0 + h_1(\alpha, \xi, \eta) v_0 + h_2(\alpha, \xi, \eta) w_0,$$

with functions $g(\alpha, \xi, \eta)$, $h_1(\alpha, \xi, \eta)$ and $h_2(\alpha, \xi, \eta)$ which are of second order in the origin (since $Z(0) = 0$ and $DZ(0) = 0$), and such that

$$\left\{ \begin{array}{l} g(\alpha, -\xi, -\eta) = g(\alpha, \xi, \eta), \\ h_1(\alpha, -\xi, -\eta) = -h_1(\alpha, \xi, \eta), \\ h_2(\alpha, -\xi, -\eta) = -h_2(\alpha, \xi, \eta), \end{array} \right. \quad \text{and} \quad \left\{ \begin{array}{l} g(\alpha, \xi, -\eta) = -g(\alpha, \xi, \eta), \\ h_1(\alpha, \xi, -\eta) = -h_1(\alpha, \xi, \eta), \\ h_2(\alpha, \xi, -\eta) = h_2(\alpha, \xi, \eta). \end{array} \right.$$

(These correspond to the conditions that Z should commute with S_0 and anti-commute with R.) In order to impose the condition (25) we can use a scalar product such that the basis $\{u_0, v_0, w_0\}$ of U is orthonormal (compare with Lemma 4.1 and the condition $R^T R = I$). Then $N_0^T(\alpha u_0 + \xi v_0 + \eta w_0) = \xi w_0$, and (25) takes the form

$$\xi \frac{\partial g}{\partial \eta}(\alpha, \xi, \eta) = 0, \quad \xi \frac{\partial h_1}{\partial \eta}(\alpha, \xi, \eta) = 0 \quad \text{and} \quad \xi \frac{\partial h_2}{\partial \eta}(\alpha, \xi, \eta) = h_1(\alpha, \xi, \eta).$$

Imposing all these conditions leads to

$$g(\alpha, \xi, \eta) \equiv 0, \qquad h_1(\alpha, \xi, \eta) \equiv 0 \qquad \text{and} \qquad h_2(\alpha, \xi, \eta) = \xi \varphi(\alpha, \xi^2),$$

with $\varphi : \mathbb{R}^2 \to \mathbb{R}$ a smooth function such that $\varphi(0,0) = 0$. The solutions of (32) are given by either $(\alpha, \xi, \eta) = (\alpha, 0, 0)$, with $\alpha \in \mathbb{R}$ small and arbitrary (since $S_0(\alpha u_0) = u_0$ these correspond to the primary branch), or by $(\alpha, \xi, \eta) = (\alpha, \xi, 0)$ with $(\alpha, \xi) \in \mathbb{R}^2$ such that

$$\varphi(\alpha, \xi^2) = 0. \tag{38}$$

We have $\varphi(0,0) = 0$, and we assume that also the *transversality condition*

$$\frac{\partial \varphi}{\partial \alpha}(0,0) \neq 0 \tag{39}$$

is satisfied. One can easily verify that the linear operator $\mathcal{N}_0 + DZ(\alpha u_0)$ restricted to U has eigenvalues $\mu = 0$ and $\mu = \pm\sqrt{\varphi(\alpha, 0)}$, corresponding to, respectively, the multiplier $\mu = 1$ and the approximate multipliers $\mu = -\exp\left(\pm\sqrt{\varphi(\alpha, 0)}\right)$ along the primary branch; the condition (39) means that as we move along the primary branch two complex conjugate multipliers move along the unit circle and with non-zero speed towards -1, and after colliding split off the unit circle into a pair of real multipliers moving away from -1 along the real axis, one inside and the other one outside the unit circle. Assuming (39) we can solve (38) for $\alpha = \alpha^*(\xi^2)$, giving us the solution branch $\{(\alpha^*(\xi^2), \xi, 0) \mid \xi \in \mathbb{R}\}$ of the determining equation. For fixed $\xi \neq 0$ the two solutions $(\alpha^*(\xi^2), \pm\xi, 0)$ correspond to the two points of a symmetric 2-periodic orbit of Φ. For the original reversible system (36) this means that a single branch of symmetric periodic orbits bifurcates from the primary branch; the limiting period along this branch is $2T_0$, and so we have period-doubling.

Using the approach of Section 5 one can also determine the stability of these bifurcating periodic solutions. Writing $\varphi(\alpha, \xi^2) = C(\alpha) + D(\alpha)\xi^2 + O(\xi^4)$ one finds that the eigenvalues of $\mathcal{N}_0 + DZ(\alpha^*(\xi^2)u_0 + \xi v_0)$ (restricted to U) are given by $\mu = 0$ and $\mu = \pm|\xi|\sqrt{2D(0)} + O(|\xi|^2)$. Taking the exponential one obtains along the bifurcating branch two critical multipliers: these multipliers are real and off the unit circle (i.e., we have instability) if $D(0) > 0$, and they lie on the unit circle (stability) if $D(0) < 0$. So the stability is determined by the sign of a third order coefficient in the normal form.

6.3   Subharmonic bifurcation with $q \geq 3$

In this final subsection we look at the case $q \geq 3$. Let $\theta_0 := 2\pi p/q$, with $q \geq 3$, $0 < p < q$ and $\gcd(p, q) = 1$. We assume that next to the simple eigenvalue $\mu = 1$ the operator $A_0 \in \mathcal{L}(V)$ has the pair $\mu = \exp(\pm i\theta_0)$ as simple eigenvalues, and that there are no further eigenvalues which are q-th roots of unity. The 2-dimensional subspace $U_q := \text{Ker}\left((A_0 - (\cos\theta_0)I)^2 + (\sin\theta_0)^2 I\right)$

is invariant under R; let $v_0 \in U_q$ be an eigenvector of R, i.e., $Rv_0 = \epsilon v_0$, with $\epsilon = \pm 1$. Setting $w_0 := (\sin \theta_0)^{-1}(A_0 - (\cos \theta_0)I)v_0$ we find

$$Rw_0 = \frac{1}{\sin \theta_0} \left(A_0^{-1} - (\cos \theta_0)I \right) Rv_0 = \frac{-\epsilon}{\sin \theta_0} \left(A_0 - (\cos \theta_0)I \right) v_0 = -\epsilon w_0.$$

So we can assume that $\epsilon = 1$ (just interchange v_0 and w_0 in the other case), resulting in a basis $\{u_0, v_0, w_0\}$ of $U = U_0 \oplus U_q$ such that $Ru_0 = u_0$, $Rv_0 = v_0$, $Rw_0 = -w_0$, $\mathcal{N}_0 u_0 = 0$, $\mathcal{N}_0 v_0 = 0$, $\mathcal{N}_0 w_0 = 0$, $S_0 u_0 = u_0$ and

$$S_0 v_0 = (\cos \theta_0)v_0 + (\sin \theta_0)w_0, \qquad S_0 w_0 = -(\sin \theta_0)v_0 + (\cos \theta_0)w_0. \quad (40)$$

Bifurcating q-periodic points can be approximated by determining the equilibria of the normal form system $\dot{u} = Z(u)$; the vectorfield $Z(u)$ must commute with S_0 and anti-commute with R. To find the form of $Z(u)$ we identify U with $\mathbb{R} \times \mathbb{C}$, via the isomorphism

$$\zeta : \mathbb{R} \times \mathbb{C} \to U, \quad (\alpha, z) \longmapsto \zeta(\alpha, z) := \alpha u_0 + \Re(z(v_0 - iw_0));$$

then $S_0(\alpha, z) = (\alpha, \exp(i\theta_0)z)$ and $R(\alpha, z) = (\alpha, \bar{z})$. Some elementary analysis shows that the system $\dot{u} = Z(u)$ must have the form

$$\begin{aligned}
\dot{\alpha} &= g(\alpha, z)\,\Im(z^q), \\
\dot{z} &= ih_1(\alpha, z)z + ih_2(\alpha, z)\bar{z}^{q-1},
\end{aligned} \qquad (41)$$

where the functions $g : \mathbb{R} \times \mathbb{C} \to \mathbb{R}$ and $h_i : \mathbb{R} \times \mathbb{C} \to \mathbb{R}$ $(i = 1, 2)$ are invariant under S_0 and R; also $h_1(0,0) = 0$, and we will assume that $\delta := h_2(0,0) \neq 0$. Setting $z = r\exp(i\theta)$ we can write (41) in the equivalent form

$$\begin{aligned}
\dot{\alpha} &= r^q g(\alpha, r\exp(i\theta))\sin(q\theta), \\
\dot{r} &= r^{q-1} h_2(\alpha, r\exp(i\theta))\sin(q\theta), \\
\dot{\theta} &= h_1(\alpha, r\exp(i\theta)) + r^{q-2} h_2(\alpha, r\exp(i\theta))\cos(q\theta).
\end{aligned} \qquad (42)$$

The α-axis forms a line of equilibria, corresponding to the primary branch (see subsection 6.1). Using (41) one finds that $DZ(\alpha u_0)$ (restricted to U) has the eigenvalues $\mu = 0$ and $\mu = \pm ih_1(\alpha, 0)$, corresponding to the multipliers $\mu = 1$ and $\mu = \exp(\pm i(\theta_0 + h_1(\alpha, 0)))$ along the primary branch. We have already observed that $h_1(0,0) = 0$, and we will assume that

$$\tau := \frac{\partial h_1}{\partial \alpha}(0, 0) \neq 0; \qquad (43)$$

this *transversality condition* means that as we move along the primary branch a pair of simple multipliers moves with non-zero speed along the unit circle, passing through $\exp(\pm i\theta_0)$ for $\alpha = 0$.

From (42) and our assumption $h_2(0,0) \neq 0$, it follows that nontrivial equilibria (with $r \neq 0$) must be such that $\sin(q\theta) = 0$, i.e. $\theta = 0$ or $\theta = \pi/q$ modulo θ_0. For $\theta = 0 \pmod{\theta_0}$ the bifurcation equation reduces to

$$h_1(\alpha, r) + r^{q-2} h_2(\alpha, r) = 0, \qquad (44)$$

and for $\theta = \pi/q \pmod{\theta_0}$ to

$$h_1(\alpha, r\exp(i\pi/q)) - r^{q-2}h_2(\alpha, r\exp(i\pi/q)) = 0. \tag{45}$$

Under the transversality condition (43) both these equations can be solved for α as a function of r, giving respectively $\alpha = \alpha_+^*(r)$ for (44) and $\alpha = \alpha_-^*(r)$ for (45). The full set of nontrivial equilibria is then given by the union of

$$B_q^+ = \{(\alpha, z) = (\alpha_+^*(r), r\exp(ik\theta_0)) \mid r > 0,\ 0 \le k \le q - 1\}$$

and

$$B_q^- = \{(\alpha, z) = (\alpha_-^*(r), r\exp(i(\pi/q + k\theta_0))) \mid r > 0,\ 0 \le k \le q - 1\}.$$

Observe that for fixed $r_0 > 0$ the intersection of B_q^+ with $r = r_0$ is invariant under S_0 and R; it therefore corresponds to a symmetric q-periodic orbit of Φ. The same holds for B_q^-, and since $\alpha_+^*(0) = \alpha_-^*(0) = 0$ we have found two branches of symmetric subharmonics bifurcating from the primary branch at the orbit κ of (36); the limiting period along these branches is equal to qT_0.

To determine the stability of these subharmonics we linearize (42) at the points of B_q^+ and B_q^- and calculate the eigenvalues of this linearization; after somewhat lengthy but straightforward calculations we find next to the trivial eigenvalue $\mu = 0$ a pair of nontrivial eigenvalues $\mu = \pm\sqrt{\Lambda_+(r)}$ along B_q^+, and another pair $\mu = \pm\sqrt{\Lambda_-(r)}$ along B_q^-. To simplify the expressions for $\Lambda_+(r)$ and $\Lambda_-(r)$ we introduce next to the (non-zero) constants δ and τ defined before also the constant $\gamma := g(0,0)$; moreover, we can expand $h_1(\alpha, z)$ as $h_1(\alpha, z) = h_1(\alpha, 0) + \tilde{h}_1(\alpha)r^2 + O(r^3)$, and we set $\nu := \tilde{h}_1(0)$. One finds then that

$$\alpha_\pm^*(r) = -\frac{\nu}{\tau}r^2 \mp \frac{\delta}{\tau}r^{q-2} + O(r^3) \tag{46}$$

and

$$\Lambda_\pm(r) = \pm q(\gamma\tau + 2\nu\delta)r^q + q(q-2)\delta^2 r^{2q-4} + O(r^{q+1}). \tag{47}$$

For $q = 3$ we have $\alpha_\pm^*(r) = \mp\delta\tau^{-1}r + O(r^2)$ and $\Lambda_\pm(r) = 3\delta^2 r^2 + O(r^3)$; therefore both bifurcating branches will be unstable (remember that we have assumed that $\delta \ne 0$). For $q = 4$ the expressions are

$$\alpha_\pm^*(r) = -\frac{\nu \pm \delta}{\tau}r^2 + O(r^3) \quad \text{and} \quad \Lambda_\pm(r) = 4(\pm\gamma\tau \pm 2\nu\delta + 2\delta^2)r^4 + O(r^5).$$

The sign of $\Lambda_\pm(r)$ (and the corresponding stability properties along the branches $B_q^\pm$) depends on all the constants involved, and a detailed analysis becomes rather messy; we leave it to the interested reader. Finally, for $q \ge 5$ we find

$$\alpha_\pm^*(r) = -\frac{\nu}{\tau}r^2 + O(r^3) \quad \text{and} \quad \Lambda_\pm(r) = \pm q(\gamma\tau + 2\nu\delta)r^q + O(r^{q+1});$$

assuming that $\gamma\tau + 2\nu\delta \ne 0$ it follows that the branches B_q^+ and B_q^- have opposite stability properties: one is stable, the other unstable. Comparing the equations (44) and (45) it is also easily seen that the two branches will be very close to each other for high values of q; more precisely, we have $\alpha_-^*(r) = \alpha_+^*(r) + O(r^{q-2})$ as $r \to 0$.

We conclude with the remark that along the stable branch of subharmonics we will have a pair of simple multipliers moving along the unit circle, starting at $\mu = 1$; these multipliers will necessarily cross roots of unity, thereby causing further subharmonic bifurcations. Repeating this scheme leads to a *cascade* of subharmonic branches, a phenomenon which is not yet fully understood.

References

[1] Ciocci, M.-C., Periodic and Quasi-periodic Bifurcations in Reversible and Symplectic Diffeomorphisms, Ph. D. Thesis, University of Gent, in preparation.

[2] Ciocci, M.-C., Bifurcation of periodic points and normal form theory in reversible diffeomorphisms. *Proceedings Equadiff10 Conference, Prague 2001*, CD-ROM to be published.

[3] Ciocci, M.-C. and Vanderbauwhede, A., On the bifurcation and stability of periodic orbits in reversible and symplectic diffeomorphisms. *Symmetry and Perturbation Theory*, A. Degasperis and G. Gaeta (Eds.), World Scientific, 1999, 159-166.

[4] Ciocci, M.-C. and Vanderbauwhede, A., Bifurcation of periodic orbits for symplectic mappings. *J. Difference Eqns. and Appl.* **3** (1998) 485-500.

[5] Vanderbauwhede, A., Subharmonic Bifurcation at Multiple Resonances. *Proceedings of the Mathematics Conference, Birzeit University, 1998* S. Elaydi et. al. (Eds.), World Scientific, 2000, 254-276.

Evolution of the Global Behavior
of a Class of Difference Equations

AARON S. CLARK and EDWARD S. THOMAS[1]

Department of Mathematics, University at Albany, SUNY
Albany, NY 12222, USA
E-mail: aron@krikkit.math.albany.edu, et392@math.albany.edu,

Abstract We consider the class of difference equations given by

$$x_{n+1} = \frac{x_n^k + a}{x_n^l x_{n-1}}$$

where k and l are real parameters, x_0 and x_1 are positive, and $a = \frac{2^{k-l-2}-1}{2^k}$. For these equations to remain well defined in the positive first quadrant, we must have $k > l + 2$. This inequality defines the half space of *admissible parameter values*.

For any particular choice of k and l, an essentially complete description of the global behavior is contained in the *phase portrait* of a canonically associated area preserving planar diffeomorphism. This phase portrait captures the local behavior of the system near its fixed points, the stability/chaos which occurs in KAM regions, as well as the existence of blocks of points whose orbits disperse exponentially to infinity. There are four different phase portraits, each corresponding to a different, explicitly defined region of parameter space.

Using computer images, it is possible to trace the continuous evolution of one phase portrait into another as (k, l) moves through parameter space. This provides a unification of the four seemingly different types of global behavior.

Keywords Difference equation, Diffeomorphism, Elliptic stability, KAM theory, Homoclinic point

AMS Subject Classification 39A10, 39A12, 39B12

1   Introduction

We consider the class of second order difference equations given by

$$x_{n+1} = \frac{x_n^k + a}{x_n^l x_{n-1}}, \qquad (1.1)$$

---

[1]Corresponding author.

where k and l are real parameters and the initial data x_0 and x_1 are positive. The constant a is defined by $a = \frac{2^{k-l-2}}{2^k}$. This restriction not only reduces the number of parameters from three to two, but it also guarantees that the system (1.1) always has an equilibrium point at $(1/2, 1/2)$, as one can easily check.

In order for the system to remain well defined in the positive first quadrant (i.e., in order to guarantee $x_n \neq 0$ for all n) it is necessary to require that $k > l + 2$. The corresponding half-space in the (k, l) plane will be called the *half-plane of admissible parameter values* or, simply, *parameter space*.

The study of these systems was begun in [2] and continued in [1]. We can quickly summarize the results in this introduction, with more details in later sections. Parameter space is divided into four regions, as pictured in Figure 1.

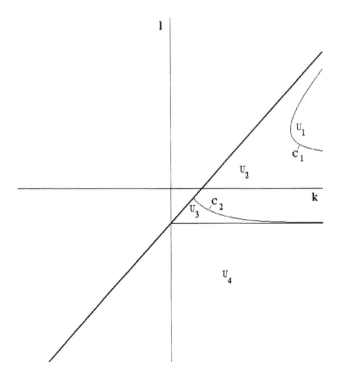

Figure 1: Regions in the global phase portrait.

Depending upon which region (k, l) lies in, the system exhibits four quite different types of *global behavior* with regard to equilibrium solutions, periodic solutions, bounded but chaotic solutions, and solutions which disperse to infinity. In this paper, we adopt a global viewpoint of the system (1.1) by studying its *evolution* as a function of k and l. This is done by assigning to each value of (k, l) a *phase portrait* of the associated system. This is, in essence, a static image which, when properly interpreted, encapsulates the behavior of the system. As (k, l) varies along a path in the parameter space crossing through all four regions, the corresponding phase portraits can be strung together, yielding an animation which shows the evolution of the global behavior of the system. This animation may be viewed at

www.math.albany.edu/~et392/dynamics/evolve.html

Our task here is to provide the interpretation of what one sees there.

Finally, we consider the issue of dispersal to infinity. In certain regions the phase portrait seems to behave quite chaotically. For example, one encounters invariant curves which form *homoclinic tangles*; these produce loops whose lengths increase without bound. We prove, however, that *most solutions are unaffected by this seeming chaos*; the solutions come from infinity, make a loop around the canonical fixed point, and disperse to infinity.

2 Associated diffeomorphisms and fixed points

As in [2] and [1], we capture the dynamics of the system (1.1) by studying the associated mapping given by

$$G(x, y) = \left(\frac{x^k + a}{x^l y} , x \right). \tag{2.1}$$

Making a change of variables $E(x, y) = ((1/2)e^x, (1/2)e^y)$ we obtain a new map $F = E^{-1} \circ G \circ E$ which is defined on the whole (x, y) plane:

$$F(x, y) = \left(\ln \left(\frac{e^{kx} + A - 1}{A} \right) - lx - y , x \right), \tag{2.2}$$

where $A = 2^{k-l-2}$.

Theorem 2.1 (*[1] Theorem 2.1*) *For all admissible parameter values ($k > l + 2$), the functions F share the following properties:*

(1) F is an area preserving diffeomorphism of $\mathbb{R}^2$;

(2) F fixes the origin; and

(3) F satisfies the symmetry condition $F^{-1} = \Delta \circ F \circ \Delta$, where $\Delta(x, y) = (y, x)$ is reflection through the diagonal and $\circ$ denotes composition.

Comments: This was proved in [2]. The word "diffeomorphism" means "C^∞ homeomorphism" and assertion (1) is obtained by showing that the Jacobian of F at any point in $\mathbb{R}^2$ is equal to 1. The second assertion follows from the fact that $E(0, 0) = (1/2, 1/2)$ and the third follows by an easy direct computation. (The symmetry condition (3) makes for simple, elegant phase portraits.)

Depending on where (k, l) lies in parameter space, our system may have either one or two fixed points. By symmetry, any fixed point must lie on the diagonal and we know that $\vec{0}$ is always fixed. For the moment, let us denote a potential second fixed point by (x^*, x^*).

Theorem 2.2 *Referring to Figure 1,*

(1) If (k, l) lies on or above the exceptional curve C_2, then there is a second fixed point lying above $\vec{0}$ on the diagonal; i.e., $x^* > 0$;

(2) If (k, l) lies in the region $\boldsymbol{U_3}$ then the second fixed point lies below $\vec{0}$, i.e., $x^* < 0$;

(3) If (k, l) lies on or below the exceptional curve $l = -2$, i.e., in region $\boldsymbol{U_4}$, then the only fixed point is $\vec{0}$.

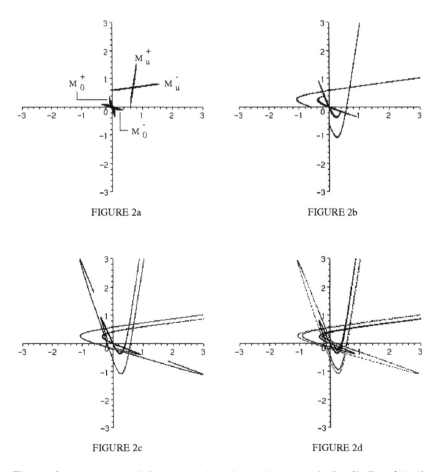

FIGURE 2a

FIGURE 2b

FIGURE 2c

FIGURE 2d

Figure 2: Formation of the interlocking homoclinic tangle for $(k, l) = (10, 3)$.

Comments: This result was established in [1] by calculating the roots of the fixed point equation $F(x^*, x^*) = (x^*, x^*)$. On a small scale, the phase portrait of F is dominated by the behavior of orbits which are close to a fixed point. We summarize the various possibilities below. The necessary calculations were carried out in [1].

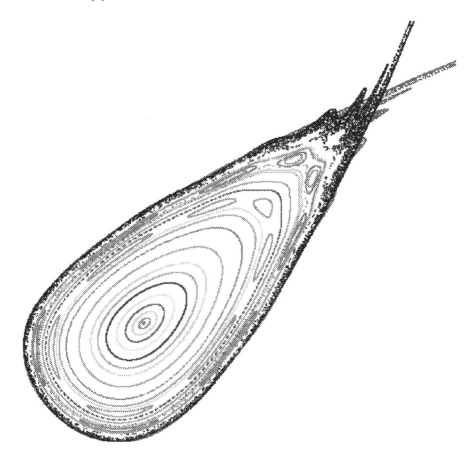

Figure 3: Invariant curves, islands, and tangles for $(k, l) = (4, 1)$ on $[-0.35, 0.8] \times [-0.35, 0.8]$.

Working down in parameter space from U_1 to U_4, the system will behave as follows: In region U_1 both fixed points are hyperbolic. A typical phase portrait is shown in Figure 2. The local stable/unstable manifold pairs of each fixed point cross transversely to produce the characteristic homoclinic tangles. Beyond this, the two tangles themselves also intersect; i.e, one has *interlocking homoclinic tangles*. Viewing this complexity near the two fixed points is difficult in a single black and white image. A color animation, available at the URL mentioned above, does a much better job. In the next section, we show that on a much larger scale, orbits are essentially unaffected by this local behavior in the sense that most orbits make a counterclockwise loop around the two fixed points and disperse to infinity in the first quadrant. As (k, l) moves across the boundary from U_1 into U_2, the system calms down to some degree. The origin becomes neutrally stable in the sense of KAM theory, developing an *invariant lobe*, while the second fixed point lying above remains hyperbolic and retains the characteristic homoclinic tangle. This is illustrated in Figure 3. The necessary *twist coefficient* computations are described in [2] and [1].

As (k, l) leaves U_2 and enters U_3 the second fixed point passes below the origin and the two fixed points adroitly *exchange behavior*; the origin becomes hyperbolic and the second fixed point elliptic. *We conjecture that, generically, the second fixed point is KAM-stable.* Every computer simulation supports this, but the necessary calculations have yet to be carried out.

The animation indicates very clearly what must happen as (k, l) continues to move down in region U_3. As l decreases, the invariant lobe around the second fixed point grows in size, stretching from the origin (whose stable/unstable manifolds form its boundary) down along the negative diagonal. As (k, l) moves into U_4 the second fixed point disappears and the boundary of the invariant lobe becomes transformed into branches of the stable/unstable manifold at the origin. Thus, in U_4, we reach the simplest possible phase portrait, as illustrated in Figure 4.

3   Dispersal to infinity

In previous sections we have discussed the local behavior of the system near its fixed points. In this section we focus on the behavior of points whose orbits are essentially unaffected by the fixed points. It is shown that, by and large, most orbits disperse to infinity at an exponential rate; in particular, the solutions of the difference equation (1.1) *do not persist*.

Definition 3.1 *Let α be a real number greater than 1. A point z is α-dispersive provided that for all n sufficiently large, $d(F^{n+1}(z)) \geq \alpha d(F^n(z))$.*

Here d denotes the distance from w to the origin in the "taxi-cab" metric $d(x, y) = |x| + |y|$. This metric is equivalent to the standard Euclidean metric [3, p. 124]. Thus, if z is α-dispersive, then its iterates tend to infinity like α^n. Our fundamental dispersion result is the following:

Theorem 3.2 *Let U be the region in the phase space given by $x > y > 0$ and $x > 3 \ln 2$. Then every point in U is α-dispersive with $\alpha = \frac{k-l+1}{3}$. In other words, the region U is uniformly α-dispersive.*

Remarks: This result was first formulated in [2] and proved under some restrictions on k and l. The proof here requires nothing more than our standing hypothesis: $k > l + 2$. For certain parameter values, this result seems quite counterintuitive. For example, if (k, l) lies in the double hyperbolic region, then, even though the system has interlocking homoclinic tangles, individual points on these tangles do not necessarily oscillate wildly as one might expect. Rather, a point may stay near the origin for a few iterates, then enter U and thereafter disperse to infinity, all the while staying in the homoclinic tangle. Thus, in a sense, Theorem 3.1 insures that there is *quite orderly behavior in the midst of chaos.*

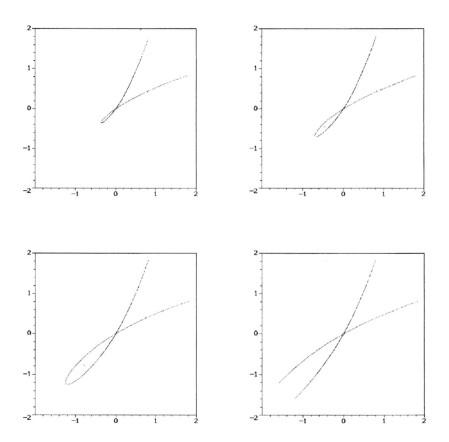

Figure 4: As (k, l) approaches the line $l = -2$, the invariant lobe expands and the lower fixed point goes to infinity on the negative diagonal.

We now turn to the proof of the theorem. This involves two things: first, we show that $F(U) \subseteq U$ and, secondly, $d(F(z)) \geq \alpha d(z)$ for all z in U.

To prove the first assertion, let $z = (x, y)$ be in U. To prove that $F(z)$ is in U, it suffices to prove that the x-coordinate of $F(z)$ is greater than x; i.e.,

$$\ln(\frac{e^{kx} + A}{2^{k-l-2}}) - lx - y > x. \tag{3.1}$$

This reduces to

$$\frac{e^{kx} + A}{2^{k-l-2}} > e^{(l+1)x} \cdot e^y. \tag{3.2}$$

Here, $A = 2^{k-l-2} - 1$ is positive and $x > y$, so it suffices to show that $e^{kx} > 2^{k-l-2} \cdot e^{(l+2)x}$, which can be rewritten, finally, as $e^{(k-l-2)x} > 2^{k-l-2}$. This reduces to $x > \ln 2$, which is certainly true in U. Thus, $F(U) \subseteq U$, whence $F^n(U) \subseteq U$ for all $n > 0$.

Now, for α-dispersion, we wish to show $d(F(z)) \geq \alpha d(z)$ for any z in U. This expands to

$$[\ln(\frac{e^{kx} + A}{2^{k-l-2}}) - lx - y] + x > \alpha(x + y),$$

which reduces to

$$\ln(\frac{e^{kx} + a}{2^{k-l-2}}) > x(\alpha + l - 1) + (\alpha + 1)y.$$

Since $A > 0$ and $x > y$ (in U), the previous inequality will follow if we can prove that

$$\ln(\frac{e^{kx} + A}{2^{k-l-2}}) > (\alpha + l - 1 + \alpha + 1)x.$$

This can be rewritten as

$$e^{(k-2\alpha-l)x} > 2^{k-l-2} - 1,$$

which will certainly follow from

$$e^{(k-2\alpha-l)x} > 2^{k-l-2}.$$

Notice that $k - 2\alpha - l$ is positive (in fact, $k - 2\alpha - l = \frac{k-l-2}{3}$), so the last inequality can be solved for x:

$$x > \frac{k - l - 2}{k - l - 2\alpha} \cdot \ln 2.$$

Now, at last, we see that, if $x > 3 \ln 2$, then our desired result will follow if $3 \geq \frac{k-l-2}{k-l-2\alpha}$. But, indeed, α was chosen precisely to yield this exact inequality! This finishes the second assertion and the theorem is proved. $\square$

So any point whose orbit lands in U will be swept to infinity at a brisk rate. What points wind up this way? One quick answer is that, *except for a* *compact neighborhood of the origin, every point in the fourth quadrant, Q_4, will be mapped into U after just one iterate of F.*

To see this, we look at images of the boundary of Q_4. If z lies on the negative y-axis, $z = (0, y)$ where $y < 0$, then $F(z) = (-y, 0)$ lies on the positive x-axis. If z lies on the positive x-axis, then by the initial computation of the theorem, $F(z)$ will lie in the sector $x > y > 0$. It follows that since F is a diffeomorphism, the preimage of U will contain all of Q_4 except for the preimage of the triangular region where $x > y > 0$ and $x \leq 4 \ln 2$.

Now we are interested in what points are *eventually* carried into U by some iterate of F. We know that almost all points in Q_4 have this property and, in view of the symmetry property of F with respect to the diagonal, it suffices to determine *which points in the third quadrant to the right of the diagonal will eventually iterate into the fourth quadrant.*

To answer this question we need a bit of *ad hoc* notation and terminology. Let Q_3^+ denote the region in the third quadrant lying to the right of the negative diagonal, Δ^-. A subset of Q_3^+ which is the intersection of Q_3^+ with a bounded neighborhood of the origin will be called a *bounded wedge*.

Theorem 3.3 *Fix (k, l) with $k > l + 2$ and $-2 < l < -1$. Then there exists an integer $N(l)$ such that F maps all of Q_3^+, except for some bounded wedge, into Q_4 in at most $N(l)$ steps.*

Reminder: Here, "F" means that F which corresponds to the particular choice of (k, l) one is looking at.

Proof. Very briefly, the idea of the proof is to show that if one looks at successive images of Δ^- under F, then *ultimately $F^n(\Delta^-)$* will be asymptotic to a line with negative slope. Since F maps the negative y-axis onto the positive x-axis, and thence into the first quadrant, this implies that all of Q_3^+, except for a neighborhood of the origin, will be mapped into Q_1 in n (or fewer) steps.

We first show that, if L is any line in $\mathbb{R}^2$, then the image of $F(L)$ is asymptotic (in the sense described below) to another line (denoted for the purpose of this proof by $F(L)^*$). Let L be the line $y = mx + b$ with $m > 0$ and let $F(L)^*$ be the line $y = \frac{-1}{l+m}x + \frac{b-\alpha}{l+m}$ where $\alpha = \ln(\frac{A}{2^{k-l-2}})$. Given $x_0 > 0$, let $L_{x_0}^-$ denote those points on L whose x-coordinates are less than $-x_0$.

We assert that, given $\epsilon > 0$, there exists an x_0 such that, if z is a point on $L_{x_0}^-$, then $F(z)$ lies within ϵ of some point on $F(L)^*$. Moreover, x_0 depends only on k, l, and ϵ (not b). To see this, write z in the form $z = (-t, -mt + b)$ where $t > 0$. Then $F(z)$ has the form

$$F(z) = (\ln(\frac{e^{-kt} + A}{2^{k-l-2}}) + (l+m)t - b, -t).$$

The horizontal line through this point hits the line $F(L)^*$ in the point

$$z^* = ((l+m)t - b + \alpha, -t).$$

The distance between these points is then given by

$$d(F(z), z^*) = \ln(\frac{e^{-kt} + A}{A}).$$

Note that, under our hypotheses, k must be positive. Hence the distance above will be less than ϵ for all $t \geq x_0$ where $x_0 = \frac{-1}{k} \ln(A(e^\epsilon - 1))$, thus proving our assertion.

There is one special case to be dealt with. If L is a vertical line, then $F(L)$ is itself a horizontal line and as the y-coordinate of a point z on L goes to $-\infty$, the x-coordinate of $F(z)$ goes to $+\infty$.

Now fix (k, l) with $k > -2$. Starting with the negative diagonal Δ^-, with slope $m_0 = +1$, consider the slope of successive asymptotes: $m_1 = $ slope of $F(\Delta^-)^*$, $m_2 = $ slope of $F(F(\Delta^-))^*$, and so on. We wish to iterate this process until some m_n is negative, at which point the process terminates and the theorem is proved; i.e., $F^n(\Delta^-)$ will be asymptotic to a line with negative slope.

The difference equation governing successive slopes is $m_n = -\frac{1}{l+m_{n-1}}$, where $m_0 = +1$. We assert that, if m_n is positive, then $m_n > m_{n-1}$. This last inequality can be rewritten as $m_{n-1} < -\frac{1}{l+m_{n-1}}$. Since the right hand side is positive by hypothesis, we can rearrange terms to obtain the equivalent desired inequality $m^2 + lm + 1 > 0$, where $m = m_{n-1}$. Completing the square and shifting terms yields $(m + \frac{l}{2})^2 > \frac{l^2-4}{4}$. But we are assuming that $-2 < l < -1$, making $l^2 < 4$, so this inequality certainly holds.

To summarize, starting with $m_0 = +1$, we iterate the sequence of slopes $m_1, m_2, \ldots$ of asymptotes. As long as $-(l + m_{n-1})$ is positive, the slopes will continue to increase.

If, for some n, m_{n-1} is greater than 2, then m_n will be positive (since $-2 < l$) and, in this case, we are done. On the other hand, if the increasing sequence $\{m_n\}$ were bounded above, then it would converge to a finite limit, m^*. But m^* would satisfy the equation $x^2 + lx + 1 = 0$ and since $l^2 - 4 < 0$, this has no real solutions.

Thus, $\{m_n\}$ is eventually positive, and we are finished. $\square$

References

[1] Clark, A. S., and Thomas, E. S., On the Global Behavior of a Class of Difference Equations, *J. Difference Equations and Appl.*, to appear.

[2] Haymond, R. E, and Thomas, E. S., Phase Portraits for a Class of Difference Equations, *J. Difference Equations and Appl.* **5** (1999), 177-202.

[3] Munkres, J. R., *Topology, A First Course*, Prentice-Hall, 1975.

A Survey of Exponential Dynamics

ROBERT L. DEVANEY

Department of Mathematics, Boston University
Boston, MA 02215 USA
E-mail: bob@bu.edu

Abstract In this paper we describe some of the interesting dynamics, topology, and geometry that arises in the iteration of the complex exponential $E_\lambda(z) = \lambda e^z$ where $\lambda > 0$. There are two quite distinct cases. When $\lambda \leq 1/e$, the Julia set for E_λ is a Cantor bouquet. When $\lambda > 1/e$, the Julia set suddenly explodes and fills the entire plane. We show that it is the appearance of indecomposable continua in the Julia set that accounts for this explosion.

Keywords Complex dynamics, Exponential function, Indecomposable continuum

AMS Subject Classification 37F10

1 Introduction

Our goal in this paper is to describe some of the interesting dynamics, topology, and geometry that arises in the iteration of entire functions such as the complex exponential $E_\lambda(z) = \lambda e^z$. We will see that the important invariant sets for this family possesses a extremely rich topological structure, including such objects as Cantor bouquets, Knaster continua, and explosion points.

For a complex analytic function E, the interesting orbits lie in the *Julia set*, which we denote by $J(E)$. This is the set on which the map is chaotic. For the exponential family, the Julia set of E_λ has three equivalent characterizations:

1. $J(E_\lambda)$ is the set of points at which the family of iterates of E_λ, $\{E_\lambda^n\}$, is not a normal family in the sense of Montel. This is the characterization that is most useful to prove theorems.

2. $J(E_\lambda)$ is the closure of the set of repelling periodic points of E_λ. This is the dynamical definition of the Julia set.

3. $J(E_\lambda)$ is the closure of the set of points whose orbits tend to ∞. This is the characterization that is most useful to compute the Julia set.

We remark that characterization 3 differs from the case of polynomial iterations, where the Julia set is the *boundary* of the set of escaping orbits. The reason for the difference is that E_λ has an essential singularity at ∞, while

0-415-31675-8/04/$0.00+$1.50

Figure 1: A Julia set for $\lambda < 1/e$.

polynomials have superattracting fixed points at ∞. The equivalence of 1 and 2 was shown by Baker, see [Ba1]. The equivalence of 1 and 3 is shown in [DT].

In this paper we will concentrate on the dynamics of E_λ where λ is real. For λ positive, the Julia set for E_λ undergoes a remarkable transformation as λ passes through $1/e$. We will show below that E_λ possesses an attracting fixed point when $0 < \lambda < 1/e$. All points in the left half plane have orbits that tend to this fixed point. Indeed, the full basin of attraction of this fixed point is open and dense in the plane.

We will show that the complement of the basin, $J(E_\lambda)$, is a *Cantor bouquet* for $0 < \lambda \leq 1/e$. Roughly speaking, a Cantor bouquet has the property that all points in the set lie on a curve (or "hair") homeomorphic to a closed half line. Each of these curves in $J(E_\lambda)$ extend to ∞ in the right half-plane.

In Figures 1 and 2 we display a computer graphics rendering of the Julia set of E_λ for a particular λ with $0 < \lambda < 1/e$. This image was computed using characterization 3 of the Julia set: Points are shaded in white and grey if their orbits ever enter the region $\operatorname{Re} z > 50$. The complement of the Julia set is displayed in black. It appears that this Julia set contains large open sets, but this in fact is not the case. The Julia set actually consists of uncountably many curves lying in the Cantor bouquet and extending to ∞ in the right half plane. These curves are packed together so tightly that the resulting set has Hausdorff dimension 2, thus giving the appearance of an open set.

At $\lambda = 1/e$, E_λ undergoes a simple saddle-node bifurcation. The attract-

Figure 2: Magnification of a Julia set for $\lambda < 1/e$.

ing fixed point merges with a repelling fixed point at this λ-value, producing a neutral fixed point. When $\lambda > 1/e$, this neutral fixed point gives way to a pair of repelling fixed points.

This apparently simple bifurcation has profound global ramifications. When $\lambda \leq 1/e$, the Julia set is a nowhere dense subset of the right half plane. However, when $\lambda > 1/e$, $J(E_\lambda)$ suddenly becomes the whole plane. No new repelling periodic points (except the two fixed points involved in the saddle-node) are born in this bifurcation; all others simply move smoothly as λ crosses through $1/e$. Yet somehow, as soon as λ exceeds $1/e$, the repelling periodic points become dense in $\mathbb{C}$.

In Figure 3 we display the Julia set for E_λ for a particular $\lambda > 1/e$. Note the striking difference between this image and that in Figure 1.

At this bifurcation, the attracting fixed point and its entire basin of attraction disappear. Most of the curves in the Cantor bouquet remain as curves in the Julia set. However, some evolve into a new and interesting set called an indecomposable continuum.

This paper is a summary of a lecture given at the International Conference on Difference Equations and Applications held in Augsburg, Germany July 30–August 3, 2001. It is a pleasure to thank the organizers of this conference for the privilege of participating.

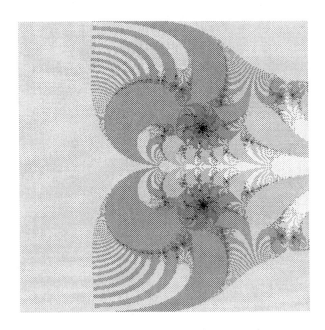

Figure 3: The Julia set for $\lambda > 1/e$.

2   Exponential Dynamics

As in the often-studied quadratic family $Q_c(z) = z^2 + c$, it is the orbit of 0 that plays a crucial role in determining the dynamics of E_λ. For the exponential family, 0 is an asymptotic value (an omitted value) rather than a critical point. Nevertheless, the orbit of 0 plays a decisive role in the determination of the structure of $J(E_\lambda)$:

Theorem 2.1 *Suppose E_λ has an attracting or rationally neutral (parabolic) periodic point. Then $E_\lambda^n(0)$ must tend to the attracting or neutral cycle. If, on the other hand, $E_\lambda^n(0) \to \infty$, then $J(E_\lambda) = \mathbb{C}$.*

The proof of the first statement in this theorem is a classical fact that goes back to Fatou. The second follows from the Sullivan No Wandering Domains Theorem [Su], as extended to the case of the exponential by Goldberg and Keen [GK] and Eremenko and Lyubich [EL].

Consider for the moment the restriction of E_λ to the real line. The exponential family undergoes a saddle node bifurcation at $\lambda = 1/e$ since, when $\lambda = 1/e$, the graph of $E_{1/e}$ is tangent to the diagonal at 1. See Figure 4. We have $E_{1/e}(1) = 1$ and $E'_{1/e}(1) = 1$. When $\lambda > 1/e$, the graph of E_λ lies above the diagonal and all orbits (including 0) tend to ∞. When $\lambda < 1/e$, the graph of E_λ crosses the diagonal twice, at an attracting fixed point a_λ and a

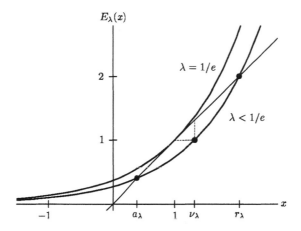

Figure 4: The graphs of E_λ for $\lambda = 1/e$ and $\lambda < 1/e$.

repeling fixed point r_λ. For later use note that $0 < a_\lambda < 1 < r_\lambda$. Note also that the orbit of 0 tends to a_λ, as it must by Fatou's theorem.

3 Cantor Bouquets

In this section, we begin the study of the dynamics of E_λ by considering the case where $\lambda \leq 1/e$. In this case $J(E_\lambda)$ is a *Cantor bouquet*. We will give a sketch of the construction of this object. For more details, see [D2]

Let $E(z) = (1/e)e^z$. We have $E(1) = 1$ and $E'(1) = 1$. If $x_0 \in \mathbb{R}$ and $x_0 < 1$, then $E^n(x_0)$ tends to the fixed point at 1. If $x_0 > 1$, then $E^n(x_0) \to \infty$ as $n \to \infty$. This can be shown using the web diagram as shown in Figure 5.

The vertical line $\operatorname{Re} z = 1$ is mapped to the circle of radius 1 centered at the origin. In fact, E is a contraction in the half plane H to the left of this line, since

$$|E'(z)| = \frac{1}{e}\exp(\operatorname{Re} z) < 1$$

if $z \in H$. Consequently, all points in H have orbits that tend to 1. Hence this half plane lies in the stable set, i.e., in the complement of the Julia set. We will try to paint the picture of the Julia set of E by painting instead its complement.

Since the half plane H is forward invariant under E, we can obtain the entire stable set by considering all preimages of this half plane. Now the first preimage of H certainly contains the horizontal lines $\operatorname{Im} z = (2k+1)\pi$, $\operatorname{Re} z \geq 1$, for each integer k, since E maps these lines to the negative real axis which lies in H. Hence there are open neighborhoods of each of these lines

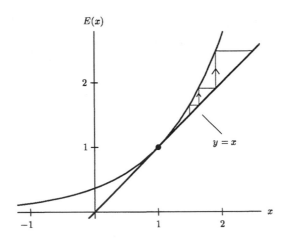

Figure 5: The graph of $E(x) = (1/e)e^x$.

that lie in the stable set. The first preimage of H is shown in Figure 6. The complement of $E^{-1}(H)$ consists of infinitely many "fingers." The fingers are $2k\pi i$ translates of each other, and each is mapped onto the complementary half plane $\mathrm{Re}\, z \geq 1$.

We denote the fingers in the complement of $E^{-1}(H)$ by C_j with $j \in \mathbb{Z}$, where C_j contains the half line $\mathrm{Im}\, z = 2j\pi$, $\mathrm{Re}\, z \geq 1$, which is mapped into the positive real axis. That is, the C_j are indexed by the integers in order of increasing imaginary part. Note that C_j is contained within the strip $-\frac{\pi}{2} + 2j\pi \leq \mathrm{Im}\, z \leq \frac{\pi}{2} + 2j\pi$.

Now each C_j is mapped in one-to-one fashion onto the entire half plane $\mathrm{Re}\, z \geq 1$. Consequently each C_j contains a preimage of each other C_k. Each of these preimages forms a subfinger which extends to the right in the half plane H. See Figure 7. The complement of these subfingers necessarily lies in the stable set.

Now we continue inductively. Each subfinger is mapped onto one of the original fingers by E. Consequently, there are infinitely many sub-subfingers which are mapped to the C_j's by E^2. So at each stage we remove the complement of infinitely many subfingers from each remaining finger.

This process is reminiscent of the construction of the Cantor set in the dynamics of polynomials when all critical points tend to ∞. In that construction, the complements of disks are removed at each stage; here we remove the complement of infinitely many fingers. As a result, after performing this operation infinitely many times, we do not end up with points. Rather, the intersection of all of these fingers, if nonempty, is a simple curve extending to ∞. See [DK].

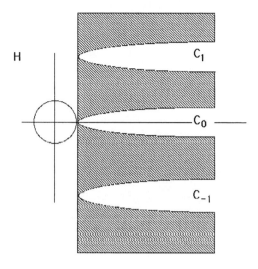

Figure 6: The preimage of H consists of H and the shaded region.

Figure 7: The second preimage of H in one of the fingers C_j.

This collection of curves forms the Julia set. E permutes these curves and each curve consists of a well-defined endpoint together with a "stem" which extends to ∞. It is tempting to think of this structure as a "Cantor set of curves," i.e., a product of the set of endpoints and the half-line. However, this is not the case as the set of endpoints is not closed.

Note that we can assign symbolic sequences to each point on these curves. To do this, we attach an infinite sequence $s_0 s_1 s_2 \ldots$ to each curve in the Julia set via the rule: $s_j \in \mathbb{Z}$ and $s_j = k$ if the jth iterate of the curve lies in C_k. The sequence $s_0 s_1 s_2 \ldots$ is called the *itinerary* of the curve.

For example, the portion of the real line $\{x \mid x \geq 1\}$ lies in the Julia set since all points (except 1) tend to ∞ under iteration, not to the fixed point. These points all have itinerary $000\ldots$.

One temptation is to say that there is a curve corresponding to every possible sequence $s_0 s_1 s_2 \ldots$ This, unfortunately, is not true, as certain sequences simply grow too quickly to correspond to orbits of E. See [DeV].

So this is $J(E)$: a "hairy" object extending toward ∞ in the right-half plane. We call this object a *Cantor bouquet*. We will see that this bouquet has some rather interesting topological properties.

We remark that the same construction works if $0 < \lambda < 1/e$. We still define the half plane H as the set $\mathrm{Re}\, z < 1$. As we saw earlier, the point 1 on the real axis sits between the attracting fixed point a_λ and the repelling fixed point r_λ, and so $E_\lambda(1) < 1$ and as a consequence $E_\lambda(H)$ is strictly contained in H. The construction of the fingers now proceeds exactly as above.

The Cantor bouquet is a remarkable object from the topological and geometric point of view. Here are just a few of its properties:

Properties.

1. There are two types of points in the Cantor bouquet: the endpoints and the points on the stem. It is known that all points on the stem have orbits that tend to ∞. Hence the set of bounded orbits is contained in the set of endpoints. In particular, the set of repelling periodic points lies in the set of endpoints. But these points are dense in $J(E_\lambda)$, so the set of endpoints accumulates on all points in J.

2. A result of Mayer [Ma] shows that the set of endpoints has the following intriguing structure: In the Riemann sphere the set of endpoints together with ∞ forms a connected set. However, the set of endpoints alone is totally disconnected! That is, removing just one point from this connected set not only disconnects the set, but also **totally** disconnects it!

3. McMullen [McM] has shown that the Hausdorff dimension of the Cantor bouquet constructed above is 2 but its Lebesgue measure is zero. This accounts for why figures 1 and 2 seem to have open regions in the Julia set.

4. Babinska has shown that the Hausdorff dimension of the set of stems is 1, but the Hausdorff dimension of the set of endpoints is 2!

4 Indecomposable Continua

We now consider the case $\lambda > 1/e$. Since the orbit of 0 tends to ∞, the Julia set is now the entire plane. For these λ values, the attracting basin for the attracting fixed point a_λ disappears. What replaces it is a collection of complicated sets known as indecomposable continua. We describe the construction of one such set in this section.

Consider the horizontal strip

$$S = \{z \mid 0 \leq \operatorname{Im} z \leq \pi\}$$

(or its symmetric image under $z \to \bar{z}$). The exponential map E_λ takes the boundary of S to the real axis and the interior of S to the upper half plane. Thus, E_λ maps certain points outside of S while other points remain in S after one application of E_λ. Our goal is to investigate the set of points whose entire orbit lie in S. Call this set Λ. The set Λ is clearly invariant under E_λ. There is a natural way to compactify this set in the plane to obtain a new set Γ. Moreover, the exponential map extends to Γ in a natural way. Our main results in this section include:

Theorem 4.1 Γ *is an indecomposable continuum.*

Moreover, we will see that Λ is constructed in similar fashion to the well known *Knaster continua* described below. Thus the topology of Λ is quite intricate. Despite this, we will show that the dynamics of E_λ on Λ is quite tame. Specifically, we will prove:

Theorem 4.2 *The restriction of E_λ to $\Lambda - \{orbit of 0\}$ is a homeomorphism. This map has a unique repelling fixed point $w_\lambda \in \Lambda$, and the α-limit set of all points in Λ is w_λ. On the other hand, if $z \in \Lambda$, $z \neq w_\lambda$, then the ω-limit set of z is either*

1. *The point at ∞, or*

2. *The orbit of 0 under E_λ together with the point at ∞.*

Thus we see that E_λ possesses an interesting mixture of topology and dynamics in the case where the Julia set is the whole plane. In the plane the dynamics of E_λ are quite chaotic, but the overall topology is tame. On our invariant set Λ, however, it is the topology that is rich, but the dynamics are tame. For more details we refer to [D1].

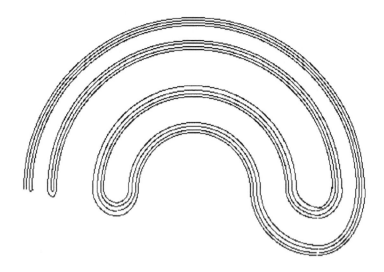

Figure 8: The Knaster Continuum.

4.1   Topological Preliminaries

In this section we review some of the basic topological ideas associated with indecomposable continua. See [Ku] for a more extensive introduction to these concepts.

Recall that a continuum is a compact, connected space. A continuum is decomposable if it is the (not necessarily disjoint) union of two proper subcontinua. Otherwise, it is indecomposable. A well known example of an indecomposable continuum is the Knaster continuum, K. One way to construct this set is to begin with the Cantor middle-thirds set. Then draw the semi-circles lying in the upper half plane with center at $(1/2, 0)$ that connect each pair of points in the Cantor set that are equidistant from $1/2$. Next draw all semicircles in the lower half plane which have for each $n \geq 1$ centers at $(5/(2 \cdot 3^n), 0)$ and pass through each point in the Cantor set lying in the interval

$$2/3^n \leq x \leq 1/3^{n-1}.$$

The resulting set is partially depicted in Figure 8.

For a proof that this set is indecomposable, we refer to [Ku]. Dynamically, this set appears as the closure of the unstable manifold of Smale's horseshoe map (see [Ba], [Sm]).

Note that the curve passing through the origin in this set is dense, since it passes through each of the endpoints of the Cantor set. It also accumulates everywhere upon itself. Such a phenomenon gives a criterion for a continuum to be indecomposable, as was shown by S. Curry.

Theorem 4.3 *Suppose X is a one-dimensional nonseparating plane continuum which is the closure of a ray that limits upon itself. Then X is indecomposable.*

We refer to [Cu] for a proof.

4.2 Construction of Λ

Recall that the strip S is given by $\{z \mid 0 \leq \text{Im}\ (z) \leq \pi\}$. Note that E_λ maps S in one-to-one fashion onto $\{z \mid \text{Im}\ z \geq 0\} - \{0\}$. Hence E_λ^{-1} is defined on $S - \{0\}$ and, in fact, E_λ^{-n} is defined for all n on $S - \{\text{orbit of } 0\}$. We will always assume that E_λ^{-n} means E_λ^{-n} restricted to this subset of S.

Define
$$\Lambda = \{z \mid E_\lambda^n(z) \in S \text{ for all } n \geq 0\}.$$

If $z \in \Lambda$ it follows immediately that $E_\lambda^n(z) \in S$ for *all* $n \in \mathbf{Z}$ provided z does not lie on the orbit of 0. Our goal is to understand the structure of Λ.

Toward that end we define L_n to be the set of points in S that leave S at precisely the nth iteration of E_λ. That is,

$$L_n = \{z \in S \mid E_\lambda^i(z) \in S \text{ for } 0 \leq i < n$$

$$\text{but } E_\lambda^n(z) \notin S\}.$$

Let B_n be the boundary of L_n.

Recall that E_λ maps a vertical segment in S to a semi-circle in the upper half plane centered at 0 with endpoints in $\mathbf{R}$. Either this semi-circle is completely contained in S or else an open arc lies outside S. As a consequence, L_1 is an open simply connected region which extends to ∞ toward the right in S as shown in Figure 9. There is a natural parameterization $\gamma_1 : \mathbf{R} \to B_1$ defined by

$$E_\lambda(\gamma_1(t)) = t + i\pi.$$

As a consequence,

$$\lim_{t \to \pm\infty} \text{Re}\ \gamma_1(t) = \infty.$$

If $c > 0$ is large, the segment $\text{Re}\ z = c$ in S meets $S - L_1$ in two vertical segments v_+ and v_- with $\text{Im}\ v_- > \text{Im}\ v_+$. E_λ maps v_- to an arc of a circle in $S \cap \{z \mid \text{Re}\ z < 0\}$ while E_λ maps v_+ to an arc of a circle in $S \cap \{z \mid \text{Re}\ z > 0\}$. As a consequence, if c is large, v_+ meets L_2 in an open interval. Since $L_2 = E_\lambda^{-1}(L_1)$, it follows that L_2 is an open simply connected subset of S that extends to ∞ in the right half plane *below* L_1.

Continuing inductively, we see that L_n is an open, simply connected subset of S that extends to ∞ toward the right in S. We may also parameterize the boundary B_n of L_n by $\gamma_n : \mathbf{R} \to B_n$ where

$$E_\lambda^n(\gamma_n(t)) = t + i\pi$$

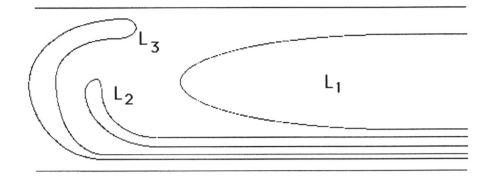

Figure 9: Construction of the L_n.

as before. Again

$$\lim_{t \to \pm\infty} \operatorname{Re} \gamma_n(t) = \infty.$$

Since each L_n is open, it follows that Λ is a closed subset of S.

Proposition 4.4 *Let $J_n = \bigcup_{i=n}^{\infty} B_i$. Then J_n is dense in Λ for each $n > 0$.*

Proof. Let $z \in \Lambda$ and suppose $z \notin B_i$ for any i. Let U be an open connected neighborhood of z. Fix $n > 0$. Since $E_\lambda^i(z) \in S$ for all i, we may choose a connected neighborhood $V \subset U$ of z such that $E_\lambda^i(V) \subset S$ for $i = 0, \dots, n$.

Now the family of functions $\{E_\lambda^i\}$ is not normal on V, since z belongs to the Julia set of E_λ. Consequently, $\bigcup_{i=0}^{\infty} E_\lambda^i(V)$ covers $\mathbf{C} - \{0\}$. In particular, there is $m > n$ such that $E_\lambda^m(V)$ meets the exterior of S. Since $E_\lambda^m(z) \in S$, it follows that $E_\lambda^m(V)$ meets the boundary of S. Applying E_λ^{-m}, we see that B_m meets V. $\square$

In fact, it follows that for any $z \in \Lambda$ and any neighborhood U of z, all but finitely many of the B_m meet V. This follows from the fact that E_λ has fixed points outside of S (in fact one such point in each horizontal strip of width 2π—see [DK]), so we may assume that $E_\lambda^m(V)$ contains this fixed point for all sufficiently large m. In particular, we have shown:

Proposition 4.5 *Let $z \in \Lambda$ and suppose that V is any connected neighborhood of z. Then $E_\lambda^m(V)$ meets the boundary of S for all sufficiently large m.*

Proposition 4.6 Λ *is a connected subset of S.*

Proof. Let G be the union of the boundaries of the L_i for all i. Since Λ is the closure of G, it suffices to show that G is connected. Suppose that this is not

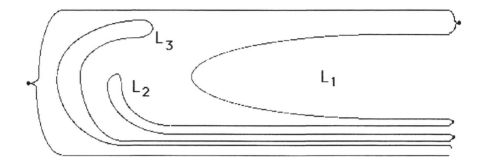

Figure 10: Embedding Γ in the plane.

true. Then we can write G as the union of two disjoint sets A and B. One of A or B must contain infinitely many of the boundaries of the L_i. Say A does. But then, if $b \in B$, the previous proposition guarantees that infinitely many of these boundaries meet any neighborhood of b. Hence b belongs to the closure of A. This contradiction establishes the result.

We can now prove:

Theorem 4.7 *There is a natural compactification Γ of Λ that makes Γ into an indecomposable continuum.*

Proof. We first compactify Λ by adjoining the backward orbit of 0. To do this we identify the "points" $(-\infty, 0)$ and $(-\infty, \pi)$ in S: this gives $E_\lambda^{-1}(0)$. We then identify the points (∞, π) and $\lim_{t \to -\infty} \gamma_1(t)$. This gives $E_\lambda^{-2}(0)$. For each $n > 1$ we identify

$$\lim_{t \to \infty} \gamma_n(t)$$

and

$$\lim_{t \to -\infty} \gamma_{n+1}(t)$$

to yield $E_\lambda^{-n-1}(0)$. This augmented space Γ may easily be embedded in the plane. See Figure 10. Moreover, if we extend the B_i and the lines $y = 0$ and $y = \pi$ in the natural way to include these new points, then this yields a curve which accumulates everywhere on itself but does not separate the plane. See the proposition above. By a theorem of S. Curry [Cu], it follows that Γ is indecomposable.

$\square$

As a consequence of this theorem, Λ must contain uncountably many composants (see [Ku], p. 213). In fact, in [DK] it is shown that Λ contains uncountably many curves.

4.3   Dynamics on Λ

In this section we describe the dynamics of E_λ on Λ.

Proposition 4.8 *There exists a unique fixed point w_λ in S if $\lambda > 1/e$. Moreover, w_λ is repelling and, if $z \in S$ − orbit of 0, $E_\lambda^{-n}(z) \to w_\lambda$ as $n \to \infty$.*

Proof. First consider the equation

$$\lambda e^{y \cot y} \sin y = y.$$

Since $y \cot y \to 1$ as $y \to 0$ and $\lambda e > 1$, we have $\lambda e^{y \cot y} \sin y > y$ for y small and positive. Since the left-hand side of this equation vanishes when $y = \pi$, it follows that this equation has at least one solution y_λ in the interval $0 < y < \pi$.

Let $x_\lambda = y_\lambda \cot y_\lambda$. Then one may easily check that $w_\lambda = x_\lambda + iy_\lambda$ is a fixed point for E_λ in the interior of S. Since the interior of S is conformally equivalent to a disk and E_λ^{-1} is holomorphic, it follows from the Schwarz Lemma that w_λ is an attracting fixed point for the restriction of E_λ^{-1} to S and that $E_\lambda^{-n}(z) \to w_\lambda$ for all $z \in S$.

Remarks.
1. Thus the α-limit set of any point in Λ is w_λ.
2. The bound $\lambda > 1/e$ is necessary for this result, since we know that E_λ has two fixed points on the real axis for any positive $\lambda < 1/e$. These fixed points coalesce at 1 as $\lambda \to 1/e$ and then separate into a pair of conjugate fixed points, one of which lies in S.

We now describe the ω-limit set of any point in Λ. Clearly, if $z \in B_n$ then $E_\lambda^{n+1}(z) \in \mathbf{R}$ and so the ω-limit set of z is infinity. Thus we need only consider points in Λ that do not lie in B_n. We will show:

Theorem 4.9 *Suppose $z \in \Lambda$ and $z \neq w_\lambda$, $z \notin B_n$ for any n. Then the ω-limit set of z is the orbit of 0 under E_λ together with the point at infinity.*

To prove this we first need a lemma.

Lemma 4.10 *Suppose $z \in \Lambda$, $z \neq w_\lambda$. Then $E_\lambda^n(z)$ approaches the boundary of S as $n \to \infty$.*

Proof. Let h be the uniformization of the interior of S taking S to the open unit disk and w_λ to 0. Recall that E_λ^{-1} is well defined on S and takes S inside itself. Then $g = h \circ E_\lambda^{-1} \circ h^{-1}$ is an analytic map of the open disk strictly inside itself with a fixed point at 0. This fixed point is therefore attracting by the Schwarz Lemma. Moreover, if $|z| > 0$ we have $|g(z)| < |z|$. As a consequence, if $\{z_n\}$ is an orbit in Λ, we have $|h(z_{n+1})| > |h(z_n)|$, and so $|h(z_n)| \to 1$ as $n \to \infty$.

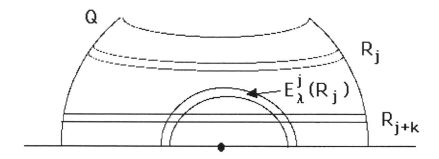

Figure 11: The return map on Q.

The remainder of the proof is essentially contained in [DK] (see pp. 45-49). In that paper it is shown that there is a "quadrilateral" Q containing a neighborhood of 0 in $\mathbf{R}$ as depicted in Figure 11. The set Q has the following properties:

1. If $z \in \Lambda - \bigcup_n B_n$ and $z \neq w_\lambda$, then the forward orbit of z meets Q infinitely often.

2. Q contains infinitely many closed "rectangles" $R_k, R_{k+1}, R_{k+2}, \ldots$ for some $k > 1$ having the property that if $z \in R_j$, then $E_\lambda^j(z) \in Q$ but $E_\lambda^i(z) \notin Q$ for $0 < i < j$.

3. If $z \in Q$ but $z \notin \bigcup_{j=k}^\infty R_j$, then $z \in L_n$ for some n.

4. $E_\lambda^j(R_j)$ is a "horseshoe" shaped region lying below R_j in Q as depicted in Figure 11.

5. $\lim_{j \to \infty} E_\lambda^j(R_j) = \{0\}$.

As a consequence of these facts, any point in Λ has orbit that meets the $\cup R_j$ infinitely often. We may thus define a return map

$$\Phi : \Lambda \cap (\cup_j R_j) \to \Lambda \cap \cup_j R_j$$

by

$$\Phi(z) = E_\lambda^j(z)$$

if $z \in R_j$. By item 4, $\Phi(z)$ lies in some R_k with $k > j$. By item 5, it follows that

$$\Phi^n(z) \to 0$$

for any $z \in \Lambda \cap Q$. Consequently, the ω-limit set of z contains the orbit of 0 and infinity.

For the opposite containment, suppose that the forward orbit of z accumulates on a point q. By the Lemma, q lies in the boundary of S. Now the orbit of z must also accumulate on the preimages of q. If q does not lie on the orbit of 0, then these preimages form an infinite set, and some points in this set lie on the boundaries of the L_n. But these points lie in the interior of S, and this contradicts the Lemma. Thus the orbit of z can only accumulate in the finite plane on points on the orbit of 0. Since the "preimage" of 0 is infinity, the orbit also accumulates at infinity.

$\square$

5   Final remarks

It is known [BD] that there are uncountably many curves in the λ-plane having the property that, if λ lies on one of these curves, then $E_\lambda^n(0) \to \infty$. Consequently, for such a λ-value, the Julia set of E_λ is again the complex plane. For these λ-values, a variant of the above construction also yields invariant indecomposable continua in the Julia set [MR]. Whether these continua are homeomorphic to any of those constructed above is an open question.

Douady and Goldberg [DoG] have shown that if $\lambda, \mu > 1/e$, then E_λ and E_μ are not topologically conjugate. Each such map possesses invariant indecomposable continua Λ_λ and Λ_μ in S, and the dynamics on each are similar, as shown above. In fact, we conjecture that each pair of these invariant sets is non-homeomorphic.

A simpler semilinear model mapping that mimics the behavior of E_λ has been constructed in [DMR]. The indecomposable continua constructed in this paper should be easier to deal with than those of E_λ, though we conjecture that they are homeomorphic to specific indecomposable continua in the exponential family.

It is also known that the set of points whose itineraries feature blocks of 0's whose length goes to ∞ quickly is an indecomposable continuum [DJ]. The exact structure of these sets, however, is far from understood.

M. Lyubich has shown that each Λ_λ is a set of measure 0 in S. Indeed, it follows from his work [Ly] that the set of points in $\mathbf{C}$ whose orbits have arguments that are equidistributed on the unit circle have full measure. In Λ_λ, the arguments of all orbits tend to 0 and/or π, and so Λ_λ has measure 0 in S.

References

[Ba]   Barge, M., Horseshoe maps and inverse limits. *Pacific J. Math.* **121** (1986),29–39.

[Ba1]   Baker, I. N., Repulsive Fixpoints of Entire Functions. *Math. Z.* **104** (1968), 252–256.

[BD]   Bhattacharjee, R. and Devaney, R. L., Tying hairs for structurally stable exponentials. Preprint.

[Cu]   Curry, S., One-dimensional nonseparating plane continua with disjoint ϵ-dense subcontinua. *Topol. and its Appl.* **39** (1991),145–151.

[D1]   Devaney, R. L., Knaster-like continua and complex dynamics. *Ergodic Theory and Dynamical Systems* **13** (1993), 627–634.

[D2]   Devaney, R. L., Se^x: dynamics, topology, and bifurcations of complex exponentials. *Topology and its Applications* **110** (2001), 133–161.

[DeV]   DeVille, R. E. Lee., Itineraries of entire functions. *J. Diff. Eq. Appl.* **7** (2001), 193–214.

[DJ]   Devaney, R. L. and Jarque, X., Indecomposable continua in exponential dynamics. Preprint.
         See http://math.bu.edu/people/bob/papers.html.

[DK]   Devaney, R. L. and Krych, M., Dynamics of Exp(z), *Ergodic Theory and Dynamical Systems* **4** (1984), 35–52

[DMR]   Devaney, R. L. and Moreno Rocha, M., A semilinear model for exponential dynamics and topology. Preprint.
         See http://math.bu.edu/people/bob/papers.html.

[DoG]   Douady, A. and Goldberg, L., The nonconjugacy of certain exponential functions. In *Holomorphic Functions and Moduli*. MSRI Publ., Springer Verlag (1988), 1–8.

[DT]   Devaney, R. L. and Tangerman, F., Dynamics of entire functions near the essential singularity, *Ergodic Thy. Dynamical Syst.* **6** (1986), 489–503.

[EL]   Eremenko, A. and Lyubich, M. Yu., Iterates of entire functions. *Dokl. Akad. Nauk SSSR* **279** (1984), 25–27. English translation in *Soviet Math. Dokl.* **30** (1984), 592–594.

[GK]   Goldberg, L. R. and Keen, L., A finiteness theorem for a dynamical class of entire functions, *Ergodic Theory and Dynamical Systems* **6** (1986), 183–192.

[Ku]   Kuratowski, K., *Topology* Vol. 2. Academic Press, New York, 1968.

[Ly]   Lyubich, M., Measurable dynamics of the exponential, *Soviet Math. Dokl.* **35** (1987), 223–226.

[Ma]   Mayer, J., An explosion point for the set of endpoints of the Julia set of $\lambda \exp(z)$. *Erg. Thy. and Dyn. Syst.* **10** (1990), 177–184.

[McM]   McMullen, C. Area and Hausdorff Dimension of Julia Sets of Entire Functions. *Trans. A.M.S.* **300** (1987), 329-342.

[MR]   Moreno Rocha, M., Existence of indecomposable continua for the unstable exponential. Preprint.

[Sm]   Smale, S., Diffeomorphisms with many periodic points. *Differential and Combinatorial Topology.* Princeton University Press, 1964, 63–80.

[Su]   Sullivan, D., Quasiconformal maps and dynamical systems I, solutions of the Fatou-Julia problem on wandering domains. *Ann. Math.* **122** (1985), 401–418.

Farey's Rule for Bifurcations of Periodic Trajectories in a Class of Multi-Valued Interval Maps

VLADIMIR V. FEDORENKO [1]

Institute of Mathematics
National Academy of Sciences of Ukraine
01601 Kiev, Ukraine
E-mail: vfedor@imath.kiev.ua

Abstract A class of dynamical systems with a two-sheeted phase space generated by two monotonic continuous maps of an interval is considered. We describe the different types of dynamical behavior and bifurcations of recurrent trajectories in this class. In particular, for stable periodic trajectories we prove the existence of "period-adding" bifurcations and bifurcations subordinate to *Farey's rule*. These results are applied to some classes of difference equations and boundary value problems.

Keywords Dynamical system, Two-sheeted phase space, Stable periodic trajectory, Bifurcation, Farey's rule, Difference equation

AMS Subject Classification 37G15, 39A11

1 Properties of a class of two-sheeted one-dimensional maps

Dynamical systems with multi-sheeted phase space often are used as mathematical models of real phenomena and processes, in particular in control theory [1]. On the other hand, such dynamical systems appear in the investigation of ordinary differential equation systems or under exact reduction of some boundary value problems to difference equations with continuous time. One such problem is considered in the Appendix.

Let f be a map defined by

$$f(x) \;=\; \begin{cases} g(x), & \text{if} \quad 0 \le x \le b, \quad \text{and} \\ h(x), & \text{if} \quad a < x \le 1 \,, \end{cases} \tag{1}$$

---

[1]Supported by the Ministry of Education and Science of Ukraine, State Fund of Fundamental Investigations, project No 01.07/00081.

with parameters a, b which belong to the domain $\Omega = \{\omega = (a, b) \in \mathbb{R}^2 \mid 0 < a \leq b < 1\}$ and g, h are monotonically increasing continuous functions defined on $I = [0, 1] \subset \mathbb{R}$ and such that $0 < h(x) < x < g(x) < 1$ for all $x \in (0, 1)$ (and, hence, $h(0) = 0, g(1) = 1$).

From the definition above one can see that f is either a single-valued (if $a = b$) or a two-valued (if $a < b$) map.

If $a < b$, we have to make additional assumptions in order to obtain a uniquely defined trajectory $x_n = f^n(x_0)$, $n = 0, 1, 2, \ldots$ for each x_0. If $x_0 \in (a, b]$ we set $x_1 = g(x_0)$; if $x_n \in (a, b]$ for $n > 0$ we use the following rule (which is the ordinary rule under existence of "hysteresis"): if $x_n = g(x_{n-1})$, then $x_{n+1} = g(x_n)$; if $x_n = h(x_{n-1})$, then $x_{n+1} = h(x_n)$, too. Under such an agreement the map generates a dynamical system with so-called two-sheeted phase space. For $n \geq 0$ it is defined by

$$x_{n+1} = \begin{cases} g(x_n), & \text{if} \quad \text{either } x_n \leq a \text{ or } a < x_n \leq b \text{ and then} \\ & \text{either } n = 0 \text{ or } n > 0 \text{ and } x_n = g(x_{n-1}), \quad (2) \\ h(x_n), & \text{otherwise.} \end{cases}$$

The dynamical system defined above can have the wide spectrum of behavior from existence of one attracting periodic trajectory to chaos. In order to be close to the main subject of this paper, namely stable periodic trajectories, we will always assume that the conditions

$$g, h \in Lip_L(I, I) \quad \text{with} \quad L < 1 \tag{3}$$

are fulfilled.

For any measurable set $B \subseteq I$, mes $f(B) < L \cdot$mes B; hence mes $f^n(B) \to 0$ as $n \to 0$. This means that the set $A(f) = \bigcap_{n=0}^{\infty} f^n(I)$ is nowhere dense; therefore, the attractor of the dynamical system (2) is contained inside a nowhere dense subset of the interval I.

Let $\mathcal{D}(f) = \{x \in (0, 1) \mid \exists n \geq 0, f^n(x) \in \{a, b\}\}$; i.e., $\mathcal{D}(f)$ is a set of points from $(0, 1)$, each of which gets mapped into a or b after a finite number of iterations. The set $\mathcal{D}(f)$ can be finite (possibly empty) or infinite. The set $\mathcal{D}(f)$ is a nowhere dense set.

The following fundamental property is obvious.

Lemma A *If J is an interval from $I \setminus \mathcal{D}(f)$ then for each $n > 0$ $f^n(J)$ is also an interval, and for any two points $x', x'' \in J$ $|f^n(x') - f^n(x'')| < L^n |x' - x''|$.*

Thus, the set $\mathcal{D}(f)$ defines a partition of the set $I \setminus \mathcal{D}(f)$ into a finite (if $\mathcal{D}(f)$ is finite) or countable number of intervals J_s, $s = 1, 2, \ldots$ with disjoint interiors which form a Markov partition of the phase space: for any interval $J_{s'}$, there exists an interval $J_{s''}$ such that $f(J_{s'}) \subseteq J_{s''}$. It allows us to reduce the study of the dynamical system (2) to the analysis of possible paths of the graph induced by the transfer matrix of the Markov partition. In particular, closed loops of the graph correspond to periodic trajectories of the dynamical system.

From the analysis of the graph one can see that the three statements

   i) $\mathcal{D}(f)$ is a finite set,

   ii) there exits a asymptotically periodic trajectory,

   iii) every trajectory is asymptotically periodic

are equivalent for the dynamical system (2) [2].

The *typical situation* for the domain Ω is that *every trajectory of the dynamical system (2) is asymptotically periodic* [2]: parameter values for which the dynamical system (2) has asymptotically periodic trajectories form a set containing an open dense subset of Ω. Almost always the dynamical system (2) has either one or two periodic trajectories. In the latter case the difference between periods equals 2.

We will call any function $\psi : \mathbf{R}^1 \to G$, where G is a finite set from $\mathbf{R}^1$, a step-function.

The following lemma is a consequence of Lemma A.

Lemma B *Assume $\mathcal{D}(f)$ is a finite set. Then there exists an integer $m > 1$ such that for any $x \in I$ the limit $\lim_{i\to\infty} f^{mi}(x)$ exists and the limiting function $f^*(x)$ is a step-function which is constant on each interval of $I \backslash \mathcal{D}(f)$. Moreover, for any $\epsilon > 0$ there are integers $m > 0$ and $N > 0$ such that $|f^{mn}(x) - f^*(x)| < \epsilon$ for any $x \in I \backslash \mathcal{D}(f)$ and $n > N$.*

1.1 Types of periodic trajectories

We need a more detailed classification of periodic trajectories based not only on their periods but also using the so-called type of the trajectories [2].

Let P be a periodic trajectory of the dynamical system (2) for some $\omega' = (a', b') \in \Omega$, and $P_1 = \{\beta \in P \mid f(\beta) = g(\beta)\}$, $P_2 = \{\beta \in P \mid f(\beta) = h(\beta)\}$. Let $\beta_1 < \beta_2 < \ldots < \beta_{n_1}$ be points of the set P_1 and $\beta_{n_1+1} < \ldots < \beta_{n_1+n_2}$ be points of the set P_2. Denote by p the number of points $\beta \in P_1$ whose presages belong to P_2. Then the number of points $\beta \in P_2$ whose presages belong to P_1 is also p.

The trajectory P generates a cyclic permutation π of length $n_1 + n_2$: $\pi(i) = j$ if $f(\beta_i) = \beta_j$, $1 \leq i, j \leq n_1 + n_2$.

For every periodic trajectory P of the dynamical system (2) there exist positive coprime integers p, q, $p < q$, and a nonnegative integer d such that the cyclic permutation generated by the periodic trajectory P has the form [2]

$$\pi(i) = \begin{cases} i+p, & \text{if } 1 \leq i \leq q - 2p + d, \\ i+p+d, & \text{if } q - 2p + d < i \leq q - p + d, \\ i-q+p-d, & \text{if } q-p+d < i \leq q+d, \\ i-p, & \text{if } q+d < i \leq q+2d, \end{cases} \tag{4}$$

if $2p \leq q$, and if $2p > q$, it equals $h \circ \pi \circ h$, where $h(i) = q + 2d + 1 - i$, $i = 1, ..., q + 2d$.

The map (4) is a cyclic permutation if and only if p and q are coprime numbers. From formula (4) one can see the meaning of p, q and d. The period of the trajectory P generating the cyclic permutation π is $q + 2d$, $d = n_2 - p$ and $q = n_1 + n_2 - 2d$.

In what follows, we always assume that p and q are coprime numbers, $p < q$, and d is a nonnegative integer. We shall call the triple (p, q, d) the *type* of the periodic trajectory.

Let $\Omega_{p,q,d} = \{\omega \in \Omega |$ the periodic trajectory of dynamical system (2) has type $(p, q, d)\}$. As it follows from [2], for every triple (p, q, d), the sets $\Omega_{p,q,d}$ and *int* $\Omega_{p,q,d}$ are nonempty connected sets with a rectangular shape. In particular, $\Omega^0_{p,q,d} = int\ \Omega_{p,q,d} \setminus (\overline{\Omega}_{p,q,d-1} \cup \overline{\Omega}_{p,q,d+1})$ is a nonempty set. If $\omega \in \Omega^0_{p,q,d}$, then the dynamical system (2) has a unique periodic trajectory which attracts all other trajectories (here $\Omega_{p,q,-1} = \emptyset$).

1.2   Properties of the bifurcation diagram

Also we will always assume the condition

$$g(0) > h(1), \quad \text{i.e.} \quad g([0, b]) \cap h((a, 1]) = \emptyset. \tag{5}$$

This property simplifies the investigation of the dynamical system(2) and the formulation of the properties of the bifurcation diagram. The bifurcation diagram of types of periodic trajectories is characterized by the following properties [2,3].

Property 1. $\Omega_{p,q,d} \cap \Omega_{p',q',d'} \neq \emptyset$ if $p = p'$, $q = q'$ and $| d - d' | \leq 1$.

Property 2. For any triple (p, q, d),

a) if $\omega \in int\ \Omega_{p,q,d} \setminus (\overline{\Omega}_{p,q,d-1} \cup \overline{\Omega}_{p,q,d+1}$, the dynamical system (2) has a unique periodic trajectory which attracts all other trajectories (here $\Omega_{p,q,-1} = \emptyset$);

b) if $\omega \in int\ \Omega_{p,q,d} \cap int\ \Omega_{p,q,d+1}$, the dynamical system(2) has two periodic trajectories which attract all other trajectories;

c) if $\omega \in \overline{\Omega}_{p,q,d} \setminus int\ \Omega_{p,q,d}$, the dynamical system(2) has either none, or one, or two periodic trajectories, and every trajectory of the dynamical system (2) is asymptotically periodic.

Property 3. $\Omega^* = \Omega \setminus \bigcup_{p,q,d} \overline{\Omega}_{p,q,d}$ is nowhere dense in Ω. For any $\omega \in \Omega^*$ every trajectory of the dynamical system (2) is asymptotically almost periodic.

Let $\Omega_{p,q} = \bigcup_d \Omega_{p,q,d}$ and $\Omega^d = \bigcup_{p,q}(\Omega_{p,q,d} \setminus \Omega_{p,q,d+1})$.

Property 4. For any p, q, $\Omega_{p,q}$ is a connected set and $\Omega \setminus \Omega_{p,q}$ is made up of two connected components. For any d the closure of Ω^d is a connected set, and if $d > 0$, then $\Omega \setminus \overline{\Omega}^d$ is made up of two connected components.

Property 5. For any $d_1 < d_2$, $\bigcup_{d=d_1}^{d_2} \Omega_{p,q,d}$ is a connected set; for any p_i, q_i, $i = 1, 2$, with $\frac{p_1}{q_1} < \frac{p_2}{q_2}$, the closure of $\bigcup_{\frac{p_1}{q_1} < \frac{p}{q} < \frac{p_2}{q_2}} \Omega_{p,q}$ is a connected set.

In order to enumerate all fractions from the interval $\left[\frac{p_1}{q_1}, \frac{p_2}{q_2}\right] \subset [0,1]$, it is convenient to use a well-known notion from number theory, namely, the *Farey sequence* [4]. This notion is based on the following property: if $p_1 q_2 - p_2 q_1 = -1$, then $\frac{p_1}{q_1} < \frac{p_1 + p_2}{q_1 + q_2} < \frac{p_2}{q_2}$. From this property it follows that there exist bifurcations of the dynamical system (2) when stable periodic trajectories with arbitrarily large periods appear under arbitrary small changes of the parameters.

2   Properties of solutions of difference equation

Consider the difference equation

$$x(t+1) = f(x(t)) \tag{6}$$

where $x(t) \in \mathbb{R}$. If t is discreet then equation (6) generates a dynamical system of the form (2) and one can formulate the properties of periodic solutions of this difference equation in the same way as for the dynamical system described in the previous section.

In the present section we consider equation (6) with a continuous argument t and describe bifurcations of periodic solutions following [3]. We will use an integral metric which is given by the norm $||x||_{[\alpha,\beta]} := \int_\alpha^\beta |x(t)|\, dt$.

Theorem 1. *If $\mathcal{D}(f)$ is a finite set, then every solution of the difference equation (6) with continuous argument converges to a periodic step-function in the integral metric as $t \to +\infty$.*

Suppose the dynamical system (2) has a unique periodic trajectory forming the cycle $\{\beta_1, ..., \beta_n\}$. If π is a cycle permutation generated by this periodic trajectory, then every limit step-function x^* has the following property: if $x^*(t_0) = \beta_{i_0}$ at $t_0 \in \mathbf{R}^1$, then $x^*(t_0 + j) = \beta_{\pi^j(i_0)}$ for any positive integer j. Thus, in this case, periodic step-functions generated by the equation $x(t+1) = f(x(t))$ can also be classified by cyclic permutations and, hence, by the triple (p, q, d) introduced above. We will call this triple the *type* of the periodic step-function, too. Moreover, it makes sense to call a triple (p, q, d) the *type* of the solution of the difference equation (6) if the limiting step-function for this solution is of type (p, q, d). This allows us to state the following properties of solutions:

Theorem 2. *Let $\omega \in \Omega^0_{p,q,d}$. For any function $\varphi : [0,1] \to [0,1]$, the solution x_φ of the difference equation (6) with initial function φ is of type (p, q, d).*

Recall that if $\omega \in \Omega^0_{p,q,d}$, then the dynamical system (2) has a unique periodic trajectory of type (p, q, d) which attracts all other trajectories. Making use of Theorem 2, the analogous properties 4 and 5 from the previous section can be stated for solutions of the difference equation (6) as follows:

Theorem 3. *Let ω_1, ω_2 be two collections of parameters from Ω^0 and let the triples (p_i, q_i, d_i), $i = 1, 2$, be the types of the corresponding periodic trajectories of the dynamical system (2). Then*

1) if $(p_1, q_1) \neq (p_2, q_2)$ and $\frac{p_1}{q_1} < \frac{p_2}{q_2}$, then for any arc $\mathcal{L}$ inside Ω^0 connecting ω_1, ω_2, and any pair (p, q) such that $\frac{p_1}{q_1} < \frac{p}{q} < \frac{p_2}{q_2}$, there exists an $\omega \in \mathcal{L}$ for which the type of solutions of the difference equation (6) is (p, q, d) with some $d \geq 0$;

2) if $(p_1, q_1) = (p_2, q_2)$ and $d_1 < d_2$, then there is an arc $\mathcal{L}$ inside Ω connecting ω_1, ω_2 such that a) for any $\omega \in \mathcal{L}$, the corresponding solution of the difference equation (6) has the type (p_1, q_1, d) with some d, $d_1 \leq d \leq d_2$, and b) for any d', $d_1 < d' < d_2$, there exists an $\omega' \in \mathcal{L}$ for which the type of solutions of equation (6) is (p_1, q_1, d').

3   Appendix

In this appendix we consider hyperbolic partial differential equations with nonlinear boundary conditions which can be reduced to scalar nonlinear difference equation [5-11].

In [8, 9] an "infinite-dimensional" generalization of Chua's circuit was considered which is obtained by replacing the LC resonant circuit by a lossless transmission line of length l terminated on its left end ($x = 0$) by a short circuit. The main objective of these papers was to study behavior as $t \to \infty$ of the voltage $v(x, t)$ and the current $i(x, t)$ along the transmission line ($0 \leq x \leq l$).

The mathematical model of the circuit is described by the linear system of partial differential equations

$$\frac{\partial v(x, t)}{\partial x} = -L \frac{\partial i(x, t)}{\partial t}, \tag{7}$$

$$\frac{\partial i(x, t)}{\partial x} = -C \frac{\partial v(x, t)}{\partial t}, \tag{8}$$

where L and C denote the inductance and capacitance, respectively, per unit length of the transmission line. The boundary conditions at $x = 0$ and $x = l$ are defined by

$$v(0, t) = 0, \tag{9}$$

$$i(l, t) = G(v(l, t) - E - Ri(l, t)) + C_1 \frac{\partial(v(l, t) - Ri(l, t))}{\partial t}, \tag{10}$$

where $G(\cdot)$ is given by

$$G(u) = \begin{cases} m_0 u, & |u| \leq 1, \\ m_1 u - (m_1 - m_0)\mathrm{sgn}\, u, & |u| \geq 1. \end{cases} \tag{11}$$

We always assume that $m_0 < 0$, $m_1 > 0$ and $E > 0$. The initial conditions for (1)-(4) are determined by the initial values (at $t = 0$) of the voltage $v_0(x)$ and the current $i_0(x)$.

We consider solutions of the boundary value problem (7)-(10) which have the form

$$v(x,t) \;=\; \alpha(t - \frac{x}{\nu}) - \alpha(t + \frac{x}{\nu}), \tag{12}$$

$$i(x,t) \;=\; \frac{1}{Z}[\alpha(t - \frac{x}{\nu}) + \alpha(t + \frac{x}{\nu})], \tag{13}$$

where $\nu = \sqrt{1/LC}$, $Z = \sqrt{L/C}$ and α is an arbitrary differentiable function.

Substituting (12), (13) into the boundary condition (10) with $C_1 = 0$ we obtain a difference equation for the function α. If we introduce new variables

$$\tau = \frac{\nu}{2l}t - \frac{1}{2}, \qquad \gamma(\tau) = \alpha\left(\frac{2l}{\nu}\tau\right), \tag{14}$$

we obtain the difference equation

$$\gamma(\tau + 1) = S(\gamma(\tau)). \tag{15}$$

Here $S(\gamma)$ is a piecewise linear function defined by linear transformations of the function $G(u)$; it has three branches. $S(\gamma)$ is either single-valued or three-valued. The analytical form of $S(\gamma)$ is in [9].

The "middle branch" of (15) does not affect the asymptotic behavior of solutions: (a) if $S(\gamma)$ is multivalued, a solution "belonging to the middle branch" will leave this branch in finite time (provided it does not start with the fixed point); (b) according to the ordinary rule of "hysteresis", any solution may not switch into the middle branch from the other two branches. Therefore we can disregard the middle branch in the sequel.

Initial values of the voltage $v_0(x)$ and the current $i_0(x)$ determine the initial conditions for the difference equation (15):

$$\gamma_0(\tau) \;=\; \begin{cases} \dfrac{v_0(-y) + Zi_0(-y)}{2}, & y = (2\tau - 1)l, \quad 0 \leq \tau < \dfrac{1}{2}, \\[2ex] \dfrac{-v_0(y) + Zi_0(y)}{2}, & \qquad\qquad\quad -\dfrac{1}{2} \leq \tau < 1. \end{cases} \tag{16}$$

References

[1] Gaushus, E. V., *Investigation of Dynamical Systems Using the Point Mapping Method*, Moscow, Nauka, 1976, 368 p. (in Russian).

[2] Sharkovsky, O. M., Ivanov, A. F., Fedorenko, V. V. and Fedorenko, O. D., Coexistence of periodic orbits for a class of discontinuous maps, *Proc. Nation. Acad. Sci. Ukraine* No.11 (1996), 20-25 (in Ukrainian).

[3] Fedorenko, A. D., Fedorenko, V. V., Ivanov, A. F. and Sharkovsky, A. N., Solution behaviour in a class of difference-differential equations, *Bulletin of the Austral. Math. Soc.* **57** (1998) , 37-48.

[4] Hardy, G. H. and Wright, E. M., *An Introduction to the Theory of Numbers*. Clarendon Press, Oxford, 1979.

[5] Witt, A. A., On the theory of the violin string, *J.Technical Physics* **6** (1936), 1459-1470 (in Russian).

[6] Nagumo, J. and Shimura, M., Self-oscillations in a transmission line with a tunnel diode, *Proc. IRE* (1961), 1281-1291.

[7] Sharkovsky, A. N., Maistrenko, Yu. L., and Romanenko, E. Yu., *Difference Equations and Their Applications*, Ser. Mathematics and Its Applications, v.250, Kluwer Academic Publishers, Dordrecht-Boston-London, 1993, 358 p.

[8] Sharkovsky, A. N., Chaos From a Time-Delayed Chua's Circuit, *IEEE Transactions on Circuits and Systems. I* **40** (1993), 781-783.

[9] Sharkovsky, A. N., Ideal turbulence in an idealized time-delayed Chua's circuit, *Intern. J. Bifurcation and Chaos* **4** (1994), 303-309.

[10] Chen, G., Hsu, S. B. and Zhou, J., Chaotic vibrations of the one-dimensional wave equation due to a self-excitation boundary condition. II: Energy injection, periodic doubling and homoclinic orbits, *Int. J. Bifurcation and Chaos* **8** (1998), 423-445.

[11] Chen, G., Hsu, S. B. and Zhou, J., Chaotic vibrations of the one-dimensional wave equation due to a self-excitation boundary condition. III: Natural hysteresis memory effects, *Int. J. Bifurcation and Chaos* **8** (1998), 447-470.

Combinatorics of Angle-Doubling: Translation Principles for the Mandelbrot set

KARSTEN KELLER

Department of Mathematics and Computer Science,
University Greifswald
D-17487 Greifswald, Germany
E-mail: keller@mail.uni-greifswald.de

Abstract Much of the structure of the Mandelbrot M set can be described from the pure combinatorial viewpoint. In particular, the naturally ordered system of hyperbolic components labeled with their periods contains much information about the shape of M. We discuss similarities between special parts of this system. The considerations rely on a tight connection of holomorphic quadratic iteration to the angle-doubling map on the circle, which was comprehensively studied in the monograph [5] of the author.

Keywords Quadratic iteration, Mandelbrot set, Angle-doubling map

AMS Subject Classification 37F45, 58F03

1 Introduction

The detailed structure of the Mandelbrot set M is extremely complicated. However, much of it can be described by different kinds of symmetry and self-similarity. For example, each neighborhood of a boundary point of M contains infinitely many topological copies of M itself, which is a consequence of the tuning results by Douady and Hubbard (see [3], [8], [2]).

The present paper deals with symmetries whose nature is a combinatorial one. To put us in a position to formulate the main statements, it is necessary to recall some facts on the Mandelbrot set. Beside the standard paper [1] by Douady and Hubbard, the references are [9, 5, 10].

Hyperbolic components. The well-known Mandelbrot set M reflects qualitative changes in the iteration of the quadratic maps $p_c : \mathbb{C} \hookleftarrow$ given by $p_c(z) = z^2 + c$ for running parameter $c \in \mathbb{C}$. Since p_c is locally invertible at points different from the critical point 0 with image c, the structure of the orbit $\{c, p_c(c), p_c^{\circ 2}(c), \ldots\}$ of c indicates how complicated the iteration of the

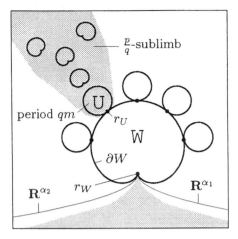
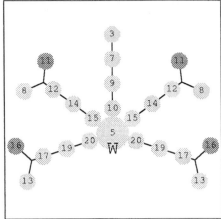

Figure 1: Anatomy of a hyp. comp.   Figure 2: Partial Transl. Principle I.

whole map p_c is. There are two special cases with simple iteration structure:

1. The orbit of c is unbounded; in this case $\lim_{n\to\infty} p_c^{\circ n}(c) = \infty$. (If considering p_c on the Riemann sphere, c is attracted by the fixed point at infinity.) The complement of the set of all those c is the *Mandelbrot set M*.

2. p_c possesses an attractive periodic orbit $\{z = p_c^{\circ m}(z), p_c(z), \ldots, p_c^{\circ m-1}(z)\}$: the *multiplier* of the orbit, i.e. the derivative of the first return map $p_c^{\circ m}$ at any point of the orbit, has absolute value less than 1. If such an orbit exists for p_c, it is unique and attracts c, i.e. for sufficiently large n the orbits of $p_c^{\circ n}(c)$ and of z are not distinguishable from the numerical viewpoint.

M is connected, and the set of all c providing an attractive periodic orbit is open and contained in M (and is conjectured to be dense in M). A connectedness component W of the latter set is called a *hyperbolic component* of M. (The Mandelbrot set is drawn in the center of Figure 3. The white connected parts indicate some of its hyperbolic components.) The period of the unique attractive periodic orbit for p_c is the same for all $c \in W$ justifying to call it the *period* of W. Note that the orbit depends analytically on $c \in W$.

From the conformal point of view each hyperbolic component W is an open disc. Namely, the *multiplier map* assigning to each $c \in W$ the multiplier of the attractive periodic orbit for p_c forms a conformal isomorphism from W onto the unit disk, and it extends continuously to a homeomorphism from the closure of W onto the closed unit disk. The unique parameter $r_W \in \partial W$ being mapped to 1 by this homeomorphism is called the *root* of W. Note that $M \setminus \{r_W\}$ splits into two connectedness components (see Figure 1).

Each parameter in ∂W being mapped to $e^{2\pi \frac{p}{q} i}$ for p, q positive and relatively prime with $p < q$ is the root r_U of a new hyperbolic component U of period qm "directly bifurcating" from W, as illustrated by Figure 1. (Each point in the attractive orbit of period m bifurcates into q points yielding a new attractive periodic orbit of period qm.) That component of $M \setminus r_U$ containing U is called the $\frac{p}{q}$-*sublimb* of W. Note that each point of the component of $M \setminus r_W$ containing W is either in the closure or in one of the sublimbs of W.

Having the shape of the Mandelbrot set in view, it suggests to say that a hyperbolic component is *behind* another one W if it lies in one of the sublimbs of W. Moreover, we need the following definition introduced in [6]:

Definition 1.1 (Visibility)
Let W, U be hyperbolic components. U is said to be visible *from W if U lies behind W and there is no hyperbolic component V of period less than the period of U between W and U, i.e., with V behind W and U behind V.*

Remark. By a statement known as Lavaurs' Lemma, between two hyperbolic components of common period there is one with less period (see [7]). So the meaning of visibility would not change if one would substitute *period less than* by *period less than or equal* in Definition 1.1.

Translation Principles. The structure of visibility behind a hyperbolic component is surprisingly regular. This was first observed by Lau and Schleicher [6] and led to a *translation principle* valid for a large class of hyperbolic components. Unfortunately, it turned out to be false for many hyperbolic components, but a *partial translation principle* is valid in general.

Let us have a look at the system of all hyperbolic components visible from a given hyperbolic component W of period m and lying in a $\frac{p}{q}$-*sublimb*. This system is finite (which will become clear below) and forms a tree: The *nodes* are the hyperbolic components, and two hyperbolic components are *connected* by an *edge* if one of them is behind the other one and no hyperbolic component of the system is between them. There is the *root*-component U of the tree which bifurcates directly from W and has period qm. Since all other visible hyperbolic components in the sublimb must be behind U (see Figure 1), by the above remark their period is less than qm.

We want to denote this tree by $\mathfrak{Vis}_{\frac{p}{q}}(W)$ and to call it a *visibility tree*. The central definition of our discussion is the following one:

Definition 1.2 (Translation-equivalence of visibility trees)
Two visibility trees $\mathfrak{Vis}_{\frac{p_1}{q_1}}(W)$ and $\mathfrak{Vis}_{\frac{p_2}{q_2}}(W)$ are called translation-equivalent *if they coincide, including the embedding into the plane, when all periods in $\mathfrak{Vis}_{\frac{p_1}{q_1}}(W)$ are increased by $(q_2 - q_1)m$.*

It turns out that for a hyperbolic component there are at most three classes of translation-equivalent visibility trees. This is the statement of the following theorem, which was stated and proved in [5] (see Theorem 1.24).

Theorem 1.3 (Partial Translation Principle)
For a given hyperbolic component W, each visibility tree other than $\mathfrak{Vis}_{\frac{1}{2}}(W)$ is translation-equivalent to $\mathfrak{Vis}_{\frac{1}{3}}(W)$ or to $\mathfrak{Vis}_{\frac{2}{3}}(W)$.

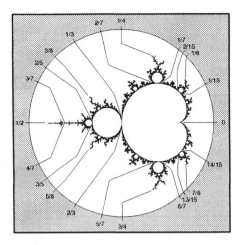
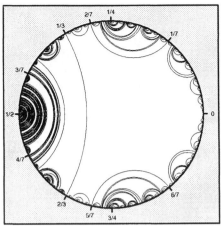

Figure 3: Mandelbrot set.   Figure 4: Abstract Mandelbrot set.

Figure 2 demonstrates Theorem 1.3. A hyperbolic component W of period 5 together with five visibility trees are shown (see also Remark 2 below Theorem 1.5). In the middle one sees $\mathfrak{Vis}_{\frac{1}{2}}(W)$. On the right and on the left there are the two translation-equivalent visibility trees $\mathfrak{Vis}_{\frac{1}{4}}(W), \mathfrak{Vis}_{\frac{1}{3}}(W)$ and $\mathfrak{Vis}_{\frac{2}{3}}(W), \mathfrak{Vis}_{\frac{3}{4}}(W)$, respectively. Note that the trees on the left coincide with those on the right, but their embeddings into the plane are different.

In Section 2 of the present paper we want to discuss translation-equivalence of visibility trees beyond the generally valid Partial Translation Principle, and Section 3 is devoted to the proof of the following supplement to Theorem 1.3:

Theorem 1.4 (Small visibility trees)
Let W be a hyperbolic component of period m and let $\mathfrak{T}$ be a visibility tree of W different from $\mathfrak{Vis}_{\frac{1}{2}}(W)$, and divide all periods of hyperbolic components in $\mathfrak{T}$ by m. Then the resulting remainders are mutually different.

Why the angle-doubling map?   The combinatorics of the system of hyperbolic components can be described within the framework of the angle-doubling map $h(\alpha) = 2\alpha \bmod 1$ on the unit circle, which we identify with the interval $[0, 1]$ according to $e^{2\pi\alpha i} \longleftrightarrow \alpha$. Here 0 and 1 are considered to be one point. Points periodic with respect to h are of a special interest. With exception of the fixed point $0('='1)$ they coincide with those rationals $\alpha \in]0, 1[$ represented by a fraction with odd denominator. The denominator can be chosen of the form $p^m - 1$, where the minimal possible m is the period of α. Thus the cardinality of periodic points for fixed period is finite.

According to Douady and Hubbard there is a conformal homeomorphism Φ mapping the complement of the Mandelbrot set M onto the complement of the unit-disk. The sets $\mathbf{R}^{\alpha} = \Phi^{-1}(\{re^{2\pi\alpha i} \mid r \in]1, \infty[\})$, $\alpha \in [0, 1]$ called *external rays* play a crucial role in understanding the combinatorial structure of M, in particular, for α periodic. Namely, there is a division of the set of periodic points into pairs $\alpha_1\alpha_2$ each corresponding to a unique hyperbolic component $W = W^{\alpha_1\alpha_2}$ and vice versa:

The external rays $\mathbf{R}^{\alpha_1}$ and $\mathbf{R}^{\alpha_2}$, together with the root of W, form a simple curve $\mathbf{R}^{\alpha_1\alpha_2}$ dividing the complex plane into two parts and separating the closure of W and its sublimbs from the rest of M (see Figure 1). The periods of α_1, α_2 and of W coincide. $\alpha_1\alpha_2 = 01$ is degenerate in the sense that the rays $\mathbf{R}^0$ and $\mathbf{R}^1$ coincide. However, they "land" at the root of the largest hyperbolic component having period 1, and the curve $\mathbf{R}^{01}$ provides "two parts" of the plane: one is $\mathbb{C} \setminus \mathbf{R}^{01}$ and the other is taken to be empty.

The curves $\mathbf{R}^{\alpha_1\alpha_2}$ obtained are mutually disjoint. Imagining a circle at infinity, each of them connects the points α_1 and α_2 (see Figure 3). This allows description of the mutual position of hyperbolic components $W^{\alpha_1\alpha_2}$ in a pure disk model: The pairs $\alpha_1\alpha_2$ can be interpreted as mutually disjoint chords drawn into the unit disk with endpoints α_1 and α_2 on its boundary (see Figure 4). Note that the situation of Figure 2 is given for $W = W^{13/31\,18/31}$.

We will prove Theorem 1.4 in the language of the chord system considered, on the base of [5]. (Note that many ideas used in [5] were taken from Thurston's unpublished paper [11]). As in [5], let $\mathcal{B}_*$ be the set of all chords (=pairs) with endpoints different from 0,1. (Beyond this we have the trivial chord 01.) The period of a chord is defined to be the period of its endpoints.

There is an obvious translation of the concepts "behind", "between" and "visible", and that what we now will call "immediately visible" means "directly bifurcating" in the language of hyperbolic components. The correspondence between hyperbolic components and chords, and the fact that the number of points of a given period in $[0, 1]$ is finite, show finiteness of all visibility trees. With the translations above, the statement to be shown is

Theorem 1.5 *Let* $S = \alpha\overline{\alpha} \in \mathcal{B}_* \cup \{01\}$ *be of period* m, *and let* $B \in \mathcal{B}_*$ *be immediately visible from* S *of period not less than* $3m$. *If* k_1 *and* k_2 *are different periods of visible chords behind* B, *then* m *does not divide* $k_1 - k_2$.

Remarks. 1. The reader may ask for the construction rules behind Figures 2, 4, 5 and 6. Without going into details, we note that $\mathcal{B}_* = \{B_1, \dots, B_n, B_{n+1}, \dots\}$, where B_{n+1} is obtained successively: If the chords $B_1, B_2, \dots, B_n$ are already given, then B_{n+1} does not intersect them and with this property has minimal possible period and minimal possible endpoints in $]0, 1[$ (see [7, 5]).

2. Figures 2, 5 and 6 show a bit more than the originally defined visibility trees. Also those branching points of M separating at least three components of the visibility trees are included. Note that branching points of M are parameters c preperiodic for p_c. For translation-equivalent visibility trees the trees obtained by adding those branching points are also translation-equivalent. This fact, which we do not prove here, can be obtained from a more detailed investigation of the combinatorical structure of M.

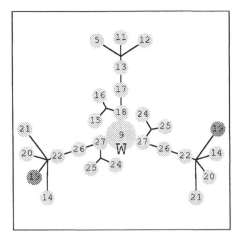 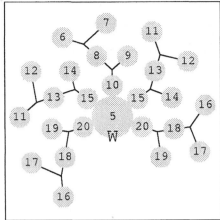

Figure 5: Partial Transl. Principle II.   Figure 6: "Complete" Transl. Princ.

2   More on translation-equivalence

As Figures 2 and 5 illustrate, the Partial Translation Principle cannot be extended to a general translation-equivalence statement. (The components given in dark-grey have no copy in the $\frac{1}{2}$-sublimb.) There is, however, the exceptional class of *narrow* hyperbolic components W – which can be characterized by the fact that all hyperbolic components behind W have period greater than W – found by Lau and Schleicher (see [6], Proposition 10.3):

Proposition 2.1 *All sublimbs of a narrow hyperbolic component are translation-equivalent.*

The periods of hyperbolic components in $\mathfrak{Vis}_{\frac{p}{q}}(W)$ for W of period m are greater than $q(m-2)$ and do not exceed qm (see [5], Proposition 3.65(i)). For narrow W there are exactly m hyperbolic components, of periods $(q-1)m+1, (q-1)m+2, \dots, qm-1$ and qm (see [6], Theorem 10.2). (An example is given by Figure 6, where $W = W^{5/31\,6/31}$ is of period 5.) There is another weakening of the Translation Principle saying the following:

Proposition 2.2 *Let W be a hyperbolic component of period m, and let r_1, r_2 be the smallest periods for a hyperbolic component in $\mathfrak{Vis}_{\frac{p_1}{q_1}}(W)$ and $\mathfrak{Vis}_{\frac{p_2}{q_2}}(W)$, respectively. Then it holds $r_1 - r_2 = (q_1 - q_2)m$.*

The statement was proved by Lau and Schleicher (see [6], Corollary 8.5), and an alternative proof was given by Kauko [4]. Let us finish our discussion by adressing two questions, which arise in relation with Theorem 1.4. One asks for a kind of completeness of visibility trees as given in the narrow case.

Question 2.3 *Do the periods of hyperbolic components in a visibility tree of W different from $\mathfrak{Vis}_{\frac{1}{2}}(W)$ provide each of the remainders $1, 2, \dots, m-1$?*

Note that a lot of computer experiments do not lead to a negative answer to Question 2.3. There are, however, negative counter-examples for $\mathfrak{Vis}_{\frac{1}{2}}(W)$ (see Figures 2 and 5), but we cannot answer

Question 2.4 *Is the statement of Theorem 1.4 true for* $\mathfrak{T} = \mathfrak{Vis}_{\frac{1}{2}}(W)$?

3   Proof of Theorem 1.4

We now prove that Theorem 1.5 is equivalent to Theorem 1.4; and for the notions, prerequisites, and statements used here, we refer to [5], in particular, to Chapter 3.4. The substantial part of the proof is the following statement:

Lemma 3.1 *Let* $S = \alpha\overline{\alpha} \in \mathcal{B}_*$, *and let* $B_1, B_2 \in \mathcal{B}_*$ *be immediately visible from* S. *Further, let* (S_1, S_3) *and* (S_2, S_4) *be two parameter-like dynamic pairs of the same STEP* n *behind* B_1 *and* B_2 *such that the kneading sequences of* S_1 *and* S_2 *have the same initial subwords of length* $n - 1$. *(The kneading sequences are well-defined by [5], Proposition 3.68.) Then the periods of* B_1 *and* B_2, *and equivalently the STEP's of* R_{B_1} *and* R_{B_2}, *coincide.*

Proof: Let $\mathbf{v} = \mathbf{v}^S, e = e^S$, and let $\widehat{S_1}|n - 1 = \widehat{S_2}|n - 1 = w_1 w_2 \ldots w_{n-1}$. Further, note that $\{h^{n-1}(S_1), h^{n-1}(S_3)\} = \{h^{n-1}(S_2), h^{n-1}(S_4)\} = \{\dot{S}, \ddot{S}\}$, and look at Proposition 2.43 in [5]. For convenience, we want to say that $i \in \{0, 1, \ldots, n - 2\}$ is related to an infinite gap G of $\eth\mathcal{B}^\alpha$ if $(h^i(S_1), h^i(S_3))$ and $(h^i(S_2), h^i(S_4))$ lie behind boundary chords of G being different and not belonging to the two longest ones of G.

Let j be maximal with $\{h^{j-1}(S_1), h^{j-1}(S_3)\} \neq \{h^{j-1}(S_2), h^{j-1}(S_4)\}$. We can assume that $j > 1$. The chords R_{B_1} and R_{B_2} have the same *STEPs* if all i with $0 \leq i < j$ are related to an infinite gap. The latter can be shown by using the following simple statements for $i = 1, 2, \ldots, n - 1$:

1. If $(h^i(S_1), h^i(S_3))$ [resp. $(h^i(S_2), h^i(S_4))$] lies behind a boundary chord $R \neq S$ of the critical value $Gap^\alpha_{\mathbf{v}*}$, then $(h^{i-1}(S_1), h^{i-1}(S_3))$ [resp. $(h^{i-1}(S_2), h^{i-1}(S_4))$] lies behind one of the boundary chords $l^\alpha_0(R), l^\alpha_1(R)$ of Gap^α_*.

2. If the dynamic pair $(h^i(S_1), h^i(S_3))$ [resp. $(h^i(S_2), h^i(S_4))$] is not behind S, then $(h^{i-1}(S_1), h^{i-1}(S_3))$ [resp. $(h^{i-1}(S_2), h^{i-1}(S_4))$] is behind $\dot{S}$ or $\ddot{S}$, and one has $(h^{i-1}(S_1), h^{i-1}(S_3)) = (l^\alpha_{w_i}(h^i(S_1)), l^\alpha_{w_i}(h^i(S_3)))$ [resp. $(h^{i-1}(S_2), h^{i-1}(S_4)) = (l^\alpha_{w_i}(h^i(S_2)), l^\alpha_{w_i}(h^i(S_4)))$].

3. If i is related to $Gap^\alpha_{\mathbf{v}*}$, then $i - 1$ is related to Gap^α_*.

4. If i is related to an infinite gap G different from $Gap^\alpha_{\mathbf{v}*}$, then $i - 1$ is related to one of the two preimage gaps of G.

Indeed, $(h^j(S_1), h^j(S_3)) = (h^j(S_2), h^j(S_4))$ must lie behind S by 2.; hence $j - 1$ is related to Gap^α_* by 1., and the rest follows by induction. ∎

In order to demonstrate how Theorem 1.5 follows from Lemma 3.1, let S, m, B, k_1 and k_2 be given as in Theorem 1.5, and let qm with $q \geq 3$ be the period of B. Assuming $k_1 - k_2 = rm, r \geq 1$, fix a visible chord $Q_1 \in \mathcal{B}_*$ of period k_1 behind $B_1 = B$, and according to Theorem 1.3 a visible chord $Q_2 \in \mathcal{B}_*$ of period k_1 behind an immediately visible chord B_2 of period $(q + r)m$.

By [5], Theorem 3.75, there exist semi-visible dynamic pairs (S_1, S_3) and (S_2, S_4) with $PChord((S_1, S_3)) = Q_1$ and $PChord((S_2, S_4)) = Q_2$ satisfying the assumptions of Lemma 3.1 for $n = k_1$ (compare [5], Definition 3.72). On the other hand, B_1 and B_2 are different. This contradicts Lemma 3.1.

Acknowledgement The author thanks S. Winter for providing the base for Figures 5 and 6.

References

[1] Douady, A., Hubbard, J., Étude dynamique des polynômes complexes. *Publications Mathématiques d'Orsay* 84-02 (1984) / 85-02 (1985).

[2] Douady, A., Hubbard, J., On the dynamics of polynomial-like mappings, *Ann. Sci. École Norm. Sup. (4)*, 18 (1985), 287-343.

[3] Haïssinsky, Modulation dans l'ensemble de Mandelbrot, In: *The Mandelbrot set, Theme and Variations* (Tan Lei, Ed.), *London Math. Soc. Lecture Note Ser.* 274, Cambridge Univ. Press. 2000, 37-65.

[4] Kauko, V., Trees of visible components in the Mandelbrot set, *Fund. Math.* 164 (2000), 41–60.

[5] Keller, K., *Invariant Factors, Julia Equivalences and the (Abstract) Mandelbrot set*, LNM 1732, Berlin-Heidelberg-NewYork 2000.

[6] Lau, E., Schleicher, D., Internal addresses in the Mandelbrot set and irreducibility of polynomials. *Stony Brook IMS Preprint 1994/19*.

[7] Lavaurs, P., Une déscription combinatoire de l'involution définie par M sur les rationnels à dénominateur impair. *C. R. Acad. Sc. Paris Série I, t. 303* (1986), 143-146.

[8] Milnor, J., Self-similarity and hairiness in the Mandelbrot set. In: *Computers in Geometry and Topology* (Tangora, Ed.), *Lect. Notes Pure Appl. Math.* 114, Dekker 1989, 211-257.

[9] Milnor, J., Periodic orbits, external rays and the Mandelbrot set: An expository account, *Stony Brook IMS Preprint 1999/3*.

[10] Schleicher, D., Rational parameter rays of the Mandelbrot set, *Astérisque* 261 (2000), 405–443.

[11] Thurston, W.P., On the combinatorics and dynamics of iterated rational maps. Preprint, Princeton 1985.

The Inflation and Perturbation of Nonautonomous Difference Equations and Their Pullback Attractors [1]

PETER KLOEDEN

FB Mathematik, Johann Wolfgang Goethe Universität
D-60054 Frankfurt am Main, Germany
E-mail: kloeden@math.uni-frankfurt.de

and

VICTOR KOZYAKIN [2]

Institute for Information Transmission Problems
Russian Academy of Sciences
Bolshoj Karetny lane, 19, 101447 Moscow, Russia
E-mail: kozyakin@iitp.ru

Abstract Nonautonomous difference equations are formulated as difference cocycles driven by an autonomous dynamical system. Pullback attractors are the appropriate generalization of autonomous attractors to cocycles and their existence follows when the difference cocycle has a pullback absorbing set. The effects of perturbing the driving autonomous system of a difference cocycle are considered here, in the weaker sense of perturbations of its input variable in the nonautonomous difference equation and in the stronger sense of the driving system being perturbed to a nearby driving system. The existence of a pullback attractor of the inflated cocycle dynamics and the shadowing of the driving system are important assumptions here.

Keywords Nonautonomous difference equation, Difference cocycle, Pullback attractor, Inflated attractors and dynamics, Driving system, Skew-product system, Total stability

AMS Subject Classification 34C35, 37B55

---

[1]This work was supported by the DFG Forschungsschwerpunkt "Ergodentheorie, Analysis und effiziente Simulation dynamischer Systeme".

[2]Supported by the Russian Foundation for Basic Research grants No. 00-15-96116 and 0001-00571.

1   Introduction

In this article we discuss the perturbation and inflation of *pullback attractors* for nonautonomous difference equations which are formulated as a skew-product system, i.e., consisting of a cocycle mapping in the state space that is driven by an autonomous dynamical system [6]. The existence of a pullback attractor for a setvalued "inflation" of the cocycle dynamics implies the total stability of the pullback attractor of the original system, i.e. the persistence of the pullback attractor under perturbations of the cocycle mapping. However, if the driving system itself is perturbed to a new driving system, then an additional property such as the shadowing of the original driving system needs also to be assumed. These concepts are introduced and developed here.

For simplicity we shall restrict attention to the Euclidean state space $\mathbb{R}^d$, though our results can be extended to more general metric or Banach state spaces with appropriate modifications to assumptions. To describe the proximity and convergence of sets, we recall that the *Hausdorff separation* $H^*(A,B)$ of nonempty compact subsets A, B of $\mathbb{R}^d$ is defined as

$$H^*(A, B) := \max_{a \in A} \operatorname{dist}(a, B) = \max_{a \in A} \min_{b \in B} \|a - b\|$$

and that $H(A, B) = \max \{H^*(A, B), H^*(B, A)\}$ defines a metric, called the *Hausdorff metric*, on the space $\mathcal{H}(\mathbb{R}^d)$ of nonempty compact subsets of $\mathbb{R}^d$. In addition, $B[A, \epsilon]$ denotes the closed neighborhood of a compact set A with radius ϵ and $B(A, \epsilon)$ denotes the open neighborhood of a compact set A with radius ϵ, i.e.,

$$B[A, \epsilon] = \{x \in \mathbb{R}^d : \operatorname{dist}(x, A) \le \epsilon\}, \qquad B(A, \epsilon) = \{x \in \mathbb{R}^d : \operatorname{dist}(x, A) < \epsilon\}.$$

2   The autonomous case

Successive iteration of an autonomous difference equation

$$x_{n+1} = f(x_n) \tag{1}$$

generates the forwards solution mapping $\phi : \mathbb{Z}^+ \times \mathbb{R}^d \to \mathbb{R}^d$ defined by

$$x_n = \phi(n, x_0) = f^n(x_0) = \underbrace{f \circ f \circ \cdots \circ f}_{n \text{ times}}(x_0),$$

which satisfies the *initial condition* $\phi(0, x_0) = x_0$ and the *semigroup property*

$$\begin{aligned}
\phi(n, \phi(m, x_0)) &= f^n(\phi(m, x_0)) = \\
&f^n \circ f^m(x_0) = f^{n+m}(x_0) = \phi(n + m, x_0)
\end{aligned} \tag{2}$$

for all $x_0 \in \mathbb{R}^d$, and integers n, $m \ge 0$. Property (2) says that the solution mapping ϕ forms a semigroup under composition; it is typically only a semigroup rather than a group since the mapping f need not be invertible. Assuming that the mapping f in the difference equation (1) is at least continuous, it follows that the mappings $\phi(n, \cdot)$ are continuous for every $n \in \mathbb{Z}^+$. The solution mapping ϕ then generates a discrete time *autonomous semidynamical system* on the state space $\mathbb{R}^d$ [1, 15].

A nonempty compact subset A of $\mathbb{R}^d$ is called *invariant* under ϕ, or ϕ-invariant, if $\phi(n, A) = A$ for all $n \in \mathbb{Z}^+$ or, equivalently, if $f(A) = A$. Simple examples are steady state solutions and periodic solutions; in the first case A consists of a single point, which must thus be a fixed point of the mapping f, and for a solution with period r it consists of a finite set of r distinct points $\{p_1, \ldots, p_r\}$ which are fixed point of the composite mapping f^r (but not for an f^j with j smaller than r). Invariant sets can also be much more complicated, for example fractal sets. Many are the ω-limit sets of some trajectory, i.e. defined by

$$\omega^+(x_0) = \left\{ y \in \mathbb{R}^d \ : \ \exists n_j \to \infty, \ \phi(n_j, x_0) \to y \right\},$$

which is nonempty, compact and ϕ-invariant when the forwards trajectory $\{\phi(n, x_0); \ n \in \mathbb{Z}^+\}$ is bounded. The asymptotic behavior of a autonomous semidynamical system is characterized by its ω-limit sets in general, and its attractors and their associated absorbing sets in particular. An *attractor* is a nonempty ϕ-invariant compact set A^* that attracts all trajectories starting in some neighborhood $\mathcal{U}$ of A^*, that is with $\omega^+(x_0) \subset A^*$ for all $x_0 \in \mathcal{U}$ or, equivalently, with $\lim_{n \to \infty} \operatorname{dist}(\phi(n, x_0), A^*) = 0$ for all $x_0 \in \mathcal{U}$. A nonempty compact subset A^* of $\mathbb{R}^d$ is a global *maximal attractor* of the discrete time autonomous semidynamical system ϕ on $\mathbb{R}^d$ if it is ϕ-invariant and attracts bounded sets; i.e.,

$$\lim_{n \to \infty} H^*\left(\phi(n, D), A^*\right) = 0 \quad \text{for any bounded subset } D \subset \mathbb{R}^d. \tag{3}$$

The existence and approximate location of such a maximal attractor follows from that of more easily found absorbing sets, which typically have a convenient simpler shape such as a ball or ellipsoid. A nonempty compact subset B of $\mathbb{R}^d$ is called an *absorbing set* of a discrete time autonomous semidynamical system ϕ on $\mathbb{R}^d$ if for every bounded subset D of $\mathbb{R}^d$ there exists a $N_D \in \mathbb{Z}^+$ such that $\phi(n, D) \subset B$ for all $n \geq N_D$ in $\mathbb{Z}^+$.

Theorem 2.1 *Suppose that a discrete time autonomous semidynamical system ϕ on $\mathbb{R}^d$ generated by a continuous mapping f has an absorbing set B. Then ϕ has a unique global maximal attractor $A^* \subset B$ given by*

$$A^* = \bigcap_{m \geq 0} \overline{\bigcup_{n \geq m} \phi(n, B)}. \tag{4}$$

A maximal attractor (but not always a local or point-attracting attractor; see [1]) is in fact uniformly Lyapunov asymptotically stable in that it is also

Lyapunov stable; i.e., for every $\epsilon > 0$ there exists a $\delta = \delta(\epsilon) > 0$ such that

$$\text{dist}\,(\phi(n, x_0), A^*) < \epsilon \quad \text{for all} \;\; n \in \mathbb{Z}^+ \;\; \text{whenever} \;\; \text{dist}\,(x_0, A^*) < \delta,$$

as well as attracting bounded sets as in (3). Uniformly Lyapunov asymptotically stable sets can be characterized by a Lyapunov functions [1, 15], which can be used to establish the existence of an absorbing set and hence that of a nearby maximal attractor in a perturbed autonomous system; i.e., if the semidynamical system generated by the autonomous difference equation (1) has a global maximal attractor A, then for any continuous $g : \mathbb{R}^d \to \mathbb{R}^d$ with $\sup_{x \in \mathbb{R}^d} \|g(x)\| \leq 1$ and $\epsilon > 0$ small enough, the semidynamical system generated by the perturbed autonomous difference equation

$$x_{n+1} = f(x_n) + \epsilon\, g(x_n)$$

has a maximal autonomous attractor $A^{\epsilon, g}$ which converges upper semicontinuously to A in the sense that

$$H^*(A^{\epsilon, g}, A) \to 0 \quad \text{as } \epsilon \to 0+ \,.$$

This property is often known as *total stability*. Similarly, for sufficiently small $\epsilon > 0$, the *setvalued* semidynamical system Φ^ϵ generated by the "inflated" difference inclusion

$$x_{n+1} \in F^\epsilon(x_n),$$

where F^ϵ is defined by $F^\epsilon(x) := B[f(x), \epsilon] = f(x) + B[0, \epsilon]$, has a maximal attractor A^ϵ, which converges continuously to A (since it contains A as well as all above $A^{\epsilon, g}$); i.e., $H(A^\epsilon, A) \to 0$ as $\epsilon \to 0+$.

In fact, in the autonomous case under discussion, the uniform asymptotic stability of the maximal attractor A, its total stability, and the existence of an "inflated" attractor A^ϵ for sufficiently small $\epsilon > 0$ all imply each other.

3   Nonautonomous difference equations

Difference equations of the form

$$x_{n+1} = f_n(x_n), \tag{5}$$

in which the mappings on the right hand side are allowed to vary with the time instant n are called *nonautonomous* difference equations. Such nonautonomous difference equations arise quite naturally in many different ways. The mappings f_n in (5) may of course vary completely arbitrarily, but often there is some relationship between them or some regularity in the way in which they are chosen. For example, the mappings may all be the same as in the very special autonomous subcase or they may vary periodically within, or be chosen irregularly from, a finite family $\{g_1, \cdots, g_r\}$, in which case (5) can be rewritten as

$$x_{n+1} = g_{k_n}(x_n), \tag{6}$$

where the $k_n \in \{1, \cdots, r\}$ and $f_n = g_{k_n}$. More generally, a difference equation may involve a parameter $q \in Q$ which varies in time by choice or randomly, giving rise to the nonautonomous difference equation

$$x_{n+1} = f(x_n, q_n), \tag{7}$$

so $f_n(x) = f(x, q_n)$ here for the prescribed choice of $q_n \in Q$. Another example, the difference equation (5) may represent a variable time-step discretization method for an autonomous differential equation. See [6, 11, 12, 14] for more details and examples.

The nonautonomous difference equation (5) has the forwards solution mapping ϕ defined through iteration by

$$x_n = \phi(n, n_0; x_{n_0}) = f_{n-1} \circ \cdots \circ f_{n_0}(x_{n_0}) \qquad \text{for all} \quad n > n_0, \tag{8}$$

for the initial value $\phi(n_0, n_0; x_{n_0}) = x_{n_0}$ at time $n = n_0 \in \mathbb{Z}$.

As in the autonomous case, the long–term or asymptotic behavior and related concepts such as asymptotic stability, limit sets and attractors are of major interest. However, the general nonautonomous case differs crucially from the autonomous case in that the starting time n_0 is just as important as the time that has elapsed since starting; i.e., $n - n_0$, and hence many of the concepts that have been developed and extensively investigated for autonomous dynamical systems in general and autonomous difference equations in particular are either too restrictive or no longer valid or meaningful. Moreover, the above formalism is often too general to allow useful assertions to be made about the dynamics of the nonautonomous system as it does not say anything explicitly about how the solution mapping changes in time. Such information can be incorporated through a driving system in the skew-product formalism of a nonautonomous dynamical system.

3.1   Skew-product formalism

Let (P, d_P) be a metric space, which we call the parameter set, and let $\theta = \{\theta_n\}_{n \in \mathbb{Z}}$ be a group of continuous mappings from P onto itself; i.e., with $\theta_0(p) = p$ and $\theta_n \circ \theta_m(p) = \theta_{n+m}(p)$ for all $p \in P$ and $n, m \in \mathbb{Z}$ (henceforth we write $\theta_n p$ instead of $\theta_n(p)$). Essentially, θ is a discrete time autonomous dynamical system on P that models the *driving mechanism* for the change in the mappings f_n on the right hand side of the nonautonomous difference equation (5), which we now write as

$$x_{n+1} = f(\theta_n p, x_n) \tag{9}$$

for $n \in \mathbb{Z}$, where $f : P \times \mathbb{R}^d \to \mathbb{R}^d$ is a continuous mapping. The corresponding solution mapping $\phi : \mathbb{Z}^+ \times P \times \mathbb{R}^d \to \mathbb{R}^d$ is now defined by

$$\phi(0, p, x) := x, \qquad \phi(j, p, x) := f(\theta_{j-1} p, \cdot) \circ \cdots \circ f(p, x), \qquad j \in \mathbb{N},$$

for each $p \in P$ and $x \in \mathbb{R}^d$. The mapping ϕ satisfies the *initial value property* $\phi(0, p, x) := x$ and the *cocycle property* with respect to the driving system θ on P, i.e.,

$$\phi(i + j, p, x) := \phi\left(i, \theta_j p, \phi\left(j, p, x\right)\right) \tag{10}$$

for all $i, j \in \mathbb{Z}^+$, $p \in P$ and $x \in \mathbb{R}^d$, and will be called a discrete time or *difference cocycle* with respect to the driving system θ on P. Note that each of the mappings $\phi(j, \cdot, \cdot) : P \times \mathbb{R}^d \to \mathbb{R}^d$ here is continuous.

Remark 3.1 If ϕ be a difference cocycle on $\mathbb{R}^d$ with respect to a group $\theta = \{\theta_n\}_{n \in \mathbb{Z}}$ of mappings of metric space P into itself. Then the mapping Π : $\mathbb{Z}^+ \times P \times \mathbb{R}^d \to P \times \mathbb{R}^d$ defined by

$$\Pi(j, p, x) := (\theta_j p, \phi(j, p, x))$$

for all $j \in \mathbb{Z}^+$, $(p, x) \in P \times \mathbb{R}^d$ forms an autonomous semidynamical system on the state space $P \times \mathbb{R}^d$; i.e., the set of mappings $\{\Pi(j, \cdot, \cdot)\}_{j \in \mathbb{Z}^+}$ of $P \times \mathbb{R}^d$ into itself is a semigroup, thus a discrete time autonomous semidynamical system, which is called a discrete time *skew-product system* [2, 15].

The above examples can be reformulated in the skew-product formalism with appropriate choices of parameter space P and θ. The nonautonomous difference equation (5) with continuous mappings $f_n : \mathbb{R}^d \to \mathbb{R}^d$ generates a difference cocycle ϕ over the parameter set $P = \mathbb{Z}$ with respect to the group of left shift mappings $\theta_j := \theta^j$ for $j \in \mathbb{Z}$, where $\theta n := n + 1$ for $n \in \mathbb{Z}$. Here ϕ is defined by

$$\phi(0, n, x) := x \quad \text{and} \quad \phi(j, n, x) := f_{n+j-1} \circ \cdots \circ f_n(x), \qquad j \in \mathbb{N},$$

for all $n \in \mathbb{Z}$ and $x \in \mathbb{R}^d$. The mappings $\phi(j, n, \cdot) : \mathbb{R}^d \to \mathbb{R}^d$ here are all continuous. (The autonomous case can be considered as a difference cocycle with respect to a singleton parameter set $P = \{p_0\}$ with θ consisting just of the identity mapping on P.)

While $\mathbb{Z}$ appears to be the natural choice for the parameter set above, in the following example the use of sequence spaces is more advantageous as such spaces are often compact. As will be seen in Theorem 3.2, stronger assertions can then be made about the dynamical behavior of the difference cocycle.

The nonautonomous difference equation (6) with continuous mappings g_k : $\mathbb{R}^d \to \mathbb{R}^d$ for $k \in \{1, \cdots, r\}$ generates a difference cocycle over the parameter set $P = \{1, \cdots, r\}^{\mathbb{Z}}$ of bi-infinite sequences $p = \{k_n, n \in \mathbb{Z}\}$ with $k_n \in \{1, \cdots, r\}$ with respect to the group of left shift operators $\theta_n := \theta^n$ for $n \in \mathbb{Z}$, where $\theta\{k_n, n \in \mathbb{Z}\} = \{k_{n+1}, n \in \mathbb{Z}\}$. The mapping ϕ defined by

$$\phi(0, p, x) := x \quad \text{and} \quad \phi(j, p, x) := g_{k_{j-1}} \circ \cdots \circ g_{k_0}(x), \qquad j \in \mathbb{N},$$

for all $x \in \mathbb{R}^d$, where $p = \{k_n, n \in \mathbb{Z}\}$, is a difference cocycle. Note that the parameter space $P = \{1, \cdots, r\}^{\mathbb{Z}}$ here is a compact metric space with the metric

$$d(p, p') = \sum_{n=-\infty}^{\infty} (r+1)^{-|n|} \, |k_n - k'_n| \, .$$

In addition, the mappings $\theta_n : P \to P$ and $\phi(j, \cdot, \cdot) : P \times \mathbb{R}^d \to \mathbb{R}^d$ here are all continuous.

3.2   Pullback attractors for difference cocycles

The concept of an autonomous maximal attractor for the discrete time skew-product system is not always appropriate as the cocycle dynamics in the state space $\mathbb{R}^d$ are often of prime importance, with the driving system dynamics in the space P being of somewhat lesser direct interest. The concept of pullback attractor provides a useful analog of an attractor for the nonautonomous cocycle dynamics. See [4, 6, 11, 12, 14].

A family $\hat{A} = \{A_p : p \in P\}$ of nonempty compact subsets of $\mathbb{R}^d$ is called a *pullback attractor* of a difference cocycle ϕ on $\mathbb{R}^d$ if it is ϕ-*invariant*, i.e., $\phi(j, p, A_p) = A_{\theta_j p}$ for all $j \in \mathbb{Z}^+$, and *pullback attracts* bounded sets, i.e.,

$$H^* (\phi(j, \theta_{-j}p, D), A_p) = 0 \qquad j \to \infty \tag{11}$$

for all $p \in P$ and all bounded subsets D of $\mathbb{R}^d$.

The pullback absorbing sets, in general, now depend on the parameter too. A family $\hat{B} = \{B_p : p \in P\}$ of nonempty compact subsets of $\mathbb{R}^d$ is called a *pullback absorbing set family* for a difference cocycle ϕ on $\mathbb{R}^d$ if for each $p \in P$ and every bounded subset D of $\mathbb{R}^d$ there exists an $N_{p,D} \in \mathbb{Z}^+$ such that $\phi(j, \theta_{-j}p, D) \subseteq B_p$ for all $j \geq N_{p,D}$ and $p \in P$. If $N_{p,D}$ is independent of p, then $\hat{B}$ is said to be *uniformly absorbing*. A proof of the following theorem can be found in [11] (see also [12]).

Theorem 3.2 *Suppose that a difference cocycle ϕ with $\phi(j, p, \cdot) : \mathbb{R}^d \to \mathbb{R}^d$ continuous for each $j \in \mathbb{Z}^+$ and $p \in P$ has a pullback absorbing set family $\hat{B} = \{B_p : p \in P\}$. Then there exists a pullback attractor $\hat{A} = \{A_p : p \in P\}$ with component sets determined uniquely by*

$$A_p = \bigcap_{n \geq 0} \overline{\bigcup_{j \geq n} \phi(j, \theta_{-j}p, B_{\theta_{-j}p})}. \tag{12}$$

If, in addition, P is a compact metric space, the θ_n are bijective and continuous, the mappings $\phi(j, \cdot, \cdot) : P \times \mathbb{R}^d \to \mathbb{R}^d$ are continuous for all $j \in \mathbb{Z}^+$, and $\hat{B}$ is uniformly absorbing, then

$$\lim_{n \to \infty} \sup_{p \in P} H^* (\phi(n, p, D), A(P)) = 0 \tag{13}$$

for any bounded subset D of $\mathbb{R}^d$, where $A(P) := \overline{\bigcup_{p \in P} A_p}$.

4   Inflated difference cocycles and pullback attractors

The "ϵ-inflation" of the of the nonautonomous difference equation (9) leads to the nonautonomous difference inclusion

$$x_{n+1} \in F^\epsilon \left(\theta_n p, x_n \right) \tag{14}$$

driven by the autonomous dynamical system $\theta = \{\theta_n\}_{n \in \mathbb{Z}}$ on P. The set

$$F^\epsilon(p, x) := B[f(p, x), \epsilon] = f(p, x) + B[0, \epsilon]$$

is compact and convex, and the setvalued mapping $(\epsilon, p, x) \mapsto F^\epsilon(p, x)$ is continuous in the variables (ϵ, p, x). The difference inclusion (14) thus generates a compact setvalued cocycle mapping $\Phi^\epsilon(n, p, x)$, which is continuous in the variables (ϵ, p, x). Φ^ϵ will be called ϵ-*inflated difference cocycle* of the single-valued cocycle ϕ of system (9). Note that $\phi(n, p, x) \in \Phi^\epsilon(n, p, x)$ for all n, p, x and $\epsilon > 0$.

Below we also consider the ϵ-*internal inflation* of a nonautonomous difference equation (9), which is a difference inclusion like (14), except now the set $F^\epsilon(p, x)$ is defined as

$$F^\epsilon(p, x) := f \left(B[0, \epsilon], x \right) \equiv \bigcup_{d_P(q, p) \le \epsilon} f(q, x).$$

The corresponding setvalued difference cocycle will be called the ϵ-*internally inflated difference cocycle*. See [8, 9, 14].

4.1   Inflated pullback attractors

Pullback attractors for a setvalued ϵ-inflated cocycle Φ^ϵ are defined analogously to the single-valued case. A family $\widehat{A}^\epsilon = \{A_p^\epsilon : p \in P\}$ of nonempty compact subsets of $\mathbb{R}^d$ is called pullback attractor of Φ^ϵ, or an ϵ-*inflated pullback attractor* of ϕ, if it is Φ^ϵ-invariant, i.e.,

$$\Phi^\epsilon(j, p, A_p^\epsilon) = A_{\theta_j p}^\epsilon \qquad \text{for all} \quad j \in \mathbb{Z}^+, p \in P,$$

and if it pullback attracts nonempty bounded subsets of $\mathbb{R}^d$, i.e.,

$$\lim_{j \to \infty} H^* \left(\Phi^\epsilon(j, \theta_{-j} p, D), A_p^\epsilon \right) = 0$$

for all $p \in P$ and nonempty bounded subsets D of $\mathbb{R}^d$.

The existence of an ϵ-inflated pullback attractor follows from that of a corresponding pullback absorbing family as in the single-valued case. However, it need not follow from the existence of a pullback attractor for the associated single-valued cocycle. Similarly, the pullback attractor (of the associated single-valued cocycle) is generally not totally stable, but the existence of an inflated pullback attractor for some $\epsilon > 0$ does imply totally stable.

The following theorem shows that if an ϵ_0-inflated pullback attractor exists for a particular value $\epsilon_0 > 0$, then an ϵ-inflated pullback attractor exists for all smaller values of ϵ, including $\epsilon = 0$, and are nested as ϵ decreases. The result also applies to internally inflated difference cocycles. The proof follows directly from definitions.

Theorem 4.1 *Suppose for some $\epsilon_0 > 0$ that the cocycle ϕ has an ϵ_0-inflated attractor $\widehat{A}^{\epsilon_0} = \{A_p^{\epsilon_0} : p \in P\}$. Then the cocycle ϕ has an ϵ-inflated attractor $\widehat{A}^{\epsilon} = \{A_p^{\epsilon} : p \in P\}$ for every $\epsilon \in [0, \epsilon_0]$ and these are related through*

$$A_p^{\epsilon} \subset A_p^{\epsilon'}, \qquad A_p^{\epsilon} = \bigcap_{\epsilon < \epsilon'} A_p^{\epsilon'}, \tag{15}$$

for any $0 \leq \epsilon < \epsilon' \leq \epsilon_0$ and each $p \in P$.

5 Perturbation of pullback attractors

There are various different ways in which a nonautonomous difference equation (9) can be perturbed. The most obvious way, given our remark on the importance of the cocycle dynamics, is by a direct perturbation to the mapping f in (9) resulting in a perturbed difference equation

$$x_{n+1} = f_{\epsilon}(\theta_n p, x_n), \tag{16}$$

where $f_{\epsilon} : P \times \mathbb{R}^d \to \mathbb{R}^d$ is a continuous mapping and the corresponding solution mapping $\phi_{\epsilon} : \mathbb{Z}^+ \times P \times \mathbb{R}^d \to \mathbb{R}^d$ is a difference cocycle with respect to the *same* driving system as in the unperturbed difference equation (9). Suppose that $f_{\epsilon}(p, x) \subset f(p, x) + B[0, \epsilon]$ for all $x \in \mathbb{R}^d$ and $p \in P$, i.e., if $\sup_{(p,x) \in P \times \mathbb{R}^d} \|f_{\epsilon}(x, p) - f(x, p)\| \leq \epsilon$, and suppose that the unperturbed difference equation (9) has an ϵ_0-inflated pullback attractor $\widehat{A}^{\epsilon_0} = \{A_p^{\epsilon_0} : p \in P\}$ for some $\epsilon_0 \geq \epsilon$. Hence (9) itself has a pullback attractor $\widehat{A} = \{A_p : p \in P\}$ with $A_p \subset A_p^{\epsilon_0}$ for each $p \in P$ and the perturbed difference equation (16) has a pullback attractor $\widehat{A}^{\epsilon,pert} = \{A_p^{\epsilon,pert} : p \in P\}$ with $A_p^{\epsilon,pert} \subset A_p^{\epsilon_0}$ for each $p \in P$. Moreover the upper semi continuous convergence of component subsets holds, i.e.,

$$\lim_{\epsilon \to 0} H^*\left(A_p^{\epsilon,pert}, A_p\right) = 0$$

for each $p \in P$.

Another type of perturbation is through the driving system and thus indirectly on the driven cocycle dynamics: This might occur in the "weak" sense of perturbations to the p-variable in the f mapping, i.e., resulting in a perturbed difference equation

$$x_{n+1} = f(q_n, x_n)$$

with the same mapping f as before and an arbitrary sequence $\{q_n : n \in \mathbb{Z}\}$ with $d_P(q_n, \theta_n p) \le \epsilon$ for all $n \in \mathbb{Z}$ with some sufficiently small ϵ. (Think of a round-off error or digitization error when inputting the driving system values $p_n = \theta_n p$ into the difference equation.) This situation is covered in that just discussed involving the inflation of the mapping f in the original nonautonomous difference equation (9).

Alternatively, the driving system θ on P itself might be perturbed, resulting in a new driving system θ^ϵ on P, and hence the perturbed difference equation

$$x_{n+1} = f\left(\theta_n^\epsilon p, x_n\right) \tag{17}$$

with the same mapping f as before. This thus perturbs the effect of the driving system in a "strong" sense. It is reasonable to assume that $d_\infty(\theta^\epsilon, \theta)$ $:= \sup_{p \in P} d_P(\theta^\epsilon p, \theta p) \to 0$ as $\epsilon \to 0$ and hence, by continuity of composite functions, that $d_\infty(\theta_n^\epsilon, \theta_n) \to 0$ as $\epsilon \to 0$ for each $n \in \mathbb{Z}$. If this convergence were uniform in $n \in \mathbb{Z}$, then we would be in the previous situation with q_n $:= \theta_n^\epsilon p$. But uniformity is too strong an assumption in most instances.

5.1   Perturbation of a shadowing driving system

An autonomous dynamical system θ on P is said to have the *shadowing property* if for any $\epsilon > 0$ there exists a $\delta = \delta(\epsilon) > 0$ such that for any bi-infinite sequence $\{q_n, n \in \mathbb{Z}\}$ in P satisfying

$$d_P\left(q_{n+1}, \theta q_n\right) < \delta, \qquad \text{for almost all } n \in \mathbb{Z}, \tag{18}$$

there exists an exact solution $\{p_n, n \in \mathbb{Z}\}$ of θ, i.e. with $p_{n+1} = \theta p_n$ for all $n \in \mathbb{Z}$, such that

$$d_P\left(p_n, q_n\right) < \epsilon \qquad \text{for all } n \in \mathbb{Z} \tag{19}$$

holds.

In our context we consider the sequence $\{q_n, n \in \mathbb{Z}\}$ in P to be a solution of some perturbed driving system θ^ϵ satisfying $d_\infty(\theta^\epsilon, \theta) \to 0$ as $\epsilon \to 0$, i.e., with $q_{n+1} = \theta^\epsilon q_n$ for all $n \in \mathbb{Z}$.

Let ϕ^ϵ be the difference cocycle generated by (17) with the perturbed driving system θ^ϵ. We will show that ϕ^ϵ has a pullback attractor when the original difference cocycle has an inflated pullback attractor and a shadowing driven system. The following lemma is needed in the proof.

Lemma 5.1 *Suppose that the driving system θ has the shadowing property and that the perturbed driving system θ^ϵ satisfies $d_\infty(\theta^\epsilon, \theta) < \delta$ for $\delta = \delta(\epsilon)$ as in the shadowing property. Then, for any nonempty compact subset D of $\mathbb{R}^d$,*

$$\phi^\epsilon(n, D, q) \subseteq \bigcup_{d_P(p,q) \le \epsilon} \Phi^\epsilon(n, D, p), \qquad n \in \mathbb{Z}^+, \tag{20}$$

and

$$\phi^\epsilon(n, D, \theta^\epsilon_{-n}q) \subseteq \bigcup_{d_P(p,q) \le \epsilon} \Phi^\epsilon(n, D, \theta_{-n}p), \qquad n \in \mathbb{Z}^+, \tag{21}$$

where Φ^ϵ is the internally ϵ-inflated cocycle solution mapping of the original difference cocycle ϕ.

Proof. Fix an $\epsilon > 0$ and $q_0 \in P$. Let $x_n := \phi^\epsilon(n, x_0, q_0)$ denote the solution of the driven equation of

$$x_{n+1} = f(\theta^\epsilon_n q_0, x_n), \qquad x(0) = x_0,$$

with the perturbed driving system θ^ϵ. Then this solution satisfies the internally inflated difference inclusion

$$x_{n+1} \in F_\epsilon(\theta_n p_0, x_n), \qquad x(0) = x_0,$$

where p_0 corresponds to q_0 under the shadowing property since

$$d_P(q_{n+1}, \theta q_n) \le d_P(q_{n+1}, \theta^\epsilon q_n) + d_P(\theta^\epsilon q_n, \theta q_n) \le d_\infty(\theta^\epsilon, \theta) < \delta(\epsilon)$$

and thus $d_P(\theta_n p_0, \theta^\epsilon_n q_0) < \epsilon$ for all $n \in \mathbb{Z}$. It then follows from the definition of the solution mapping for the internally inflated system that

$$\phi^\epsilon(n, x_0, q_0) = x_n \in \Phi^\epsilon(n, x_0, p_0)$$

and the required forwards inclusion (20) is an immediate consequence of the fact that $d_P(p_0, q_0) < \epsilon$. The backwards inclusion (21) is proved analogously. $\square$

Remark 5.2 Although quite simple, Lemma 5.1 is important because it clearly shows the differing influences of the weak and strong forms of perturbations of the driving system on the behavior of the system. The former manifests itself through the second ϵ in (20) and (21), i.e., in the Φ^ϵ term, and the latter through the first ϵ in (20) and (21), i.e., under the set union symbol. $\square$

Theorem 5.3 *Suppose that the original system 16, that the driven cocycle system ϕ possesses a uniform internally ϵ-inflated attractor $\widehat{A}^\epsilon := \{A^\epsilon_p : p \in P\}$ for each $\epsilon \in [0, \epsilon_0]$ for some $\epsilon_0 > 0$, and that the driving system θ on P has the shadowing property. In addition, suppose that the perturbed driving system θ^ϵ satisfies $d_\infty(\theta^\epsilon, \theta) \le \delta$ for $\delta = \delta(\epsilon)$ of the shadowing property for some $\epsilon \in (0, \epsilon_0]$.*

Then the perturbed difference cocycle ϕ^ϵ has a pullback attractor $\widehat{A}^{\epsilon, pert} := \{A^{\epsilon, pert}_q : q \in P\}$ such that

$$A^{\epsilon, pert}_q \subseteq \bigcup \{A^\epsilon_p : d_P(p, q) \le \epsilon\}, \qquad q \in P. \tag{22}$$

Proof of Theorem 5.3. Let $\epsilon \in (0, \epsilon_0]$ correspond to the $\delta(\epsilon)$ of the shadowing property. Then by the uniformity assumption on the internally ϵ-inflated attractor $\widehat{A}^\epsilon$, for any $\sigma > 0$ and nonempty compact subset D of $\mathbb{R}^d$ there exists a $N = N(\epsilon, \sigma, D) \geq 0$ such that

$$\Phi^\epsilon(n, D, p) \subset B\left(A^\epsilon_{\theta_n p}, \sigma\right) \qquad \text{for all } n \geq N(\epsilon, \sigma, D), \ p \in P.$$

It follows immediately from this inclusion and from Lemma 5.1 that

$$\phi^\epsilon(n, D, q) \subseteq B\left(\bigcup\{A^\epsilon_{\theta_n p} : d_P(p, q) \leq \epsilon\}, \sigma\right) \qquad (23)$$

and

$$\phi^\epsilon(n, D, \theta^\epsilon_{-n} q) \subseteq B\left(\bigcup\{A^\epsilon_p : d_P(p, q) \leq \epsilon\}, \sigma\right) \qquad (24)$$

for all $n \geq N(\epsilon, \sigma, D)$ and $q \in P$. Now fix an arbitrary $\sigma > 0$ and define

$$B_\sigma := B\left[\bigcup\{A^{\epsilon_0}_p : p \in P\}, \sigma\right],$$

where $\widehat{A}^{\epsilon_0} := \{A^{\epsilon_0}_p : p \in P\}$ is the uniform internally ϵ_0-inflated attractor. This set B_σ is compact since the set P and the sets $A^{\epsilon_0}_p$ are compact and the mapping $p \mapsto A^{\epsilon_0}_p$ is upper semicontinuous. Moreover, by Theorem 4.1, B_σ contains any set A^ϵ_p with $\epsilon \in [0, \epsilon_0)$ and $p \in P$. Hence by (23) and (24), respectively,

$$\phi^\epsilon(n, D, q) \subseteq B\left(\bigcup\{A^\epsilon_{\theta_n p} : d_P(p, q) \leq \epsilon\}, \sigma\right) \subseteq B_\sigma$$

and

$$\phi^\epsilon(n, D, \theta^\epsilon_{-n} q) \subseteq B\left(\bigcup\{A^\epsilon_p : d_P(p, q) \leq \epsilon\}, \sigma\right) \subseteq B_\sigma \qquad (25)$$

for all $n \geq N(\epsilon, \sigma, D)$ and $q \in P$.

The existence of an attractor (forwards and pullback) $\widehat{A}^{\epsilon, pert} := \{A^{\epsilon, pert}_p : p \in P\}$ of the perturbed difference cocycle ϕ^ϵ follows from the above inclusions by Theorem 3.2; see also Theorems 2.8 or 2.9 of [2]. In particular, for $(\phi^\epsilon, \theta^\epsilon)$ instead of (ϕ, θ) with the pullback absorbing system consisting of the same subset B_σ gives

$$A^{\epsilon, pert}_q = \bigcap_{\tau \geq 0} \overline{\bigcup_{n \geq \tau} \phi^\epsilon(n, B_\sigma, \theta^\epsilon_{-n} q)} \subseteq B\left(\bigcup\{A^\epsilon_p : d_P(p, q) \leq \epsilon\}, \sigma\right), \qquad q \in P,$$

where the set inclusion follows from (25) with $D = B_\sigma$. The desired inclusion (22) then follows since $\sigma > 0$ can be chosen arbitrarily small. $\qquad \square$

References

[1] N. Bhatia and G.P. Szegö, *Stability Theory of Dynamical Systems.* Springer–Verlag, Berlin, 1970.

[2] D. Cheban, P. E. Kloeden and B. Schmalfuß, The relationship between pullback, forwards and global attractors of nonautonomous dynamical systems. *Nonlinear Dynamics & Systems Theory.* (to appear).

[3] H. Crauel and F. Flandoli, Attractors for random dynamical systems, *Probab. Theory Relat. Fields,* **100** (1994), 365–393.

[4] L. Grüne and P.E. Kloeden, Discretization, inflation and perturbation of attractors, in *Ergodic Theory; Analysis and Efficient Simulation of Dynamical Systems.* (Editor: B. Fiedler), Springer Verlag, 2001, pp. 399–416.

[5] P.E. Kloeden, Lyapunov functions for cocycle attractors in nonautonomous difference equation, *Izvestiya Akad Nauk RM. Mathematika* **26** (1998), 32–42.

[6] P.E. Kloeden, Pullback attractors in nonautonomous difference equations. *J. Difference Eqns. Applns.* **6** (2000), 33–52.

[7] P. E. Kloeden, H. Keller and B. Schmalfuß, Towards a theory of random numerical dynamics. In *Random Dynamical Systems. A Festschrift in Honour of Ludwig Arnold.* Editors: F. Colonius, M. Gundlach and W. Kliemann. Springer-Verlag, 1999, pp. 259–282.

[8] P.E. Kloeden and V.S. Kozyakin, The inflation of attractors and discretization: the autonomous case, *Nonlinear Anal. TMA* **40** (2000), 333–343.

[9] P.E. Kloeden and V.S. Kozyakin, The inflation of nonautonomous systems and their pullback attractors, *Transactions of the Russian Academy of Natural Sciences, Series MMMIU.* **4**, No. 1-2, (2000), 144-169.

[10] P.E. Kloeden and V.S. Kozyakin, The perturbation of attractors of skew-product flows with a shadowing driving system. *Discrete and Continuous Dynamical Systems.* **7** (2001), 883–893.

[11] P.E. Kloeden and B. Schmalfuß, Nonautonomous systems, cocycle attractors and variable time–step discretization, *Numer. Algorithms* **14** (1997), 141–152.

[12] P.E. Kloeden and B. Schmalfuß, Asymptotic behavior of nonautonomous difference inclusions. *Systems & Control letters* **33** (1998), 275–280.

[13] M.A. Krasnosel'skii, *The Operator of Translation along Trajectories of Differential Equations*, Translations of Mathematical Monographs, Volume 19. American Math. Soc., Providence, R.I., 1968.

[14] G. Ochs, Random attractors: robustness, numerics and chaotic dynamics, in *Ergodic Theory; Analysis and Efficient Simulation of Dynamical Systems*. (Editor: B. Fiedler), Springer-Verlag, 2001, pp. 1–30.

[15] K.S. Sibirsky, *Introduction to Topological Dynamics*. Noordhoff, Leyden, 1975.

A Short Introduction to Asynchronous Systems

VICTOR KOZYAKIN [1]

Institute for Information Transmission Problems
Russian Academy of Sciences
101447 Bolshoj Karetny lane, 19, Moscow, Russia
E-mail: kozyakin@iitp.ru

Abstract Looking at the dynamics of the system described by the equation

$$x(n + 1) = f(x(n))$$

one may say that coordinates of the vector $x = \{x_1, x_2, \ldots, x_N\}$ are updated *synchronously*. What happens with the system if coordinates of the vector x are updated *asynchronously*, i.e., if at a given moment n only coordinates with indices i from some set $\omega(n) \subseteq \{1, 2, \ldots, N\}$ are changed in accordance with the law

$$x_i(n + 1) = f_i(x(n))$$

while others remain intact? This is the main topic which is discussed below.

Examples of asynchronous systems are multiprocessor systems, distributed digital networks, discrete-time models of market economy, etc.

The main attention will be paid to discussion of the problem how asynchronism affects stability of the system. Examples show that all possible combinations of stability/instability for the pair synchronous/asynchronous system may occur. And also, simple examples demonstrate that the problem of investigation of stability for asynchronous systems is more complicated than for synchronous ones, even in the linear case. Nevertheless, in some situations asynchronous systems possess more robust properties than synchronous ones.

Formal explanation of this fact will be presented, and various methods of stability investigation for asynchronous systems will be discussed.

Keywords Asynchronous systems, Stability theory, Complexity

AMS Subject Classification 93A10, 93D05, 93D15

---

[1]Supported by the Russian Foundation for Basic Research grants No. 00-15-96116 and 0001-00571.

1   Introduction

The theory of asynchronous systems took its rather distinctive shape about 20 years ago. It was grounded on quite practical problems concerning functioning of distributed computational networks and from the very beginning demonstrated plenty of mathematically difficult, though easily formulated, problems. So, from different points of view it should be very attractive field of investigation for mathematicians. Nevertheless, until now only a few mathematicians are familiar with asynchronous systems. To overcome this, on the ICDEA-2001 there was delivered an educational talk "Asynchronous systems: an intersection point of 'easy questions' with difficult solutions", essential moments of which are presented below.

The simplest, though at the same time the most principal, object of investigation in the theory of difference equations is the autonomous equation

$$x(n+1) = f(x(n))$$

which is often used to describe the dynamics of a system

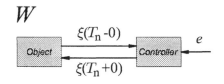

Figure 1: General dynamical system.

where $\{T_n\}$ are discrete moments at which the state $\xi(t)$ of the object is *instantaneously* updated by the controller in accordance to the law

$$\xi(T_n + 0) = f(\xi(T_n - 0)) \tag{1}$$

and the connection between "physical" state vector $\xi(t)$ and "abstract" one $x(n)$ is established by the relation $x(n) := \xi(T_n - 0)$.

Clearly, Fig. 1 is very simplified and schematic. A more adequate situation is represented on Fig. 2.

Fig. 2 reflects the fact that usually practical systems consist of more than one subsystem (components) the state of which are updated *synchronously* (say, when an updating mechanism is triggered by a time impulse coming from an external clock). In this case equation (1) also takes the more detailed form

$$\begin{aligned}
\xi_1(T_n + 0) &:= f_1(\xi(T_n - 0)), \\
\xi_2(T_n + 0) &:= f_2(\xi(T_n - 0)).
\end{aligned} \tag{2}$$

Now one may try to go further and to consider even more realistic situations when different components of the system W on Fig. 2 are updated

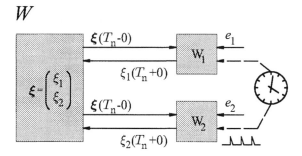

Figure 2: Two-component system.

non-synchronously with each other. Such systems are plotted on Fig. 3 where the so-called *phase-asynchronous* mode of updating is demonstrated and on Fig. 4 where the so-called *frequency-asynchronous* mode of updating is demonstrated.

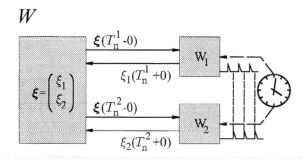

Figure 3: Two-component system with asynchronous updating moments $T_n^1 := n\tau + \varphi_1$, $T_n^2 := n\tau + \varphi_2$ where $\varphi_1 \neq \varphi_2$.

And now, one may formulate the key problem:

> What happens with the dynamics of the system W if updating moments for different components will not coincide?

2 Asynchronous systems

While for a two-component system with synchronous updating mode (*synchronous system*) we have the dynamic's equation (2), for a two-component

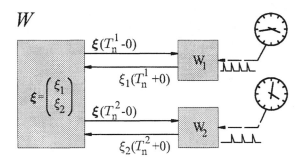

Figure 4: Two-component system with asynchronous updating moments: $T_n^1 := n\tau_1 + \varphi_1$, $T_n^2 := n\tau_2 + \varphi_2$ where $\tau_1 \neq \tau_2$.

system with asynchronous updating mode (*asynchronous system*) we obtain outwardly similar, but nevertheless different equations of dynamics:

$$\begin{aligned}
\xi_1(T_n^1 + 0) &:= f_1(\xi(T_n^1 - 0)), \\
\xi_2(T_n^2 + 0) &:= f_2(\xi(T_n^2 - 0)).
\end{aligned} \tag{3}$$

This means, in fact, that for a given updating moment $t = T_n \in \{T_n^1\} \cup \{T_n^2\}$ the state vector $(\xi_1(t), \xi_2(t))$ of the system W is changed in accordance to the law

$$\begin{array}{lclclcl}
\xi_1(T_n + 0) &:=& \xi_1(T_n - 0), & & &:=& f_1(\xi(T_n - 0)), \\
\xi_2(T_n + 0) &:=& f_2(\xi(T_n - 0)), & \text{or} & &:=& \xi_2(T_n - 0),
\end{array} \tag{4}$$

depending on whether $T_n \in \{T_k^2\}$ while $T_n \notin \{T_k^1\}$ or vise versa $T_n \in \{T_k^1\}$ while $T_n \notin \{T_k^2\}$. And only in the case when $T_n = T_m^1 = T_k^2$ or, what is the same $T_n \in \{T_k^1\} \cap \{T_k^2\} \neq \emptyset$, updating law for the system W at the moment $t = T_n$ is described by equations (2).

Unfortunately, description of the dynamics of the system W in terms of "impulse" equations (2) or (3) is not very convenient. Thus, it is natural to describe the dynamics of W in terms of more habitual difference equations.

Clearly, as was mentioned earlier, in the simplest situation the dynamics of synchronous system W is covered by the vector equation (1) which can be rewritten in the following form:

$$x(n+1) := f(x(n)) = \begin{pmatrix} f_1(x(n)) \\ f_2(x(n)) \end{pmatrix}.$$

For asynchronous systems by analogy with (4) one can get:

$$x(n+1) := \begin{pmatrix} x_1(n) \\ f_2(x(n)) \end{pmatrix} \quad \text{or} \quad := \begin{pmatrix} f_1(x(n)) \\ x_2(n) \end{pmatrix} \tag{5}$$

depending on whether $T_n \in \{T_k^2\}$ or $T_n \in \{T_k^1\}$.

3   More formal look

Summarize now, what one needs to describe the dynamics of asynchronous systems?

- For each n, one should know the set $w(n)$ of indices of the coordinates of the state vector x updated at the moment n. In the case of two-dimensional systems the set $w(n)$ may be $\{1\}$ or $\{2\}$ or $\{1,2\}$; in the general case of N-component systems the set $w(n)$ is a non-empty subset of the set $\{1, 2, \ldots, N\}$.
- For each set of indices w, one should define the mapping $f_w(x)$ (*the $w-$ mixture of the mapping f*) i-th coordinate of which is defined in accordance with the rule

$$f_{w,i}(x) :- \begin{cases} f_i(x), & \text{if } i \in w, \\ x_i, & \text{if } i \notin w; \end{cases} \tag{6}$$

- Finally, one should write the "asynchronous" equation of dynamics

$$x(n+1) := f_{w(n)}(x(n)). \tag{7}$$

So, to describe the dynamics of the asynchronous version of the system W one should replace "synchronous" equation (1) by its "asynchronous" counterpart (7).

As is seen, the right-hand term of "asynchronous" equation (7) essentially depends on the procedure of "taking the mixture" of the mapping f. So, look at some examples of mixtures for different mappings.

Example 3.1

$$\text{Let}\quad f(x) = \begin{pmatrix} f_1(x) \\ f_2(x) \\ f_3(x) \\ f_4(x) \end{pmatrix}, \quad \text{then}\quad f_{\{2\}}(x) = \begin{pmatrix} x_1 \\ f_2(x) \\ x_3 \\ x_4 \end{pmatrix}.$$

Example 3.2 *Let $f(x)$ be a linear mapping, $f(x) = Ax$, where A is the matrix*

$$A = \begin{pmatrix} a_{11} & a_{12} & a_{13} & a_{44} \\ a_{21} & a_{22} & a_{23} & a_{44} \\ a_{31} & a_{32} & a_{33} & a_{44} \\ a_{41} & a_{42} & a_{43} & a_{44} \end{pmatrix}$$

then $f_w(x) = A_w x$ where

$$A_{\{2\}} = \begin{pmatrix} 1 & 0 & 0 & 0 \\ a_{21} & a_{22} & a_{23} & a_{44} \\ 0 & 0 & 1 & 0 \\ 0 & 0 & 0 & 1 \end{pmatrix}, \quad A_{\{1,3\}} = \begin{pmatrix} a_{11} & a_{12} & a_{13} & a_{14} \\ 0 & 1 & 0 & 0 \\ a_{31} & a_{32} & a_{33} & a_{34} \\ 0 & 0 & 0 & 1 \end{pmatrix}.$$

Another thing that we need to pay attention to when considering asynchronous equation (7) is the *index* sequence $\omega(n)$, as the properties of this sequence dramatically affect properties of the right-hand term of equation (7). These properties will be discussed in more details later, while now we shall make only one important remark.

Remark 3.3 From physical considerations it is natural to suppose that each coordinate of the vector x is updated infinitely many times, i.e., each $i \in \{1, 2, \ldots, N\}$ belongs to infinitely many sets $\omega(n)$, $n \geq 0$ (the sequences $\omega(n)$ possessing this property will be called *admissible*). The meaning of such a requirement is that each component is "alive" forever as in the opposite case the state vector of the corresponding component will be constant starting from some moment. When considering the long-term dynamics, this will allow us to exclude this component from further considerations and to reduce the dimension of the state vector of the system under consideration.

4   Short historical survey

Now, that we have become acquainted with the notion of asynchronous system, we present a short survey of basic stages in forming the idea of asynchronous systems.

Long ago (we may say "in prehistoric times"), the method of Simple Iterations and the Gauss-Seidel-method of solving linear or nonlinear equations were known. The method of Gauss–Seidel can be treated as the asynchronous version of the method of Simple Iterations with choosing

$$\omega(n) := \{n - 1 \pmod{N} + 1\}.$$

It is known (e.g., [1]) that the method of Simple Iterations and the method of Gauss–Seidel may converge or diverge independently of each other; thus, all combinations of stability/instability for the pair of synchronous system and its asynchronous counterpart may occur. In terms of asynchronous systems this leads to the principal conclusion: *stability/instability properties of a system may change dramatically depending on whether the system operates in synchronous or asynchronous mode.*

It seems that the first distinctive formulations of the idea of asynchronism, as applied to investigation of dynamic properties of control systems, appeared sometime in 1950s. They were mainly connected with attempts to consider impulse systems in control theory with asynchronously interacting components (see, e.g., Sklansky & Raggaciny [13, 14], Kranc [8, 9], Fan Chung Wuy [7]). Principal attention was paid to investigation of the so-called multirate impulse systems which are essentially (in our terminology) the systems with "frequency" updating components. The main conclusion that may be drawn from these works is that theoretical investigation of asynchronous systems, even in linear case, is much more complicated compared with that for usual, synchronous, systems.

In the 1960-70s the main interest to asynchronous systems was motivated by necessity to develop tools for the so-called parallel methods of computation.

In 1969 the work of Chazan & Miranker [6] was published, in which the stability of linear asynchronous systems with positive matrices was *fully* investigated. The word *fully* means that the principal necessary and sufficient condition of stability has been formulated in terms of *all possible asynchronous systems with a given matrix*, but not in terms of an individual asynchronous system with a given updating law $\{\omega(n)\}$.

In the 1990s an accumulated knowledge in the theory of asynchronous systems was summarized to some extent in monographs by Bertsekas & Tsitsiklis [2], Asarin, Kozyakin et. al. [1] and Bhaya & Kaszkurewicz [3]. These three monographs precisely reflect three directions of investigation of asynchronous systems formed up to now. The first direction is originated from the needs of computational mathematics, and its principal problem is how to organize a computational procedure to obtain most efficiently (fast, with low memory consumption or processor load) the converging iteration algorithm. The second direction is originated from considerations of control theory and its principal problem whether the system under consideration remains stable or becomes unstable upon a perturbation with asynchronous data transmission between different components. The third direction is the attempt to develop "robust" linear algebraic conditions which enable stability of a system independent of whether this system operates in synchronous or asynchronous mode.

5   First set of problems

Let us formulate some natural problems arising in investigation of stability of asynchronous systems. It is worthwhile to stress that, as will be seen below, the main problems arising in investigation of asynchronous systems originate not from the fact of linearity or non-linearity of the system under consideration. So, below only linear systems will be considered.

Given an $N \times N$ real matrix A, consider the usual synchronous linear system

$$x(n+1) = Ax(n) \qquad (W)$$

and its asynchronous counterpart

$$x(n+1) = A_{\omega(n)}x(n) \qquad (W_a)$$

Question 5.1 *What kind of conditions should be imposed on A under which stability of (W) implies stability of (W$_a$)*

- *for all admissible (see definition in Remark 3.3) sequences $\{\omega(n)\}$?*

- *for sequences $\{\omega(n)\}$ from some class (corresponding, e.g., to phase or frequency updating mode, some stochastic low of updating of coordinates, etc.)?*
- *for some individual sequence $\{\omega(n)\}$?*

Another set of problems covers perturbation properties of asynchronous equation (W_a) – more precisely, given an $N \times N$ real matrix A and a sequence $\{\omega(n)\}$, such that the asynchronous equation (W_a) be asymptotically stable.

Question 5.2 *How will the stability of (W_a) be changed if we perturb the matrix A?*

Question 5.3 *How will the stability of (W_a) be changed if we slightly perturb the updating sequence $\{\omega(n)\}$? And what is the meaning of the term* slight *perturbation of $\{\omega(n)\}$ from an application point of view?*

In connection with the last question one should take into account that a "small" perturbation from the physical point of view of updating moments $\{T_n^i\}$ as a rule results in a "big" perturbation of the right-hand terms of asynchronous equations. For example, in the case of phase-frequency updating moments $T_n^i := n\tau_i + \varphi_i$ it is natural from the physical point of view to treat perturbation of moments T_n^i as "slight" or "small" if the parameters τ_i or φ_i are slightly perturbed. But such a slight perturbation of T_n^i will result in a rather 'big' perturbation in any reasonable metric of the set-valued sequence $\omega(n)$.

Even in physically "natural" situations the behavior of the sequence $\{\omega(n)\}$ is rather complicated, e.g., it is not periodic. As a result, it is more difficult to investigate individual asynchronous equations rather than classes of such equations. So, there naturally arises the following general question.

Question 5.4 *How do the right-hand terms of asynchronous equations depend on $\omega(n)$?*

Another set of questions arises when one is interested in the problem of synthesis. Given an $N \times N$ real matrix A, again consider the asynchronous equation (W_a).

Question 5.5 *Is it possible to choose such a sequence $\{\omega(n)\}$ which makes equation (W_a) stable?*

Question 5.6 *Is it possible to choose such a non-degenerate matrix Q and a sequence $\{\omega(n)\}$ which makes stable the "equivalent" equation*

$$x(n+1) = \left(QAQ^{-1}\right)_{\omega(n)} x(n).$$

Finally, we formulate a very simple question. Given an $N \times N$ real matrix A and equations

$$x(n+1) \quad = \quad Ax(n), \qquad\qquad (W)$$
$$x(n+1) \quad = \quad A^*x(n). \qquad\qquad (W^*)$$

It is well known that these equations are simultaneously either stable or unstable. Now, consider asynchronous versions of equations (W) and (W^*):

$$x(n+1) \quad = \quad A_{\omega(n)}x(n), \qquad\qquad (W_a)$$
$$x(n+1) \quad = \quad (A^*)_{\omega(n)}x(n). \qquad\qquad (W_a^*)$$

Question 5.7 *Is it true that stability of equation (W_a) implies stability of equation (W_a^*) and vise versa?*

The questions formulated above are gathered here not by a difficulty principle but simply to demonstrate that even quite naturally formulated questions which have evident answers in a synchronous setting become not-so-evident when one starts to investigate them in the asynchronous formulation. In more details these and other questions are discussed in [1]. Some of them have quite natural answers that can be obtained relatively easy; others also have natural answers, but in order to prove the corresponding statements a special technique has to be developed; and, finally, some of these questions are unresolved until now.

6   Chazan–Miranker theorem

Partial answers to some questions formulated in the previous section will be given below.

Theorem 6.1 (Chazan & Miranker, [6]) *Let $A = (a_{ij})$ be a real matrix with positive entries, $a_{ij} > 0$. If $\rho(A) < 1$ then any asynchronous equation (W_a) is asymptotically stable. If $\rho(A) \geq 1$ then such a sequence $\{\omega(n)\}$ can be found that the corresponding asynchronous equation will be not asymptotically stable.*

In control theory, instead of saying "any equation (from some class) is stable", one usually says "the equation is *absolutely stable in a class of all asynchronous equations (W_a).*" This remark makes it possible to reformulate Theorem 6.1 in the following way.

Theorem 6.2 *Let $A = (a_{ij})$ be a real matrix with positive entries, $a_{ij} > 0$. Then the asynchronous equation (W_a) is absolutely asymptotically stable in the class of all admissible updating sequences $\{\omega(n)\}$ (see definition in Remark 3.3) if and only if $\rho(A) < 1$.*

The following theorem allows use of the Chazan–Miranker criterium for obtaining sufficient conditions for stability of asynchronous systems with arbitrary matrices.

Theorem 6.3 (Majorization Principle) *Let $B = (b_{ij})$ be a real matrix with positive entries satisfying $\rho(B) < 1$, and let the matrix $A = (a_{ij})$ be such that $|a_{ij}| \leq b_{ij}$. Then the asynchronous equation (W_a) is absolutely asymptotically stable in the class of all admissible updating sequences $\{\omega(n)\}$.*

The following criterium for stability is again a word-to-word reformulation of the fact well known in the synchronous setting.

Theorem 6.4 ([1]) *If the matrix A is symmetric, $A = A^*$, then the asynchronous equation (W_a) is absolutely asymptotically stable in the class of all admissible updating sequences $\{\omega(n)\}$ if and only if $\rho(A) < 1$.*

In [1] a criterium for absolute asymptotic stability of equation (W_a) in the class of all admissible updating sequences $\{\omega(n)\}$ for the case of 2×2 matrices A is obtained.

Unfortunately, theorems presented in this section almost exhaust the set of easily formulated statements concerning investigation of the problem of absolute stability of asynchronous systems known to date. Another frustrating thing is that proofs of the theorems formulated above are much more complicated than proofs of their synchronous analogs.

The proof of the sufficiency parts of the above theorems is based on the following remark: it suffices to find a norm $\| \cdot \|$ in $\mathbb{R}^N$ (called *joint strongly contracting norm*) such that

$$\|A_\omega\| \leq 1, \tag{8}$$

$$\|A_{\omega_1} A_{\omega_2} \cdots A_{\omega_k}\| \leq \gamma < 1, \quad \text{as soon as} \quad \bigcup_{i=1}^{k} \omega_i = \{1, 2, \ldots, N\}, \tag{9}$$

in order to guarantee absolute asymptotic stability of the asynchronous equation (W_a) with the matrix A.

With this remark, the sufficiency part of Theorem 6.1 is obtained with the choice of the norm $\|x\| = \max_i\{\lambda_i|x_i|\}$ and appropriate λ_i, $i = 1, 2, \ldots, N$. To prove the sufficiency part of Theorem 6.4 it suffices to set $\|x\| = \sqrt{((I - A)x, x)}$. For the case $N = 2$ the definition of corresponding norm is more complicated.

What is important is that, in fact, conditions (8), (9) are not only sufficient for stability of asynchronous systems, but also necessary.

Theorem 6.5 ([1]) *The existence of a norm $\| \cdot \|$ satisfying condition (8) is necessary and sufficient for absolute stability of asynchronous equation (W_a).*

The existence of a norm $\| \cdot \|$ satisfying conditions (8), (9) for some $\gamma < 1$ is necessary and sufficient for absolute asymptotic stability of asynchronous equation (W_a).

7   Complexity issues

Note that the usual spectral criterium $\rho(A) < 1$ for asymptotic stability of the synchronous equation (W) is well known to be equivalent with the condition

$$\exists \, \| \cdot \|, \gamma < 1 : \qquad \|A\| \leq \gamma < 1, \tag{10}$$

which is exactly the condition (9) presented in Theorem 6.5. The spectral condition $\rho(A) < 1$ can be treated as simple from various points of view – it is algorithmic, it is semialgebraic in terms of entries of the matrix A, i.e., it can be written as a finite set of algebraic equalities and inequalities over entries of the matrix A, etc. So, condition (10) also may be qualified as simple, which gives us hope that conditions (8), (9) are also rather simple for use.

Unfortunately, this is not the case. Given a set of $N \times N$ matrices A_1, $A_2, \ldots, A_k$, the norm $\| \cdot \|$ in $\mathbb{R}^N$ will be called *joint contracting (nonexpanding)* if

$$\|A_1\|, \|A_2\|, \ldots, \|A_k\| < 1 \quad (\leq 1).$$

Clearly, this condition is of the same type as conditions (8), (9).

Theorem 7.1 (Kozyakin, [1]) *If $N, k \geq 2$ then the problem of existence of the joint contracting (nonexpanding) norm is not semialgebraic.*

Blondel & Tsitsiklis [2] show that an analogous problem of computing and approximating the so-called joint spectral radius of a set of matrices A_1, $A_2, \ldots, A_k$ is NP-hard, when not impossible.

8   Robustness of stability

In spite of the fact that Theorem 6.5, as indicated in the previous section, is not too constructive, it nevertheless may help to derive quite strong properties of asynchronous systems (details see, e.g., in [1]).

Theorem 8.1 (Robustness of Stability) *Let asynchronous equation (W_a) be absolutely asymptotically stable and B be a matrix with sufficiently small norm $\|B\|$. Then the perturbed asynchronous equation*

$$x(n+1) = (A + B)_{\omega(n)} x(n)$$

is also absolutely asymptotically stable.

Theorem 8.2 (First Approximation Principle) *Let asynchronous equation (W_a) be absolutely asymptotically stable and function $f(x)$ satisfies "first approximation condition" $f(x) = Ax + o(\|x\|)$. Then the zero equilibrium of the nonlinear asynchronous equation*

$$x(n+1) = f_{\omega(n)}(x(n))$$

is absolutely asymptotically stable.

9   Finiteness conjecture

Define, for a set of matrices $\mathcal{A} := \{A_1, A_2, \ldots, A_k\}$, the *greatest Liapunov exponent* as

$$\lambda^+(\mathcal{A}) := \sup_{n \geq 1} \frac{1}{n} \log \left(\max_{A_{i_j} \in \mathcal{A}} \rho\left(A_{i_1} A_{i_2} \cdots A_{i_n}\right) \right)$$

Lagarias & Wang [10] conjectured that here, in fact, sup may be replaced by max. Recently, Bousch & Mairesse [5] built a counterexample. One of the key points in their complicated proof was the construction of a special norm which, as above, cannot be expressed explicitly by formula, but which possesses some important properties. The idea of the construction of such a norm is taken from Mañé [11, 12], and so below it will be called *Mañé norm*.

More precisely, let $A = (a_{ij})$ be a 2×2 matrix with positive entries, $a_{ij} > 0$. Then by [5] there exist a norm $\|\cdot\|$ and a number $\lambda > 0$ such that

$$\max\left\{\|A_{\{1\}} x\|, \|A_{\{2\}} x\|\right\} = \lambda \|x\| \quad \text{for} \quad x \geq 0.$$

Existence of this norm allows us to describe explicitly the fastest growing trajectory of asynchronous equation (W_a) with the matrix A and to show that $\lambda^+(\{A_{\{1\}}, A_{\{2\}}\}) = \lambda$.

10   Concluding remarks

Of course, in a short paper it is impossible to give a detailed survey of all results obtained in the theory of asynchronous systems. So, below are some of the topics which would be interesting for discussion but which were skipped here due to lack of space.

- Robustness of *instability* of asynchronous systems and its connection with the so-called overshooting effect and notion of quasi-controllability.
- Problems specific to nonlinear asynchronous systems. Here, plenty of interesting and natural examples were found in the field of neural networks.
- To investigate stability of frequency asynchronous systems there were developed quite specific set of methods which are essentially based on the symbolic dynamics methods and especially on the properties of the so-called "sturmian sequences". Here, there are known a few robustness results (unfortunately, only for 2–component systems).
- Stochastic asynchronous systems – in spite of very acute problems only some first results are known.
- Asynchronous analogs for differential equations – again, in spite of very natural statements of problems only some preliminary results were obtained in last years.
- From the applications point of view it would be very important to develop approximate methods of investigation of individual asynchronous equations.

References

[1] Asarin E.A., Kozyakin V.S., Krasnosel'skii M.A. and Kuznetsov N.A., *Stability Analysis of Desynchronized Discrete-Event Systems*, Nauka, Moscow, 1992 (in Russian).

[2] Bertsekas D.P. and Tsitsiklis J.N., *Parallel and distributed computation – Numerical methods*, Prentice Hall, Englewood Cliffs, New Jersey, 1989.

[3] Bhaya A. and Kaszkurewicz E., *Matrix stability in systems and computations*, Birkhauser, Boston, 1999.

[4] Blondel V.D. and Tsitsiklis J.N., *A survey of computational complexity results in systems and control*, Automatica **36**, 9 (2000), 1249–1274.

[5] Bousch T. and Mairesse J., *Asymptotic height optimization for topical IFS, Tetris heaps, and the finiteness conjecture*, (2000), preprint, http://topo.math.u-psud.fr/~bousch/preprints/

[6] Chazan D. and Miranker W., *Chaotic relaxation*, Linear Algebra and Appl. **43**, 2 (1969), 213–231.

[7] Fan Chung Wuy, *On tracing systems having two impulse elements with nonequal rates of sampling*, Avtomatika i Telemekhanika **19**, 10 (1958), 917–930 (in Russian).

[8] Kranc G.M., *Compensation of an error sampled system by a multirate controller*, AIEE Trans. **76**, (1957), 149–158.

[9] Kranc G.M., *Input-output analysis of multirate feedback systems*, IRE Trans. Automat. Contr. **3**, (1957), 21–28.

[10] Lagarias J.C. and Wang Y., *The finiteness conjecture for the generalized spectral radius of a set of matrices*, Linear Algebra Appl. **214**, (1995), 17–42.

[11] Mañé R., *Introdução à teoria ergódica*, IMPA, Rio (1983).

[12] Mañé R., *Generic properties and problems of minimizing measures of lagrangian systems*, Nonlinearity. **9**, (1996), 273–310.

[13] Sklansky J., *Network compensation of error-sampled feedback systems*, Ph.D diss. N.Y., (1955), 217 p.

[14] Sklansky J. and Ragazzini J.R., *Analysis of errors in sampled-data feedback systems*, AIEE Trans. **74**, (1955), 65–71.

A Local-Global Stability Principle for Discrete Systems and and Difference Equations

ULRICH KRAUSE

Fachbereich Mathematik und Informatik
Universität Bremen, 28343 Bremen, Germany
E-mail: krause@math.uni-bremen.de

Abstract A local-global stability principle for a specified class of mappings means that for every mapping in this class a fixed point which is locally asymptotically stable must be already globally asymptotically stable. Such a principle is proved for power lipschitzian selfmappings of metric spaces which yields in particular such a principle for certain selfmappings of convex cones. The latter principle is applied to discrete systems and to difference equations to establish the global asymptotic stability of an equilibrium by proving that it is locally attractive.

Keywords Local and global asymptotic stability, Non-expansive mappings, Convex cones, Part metric, Positive discrete dynamical systems, Difference equations

AMS Subject Classification 39A10, 39A11

1 Introduction

By a local-global principle for a specified class of mappings we mean the statement that for any mapping in this class a fixed point which is locally attractive (asymptotically stable) must be already globally attractive (globally asymptotically stable). In the present paper we prove such a principle at various stages of generality with an emphasis on difference equations.

Section 2 presents a general local-global principle for selfmappings of a metric space which are power lipschitzian.

Section 3 introduces cone mappings which form the class of functions we are mainly dealing with. This is a rich class of self mappings of a cone which contains, e.g., concave mappings but contains also mappings that are neither monotone increasing nor monotone decreasing.

Section 4 combines the results of the two preceding sections. The metric space considered is the interior of a cone in a Banach space which is equipped with the part metric. Cone mappings are non-expansive and, hence, power lipschitzian with respect to the part metric. The general local-global principle of Section 1 yields a local-global principle for cone mappings. This in turn leads to a local-global principle for discrete systems and difference equations, in particular, where the cone is the standard cone in finite dimensional real space.

Section 5 To prove for a given difference equation the global asymptotic stability of an equilibrium by the local-global principle one has to make sure that the equilibrium is locally attractive. In this respect, Lemma 5.1 gives conditions under which all roots of the characteristic polynomial have a modulus strictly less than one.

Section 6 illustrates the local-global principle for difference equations by examples coming from one dimensional population dynamics, Cobb-Douglas functions, multidimensional Pielou functions and nonlinear versions of the Fibonacci function.

2 A general local-global stability principle in metric spaces

Let (X, d) be a metric space. A selfmapping f of X is called *power lipschitzian* if there exists a constant $c > 0$ such that for any $x, y \in X$ there exists $N(x, y) \in \mathbb{N}(= \{0, 1, 2, \ldots\})$ with

$$d(f^n(x), f^n(y)) \leq cd(x, y) \text{ for } n \geq N(x, y).$$

Obviously, any non-expanding mapping f, i.e., $d(f(x), f(y)) \leq d(x, y)$ for all $x, y \in X$, is power lipschitzian with $c = 1$. The converse, however, does not hold, as can be seen from simple examples. If a mapping is power lipschitzian for a certain metric d then it has this property also for every metric equivalent to d. This invariance does not hold for the property of non-expansiveness.

Obviously, any fixed point of a power lipschitzian mapping must be stable. Therefore, considering asymptotic stability (local or global) it suffices to consider attractivity.

A metric space (X, d) is called *metrically convex*, if for any two distinct points $x, y \in X$ there exists $z \in X$ distinct from x and y such that $d(x, z) + d(z, y) = d(x, y)$.

Theorem 2.1. *Let (X, d) be a complete metric space which is metrically convex. For every power lipschitzian selfmapping of (X, d) it holds that a locally attractive fixed point is also globally attractive and, hence, globally asymptotically stable.*

Proof. Let f be power lipschitzian with $c > 0$ and let x^* be a fixed point of f that is locally attractive. It suffices to show that x^* is globally attractive.

Let $B(x^*, r)$ denote the open ball in (X, d) with center x^* and radius r. Since x^* is locally attractive there exists $\delta > 0$ such that

$$\lim_{n \to \infty} d(f^n(x), x^*) = 0 \text{ for all } x \in B(x^*, \delta). \tag{*}$$

We shall show that this property (*) holds also if δ is replaced by $\delta(1 + c^{-1})$. By iterating this argument property (*) holds if δ is replaced by $\delta(1 + c^{-1})^k$ for every $k \in \mathbb{N}$. This then proves that x^* is globally attractive.

Therefore, pick $x \in B(x^*, \delta(1 + c^{-1}))$ and let $0 < r = d(x, x^*) < \delta(1 + c^{-1})$. Since (X, d) is metrically convex, by Menger's Theorem (see [5]) there exists a mapping $\varphi \colon [0, r] \to X$ with $\varphi(0) = x^*$ and $\varphi(r) = x$ such that $d(\varphi(s), \varphi(t)) = |s - t|$ for all $s, t \in [0, r]$. Choose $0 < s_0 < r$ such that $r - \delta c^{-1} < s_0 < \delta$ and let $y = \varphi(s_0)$. It follows that

$$
\begin{aligned}
d(x^*, y) &- d(\varphi(0), \varphi(s_0)) &= s_0 < \delta \quad \text{and} \\
d(y, x) &- d(\varphi(s_0), \varphi(r)) &= r - s_0 < \delta c^{-1}.
\end{aligned}
$$

Since f is power lipschitzian with c there exists $N(x, y)$ such that

$$d(f^n(x), f^n(y)) \leq cd(x, y) < \delta \quad \text{for all} \quad n \geq N(x, y).$$

Since $d(x^*, y) < \delta$ property (*) implies that there exists $N(y) \in \mathbb{N}$ with $d(f^n(y), x^*) < \delta - cd(x, y)$ for all $n \geq N(y)$. Putting all of this together we obtain for $n \geq N(y), \ N(x, y)$

$$d(f^n(x), x^*) \leq d(f^n(x), f^n(y)) + d(f^n(y), \ x^*) < cd(x, y) + \delta - cd(x, y)$$

and, hence, $f^n(x) \in B(x^*, \delta)$ for $n \geq \max\{N(y), \ N(x, y)\}$. Property (*) implies that $\lim_{n \to \infty} d(f^n(x), x^*) = 0$ for $x \in B(x^*, \delta(1 + c^{-1}))$ which proves the theorem. $\qquad \square$

3   Cone mappings

Let $(E, \| \cdot \|)$ be a real Banach space and let K be a *convex cone* in E, i.e., $K + K \subset K$ and $\lambda K \subset K$ for all $\lambda > 0$. K induces a *partial ordering* $\leq$ on E by $x \leq y$ iff $y - x \in K$. Admitting that $0 \notin K$ the relation $\leq$ need not be reflexive in general. K is *normal* if there exists an equivalent norm on E that is monotone on K. Thus, for K normal we may assume that $\| \cdot \|$ is monotone on K. Define for $x, y \in K$ an equivalence relation $x \sim y$ iff $\mu x \leq y$ and $\lambda y \leq x$ for $\lambda, \mu > 0$. The corresponding equivalence classes are called the *parts* of K.

A selfmapping T of a cone K is called a *cone mapping* if for every $x, y \in K$ and $1 \leq \lambda$ from $\lambda^{-1} x \leq y \leq \lambda x$ it follows that $\lambda^{-1} Tx \leq Ty \leq \lambda Tx$, that is a cone mapping T maps any interval $[\lambda^{-1} x, \lambda x]$ into $[\lambda^{-1} Tx, \lambda Tx]$. By $\mathcal{K}$ we denote the set of all cone mappings of cone K. This set has the following general properties:

Proposition 3.1.   (i) $\mathcal{K}$ is a convex cone which contains id_K and the constant mappings.

(ii) $\mathcal{K} \circ \mathcal{K} \subset \mathcal{K}$.

(iii) Let the selfmapping T of K be *monotone*; i.e., $x \leq y$ implies $Tx \leq Ty$. Then $T \in \mathcal{K}$ iff T is *subhomogeneous*, i.e.,

$$\lambda Tx \leq T(\lambda x) \text{ for all } x \in K,\ 0 < \lambda \leq 1.$$

(iv) T maps parts into parts.

(v) If $T \in \mathcal{K}$ then T is non-expansive for the *part metric* p on K, where the latter is defined by $p(x, y) = \inf\{\log \lambda \mid \lambda^{-1}x \leq y \leq \lambda x\}$ for $x \sim y$ and $p(x, y) = +\infty$, otherwise. For K closed, $T \in \mathcal{K}$ iff T is non-expansive for the part metric.

Proof.   (i) Follows directly from the definitions $(S + T)(x) = Sx + Tx$ and $(\lambda T)(x) = \lambda Tx$ for $\lambda > 0$.

(ii) Immediate from the definition.

(iii) Obviously, any cone mapping must be subhomogeneous. For the remaining implication note that for a subhomogeneous mapping one has that $T(\lambda x) \leq \lambda Tx$ for $1 \leq \lambda$.

(iv) Follows from the definitions.

(v) Obviously, if $T \in \mathcal{K}$ then T is non-expansive for p. If T is non-expansive for p and K is closed, then for $x \sim y$ and $p(Tx, Ty) = \log \lambda_0$ one has that $\lambda_0^{-1}Tx \leq Ty \leq \lambda_0 Tx$. If $\lambda^{-1}x \leq y \leq \lambda x$, then $p(x, y) \leq \log \lambda$ and, hence, $\lambda_0 \leq \lambda$. This shows $\lambda^{-1}Tx \leq Ty \leq \lambda Tx$, and T is a cone mapping. $\square$

In finite dimensions one has the following properties in addition.

Proposition 3.2. Let $K = \mathrm{int}\mathbb{R}^n_+$ be the interior of the standard cone in $\mathbb{R}^n$ and let $\mathcal{K}$ be the set of cone mappings with respect K.

(i) Any concave selfmapping of K is in $\mathcal{K}$ and it is monotone and continuous.

(ii) If S and T are in $\mathcal{K}$ then $\max\{S, T\}$ and $\min\{S, T\}$ (taken componentwise) are in $\mathcal{K}$.

(iii) If $T \in \mathcal{K}$ then $\frac{1}{T}$ (taken componentwise) is in $\mathcal{K}$.

(iv) $T \in \mathcal{K}$ if and only if $T \colon K \to K$ and T is a cone mapping for $\mathbb{R}^n_+$ (on K).

(v) Suppose, $T \colon K \to K$ is differentiable and the following inequalities hold for the component mappings $T_i \colon K \longrightarrow \mathbb{R}_+$, $1 \leq i \leq n$:

$$\sum_{j=1}^{n} \left| \frac{\partial T_i}{\partial x_j}(x) \right| \leq T_i x \text{ on } K.$$

Then $T \in \mathcal{K}$.

Proof. (i) From the definition of concavity, i.e., $T_i(\alpha x + (1 - \alpha)y) \geq \alpha T_i x + (1-\alpha)T_i y$ for all $x, y \in K$, $\alpha \in [0, 1]$ and all component mappings $T_i \colon K \to \mathbb{R}_+$ of T, it follows that $T \in \mathcal{K}$. Furthermore, it is well-known that T_i is continuous on K and easily seen that T_i must be monotone.

(ii) and (iii) follow immediately.

(iv) Suppose $T \in \mathcal{K}$. If $\lambda^{-1}x \leqq y \leqq \lambda x$ for $x, y \in K$, $1 \leq \lambda$ and $\leqq$ the partial ordering induced by $\mathbb{R}_+^n$ then there exists $\epsilon > 0$ such that $(\lambda + \epsilon)^{-1}x \leq y \leq (\lambda + \epsilon)x$, $\leq$ the ordering relation induced by K. Therefore, $(\lambda + \epsilon)^{-1}Tx \leq Ty \leq (\lambda + \epsilon)Tx$. Since $\epsilon > 0$ is arbitrary it follows that $\lambda^{-1}Tx \leqq Ty \leqq \lambda Tx$. Similarly, the reverse implication follows by perturbing λ to $\lambda - \epsilon$.

(v) From the inequalities it follows by [11, Theorem 4.1] that T is non-expansive with respect to the part metric p. As in Proposition 3.2 (v) it follows that T is a cone mapping for the closure $\mathbb{R}_+^n$ of K (restricted on K) and by (iv) above $T \in \mathcal{K}$.   □

To check if a given mapping is a cone mapping, property (v) of Proposition 3.2 is particularly useful. For $n - 1$ the inequalities in (v) reduce to just one inequality $x|T'(x)| \leq Tx$ for all $x > 0$ which has a simple geometric interpretation. The inequality means that at every point $x > 0$ the growth of T is bounded above by $\frac{Tx}{x}$, the average growth up to this point in case of $T0 = 0$, and bounded below by the negative of $\frac{Tx}{x}$. In case the mapping T is monotone (increasing) for $n = 1$, the inequality means that $xT'(x) \leq Tx$ which in turn means that $\frac{Tx}{x}$ is not increasing. (For this property see [8, p. 289].) Actually, if T is monotone, it is clear by Proposition 3.1 (iii) that T is a cone mapping iff it is subhomogeneous. In [6] a monotone and subhomogeneous mapping is called increasing and co-radiant. But notice that even for $n - 1$ a cone mapping need neither be monotone increasing nor monotone decreasing. For example, $T \colon \mathrm{int}\mathbb{R}_+ \to \mathrm{int}\mathbb{R}_+$ given by $Tx = x + \frac{1}{x}$ is neither monotone increasing nor decreasing and, hence, not concave, but it is a cone mapping for $\mathrm{int}\mathbb{R}_+$ by Proposition 3.2 (v) because of

$$x|T'x| = |x - \frac{1}{x}| \leq x + \frac{1}{x} = Tx.$$

There is a plenitude of cone mappings. Actually, a simple application of the Stone–Weierstraß theorem for $n = 1$ shows that any continuous selfmapping of $\mathrm{int}\mathbb{R}_+$ can be approximated by differences of cone mappings for $\mathrm{int}\mathbb{R}_+$. We give some examples for cone mappings T which arise from difference equations we will consider later on.

Examples 3.3. Let $T \colon \mathrm{int}\mathbb{R}_+^n \to \mathrm{int}\mathbb{R}_+^n$ be given by

$$Tx = (x_2, \ldots, x_n, \ f(x_1, \ldots, x_n))$$

for $f \colon \mathrm{int}\mathbb{R}_+^n \to \mathrm{int}\mathbb{R}_+$.

(a) **Cobb-Douglas function:** $f(x_1, \ldots, x_n) = \prod\limits_{j=1}^{n} x_j^{r_j}$ with

$\sum\limits_{j=1}^{n} |r_j| \leq 1$. Because of $\frac{\partial f}{\partial x_j} = r_j x_j^{r_j - 1} \prod\limits_{i \neq j} x_i^{r_i}$ one has that

$$\sum_{j=1}^{n} x_j \left| \frac{\partial f}{\partial x_j}(x) \right| = \left(\sum_{j=1}^{n} |r_j| \right) f(x) \leq f(x).$$

Since trivially $\sum\limits_{j=1}^{n} x_j |\frac{\partial T_i}{\partial x_j}(x)| \leq T_i x$ for $1 \leq i \leq n-1$ it follows by Proposition 3.2 (v) that T is a cone mapping for $\mathrm{int}\mathbb{R}_+^n$.

(b) **Nonlinear Fibonacci function (cf. [9]):** $f(x_1, \ldots, x_n) = \sum\limits_{j=1}^{n} a_j x_j^{r_j}$,

$|r_j| \leq 1$, $a_j \geq 0$ and not all $a_j = 0$. Because of $\frac{\partial f}{\partial x_j} = a_j r_j x_j^{r_j - 1}$ one has that

$$\sum_{j=1}^{n} x_j \left| \frac{\partial f}{\partial x_j}(x) \right| = \sum_{j=1}^{n} |r_j| a_j x_j^{r_j} \leq \sum_{j=1}^{n} a_j x_j^{r_j} = f(x).$$

(c) **Multidimensional Pielou function (cf. [3] for $n = 1$):**
$f(x_1, \ldots, x_n) = \sum\limits_{j=1}^{n} \frac{a_j x_j + b_j}{c_j x_j + d_j}$, where $a_j, b_j, c_j, d_j \geq 0$ and $a_j + b_j > 0$, $c_j + d_j > 0$ for all j.

Let $g \colon \mathrm{int}\mathbb{R}_+ \to \mathrm{int}\mathbb{R}_+$, $g(y) = \frac{ay+b}{cy+d}$, where $a, b, c, d \geq 0$ and $a + b > 0$, $c + d > 0$. For applying Proposition 3.2 (v) it suffices to show that $y|g'(y)| \leq g(y)$ for all $y > 0$. Now, $y|g'(y)| \leq g(y)$ is equivalent to $\frac{y|ad-bc|}{(cy+d)^2} \leq \frac{ay+b}{cy+d}$ which is true because of
$y|ad - bc| \leq y(ad + bc) \leq (ay + b)(cy + d)$.

4 A local-global stability principle for discrete systems

Now we combine the general local-global principle of Section 2 with the analysis of cone mappings in Section 3. As a first result we obtain the following *local-global principle for cone mappings in Banach spaces.*

Theorem 4.1. *Let K be a normal and closed convex cone in a real Banach space and let T^k for some $k \geq 1$ be a cone mapping of $\mathrm{int}K \neq \emptyset$. Then any locally attractive fixed point of T is also globally attractive.*

Proof. $X = \mathrm{int}K$ equipped with the part metric p is a complete metric space ([15, Lemma 3]). Moreover, $(\mathrm{int}K, p)$ is metrically convex ([13, Proposition 1.12]). By Proposition 3.1 (v) the iterate T^k is non-expansive for p and, hence, power lipschitzian. Thus, Theorem 2.1 applies to the metric space (X, p) and its selfmapping T^k. Suppose $x^* \in X$ is a locally attractive fixed point of T. Obviously, x^* is a locally attractive fixed point also for T^k. Theorem 2.1 implies that x^* is globally attractive for T^k and, hence, for T, too. □

Remark 4.2. If in Theorem 4.1 especially $k = 1$ then by Theorem 2.1 it follows that a locally attractive fixed point of T is globally asymptotically stable.

Remark 4.3. Consider in particular a selfmapping T of $\mathrm{int}\,K$ that is monotone and subhomogeneous (for the partial ordering $\leq$ induced by K; K a normal and closed convex cone in a Banach space). Suppose in addition that T has a fixed point x^* which satisfies for some $0 < \delta < 1$ and all $0 < \alpha < 1$ the inequalities

$$\alpha^\delta F(x^*) \leq F(\alpha x^*) \leq F(\alpha^{-1} x^*) \leq \alpha^{-\delta} F(x^*).$$

Then T is a cone mapping and x^* is a locally attractive fixed point of T. By Theorem 4.1, the fixed point x^* is globally attractive. For the special case $K = \mathbb{R}^n_+$ see [6, Theorem 3.7].

Specializing further to finite dimensions and employing the analysis of cone mappings for that case we obtain the following *local-global principle for discrete systems and difference equations.*

Theorem 4.4.   (i) *Let T be a differentiable selfmapping of* $\mathrm{int}\mathbb{R}^n_+$ *such that for an iterate T^k it holds that*

$$\sum_{j=1}^{n} x_j \left| \frac{\partial T_i^k}{\partial x_j}(x) \right| \leq T_i^k x \text{ on } \mathrm{int}\mathbb{R}^n_+ \text{ for all } 1 \leq i \leq n.$$

Then any locally attractive fixed point of T is also globally attractive.

(ii) *Let $u(t + n) = f(u(t), u(t + 1), \ldots, u(t + n - 1))$ be a scalar difference equation of order n with $f \colon \mathrm{int}\mathbb{R}^n_+ \to \mathrm{int}\mathbb{R}_+$ differentiable and*

$$\sum_{j=1}^{n} x_j \left| \frac{\partial f}{\partial x_j}(x) \right| \leq f(x) \text{ on } \mathrm{int}\mathbb{R}^n_+.$$

Then any locally attractive equilibrium $x^* = (\xi, \ldots, \xi) \in \mathrm{int}\mathbb{R}^n_+$ of the difference equation is also globally asymptotically stable.

Proof.   (i) By Proposition 3.2 (v) the iterate T^k is a cone mapping for $\mathrm{int}\mathbb{R}^n_+$. Since $K = \mathbb{R}^n_+$ is a normal and closed convex cone in the Banach space $\mathbb{R}^n$, Theorem 4.1 implies (i).

(ii) Consider the selfmapping T of $\mathrm{int}\mathbb{R}^n_+$ defined by $T(x_1, \ldots, x_n) = (x_2, \ldots, x_n, f(x_1, \ldots, x_n))$. By assumption of f, part (i) applies to T for $k = 1$. By Remark 4.2, x^* is stable. □

Before turning to examples in Section 6 we establish a lemma in the next section which we need to check local attractivity for equilibria of difference equations.

5   A lemma on roots of certain polynomials

For applying Theorem 4.4 (ii) to a given difference equation one needs to know the existence of a locally attractive equilibrium $x^* = (\xi, \dots, \xi)$ with $\xi > 0$. For $f \colon \mathrm{int}\,\mathbb{R}^n_+ \to \mathrm{int}\,\mathbb{R}_+$ differentiable let $\alpha_{j-1} = -\frac{\partial f}{\partial x_j}(x^*)$ for an equilibrium x^*. The characteristic equation then is $X^n + \sum_{j=0}^{n-1} \alpha_j X^j = 0$ and x^* is locally attractive iff $|\lambda| < 1$ for all roots of the characteristic equation. To check this it is possible in principle to apply the well-known Schur–Cohn conditions which, however, may be difficult to handle. (For these conditions see [7], [3].) Taking the particular situation of Theorem 4.4 (ii) into account, that is

$$\sum_{j=1}^{n} x_j \left| \frac{\partial f}{\partial x_j}(x) \right| \leq f(x) \quad \text{on } \mathrm{int}\,\mathbb{R}^n_+,$$

we proceed differently. For the equilibrium $x^* = (\xi, \dots, \xi)$ we obtain that $\sum_{j=1}^{n-1} \xi |\alpha_{j-1}| \leq f(x^*) = \xi$, and, hence, $\sum_{j=0}^{n-1} |\alpha_j| \leq 1$. In case the α_j satisfy certain monotonicity properties a nice theorem of Kakeya or of Eneström–Kakeya or a generalization of those by Fujimoto ([4]) can be applied. The following lemma uses $\sum_{j=0}^{n-1} |\alpha_j| \leq 1$ but no such monotonicity conditions.

Lemma 5.1. *Consider the polynomial*

$$X^n + \sum_{j-s}^{n-1} \alpha_j X^j = 0 \text{ for a fixed } 0 \leq s \leq n-1 \qquad (*)$$

with $\alpha_j = \alpha_j^1 - \alpha_j^2 \in \mathbb{R}$, $\alpha_j^1, \alpha_j^2 \geq 0$ for all j and $\alpha_s^1 > 0$. For all roots λ of (*) one has $|\lambda| < 1$ provided one of the following two conditions is met:

(a) $\displaystyle\sum_{j=s}^{n-1} (\alpha_j^1 + \alpha_j^2) < 1.$

(b) $\displaystyle\sum_{j=s}^{n-1} \alpha_j^1 + \alpha_j^2 = 1.$ *Furthermore, there exists $s < k \leq n-1$ with $\alpha_k^1 > 0$
such that $k - s$ is not divided by a higher power of 2 then $n - s$; or,
alternatively, $\alpha_s^2 > 0$ or $\alpha_k^2 > 0$ for some $s < k \leq n-1$ such that the
highest power of 2 that divides $0 < k - s$ is not the same as for $n - s$.*

Proof. (a) Obviously, $\alpha_j = \alpha_j^1 - \alpha_j^2 \leq \alpha_j^1 + \alpha_j^2$ and $-\alpha_j = \alpha_j^2 - \alpha_j^1 \leq \alpha_j^1 + \alpha_j^2$ and, hence, $|\alpha_j| \leq \alpha_j^1 + \alpha_j^2$. Suppose λ is a root of (*) such that $|\lambda| \geq 1$. Then

$$-1 = \sum_{j=s}^{n-1} \alpha_j \lambda^{j-n}, \quad \text{and, hence,} \quad 1 \leq \sum_{j=s}^{n-1} |\alpha_j|\, |\lambda|^{j-n} \leq \left(\sum_{j=s}^{n-1} |\alpha_j| \right) \cdot |\lambda|^{-1}.$$

If $\sum_{j=s}^{n-1}(\alpha_j^1 + \alpha_j^2) < 1$ then it follows $1 < |\lambda|^{-1}$, that is $|\lambda| < 1$. Therefore, $|\lambda| < 1$ for all roots λ.

(b) If $\sum_{j=s}^{n-1}(\alpha_j^1 + \alpha_j^2) = 1$ then $\sum_{j=s}^{n-1}|\alpha_j| \le 1$, and, as above, $1 \le |\lambda|^{-1}$ that is $|\lambda| \le 1$. Therefore $|\lambda| \le 1$ for all roots λ. Suppose, there is a root λ with $|\lambda| = 1$. From (*) we obtain

$$\sum_{j=s}^{n-1}\alpha_j\lambda^{j-s} = -\lambda^{n-s}, \quad \text{hence} \quad \sum_{j=s}^{n-1}\alpha_j^1\lambda^{j-s} + \sum_{j=s}^{n-1}\alpha_j^2(-\lambda^{j-s}) = -\lambda^{n-s}. \quad (**)$$

Now, λ^{j-s} and $-\lambda^{j-s}$ and λ^{n-s} are all boundary points of the unit circle in $\mathbb{C}$. Since α_j^1, $\alpha_j^2 \ge 0$ and $\sum_{j=s}^{n-1}\alpha_j^1 + \alpha_j^2 = 1$ by assumption, equation (**) means that a certain convex combination of boundary points yields a boundary point, too. A proper convex combination of different boundary points, however, cannot yield a boundary point. Since $\alpha_s^1 > 0$ and $\lambda^{j-s} = 1$ for $j = s$ it follows that $\alpha_j^2 > 0$ implies $-\lambda^{j-s} = 1$. Therefore,

$$1 = \sum_{j=s}^{n-1}\alpha_j^1 + \sum_{j=s}^{n-1}\alpha_j^2 = -\lambda^{n-s}$$

which means that $\lambda = \exp(\frac{(2l+1)\pi i}{n-s})$ for some $l \in \mathbb{Z}$. Let $\alpha_k^1 > 0$ for some $s < k \le n-1$. By the above $\lambda^{k-s} = 1$ and, hence,

$$1 = \lambda^{k-s} = \exp\left(\frac{(2l+1)(k-s)\pi i}{n-s}\right).$$

This implies that $(2l+1)(k-s) = 2(n-s)m$ for some $m \in \mathbb{Z}$. In particular, $k - s$ must be divided by a higher power of 2 than $n - s$ which contradicts the assumption. Thus we must have $|\lambda| < 1$ for all roots in this case. For the alternative condition suppose $\alpha_k^2 > 0$ for some $s \le k \le n-1$. By the above $\lambda^{k-s} = -1$ and we get a contradiction to the assumption for $k = s$. Therefore consider the case $s < k \le n-1$. From $\lambda = \exp\left(\frac{(2l+1)\pi i}{n-s}\right)$ it follows that $-1 = \lambda^{k-s} = \exp\left(\frac{(2l+1)(k-s)\pi i}{n-s}\right)$. This implies that $(2l+1)(k-s) = (2m+1)(n-s)$ for some $m \in \mathbb{Z}$. Therefore, the highest power of 2 that divides $k - s$ is the same as for $n - s$. This contradicts the assumption and we must have $|\lambda| < 1$ for all roots λ also in this case. $\square$

6 Examples

We shall illustrate the local-global principle for difference equations, that is part (ii) of Theorem 4.4, by several examples. Thereby, we will employ also Lemma 5.1 for checking local asymptotic stability of the equilibrium.

A.   Population dynamics

Consider the scalar population model given by $x(t+1) = f(x(t))$, $t \in \mathbb{N}$ where $f \colon \mathbb{R}_+ \to \mathbb{R}_+$, $f(x) = \frac{\lambda x}{(1+x)\beta}$ with constants $\lambda > 0$, $\beta > 0$. (Cf. [12] for an analysis of this model.) Applying Proposition 3.2 (v) one easily verifies that f is a cone mapping for $\beta \leq 2$. If $\lambda > 1$ then f has a unique positive equilibrium $x^* = \lambda^{\frac{1}{\beta}} - 1$. One obtains that $|f'(x^*)| < 1$, that is x^* is locally asymptotic stable. Therefore, by Theorem 4.4 (ii), if $\lambda > 1$ and $0 < \beta \leq 2$ then there exists a unique positive equilibrium x^* which is globally asymptotic stable.

This example also demonstrates that in the formula of Theorem 4.4 (ii) the modulus on the left hand side cannot be omitted. For one has $xf'(x) \leq f(x)$ on $\mathrm{int}\mathbb{R}_+$ for all $\beta > 0$ in the example. For $\lambda > 1$, $\beta > 0$ big enough the equilibrium x^* is locally asymptotic stable, but not globally asymptotic stable.

For the next examples we know already that the selfmappings T of $\mathrm{int}\mathbb{R}^n_+$ given by $Tx = (x_2, \ldots, x_n, f(x))$ for $f \colon \mathrm{int}\mathbb{R}^n_+ \to \mathrm{int}\mathbb{R}_+$ are cone mappings. In particular, every fixed point of T must be stable.

B.   Cobb-Douglas function

$$f(x_1, \ldots, x_n) = \prod_{j=1}^{n} x_j^{r_j}, \ \sum_{j=1}^{n} |r_j| \leq 1.$$

A unique positive equilibrium x^* exists iff $\sum_{j=1}^{n} r_j \neq 1$. In this case $x^* = (1, \ldots, 1)$. For checking local asymptotic stability of x^* by Lemma 5.1 we compute $\alpha_{j-1} = -\frac{\partial f}{\partial x_j}(x^*)$ to obtain $\alpha_{j-1} = -r_j$ for $1 \leq j \leq n$. Putting $\alpha_j^1 = \max\{\alpha_j, 0\}$ and $\alpha_j^2 = -\min\{\alpha_j, 0\}$ we obtain local, and hence global, stability for $\sum_{j=1}^{n} |r_j| = \sum_{j=0}^{n-1} \alpha_j^1 + \alpha_j^2 < 1$. For the other cases, suppose $\sum_{j=1}^{n} |r_j| = 1$. Assume that there is at least one r_j positive and let $r_k > 0$. Because of $\sum_{j=1}^{n} r_j \neq 1$ there must be some r_j negative and let $r_1 < 0$. Assume that k can be chosen odd if n is odd, and even if n is even. It follows that $\alpha_0 = -r_1 > 0$ and, hence, $\alpha_0^1 = -r_1 > 0$. Also $\alpha_{k-1} = -r_k < 0$ and, hence $\alpha_{k-1}^2 = -\alpha_{k-1} = r_k > 0$. By the choice of k, the highest power that divides $k - 1$ cannot be the same as for n. Therefore, by Lemma 5.1 (b), x^* is locally asymptotically stable, and by Theorem 4.4 (ii), x^* is globally asymptotic stable. In a similar manner, other conditions on the r_j's for global asymptotic stability can be obtained from Lemma 5.1 (b).

C.   Multidimensional Pielou function

$$f(x_1, \ldots, x_n) = \sum_{j=1}^{n} \frac{a_j x_j + b_j}{c_j x_j + d_j}; \ a_j, b_j, c_j, d_j \geq 0$$

and, of course, $c_j + d_j > 0$ for all j. A quotient $\frac{ax+b}{cx+d}$ we call proper if it is not zero (i.e., not $a = b = 0$) and $ac > 0$ or $bd > 0$. If such a quotient is not proper it may disappear in the summation for f or it may be of the form rx or $\frac{r}{x}$. Since these cases will be handled in Example D we will assume here that at least one of the summands for f is proper. Also, we assume that there exists a positive equilibrium $x^* = (\xi, \ldots, \xi)$, $\xi > 0$. One obtains that

$$\alpha_{j-1} = -\frac{\partial f}{\partial x_j}(x^*) = \frac{a_j d_j - b_j c_j}{(c_j \xi + d_j)^2}.$$

Obviously,

$$\xi \frac{|a_j d_j - b_j c_j|}{(c_j \xi + d_j)^2} \leq \frac{a_j \xi + b_j}{c_j \xi + d_j} \quad \text{for all } j$$

and strict inequality holds for a proper right hand side. Therefore,

$$\xi \sum_{j=0}^{n-1} |\alpha_j| < \sum_{j=1}^{n} \frac{a_j \xi + b_j}{c_j \xi + d_j} = f(\xi, \ldots, \xi) = \xi, \quad \text{that is } \sum_{j=0}^{n-1} |\alpha_j| < 1.$$

By Lemma 5.1 (i), x^* is locally asymptotic stable and, by Theorem 4.4 (ii), also globally asymptotic stable.

D.   Nonlinear Fibonacci function

$$f(x_1, \ldots, x_n) = \sum_{j=1}^{n} a_j x_j^{r_j}, \quad |r_j| \leq 1 \text{ for all } j, \ a_j \geq 0 \text{ and not all } a_j = 0, \ n \geq 2.$$

In the following we shall discuss various cases for the exponents r_j.

(a) **Common Fibonacci and linear generalization**

   The common difference equation which generates the Fibonacci numbers is given by $f(x_1, x_2) = x_1 + x_2$. Consider the generalization $f(x_1, \ldots, x_n) = \sum_{j=1}^{n} x_j$ for $n \geq 2$. The only (nonnegative) equilibrium is $x^* = (0, \ldots, 0)$ and every orbit with $x(0) \geq 0$ tends to infinity.

(b) **Fibonacci with exponents in $[0, 1]$**

   Consider $f(x_1, \ldots, x_n) = \sum_{j=1}^{n} x_j^{r_j}$ with $r_j \in [0, 1]$. One verifies easily that there exists a positive equilibrium $x^* = (\xi, \ldots, \xi)$, $\xi > 0$, iff $\max_{1 \leq j \leq n} r_j < 1$. It follows that $\alpha_{j-1} = -\frac{\partial f}{\partial x_j}(x^*) = -r_j \xi^{r_j - 1}$ and, hence,

$$\sum_{j=0}^{n-1} |\alpha_j| = \sum_{j=1}^{n} r_j x^{r_j - 1} \leq \max_{1 \leq j \leq n} r_j < 1.$$

   By Theorem 4.4 (ii) and Lemma 5.1 one has that x^* is globally asymptotically stable. That is, in this case any positive equilibrium must be globally asymptotically stable (cf. also [9, Example 1]).

(c) **Fibonacci with exponents in $[-1, +1]$**

   Now we admit coefficients $a_j \geq 0$ and assume that $f(x) = \sum_{j=1}^{n} a_j x_j^{r_j}$, $r_j \in [-1, +1]$, has a positive equilibrium $x^* = (\xi, \ldots, \xi), \xi > 0$. It follows that $\alpha_{j-1} = -\frac{\partial f}{\partial x_j}(x^*) = -a_j r_j \xi^{r_j - 1}$ and, hence, $\sum_{j=0}^{n-1} |\alpha_j| = \sum_{j=1}^{n} |r_j| w_j$ with $w_j = a_j \xi^{r_j - 1}$. Assume first that $a_{j_0} > 0$ and $|r_{j_0}| < 1$

for some j_0. Since $|r_j| \leq 1$ for all j and $\sum_{j=1}^{n} w_j = 1$ we have for the convex combination that $\sum_{j=1}^{n} |r_j| w_j < 1$. Therefore, $\sum_{j=0}^{n-1} |\alpha_j| < 1$ and by Theorem 4.4 (ii) and Lemma 5.1 the equilibrium x^* is globally asymptotically stable.

It remains to consider the case that $|r_j| = 1$ for all j with $a_j > 0$, in which case $\sum_{j=0}^{n-1} \alpha_j| = 1$. This case is covered by the following more general situation.

(d) **Mixture of linear and inverse Fibonacci**

Consider $f(x_1, \ldots, x_n) = \sum_{j=1}^{n} (a_j x_j + b_j x_j^{-1})$, $n \geq 1$ with $a_j \geq 0, b_j \geq 0$ and at least one coefficient $\neq 0$. Let $a = \sum_{j=1}^{n} a_j$, $b = \sum_{j=1}^{n} b_j$. There exists a positive equilibrium $x^* = (\xi, \ldots, \xi)$, $\xi > 0$, iff $a < 1$ and $b > 0$ and in this case ξ is uniquely given by $\xi = \sqrt{b/(1 - ba)}$. In the following we assume that $a < 1$, $b > 0$.

It follows that $\frac{\partial f}{\partial x_j} = a_j - b_j x_j^{-2}$ and, hence,

$$\sum_{j=1}^{n} x_j \left| \frac{\partial f}{\partial x_j}(x) \right| \leq \sum_{j=1}^{n} (a_j x_j + b_j x_j^{-1}) = f(x).$$

By Theorem 4.4 (ii) the equilibrium x^* is globally asymptotically stable if it is locally attractive. For applying Lemma 5.1 we compute $\alpha_{j-1} = -\frac{\partial f}{\partial x_j}(x^*) = b_j \xi^{-2} - a_j = \frac{b_j}{b}(1 - a) - a_j$. Let $\alpha_{j-1}^1 = \frac{b_j}{b}(1 - a)$ and $\alpha_{j-1}^2 = a_j$, $1 \leq j \leq n$. Obviously, $\sum_{j=0}^{n-1} \alpha_j^1 + \alpha_j^2 = 1$. From Lemma 5.1 it follows that x^* is globally asymptotically stable provided the following conditions are met: There exists a smallest index $j_1 \in \{1, \ldots, n\}$ such that $b_{j_1} > 0$,

and there exists a second index $j_1 < j_2$ with $b_{j_2} > 0$ such that $j_2 - j_1$ is not divided by a higher power of 2 than $n + 1 - j_1$

or $a_{j_1} > 0$ or $a_{j_3} > 0$ for some index $j_1 < j_3$ such that the highest power of 2 that divides $j_3 - j_1$ is not the same as for $n + 1 - j_1$.

In particular, the equilibrium is globally asymptotically stable if for some j one has $a_j > 0$ and $b_j > 0$. Consider on the opposite, the case that for each j either a_j or b_j is (strictly) positive. Suppose in addition that $n \geq 2$ is even and $a_j > 0$ for j even and $b_j > 0$ for j odd. Then one can choose $j_1 = 1$ together with $j_2 = 3$ (for $n \geq 3$) or $j_1 = 1$ together with $j_3 = 2$ (for $n = 2$) to conclude that x^* is globally asymptotically stable. (Cf. [14, Theorem 6] for this and similar cases. In [14] the nonlinear Fibonacci difference equation is analyzed also for time dependent coefficients.)

(e) **Inverse Fibonacci**

As a special case of (d) we consider the inverse Fibonacci function that is

$f(x_1, \ldots, x_n) = \sum_{j=1}^{n} b_j x_j^{-1}$, $n \geq 1$, with $b_j \geq 0$ and $b = \sum_{j=1}^{n} b_j > 0$. There is a unique positive equilibrium $x^* = (\xi, \ldots, \xi)$ with $\xi = \sqrt{b}$. From part (d) it follows that this equilibrium is globally asymptotically stable provided there exist two indices j_1 and j_2 such that $b_{j_1} > 0$, $b_{j_2} > 0$ and $0 < j_2 - j_1$ is not divided by a higher power of 2 than $n+1-j_1$. For example, let $n = 3$ and $b_1 > 0$, $b_2 > 0$, $b_3 = 0$. Then by the above x^* is globally asymptotically stable. This is not true for $b_1 > 0$, $b_2 = 0, b_3 > 0$ (see (f) below).

The above type of difference equation has been analyzed in detail in [2] where it is shown more generally that every solution converges to a periodic solution and that the period p can be beautifully computed from the indices j with $b_j > 0$. This result yields in particular a characterization of global asymptotic stability. ([2, Theorem 2.3]. The paper contains a lengthy proof that the equilibrium x^* is locally stable [2, Theorem 2.1]. This, however, can be immediately concluded, even for the mixed case (d), from the fact that f is a cone function.)

(f) **Some special cases**

The general statements in (d) and (e), respectively, we illustrate by cases where the assumptions are not met and the conclusions fail. As a mixture of a linear and an inverse Fibonacci function consider $f(x_1, x_2, x_3) = x_1^{-1} + \frac{1}{2} x_2$. To apply the criterion in (d) choose $j_1 = 1$ and $j_3 = 2$. This gives $j_3 - j_1 = 1$ and $n + 1 - j_1 = 3$ and, therefore, the highest power of 2 that divides $j_3 - j_1$ is the same as for $n+1 - j_1$. Indeed, for the equilibrium x^* given by $\xi = \sqrt{2}$ one obtains that $\alpha_0 = \frac{1}{2}$, $\alpha_1 = -\frac{1}{2}$, $\alpha_2 = 0$. The characteristic equation as in Lemma 5.1 becomes $X^3 - \frac{1}{2} X + \frac{1}{2} = 0$ and has the root $\lambda = -1$. Since $x = (1, 2, 1) \neq x^*$ is a periodic point for $Tx = (x_2, x_3, f(x_1, x_2, x_3))$, the equilibrium x^* is not globally asymptotically stable.

As a purely inverse Fibonacci function consider $f(x_1, x_2, x_3) = x_1^{-1} + x_3^{-1}$. To apply the criterion in (e) choose $j_1 = 1$, $j_2 = 3$. This gives $j_2 - j_1 = 2$ which is divided by 2 and $n + 1 - j_1 = 3$ which is not divided by 2. Indeed, for the equilibrium x^* given by $\xi = \sqrt{2}$ one obtains that $\alpha_0 = \frac{1}{2}$, $\alpha_1 = 0$ and $\alpha_2 = \frac{1}{2}$. The characteristic equation becomes $X^3 + \frac{1}{2} X^2 + \frac{1}{2} = 0$ and has the root $\lambda = -1$. Here, too, $x = (1, 2, 1) \neq x^*$ is a periodic point for the mapping T induced by f, the equilibrium x^* is not globally asymptotically stable.

References

[1] R. DeVault, G. Ladas, and S.W. Schultz. On the recursive sequence $x_{n+1} = \frac{A}{x_n} + \frac{1}{x_{n-2}}$. Proc. Am. Math. Soc., 126:3257–3261, 1998.

[2] H. El-Metwally, E.A. Grove, G. Ladas, and H.D. Voulov. On the global attractivity and the periodic character of some difference equations, unpublished typescript, 2001.

[3] S.N. Elaydi. *An Introduction to Difference Equations*. Springer, New York, second edition, 1999.

[4] T. Fujimoto. Generalizations of Kakeya's theorem. *Kagawa University Economic Review*, 60:73–79, 1988.

[5] K. Goebel and W.A. Kirk. *Topics in Metric Fixed Point Theory*. Cambridge University Press, Cambridge, 1990.

[6] P.E. Kloeden and A.M. Rubinov. Attracting sets for increasing co-radiant and topical operators, to be published, *Mathematische Nachrichten*, 2002.

[7] V.L. Kocic and G. Ladas. *Global Asymptotic Behavior of Nonlinear Difference Equations of Higher Order with Applications*. Kluwer Academic Publishers, Dordrecht, 1993.

[8] M.A. Krasnoselskii and P.P. Zabreiko. *Geometrical Methods of Nonlinear Analysis*. Springer-Verlag, Berlin, 1984.

[9] U. Krause. Stability trichonomy, path stability, and relative stability for positive nonlinear difference equations of higher order. *J. Difference Equ. Appl.*, 1:323–346, 1995.

[10] U. Krause and T. Nesemann. *Differenzengleichungen und diskrete dynamische Systeme*. B.G. Teubner, Stuttgart, 1999.

[11] U. Krause and R.D. Nussbaum. A limit set trichotomy for self–mappings of normal cones in Banach spaces. *Nonlinear Analysis (TMA)*, 20:855–870, 1993.

[12] R.M. May. Simple mathematical models with very complicated dynamics. *Nature*, 261:459–467, 1976.

[13] R.D. Nussbaum. Hilbert's projective metric and iterated nonlinear maps. *Memoirs Amer. Math. Soc.*, No. 391, 1988.

[14] B. Scherenberger. On the behavior of the recursive sequence $u(t + n) = \sum_{i=1}^{n}(a_i(t)/u(t + i - 1)^{r_i})$. *J. Difference Equ. Appl.*, 6:625–639, 2000.

[15] A.C. Thompson. On certain contraction mappings in partially ordered vector spaces. *Proc.Am.Math.Soc.*, 14:438–443, 1963.

Stability Implications of Bendixson Conditions for Difference Equations

C. CONNELL McCLUSKEY

Department of Mathematics and Statistics, University of Victoria
PO Box 3045 STN CSC, Victoria BC, Canada, V8W 3P4

and

JAMES. S. MULDOWNEY [1]
Department of Mathematical and Statistical Sciences
University of Alberta, Edmonton AB, Canada, T6G 2G1
E-mail: jim.muldowney@ualberta.ca

Abstract It is known that certain conditions which preclude the existence of non-constant periodic solutions to autonomous differential equations have strong implications for the asymptotic behavior of individual solutions and for global stability questions. Here we investigate similar questions for difference equations.

Keywords Periodic orbit, Equilibrium, Stability, Hyperbolicity

AMS Subject Classification 39A11, 37D05

1 Differential equations in $\mathbb{R}^n$

Let $f \in C^1(\mathbb{R}^n \to \mathbb{R}^n)$ and consider the flow $x(t) = \phi_t(x)$, $x(0) = x$ determined by the differential equation

$$\dot{x} = f(x). \tag{1}$$

When $n = 2$ the existence of non-trivial periodic orbits of the flow is precluded in any simply connected set in which the familiar Bendixson Condition

$$\operatorname{div} f \neq 0 \tag{2}$$

is satisfied. We will briefly recall three different proofs of this statement.

---

[1]Supported by the Natural Sciences and Engineering Research Council, Grant A7197

Let $C = \{x = x(t), 0 \le t \le \omega\}$, be a periodic orbit of (1), that bounds a region D in which (2) holds. The first, which is the elegant traditional proof, uses Green's Theorem, $\int_D \operatorname{div} f = \int_C f^1 dx^2 - f^2 dx^1$. This yields a contradiction as the left-hand side is non-zero since $\operatorname{div} f$ is of one sign in D from (2) and the right-hand side is zero from (1), for example see [2]. The second proof, see [7], uses the fact that, if $\operatorname{div} f < 0$, then areas in $\mathbb{R}^2$ decrease under the map $x \mapsto \phi_t(x)$ which would contradict the fact that D is invariant. The same contradiction is reached if $\operatorname{div} f > 0$. The third proof, also from [7], observes that, if the differential equations

$$\dot{y} = \operatorname{div} f(\phi_t(x)) y \tag{3}$$

are equi-uniformly asymptotically stable with respect to x in a simply connected set D, then areas in D tend to zero under the map $x \mapsto \phi_t(x)$, as $t \to \infty$, precluding the invariance of any such set with finite positive area.

It is well known that (2) does not preclude periodic solutions when $n > 2$. There have been conditions developed from Stokes' Theorem, in a manner related to the first proof above, that place restrictions on the orientation of periodic orbits; see for example [1]. The second and third proofs have generalizations (see [7] and [3]) that rule out the existence of periodic orbits in higher dimensions and reduce to the Bendixson Condition when $n = 2$. This is possible through the use of the second compound equations

$$\dot{y} = \left[\frac{\partial f}{\partial x}(\phi_t(x))\right]^{[2]} y, \tag{4}$$

linear systems of order $\binom{n}{2} \times \binom{n}{2}$ that reduce to (3) when $n = 2$ and, when $n > 2$, play the same role as (3) in the evolution of infinitesimal 2-dimensional areas in the dynamics of (1).

The second proof of Bendixson's Condition when $n = 2$ has the following generalization to $n \ge 2$. Equation (1) has no non-equilibrium periodic solutions if, for each $x \in \mathbb{R}^n$ and each solution $y(t)$ of (4), $\|y(t)\|$ decreases; here $\|\cdot\|$ is any absolute norm on $\mathbb{R}^{\binom{n}{2}}$. This ensures that a measure of 2-dimensional surface area decreases in the dynamics of (1) and the assumption of the existence of a non-trivial periodic orbit C is contradicted by considering that a minimum surface with C as its boundary evolves as a surface also with boundary C and lower area. Note that $\operatorname{div} f < 0$ is just such a condition when $n = 2$. Any one of the following is a Bendixson condition when $n > 2$ and reduces to $\operatorname{div} f < 0$ when $n = 2$:

$$\lambda_1 + \lambda_2 \;<\; 0, \tag{5}$$

$$\max_{r \ne s} \left\{ \frac{\partial f_r}{\partial x_r} + \frac{\partial f_s}{\partial x_s} + \sum_{q \ne r,s} \left(\left|\frac{\partial f_r}{\partial x_q}\right| + \left|\frac{\partial f_s}{\partial x_q}\right| \right) \right\} \;<\; 0,$$

$$\max_{r \ne s} \left\{ \frac{\partial f_r}{\partial x_r} + \frac{\partial f_s}{\partial x_s} + \sum_{q \ne r,s} \left(\left|\frac{\partial f_q}{\partial x_r}\right| + \left|\frac{\partial f_q}{\partial x_s}\right| \right) \right\} \;<\; 0.$$

The condition $\operatorname{div} f > 0$ when $n = 2$ has the following generalizations when $n > 2$:

$$\lambda_{n-1} + \lambda_n > 0, \tag{6}$$

$$\min_{r \neq s} \left\{ \frac{\partial f_r}{\partial x_r} + \frac{\partial f_s}{\partial x_s} - \sum_{q \neq r,s} \left(\left| \frac{\partial f_r}{\partial x_q} \right| + \left| \frac{\partial f_s}{\partial x_q} \right| \right) \right\} > 0,$$

$$\min_{r \neq s} \left\{ \frac{\partial f_r}{\partial x_r} + \frac{\partial f_s}{\partial x_s} - \sum_{q \neq r,s} \left(\left| \frac{\partial f_q}{\partial x_r} \right| + \left| \frac{\partial f_q}{\partial x_s} \right| \right) \right\} > 0.$$

In (5) and (6), $\lambda_1(x) \geq \lambda_2(x) \geq \cdots \geq \lambda_{n-1}(x) \geq \lambda_n(x)$ are the eigenvalues of the matrix $\frac{1}{2} \left(\frac{\partial f}{\partial x}^* + \frac{\partial f}{\partial x} \right)(x)$. The condition $\lambda_1 + \lambda_2 < 0$ was first established by R.A. Smith [10] as a Bendixson condition for systems (1) that have global attractors. Details of the argument outlined here are given in [7].

The third proof above generalizes to the following statement. If the equations (4) are equi-uniformly asymptotically stable with respect to x in a simply connected set D, then 2-dimensional areas in D tend to zero under the map $x \mapsto \phi_t(x)$, as $t \to \infty$, precluding the invariance of any such set with finite positive area. This can be proved by considering any 2-dimensional surface, whose boundary is a periodic orbit C, and showing that the area of any such a surface has a positive lower bound dependent only on C. The surface evolves as a surface also with boundary C and whose area tends to zero in contradiction of the positive lower bound.

The implications of Bendixson's conditions are stronger than the preclusion of periodic orbits. The conditions outlined here imply that the only non-wandering points in the system are equilibria. In particular, they imply that a non-empty alpha or omega limit set of an orbit is a single equilibrium. They thus provide a powerful tool for the study of global stability questions. A general proof in [4] is non-elementary and depends on the Pugh Closing Lemma; an elementary proof in the case $n = 2$ is given in [6].

2   Difference equations in $\mathbb{R}^n$

Again let $f \in C^1(\mathbb{R}^n \to \mathbb{R}^n)$ and consider the recurrence

$$x_{t+1} = f(x_t), \qquad t = 0, 1, 2, \cdots. \tag{7}$$

A *solution* is the sequence

$$x_t = f^t(x_0), \qquad t = 0, 1, 2, \cdots, \tag{8}$$

as long as the iterate f^t of f exists. The set $\gamma_+(x_0) = \{ f^t(x_0) : t = 0, 1, 2, \cdots \}$ is the *orbit* of x_0. If $0 < \omega$ and $x = f^\omega(x)$ then x is a ω-*periodic point* and it is a *proper* ω-*periodic point* if ω is its least period: $x \neq f^\tau(x)$ for each τ, $0 < \tau < \omega$. A 1-periodic point is an *equilibrium* and a periodic point is *non-trivial* if it is not an equilibrium. The second and third approaches of Section 1 to the problem of excluding the possibility of periodic solutions to differential equations have straightforward analogues for difference equations of the form (7) with, frequently, similar stability implications.

First, in analogy to the surfaces of Section 1 with a periodic orbit as their boundary, suppose that x_0 is a proper 2-periodic point. If the map f diminishes distance, then we have a contradiction if we consider the line segment $[x_0, x_1]$ which is mapped to a curve of lesser length but also with the end-points x_1, x_0 contradicting the minimality of the length of the line segment. Conditions of this type may be expressed in the form of the following inequality on $\mathbb{R}^n$:

$$\left\| \frac{\partial f}{\partial x} \right\| < 1 \tag{9}$$

where $\|\cdot\|$ denotes a norm on $\mathbb{R}^n$ and the matrix norm that it induces. The norm defines a measure of length $\|\lambda\|$ for a smooth curve $t \mapsto \lambda(t) \in \mathbb{R}^n$, $0 \leq t \leq 1$, by

$$\|\lambda\| = \int_0^1 \|\lambda'(t)\| \, dt. \tag{10}$$

Consider the line segment $\lambda(t) = tx_1 + (1-t)x_0$, $0 \leq t \leq 1$, which is mapped to the curve $f \circ \lambda(t)$ which also has end-points $x_1 = f(x_0)$ and $x_0 = f(x_1)$ and whose length is

$$\|f \circ \lambda\| = \int_0^1 \left\| \frac{\partial f}{\partial x}(\lambda(t)) \lambda'(t) \right\| dt \tag{11}$$

$$\leq \int_0^1 \left\| \frac{\partial f}{\partial x}(\lambda(t)) \right\| \|\lambda'(t)\| \, dt < \int_0^1 \|\lambda'(t)\| \, dt = \|\lambda\| ,$$

by (9) and (10). It is an easy exercise in the triangle inequality to show that the line segment λ minimizes the length of curves with end-points x_0, x_1, but not necessarily uniquely for some $\|\cdot\|$. This is contradicted by (11) if x_0 is 2-periodic. The choices of the l^2, l^1 and l^∞ norms give the following examples of (9)

$$\sigma_1 < 1, \quad \max_r \left\{ \sum_{s=1}^n \left| \frac{\partial f_r}{\partial x_s} \right| \right\} < 1, \quad \max_s \left\{ \sum_{r=1}^n \left| \frac{\partial f_r}{\partial x_s} \right| \right\} < 1, \tag{12}$$

where σ_1 is the largest singular value of $\frac{\partial f}{\partial x}$. Here it is not required that solutions of (7) exist for all t. Indeed, the preceding argument shows that $\{x_0, x_1\}$ cannot be a proper 2-periodic orbit if the line segment $[x_0, x_1]$ (or any other curve that minimizes the measure $\|\cdot\|$ of arc length) lies in the set where (9) is satisfied. Conditions that rule out ω-periodic orbits are given by replacing f by $f^{\omega-1}$ in the foregoing discussion.

If we have the stronger requirement that solutions of (7) exist for all t then we no longer need to restrict our attention to sets in which the length $\|\lambda\|$ is minimized as described. The analogue of the third approach yields the result that no arcwise connected set D can contain two distinct points x_j, x_k of a non-trivial periodic orbit if the linearizations of (7),

$$y_{t+1} = \frac{\partial f}{\partial x}(x_t) y_t \tag{13}$$

are equi-uniformly asymptotically stable with respect to $x_0 = x \in D$. This means that there exist constants K and σ such that, for all $x \in D$,

$$\left\| \frac{\partial}{\partial x} f^t(x) \right\| \leq K\sigma^t, \ \sigma < 1. \tag{14}$$

Condition (14) implies that the orbit of x attracts all nearby orbits exponentially. If x is a periodic point, then its orbit is said to be *stable hyperbolic* if (14) is satisfied.

LEMMA 1. *Suppose that* $\overline{\gamma_+(x_0)}$ *is compact. Then the omega limit set of* $\gamma_+(x_0)$ *is a stable hyperbolic periodic orbit if there exist constants* K, σ *such that* (14) *is satisfied for all* $x \in \gamma_+(x_0)$.

Lemma 1 is Theorem 2.1 of [8]. The condition (14) is also necessary for this conclusion.

LEMMA 2. *Suppose that* D *is arcwise connected and that there exist constants* K, σ *such that* (14) *is satisfied for all* $x \in D$. *Then there is no non-trivial periodic orbit of* (7) *that intersects* D *in two or more points.*

PROOF: Suppose on the contrary that a non-trivial ω-periodic orbit exists and that it intersects D in at least two points x_j, x_i. Let $t \mapsto \lambda(t) \in D$ be a C^1 function on $[0, 1]$ such that $\lambda(0) = x_j$ and $\lambda(1) = x_i$. Then, replacing f by $f^{k\omega}$ in (11), we find that

$$\left\| f^{k\omega} \circ \lambda \right\| \leq K\sigma^{k\omega} \|\lambda\| \to 0, \ \text{as } k \to \infty, \tag{15}$$

contradicting $0 < \|x_i - x_j\| \leq \left\| f^{k\omega} \circ \lambda \right\|$.

Combining Lemmas 1 and 2, we find the following theorem.

THEOREM. *Suppose that* D *is an arcwise connected set in* R^n *and that there exist constants* K, σ *such that* (14) *is satisfied for all* $x \in D$. *If there is at least one orbit* $\gamma_+(x_0)$ *whose closure is a compact subset of* D, *then there is a unique stable hyperbolic equilibrium* $y \in D$ *such that for all* $x \in D$

$$\lim_{t \to \infty} f^t(x) = y. \tag{16}$$

PROOF: First, Lemma 1 implies that the omega limit set is a stable hyperbolic periodic orbit in D. Then Lemma 2 rules out non-trivial periodic orbits so the omega limit set is an equilibrium y. Now (16) follows since, if $x \in D$, choose $t \mapsto \lambda(t) \in D$ such that $\lambda \in C^1[0,1]$, $\lambda(0) = x$, $\lambda(1) = y$ and so from (14), as in (11), (15),

$$\left\| f^t(x) - y \right\| = \left\| f^t(x) - f^t(y) \right\| \le \left\| f^t \circ \lambda \right\| \le K\sigma^t \left\| \lambda \right\| \to 0, \text{ as } t \to \infty,$$

completing the proof.

3   Other Bendixson conditions

We conclude with a brief sampling of Bendixson conditions established in [5] for which the stability implications are unclear. Broadly, these are based on analogues of the first proof of the 2-dimensional Bendixson criterion, consideration of the action of a 2-form on an oriented area in the plane the boundary of which is a periodic orbit. We instead consider actions of 1-forms on oriented arcs. The simplest case of this type of result is as follows. Suppose $n = 1$ and the condition

$$1 + f'(x) \ne 0 \tag{17}$$

is satisfied on an interval $I \subset \mathbb{R}$; then (7) has no non-trivial 2-periodic orbits in I. This follows from the observation that, with $x_1 = f(x_0)$, $x_2 = f(x_1)$,

$$x_2 - x_0 = (x_1 - x_0) + (x_2 - x_1) = \int_{x_0}^{x_1} [1 + f'(x)]\, dx \ne 0 \tag{18}$$

if $x_0, x_1 \in I$ so that $f^2(x_0) \ne x_0$. This is considerably less restrictive than the requirement of the preceding section that f decreases arc length; for example, condition (9) is $|f'(x)| < 1$ in this case. In fact (17) precludes the existence of any non-trivial periodic orbit $\{x_t\}$ of (7) by Sharkovsky's Theorem [9] page 66, since the existence of a ω-periodic orbit implies the existence of a τ-periodic orbit if τ is less than ω in the Sharkovsky order. But 2 is less than $\omega \ne 1$, 2. More generally, it is easy to see that the same conclusion can be drawn if there exists a C^1 function $t \mapsto \alpha(t)$ such that

$$\alpha'(x) + \alpha'(f(x))\, f'(x) \ge 0 \tag{19}$$

with strict inequality if $f(x) \ne x$. In [5], it is shown that, for a diffeomorphism f, the condition (19) is both necessary and sufficient for the non-existence of non-trivial periodic orbits. In this vein, Theorem 2 of [5] discusses generalizations of (19) of the form

$$\frac{d}{dx}\left[\alpha(x) + \alpha(f(x)) + \cdots + \alpha\left(f^{\omega-1}(x)\right)\right] \ge 0$$

for the exclusion of ω-periodic orbits.

Some higher dimensional results analogous to the foregoing examples are as follows. There are no proper 2-periodic solutions of (7) if

$$\lambda_n > 0 \text{ or } \lambda_1 < 0,$$

where $\lambda_1 \geq \lambda_2 \geq \cdots \geq \lambda_n$ are the eigenvalues of $I + \frac{1}{2}\left(\frac{\partial f}{\partial x}^* + \frac{\partial f}{\partial x}\right)$. The same conclusion may be drawn if the matrix $I + \frac{\partial f}{\partial x}$ is diagonally dominant either by rows or by columns. Again these are considerably weaker conditions than (12). We are not aware of the stability implications of these weaker conditions.

References

[1] Busenberg, S. N. and van den Driessche, P., A method for proving the nonexistence of limit cycles, *Journal of Mathematical Analysis and Applications* **172** (1993), 463-479.

[2] Jordan, D. W. and Smith, P., *Nonlinear Ordinary Differential Equations* (second edition), Clarendon Press, Oxford, 1987.

[3] Li, Y. and Muldowney, J. S., On Bendixson's criterion, *Journal of Differential Equations* **106** (1993), 27-39.

[4] Li, M. Y. and Muldowney, J. S. On R. A. Smith's autonomous convergence theorem, *Rocky Mountain Journal of Mathematics* **25** (1995), 365-379.

[5] McCluskey, C. C. and Muldowney, J. S., Bendixson Dulac criteria for difference equations, *Journal of Dynamics and Differential Equations* **10** (1998), 567-575.

[6] McCluskey, C. C. and Muldowney, J. S., Stability implications of Bendixson's criterion, *SIAM Review* **40** (1998), 931-934.

[7] Muldowney, J. S., Compound matrices and ordinary differential equations, *Rocky Mountain Journal of Mathematics* **20** (1990), 857-871.

[8] Muldowney, J. S., Stable hyperbolic limit cycles, *Differential Equations and Applications to Biology and Industry* (Martelli, M., Cooke, K. Cumberbatch, E. Tang, B. and Thieme, H., editors) World Scientific, Singapore, 1995.

[9] Robinson, C., *Dynamical Systems*, CRC Press, Boca Raton, 1985.

[10] Smith, R. A., Some applications of Hausdorff dimension inequalities for ordinary differential equations, *Proceedings of the Royal Society of Edinburgh Section A* **104** (1986), 235-259.

Optimal Topological Chaos
in Dynamic Economies [1]

KAZUO NISHIMURA and TADASHI SHIGOKA

Institute of Economic Research, Kyoto University
Sakyoku, Kyoto 606-8501, Japan
E-mail: nishimura@kier.kyoto-u.ac.jp, shigoka@pop.nifty.com

and

MAKOTO YANO

Department of Economics, Keio University
2-15-45 Mita, Minato-ku, Tokyo 108-8345, Japan
E-mail: myano@tkd.att.ne.jp

Abstract The present study deals with a parameterized family of economies that is based on a two-sector optimal growth model. This study demonstrates that no matter how close to 0 the discount rate of future utilities is set, there exists a topologically chaotic policy function that lies in the interior of the technology set. The result is proved by appealing to the theorem of Block (1981) which shows that some stability exists in Sharkovskii's (1964) theorem with C^0 perturbations.

Keywords Chaos, Sharkovskii, Optimal growth, Interior solutions

AMS Subject Classification 91B62, 91B66

1 Introduction

Since the work of Boldrin and Montrucchio (1986) and Deneckere and Pelikan (1986), a major issue in the literature on the non-linear optimal dynamics has been on the discount rate of future utilities at which a chaotic optimal path may appear. In the standard model of optimal growth, the discount rate is equal to the long-run interest rate for a unit period. If, therefore, the long-run interest rate per annum is set at an empirically reasonable level, the discount rate of future utilities in a particular model is positively correlated to the length of a unit period of that model; if, for example, the discount rate is 50%, it is unreasonable to assume that the length of a unit period is less than

[1] We are grateful to Alexander N. Sharkovskii, Yoichiro Takahashi and, in particular, Keiichiro Iwai who provided an invaluable help at the first stage of the research.

a decade. This implies that in a model with too high a discount rate, the chaotic behavior may not be observable within a practical length of time even if an optimal path is chaotic.

As the series of turnpike theorems suggests, the lower the discount rate, the less likely optimal paths are chaotic (Brock and Scheinkman, 1976, Cass and Shell, 1976, Scheinkman, 1976, McKenzie, 1983, 1986).

Despite these results, it has also been demonstrated that optimal paths can behave chaotically even if the discount rate is arbitrarily close to 0 (Nishimura and Yano, 1993 and 1995, and Nishimura, Sorger and Yano, 1994). These results are, however, based on the case in which a part of the graph of a policy function lies on the boundary of the technology set of the model (boundary policy function).[2]

The present study intends to extend this line of research by demonstrating that no matter how close to 0 the discount rate of future utilities is set, there exists a topologically chaotic policy function that lies in the interior of the technology set. That an economic activity lies on the boundary of the technology set in a particular period implies that the consumption level in that period is on the lower boundary of the consumption set; in other words, the consumers cannot survive at a lower level of consumption. As is well known in the conventional literature on optimal growth, such a pathological case can be excluded by imposing the Inada condition on the utility function, which the models of Nishimura and Yano (1993, 1995) and Nishimura, Sorger and Yano (1994) do not satisfy. This study demonstrates a way in which the chaotic optimal dynamics captured by those studies may be preserved even in the case in which the Inada condition is satisfied.

2   Mathematical note

In this section, we explain the mathematical concepts and results that are related to our result.

2.1   Periodic orbit and topological entropy

Let $f : X \to X$ be a continuous map from a subset X of a topological space into itself. By a dynamical system, we mean a pair (X, f). A *trajectory* is a sequence $\{f^n(x)\}_{n \geq 0}$ for a given $x \in X$, where $f^0(x) = x$ and $f^n(x) = f(f^{n-1}(x))$ for $n \geq 1$. The set $\{y \in X| \ y = f^n(x), \ n \geq 0\}$ is called a *orbit* from $x \in X$.

If $x \in X$ satisfies $f^k(x) = x$ for some integer $k > 1$, but $f^n(x) \neq x$ for $n = 1, \ldots, k-1$, then x is called a *periodic point* of (X, f) *with period* k. The

---

[2]In the recent literature on dynamics, there has been a large number of studies that are concerned with the complex behavior of economic variables. In addition to those studies above, Nishimura and Yano (1994) and Majumdar and Mitra (1994) investigate the non-linear dynamics that emerges in deterministic dynamic optimization models.

orbit from a periodic point with period k is called a *periodic orbit of period k*. In this case, we say that a dynamical system (X, f) has a *cycle of period k*.

Assume that X is compact. Let α be a cover of X consisting of open subsets. By the compactness of X, it is possible to find the minimum number of sets in α that covers X. Let $N(\alpha)$ be that minimum number (finite). For two open covers α and β of X, we denote by $\alpha \vee \beta$ the open cover of X consisting of all sets of the form $U \cap V$ for $U \in \alpha$ and $V \in \beta$. Let $f^{-1}\alpha$ be the set of inverse images of U in cover α by the mapping f,

$$f^{-1}\alpha = \{f^{-1}(U)|\, U \in \alpha\}, \tag{1}$$

which is also a cover of X. It may be demonstrated that $\lim_{n\to\infty} \frac{1}{n} \log N(\alpha \vee f^{-1}\alpha \vee \ldots \vee f^{-(n-1)}\alpha)$ is well defined (see Walters, 1975, for a proof). The *topological entropy* of a dynamical system (X, f) is defined as

$$\mathrm{ent}(X, f) = \sup_{\alpha} \left\{ \lim_{n\to\infty} \frac{1}{n} \log N(\alpha \vee f^{-1}\alpha \vee \ldots \vee f^{-(n-1)}\alpha) \right\}, \tag{2}$$

where the supremum is taken over all open covers α of X. By construction, topological entropy is non-negative, i.e., $\mathrm{ent}(X, f) \geq 0$. If a dynamical system (X, f) has *positive topological entropy*, i.e., $\mathrm{ent}(X, f) > 0$, it is said that (X, f) exhibits *topological chaos*.

2.2 Sharkovskii's Theorem

In what follows, we focus on the case in which an underlying space is a compact segment in the set of real numbers. Let I denote a closed, bounded interval on the real line R, and let $C^0(I, I)$ denote the space of continuous mappings of I into I endowed with the topology of uniform convergence. Consider the following total ordering $\succ$ of the positive integers.

$$3 \succ 5 \succ \ldots \succ 2 \cdot 3 \succ 2 \cdot 5 \succ \ldots$$
$$\ldots \succ 2^2 \cdot 3 \succ 2^2 \cdot 5 \succ \ldots \succ 2^2 \succ 2 \succ 1.$$

We will refer to this ordering $\succ$ as *Sharkovskii's ordering*. If $n \succ m$, it is said that n (resp. m) is greater (resp. less) than m (resp. n) in the Sharkovskii's ordering. Sharkovskii (1964) has proved the following fundamental result.[3]

Proposition 1 *Let $f \in C^0(I, I)$. If a dynamical system (I, f) has a cycle of period n and if m is less than n in the Sharkovskii's ordering, i.e., $n \succ m$, then it has a cycle of period m.*

Block (1981) has proved the following result which shows that some stability exists in Sharkovskii's theorem with C^0 perturbations.

---

[3]Sharkovskii (1964) is written in Russian. Sharkovskii (1995) is the English translation of Sharkovskii (1964). This translation is also available in Alsedà et al. (1996, pp. 1–11).

Proposition 2 *Let $f \in C^0(I, I)$ and suppose that a dynamical system (I, f) has a cycle of period n. There is a neighborhood N of f in $C^0(I, I)$ such that for each $g \in N$ and every positive integer m with m less than n in the Sharkovskii's ordering, i.e., $n \succ m$, a dynamical system (I, g) has a cycle of period m.*

Our main result is concerned with a perturbation in a parameter determining a dynamical system. In order to deal with this case, let Λ be a topological space of parameter values. Let $\lambda \to f_\lambda$ be a continuous mapping on the parameter space Λ into $C^0(I, I)$. If for some $\lambda_0 \in \Lambda$, a dynamical system (I, f_{λ_0}) has a cycle of period $3 \cdot 2^n$ for some $n \geq 0$, then, by the theorem of Block (1981), there exists an open neighborhood $U \subset \Lambda$ of λ_0 such that the dynamical system (I, f_λ) has a cycle of period $5 \cdot 2^n$ for any $\lambda \in U$. We will demonstrate our main result based on this fact.

By Sharkovskii's theorem, a dynamical system (I, f) has a cycle of period $3 \cdot 2^n$ for some $n \geq 0$, if and only if it has a cycle of period which is not a power of 2. The next result relates the existence of a cycle of period $3 \cdot 2^n$ to that of topological chaos (see Alsedà et al., 1993, for a proof).

Proposition 3 *Let $f \in C^0(I, I)$. A dynamical system (I, f) has positive topological entropy, i.e., $\text{ent}(I, f) > 0$, if and only if it has a cycle of period which is not a power of 2.*

3   Model

An economy is described by a triplet (Ω, v, ρ) consisting of a *technology set* $\Omega \subset R_+ \times R_+$, a *reduced-form utility function* $v : \Omega \to R_+$ and the discount factor of future utilities $\rho \in (0, 1)$. Given an initial condition $k \in R_+$, the optimal growth model of an economy (Ω, v, ρ) can be described by the following optimization problem.

$$\max_{\{k_t\}} \sum_{t=0}^{\infty} \rho^t v(k_t, k_{t+1}) \text{ s.t. } k_0 = k. \tag{3}$$

The discount rate of future utilities is $\rho^{-1} - 1$. A *feasible path* is a sequence $(k_0, k_1, \ldots)$ such that $(k_t, k_{t+1}) \in \Omega$ for $t \geq 0$. Denote by K the set of feasible paths and by $K(x)$ the set of feasible paths from x. If a feasible path satisfies $(k_t, k_{t+1}) \in \text{int } \Omega$ for all $t \geq 0$, then it is called an *interior path*.

The economic importance of an interior path may be explained as follows. Suppose that the economy is endowed with stock k_t at time t and is to achieve stock k_{t+1} at time $t + 1$. If the economy chooses k_{t+1} in such a way that (k_t, k_{t+1}) lies on the upper boundary of Ω, it may be considered that the consumption level is at the minimum possible level for the economy, in other words, that the consumers cannot survive with a consumption level below that minimum level. If, instead, the economy chooses k_{t+1} in such a way that (k_t, k_{t+1}) lies on the lower boundary of Ω, the end-of-a-period stock level becomes 0. These pathological cases do not appear along an interior path.

We impose the following assumptions on the technology set Ω.

A1: $(0,0) \in \Omega$.
A2: Ω *is closed and convex.*
A3: There is $\bar{x} > 0$ *such that* $x > \bar{x}$ *and* $(x,y) \in \Omega$ *imply* $y < x$, *and* $0 < x < \bar{x}$ *implies* $y > x$ *for some* y *satisfying* $(x,y) \in \Omega$.
A4: If $(x,y) \in \Omega$, $x' \geq x$ *and* $0 \leq y' \leq y$, *then* $(x',y') \in \Omega$.

We impose the following assumptions on the utility function v.

A5: $v : \Omega \to R_+$ *is continuous.*
A6: If $(x,y) \in \Omega$, $x' \geq x$ *and* $0 \leq y' \leq y$, *then* $v(x',y') \geq v(x,y)$.

A feasible path $(k_0, k_1, \ldots)$ from $k_0 = k$ is an *optimal path* from k if it solves the maximization problem in (3). Under A1–A6, there exists an optimal path from every $k \in R_+$.

Let $I = [0, \bar{x}]$ be the closed interval between 0 and $\bar{x}$, where $\bar{x}$ is that of Assumption 3. We call $h : I \to I$ the *policy function* of an economy (Ω, v, ρ) if for any $k \in I$, the optimal path k_t from k satisfies $k_{t+1} = h(k_t)$ for $t \geq 0$. In particular, $h : I \to I$ is called an *interior policy function* if $(k, h(k))$ lies in the interior of the technology set Ω for any $k \in \text{int } I$.

In the next section, we will demonstrate that there exists an economy that has an interior and topologically chaotic policy function no matter how close to 1 the discount factor of the economy, ρ, is.

4 Interior topological chaos

We will deal with parameterized economies that are based on a two-sector optimal growth model. To this end, take $\lambda \in [0,1)$, and think of the following functions:

$$u_\lambda(c_t) = c_t^{1-\lambda} \geq 0 ; \tag{4}$$

$$F_{1\lambda}^\pi(K_1, L_1) = \left[\frac{1}{2} K_1^{-\frac{1}{\lambda}} + \frac{1}{2} \left(\frac{L_1}{a} \right)^{-\frac{1}{\lambda}} \right]^{-\lambda} ; \tag{5}$$

$$F_{2\lambda}^\pi(K_2, L_2) = \mu_\lambda \left[\frac{1}{2} K_2^{-\frac{1}{\lambda}} + \frac{1}{2} \left(\frac{L_2}{b} \right)^{-\frac{1}{\lambda}} \right]^{-\lambda} , \tag{6}$$

where $\mu_\lambda = \left[\frac{1}{2} \mu^{\frac{1}{\lambda}} + \frac{1}{2} \right]^\lambda$, and where $\pi = (\mu, a, b)$ is in the set

$$\Pi = \{(\mu, a, b) | b > a > 0, \ 1/a > \mu > 1/\rho, \ \mu > (b-a)/a\}. \tag{7}$$

With these functions, define

$$
\begin{aligned}
v_\lambda^\pi(k_{t-1}, k_t; \pi) = \quad &\max_{(c_t, K_1, L_1, K_2, L_2) \geq 0} u_\lambda(c_t) \\
&\text{s.t. } c_t \leq F_{1\lambda}^\pi(K_1, L_1) \\
&\quad k_t \leq F_{2\lambda}^\pi(K_2, L_2) \\
&\quad K_1 + K_2 \leq k_{t-1} \\
&\quad L_1 + L_2 \leq 1 \quad .
\end{aligned}
\tag{8}
$$

In the above maximization problem, u_λ may be thought of as the representative consumer's utility function, and $F_{1\lambda}^\pi$ and $F_{2\lambda}^\pi$ may be thought of as the production functions of sectors C and K, which produce a pure consumption good C and a pure capital good K, respectively. Arguments c_t and k_t, respectively, represent the outputs of goods C and K. Each sector uses both the capital good K and labor L as input. Arguments k_{t-1} and 1, respectively, represent the total amount of capital and that of labor. They are divided between sectors C and K to produce outputs k_t and c_t. The capital good input must be made one period prior to the period in which output is produced. Given the above setting, the technology set Ω_λ^π is given by

$$
\Omega_\lambda^\pi = \{(x, y) \in R_+^2 \mid y \leq \mu_\lambda \left[\frac{1}{2} x^{-\frac{1}{\lambda}} + \frac{1}{2} b^{\frac{1}{\lambda}}\right]^{-\lambda} \}.
\tag{9}
$$

Given the above notation, $(\Omega_\lambda^\pi, v_\lambda^\pi, \rho)$ is a parameterized economy, which satisfies Assumptions A1 through A6. Note that if (k_{t-1}, k_t) lies on the upper boundary of technology set Ω_λ^π, no consumption good is produced, i.e., $c_t = 0$. Moreover, the $\bar{x}$ of Assumption A3 is independent of λ and equal to $\bar{x} = \mu/b$. This implies $I = [0, \mu/b]$.

Our main theorem is as follows.

Theorem 1 *For any $\rho' \in (0, 1)$, there are $\pi \in \Pi$, $\delta \in (0, 1]$ and $\rho \in (\rho', 1)$ such that if $\lambda \in (0, \delta)$, the policy function of an economy $(\Omega_\lambda^\pi, v_\lambda^\pi, \rho)$ is interior and topologically chaotic.*

5 Boundary topological chaos

Focusing on the case of $\lambda = 0$, Nishimura and Yano (1995) demonstrate that the economy $(\Omega_0^\pi, v_0^\pi, \rho)$ may have a topologically chaotic policy function no matter how close the discount factor ρ is to 1. However, their result involves a policy function the graph of which partly lies on the boundary of technology set Ω_0^π. In this study, we will demonstrate Theorem 1 by extending the result of Nishimura and Yano (1995).

In order to explain the result of Nishimura and Yano (1995), take the case of $\lambda = 0$. In this case, the utility function becomes linear, and $u_0(c_t) = c_t$. The production functions become of the Leontief type:

$$
c_t = \min \left[K_1, \frac{L_1}{a}\right];
\tag{10}
$$

$$k_t = \mu \min \left[K_2, \frac{L_2}{b} \right] . \tag{11}$$

The reduced form utility function is the same as the social production function. And it has the following form.

$$v_0^\pi(k_{t-1}, k_t) = \left[\begin{array}{ll} k_{t-1} - \frac{k_t}{\mu} & \text{for } k_t \leq -\frac{a\mu}{b-a} k_{t-1} + \frac{\mu}{b-a} \\ \frac{1}{a} - \frac{b k_t}{a\mu} & \text{for } k_t > -\frac{a\mu}{b-a} k_{t-1} + \frac{\mu}{b-a} \end{array} \right. . \tag{12}$$

The result of Nishimura and Yano (1995) is concerned with the condition under which the policy function of the economy (Ω_0, v_0, ρ) coincides with the following function:

$$f^\pi(k) = \left[\begin{array}{ll} \mu k & \text{if } 0 \leq k \leq 1/b \\ -\frac{a}{b-a} \mu k + \frac{1}{b-a} \mu & \text{if } 1/b \leq k \leq \mu/b \end{array} \right. \tag{13}$$

Proposition 4 *For each non-negative integer, N, define $\Pi(N) \subset \Pi$ as the set of parameters π such that $1/b$ is a periodic point of dynamical system $f^\pi : I \to I$ with period 3×2^N. Then, for each N, there is an open interval $(\rho''(N), \rho'(N))$ such that the following two conditions hold. (a) For any $\rho \in (\rho''(N), \rho'(N))$, there is $\pi' \in \Pi(N)$ such that $f^{\pi'} : I \to I$ is the unique optimal policy function of the economy $(\Omega_0^{\pi'}, v_0^{\pi'}, \rho)$. (b) As $N \to \infty$, $\rho''(N) \to 1$.*

Since $1/b$ is a periodic point of f^π with period 3×2^N, Propositions 3 and 4 imply the existence of a topologically chaotic policy function no matter how close the discount factor is to 1. However, the policy function lies on the boundary of technology set Ω_0 on interval $[0, 1/b]$, since $f^\pi(k) = \mu k$ on $[0, 1/b]$.

6 Proof of Theorem 1

In this section, we demonstrate the existence of an interior topologically chaotic policy function by perturbing the economy $(\Omega_0^\pi, v_0^\pi, \rho)$ to $(\Omega_\lambda^\pi, v_\lambda^\pi, \rho)$ with $\lambda > 0$. In order to prove Theorem 1, first, we prove several lemmas.

Lemma 1 *A reduced form utility function $v_\lambda^\pi(k_{t-1}, k_t)$ is continuous in $(k_{t-1}, k_t, \lambda, \pi)$.*

Proof: Let $B(k_{t-1}, k_t, \lambda, \pi) = \{(K_1, L_1) \geq 0 | k_t \leq F_{2\lambda}^\pi(k_{t-1} - K_1, 1 - L_1)\}$. Then, $v_\lambda^\pi(k_{t-1}, k_t) = \max u_\lambda(F_{1\lambda}^\pi(K_1, L_1))$ s.t. $(K_1, L_1) \in B(k_{t-1}, k_t, \lambda, \pi)$. Since $B(k_{t-1}, k_t, \lambda, \pi)$ is lower-hemi continuous in $(k_{t-1}, k_t, \lambda, \pi)$, $v_\lambda^\pi(k_{t-1}, k_t)$ is continuous. **Q.E.D**.

Lemma 2 *For any $\lambda \in (0, 1)$, the policy function of an economy $(\Omega_\lambda^\pi, v_\lambda^\pi, \rho)$, $h_\lambda^\pi : I \to I$, exists and is unique and continuous.*

Proof: Define

$$V_\lambda^\pi(k) = \max_{\{k_t\}} \sum_{t=0}^\infty \rho^t v_\lambda^\pi(k_t, k_{t+1}) \text{ s.t. } k_0 = k \tag{14}$$

and

$$H_\lambda^\pi(k) = \arg\max_h [v_\lambda^\pi(k, h) + \rho V_\lambda^\pi(h)], \tag{15}$$

where $\arg\max_x f(x)$ denotes the set of x that maximizes $f(x)$. By the Bellman principle, any optimal path $\{k_t\}$ satisfies $k_{t+1} \in H_\lambda^\pi(k_t)$ for $t \geq 0$. Since $u_\lambda(c)$ is strictly concave in c, for any given k, $v_\lambda^\pi(k, h)$ is strictly concave in h. Thus, $H_\lambda^\pi(k_t)$ is a singleton for any $k_t \geq 0$. Thus, $\{h_\lambda^\pi(k)\} = H_\lambda^\pi(k)$ exists and is uniquely determined.

Since $v_\lambda^{\pi^*}$ is continuous, the correspondence H_λ^π is upper-hemi continuous. Since $H_\lambda^\pi(k)$ is a singleton for any $k \in I$, $h_\lambda^\pi(k)$ is continuous in k. **Q.E.D**.

Lemma 3 *For any $\lambda \in (0, 1)$ and $\pi \in \Pi$, $(k, h_\lambda^\pi(k))$ does not lie on the upper boundary of Ω_λ^π if $k > 0$.*

Proof: Let $k_1 = h_\lambda^\pi(k_0)$. Given $k_0 > 0$, suppose that (k_0, k_1) lies on the upper boundary of Ω_λ^π. Since $k_0 > 0$, by (9), $k_1 = \mu_\lambda \left[\frac{1}{2} k_0^{-\frac{1}{\lambda}} + \frac{1}{2} b^{\frac{1}{\lambda}}\right]^{-\lambda} > 0$. This implies, by (15), that for any $\varepsilon, k_1 > \varepsilon > 0$,

$$v_\lambda^\pi(k_0, k_1 - \varepsilon) - v_\lambda^\pi(k_0, k_1) \leq \rho[V_\lambda^\pi(k_1) - V_\lambda^\pi(k_1 - \varepsilon)]. \tag{16}$$

Since V_λ^π is concave on R_+ and $k_1 > 0$, there is $\beta > 0$ such that for any $\varepsilon, k_1 > \varepsilon > 0$,

$$[V_\lambda^\pi(k_1) - V_\lambda^\pi(k_1 - \varepsilon)]/\varepsilon < \beta. \tag{17}$$

However, since (4) implies that $u_\lambda(c)$ satisfies the Inada condition at $c = 0$, it follows that as $\varepsilon \to 0$,

$$[v_\lambda^\pi(k_0, k_1 - \varepsilon) - v_\lambda^\pi(k_0, k_1)]/\varepsilon \to \infty. \tag{18}$$

Since (17) and (18) contradict (16), the lemma is proved. **Q.E.D.**

Lemma 4 *Let ρ', ρ'' and π' be those of Proposition 4. Given $\rho \in (\rho'', \rho')$, let $h_\lambda^{\pi'} : I \to I$ be the policy function of an economy $(\Omega_\lambda(\pi'), v_\lambda(\cdot; \pi'), \rho)$. Then, as $\lambda \to 0$, $h_\lambda^{\pi'}(k)$ converges to $h_0^{\pi'}(k)$ uniformly in k; i.e., for any $\varepsilon > 0$, there is $\delta > 0$ such that if $0 < \lambda < \delta$, $|h_\lambda^{\pi'}(k) - h_0^{\pi'}(k)| < \varepsilon$ for any $k \in I$.*

Proof: Let $k_1(\lambda) = h_\lambda^{\pi'}(k_0)$. By (15), it holds that

$$v_\lambda^{\pi'}(k_0, k_1) + \rho V_\lambda^{\pi'}(k_1) \geq v_\lambda^{\pi'}(k_0, k) + \rho V_\lambda^{\pi'}(k)$$

for any k such that $v_\lambda^{\pi'}(k_0, k)$ is well defined. It is possible to find a subsequence λ_i such that as $\lambda_i \to 0$, $k_1(\lambda_i) \to k_1$ for some k_1. Then, by the continuity of v_λ^π and V_λ^π in λ, $k_1 \in H_0^{\pi'}(k)$. By Proposition 4, therefore, $k_1 = h_0^{\pi'}(k_0)$ for any $k_0 \in I$. Since both $h_0^{\pi'}(k)$ and $h_\lambda^{\pi'}(k)$ are continuous in k, the uniform convergence follows. **Q.E.D.**

Proof of Theorem 1: Since $\pi' \in \Pi(N)$, $h_0^{\pi'}$ has a cycle of period 3×2^N. By Proposition 2 and Lemma 4, there is a constant $\delta'(N) \in (0, 1]$ such that for any $\lambda \in [0, \delta'(N))$, $(I, h_\lambda^{\pi'})$ has a cycle of period 5×2^N. By Proposition 3, therefore, $(I, h_\lambda^{\pi'})$ is topologically chaotic for any $\lambda \in [0, \delta'(N))$. By Lemma 4, there is a constant $\delta''(N) \subset (0, 1]$ such that for any $\lambda \in [0, \delta''(N))$, $h_\lambda^{\pi'}(k) > 0$ for any $k \in \text{int } I$. By Lemma 3, therefore, for any $\lambda \in (0, \delta''(N))$, $h_\lambda^{\pi'} : I \to I$ is an interior policy function. Let $\delta(N) = \min[\delta'(N), \delta''(N)]$. Then, for any $\lambda \in (0, \delta(N))$, $h_\lambda^{\pi'} : I \to I$ is a topologically chaotic interior policy function. Since, by Proposition 4, $\rho''(N) \to 1$, as $N \to \infty$, Theorem 1 is proved. **Q.E.D.**

References

[1] Alsedà, L., J., F. Balibrea and J. Llibre eds., (1996): *Thirty Years after Sharkovskii's Theorem: New Perspectives: June 13-18, 1994 Murcia, Spain*, World Scientific, Singapore.

[2] Alsedà, L., J. Llibre and M. Misurewicz (1993): *Combinatorial Dynamics and Entropy in Dimension One*, World Scientific, Singapore.

[3] Block, L (1981): Stability of Periodic Orbits in the Theorem of Šarkovskii, *Proceedings of the American Mathematical Society* 81, 333–336.

[4] Boldrin, M., and L. Montrucchio (1986): On the Indeterminacy of Capital Accumulation Paths, *Journal of Economic Theory*, 40, 26–39.

[5] Brock, W., and J. Scheimkman (1976): Global Asymptotic Stability of Optimal Control Systems with Applications to the Theory of Economic Growth, *Jounal of Economic Theory* 12, 164–190.

[6] Cass, D., and K. Shell (1976): The Structure and Stability of Competitive Dynamical Systems, *Journal of Economic Theory* 12, 31–70.

[7] Deneckere, R. and S. Pelikan (1986): Competitive Chaos, *Journal of Economic Theory* 40, 13–25.

[8] Majumdar, M., and T. Mitra (1994): Periodic and Chaotic Programs of Optimal Intertemporal Allocation in an Aggregative Model with Wealth Effects, *Economic Theory* 4, 649–676.

[9] McKenzie, L. (1983): Turnpike Theory, Discounted Utility and the Von Neumann Facet, *Journal of Economic Theory* 30, 330–352.

[10] McKenzie, L. (1986): Optimal Economic Growth, Turnpike Theorems and Comparative Dynamics, in: K. J. Arrow and M. D. Intriligator, eds., *Handbook of Mathematical Economics*, Vol. III, North-Holland, Amsterdam.

[11] Nishimura, K., G. Sorger and M. Yano (1994): Ergodic Chaos in Optimal Growth Models with Low Discount Rates, *Economic Theory* 4, 705–717.

[12] Nishimura, K., and M. Yano (1993): Optimal Chaos when Future Utilities are Discounted Arbitrarily Weakly, in *Nonlinear Analsis and Mathematical Economics*, Lecture Note Series, Institue of Mathematical Analysis, Kyoto University.

[13] Nishimura, K., and M. Yano (1994): Optimal Chaos, Non-linearity and Feasibility Conditions, *Economic Theory* 4, 689–704.

[14] Nishimura, K., and M. Yano (1995): Non-linear Dynamics and Chaos in Optimal Growth: an Example, *Econometrica* 63, 981–1001.

[15] Sharkovskii, A. N. (1964): Coexistence of Cycles of a Continuous Map of the Line into Itself, *Ukrainian Mathematical Journal* 16, 61–71. [In Russian]

[16] Sharkovskii, A. N. (1995): Coexistence of Cycles of a Continuous Map of the Line into Itself, *International Journal of Bifurcation and Chaos* 5, 1263–1273. [The English Translation of Sharkovskii (1964)]

[17] Scheinkman, J. (1976): On Optimal Steady States of N-Sector Growth Models when Utility is Discounted, *Journal of Economic Theory* 12, 11–30.

[18] Walters, P. (1975): *Ergodic Theory: Introductory Lectures*, Lecture Notes in Mathematics 456, Springer-Verlag, Berlin.

Dynamics of the Tangent Map

H. OLIVEIRA and J. SOUSA RAMOS

Departamento de Matemática, Instituto Superior Técnico
Av. Rovisco Pais 1049-001 Lisboa, Portugal
E-mail: holiv@math.ist.utl.pt, sramos@math.ist.utl.pt

Abstract When we iterate the tangent map of the family $\lambda \tan(z)$ with initial condition and the parameter both restricted to the imaginary axis, the dynamics of the even iterates can be viewed as real and one dimensional. We study the bifurcations which occur when we change the parameter. We prove that there exists an isomorphism between a subtree of this map and the combinatorial tree of the quadratic map. We obtain as a corollary that the Sharkovsky ordering applies to this map.

Keywords Symbolic dynamics, Combinatorial dynamics, Maps of the interval, Tangent map

AMS Subject Classification 37B10, 37E05, 37E15

1 Introduction

In this paper we study some particular properties of the dynamics generated by the iterates of the complex trigonometric family of maps f_λ, depending on one complex parameter, λ

$$f_\lambda = \lambda \tan(z),$$

studied by Devaney and Keen [1] and by Keen and Kotus [2]. In this paper, divided in three sections, we present first a restriction of this map to the imaginary axis and explain the general framework of our study, including the tools of symbolic dynamics. In the second section we obtain some results about the bifurcations which can occur. In the last section we discuss the organization of the periodic structure of the orbits and the Sharkovsky ordering. More results can be found in Oliveira and Sousa Ramos [5].

Let x and β be real numbers, when $\lambda = i\beta$ and $z = ix$ the first iterate $f_\lambda(z) = f_{i\beta}(ix) = -\beta \tanh(x)$ is real, the second iterate $f_\lambda(f_\lambda(z)) = f_{i\beta}^2(ix) = -i\beta \tan(\beta \tanh(x))$ is imaginary. So, in this case every odd iterate is real and every even iterate is imaginary. Trying to understand the structure of the "Mandelbrot" set of the family f_λ on the complex plane, we use as a first partial picture this return function to the imaginary axis.

In the line of reasoning exposed before we devote our efforts to the real family

$$g_\beta = -\beta \tan\left(\beta \tanh(x)\right),$$

which describes the even iterates of f_λ when both the initial condition and the parameter are imaginary numbers.

The geometric behavior of the map g_β[1] depends on the parameter β. When $\frac{\pi}{2} < \beta < \frac{3\pi}{2}$ the graph of g_β has two vertical asymptotes (as is shown in the figure 1). This map is symmetric

$$g_\beta(x) = -g_\beta(-x). \tag{1}$$

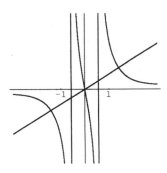

Figure 1: Graph of g_β when $\beta = 3.1$.

When $\frac{3\pi}{2} < \beta < \frac{5\pi}{2}$ the graph of g_β has four vertical asymptotes. Generally for $\frac{(2k-1)\pi}{2} < \beta < \frac{(2k+1)\pi}{2}$, $k = 1, 2, \ldots$ we have $2k$ asymptotes, a fact related to the strips studied by Devaney and Keen [1].

In our work we focus our attention only on the strip $\frac{\pi}{2} \le \beta \le \frac{3\pi}{2}$.

We define the orbit of a real point x_0 as a sequence of numbers $O(x_0) = \{x_j\}_{j=0,1,\ldots}$ such that $x_j = g_\beta^j(x_0)$ where g_β^j is the composition of order j of g_β with itself. Any point x is periodic with period $n > 0$ if the condition $g_\beta^n(x) = x$ is fulfilled with n minimal. Because of condition 1 the orbit of any point x is symmetric to the orbit of $-x$.

We denote the points where g_β has vertical asymptotes by

$$a(\beta) = -\tanh^{-1}\left(\frac{\pi}{2\beta}\right) \text{ and } b(\beta) = \tanh^{-1}\left(\frac{\pi}{2\beta}\right).$$

We denote the value of the image of $+\infty$ by $d(\beta) = -\beta \tan(\beta)$ and the image of $-\infty$ by $-d(\beta) = \beta \tan(\beta)$.

Consider the alphabet $\mathcal{A} = \{L, A, M, B, R\}$. The address $\mathbf{A}(x)$ of a real point x is defined to be L if $x_j < a$, A if $x_j = a$, M if $a < x_j < b$, B

---

[1]This map doesn't have a fixed sign of the schwarzian derivative $Sg_\beta(x)$. When $\beta > 1$, $Sg_\beta(x)$ is positive for $x \in \,]-\cosh^{-1}(\sqrt{\beta}), \cosh^{-1}(\sqrt{\beta})[$, with maximum at the origin which is $2(\beta^2 - 1)$ and is nonpositive in the complement of the previous set in $\mathbb{R}$, with the horizontal asymptote $y = -2$.

if $x_j = b$, and R if $x_j > b$. We can apply this function to an orbit of a given real point x_0; we associate to that orbit one infinite symbolic sequence $I(x_0) = \mathbf{A}(x_0)\mathbf{A}(x_1)\mathbf{A}(x_2)...\mathbf{A}(x_n)...$ the symbolic itinerary of x_0. Take the order naturally induced from the order in the real axis $L \prec A \prec M \prec B \prec R$. We introduce a parity function $\rho(S)$, for any finite sequence S, as $+1$ if S has an even number of symbols, and -1 otherwise. Let Σ denote the set of all sequences written with the alphabet $\mathcal{A}$. We can define an ordering $\prec$ on the set Σ such that given two symbolic sequences $P = P_0 P_1 P_2...$ and $Q = Q_0 Q_1 Q_2...$ let n be the first integer such that $P_n \neq Q_n$. Denote by $S = S_0 S_1 S_2...S_{n-1}$ the common first subsequence of both P and Q. Then, we say that $P \prec Q$ if $P_n \prec Q_n$ and $\rho(S) = +1$ or $Q_n \prec R_n$ if $\rho(S) = -1$. If no such n exists we say that $P = Q$. This ordering is originated by the fact that when $x < y$ we have $I(x) \prec I(y)$ or $I(x) = I(y)$.

Here the orbits determining the ordering of the combinatorial structure of the dynamic of g_β are the itineraries of $+\infty$ and $-\infty$, these orbits are "superstable" in the sense that $\lim_{x \to \pm\infty} g'_\beta(x) = 0$. We can use also the itineraries of $a(\beta)^-$ (or $b(\beta)^-$) and $b(\beta)^+$ (or $a(\beta)^+$) with almost the same results. Any itinerary of $|\infty$ (resp. $-\infty$) is also a maximal orbit (minimal) in the ordering defined in this section. As we know, every orbit with initial condition x_0 is the symmetric of the orbit with initial condition $-x_0$, so any orbit initiated in $+\infty$ has as counterpart a symmetric orbit started in $-\infty$; therefore we shall focus the admissibility rules only on the orbits of $+\infty$. To state these rules we will use the shift operator σ defined with the usual meaning.

Given a particular sequence Q we define another operator τ acting on Q; this operator interchanges the symbols L and R in Q leaving the symbols M unchanged, for instance $\tau(RLMR) = LRML$. Given any itinerary of $+\infty$ which we call S, the corresponding itinerary of $-\infty$ is $\tau(S)$. The kneading pair is in our case $(S, \tau(S))$. The concept of kneading was introduced by Milnor and Thurston [4].

Admissibility rules: Let S be a given sequence of symbols and $(S, \tau(S))$ be a pair of sequences. $(S, \tau(S))$ is a kneading pair and S is a kneading sequence, if S satisfies the admissibility condition $\tau(S) \prec \sigma^k(S) \prec S$ for every integer k.

A bistable periodic orbit attracts both the orbit of $+\infty$ and the orbit of $-\infty$. Any bistable orbit has an itinerary $S = P\tau(P)$; as a consequence this type of orbit must have even period.

When the parameter satisfies $\frac{\pi}{2} < \beta < \beta_1$, where β_1 is the solution of $b(\beta) = \tanh^{-1}\left(\frac{\pi}{2\beta}\right) = -b\tan(\beta) = d(\beta)$ (approximately 2.941812), the orbit of any initial condition in $\left]\tanh^{-1}\left(\frac{\pi}{2\beta}\right), +\infty\right[$ is invariant in that interval. In particular the symbolic orbit of $+\infty$ is R^∞. This happens because when $\beta < \beta_1$ the image of $+\infty$ is greater than $b(\beta)$. The corresponding situation holds for initial condition in the interval $\left]-\infty, -\tanh^{-1}\left(\frac{\pi}{2\beta}\right)\right[$.

Lemma 1.1 *When the parameter $\beta_1 < \beta < \beta_2$ where β_2 is the solution of $b(\beta) = \tanh^{-1}\left(\frac{\pi}{2\beta}\right) = \beta\tan(\beta) = -d(\beta)$ the blocks LR and RL do not occur in the symbolic orbit (β_2 is approximately 3.29749).*

Proof The image of any point $x > b(\beta)$ is greater than $a(\beta)$, so after R we must have R or M. Similarly the image of any point $x < a(\beta)$ is always less than $b(\beta)$, so after an L we cannot have any R. $\square$

2   Types of bifurcations

In this section we enumerate briefly the types of bifurcations which can occur when we change the parameter β. The structure of the bifurcations will be the basis of the organization of the combinatorial tree enumerating the admissible periodic orbits discussed in the next section.

Type $B1$: Period doubling. When we increase the parameter β from $\beta = \frac{\pi}{2}$ we have the usual bifurcation of period doubling when $g_\beta(x) = x$ and $g'_\beta(x) = -1$; a stable orbit with period 1 turns unstable and a new orbit with period 2 is generated. We have this type of bifurcation both for x positive and negative ($x \simeq \pm 1.84475, \beta_d \simeq 2.66633$). Let us focus on the orbit on the right side. Symbolically, when we follow the orbit of $+\infty$, we see only R^∞. When we analyze the numeric orbit of $+\infty$, however, in case $\beta < \beta_d$ we see that this orbit is stable and asymptotic to a fixed point, while in case $\beta > \beta_d$ this orbit is asymptotic to a stable cycle of period 2. Symbolically we see this orbit only when the parameter β approaches β_1, obtaining there $(RB)^\infty$. We have also this type of bifurcation for higher periods.

Type $B2$: This happens when one orbit crosses a discontinuity when we vary the parameter β. In that case the orbits of $-\infty$ and $+\infty$ collide and a new orbit (bistable) results from the two previous orbits.

Type $B3$: Pitchfork bifurcation. We shall see in Lemma 2.2 how this bifurcation occurs.

Type $B4$: Saddle node bifurcation. When we increase the parameter β from $\beta = \frac{\pi}{2}$ we have the bifurcation characterized by $g_\beta^n(x) = x$ and $g_\beta^{n'}(x) = 1$, with solutions both for x positive and negative. When this type of bifurcation occurs we have a creation of a hyperbolic pair of orbits, one stable and the other unstable, each one of period n. It is easy to see that this type of bifurcation occurs only with $n = 2(2k+1)$ because only the even iterates can have a positive derivative. The orbit generated by this type of bifurcation must be an admissible bistable orbit; see Oliveira and Sousa Ramos [5]. For instance, the symbolic orbit $(RMLLMR)^\infty$ with period 6 is generated by this type of bifurcation.

We first give some instances of the less usual types of bifurcation, the nonclassical period doubling $B2$ and the pitchfork $B3$. We note that each bifurcation ($B1$, $B2$, $B3$ and $B4$) occurs many times for different values of the parameter β.

Lemma 2.1 *Type B2. When $\beta = \beta_1$ there occurs a bifurcation; the symbolic itinerary of $+\infty$ changes from $(RR)^\infty$ to $(RB)^\infty$ and then to $(RMLM)^\infty$. The block of the first orbit is RR, the last R changes to M in the bifurcation process, and the new itinerary is $(RM\tau(RM))^\infty$. This itinerary results from the collision of the two orbits one of $-\infty$ and the other of $+\infty$.*

Proof As we know $g_\beta(+\infty) = d(\beta)$, when β approaches β_1 the image of $+\infty$ approaches the right discontinuity $\lim_{\beta \to \beta_1^-} g_\beta(+\infty) = b(\beta_1)^+$ and $g_\beta\left((b(\beta_1))^+\right) = +\infty$, so the symbolic itinerary for $\beta = \beta_1^-$ can be written as $(RB)^\infty$. When β crosses the discontinuity we have $\beta = \beta_1^+$, the image of $+\infty$ is now $b(\beta_1)^-$ but $g_{\beta_1}\left(b(\beta_1)^-\right) = -\infty$, and the symbolic itinerary of $+\infty$ begins now by RB^-. The new orbit contains also the orbit of $-\infty$ (the second iterate of $+\infty$), is bistable and symmetric: the orbit of $-\infty$ begins now by $\tau(RB^-) = LA^+$. The second iterate of $-\infty$ is $+\infty$ and the new orbit, for β_1^+, has period 4, resulting from the fusion of the two orbits of $-\infty$ and $+\infty$, for β_1^-, each one with period 2. The new orbit has the periodic block $RB^-\tau(RB^-) = RB^-LA^+$ which we can write like $Q = RMLM$, the new orbit is Q^∞. $\square$

Lemma 2.2 *A pitchfork bifurcation (type B3) occurs when $\beta \simeq 3.060687$ and we increase β. Before the bifurcation there exists only one bistable orbit attracting both the orbits of $+\infty$ and $-\infty$. After the bifurcation this unique stable orbit splits in two different stable orbits – one attracting the orbit of $+\infty$, the other the orbit of $-\infty$ – and one unstable orbit. The symbolic itinerary of the stable orbit is $RMLM$, and the three resulting orbits have exactly the same symbolic itineraries immediately after the bifurcation process.*

Proof This bifurcation occurs after the bifurcation described in Lemma 2.1. We know that after bifurcations of type $B2$ we have only one stable orbit resulting from the collapse of the orbits of $+\infty$ and $-\infty$. This orbit has symbolic itinerary $RMLM$. The numeric orbit is asymptotic to $\{x_0, x_1, x_2, x_3\}$, this last orbit must be symmetric because is bistable, so $x_0 = -x_2$ and $x_1 = -x_3$ with $x_0, x_1 > 0$. Let us focus on this numeric orbit. The derivative of $g_\beta^4(x_0)$, is

$$\lambda_\beta = \frac{d}{dx}g_\beta^4(x)\bigg|_{x_0} = g_\beta'(x_0)\,g_\beta'(x_1)\,g_\beta'(x_2)\,g_\beta'(x_3) = \left(g_\beta'(x_0)\,g_\beta'(x_1)\right)^2$$

and must be positive. When we increase β from β_1, λ_β increases from values near 0 reaching $\lambda_\beta = 1$, for some value of β (actually near $\beta \simeq 3.060687$); we get $\frac{d}{dx}g_\beta^4(x) = 1$, but in that case the second derivative is

$$\frac{d^2}{dx^2}g_\beta^4(x)\bigg|_{x_0} = g_\beta''(x_0)\,g_\beta'(x_1)\,g_\beta'(x_2)\,g_\beta'(x_3) + g_\beta''(x_0)\,g_\beta'(x_1)\,g_\beta'(x_2)\,g_\beta'(x_3)$$

$$+ g_\beta'(x_0)\,g_\beta'(x_1)\,g_\beta''(x_2)\,g_\beta'(x_3) + g_\beta'(x_0)\,g_\beta'(x_1)\,g_\beta'(x_2)\,g_\beta''(x_3).$$

This second derivative must be 0, because $g''_\beta (x_0) = -g''_\beta (x_2)$ and $g''_\beta (x_1) = -g''_\beta (x_3)$. Finally $\frac{d}{d\beta} g^4_\beta (x) \Big|_{x_0} = 0$ and

$$\frac{d^3}{dx^3} g^4_\beta (x) \Big|_{x_0} = S g^4_\beta (x_0) = 2 \left(\left(g'_\beta (x_0) \right)^2 S g_\beta (x_1) + S g_\beta (x_0) \right) < 0,$$

because $0 < \left(g'_\beta (x_0) \right)^2 << 1$, $-2 < S g_\beta (x_1) < 2 \left(\beta^2 - 1 \right) < 18$ and $S g_\beta (x_0) \simeq -2$. We have the conditions of the classical pitchfork bifurcation. Now the stable orbits are different; we don't have symmetry in the orbits of $+\infty$ and $-\infty$ and those orbits must split apart. After the bifurcation there are two numeric stable orbits $\{x^+_0, x^+_1, x^+_2, x^+_3\}$ and $\{x^-_0, x^-_1, x^-_2, x^-_3\}$; the first one is the orbit associated with $+\infty$, and the second is the orbit associated with $-\infty$. Finally we notice, for instance, that x^+_3 and x^-_1 were originated from x_3 and must change in the opposite direction when we change the parameter β; the same happens to the other values of the two orbits. $\square$

Theorem 2.3 *When β increases from $\frac{\pi}{2}$, we have a cascade of period doubling bifurcations, in fact a sequence of bifurcations of the type $B1 \to B2 \to B3 \to B2 \to B3 \to \dots$. Intercalated with these cascades we have sequences of bifurcations of the type $B4 \to B2$, where $B2$ is in the reverse order than previously; a bistable orbit splits apart in two stable orbits, each one with period halved.*

The complete proof of this result can be seen in Oliveira and Sousa Ramos [5]. We describe here how the proof is outlined. Let P be the block of a periodic symbolic orbit $(P)^\infty$ of $+\infty$. Now suppose that this block ends by a M or R. For each bifurcation of the type $B2$, if the last symbol is R, it changes to M; if the last symbol is M, it changes to L. The block P after each bifurcation changes to P' and the new periodic block is now $P'\tau(P') = Q$ (which is bistable), with period doubled. The new orbit, Q^∞, results from the crossing of the discontinuity and the fusion of two periodic orbits as in the Lemma 2.1. Then occurs a pitchfork splitting $B3$ and the process reinitiates accumulating in a non periodic orbit. After each one of these cascades we have a saddle node bifurcation $B4$ generating a bistable orbit (and one unstable orbit) with even periods $2(2k+1)$ which splits in two odd period stable orbits by a reverse $B2$ bifurcation. This process never produces an L at the end of the sequence associated to $+\infty$ for $\beta < \beta_2$.

3   Combinatorial trees

We consider now the trees that organize the periodic structure of the orbits of real mappings depending on one parameter; examples of this procedure can be found in [3] for real maps or in [6] for complex families. These trees are, in general, indexed by the parameter. The next theorem summarizes the information of the combinatorial tree associated with g_β for $\frac{\pi}{2} \leq \beta \leq \pi$, where the topological entropy has the same variation as in the quadratic case. We use the capitals L, M and R related to the symbolic dynamics of g_β and r, l in the case of the quadratic map. These last symbols have the usual meaning in unimodal maps; see for instance [3].

Let us introduce the combinatorial tree T_1 associated to g_β. In this tree, each vertex represents an ordered kneading pair

$$(S, \tau(S)) = (P_1 P_2 ... P_k, \tau(P_1 P_2 ... P_k)) .$$

The first element of the pair is the symbolic itinerary of an orbit with initial condition $b(\beta)^+$ or $a(\beta)^+$ (here we omit the initial condition from the symbolic itinerary) with the first symbol $P_1 = R$ (the address of $+\infty$). The second element is the symbolic itinerary of an orbit with initial condition $a(\beta)^-$ or $b(\beta)^-$ and first symbol L. The last symbol P_k can be a B or an A. In this tree, we represent every admissible symbolic orbit. To avoid overloading the notation we only represent S in the tree presented at the end of this section.

Theorem 3.1 *There is an isomorphism between the symbolic tree T_1 describing the combinatorial structure of the periodic orbits of the critical points (kneading sequences) of the family of maps g_β (with $\frac{\pi}{2} \leq \beta \leq \pi$) and the symbolic tree T_2 of the quadratic family $q_\mu = 4\mu x (1 - x)$ (with $0 < \mu \leq 1$).*

Proof First we prove that given a combinatorial tree of g_β we can write the tree of the map q_μ, and then the converse. We use lemma 1.1 and simple analytical and geometrical reasonings of the same type used in that proof.

We see that below each vertex of the tree T_1 we can have only two edges: a vertex ending in B can go to MA or RB, a vertex ending in A can go to LA or MB. Using these rules it is easy to see that if we substitute R and L by r, and M by l, we obtain the desired correspondence.

We prove now that we can translate uniquely the tree of q_μ in terms of the tree of g_β. We change the symbols l by symbols M. The problem consists of choosing an R or an L for each r. Every kneading sequence represented on T_1 begins by R, so we have a constructive rule to translate the sequences of q_μ in terms of g_β. We see easily that the only admissible blocks of the kneading sequences of g_β are of the type $RM^{2k}R$, $LM^{2k}L$, $RM^{2k+1}L$ and $LM^{2k+1}R$ with $k \geq 0$. So, after a block l^{2k} (translated by M^{2k}) we choose the same symbol as the precedent of this block and the opposite symbol if we have a block l^{2k+1}. More details can be found in [5]. $\square$

Remark 3.2 *We note that for $\beta \leq \pi$, the combinatorial structure of the kneading sequences of our map and the quadratic map are the same, but the nature of the bifurcations is quite different; only the organization of the sequence of bifurcations has a parallelism in the two families.*

Corollary 3.3 *The Sharkovsky Theorem [7] applies to the map g_β.*

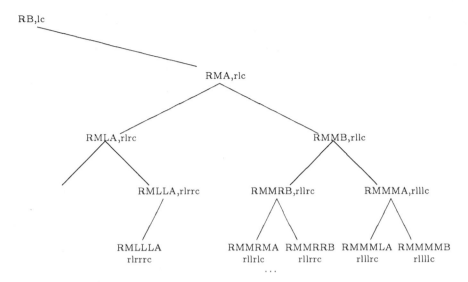

References

[1] Devaney, R. and Keen, L. [1989] "Dynamics of tangent," in *Dynamical Systems: Proc. Univ. of Maryland 1986–1987, Lecture Notes in Mathematics* vol. 1342 (Springer, Berlin, New York), pp. 105–111.

[2] Keen, L. and Kotus, J. [1997] "Dynamics of the family $\lambda \tan z$," *Conformal Geometry and Dynamics*, Vol I, 28–57.

[3] Lampreia, J., Rica da Silva, A. and Sousa Ramos, J. [1986] "Subtrees of the unimodal maps tree", *Boll. Un. Mat. Ital.* C(6) 5, no 1, 159–167.

[4] Milnor, J. and Thurston, W. [1988], "On iterated maps of the interval," in *Dynamical Systems: Proc. Univ. of Mariland 1986-1987, Lecture Notes in Mathematics vol. 1342* (Springer, Berlin, New York), 465–563.

[5] Oliveira, H. and Sousa Ramos, J. [2002] "Topological entropy of the tangent map," to appear.

[6] Sarreira, M. and Sousa Ramos, J. [1991] "Symbolic dynamics of iteration of cubic complex maps," *European Conference on Iteration Theory* (ECIT 1989) (World Scientific, Singapore), pp. 217–229.

[7] Sharkovsky, A. N. [1964] "Coexistence of cycles of a continuous map of the line into itself," *Ukrain. Math. Zh.* 16, 61–71.

Thresholds, Mode Switching and Complex Dynamics

HASSAN SEDAGHAT

Department of Mathematics
Virginia Commonwealth University
Richmond, VA 23284-2014, USA
E-mail: hsedagha@vcu.edu

Abstract In this note, mode switching in threshold systems is considered in the special case of ejector cycles. The existence and the global properties of such a cycle (from periodic behavior to chaos) are studied in a model from combat theory.

Keywords Threshold, Mode switching, Ejector cycle, Polymodal system

AMS Subject Classification 39A11

1   Introduction

Systems of difference equations that possess thresholds and discontinuities are quite commonly encountered in scientific models in general, and models in the social sciences in particular; see, e.g., [1–4] and the references therein. A few specific examples appear below. In spite of this, there are few mathematical tools that are broadly applicable to these systems. In this note, we consider briefly first a general classification of threshold models as polymodal systems, and then consider the concept of *ejector cycles* that pertains to such systems. These cycles contain global information about the system as each goes through its mode-switching sequence. We illustrate this behavior by analyzing equations from a model of ground combat.

2   Polymodal systems

Defintion 1 *Let $D \subset \mathbb{R}^m$ be a nonempty set, and let $F \in C(D, \mathbb{R}^m)$. A point $\widehat{x} \in D$ is an* ejection point *of F if $F(\widehat{x}) \notin \overline{D}$. The set E of all ejection points of F in D is the* ejector *of D relative to F; i.e.,*

$$E \doteq \left\{ x \in D : F(x) \notin \overline{D} \right\}.$$

0-415-31675-8/04/$0.00+$1.50

Defintion 2 *A polymodal system in $\mathbb{R}^m$ with k modes is a function-set collection*

$$\{(F_i, D_i): i = 1, \ldots, k\}, \ k \geq 2$$

where for each i all of the following are true:

(a) $D_i \subset \mathbb{R}^m$ *is nonempty and disjoint from D_j for $j \neq i$;*

(b) $F_i \in C(D_i, \overline{D})$ *where $D \doteq \bigcup_{j=1}^{k} D_j$;*

(c) D_i *contains a nonempty ejector E_i relative to F_i.*

Each pair (F_i, D_i) is a *mode* of the system. We also define the usually, though not necessarily, discontinuous *join* of the maps F_i as

$$F \doteq \sum_{i=1}^{k} \chi_{D_i} F_i : D \to \overline{D},$$

where χ_S is the characteristic function of the set S; i.e., $\chi_S(x) = 1$ if $x \in S$ and $\chi(x) = 0$ if $x \notin S$.

Next, we give some examples of polymodal systems from the social sciences. For additional examples, and a more detailed discussion of polymodal systems, see [6].

Examples 1 (Addiction, duopoly, arms race) G. Feichtinger proposes two sets of equations in [4] (and the references given therein) that involve thresholds (and are thus polymodal). The first set

$$
\begin{aligned}
x_{n+1} &= ax_n + b\chi_{\{x_n > y_n\}} \\
y_{n+1} &= y_n + c(x_n - y_n)
\end{aligned}
$$

models "habit formation" in use of addictive substances (e.g., tobacco, alcohol, drugs) where x_n is the habit's (e.g., smoking) consumption capital in period n and y_n is the "threshold in the habit stock" so that consumption takes place only if x_n exceeds y_n. Also, $0 < a < 1$, $b, c > 0$. It is easy to see that this is a bimodal system with D_1 and D_2 the opposite sides of the diagonal $y = x$.

The second set of equations model dynamic interaction in a simple duopoly with "asymmetries." If x_n and y_n denote the sizes (as measured by sales or market shares) of the two firms in the duopoly in period n, then

$$
\begin{aligned}
x_{n+1} &= (1 - \alpha)x_n + a\chi_{\{x_n > y_n\}} \\
y_{n+1} &= (1 - \beta)x_n + b\chi_{\{x_n > y_n\}}
\end{aligned}
$$

where $\alpha, \beta \in (0, 1)$ and $a, b > 0$. These equations may also be used as a nonlinear extension of Richardson-type model of arms race.

Example 2 (A model of ground combat) J. Epstein [3] proposes a simple deterministic model in order to illustrate the role of a defender's withdrawal as a feedback mechanism that can substantially affect the outcome of combat. A special case of this model involves the following equations:

$$x_{n+1} = x_n + \frac{1}{a}(a - x_n)\left[a - x_n(1 - y_n)\right] \tag{1}$$

$$y_{n+1} = \left\{y_n + \frac{1 - y_n}{1 - d}[x_n(1 - y_n) - d]\right\}\chi_{\{x_n(1-y_n)\geq d\}}$$

These equations are taken from [5] where the ground-combat version of Epstein's model is treated in a rigorous way. There are two combatants, an "attacker" and a "defender". The latter will withdraw if its attrition level exceeds a prescribed level (losses measured in terms of standard military "scores"). The variables and constants have the following meanings:

a: attacker's attrition rate threshold, $a \in (0,1)$
d: defender's attrition rate threshold, $d \in (0,1)$
y_n: defender's withdrawal rate in period (e.g., day) n
x_n: attacker's prosecution rate of combat in period n
$x_n(1 - y_n)$: defender's attrition rate

Also, assumed are the initial value restrictions: $x_0 > 0$, $y_0 = 0$. Equations (1) describe a threshold model, which can be expressed as a bimodal system:

$$
\begin{aligned}
D_1 &= \{(x,y) \in [0,\infty)^2 : x(1 - y) \geq d\} \\
D_2 &= \{(x,y) \in [0,\infty)^2 : x(1 - y) < d\} \\
F_1(x,y) &= [f(x,y), g(x,y)], \ (x,y) \in D_1 \\
F_2(x,y) &= [f(x,y), 0], \ (x,y) \in D_2
\end{aligned}
$$

where

$$
\begin{aligned}
f(x,y) &= x + (a - x)[a - x(1 - y)]/a \\
g(x,y) &= y + (1 - y)[x(1 - y) - d]/(1 - d).
\end{aligned}
$$

Sets D_1 and D_2 represent the regions in $[0,\infty)^2$ that lie, respectively, below and above the curve $y = 1 - d/x$.

Examples 3 (One dimensional, multi-regime economic systems) Economic models often involve thresholds that range from physical constraints (since negative values are not permissible either for functions or for variables) to different "regimes" that a system can exist in. Piecewise linear equations provide the simplest examples of these types of systems. Several polymodal models are discussed in [2] together with a discussion of these models as "multiple phase systems." In all the polymodal systems in [2], the join map $\sum_{i=1}^{k} \chi_{Di} F_i$ is continuous (though not necessarily smooth). More economic examples (too many to list specifically) appear in other sources, e.g., [1].

3   Ejector cycles and an application

Each ejector E_i by itself is anti-invariant, where points of D_i move out by the action of F. In some cases, a collection of two or more ejectors can form an *ejector chain* that leads trajectories through the sets D_i. When such a chain is closed, trajectories return to the ejector from which they started, and the initial ejector will contain an invariant subset under the action of several of the F_i composed with each other.

Definition 3   *Let (F_i, D_i), $i = 1, \ldots, k$, $k \geq 2$ be a polymodal system with k components. Let $2 \leq l \leq k$ and assume that $E_{i_1}, \ldots, E_{i_l}$ are ejectors in $D_{i_1}, \ldots, D_{i_l}$ respectively. Then the collection $\mathcal{E} = \{E_{i_1}, \ldots, E_{i_l}\}$ is an ejector cycle of length l if for each $j = 1, \ldots, l$ there is a nonempty subset $E'_{i_j} \subset E_{i_j}$ such that*

$$F_{i_j}(E'_{i_j}) \subset E'_{i_{j+1}}, \ 1 \leq j \leq l - 1, \ F_{i_l}(E'_{i_l}) \subset E'_{i_1}.$$

We call the continuous mapping

$$\psi \doteq F_{i_l} \circ \cdots \circ F_{i_1} : E'_{i_1} \rightarrow E'_{i_1}$$

a cycle map *of the polymodal system corresponding to the ejector cycle $\mathcal{E}$.*

Note that a cycle map is a standard continuous mapping on an invariant region, namely E'_{i_1}. As such, the standard theory of continuous maps applies to ψ and the various properties of ψ provide information on the behavior of the polymodal system. For example, if ψ has a cycle of length l, then there is a cycle of length kl in the polymodal system. Similarly, an aperiodic trajectory of ψ gives rise to an aperiodic trajectory in the system. It is therefore of great interest to identify ejector cycles, whenever they exist.

It is not difficult to give specific examples of polymodal systems in dimensions 1 and 2 that possess ejector cycles and exhibit complex behavior. Here, however, we study the system described by equations (1) above. Our aim is to show that for various ranges of parameter values, *there are ejector cycles whose cycle maps are topologically conjugate to maps of the interval*. This fact is then used to draw conclusions about the behavior of equations (1). We begin with two lemmas whose proofs are given in [5].

Lemma 1   *Let ξ be the largest real root of the cubic polynomial*

$$C(t) = -(1 - t)(t^2 - at + a^2) + ad(1 - d).$$

Then $\xi < 1$. If $a > 1/2$, then C is strictly increasing and ξ is its only real root. Further, $\xi \in (d, a)$ if $a > d \geq 1/2$ and $\xi \in (1 - d, a)$ if $1 - a < d < 1/2$.

Lemma 2   *Let $d \leq a$ and consider the quintic polynomial*

$$Q(t) = a + t(1 - t)(t - a)\frac{t^2 - at + a^2}{a^3(1 - d)}, \quad t \geq 0.$$

(a) *Q has a unique real root ζ, and $\zeta \in (1, 1+a)$; in fact, there is $\varepsilon > 0$ such that Q is strictly decreasing on the interval $(1 - \varepsilon, \infty)$, and Q maps the interval $[1, \zeta]$ homeomorphically onto $[0, a]$;*

(b) *Assume that $d < a$. Then all fixed points of Q that exceed d are in the interval $[a, 1)$. If $a \geq 1/2$, then a is the only fixed point of Q that is larger than d. On the other hand, if $a < 1/2$ and*

$$d \geq 1 - \frac{1}{4a} \tag{2}$$

then Q has a fixed point in (a, β^-) and another in $(\beta^+, 1)$, where

$$\beta^\pm = \frac{1 \pm \sqrt{1 - 4a(1 - d)}}{2}.$$

Theorem 1 (a) *Let $1 - a < d < 1/2$ and let $\xi \in (1 - d, a)$ be the unique zero of $C(t)$ in Lemma 1. If $\{(x_n, y_n)\}$ is a solution of (1) with $y_0 = 0$ and $x_0 \in (\xi, a]$ then $\{x_n\}$ increases monotonically to a, but for all n,*

$$y_{2n} = 0, \ y_{2n+1} = \frac{x_{2n} - d}{1 - d}$$

In particular, the trajectory $\{(x_n, y_n)\}$ converges to the 2-cycle

$$\Gamma = \{(a, 0), (a, y_\infty)\}$$

where $y_\infty = (a - d)/(1 - d)$.

(b) *Let $a > d \geq 1/2$, and let $\xi \in (d, a)$ be the unique zero of $C(t)$ in Lemma 1. Then the same behavior as in Part (a) is obtained.*

Proof (a) We first show that system (1) has an ejector cycle. For $x > \xi$,

$$
\begin{aligned}
\psi(x, 0) &= F_2 \circ F_1(x, 0) \\
&= F_2(x + (a - x)^2/a, (x - d)/(1 - d)) \\
&= (Q(x), 0)
\end{aligned}
$$

where Q is defined in Lemma 2. The action of F_2 is well defined because if $\tilde{x}_1$ and $\tilde{y}_1$ denote the two coordinates of $F_1(x, 0)$ above, then $\tilde{x}_1(1 - \tilde{y}_1) < d$ if and only if $C(x) > 0$. This last inequality is true by Lemma 1 if $x > \xi$. Further, if $x \in (\xi, a]$, then

$$a - Q(x) = \frac{\tilde{x}_1(1 - \tilde{y}_1)}{a}(a - \tilde{x}_1) < a - \tilde{x}_1$$

so that

$$Q(x) > \tilde{x}_1 > x. \tag{3}$$

Since $Q(a) = a$, it follows that $Q(I) \subset I$ where $I = (\xi, a]$. Hence, ψ is the cycle map of an ejector cycle with domain $E'_1 = I \times \{0\}$, and ψ is topologically isomorphic to Q on I. In addition, the inequalities (3) imply that $\{x_n\}$ increases to a if $x_0 \in (\xi, a]$ (the even terms are $x_{2n} = Q^n(x_0)$ and the odd terms are given by $x_{2n+1} = \tilde{x}_1(x_{2n})$). Also, $\{y_n\}$ behaves as claimed, since $y_{2n} = 0$ (being the second coordinate of $F_2 \circ F_1(x_{2n-2}, 0)$) and $y_{2n+1} = (x_{2n} - d)/(1 - d)$, which converges to y_∞.

(b) This is done in essentially the same way as (a). ∎

Remark   The image $F_1(I, 0)$ of I is the locus of all odd terms (x_{2n+1}, y_{2n+1}). These are the coordinates of $F_1(x_{2n}, 0)$, i.e.,

$$x_{2n+1} = x_{2n} + (a - x_{2n})^2/a, \quad y_{2n+1} = (x_{2n} - d)/(1 - d).$$

Eliminating x_{2n} from these equations shows that $F_1(I, 0)$ is a connected segment of the parabola

$$x = l(y) = d + (1 - d)y + \frac{1}{a}[a - d - (1 - d)y]^2. \tag{4}$$

Theorem 2 below looks at the situation where $x_0 > a$. Again, we find an ejector cycle, but the dynamics are not as simple as in Theorem 1.

Theorem 2   *Assume that one of the following conditions hold:*

(i) $a > d \geq 1/2$;
(ii) $1 - a < d < 1/2$.

Then every trajectory with $x_0 \in (a, 1)$ and $y_0 = 0$ converges to the cycle Γ of Theorem 1 from the right, with $\{y_n\}$ having the same behavior as in Theorem 1(a) but now $\{x_n\}$ converges non-monotonically to a from the right in the manner $x_{2n+2} < x_{2n} < x_{2n+1}$ for every n.

Proof   As in the proof of Theorem 1, we have

$$\psi(x, 0) = F_2(\tilde{x}_1, \tilde{y}_1) = (Q(x), 0). \tag{5}$$

If (i) or (ii) hold, and ξ is as defined in Lemma 1, then $\xi < a$ so $\tilde{x}_1(1 - \tilde{y}_1) > d$, and the action of F_2 is well defined for $x > a$. Also by Lemma 2, Q has no fixed points (except a). It follows that if $x \in (a, 1)$ and $y = 0$, then $\tilde{x}_1 > x$. Further,

$$0 < Q'(a) = \frac{1 - a}{1 - d} < 1$$

so that $a < Q(x) < x < \tilde{x}_1$ for $x \in (a, 1)$. These inequalities and (5) establish the pattern described in the statement of the theorem. ∎

As in the previous case, it is evident that the odd terms (x_{2n+1}, y_{2n+1}) fall on a connected segment of the parabola (4) and the even terms (x_{2n}, y_{2n})

fall in the interval $I = (a, 1)$ on the x-axis. We quote one more result that gives sufficient conditions for the occurrence of chaotic behavior. The proof is self-evident, since ψ is topologically conjugate to Q on a suitable interval I.

Theorem 3 *Assume that the polynomial Q has a fixed point in the interval $(a, 1)$, and that p is the larger fixed point with $p > \xi$. If the interval $(\xi, 1)$ contains a subinterval I with $p \in Q(I) \subset I$, and if $x_0 \in I$, $y_0 = 0$, then for $n \geq 1$, the following are true:*

(a) If p is attracting (e.g., $|Q'(p)| < 1$), then $\{(x_n, y_n)\}$ converges to the 2-cycle

$$\Psi = \left\{ (p, 0), \left(p + \frac{(p-a)^2}{a}, \frac{p-d}{1-d} \right) \right\}.$$

(b) If p is unstable (e.g., $Q'(p) < -1$), and Q has a limit cycle $\{c_1, \ldots, c_k\}$ in I, then $\{(x_n, y_n)\}$ converges to a $2k$-cycle whose even-indexed terms are $(c_i, 0)$, $i = 1, \ldots, k$.

(c) If Q is chaotic on the invariant interval I (e.g., it a has snap-back repeller in I) then F is chaotic and has periodic points of all possible even periods. The even indexed terms are in $I \times \{0\}$ and the odd indexed terms are on the parabola (4).

To see that the last case is in fact possible, consider a special case: $a = 0.465$, $d = 0.455$. In this case, direct computation shows that Q has a snap-back repeller in $I = [0.73, 0.95]$ and chaos obtains. See [6] for more details.

References

[1] Benhabib, J., (Ed.) *Cycles and Chaos in Economic Equilibirum*, Princeton U. Press, Princeton, 1992.

[2] Day, R.H., *Complex Economic Dynamics*, MIT Press, Cambridge, 1994.

[3] Epstein, J.M., *The Calculus of Conventional War*, Brookings Inst., Washington, DC, 1985.

[4] Feichtinger, G., Nonlinear threshold dynamics: Further examples for chaos in social sciences, in: *Economic Evolution and Demographic Change*, G. Haag, et al. (Eds.), Springer-Verlag, 1992.

[5] Sedaghat, H., Convergence, oscillations and chaos in a discrete model of combat, *SIAM Review*, 44, 74-92, 2002.

[6] Sedaghat, H., *Nonlinear Difference Equations: Theory with Applications to Social Science Models*, Kluwer, Dordrecht, 2003.

Computation of Nonautonomous Invariant Manifolds

STEFAN SIEGMUND

Department of Mathematics
University of California
Berkeley, CA 94720, USA
E-mail: siegmund@math.berkeley.edu

Abstract We use a cell mapping algorithm to approximate the local unstable manifold of a nonautonomous difference equation which could be either deterministic or random. The proof of convergence is carried out by showing that the unstable manifold is pullback attracting and that the algorithm approximates this pullback attracting set. We make use of the general notion of a nonautonomous dynamical system. Explicit examples are given to illustrate the results.

Keywords Invariant manifolds, Nonautonomous difference equations, Pullback attractor, Continuation technique, Cell mapping algorithm

AMS Subject Classification 39A10, 37B55

1 Introduction

We consider nonautonomous difference equations which could be either *deterministic*

$$x_{n+1} = f_n(x_n) \tag{1}$$

with $n \in \mathbb{Z}$, $x_n \in \mathbb{R}^d$, or *random*

$$x_{n+1} = f_{\omega_n}(x_n) \tag{2}$$

where $\omega = (\ldots, \omega_{-1}, \omega_0, \omega_1, \ldots)$ could be an element of a probability space Ω consisting of sequences of random numbers. If we fix a sequence ω then the random difference equation can be interpreted as a deterministic nonautonomous difference equation. The random sequence (ω_n) chooses randomly an index of f. One also says that pathwise the random equation is a deterministic equation. And this pathwise interpretation of the random case enables us to do the same numerics for nonautonomous deterministic (1) and random (2) difference equations.

0-415-31675-8/04/$0.00+$1.50

One of the main differences between autonomous and nonautonomous dynamics is that a solution of a nonautonomous difference equation which is shifted in time is again a solution, possibly with another starting value at time 0, but is again a solution parameterizing the same trajectory. For nonautonomous equations shifted solutions are not solutions anymore, in general, and for that reason we cannot project solutions into the state space $\mathbb{R}^d$ to get trajectories but instead have to visualize them as graphs over time.

We use, extend and combine ideas and techniques of Ludwig Arnold, Peter Kloeden, Gunter Ochs, Hannes Keller, Michael Dellnitz and Oliver Junge to formulate a cell mapping algorithm which approximates the unstable manifold, and prove its convergence in the Hausdorff distance. The rigorous mathematical formulation of the algorithm and the proof of convergence are carried out using the general notion of a nonautonomous dynamical system.

Moreover we discuss the relationship between the (local) unstable manifold and the (local) nonautonomous attractor, especially the so-called pullback attractor.

It is somehow natural to compute the unstable manifold. Consider this autonomous linear 2-dimensional example

$$x_{n+1} = \frac{1}{2}x_n, \qquad y_{n+1} = 2y_n.$$

The two equations are decoupled with eigenvalues $\frac{1}{2}$ and 2, the stable manifold is the x-axis, the unstable manifold is the y-axis. Now think of all the solutions starting at time 0 in the hatched set. At time 1 this set is contracted in x-direction and expanded in y-direction, and as time continues the hatched set somehow converges to the unstable manifold.

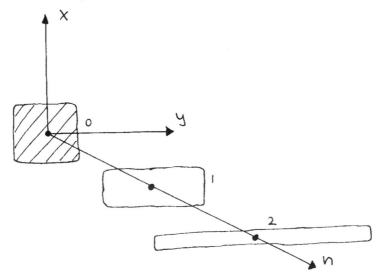

Figure 1: The hatched box at time 0 converges to the y-axis.

Note that since we have an autonomous equation, it does not matter whether we start at time 0 or at some other time; the picture is the same due to the translation invariance of the autonomous equation.

2   Preparations

Consider again our nonautonomous deterministic (or random) difference equation together with an arbitrary but fixed *reference solution* $\mu : \mathbb{Z} \to \mathbb{R}^d$. We want to compute the unstable manifold of μ. μ can be a constant solution or a periodic solution or an arbitrary solution which exists for all times. For convenience we do assume unique existence of backwards solutions; for the noninvertible case see Rasmussen [9].

Now we apply two preparatory steps to the nonautonomous difference equation to get a simplified standard situation. First we transform the reference solution $\mu : \mathbb{Z} \to \mathbb{R}^d$ into the zero solution with the transformation $x \mapsto x - \mu_n$. This transformation changes our original equation to this new equation which is written as linear part A_n plus nonlinearity F_n

$$x_{n+1} = Df_n(\mu_n)x_n + [f_n(x_n + \mu_n) - f_n(\mu_n) - Df_n(\mu_n)x_n] = A_n x_n + F_n(x_n).$$

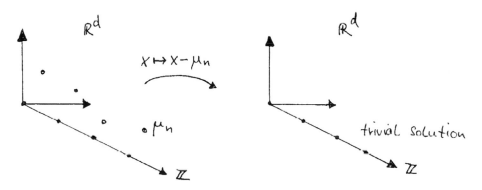

Figure 2: The transformed equation has the (trivial) zero solution.

Note that this simple transformation is useless in an autonomous framework, since the transformed equation with the zero solution is nonautonomous in general.

As a second preparatory step we globalize the system with a cutoff procedure which does not change the dynamics in a cylinder containing the zero solution. Thereto let $\chi : \mathbb{R} \to \mathbb{R}^+$ be a C^∞ bump function which equals 1 on an interval $[-r, r]$, $r > 0$. We replace our equation with the new equation

$$x_{n+1} = A_n x_n + F_n(x_n) \cdot \chi(\|x_n\|)$$

which has the same linear part but a modified nonlinearity which equals the old one in a cylinder and is zero outside. The effect of this cutoff procedure is that the influence of the nonlinearity is restricted to the area of interest, the cylinder, whereas the dynamics outside are those of the linear system.

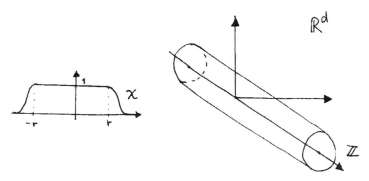

Figure 3: The bump function χ cuts off the nonlinearity outside the cylinder.

Instead of introducing new letters for the modified nonlinearity we will write again F_n instead of $F_n \cdot \chi$.

The solution of a nonautonomous equation is not a dynamical system. It is what I call a *nonautonomous dynamical system*. This is a relatively new notion and its birth was prepared by Ludwig Arnold, Peter Kloeden, Björn Schmalfuß, George Sell and many others.

Definition 2.1 (Nonautonomous Dynamical System (NDS)) *A (discrete) nonautonomous (semi)-dynamical system (NDS) on a metric space X with base set P consists of two ingredients:*

(i) a driving system $\theta : \mathbb{Z} \times P \to P$ modelling the nonautonomy; i.e., the family $\theta(n, \cdot) = \theta(n) : P \to P$ of self-mappings of P satisfies the group property

$$\theta(0) = \mathrm{id}_P \,, \quad \theta(n + l) = \theta(n) \circ \theta(l)$$

for all $n, l \in \mathbb{Z}$.

(ii) a cocycle $\varphi : \mathbb{Z}_0^+ \times P \times X \to X$ over θ; i.e., the family $\varphi(n, p, \cdot) = \varphi(n, p) : X \to X$ of self-mappings of X satisfies the cocycle property

$$\varphi(0, p) = \mathrm{id}_X \,, \quad \varphi(n + l, p) = \varphi(n, \theta(l)p) \circ \varphi(l, p)$$

for all $n, l \geq 0$ and $p \in P$. Moreover $x \mapsto \varphi(n, p, x)$ is continuous.

Remark 2.2 *(i) The set P is called* base *and is an arbitrary set, e.g., a compact topological space or a probability space. The driving system usually has additional regularity, e.g. it is continuous or ergodic.*

(ii) The pair of mappings

$$(\theta, \varphi) : \mathbb{Z}_0^+ \times P \times X \to P \times X \,, \quad (n, p, x) \mapsto (\theta(n, p), \varphi(n, p, x))$$

is a special semi-dynamical system, a so-called skew product flow *(usually one requires additionally that P is a topological space and that (θ, φ) is continuous). If $P = \{p\}$ consists of one point then the cocycle φ is a semi-dynamical system.*

(iii) We use the abbreviations $\theta^n p$ for $\theta(n, p)$ and $\varphi(n, p)x$ for $\varphi(n, p, x)$. We also say that φ is an NDS to abbreviate the situation of Definition 2.1.

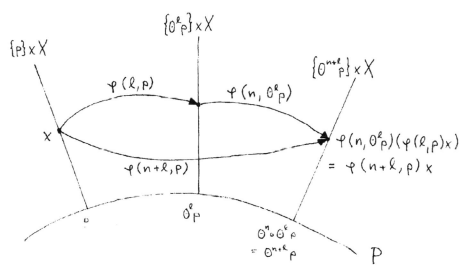

Figure 4: The cocycle φ over the driving system θ acts on the fibers.

The cocycle property is easy to understand with this figure. Think of the cocyle φ acting on the bundle $P \times X$.

How does a nonautonomous difference equation generate an NDS? By choosing different base spaces P and driving systems θ we have the choice to generate different NDS. I explain the simplest construction here which is sufficient for our purpose. Starting with the deterministic situation we take time, i.e., the integers as the base set $P := \mathbb{Z}$ and the driving system which models the nonautonomy is just the time shift $\theta : \mathbb{Z} \times P \to P$, $\theta^n l = n + l$. This driving system corresponds to the trick of making a nonautonomous difference equation autonomous by introducing a new variable for the time. To define the cocycle let $y(\cdot, l, x)$ be the solution starting at time l in $x \in \mathbb{R}^d$. Then our state space X is $\mathbb{R}^d$ and φ is just the solution with one difference: the first argument of φ is not the absolute time but it is the time which elapsed since we started at time l in x

$$\varphi : \mathbb{Z}_0^+ \times P \times X \to X, \qquad \varphi(n, l, x) := y(n + l, l, x).$$

In the random case the index ω_n could e.g. range between 1 and m and is an entry of a doubly-infinite sequence ω. For our numerical considerations it is not important but to generate a random dynamical system (RDS) which is an NDS with additional structure, one has to make the set Ω of doubly-infinite sequences ω a probability space. Thereto we could, e.g., give equal probability to each selection of a number between 1 and m and then extend this probability by Kolmogorov's theorem to doubly-infinite sequences to get the probability measure $(\delta_{1/m}(1) + \cdots + \delta_{1/m}(m))^{\mathbb{Z}}$. In any case the random difference equation generates an NDS in the following way. Take $\Omega := \{1, \ldots, m\}^{\mathbb{Z}}$ as the base set P consisting of all doubly-infinite sequences $\omega = (\ldots, \omega_{-1}, \omega_0, \omega_1, \ldots)$. The driving system $\theta : \mathbb{Z} \times P \to P$, $\theta^n \omega = (\ldots, \omega_{n-1}, \omega_n, \omega_{n+1}, \ldots)$ models the nonautonomy in the way that it chooses a new random number at every time step, and this is done by shifting the sequence to the left each time step, hence the entry of $\theta^n \omega$ at position 0 is ω_n. To define the cocycle let $z(\cdot, \omega, x)$ be the solution of (2) starting at time 0 in $x \in \mathbb{R}^d$. Then the state space is $X := \mathbb{R}^d$ and the cocycle

$$\varphi : \mathbb{Z}_0^+ \times P \times X \to X, \qquad \varphi(n, \omega, x) := z(n, \omega, x)$$

equals the solution. Having these constructions in mind we say that "the solution of a nonautonomous difference equation is a nonautonomous dynamical system".

An invariant manifold of a dynamical system is a subset of the state space. An invariant manifold of an NDS is a subset of the extended state space $P \times X$, a so-called *nonautonomous set*.

Definition 2.3 (Nonautonomous Set) *A family* $M = (M(p))_{p \in P}$ *of non-empty sets* $M(p) \subset X$ *is called a* nonautonomous set *and* $M(p)$ *is called the p-fiber of M or the* fiber *of M over p. We say that M is* closed, open, bounded, *or* compact, *if every fiber has the corresponding property. For notational convenience we use the identification* $M \simeq \{(p, x) : p \in P, x \in M(p)\} \subset P \times X$.

Definition 2.4 (Invariance of Nonautonomous Set) *A nonautonomous set M is called* forward invariant *under the NDS φ, if $\varphi(t, p)M(p) \subset M(\theta_t p)$ for $t \geq 0$ and $p \in P$. It is called* invariant, *if $\varphi(t, p)M(p) = M(\theta_t p)$ for $t \geq 0$ and $p \in P$.*

An invariant manifold of a difference equation that satisfies our assumptions is an invariant nonautonomous set which can be dynamically characterized as the set of solutions converging exponentially fast to 0 as time n tends to ∞ (stable manifold) or of solutions converging exponentially to 0 as n tends to $-\infty$ (unstable manifold), etc. In the deterministic situation the existence of the unstable manifold $W^u = (W^u(p))_{p \in P}$ is guaranteed (see, e.g., Aulbach, Pötzsche and Siegmund [2]) if the linear part A_n is hyperbolic, i.e., admits an exponential dichotomy, and the nonlinearity is Lipschitz

$$\|F_n(x) - F_n(y)\| \leq L\|x - y\|$$

with L small in terms of the hyperbolicity constants of A_n. Since we applied a cutoff procedure, the Lipschitz condition is in fact a local condition for our original equation.

In the random case the existence of the unstable manifold $W^u = (W^u(p))_{p \in P}$ follows from L. Arnold [1, Thm. 7.3.1] if the linear part A_{ω_n} satisfies the integrability condition of the multiplicative ergodic theorem and the nonlinearity is again Lipschitz.

A picture of an invariant manifold is a picture in the extended state space.

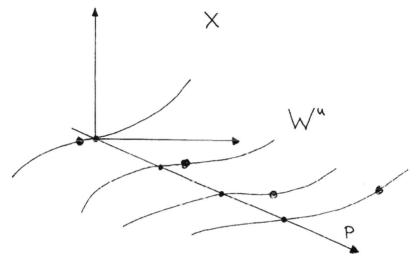

Figure 5: The unstable manifold consists of solutions which decay for $n \to -\infty$.

The unstable manifold W^u consists of fibers $W^u(p)$ indexed by the base points; i.e., in the deterministic case they are indexed by the integers and in the random case they are indexed by the random sequences, where the shifts of one fixed sequence ω can again be identified with $\mathbb{Z}$.

Recall that if D and A are nonempty closed sets in $\mathbb{R}^d$, the Hausdorff semi-distance $d(D|A)$ is defined by

$$d(D|A) := \sup_{x \in D} d(x, A), \quad d(x, A) := \inf_{y \in A} d(x, y) = \inf_{y \in A} \|x - y\|,$$

and the Hausdorff distance is $d_H(D, A) = d(D|A) + d(A|D)$.

Now we want to generalize the notion of an attracting set from dynamical systems to nonautonomous dynamical systems. A natural generalization of convergence to a nonautonomous set $A = (A(p))_{p \in P}$ seems to be the *forwards running convergence* defined by

$$d(\varphi(n, p)x, A(\theta^n p)) \to 0 \quad \text{for } n \to \infty.$$

However, this does not ensure convergence to a specific component set $A(p)$ for a fixed p. For that one needs to start "progressively earlier" at $\theta^{-n}p$ in order to "finish" at p. This leads to the concept of *pullback convergence* defined by

$$d(\varphi(n, \theta^{-n}p)x, A(p)) \to 0 \quad \text{for } n \to \infty.$$

Using this we can define a *pullback attractor*.

Definition 2.5 (Pullback Attractor) *Let φ be an NDS and A be a compact invariant nonautonomous set. Then A is called (global)* pullback attractor *if for every bounded set $D \subset X$ and $p \in P$*

$$\lim_{n \to \infty} d\left(\varphi(n, \theta^{-n}p)D \mid A(p)\right) = 0.$$

The concept of pullback convergence was introduced in the mid 1990s in the context of random dynamical systems (see Crauel and Flandoli [4], Flandoli and Schmalfuss [5], and Schmalfuss [10]) and has been used e.g. in numerical dynamics by Peter Kloeden (see e.g. [7]). Note that a similar idea had already been used in the 1960s by Mark Krasnoselski [8] to establish the existence of solutions that exist and remain bounded on the entire time set.

3   Box algorithm

The box algorithm is the idea of the numerics we want to do. To understand it take a step size $K \in \mathbb{N}$ and a compact set $Q \subset \mathbb{R}^d$; let us call it *main box*. It is partitioned into smaller *single boxes*. Interpret Q as a set in the p-fiber of the extended state space $P \times \mathbb{R}^d$. Take copies of Q in the $\theta^K p$-fiber, the $\theta^{2K} p$-fiber and so on. Then we start not with a single point in the p-fiber but with a collection B_I of single boxes in the main box Q. We map all the points in the selected single boxes K time steps forward with the cocycle $\varphi(K, p)$. The image has a (possibly empty) intersection with a minimal number of single boxes in the θ^K-fiber, and the collection B_J of these boxes is defined to be the image of the so-called *box NDS* $\hat\varphi$.

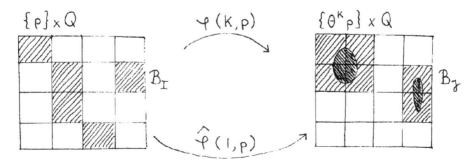

Figure 6: The box NDS $\hat\varphi$ maps a collection B_I of single boxes to B_J.

More precisely, we construct $\hat\varphi$ as an NDS in the following way.

Definition 3.1 (Box NDS) *Let* $\varphi : \mathbb{Z}_0^+ \times P \times \mathbb{R}^d \to \mathbb{R}^d$ *be an NDS. Choose a compact set* $Q \subset \mathbb{R}^d$ *(called* main box*), a step size* $K \in \mathbb{N}$ *and a finite collection* $\mathcal{B} = (B_i)_{i=1}^m$ *of connected, closed subsets* B_i *of* Q *(called* single boxes*) with:*

1. $\bigcup_{i=1}^m B_i = Q$,

2. $\text{int } B_i \cap \text{int } B_j = \emptyset$ *if* $i \neq j$, $1 \leq i, j \leq m$.

Then $\hat{\varphi} : \mathbb{Z}_0^+ \times P \times \mathcal{P}(\mathcal{B}) \to \mathcal{P}(\mathcal{B})$ *is the* box NDS *over* $\hat{\theta} = \theta^K$, *where*

$$\mathcal{P}(\mathcal{B}) := \{B_I = \cup_{i \in I} B_i : I \subset \{1, \ldots, m\}\}$$

and $\hat{\varphi}(n, p) = \hat{\varphi}(1, \theta^{(n-1)K} p) \circ \cdots \circ \hat{\varphi}(1, p)$ *is defined by*

$$\hat{\varphi}(1, p) B_I := B_J = \bigcup_{j \in J} B_j$$

with $J = \{j \in \{1, \ldots, m\} : \varphi(K, p) B_I \cap B_j \neq \emptyset\}$.

4 Main result

Theorem 4.1 (Box NDS approximates Unstable Manifold) *Let* $\varphi :$ $\mathbb{Z}_0^+ \times P \times \mathbb{R}^d \to \mathbb{R}^d$ *be an NDS with unstable manifold* $W^u = (W^u(p))_{p \in P}$ *generated by a nonautonomous deterministic or random difference equation which satisfies our assumptions. Let* $\hat{\varphi}$ *be a corresponding box NDS with main box* $Q \subset \mathbb{R}^d$. *Then for every accuracy* $\varepsilon > 0$ *and* $p \in P$ *there exists a box diameter* $\delta > 0$, *a step size* $K \in \mathbb{N}$ *and a minimal number* $n_0 \in \mathbb{N}$ *of computation steps such that for* $\text{diam} \mathcal{B} < \delta$ *and* $n \geq n_0$

$$d_H(\hat{\varphi}(n, \theta^{-n} p) Q, W^u(p) \cap Q) < \varepsilon.$$

Before we give the proof, let us look at an example.

Example 4.2 *Martin Rasmussen [9] extended the software package* GAIO *by Michael Dellnitz and Oliver Junge to the nonautonomous situation and he implemented the Box NDS* $\hat{\varphi}$. *The following series of pictures shows the approximations of the unstable manifold for the nonautonomous difference equation which we get from*

$$\begin{aligned} x_{n+1} &= .5x_n - .7y_n^2 \\ y_{n+1} &= 2y_n \end{aligned}$$

by transforming it with $\binom{x}{y} \mapsto T_n^{-1}\binom{x}{y}$ *where* $T_n = \begin{pmatrix} \cos .1n & -\sin .1n \\ \sin .1n & \cos .1n \end{pmatrix}$. *We start at time* -7 *and approximate the unstable manifold in the fiber at time* 7 *in the main box* $Q = [-4, 4]^2$ *with* $|\mathcal{B}| = 2^{18}$ *boxes of diameter* $\delta = 2.2 \cdot 10^{-2}$.

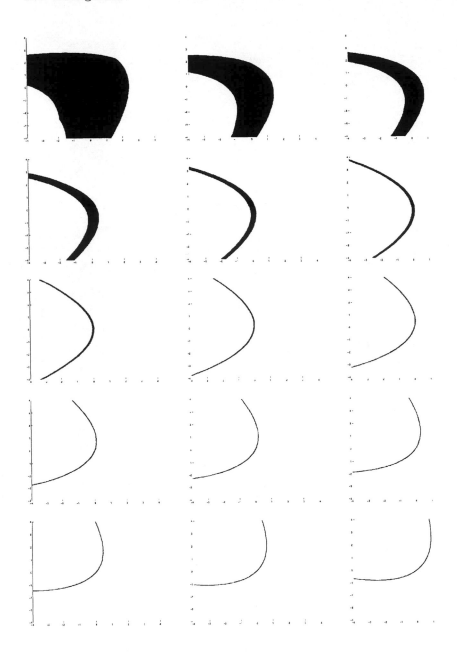

Figure 7: The iterates of Q under the box NDS $\hat{\varphi}$ from time -7 to 7.

Proof of Theorem 4.1. The first step of this proof is due to Martin Rasmussen and the details are part of his diploma thesis [9].

Step 1: The unstable manifold is pullback attracting

$$\lim_{n \to \infty} d\left(\varphi(n, \theta^{-n}p)Q|W^u(p)\right) = 0.$$

To see this one starts with an arbitrary x and shows that it approaches W^u uniformly in P. Now the main idea is to use the hyperbolicity of the equation in the following way: If x lies on the stable manifold then the whole cocycle through x also lies in the stable manifold; it converges to 0 and we have explicit, uniform estimates for the rate of convergence. If x does not lie in the stable manifold, it does not converge to 0 but we have the so-called stable foliation of the unstable manifold which yields a nonautonomous set consisting of all starting points such that the cocycle converges to the cocycle through x; this set is nothing other than the stable manifold of the cocycle through x and one can show that it has a unique intersection with the unstable manifold in each fiber (for this result see, e.g., Siegmund [11] or Aulbach and Wanner [3]). Hence the cocycle through x converges to the unstable manifold and we have explicit, uniform estimates for the rate of convergence. This leads to the pullback attractivity of the unstable manifold.

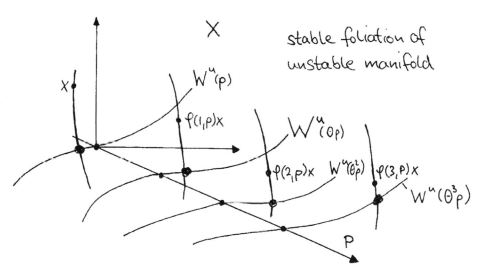

Figure 7: The stable foliation hits the unstable manifold in the fibers in exactly one point.

Step 2: For every $\varepsilon > 0$, $p \in P$ and $K \in \mathbb{N}$, there exist a $\delta > 0$ and $n_0 \in \mathbb{N}$ such that if diam $\mathcal{B} < \delta$ and $n \geq n_0$, then

$$d(\hat\varphi(n, \theta^{-nK}p)Q|W^u(p) \cap Q) < \varepsilon.$$

This result follows from H. Keller and G. Ochs [6] using the fact that W^u is pullback attracting.

Step 3: For $j = 1, \ldots, n_0$ we have

$$\varphi(K, \theta^{-jK} p)(A(\theta^{-jK} p) \setminus Q) \subset A(\theta^{-(j-1)K} p) \setminus Q. \tag{3}$$

This condition simply means that if the part of the fiber $W^u(\theta^{-jK} p)$ which is outside of Q is mapped K time steps forward by φ, then the image should remain outside the main box Q. It is fulfilled because the unstable manifold W^u consists of solutions which grow exponentially. Possibly after enlarging K (which does not affect the constants chosen in Step 2), we can always ensure that (3) is satisfied.

Now we show that under the condition (3)

$$d(W^u(p) \cap Q | \hat{\varphi}(n_0, \theta^{-n_0 K} p) Q) = 0$$

holds, since this implies that the *Hausdorff distance* between the numerical approximation and the unstable manifold is less than ε and we are finished.

To this end we prove for $l \in \{0, \ldots, n_0\}$:

$$W^u(\theta^{-(n_0-l)K} p) \cap Q \subset \hat{\varphi}(l, \theta^{-n_0 K} p) Q.$$

For $l = 0$ obviously $W^u(\theta^{-n_0 K} p) \cap Q \subset Q$. Now we assume the assertion for some $l \in \{0, \ldots, n_0 - 1\}$ and prove it for $l + 1$. Using the invariance of W^u and the fact that $f(A \cup B) = f(A) \cup f(B)$ we obtain

$$
\begin{aligned}
& W^u(\theta^{-(n_0-l-1)K} p) \cap Q \\
=\ & \varphi(K, \theta^{-(n_0-l)K} p) W^u(\theta^{-(n_0-l)K} p) \cap Q \\
=\ & \varphi(K, \theta^{-(n_0-l)K} p)[(W^u(\theta^{-(n_0-l)K} p) \cap Q) \cup (W^u(\theta^{-(n_0-l)K} p) \setminus Q)] \cap Q \\
=\ & [\varphi(K, \theta^{-(n_0-l)K} p)(W^u(\theta^{-(n_0-l)K} p) \cap Q) \\
& \cup \varphi(K, \theta^{-(n_0-l)K} p)(W^u(\theta^{-(n_0-l)K} p) \setminus Q)] \cap Q \\
\subset\ & [\varphi(K, \theta^{-(n_0-l)K} p)(W^u(\theta^{-(n_0-l)K} p) \cap Q) \cup (W^u(\theta^{-(n_0-l-1)K} p) \setminus Q)] \cap Q \\
=\ & \varphi(K, \theta^{-(n_0-l)K} p)(W^u(\theta^{-(n_0-l)K} p) \cap Q) \cap Q \\
\subset\ & \varphi(K, \theta^{-(n_0-l)K} p) \hat{\varphi}(l, \theta^{-n_0 K} p) Q \cap Q \\
\subset\ & \hat{\varphi}(1, \theta^{-(n_0-l)K} p) \hat{\varphi}(l, \theta^{-n_0 K} p) Q \cap Q \\
=\ & \hat{\varphi}(l+1, \theta^{-n_0 K} p) Q
\end{aligned}
$$

proving the assertion for $l+1$. For $l = n_0$ we get $W^u(p) \cap Q \subset \hat{\varphi}(n_0, \theta^{-n_0 K} p) Q$ and this implies

$$d(W^u(p) \cap Q | \hat{\varphi}(n_0, \theta^{-n_0 K} p) Q) = 0.$$

■

References

[1] L. Arnold, *Random Dynamical Systems*, Springer-Verlag, Berlin, Heidelberg, New York, 1998.

[2] B. Aulbach, C. Pötzsche and S. Siegmund, A smoothness theorem for invariant fiber bundles, *Journal of Dynamics and Differential Equations* **14** (2002), 519–547.

[3] B. Aulbach and T. Wanner, Invariant foliations for Carathédory-type differential equations in Banach spaces, in *Advances of Stability Theory at the End of 20th Century* (A. A. Martynyuk, Ed.) Taylor and Francis, London, 2003, 1–14.

[4] H. Crauel and F. Flandoli, Attractors for random dynamical systems, *Probab. Theory Relat. Fields* **100** (1994), 365–393.

[5] F. Flandoli and B. Schmalfuss, Random attractors for the 3d stochastic Navier Stokes equation with multiplicative white noise, *Stochastics and Stochastic Reports* **59** (1996), 12–45.

[6] H. Keller and G. Ochs, Numerical Approximation of Random Attractors, in: H. Crauel, M. Gundlach (Eds.), *Stochastic Dynamics*, Springer 1999.

[7] P. Kloeden, H. Keller and B. Schmalfuss Towards a Theory of Random Numerical Dynamics, in: H. Crauel, M. Gundlach (Eds.), *Stochastic Dynamics*, Springer 1999.

[8] M.A. Krasnoselski, The operator of translation along trajectories of differential equations, *Translations of Mathematical Monographs*, vol. 19, American Math. Soc., Providence, R.I., 1968.

[9] M. Rasmussen, Approximation von Attraktoren und invarianten Mannigfaltigkeiten nichtautonomer Systeme. Diploma Thesis. Universität Augsburg, 2002.

[10] B. Schmalfuss, The stochastic attractor of the stochastic Lorenz system in nonlinear dynamics: Attractor approximation and global behaviour, *Nonlinear Dynamics: Attractor Approximation and Global Behaviour.* ISAM '92 (N. Koksch, V. Reitmann, and T. Riedrich, Eds.), Technical University Dresden, 1992, 27–52.

[11] S. Siegmund, Spektraltheorie, glatte Faserungen und Normalformen für Differentialgleichungen vom Carathéodory-Typ. (German). *Augsburger Mathematisch-Naturwissenschaftliche Schriften* **30**, Wißner Verlag (1999).

Dynamic Equations on Time Scales

Exponential Functions and Laplace Transforms
for Alpha Derivatives 231
E. Akin-Bohner and M. Bohner

Integration on Measure Chains 239
B. Aulbach and L. Neidhart

Asymptotic Formulae for Dynamic Equations on Time Scales
with a Functional Perturbation 253
S. Castillo and M. Pinto

Oscillation of a Matrix Dynamic Equation on a Time Scale 267
L. Erbe

Continuous Dependence in Time Scale Dynamics 279
B. M. Garay, S. Hilger and P. E. Kloeden

On the Riemann Integration on Time Scales 289
G. Sh. Guseinov and B. Kaymakçalan

Cauchy Functions and Taylor's Formula for Time Scales $\mathbb{T}$ 299
R. J. Higgins and A. Peterson

Embedding a Class of Time Scale Dynamics into
O.D.E. Dynamics with Impulse Effect 309
J. López Fenner

An Oscillation Criterion for a Dynamic Sturm-Liouville Equation 317
Z. Opluštil and Z. Pospíšil

Two Perturbation Results for Semi-Linear Dynamic
Equations on Measure Chains 325
C. Pötzsche

Exponential Functions and Laplace Transforms for Alpha Derivatives

ELVAN AKIN–BOHNER and MARTIN BOHNER

Department of Mathematics, Florida Institute of Technology
Melbourne, Florida 32901, USA
E-mail: eakin@math.unl.edu, bohner@umr.edu

Abstract We introduce the exponential function for alpha derivatives on generalized time scales. We also define the Laplace transform that helps to solve higher order linear alpha dynamic equations on generalized time scales. If $\alpha = \sigma$, the Hilger forward jump operator, then our theory contains the theory of delta dynamic equations on time scales as a special case. If $\alpha = \rho$, the Hilger backward jump operator, then our theory contains the theory of nabla dynamic equations on time scales as a special case. Hence differential equations, difference equations (using the forward or backward difference operator), or q-difference equations (using the forward or backward q-difference operator) can be accommodated within our theory. We also present various properties of the Laplace transform and offer some examples.

Keywords Alpha derivative, Generalized time scale, Exponential function, Laplace transform

AMS Subject Classification 39A12, 39A13

1 Introduction

We consider *generalized time scales* $(\mathbb{T}, \alpha)$ as introduced in [1], i.e., $\mathbb{T} \subset \mathbb{R}$ is a nonempty set such that every Cauchy sequence in $\mathbb{T}$ converges to a point in $\mathbb{T}$ (with the possible exception of Cauchy sequences that converge to a finite infimum or supremum of $\mathbb{T}$), and α is a function that maps $\mathbb{T}$ into $\mathbb{T}$. A function $f : \mathbb{T} \to \mathbb{R}$ is called *alpha differentiable* at a point $t \in \mathbb{T}$ if there exists a number $f_\alpha(t)$, the so-called *alpha derivative* of f at t, with the property that for every $\varepsilon > 0$ there exists a neighborhood U of t such that

$$|f(\alpha(t)) - f(s) - f_\alpha(t)(\alpha(t) - s)| \le \varepsilon |\alpha(t) - s|$$

is true for all $s \in U$. If $\mathbb{T}$ is closed and $\alpha = \sigma$, the Hilger forward jump operator, then $f_\alpha = f^\Delta$ is the usual delta derivative (see [4, 6, 7]), which contains as special cases derivatives f' (if $\mathbb{T} = \mathbb{R}$) and differences Δf (if

$\mathbb{T} = \mathbb{Z}$). If $\mathbb{T}$ is closed and $\alpha = \rho$, the Hilger backward jump operator, then $f_\alpha = f^\nabla$ is the nabla derivative (see [3] and [4, Section 8.4]).

In this paper we consider linear alpha dynamic equations of the form

$$y_\alpha = p(t)y \quad \text{with} \quad 1 + p(t)\mu_\alpha(t) \neq 0,$$

where $\mu_\alpha(t) = \alpha(t) - t$ is the *generalized graininess*. If the initial value problem

$$y_\alpha = p(t)y, \quad y(t_0) = 1$$

has a unique solution, we denote it by $e_p(t, t_0)$ and call it the *generalized exponential function*. Note that e_p also depends on α, but we choose not to indicate this dependence as it should be clear from the context. The exponential function satisfies some properties, which are presented in the next section of this paper. Similarly as in [5], the exponential function may be used to define a *generalized Laplace transform*, which is helpful when solving higher order linear alpha dynamic equations with constant coefficients. We illustrate this technique with an example in the last section. This example features an α which neither satisfies $\alpha(t) \geq t$ for all $t \in \mathbb{T}$ nor $\alpha(t) \leq t$ for all $t \in \mathbb{T}$, and hence this example can not be accommodated in the existing literature on delta and nabla dynamic equations.

2   Alpha derivatives, exponentials, and Laplace transforms

For a function $f : \mathbb{T} \to \mathbb{R}$ we denote by f_α the alpha derivative as defined in the introductory section, and we also put $f^\alpha = f \circ \alpha$. Then the following rules (see [4, Section 8.3]) are valid:

- $f^\alpha = f + \mu_\alpha f_\alpha$;

- $(fg)_\alpha = fg_\alpha + f_\alpha g^\alpha$ ("Product Rule");

- $\left(\dfrac{f}{g}\right)_\alpha = \dfrac{f_\alpha g - f g_\alpha}{g g^\alpha}$ ("Quotient Rule").

We may use these rules to find

$$e_p(\alpha(t), t_0) = e_p^\alpha(t, t_0) = e_p(t, t_0) + \mu_\alpha(t)p(t)e_p(t, t_0),$$

i.e.,

- $e_p(\alpha(t), t_0) = [1 + p(t)\mu_\alpha(t)]\, e_p(t, t_0)$,

by putting $y(t) = e_p(t, t_0)e_q(t, t_0)$,

$$
\begin{aligned}
y_\alpha(t) &= e_p(t, t_0)q(t)e_q(t, t_0) + p(t)e_p(t, t_0)e_q(\alpha(t), t_0) \\
&= [p(t) + q(t) + \mu_\alpha(t)p(t)q(t)]\, y(t),
\end{aligned}
$$

i.e.,

- $e_p e_q = e_{p \oplus q}$, where $p \oplus q := p + q + \mu_\alpha pq$,

and by putting $y = e_p(t, t_0)/e_q(t, t_0)$,

$$y_\alpha(t) = \frac{p(t)e_p(t, t_0)e_q(t, t_0) - e_p(t, t_0)q(t)e_q(t, t_0)}{e_q(t, t_0)e_q(\alpha(t), t_0)}$$

$$= \frac{p(t) - q(t)}{1 + \mu_\alpha(t)q(t)}y(t),$$

i.e.,

- $\dfrac{e_p}{e_q} = e_{p \ominus q}$, where $p \ominus q := \dfrac{p - q}{1 + \mu_\alpha q}$.

Note also that $\ominus q := 0 \ominus q = -q/(1 + \mu_\alpha q)$ satisfies $q \oplus (\ominus q) = 0$ and that $p \ominus q = p \oplus (\ominus q)$. Again, $\oplus$ and $\ominus$ depend on α, but in order to avoid many subscripts we choose not to indicate this dependence as it should be clear from the context. We also remark that the set of *alpha regressive* functions

$$\mathcal{R}_\alpha = \{p : \mathbb{T} \to \mathbb{R}| \ 1 + p(t)\mu_\alpha(t) \neq 0 \text{ for all } t \in \mathbb{T}\}$$

is an Abelian group under the addition $\oplus$, and $\ominus p$ is the additive inverse of $p \in \mathcal{R}_\alpha$.

Now, similarly as in [5], the Laplace transform for functions $x : \mathbb{T} \to \mathbb{R}$ (from now on we assume that $\mathbb{T}$ is unbounded above and contains 0) may be introduced as

$$\mathcal{L}\{x\}(z) = \int_0^\infty x(t)e^\alpha_{\ominus z}(t, 0)d_\alpha t \quad \text{with} \quad z \in \mathcal{R}_\alpha \cap \mathbb{R},$$

whenever this Cauchy alpha integral is well defined. As an example, we calculate $\mathcal{L}\{e_c(\cdot, 0)\}$, where $c \in \mathcal{R}_\alpha$ is a constant such that $\lim_{t \to \infty} e_{c \ominus z}(t, 0) = 0$. Then

$$\mathcal{L}\{e_c(\cdot, 0)\}(z) = \int_0^\infty e_c(t, 0)e^\alpha_{\ominus z}(t, 0)d_\alpha t$$

$$= \int_0^\infty [1 + \mu_\alpha(t)(\ominus z)(t)] \, e_c(t, 0)e_{\ominus z}(t, 0)d_\alpha t$$

$$= \int_0^\infty \left[1 - \frac{\mu_\alpha(t)z}{1 + \mu_\alpha(t)z}\right] e_{c \ominus z}(t, 0)d_\alpha t$$

$$- \int_0^\infty \frac{1}{1 + \mu_\alpha(t)z}e_{c \ominus z}(t, 0)d_\alpha t$$

$$= \frac{1}{c - z}\int_0^\infty (c \ominus z)(t)e_{c \ominus z}(t, 0)d_\alpha t$$

$$= \frac{1}{c - z}\int_0^\infty (e_{c \ominus z}(\cdot, 0))_\alpha d_\alpha t$$

$$= \frac{1}{z - c}.$$

Under appropriate assumptions we can also show

- $\mathcal{L}\{x_\alpha\}(z) = z\mathcal{L}\{x\}(z) - x(0);$
- $\mathcal{L}\{x_{\alpha\alpha}\}(z) = z^2\mathcal{L}\{x\}(z) - zx(0) - x_\alpha(0);$
- $\mathcal{L}\{X\}(z) = \dfrac{1}{z}\mathcal{L}\{x\}(z),$ where $X(t) = \displaystyle\int_0^t x(\tau)d_\alpha\tau.$

Further results can be derived as in [4, Section 3.10].

3   An example

To illustrate the use of our Laplace transform, we consider the initial value problem

$$y_{\alpha\alpha} - 5y_\alpha + 6y = 0, \quad y(0) = 1, \quad y_\alpha(0) = 5.$$

By formally taking Laplace transforms, we find

$$\begin{aligned} 0 &= z^2\mathcal{L}\{y\}(z) - zy(0) - y_\alpha(0) - 5\left[z\mathcal{L}\{y\}(z) - y(0)\right] + 6\mathcal{L}\{y\}(z) \\ &= (z^2 - 5z + 6)\mathcal{L}\{y\}(z) - z \\ &= (z-2)(z-3)\mathcal{L}\{y\}(z) - z \end{aligned}$$

so that

$$\mathcal{L}\{y\}(z) = \frac{z}{(z-2)(z-3)} = \frac{3}{z-3} - \frac{2}{z-2} = \mathcal{L}\{3e_3(\cdot,0) - 2e_2(\cdot,0)\}(z).$$

Hence, if $e_2(\cdot,0)$ and $e_3(\cdot,0)$ exist, we let

$$y = 3e_3(\cdot,0) - 2e_2(\cdot,0),$$

and then

$$y_\alpha = 9e_3(\cdot,0) - 4e_2(\cdot,0) \quad \text{and} \quad y_{\alpha\alpha} = 27e_3(\cdot,0) - 8e_2(\cdot,0)$$

so that indeed $y(0) = 3 - 2 = 1$, $y_\alpha(0) = 9 - 4 = 5$, and $y_{\alpha\alpha} - 5y_\alpha + 6y = 0$. Let us now consider several special cases of this example.

(a) $\mathbb{T} = \mathbb{R}$ and $\alpha(t) = t$ for all $t \in \mathbb{T}$. Then $e_c(t,0) = e^{ct}$ for any constant $c \in \mathbb{R}$, and the solution is given by

$$y(t) = 3e^{3t} - 2e^{2t}.$$

(b) $\mathbb{T} = \mathbb{N}_0$ and $\alpha(t) = 2t + 1$ for all $t \in \mathbb{T}$. Note that $e_c(\cdot,0)$ is only defined on

$$\{t_m = 2^m - 1 \mid m \in \mathbb{N}_0\} \subset \mathbb{T}.$$

Since $\mu_\alpha(t) = t + 1$, we find that e_c satisfies

$$e_c(t_{k+1},0) = (1 + c\mu_\alpha(t_k))e_c(t_k,0) = (1 + c2^k)e_c(t_k,0),$$

Table 1: $y = 3e_3(\cdot, 0) - 2e_2(\cdot, 0)$ for (b)

t	$e_2(t,0)$	$e_3(t,0)$	$y(t)$	$y_\alpha(t)$	$y_{\alpha\alpha}(t)$
0	1	1	1	5	19
1	3	4	6	24	84
3	15	28	54	192	636
7	135	364	822	2736	8748
15	2295	9100	22710	72720	
31	75735	445900	1186230		

Table 2: $y = 3e_3(\cdot, 0) - 2e_2(\cdot, 0)$ for (c)

t	$e_2(t,0)$	$e_3(t,0)$	$y(t)$	$y_\alpha(t)$	$y_{\alpha\alpha}(t)$
-6	-27	-125	-321		
-4	9	25	57	189	
-2	-3	-5	-9	-33	-111
0	1	1	1	5	19
2	$-0.\bar{3}$	-0.2	$0.0\bar{6}$	$-0.4\bar{6}$	$-2.7\bar{3}$
4	$0.\bar{1}$	0.04	$-0.10\bar{2}$	$-0.08\bar{4}$	$0.19\bar{1}$

and hence we obtain for constant $c \in \mathcal{R}_\alpha$

$$e_c(t_m, 0) = \prod_{k=0}^{m-1} (1 + c2^k).$$

See Table 1 for some numeric values. Note that $y_{\alpha\alpha} - 5y_\alpha + 6y = 0$ in each row.

(c) $\mathbb{T} = \mathbb{Z}$ and $\alpha(t) = t - 2$ for all $t \in \mathbb{T}$. Note that $e_c(t, 0)$ is only defined for all even integers. Since $\mu_\alpha(t) \equiv -2$, we find that e_c satisfies

$$e_c(\alpha(t), 0) = (1 + c\mu_\alpha(t))e_c(t, 0) = (1 - 2c)e_c(t, 0),$$

and hence we obtain for constant $c \neq 1/2$

$$e_c(t, 0) = (1 - 2c)^{-t/2}.$$

See Table 2 for some numeric values.

(d) $\mathbb{T} = \mathbb{Z}$ and $\alpha(t) = t + 1 + 2(-1)^t$ for all $t \in \mathbb{T}$. Note that while in the previous two examples $\alpha(t) \geq t$ for all $t \in \mathbb{T}$ and $\alpha(t) \leq t$ for all $t \in \mathbb{T}$, respectively, none of these properties hold in the current example. This time $\alpha : \mathbb{T} \to \mathbb{T}$ is additionally a bijection and hence $e_c(t, 0)$ is defined on the entire set $\mathbb{T}$ when $c \notin \{-1/3, 1\}$. We have

$$\mu_\alpha(t) = 1 + 2(-1)^t = \begin{cases} 3 & \text{if } t \text{ is even} \\ -1 & \text{if } t \text{ is odd.} \end{cases}$$

Table 3: $y = 3e_3(\cdot, 0) - 2e_2(\cdot, 0)$ for (d)

t	$e_2(t,0)$	$e_3(t,0)$	$y(t)$	$y_\alpha(t)$	$y_{\alpha\alpha}(t)$
0	1	1	1	5	19
1	-1	-0.5	0.5	-0.5	-5.5
2	-7	-20	-46	-152	-484
3	7	10	16	62	214
4	49	400	1102	3404	10408
5	-49	-200	-502	-1604	-5008
6	-343	-8000	-23314	-70628	
7	343	4000	11314	34628	
8	2401	160000	475198		
9	-2401	-80000	-235198		

As an example we calculate $e_2(t, 0)$ for some values of t. Since $e_2(0,0) = 1$ and $\alpha(0) = 3$, we find

$$e_2(3, 0) = e_2(\alpha(0), 0) = (1 + 2\mu_\alpha(0))e_2(0, 0) = 7.$$

Next,

$$e_2(2, 0) = e_2(\alpha(3), 0) = (1 + 2\mu_\alpha(3))e_2(3, 0) = -7,$$

and similarly,

$$e_2(5, 0) = -7^2, \quad e_2(4, 0) = 7^2, \quad e_2(7, 0) = 7^3, \quad e_2(6, 0) = -7^3$$

and so on. In general, we find

$$e_c(t, 0) = \begin{cases} \{(1-c)(1+3c)\}^{t/2} & \text{if } t \text{ is even} \\ (1-c)^{(t-3)/2}(1+3c)^{(t-1)/2} & \text{if } t \text{ is odd,} \end{cases}$$

which can be written in closed formula as

$$e_c(t, 0) = \frac{\{(1-c)(1+3c)\}^{\lfloor t/2 \rfloor}}{(1-c)^{\chi(t)}} = \frac{\{(1-c)(1+3c)\}^{\lfloor t/2 \rfloor}}{(1-c)^{\lceil t/2 \rceil - \lfloor t/2 \rfloor}},$$

where $\chi = \chi_{2\mathbb{Z}+1}$ is the characteristic function for the odd integers. Again we refer to Table 3 for some numeric values.

References

[1] C. D. Ahlbrandt, M. Bohner, and J. Ridenhour. Hamiltonian systems on time scales. *J. Math. Anal. Appl.*, 250:561–578, 2000.

[2] E. Akın, L. Erbe, B. Kaymakçalan, and A. Peterson. Oscillation results for a dynamic equation on a time scale. *J. Differ. Equations Appl.*, 7:793–810, 2001.

[3] F. M. Atıcı and G. Sh. Guseinov. On Green's functions and positive solutions for boundary value problems on time scales. *J. Comput. Appl. Math.*, 2002. Special Issue on "Dynamic Equations on Time Scales", edited by R. P. Agarwal, M. Bohner, and D. O'Regan. To appear.

[4] M. Bohner and A. Peterson. *Dynamic Equations on Time Scales: An Introduction with Applications*. Birkhäuser, Boston, 2001.

[5] M. Bohner and A. Peterson. Laplace transform and Z-transform: Unification and extension. *Methods Appl. Anal.*, 2002. To appear.

[6] S. Hilger. Analysis on measure chains — a unified approach to continuous and discrete calculus. *Results Math.*, 18:18–56, 1990.

[7] B. Kaymakçalan, V. Lakshmikantham, and S. Sivasundaram. *Dynamic Systems on Measure Chains*, volume 370 of *Mathematics and its Applications*. Kluwer Academic Publishers, Dordrecht, 1996.

[8] W. G. Kelley and A. C. Peterson. *Difference Equations: An Introduction with Applications*. Academic Press, San Diego, second edition, 2001.

Integration on Measure Chains

BERND AULBACH and LUDWIG NEIDHART

Department of Mathematics, University of Augsburg

D-86135 Augsburg, Germany

E-mail: aulbach@math.uni-augsburg.de

neidhart@math.uni-augsburg.de

Abstract In its original form the calculus on measure chains is mainly a differential calculus. The notion of integral being used, the so-called Cauchy integral, is defined by means of antiderivatives and, therefore, it is too narrow for the development of a full infinitesimal calculus. In this paper, we present several other notions of integral such as the Riemann, the Cauchy-Riemann, the Borel and the Lebesgue integral for functions from a measure chain to an arbitrary real or complex Banach space. As in ordinary calculus, of those notions only the Lebesgue integral provides a concept which ensures the extension of the original calculus on measure chains to a full infinitesimal calculus including powerful convergence results and complete function spaces.

Keywords Measure chain, Time scale, Cauchy integral, Riemann integral, Cauchy-Riemann integral, Borel integral, Lebesgue integral, Bochner integral

AMS Subject Classification 26A42, 28A25, 28B05, 39A10, 39A12

1 Introduction

The creation of *measure chains* was motivated by the longstanding desire to have some means that make it possible to treat problems arising in the theory of differential and/or difference equations in a unified way. With this aim in mind, Stefan Hilger introduced in his PhD thesis [7] (see also [8]) the concept of a *measure chain* and developed a rather complete theory of differentiation for functions which are defined on a measure chain (or a subset of it) and have their values in an arbitrary real or complex Banach space. As far as integration of those functions is concerned, he mentioned the possibility of a measure theoretic approach; however, in favor of an application of his theory to a prototype problem on invariant manifolds, he confined his study to a basic notion of integral which is simply defined by means of antiderivatives. The corresponding notion of integral, the so-called *Cauchy integral*, turned out to be sufficient for the achievement of the original goals, and, even in the long run, it has proved to be a successful concept leading to many research activities providing interesting new results in the field of dynamic equations on measure chains (among many others, ten papers in this volume).

While most of these activities focus on *applications* of the calculus on measure chains to dynamic equations, the *foundations* of this calculus have obtained only minor attention. In some papers on dynamic equations, the calculus itself has been promoted, but only as much as it was needed for a certain purpose (for the *chain rule*, e.g., see Keller [10] and Pötzsche [13]). As to our knowledge, the only papers (or theses) dealing with the *theory of integration* on measure chains are Sailer [14], Neidhart [12], Guseinov and Kaymakçalan [9] and Bohner and Peterson [4, Chapter 5].

In this paper, we give a survey of the main results contained in the diploma thesis [12] of the second author. In fact, we outline the definitions of the Cauchy, the Riemann, the Cauchy-Riemann, the Borel and the Lebesgue integral for functions from a measure chain into an arbitrary real or complex Banach space; and we briefly sketch their mutual interrelations. For further information on this topic we refer the reader to Neidhart [12].

2   Basic fact about measure chains

For the reader's convenience we briefly state some facts on measure chains which are used in this paper. For more details we refer to Hilger [8] and Bohner and Peterson [3, Section 8.1].

A *measure chain* is a triple $(\mathbb{T}, \leq, \mu)$ consisting of a set $\mathbb{T}$, a relation $\leq$ on $\mathbb{T}$ and a function $\mu : \mathbb{T} \times \mathbb{T} \to \mathbb{R}$, the so-called *growth calibration*, such that the following axioms hold:

Axioms on $\leq$:

(Reflexivity)	$\forall\, x \in \mathbb{T}:$	$x \leq x$
(Antisymmetry)	$\forall\, x, y \in \mathbb{T}:$	$x \leq y \leq x \;\Rightarrow\; x = y$
(Transitivity)	$\forall\, x, y, z \in \mathbb{T}:$	$x \leq y \leq z \;\Rightarrow\; x \leq z$
(Totality)	$\forall\, x, y, z \in \mathbb{T}:$	$x \leq y \;$ or $\; y \leq x$
(Completeness)	Any non-void subset of $\mathbb{T}$ which is bounded above has a least upper bound	

Axioms on μ :

(Cocycle property)	$\forall\, x, y, z \in \mathbb{T}:$	$\mu(x, y) + \mu(y, z) = \mu(x, z)$
(Strong isotony)	$\forall\, x, y \in \mathbb{T}:$	$x > y \;\Rightarrow\; \mu(x, y) > 0$
(Continuity)	μ is continuous	

In order to fix (possibly ambiguous) notation we denote by $\sigma : \mathbb{T} \to \mathbb{T}$ the *forward jump operator*, i.e., $\sigma(t) := \inf\{s \in \mathbb{T} : s > t\}$ on $\mathbb{T}$, and by $\rho : \mathbb{T} \to \mathbb{T}$ the *backward jump operator*, i.e., $\rho(t) := \sup\{s \in \mathbb{T} s < t\}$. Thus, a point $t \in \mathbb{T}$ which is not a maximum of $\mathbb{T}$ is called *right-scattered* if $\sigma(t) > t$, *right-dense* if $\sigma(t) = t$, *left-scattered* if $\rho(t) < t$ and *left-dense* if $\rho(t) = t$. Finally, by $\mathbb{T}^\kappa$ we denote the set $\{t \in \mathbb{T} : t$ is no isolated maximum of $\mathbb{T}\}$ and by $\mu^*(t) := \mu(\sigma(t), t)$ the *graininess* $\mu^* : \mathbb{T} \to [0, \infty)$ of $\mathbb{T}$.

Looking for examples of measure chains it is apparent that subsets of the real line together with the usual ordering $\leq$ and $\mu(s, r) := s - r$ are good candidates. The question of which of those sets are in fact measure chains is answered as follows (see Hilger [8, Theorem 1.5.1]):

Theorem 2.1 *A subset $\mathbb{T}$ of $\mathbb{R}$ is a measure chain (with respect to the usual ordering $\leq$ of $\mathbb{R}$ and $\mu(s, r) = s - r$) if and only if $\mathbb{T}$ has the form $\mathbb{T} = I \setminus O$ where I is an interval and O an open subset of $\mathbb{R}$.*

Thus, real *intervals* and *closed subsets* of $\mathbb{R}$ are important examples of measure chains. On the other hand, they are more than just *examples*; they are to some extent representative. In order to see this, the notion of an isomorphism between measure chains can be introduced by saying that two measure chains T_1 and T_2 with respective growth calibrations μ_1 and μ_2 are *isomorphic* if there exists a bijection $f : \mathbb{T}_1 \to \mathbb{T}_2$ such that $\mu_2(f(s), f(r)) = \mu_1(s, r)$ for all $r, s \in \mathbb{T}_1$. With this notion at hand, the following can be said (see Hilger [8, Theorem 2.1]).

Theorem 2.2 *Each measure chain is isomorphic to a set of the form $I \setminus O$ where I is a real interval and O an open subset of $\mathbb{R}$.*

It is due to this theorem that measure chains are usually considered to be subsets of $\mathbb{R}$ and, even more, that they are *closed* subsets of $\mathbb{R}$. In fact, in most of the literature a measure chain is by definition a closed subset of $\mathbb{R}$ and, as such, it is called a *time scale*. While this is legitimate and (perhaps) helpful for visualizing measure chains, from the point of view of mathematical aesthetics the original definition seems more appropriate.

Before we dwell on the topic of this paper we introduce a means to classify measure chains according to their orders, being dense or not (see Neidhart [12, Definition 19]). This distinction is needed in Section 5 on the Cauchy-Riemann integral.

Definition 2.3 *A measure chain $(\mathbb{T}, \leq, \mu)$ is called* densely ordered *if for any two points $x, z \in \mathbb{T}$ there exists a point $y \in \mathbb{T}$ such that $x \leq y \leq z$ and $x \neq y$ as well as $y \neq z$.*

The following theorem is proved in Neidhart [12, Theorem 80].

Theorem 2.4 *A measure chain is densely ordered if and only if it is isomorphic to a real interval.*

This theorem tells us that the densely ordered measure chains are, in a way, trivial. The reason is that the calculus on measure chains reduces to the ordinary calculus of real numbers if the underlying measure chain is isomorphic to an interval.

3   The Cauchy integral

The notion of integral which is commonly used in the literature on measure chains or time scales is the one introduced by Hilger [7], so-called Cauchy integral. In order to recall its definition we have to introduce two notions. To this end let $\mathbb{T}$ be an arbitrary measure chain and $\mathcal{Y}$ an arbitrary real or complex Banach space.

Definition 3.1 *A function $f : \mathbb{T} \to \mathcal{Y}$ is called* regulated *if its left-sided limits exist in all left-dense points of $\mathbb{T}$ and its right-dense limits exist in all right-dense points of $\mathbb{T}$.*

Definition 3.2 *A function $F : \mathbb{T} \to \mathcal{Y}$ is a called a* pre-antiderivative *of a function $f : \mathbb{T} \to \mathcal{Y}$ if F is continuous and if there exists a set $D \subseteq \mathbb{T}^\kappa$ such that $\mathbb{T}^\kappa \setminus D$ is countable, contains no right-scattered points and has the property that the restriction of F to D is differentiable with derivative f.*

The crucial relation between regulated functions and pre-antiderivatives is described in the following theorem (see Hilger [8, Theorem 4.2]).

Theorem 3.3 *If $f : \mathbb{T} \to \mathcal{Y}$ is a regulated function, then there exists at least one pre-antiderivative $F : \mathbb{T} \to \mathcal{Y}$ of f. Moreover, for any two $a, b \in \mathbb{T}$ the difference $F(a) - F(b)$ does not depend on the choice of F.*

With this theorem at hand it is straightforward to define an integral for regulated functions.

Definition 3.4 *For any regulated function $f : [a, b] \to \mathcal{Y}$ from an interval of a measure chain $\mathbb{T}$ to a Banach space $\mathcal{Y}$, the* Cauchy integral *is defined by*

$$\int_a^b f(x) \, \Delta x := F(b) - F(a)$$

where $F : [a, b] \to \mathcal{Y}$ is any pre-antiderivative of f on $[a, b]$.

Since the Cauchy integral is widely known and extensively used in the literature we do not dwell on it any further. We rather want to mention that the main *advantage* of this integral is its simplicity, i.e., its simple derivation from the concept of differentiation. No "construction" of the integral by means of some limiting process is necessary. On general measure chains, however, this advantage is somewhat obscured by the fact that pre-antiderivatives are involved, objects which are quite subtle and not easy to handle.

The main *disadvantage* of the Cauchy integral is, that the set of functions which are integrable, i.e., the set of regulated functions, is too small. In fact, as known from ordinary calculus, the set of regulated functions is even a proper subset of the set of Riemann integrable functions, not to mention the set of Lebesgue integrable functions.

4   The Riemann integral

In ordinary calculus, the Riemann integral of a real-valued function is usually defined by either using *Riemann* sums or upper and lower *Darboux* sums. If the values of the function under consideration lie in an arbitrary Banach space, however, the Darboux sum approach cannot be used because of the lacking order structure of a general Banach space. Since, in this paper, we deal with Banach space-valued functions, we therefore mimic the Riemann sum approach.

To this end we first explain what we mean by saying that, for some $\delta > 0$, a *partition* $Z = (a_0, \ldots, a_n)$ of the interval $[a, b]$ *is finer than* δ. In fact, we mean that for each $i = 1, \ldots, n$ we have

either $\mu(a_i, a_{i-1}) \leq \delta$ or both $\mu(a_i, a_{i-1}) > \delta$ and $a_i = \sigma(a_{i-1})$.

With this notion at hand we can define the integrability of a Banach space-valued function in the sense of Riemann.

Definition 4.1 *A function $f : [a, b] \to \mathcal{Y}$ from an interval $[a, b]$ of a measure chain $\mathbb{T}$ to Banach space $\mathcal{Y}$ is called* Riemann integrable *if there exists a $y \in \mathcal{Y}$ such that for any $\varepsilon > 0$ there is a $\delta > 0$ with the following property: For any partition $Z = (a_0, \ldots, a_n)$ of $[a, b]$ which is finer than δ and any set of points $y_1, \ldots, y_n \in \mathcal{Y}$ with $y_j \in [a_{j-1}, a_j)$ for $j = 1, \ldots, n$ one has*

$$\left\| y - \sum_{j=1}^{n} f(y_j)\, \mu(a_j, a_{j-1}) \right\| \leq \varepsilon.$$

Using the fact that for any $\delta > 0$ there always exists a partition of $[a, b]$ which is finer than δ, we easily get the following result.

Theorem 4.2 *If $f : [a, b] \to \mathcal{Y}$ is Riemann integrable according to Definition 4.1, then the $y \in \mathcal{Y}$ appearing in this definition is uniquely determined, and it is called the* Riemann integral *of f, in signs*

$$y = \int_a^b f(x)\, dx.$$

The *advantage* of the Riemann integral over the Cauchy integral is that the set of Riemann integrable functions is definitively larger than the set of Cauchy integrable functions, i.e. the set of regulated functions. In fact, from ordinary calculus we know (see, e.g., Elstrodt [6, IV, Theorem 6.1]) that a real- or complex-valued bounded function is Riemann integrable if and only if it is Lebesgue-almost everywhere continuous.

The various *disadvantages* of the Riemann integral are well known from ordinary calculus. Apart from the fact that the set of Riemann integrable functions is too small (compared to the set of Lebesgue integrable functions) we just mention the lack of reasonable convergence results.

For more on the Riemann integral on time scales we refer to Sailer [14], Guseinov and Kaymakçalan [9], Bohner and Peterson [4, Chapter 5].

5   The Cauchy-Riemann integral

In ordinary calculus, the Cauchy-Riemann integral (cf. Amann and Escher [1, Chapter VI.3]) is defined for functions $f : I \to \mathcal{Y}$ where I is a compact interval and $\mathcal{Y}$ a Banach space. The idea underlying this concept of integral is first to assign to any step function (more precisely, to any function which is constant on any *open* interval of a partition of I; we call such a function therefore an *o-step function*) an integral value, and then to extended the set of integrable functions from the Banach space $St_o(I, \mathcal{Y})$ of o-step functions to the Banach space $\mathcal{R}(I, \mathcal{Y})$ of regulated functions. This extension is based on the *Linear Extension Theorem* of Functional Analysis (cf. [1, Theorem 2.6] or [11, Theorem 3.1]) which reads as follows:

Theorem 5.1 *Let $\mathcal{X} \neq \{0\}$ and $\mathcal{Y}$ be normed linear spaces and $\mathcal{D} \neq \{0\}$ a subspace of $\mathcal{X}$ which is dense in $\mathcal{X}$. Then, for any bounded linear operator $A : \mathcal{D} \to \mathcal{Y}$ there is exactly one bounded linear operator $\overline{A} : \mathcal{X} \to \mathcal{Y}$ which is an extension of A. Moreover, for all $x_0 \in \mathcal{X}$ one has $\overline{A}(x_0) = \lim_{x \to x_0} A(x)$.*

This theorem can be applied to the setting where (within the Banach space $\mathcal{B}(I, \mathcal{Y})$ of bounded functions from I to $\mathcal{Y}$) one has $\mathcal{D} = St_o(I, \mathcal{Y})$, $\mathcal{X} = \mathcal{R}(I, \mathcal{Y})$ and A is the integral operator which assigns to each o-step function its canonical integral value. This situation is depicted in the following diagram.

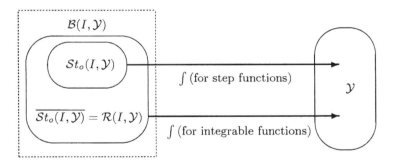

In order to carry over this situation from ordinary to the measure chain calculus we call a function f from an interval $I = [a, b]$ of a measure chain $\mathbb{T}$ to a Banach space $\mathcal{Y}$ an *o-step function* if there exists a partition $Z = (a_0, \ldots, a_n)$ of $[a, b]$ such that the restriction of f to any of the *open* intervals (a_{i-1}, a_i), $i = 1, \ldots, n$, is constant. The set of o-step functions $f : I \to \mathcal{Y}$ is denoted by $St_o(I, \mathcal{Y})$, and the canonical integral operator $\int_{(Z)} : St_o(I, \mathcal{Y}) \to \mathcal{Y}$ is defined by means of

$$\int_{(Z)} f := \sum_{i=1}^{n} m_i \, \mu(a_i, a_{i-1}) \tag{1}$$

where m_i is the constant value of f on the open interval (a_{i-1}, a_i) and $\mu(\cdot, \cdot)$ the growth calibration of $\mathbb{T}$.

If $\mathbb{T}$ is *densely ordered*, it can be shown along the lines of ordinary calculus that the value (1) is independent of the choice of the partition Z. However, if $\mathbb{T}$ is *not densely ordered*, the integral value (1) may change with the partition, and even may fail to be well defined (if at least one of the open intervals (a_{i-1}, a_i) is empty). That, in fact, one may have $\int_{(Z)} f \neq \int_{(U)} f$ for different partitions Z and U can be seen by means of the following simple example.

Example 5.2 On the measure subchain $\{0, 1, 2, 3, 4, 5\}$ of $\mathbb{R}$ we consider the real-valued function f whose values are $f(0) = f(1) = f(2) = 10$ and $f(3) = f(4) = f(5) = 1$. Then f is an o-step function with respect to the two partitions $Z := (0, 2, 5)$ and $U := (0, 3, 5)$, and the corresponding integral values $\int_{(Z)} f = 2 \cdot 10 + 3 \cdot 1 = 23$ and $\int_{(U)} f = 3 \cdot 10 + 2 \cdot 1 = 32$ are different.

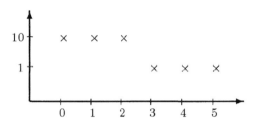

If, in contrast to the previous example, a measure chain is densely ordered, then the Cauchy-Riemann integral can be introduced as in ordinary calculus using the Extension Theorem 5.1. A summary of this approach using o-step functions is as follows (see Neidhart [12, Section 8.1]).

Theorem 5.3 *If $\mathbb{T}$ is a densely ordered measure chain and $f : [a, b] \rightarrow \mathcal{Y}$ an o-step function from an interval $[a, b] \subseteq \mathbb{T}$ to a Banach space $\mathcal{Y}$, then the integral (1) is well-defined and independent of the partition Z of $\mathbb{T}$. Moreover, for any function $f \in \mathcal{S}t_o(I, \mathcal{Y})$ the Cauchy-Riemann integral is defined as*

$$\int_a^b f(x)\,dx = \lim_{n \to \infty} \int_{(Z)} f_n \tag{2}$$

where $(f_n)_{n \in \mathbb{N}}$ is any sequence of o-step functions which converges to f uniformly on $[a, b]$.

On the other hand, if $\mathbb{T}$ is not densely ordered, the integral (1) may not exist or may depend on the choice of the partition Z, and hence the limit (2) may not exist.

The difficulties with the o-step functions (whose use was motivated from ordinary calculus) can be avoided if one modifies the kind of step functions being used. To this end we introduce the notion of an *h-step function* for any function f from an interval $I = [a, b]$ of a measure chain $\mathbb{T}$ to a Banach space $\mathcal{Y}$, if there exists a partition $Z = (a_0, \ldots, a_n)$ of $[a, b]$ such that f is constant on any *half-open* interval $[a_{i-1}, a_i)$, $i = 1, \ldots, n$, of the partition Z.

The set of h-step functions $f : I \to \mathcal{Y}$ is denoted by $St_h(I, \mathcal{Y})$, and the canonical integral operator $\int_{[Z]} : St_h(I, \mathcal{Y}) \to \mathcal{Y}$ is defined by

$$\int_{[Z]} f := \sum_{i=1}^{n} m_i \, \mu(a_i, a_{i-1}) \tag{3}$$

where, again, m_i is the constant value of f on $[a_{i-1}, a_i)$ and $\mu(\cdot, \cdot)$ the growth calibration of $\mathbb{T}$.

At first glance, the advantage of h-step functions over o-step functions is not apparent. However, one can show that for any measure chain $\mathbb{T}$ (densely ordered or not) the integral (3) is well defined and independent of the partition Z of I. Moreover, any h-step function is also an o-step function, and if the integral $\int_{(Z)} f$ exists, it coincides with $\int_{[Z]} f$. In any case, an application of Theorem 5.1 yields the following result (see Neidhart [12, Section 8.1]).

Theorem 5.4 *For any measure chain $\mathbb{T}$ and any h-step function f from an interval $I = [a, b] \subseteq \mathbb{T}$ to a Banach space $\mathcal{Y}$ the integral (3) is well-defined and independent of the partition Z of I. Moreover, for any function $f \in \overline{St_h(I, \mathcal{Y})}$ the Cauchy-Riemann integral is defined as*

$$\int_a^b f(x) \, dx = \lim_{n \to \infty} \int_{(Z)} f_n \tag{4}$$

where $(f_n)_{n \in \mathbb{N}}$ is any sequence of h-step functions which converges to f uniformly on $[a, b]$.

Comparing Theorems 5.3 and 5.4 one might come to the conclusion that the o-step functions approach to the Cauchy-Riemann integral could be dismissed in favor of the h-step functions approach, because the former requires a densely ordered measure chain while the latter one works for *all* measure chains. This guess, however, is not quite true. The reason is that in either case the set of integrable functions is determined by an application of the Linear Extension Theorem 5.1 (where $\mathcal{X}$ is the Banach space $\mathcal{B}(I, \mathcal{Y})$ of bounded functions from I to $\mathcal{Y}$, equipped with the sup-norm). In the first case this extension leads from the normed linear space $\mathcal{D} := St_o(I, \mathcal{Y})$ to its closure $\overline{St_o(I, \mathcal{Y})}$, and in the second case from $\mathcal{D} := St_h(I, \mathcal{Y})$ to $\overline{St_h(I, \mathcal{Y})}$. The point now is, that $\overline{St_o(I, \mathcal{Y})}$ is identical with the Banach space $\mathcal{R}(I, \mathcal{Y})$ of regulated functions, while $\overline{St_h(I, \mathcal{Y})}$ can be a proper subspace of $\mathcal{R}(I, \mathcal{Y})$. In any case, however, $\overline{St_h(I, \mathcal{Y})}$ contains all rd-continuous (and hence all continuous) functions (see Neidhart [12, Theorem 149]).

In summary, we can say that on *densely ordered* measure chains one should use the o-step function approach. This yields the full set of regulated functions $\mathcal{R}(I, \mathcal{Y})$ as the set of integrable functions while the h-step function approach may yield a proper subset. On measure chains which are *not densely ordered*, however, the h-step function approach must be used, because the o-step function approach may fail. But then the set of integrable functions may be a proper subset of $\mathcal{R}(I, \mathcal{Y})$.

The overall picture of this situation is given in the following diagram.

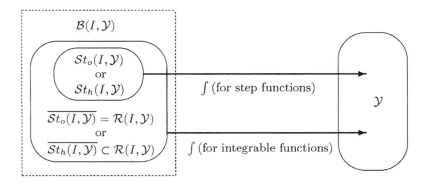

The main *advantage* of the Cauchy-Riemann integral is its quick and elegant introduction. In the context of general measure chains, however, this advantage is obscured by the fact that one has to distinguish between densely ordered measure spaces and those which are not densely ordered, and such a distinction contradicts the general philosophy of the calculus of measure chains. The main *disadvantage* of the Cauchy integral, however, is, that the set of integrable functions is too small. In fact, it is a (possibly even proper) subset of the set of regulated functions, which in turn is definitively smaller than the set of Riemann integrable functions, not to mention the Lebesgue integrable functions.

6   Measure and integral on measure chains

As the term "measure chain" already indicates, and as Hilger mentioned in [7, page 12] (see also [8, page 25]), the growth calibration of a measure chain $\mathbb{T}$ induces a measure on $\mathbb{T}$ in a canonical way. On the other hand, knowing a measure on $\mathbb{T}$, the construction of an integral for functions $f : \mathbb{T} \to [-\infty, +\infty]$ is a straightforward task of measure theory. One may wonder that these facts have not been observed and picked up at the early states of the usage of measure chains, and that still today the Cauchy integral is the standard integral on measure chains.

In the remainder of this paper we sketch how the construction of the measure theoretic integral indeed works for functions from an arbitrary measure chain $\mathbb{T}$ to an arbitrary real or complex Banach space $\mathcal{Y}$.

To this end we first generate a suitable σ-Algebra over $\mathbb{T}$ by noticing that each measure chain is a topological space, generated by the open intervals, and that in each topological space the set of open sets generates a σ-Algebra, the so-called *Borel σ-Algebra* $\mathfrak{B}$. In order to construct a measure on $\mathfrak{B}$ which is compatible with the growth calibration $\mu : \mathbb{T} \times \mathbb{T} \to \mathbb{R}$, we introduce a suitable generator $\mathfrak{J}$ of $\mathfrak{B}$ (see Neidhart [12, Theorems 332 and 333]).

Theorem 6.1 *The system of sets* $\mathfrak{J} := \big\{ [a,b) : a,b \in \mathbb{T},\ a \leq b \big\} \cup M$, *where*

$$M = \left\{ \begin{array}{ll} \varnothing & , \quad \textit{if } \mathbb{T} \textit{ has no Maximum,} \\ \{\,\max \mathbb{T}\,\} , & \textit{otherwise,} \end{array} \right.$$

is a semi-ring over $\mathbb{T}$ *and a generator of* $\mathfrak{B}$.

That the semi-ring $\mathfrak{J}$ is indeed the right choice for our purpose can be seen from the following result (see Neidhart [12, Theorem 338]):

Theorem 6.2 *The mapping* $\nu : \mathfrak{J} \to \mathbb{R}$ *defined by*

$$\nu(A) := \left\{ \begin{array}{ll} \mu(r,s), & \textit{if } A = [s,r), \textit{ where } s,r \in \mathbb{T} \textit{ and } s \leq r, \\ 0 & , \quad \textit{if } \max \mathbb{T} \textit{ exists and } A = \{\max \mathbb{T}\}, \end{array} \right.$$

is a σ-*finite pre-measure on* $\mathfrak{J}$.

In view of the Extension Theorems of measure theory (see, e.g., Cohn [5], Elstrodt [6]), the pre-measure ν can be extended to a uniquely determined measure on $\mathfrak{B}$, the so-called *Borel measure* β. Moreover, if $\mathfrak{L}$ denotes the measure theoretic completion of $\mathfrak{B}$ (i.e., the union of $\mathfrak{B}$ with the set of all subsets of null-sets of $\mathfrak{B}$), then $\mathfrak{L}$ is a σ-Algebra as well, the so-called *Lebesgue* σ-*Algebra*, and there exists exactly one extension of β (and thus of ν) to a measure λ on $\mathfrak{L}$, the so-called *Lebesgue measure*.

The situation we have gained so far is depicted in the following diagram where $\mathfrak{P}(\mathbb{T})$ denotes the set of all subsets of $\mathbb{T}$.

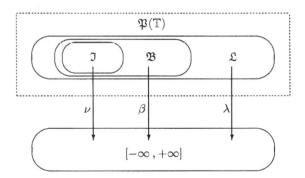

Having the measure spaces $(\mathbb{T}, \mathfrak{B}, \beta)$ and $(\mathbb{T}, \mathfrak{L}, \lambda)$ at hand we are now in a position to introduce the Borel and the Lebesgue integral for functions from a measure chain to a Banach space by simply employing the standard procedure from measure theory. Since this is commonly known only for *real-valued* functions, we first make a brief excursion to the so-called *Bochner integral* which is the proper notion of integral for *Banach space-valued* functions. For more details on the construction of this kind of integral we refer to Cohn [5, Appendix E] (see also Aulbach and Wanner [2, Appendix A]).

Let $(X, \mathfrak{A}, \alpha)$ be a measure space over an arbitrary set X and $\mathfrak{B}(\mathcal{Y})$ the Borel σ-Algebra of a Banach space $\mathcal{Y}$. Then a function $f : X \to \mathcal{Y}$ is called *measurable* if $f^{-1}(B) \in \mathfrak{A}$ for all $B \in \mathfrak{B}(\mathcal{Y})$, and a measurable function is called *simple* if it attains only finitely many values. Roughly speaking, the introduction of an integral for measurable functions $f : X \to \mathcal{Y}$ is to first define the integral for simple functions (in the obvious way) and then to define the integral of f by suitably approximating f by a sequence $(f_n)_{n \in \mathbb{N}}$ of simple functions and to define the integral of f as the limit of the integrals of the f_n.

In case $\mathcal{Y}$ equals $\mathbb{R} \cup \{\pm\infty\}$ or $\mathbb{C}$, this is the standard procedure which can be found in any textbook on measure theory. If $\mathcal{Y}$ is a (real or complex) Banach space with *finite* dimension, one may choose a basis $\{y_1, \ldots, y_d\}$ of $\mathcal{Y}$ and define the (real- or complex-valued, respectively) coordinate mappings f_i of f through the relation $f(x) =: \sum_{i=1}^{d} f_i(x) \cdot y_i$ for all $x \in X$. By means of the relation

$$\int_X f(x)\, d\mu := \sum_{i=1}^{d} \left(\int_X f_i(x)\, d\mu \right) \cdot y_i$$

the definition of the integral of f is then reduced to the well known integral for real- or complex-valued functions. If the Banach space $\mathcal{Y}$ is *infinite-dimensional*, however, the general construction of the integral breaks down if the image of X under f is not *separable*, i.e., if $f(X)$ does not contain a dense countable subset. This can be seen as follows:

- If $(f_n)_{n \in \mathbb{N}}$ is a sequence of simple functions $f_n : X \to \mathcal{Y}$ converging to $f : X \to \mathcal{Y}$, then $f(X)$ is contained in the closure of the set $\bigcup_{n=1}^{\infty} f_n(X)$. And since the functions f_n are simple, the sets $f_n(X)$ are finite, thus $\bigcup_{n=1}^{\infty} f_n(M)$ is countable. Consequently, $f(X)$ is necessarily separable.

In addition, in the infinite-dimensional context the following problem arises:

- The sum of two measurable functions from X to $\mathcal{Y}$ is not necessarily measurable (see Cohn [5, Appendix E, Exercise 2]).

In order to overcome these complications one has to replace the class of measurable functions by a more adequate class. In fact, to suitably strengthen the notion of measurability, a function $f : X \to \mathcal{Y}$ is called *strongly measurable* if it is measurable and if $f(X)$ is separable. This definition immediately implies that every simple and every measurable function is strongly measurable. On the other hand, if $\mathcal{Y}$ is a separable Banach space (which is particularly true if $\mathcal{Y}$ is finite-dimensional) then strong measurability is the same as measurability.

Employing the concept of strong measurability, the introduction of an integral for functions from a measure space $(X, \mathfrak{A}, \alpha)$ to a Banach space $(\mathcal{Y}, |\cdot|)$ is almost straightforward. In fact, the set of strongly measurable functions is a linear space (with the usual operations) which is closed under the formation of pointwise limits, and any strongly measurable function is the pointwise limit of a sequence of simple functions. A strongly measurable function $f : X \to \mathcal{Y}$

is then called *(Bochner) integrable*, if the function $|f| : X \to \mathbb{R}$ is integrable with respect to the measure α, and the integral of f is defined by means of an approximating sequence $(f_n)_{n \in \mathbb{N}}$ of simple functions as follows: Each f_n can be written in the canonical form $\sum_{j=1}^{k} a_j \chi_{A_j}$, and hence its integral is defined as $\int_X f_n(x) \, d\alpha := \sum_{j=1}^{k} a_j \alpha(A_j)$. Then, for an arbitrary strongly measurable function $f : X \to \mathcal{Y}$ the integral is defined as

$$\int_X f(x) \, d\alpha := \lim_{n \to \infty} \int_X f_n(x) \, d\alpha \,,$$

where $(f_n)_{n \in \mathbb{N}}$ is a sequence of simple functions converging to f. That this integral is indeed well defined follows as in the case of real-valued functions. Of the other properties of this integral we just mention that Lebesgue's Dominated Convergence Theorem looks the same as for real-valued functions, while Beppo Levi's Monotone Convergence Theorem is, of course, not available due to the lacking order structure of general Banach spaces. For more properties of the Bochner integral we refer to Cohn [5, Appendix E].

Returning from general measure spaces $(X, \mathfrak{A}, \alpha)$ to the measure spaces $(\mathbb{T}, \mathfrak{B}, \beta)$ and $(\mathbb{T}, \mathfrak{L}, \lambda)$ appearing in the calculus of measure chains, for any function f from an arbitrary measure chain $\mathbb{T}$ to an arbitrary real or complex Banach space $\mathcal{Y}$ we immediately obtain the *Borel* and the *Lebesgue integral*

$$\int_{\mathbb{T}} f(x) \, d\beta \quad \text{and} \quad \int_{\mathbb{T}} f(x) \, d\lambda$$

by simply applying the above-mentioned general result. As a first relation to the previously described integrals we get (see Neidhart [12, Theorem 349]):

Theorem 6.3 *Any regulated function $f : \mathbb{T} \to \mathcal{Y}$ is Borel and Lebesgue integrable.*

Basically, the definition of the Borel and Lebesgue integral applies to functions which are defined throughout the *whole* measure chain under consideration, and so the (seemingly redundant) question arises of how to define the Borel and the Lebesgue integral for functions which are defined on subsets, in particular intervals, of a measure chain only. In order to answer this question we take two points a and b of a measure chain $\mathbb{T}$ with $a < b$ and notice that the intervals $[a, b]$, (a, b), $[a, b)$ and $(a, b]$ are Borel and Lebesgue measurable. Thus, using the suitable restrictions of f the following integral values are well defined:

$$\int_{[a,b]} f(x) \, d\beta, \quad \int_{(a,b)} f(x) \, d\beta, \quad \int_{[a,b)} f(x) \, d\beta, \quad \int_{(a,b]} f(x) \, d\beta,$$

$$\int_{[a,b]} f(x) \, d\lambda, \quad \int_{(a,b)} f(x) \, d\lambda, \quad \int_{[a,b)} f(x) \, d\lambda, \quad \int_{(a,b]} f(x) \, d\lambda.$$

While in real calculus those terms coincide (if f is Lebesgue integrable), this is not the case in general measure chains. This is due to the following result (see Neidhart [12, Theorem 341]).

Theorem 6.4 *Any singleton* $\{t\} \subseteq \mathbb{T}$ *is Lebesgue measurable, and we have*

$$\lambda(\{t\}) = \beta(\{t\}) = \mu^*(t).$$

Hence, the Lebesgue and the Borel measure of a singleton $\{t\}$ *is* 0 *for right-dense points* t, *while for right-scattered* t *it has the positive value of the graininess* $\mu^*(t)$.

Due to this theorem the question arises which of the four Borel and which of the four Lebesgue integrals is suitable for the definition of the respective integral between a and b. It turns out (see Neidhart [12, Section 10.2]) that in both cases the use of the half-open interval $[a, b)$ is the choice which leads to the desired result (Theorem 6.5 below). We thus define for any $a, b \in \mathbb{T}$ with $a < b$ and any Borel or Lebesgue integrable function $f : \mathbb{T} \to \mathcal{Y}$ the *Borel* and the *Lebesgue integral* by

$$\int_a^b f(x)\, d\beta := \int_{[a,b)} f(x)\, d\beta \quad \text{and} \quad \int_a^b f(x)\, d\lambda := \int_{[a,b)} f(x)\, d\lambda,$$

respectively. With these notions at hand we finally get the following result which relates the various notions of integrals considered in this paper (see Neidhart [12, Theorem 350]).

Theorem 6.5 *Suppose* $\mathbb{T}$ *is any measure chain,* $\mathcal{Y}$ *any real or complex Banach space and* $f : \mathbb{T} \to \mathcal{Y}$ *strongly measurable. Then, if for some* $a, b \in \mathbb{T}$ *with* $a < b$ *the restriction of* f *to the interval* $[a, b)$ *is Lebesgue integrable, it is also integrable in the sense of Cauchy, Riemann, Cauchy-Riemann and Borel, and the corresponding integrals have the same value as* $\int_a^b f(x)\, d\lambda$.

We close this section by noticing that the Lebesgue integral for real-valued functions on time scales has recently been considered by Bohner and Guseinov in Bohner and Peterson [4, Chapter 5].

7   Conclusion: The best is (almost) for free

We summarize our previous considerations by stating that – as in ordinary calculus – the Lebesgue integral is by far superior to all other notions of integrals which are possible on measure chains. The Lebesgue integral not only provides the largest set of integrable functions, its derivation is even simpler than the construction of the other types of integrals, because most of the technical details can be avoided by simply quoting standard results from measure theory. In fact, once it is observed that the growth calibration of a measure chain canonically generates a σ-finite pre-measure on the semi-ring of half-open intervals, the rest of the work is done by standard Extension Theorems. We therefore come to the astonishing conclusion that in the context of integration on measure chains the best is (almost) for free.

References

[1] Amann, H. and Escher, J., *Analysis II* (in German). Birkhäuser, Boston, 1999.

[2] Aulbach, B. and Wanner, T., Integral manifolds for Carathéodory type differential equations in Banach spaces. In: Aulbach, B. and Colonius, F. (Eds.), *Six Lectures on Dynamical Systems*. World Scientific, Singapore, 1996.

[3] Bohner, M. and Peterson, A., *Dynamic Equations on Time Scales*. Birkhäuser, Boston, 2001.

[4] Bohner, M. and Peterson, A. (Eds.), *Advances in Dynamic Equations on Time Scales*. Birkhäuser, Boston, 2003.

[5] Cohn, D. L., *Measure Theory*. Birkhäuser, Boston, 1980.

[6] Elstrodt, J., *Maß– und Integrationstheorie* (in German). Springer-Verlag, New York, Second Edition 1999.

[7] Hilger, S., *Ein Maßkettenkalkül mit Anwendungen auf Zentrumsmannigfaltigkeiten* (in German). PhD thesis, University of Würzburg, 1988.

[8] Hilger, S., Analysis on Measure Chains. A Unified Approach to Continuous and Discrete Calculus. *Results in Mathematics* **10** (1990), 18–56.

[9] Guseinov, G. Sh. and Kaymakçalan, B., On the Riemann integration on time scales, this book, 289–298.

[10] Keller, S., Asymptotisches Verhalten invarianter Faserbündel bei Diskretisierung und Mittelwertbildung im Rahmen der Analysis auf Zeitskalen (in German). PhD thesis, University of Augsburg, 1999.

[11] Lang, S., *Real and Functional Analysis*. Springer-Verlag, New York, Third edition 1993.

[12] Neidhart, L., Integration im Rahmen des Maßkettenkalküls (in German). Diploma thesis, University of Augsburg, 2001.

[13] Pötzsche, C., Chain rule and invariance principle on measure chains. *J. Comput. Appl. Math.* **141** (2002), 249–254.

[14] Sailer, S., Riemann-Stieltjes-Integrale auf Zeitmengen (in German). Final examination thesis, University of Augsburg, 1992.

Asymptotic Formulae for Dynamic Equations on Time Scales with a Functional Perturbation [1]

SAMUEL CASTILLO

Departamento de Matemática, Facultad de Ciencias
Universidad del Bío-Bío, Casilla 5-C, Concepción, Chile
E-mail: scastill@ubiobio.cl

and

MANUEL PINTO

Departamento de Matemática, Facultad de Ciencias
Universidad de Chile, Casilla 653, Santiago, Chile
E-mail: pintoj@uchile.cl

Abstract In this work we give a unification of discrete and continuous results about asymptotic formulae for solutions of dynamic systems on time scales with a functional perturbation.

Keywords Asymptotic formula, Functional perturbation, Levinson theorem

AMS Subject Classification 39A12

1 Introduction

In their well known book, Coodington and Levinson (see [6]) study the differential system

$$y' = (A + B(t) + R(t))y, \ t \geq 0 \tag{1}$$

where A, $B(t)$ and $R(t)$ are $N \times N$ matrices with complex entries. They assume that all eigenvalues of A : $\mu_1, ..., \mu_N$ are different and simple; $B(t) \to 0$ as $t \to +\infty$, all eigenvalues of $A + B(t)$: $\lambda_1(t), ..., \lambda_N(t)$ satisfy $\lambda_j(t) \to \mu_j$ as $t \to +\infty$, there is $k : 1 \leq k \leq N$ such that $\int^t \mathrm{Re}[\lambda_j(\tau) - \lambda_k(\tau)]d\tau \to -\infty$ as $t \to +\infty$ for $j > k$, and there is $\eta \in \mathbf{R}$ such that $\int_s^t \mathrm{Re}[\lambda_j(\tau) - \lambda_k(\tau)]d\tau \geq \eta$ if

---

[1]Research supported by Fondecyt 8990013.

$t \geq s$ for $j \leq k$ and $B', R \in L^1$. Their conclusion is the existence of a solution $y = y_k(t)$ of (1) such that

$$y_k(t) = \exp\left(\int_0^t \lambda_k(\tau)d\tau\right)(v + o(1)),$$

as $t \to +\infty$. Versions of this result for functional differential and difference equations were studied by S. Castillo and M. Pinto [4]. Discrete and continuous results can be unified and generalized by means of the theory of "time scales" which was proposed and developed by B. Aulbach and S. Hilger (see [1, 8]). A version of the Levinson Theorem proposed above for time scales with nonlinear and functional perturbation is studied here. A finite dimensional version for linear systems on time scales is given by Bohner and Lutz [2].

This work is organized as follows: Section 2 contains some preliminary facts about time scales. Section 3 contains the main results of this work. Section 4 is devoted to the proofs.

2   Basic notions in time scales

For detailed information about time scales, the reader could see Aulbach and Hilger [1] and Hilger [8].

A *time scale* is a closed subset $\mathbf{T}$ of $\mathbf{R}$. Let $\mathbf{B}$ be a complex Banach algebra. For a function $f : \mathbf{T} \to \mathbf{B}$, the operator $^\Delta$ is defined by

$$f^\Delta(t) = \lim_{s \to t, s \in \mathbf{T} - \{\sigma(t)\}} \frac{f(\sigma(t)) - f(s)}{\sigma(t) - s}, \tag{2}$$

where $\sigma(t) = \inf\{s \in \mathbf{T} : s > t\}$ is *a jump function*. Another jump function is given by $\rho(t) = \sup\{s \in \mathbf{T} : s < t\}$, for all $t \in \mathbf{T}$.

Examples:

1. If $\mathbf{T} = \mathbf{R}$, then $f^\Delta(t) = \frac{df}{dt}$.

2. If $\mathbf{T} = \mathbf{Z}$, then $f^\Delta(n) = f(n+1) - f(n)$.

3. If $\mathbf{T} = h\mathbf{Z}$, with $h > 0$, then $f^\Delta(n) = \frac{1}{h}(f(t+h) - f(t))$.

So, difference and differential equations can be generalized by means of the operator $^\Delta$. The *graininess* is defined by $\mu(t) = \sigma(t) - t$ and $f^\sigma(t) = f(\sigma(t))$ for all $t \in \mathbf{T}$. If the limit (2) exists for $t = t_0$, then f is called *differentiable at* $t = t_0$. If f is differentiable at every $t \in \mathbf{T}$, f is called *differentiable*.

Since we are interested in asymptotic formulas, we will assume that $\mathbf{T}$ has no real upper bound. So, from now on, we will ask that $\lim_{n \to +\infty} t_m = +\infty$ for some sequence $(t_m)_{m \in \mathbf{N}} \subseteq \mathbf{T}$.

The following lemma gives some basic properties of the operator $^\Delta$.

Lemma 1 *(See [1], Theorems 3 and 4)*

   i) If f is differentiable, then it is continuous in t;

   ii) If f is continuous at t_0 and t_0 is right-scattered, then f is differentiable at t_0;

   iii) If $f^\Delta(t)$ exists, then $f(\sigma(t)) = f(t) + \mu(t)f^\Delta(t)$;

   iv) If $\sigma(t) > t$ for all $t \in \mathbf{T}$, then $f^\Delta(t) = \frac{f(\sigma(t)) - f(t)}{\mu(t)}$;.

   v) If f and g are differentiable in t, then fg is differentiable and $(fg)^\Delta = f^\sigma g^\Delta + f^\Delta g$.

   vi) If f is differentiable in t and invertible on $\mathbf{T}$, then f^{-1} is differentiable in $\mathbf{T}$ and $(f^{-1})^\Delta(t) = -(f^\sigma)^{-1}(t)f^\Delta(t)f^{-1}(t)$.

A function is called *rd-continuous*, if

   i) $\lim_{s \to t^-} f(s)$ exists when $\rho(t) = t$;

   ii) $\lim_{s \to t} f(s) = f(t)$ when $\sigma(t) = t$.

A function $F : \mathbf{T} \to \mathbf{B}$ is called an *antiderivative* of f iff $F^\Delta(t) = f(t)$ for all $t \in \mathbf{T}$. We define

$$\int_s^t f(\tau)\Delta\tau := F(s) - F(r). \tag{3}$$

We will need the following result:

Lemma 2 *(See [1], Theorems 6 and 7). Under the above assumptions, let $f : \mathbf{T} \to \mathbf{B}$ be a rd-continuous function. Then we have:*

   i) f has an antiderivative;

   ii) $\int_t^{\sigma(t)} f(\tau)\Delta\tau = \mu(t)f(t)$ for all $t \in \mathbf{T}$;

   iii) $|\int_s^t f(\tau)\Delta\tau| \leq \int_s^t |f(\tau)|\Delta\tau$ for $s, t \in \mathbf{T} : s < t$.

Examples:

   1. If $\mathbf{T} = \mathbf{R}$ then

$$\int_a^b f(\tau)\Delta\tau = \int_a^b f(\tau)d\tau.$$

   2. If $\mathbf{T} = h\mathbf{Z}$ then

$$\int_a^b f(\tau)\Delta\tau = h \sum_{j=0}^{N-1} f(a + jh),$$

   where $h > 0$, $a \in \mathbf{T}$ and $b = a + jN$.

3. If $\mathbf{T} = \dot{\cup}_{j=1}^{+\infty}[a_j, b_j]$ then

$$\int_s^t f(\tau)\Delta\tau = \sum_{j=n_a+1}^{n_b} (a_j - b_{j-1})f(b_{j-1}) + \sum_{j=n_a}^{n_b} \int_{a_j}^{b_j} f(\tau)d\tau + \int_{n_b}^t f(\tau)d\tau,$$

where $n_t \in \mathbf{N}$ is such that $n_t \leq t < n_t + 1$, for all $t \in \mathbf{T}$.

We define $L_{\mathbf{T}}^p(I, \mathbf{B})$ for $p \geq 1$, $I \subseteq \mathbf{R}$ is an interval, as the set of the functions $f : I \cap \mathbf{T} \to \mathbf{C}$ such that

$$\int_a^b |f(\tau)|^p \Delta\tau < +\infty,$$

with $a = \inf I$ and $b = \sup I$.

Remark: Under a clear context the set $L_{\mathbf{T}}^p(I, \mathbf{C})$ will be denoted by L_T^p.

Lemma 3 *Under the above definitions, let $H(t) : E \to E$ be bounded linear operator-valued matrices for all $t \in \mathbf{T}$. Moreover, suppose that H is regressive, i.e.,*

$$I + \mu(t)H(t) \text{ is invertible for each } t \in \mathbf{T},$$

where I denotes the identity-operator on E. Then for $t_0 \in \mathbf{T}$, there is an invertible bounded linear operator $T : E \to E$ such that

$$T^\Delta(t) = H(t)T(t) \text{ and } T(t_0) = I.$$

A particular example is the scalar case. For a function $p : \mathbf{T} \to \mathbf{C}$ such that $1 + \mu(t)p(t) \neq 0$ for all $t \in \mathbf{T}$ (regressive) we consider the solution of the initial value problem

$$y^\Delta = p(t)y \text{ and } y(t_0) = 1$$

and denote it as $y(t) = e_p(t, t_0)$ for $t \in \mathbf{T}$. Then,

$$e_p(t, s) = \exp\left(\int_s^t \xi_{\mu(\tau)}(p(\tau))\Delta\tau\right),$$

where

$$\xi_h(z) = \begin{cases} \frac{\ln(1+hz)}{h} & \text{if } h \neq 0 \\ z & \text{if } h = 0, \end{cases}$$

for $z \in \mathbf{C}_h := \mathbf{C} - \{-\frac{1}{h}\}$.

Remark: For $z \in \mathbf{C}$ we have that

$$\lim_{h \to 0+} \xi_h(z) = z.$$

In [1] the operations $\oplus$ and $\ominus$ are defined by $p \oplus q := p + q + \mu p q$ and $\ominus p := -\frac{p}{1+\mu p}$, where p and q are regressive functions. The subtraction $p \ominus q$ is defined by $p \ominus q = p \oplus (\ominus q)$. Note that

$$\xi_h(p \ominus q) = \begin{cases} p - q, & \text{if } h = 0 \\ \frac{1}{h} \ln\left(\frac{1+hp}{1+hq}\right) & \text{if } h \neq 0 \end{cases}$$

The set of the regressive functions with the operation $\oplus$ forms an abelian group. So, we obtain the following result:

Lemma 4 *(see [1] Theorem 9 and [3]) If p, q are regressive rd-continuous functions, then we have*

i) $e_0(t, s) = 1$ and $e_p(t, t) = 1$;

ii) $e_p(\sigma(t), s) = (1 + \mu(t)p(t))e_p(t, s)$;

iii) $e_p(s, t) = \frac{1}{e_p(t,s)} = e_{\ominus p}(t, s)$;

iv) $e_p(t, \tau)e_p(\tau, s) = e_p(t, s)$;

v) $e_p(t, s)e_q(t, s) = e_{p \oplus q}(t, s)$;

vi) $e_{p \ominus q}(t, s) = \frac{e_p(t,s)}{e_q(t,s)}$.

Examples:

1. If $\mathbf{T} = \mathbf{R}$, then $e_p(t, s) = \exp\left(\int_s^t p(\tau)d\tau\right)$.

2. If $\mathbf{T} = h\mathbf{Z}$, then $e_p(t, s) = \prod_{\tau=s}^{t-1}(1 + hp(\tau))$.

3. If $\mathbf{T} = \dot{\cup}_{j=1}^{+\infty}[a_j, b_j]$ then

$$e_p(t, s) = \left[\prod_{j=n_s}^{n_t}[1 + (a_j - b_{j-1})f(b_{j-1})]\right] \times$$

$$\times \exp\left(\sum_{j=n_s+1}^{n_t}\int_{a_j}^{b_j} p(\tau)d\tau\right)\exp\left(\int_{n_t}^t p(\tau)d\tau\right),$$

where $n_t \in \mathbf{N}$ is such that $n_t \leq t < n_t + 1$, for all $t \in \mathbf{T}$.

3   Main result

We propose the following generalizations of the Levinson's asymptotic theorem.

Theorem 1 *Let* **T** *be a time scale,* $A : \mathbf{C}^N \to \mathbf{C}^N$ *a regressive linear operator and* $\{V(t)\}_{t \in \mathbf{T} \cap [0,+\infty[}$ *a family of linear operators* $\mathbf{C}^N \to \mathbf{C}^N$. *Let* $\mathcal{B}$ *be the set of bounded functions from* **T** *into* $\mathbf{C}^N$ *and* $\{F(t, \cdot)\}_{t \in \mathbf{T} \cap [0,+\infty[}$ *a family of not necessarily linear functional operators from* $\mathcal{B}$ *into* $\mathbf{C}^N$. *Suppose that for the equation*

$$y^{\Delta}(t) = (A + V(t))y(t) + F(t, y), \tag{4}$$

the following is true:

    i) $A + V(t)$ *is regressive for* $t \in \mathbf{T}$ *and* $\sigma(A) = \{\mu_j\}_{j=1}^N$, $\sigma(A + V(t)) = \{\mu_j + \varepsilon_j(t)\}_{j=1}^N$.

    ii) *There is* $k : 1 \le k \le N$, *such that*

        a) *There is* $\delta > 0$ *such that* $\mathrm{Re}\,\xi_{\mu(\tau)}(\mu_j \ominus \mu_k) < -\delta$ *for all* $j < k$ *and* $t \in \mathbf{T}$,

        b) μ_j *is simple and* $\mathrm{Re}\,\xi_{\mu(t)}(\mu_j \ominus \mu_k) \ge 0$ *or all* $j \ge k$ *and* $t \in \mathbf{T}$.

    iii) $V(\cdot)$ *is differentiable,* $V(t) \to 0$ *as* $t \to +\infty$, $V^{\Delta}(\cdot) \in L^1_{\mathbf{T}}$,

    vi) $\frac{1}{1+\mu(t)(\mu_k+\varepsilon_k(t))}|F(t,\varphi_1) - F(t,\varphi_2)| \le \gamma(t)\|\varphi_1 - \varphi_2\|$, *where* $\gamma : [0,+\infty[\to \mathbf{R}$ *is in* $L^1_{\mathbf{T}}$ *and* $\tilde{F}(t, 0) = 0$.

    Then, ε_j *could be chosen such that* $\varepsilon_j(t) \to 0$ *as* $t \to +\infty$ *for* $j = 1, ..., N$ *and equation (4) has a solution* $y_k(t)$, *defined for* $t \ge t_1$ *and* t_1 *large enough such that*

$$y_k(t) = e_{\mu_k + \varepsilon_k(\tau)}(t, t_1)(\hat{e} + o(1)), \tag{5}$$

as $t \to +\infty$, *where* $\hat{e}$ *is the eigenvector of* A *associated to* μ_k.

Remark: *Since* A *and* $A + V(t)$ *are regressive for* $t \in \mathbf{T}$, *if* $\mu(t) \ne 0$, *then* $\sigma(A) \cup \sigma(A + V(t)) \subseteq \mathbf{C}_{\mu(t)}$.

    When $\mathbf{T} = \mathbf{R}$ *we obtain the following corollary:*

Corollary 1 *Let* $A : \mathbf{C}^N \to \mathbf{C}^N$ *be a regressive linear operator such that* $\sigma(A) = \{\mu_1, \mu_2, ..., \mu_N\}$, *there are* $\delta > 0$ *and* $k : 1 \le k \le N$ *such that*

    a) $\mathrm{Re}(\mu_j - \mu_k) < -\delta$, *for* $j < k$,

    b) $\mathrm{Re}\,\mu_j \ge \mathrm{Re}\,\mu_k$ *and* μ_j *are simple for* $j \ge k$.

Let $\{V(t)\}_{t \ge 0}$ *be a family of linear operators on* $\mathbf{C}^N$ *such that* $\lim\limits_{t \to +\infty} |V(t)| = 0$ *and*

$$\int_0^{+\infty} |V'(\tau)|d\tau < +\infty.$$

Let $\{R(t)\}_{t\geq 0}$ be a family of matrices such that

$$\int_0^t |R(\tau)|d\tau < +\infty.$$

Then, system

$$y'(t) = (A + V(t))y(t) + R(t)y(t - r),$$

has a solution $y = y_k(t)$ *such that*

$$y_k(t) = \exp\left(\mu_k t + \int_0^t \varepsilon(\tau)d\tau\right)(\hat{e} + o(1)),$$

as $t \to +\infty$, *where* $A\hat{e} = \mu_k \hat{e}$, *and* $\varepsilon = \varepsilon(t)$ *is a solution of*

$$\det[(\mu_k + \varepsilon)I - A - V(t)] = 0,$$

such that $\lim_{t \to +\infty} \varepsilon(t) = 0$.

Another corollary, when $\mathbf{T} = h\mathbf{Z}$, is:

Corollary 2 *Let* $A : \mathbf{C}^N \to \mathbf{C}^N$ *be a linear operator such that* $\sigma(A) = \{\mu_1, \mu_2, ..., \mu_N\}$, *there are* $h > 0$ *such that* $I + hA$ *is invertible,* $\delta > 0$ *and* $k : 1 \leq k \leq N$ *such that*

a) $|\frac{1+h\mu_j}{1+h\mu_k}| < e^{-\delta h}$, *for* $j < k$,

b) $|\frac{1+h\mu_j}{1+h\mu_k}| \geq 1$ *and* μ_j *are simple for* $j \geq k$

Let $\{V(t)\}_{t \in h\mathbf{N}_0}$ *be a family of linear operators on* $\mathbf{C}^N$ *such that* $\lim_{t \to +\infty} |V(t)| = 0$ *and*

$$\sum_{j=0}^{+\infty} |V(h(j + 1)) - V(hj)| < +\infty.$$

Let $\{R(t)\}_{t\geq 0}$ *be a family of matrices such that*

$$\sum_{j=0}^{+\infty} |R(hj)| < +\infty.$$

Then, system

$$\frac{y(t + h) - y(t)}{h} = (A + V(t))y(t) + R(t)y(t - r),$$

has a solution $y = y_k(t)$ *such that*

$$y_k(hn) = \prod_{j=0}^{n-1}(1 + h[\mu_k + \varepsilon(hj)])(\hat{e} + o(1)),$$

as $n \rightarrow +\infty$, where $A\hat{e} = \mu_k \hat{e}$, and $\varepsilon = \varepsilon(n)$ is a solution of

$$\det[(\mu_k + \varepsilon)I - A - V(t)] = 0,$$

such that $\lim_{n \rightarrow +\infty} \varepsilon(n) = 0$.

The following result is a more general and simple version of Theorem 1.

Theorem 2 Let $B(t) : \mathbf{C}^N \rightarrow \mathbf{C}^N$ be a family of linear operators, for $t \geq 0$. Let $\mathcal{B}$ be the set of bounded functions from $\mathbf{T}$ into $\mathbf{C}^N$, and $\{\tilde{F}(t, \cdot)\}_{t \geq 0}$ a family of not necessarily linear functional operators from $\mathcal{B}$ into $\mathbf{C}^N$. Consider the $N \times B$ linear differential system

$$w^{\Delta} = B(t)w, \tag{6}$$

for $t \geq 0$ such that $w(t_0) = w_{t_0} \in \mathbf{C}^N$. Let $W(t, s)$ the Cauchy matrix of the system (6). Suppose that there are:

a) A projection $P : E \rightarrow E$ which commutes with $B(t)$.

b) A locally integrable function $\lambda : [0, \infty[\rightarrow \mathbf{C}$, a function $H : [0, \infty[\times [0, \infty[\rightarrow \mathbf{R}$ satisfying $H(t, s) \leq H(t, T)H(T, s)$ and $\lim_{t \rightarrow \infty} H(t, s) = 0$ for all $s \geq 0$, a vector $v \in E$ and a positive real constant K such that:

    i) $|W(t, s)P| \leq H(t, s)e_{\lambda(\tau)}(t, s)|$, for $t \geq s \geq 0$.

    ii) $W(t, s)v = e_{\lambda(\tau)}(t, s)v$ for all $t, s \geq 0$.

    iii) $|W(s, t)(I - P)^{-1}| \leq K|e_{\lambda(\tau)}(t, s)|$ for all $s \geq t \geq 0$.

Let $\{\tilde{F}(t, \cdot)\}_{t \geq 0}$ a family of not necessarily linear operators from a $\mathbf{T}$ into $\mathbf{C}^N$ such that

$$\frac{1}{1 + \mu(t)\lambda_k(t)}|\tilde{F}(t, e_{\lambda(\tau)}(s + \cdot, s)\varphi_1) - \tilde{F}(t, e_{\lambda(\tau)}(\cdot, s)\varphi_2)| \leq \gamma(t)\|\varphi_1 - \varphi_2\|, \tag{7}$$

where $\gamma : [0, +\infty[\rightarrow \mathbf{R}$ is in $L^1_{\mathbf{T}}$ and $\tilde{F}(t, 0) = 0$. Then, the equation

$$\zeta^{\Delta}(t) = B(t)\zeta(t) + \tilde{F}(t, \zeta) \tag{8}$$

has a solution $\zeta = \zeta(t)$, defined for $t \geq t_1$ and t_1 large enough such that

$$\zeta(t) = e_{\lambda(\tau)}(t, t_1)(v + o(1)), \tag{9}$$

as $t \rightarrow \infty$.

4   Some lemmas and proofs

Proof of Theorem 2: Let $G(t,s) = \begin{cases} W(t,s)P, & \text{if } t \geq s \geq 0 \\ W(t,s)(I-P)^{-1}, & \text{if } s > t \geq 0 \end{cases}$

Let $t_0 \geq 0$ and $C = C_{\mathrm{rd}}(\mathbf{T}, E)$ and let $\mathcal{N}$ be the operator defined on C by

$$(\mathcal{N}\zeta)(t) = e_{\lambda(\tau)}(t, t_0)v + \int_{t_0}^{+\infty} G(t, \sigma(s))\tilde{F}(s, \zeta)\Delta s \tag{10}$$

It is easy to prove that any fixed point ζ_0 of $\mathcal{N}$ is a solution of the equation (8). Then, we will prove that the operator $\mathcal{N}$ has a fixed point satisfying (9). Let us define the norm

$$\|\zeta\|_\lambda = \sup_{t \geq t_0} |e_{\ominus \lambda(\tau)}(t, t_0)\zeta(t)|$$

and the Banach space $C_\lambda = \{\zeta \in C_\sigma : \|\zeta\|_\lambda < +\infty\}$. We have that $\mathcal{N}(C_\lambda) \subseteq C_\lambda$, $\mathcal{N}$ has a unique fixed point ζ_0 in C_λ and ζ_0 satisfies (9). In fact, for a constant large enough $\tilde{K}$, we define $H_1^{t_0}(t) = \int_\sigma^t H(t,s)\tilde{\gamma}(s)\Delta s$ and $H_2^{t_0}(t) = K\int_t^{+\infty} \tilde{\gamma}(s)\Delta s$, for all $t \in \mathbf{T} \cap [t_0, +\infty[$, where $\tilde{\gamma} = \frac{\gamma}{1 + \mu\lambda_k}$. Then we obtain,

$$|\int_{t_0}^t W(t, \sigma(s))P\tilde{F}(s, \zeta)\Delta s| \quad \leq \quad |e_{\lambda(\tau)}(t, t_0)|H_1^{t_0}(t)\|\zeta\|_\lambda,$$

$$|\int_{t_0}^t W(t, \sigma(s))(I-P)^{-1}\tilde{F}(s, \zeta)\Delta s| \quad \leq \quad |e_{\lambda(\tau)}(t, t_0)|H_2^{t_0}(t)\|\zeta\|_\lambda,$$

for all $\zeta \in C_\lambda$ and $t \geq t_0$ with t_0 large enough. Moreover, if $H^{t_0} = H_1^{t_0} + H_2^{t_0}$, from (10), we have

$$|\mathcal{N}\zeta_1(t) - \mathcal{N}\zeta_2(t)| \leq |e_{\lambda(\tau)}(t, t_0)|H^\sigma(t)\|\zeta_1 - \zeta_2\|_\lambda \tag{11}$$

for all ζ_1, ζ_2 in C_λ. If we fix T and $t \geq T \geq t_0$ then

$$|H^{t_0}(t)| \leq H(t, T)\int_{t_0}^T H(T, s)|\tilde{\gamma}(s)|\Delta s + M \int_T^{+\infty} \tilde{\gamma}(s)\Delta s,$$

where $M = \max\{1, \sup_{t \geq s} |H(t,s)|\}$. Since $\tilde{\gamma} \in L_{\mathbf{T}}^1$, $H^{t_0}(t) \to 0$ as $t \to +\infty$. From (11) with $\zeta_2 = 0$, there is a constant $c > 0$ such that

$$|\mathcal{N}\zeta(t) - e_{\lambda(\tau)}(t, t_0)v| \leq c|e_{\lambda(\tau)}(t, t_0)|\|\zeta\|_\lambda$$

and

$$\lim_{t \to +\infty} |e_{\ominus\lambda(\tau)}(t, t_0)\mathcal{N}\zeta(t) - v| = 0. \tag{12}$$

Hence, $\mathcal{N}\zeta \in C_\lambda$ for all $\zeta \in C_\lambda$. If we take, in (11), t_0 so large that $\sup_{t \geq t_0} H^{t_0}(t) < 1$, we obtain that $\mathcal{N}$ is a contractive operator and by Banach Fixed Point Theorem, $\mathcal{N}$ has a unique fixed point $\zeta_0 \in C_\lambda$. The limit (12) proves our asymptotic result.

Prior to the proof of Theorem 1, some lemmas will be needed.

Lemma 5 *(Gronwall, see [8], page 54). Let* $\mathbf{T}$ *be a time scale,* $u, g : \mathbf{T} \to \mathbf{R}$ *be rd-continuos functions and* K *a positive constant such that*

$$u(t) = K + \int_{\tau}^{t} g(s)u(s)\Delta s,$$

for all $t \geq s$. *Then,* $u(t) \leq e_g(t, \tau)$ *for all* $t \geq \tau$.

Proof: Let $\omega(t) = K + \int_{\tau}^{t} g(s)u(s)\Delta s$. Then,

$$\omega^{\Delta}(t) = g(t)u(t) \leq g(t)\omega(t).$$

So, $\omega(t) \leq K e_g(t, \tau)$.

Lemma 6 *Let* $\mathbf{T}$ *be a time scale and* A *a* $N \times N$ *matrix such that* $I + \mu(t)A$ *is invertible and there are* $\delta > 0$ *and* $p \in \mathbf{C}$ *such that* $Re\xi_{\mu(t)}(\lambda) < -\delta + Re\xi_{\mu(t)}(p)$ *for all* $\lambda \in \sigma(A)$ *and* $t \in \mathbf{T}$. *If* X *is the fundamental solution of*

$$X^{\Delta} = AX,$$

then there is a positive constant K *such that*

$$|X(t, \tau)| \leq K e^{-\delta(t-\tau)} |e_p(t, \tau)|, \ t \geq \tau \geq 0. \tag{13}$$

Proof: Let γ be a rectifiable Jordan curve such that $\sigma(A) \subseteq \mathrm{Int}\gamma$. Then,

$$X(t, \tau) = \frac{1}{2\pi i} \int_{\gamma} (\lambda I - A)^{-1} e_{\lambda}(t, \tau) d\lambda,$$

for all $t \geq \tau \geq 0$. So (13) is obtained for $K = \frac{1}{2\pi} \sup_{\lambda \in \mathrm{Range}(\gamma)} |(\lambda I - A)^{-1}|$.

Lemma 7 *Let* $\mathbf{T}$ *be a time scale and* A *a* $N \times N$ *matrix such that* $I + \mu(t)A$ *is invertible and there are* $\delta > 0$ *and* $p \in \mathbf{C}$ *such that* $Re\xi_{\mu(t)}(\lambda) < -\delta + Re\xi_{\mu(t)}(p)$ *for all* $\lambda \in \sigma(A)$ *and* $t \in \mathbf{T}$. *Let* $\{B(t)\}_{t \geq 0}$ *a family of* $N \times N$ *matrices such that* $\xi_{\mu(t)}(|(I+\mu(t)A)^{-1}B(t)|) < \delta - \epsilon + Re(p)$, *for some* $\epsilon \in]0, \delta[$. *If* Y *is the fundamental solution of*

$$Y^{\Delta} = (A + B(t))Y,$$

If, $I + \mu(t)(A + B(t))$ *is invertible then there is a positive constant* K *such that*

$$|Y(t, \tau)| \leq K e^{-\epsilon(t-\tau)} |e_p(t, \tau)|, \ t \geq \tau \geq 0. \tag{14}$$

Proof: Let X be the fundamental solution of $X^{\Delta} = AX$. By Lemma 6, there is a positive constant K such that

$$|X(t, s)| \leq K e^{-\delta(t-\tau)} |e_p(t, s)|, \ t \geq \tau \geq 0.$$

By the Variation of Constants Formula (see [1], Theorem 8),

$$Y(t,\tau) = X(t,\tau) + \int_{\tau}^{t} X(t,\sigma(s))B(s)Y(s,\tau)\Delta s, \tag{15}$$

for $t \geq \tau \geq 0$. Then,

$$|Y(t,\tau)| = |X(t,\tau)| + \int_{\tau}^{t} |X(t,\sigma(s))||B(s)||Y(s,\tau)|\Delta s.$$

So,

$$u(t) \leq K + \int_{\tau}^{t} g(s)u(s)\Delta s,$$

where $u(t) = e^{\delta(t-\tau)}|e_{\ominus p}(t,\tau)||Y(t,\tau)|$ and $g(t) = [I + \mu(t)A]^{-1}B(t)$. By Lemma 5, $w(t) \leq Ke_g(t,\tau)$. Since $\xi_{\mu(t)}(g(t)) < \delta - \epsilon 0$, where $\epsilon \in]0, \delta[$, we have that

$$w(t) \leq Ke^{(\delta-\epsilon)(t-\tau)},$$

for all $t \geq \tau$. Therefore, K can be taken large enough that (14) be fulfilled.

Proof of the Theorem 1: Since the eigenvalues $\mu_k, \mu_{k+1}..., \mu_N$ of A are simple, we can define the projections to the eigenspaces relative to μ_j, by

$$Q_j(\infty) = \frac{1}{2\pi i} \int_{\gamma_j} (\lambda I - A)^{-1}d\lambda, \quad j = k, ..., N,$$

where γ_j's are rectificable Jordan's curves such that $\mu_j \in \mathrm{Int}\gamma_j$ and $\mathrm{Int}\gamma_j \cap \mathrm{Int}\gamma_l = \phi$ if $j \neq l$.

Let $\lambda_j = \mu_j + \varepsilon_j$. We can take t_0 so large that $\lambda_j(t) \in \mathrm{Int}\gamma_j$ for $j = k, ..., N$ and $t \geq t_0$. So, we can define projections to the eigenspaces of $A + V(t)$, relative to $\lambda_j(t)$, by

$$Q_j(t) = \frac{1}{2\pi i} \int_{\gamma_j} (\lambda I - A - V(t))^{-1}d\lambda, \quad j = k, ..., N, \ t \geq t_0.$$

By Bounded Convergence Theorem, $Q_j(\infty) = \lim_{t\to+\infty} Q_j(t)$.

Let $P_2(t) = Q_N(t) + \cdots + Q_{k+1}(t)$, $P_0(t) = Q_k(t)$, $P_1(t) = I - P_0(t) - P_2(t)$ and denote $P_j(\infty) = \lim_{t\to+\infty} P_j(t)$ for $j = 0, 1, 2$. Then

$$P_1(t)(\mathbf{C}^N) = \oplus_{j=1}^{k-1} M_{\lambda_j(t)}(A + V(t)), P_0(t)(\mathbf{C}^N) =$$
$$= M_{\lambda_k(t)}(A + V(t)), P_2(t)(\mathbf{C}^N) = \oplus_{j=k+1}^{N} M_{\lambda_j(t)}(A + V(t)),$$

where $M_{\lambda_j(t)}(A + V(t))$ is the eigenspace of $A + V(t)$ associated to the eigenvalue $\lambda_j(t)$ and

$$P_1(\infty)(\mathbf{C}^N) = \oplus_{j=1}^{k-1} M_{\mu_j}(A), P_0(\infty)(\mathbf{C}^N) =$$
$$= M_{\mu_k}(A), P_2(\infty)(\mathbf{C}^N) = \oplus_{j=k+1}^{N} M_{\mu_j}(A).$$

Let $S(t)$ be a solution of

$$S^{\Delta}(t) = G(t)S(t), \tag{16}$$

where $G(t) = P_1^{\Delta}(t)P_1(t) + P_0^{\Delta}(t)P_0(t) + P_2^{\Delta}(t)P_2(t)$. Since the projections $P_j(t)$ are bounded, there is $c_1 > 0$ such that

$$|G(t)| \leq \frac{c_1}{2\pi}|(\lambda I - A - V(t))^{-1}|^2 \max_{i=1,\dots,k} |\text{Var}(\gamma_i)||V'(t)|.$$

Since the resolvent $(\lambda I - A - V(t))^{-1}$ is bounded for $\lambda \in \cup_{j=k}^{N} \text{Range}\gamma_j$, there is $c_2 > 0$ such that $|G(t)| \leq c_2|V^{\Delta}(t)|$. Hence, by iii), we have that $|G(t)| \in L_{\mathbb{T}}^1$ and so $S(\infty) := \lim_{t \to +\infty} S(t)$ exists. Since $(S^{-1})^{\Delta}(t) = -\frac{\mu(t)}{1+\mu(t)}S^{-1}(t)G(t)$, we have that $S^{-1}(\infty) := \lim_{t \to +\infty} S(t)^{-1}$ also exists. So, it can be considered $S(\infty) = I$. We have that

$$S(t)P_j(\infty) = P_j(t)S(t), \tag{17}$$

for $j = 0, 1, 2$. In fact, $S_j(t) = S(t)P_j(\infty)$ is a solution of the system

$$\begin{aligned} S_j^{\Delta}(t) &= G(t)S_j(t) \\ S_j(\infty) &= P_j(\infty), \end{aligned} \tag{18}$$

for $j = 0, 1, 2$. On the other hand, if we take $S_j(t) = P_j(t)S(t)$ then,

$$S_j^{\Delta}(t) = P_j^{\sigma}(t)G(t)S(t) + P_j^{\Delta}(t)S(t).$$

Since $P_j^{\Delta}(t) = P_j^{\sigma}(t)P_j^{\Delta}(t) + P_j^{\Delta}(t)P_j(t)$,

$$S_j^{\Delta}(t) = [P_j^{\sigma}(t)G(t) + P_j^{\sigma}(t)P_j^{\Delta}(t) + [P_j^{\Delta}(t)]P_j(t)]S(t).$$

Hence,

$$S_j^{\Delta}(t) = [-P_j^{\sigma}(t)P_j^{\Delta}(t) + P_j^{\sigma}(t)P_j^{\Delta}(t) + P_j^{\Delta}(t)P_j(t)]S(t).$$

So, $S_j^{\Delta}(t) = P_j^{\Delta}(t)P_j(t)S(t)$, i.e.,

$$S_j^{\Delta}(t) = G(t)S_j(t).$$

Moreover, $S_j(\infty) = P_j(\infty)$. By uniqueness of solutions of system (18), we have that (17) is satisfied.

If we make, in (4), the change of variable $y(t) = S(t)\zeta(t)$, we have

$$\zeta^{\Delta}(t) = (A + \tilde{V}(t))\zeta(t) + \tilde{F}(t, z), \tag{19}$$

where $\tilde{V}(t) = S(t)^{-1}(A + V(t))S(t) - A$ and

$$\tilde{F}(t, \cdot) = -\frac{\mu(t)}{\mu(t) + 1}\left[S(t)^{-1}S^{\Delta}(t) - S(t)^{-1}F(t, S(t)\cdot)\right].$$

Since $S(t)$ and $S(t)^{-1}$ are bounded, $\tilde{\gamma} : [0, +\infty[\to \mathbf{R}$ defined by

$$\tilde{\gamma}(t) = |(S^\sigma(t))^{-1}||G(s)||S(s)| + \gamma(t)|(S^\sigma(t))^{-1}||S(t)| + \mu(t)|G(t)||A + V(t)|$$

is in $L^1_{\mathbf{T}}$, $|\tilde{F}(t, \varphi_1) - \tilde{F}(t, \varphi_2)| \le \tilde{\gamma}(t)\|\varphi_1 - \varphi_2\|$ and $\tilde{F}(t, 0) = 0$. Note that $P_i(\infty)(\mathbf{C}^N) \subseteq \mathbf{C}^N$ and it is invariant under $B(t) := A + \tilde{V}(t)$, i.e.,

$$B(t)P_j(\infty)(\mathbf{C}^N) \subseteq P_j(\infty)(\mathbf{C}^N),$$

for $j = 1, 0, 2$ and satisfy

$$\sigma(B(t)/P_1(\infty)\mathbf{C}^N) = \{\lambda_j(t)\}_{j=1}^{k-1},$$

$\sigma(B(t)/P_0(\infty)\mathbf{C}^N) = \{\lambda_k(t)\}$ and $\sigma(B(t)/P_2(\infty)\mathbf{C}^N) = \{\lambda_j(t)\}_{j=k+1}^{N}$. So, system $w^\Delta = B(t)w$ can be considered as

$$w_j^\Delta(t) = B_j(t)w_j(t), \tag{20}$$

where $B_i = B/P_j(\infty)\mathbf{C}^N$ y $w_j = P_j(\infty)w$ for $j = 0, 1, 2$.

- We will prove that the above splitting satisfies hypotheses i), ii), iii) of Theorem 2. More exactly, denote W_j, the evolution operator of (20),for $j = 0, 1, 2$, we will prove that given $\varepsilon > 0$, small enough, there is a constant $K > 0$ such that

A) $|W_1(t, s)| \le Ke^{-\delta_1(t-s)}|e_{\lambda_k(\tau)}(t, s)|$ for $t_0 \le s \le t$, where $\delta_1 \in]0, \delta[$,

B) $W_0(t, s)P_0(\infty) = e_{\lambda_k(\tau)}(t, s)$ for all $t, s \ge t_0$,

C) $|W_2(t, s)| \le \eta|e_{\lambda_k(\tau)}(t, s)|$, for $t_0 \le t \le s$.

In fact, to prove B) note that $\lambda_k(t)$ is the only eigenvector of $B_0(t)$, $B_0(t) = \lambda_k(t)P_0(\infty)$. Since P_0 is one dimensional, B) is easily obtained. To prove A), we observe that

$$w_1^\Delta = (A + \tilde{V}(t))w_1.$$

By Lemma 6, $|X(t, s)| \le Ke^{-\delta(t-s)}e_{\mu_k}(t, s)$ if $X^\Delta = AX$, for all $t \ge s \ge 0$. Since $V(t) \to 0$ as $t \to +\infty$, by Lemma 7, given $\epsilon \in]0, \delta[$, there is t_0 large enough such that

$$\begin{aligned} |W_1(t, s)| &\le Ke^{-\epsilon(t-s)}|e_{\mu_k}(t, s)| \\ &= Ke^{-\epsilon(t-s)}|e_{\lambda_k(\tau)}(t, s)||e_{\ominus\epsilon_k(\tau)}(t, s)|. \end{aligned}$$

Take $\delta_1 \in]0, \epsilon[$. Since $\ominus\epsilon_k(t) \to 0$ as $t \to +\infty$, we can take t_0 large enough such that $Re\xi_{\mu(t)}(\ominus\epsilon_k(t)) < \epsilon - \delta_1$ for $t \ge t_0$. Then,

$$|W_1(t, s)| \le Ke^{-\epsilon(t-s)}|e_{\lambda_k(\tau)}(t, s)|,$$

for all $t \geq s \geq t_0$ which proves A). To prove C), note that

$$
\begin{aligned}
|W_2(t,s)| &\leq \max_{j=k,\ldots,N} |e_{\lambda_j(\tau) \ominus \lambda_k(\tau)}(t,s)| e_{\lambda_k(\tau)}(t,s)| |P_2(\infty)| \\
&\leq \eta |e_{\lambda_k(\tau)}(t,s)| \, |P_2(\infty)|,
\end{aligned}
$$

for $t \leq s$. By Theorem 2, equation (19) has a solution $\zeta_k(t)$ defined for $t \geq \sigma$ and σ large enough such that

$$
\zeta_k(t) = \exp\left(\int_\sigma^t \lambda_k(\xi) d\xi \right) (\hat{e} + o(1)),
$$

as $t \to +\infty$. Since $S(t) \to I$ as $t \to +\infty$, $y_k(t) = S(t)\zeta_k(t)$ is a solution of equation (4) which satisfies (5).

References

[1] B. Aulbach and S. Hilger, Linear dynamics processes with homogeneous time scales. *Nonlinear Dynamics and Quantum Dynamical Systems.* Akademie Verlag, Berlin, (1990), 9–20.

[2] M. Bohner and D. Lutz, Asymptotic behavior of dynamic equations on Time Scale. *J. of Diff. Eqs. and Appl.* 7(1) (2001), 21–50.

[3] M. Bohner and A. Peterson, A survey of exponential functions in time scales. *Revista Cubo Matemática Educacional* 3(2) (2001), 285–301.

[4] S. Castillo and M. Pinto, Asymptotic formulae for nonlinear functional difference equations. *Computer and Math. with Appl.* Special Issue on Difference Equations III. Issue 3–5 (2001), 551–559.

[5] S. Castillo and M. Pinto, Levinson Theorem for functional differential equations. *Nonlinear Analysis T.M.A.* Vol 47/6 (2001), 3963–3965.

[6] E. Coddington and N. Levinson, *Theory of Ordinary Differential Equations.* Mc Graw Hill. New York 1955.

[7] M. Eastham, *The Asymptotic Solution of Linear Differential Systems: Applications of the Levinson Theorem.* Claremdon Press, Oxford 1989.

[8] S. Hilger, Analysis on measure chains – a unified approach to continuous and discrete calculus. *Results Math.*, *18* (1990), 18–56.

Oscillation of a Matrix Dynamic Equation on a Time Scale

LYNN ERBE[1]

Department of Mathematics and Statistics
University of Nebraska-Lincoln
Lincoln, NE 68588-0323, USA
E-mail: lerbe@math.unl.edu

Abstract We apply averaging techniques to establish an oscillation result for a matrix dynamic equation $LX(t) := [P(t)X^\Delta(t)]^\Delta + Q(t)X^\sigma(t) = 0$ on a time scale (measure chain) $\mathbb{T}$. This may be viewed as an extension of some earlier results for the differential and difference equations case.

Keywords Measure chains, Time scales, Riccati equation, Oscillation, Nonoscillation

AMS Subject Classification 34B10,39A10

1 Introduction

In this paper we shall study the second order matrix dynamic equation

$$LX(t) := \left[P(t)X^\Delta(t)\right]^\Delta + Q(t)X^\sigma(t) = 0 \tag{1}$$

on a measure chain (time scale) $\mathbb{T}$, by which we mean a nonempty closed subset of $\mathbb{R}$. Further we will assume throughout that

$$\sup \mathbb{T} = \infty,$$

since we shall be primarily interested in extending oscillation and nonoscillation criteria for the corresponding continuous and discrete cases, namely

$$[P(t)X'(t)]' + Q(t)X(t) = 0 \tag{2}$$

where $\mathbb{T}$ is the real interval $[a, \infty)$ and

$$\Delta(P_n \Delta X_n) + Q_n X_{n+1} = 0 \tag{3}$$

[1] Research was supported by NSF Grant 0072505.

where $\mathbb{T}$ is the set of nonnegative integers $\mathbb{N}_0$. For completeness, we introduce the following concepts related to the notion of time scales.

Definition *Let* $\mathbb{T}$ *be a time scale and define the* forward jump operator $\sigma(t)$ *at* t, *for* $t \in \mathbb{T}$, *by*

$$\sigma(t) := inf\{\tau > t : \tau \in \mathbb{T}\},$$

and the backward jump *operator* $\rho(t)$ *at* t, *for* $t \in \mathbb{T}$, *by*

$$\rho(t) := sup\{\tau < t : \tau \in \mathbb{T}\}.$$

We assume throughout that $\mathbb{T}$ *has the topology that it inherits from the standard topology on the real numbers* $\mathbb{R}$. *If* $\sigma(t) > t$, *we say* t *is* right-scattered, *while if* $\rho(t) < t$ *we say* t *is* left-scattered. *If* $\sigma(t) = t$ *we say* t *is* right-dense, *while if* $\rho(t) = t$ *we say* t *is* left-dense. *A function* $f : \mathbb{T} \to \mathbb{R}$ *is said to be* right-dense continuous *provided* f *is continuous at right-dense points in* $\mathbb{T}$ *and at left-dense points in* $\mathbb{T}$, *left hand limits exist and are finite. We shall also use the notation* $\mu(t) := \sigma(t) - t$ *which is called the graininess function.*

Our main concern is to extend to time scales some averaging techniques which have been applied to the study of (2) and (3) and see explicitly how the graininess function $\mu(t)$ enters the picture. In Erbe and Peterson [15] some general averaging criteria were developed for the case when the graininess is bounded. The result here gives a criterion for the situation when the graininess may be unbounded. In Erbe [10, 11] were developed some general results similar to those presented here for the scalar case.

Definition *Throughout this paper we make the blanket assumption that* $a \in \mathbb{T}$ *and we define the* interval *in* $\mathbb{T}$

$$[a, \infty) := \{t \in \mathbb{T} \ such \ that \ t \geq a\}.$$

The notion of a measure chain was introduced by S. Hilger [16]. Related work on the calculus of measure chains may be found in Agarwal and Bohner [1, 3], Agarwal, Bohner, and Wong [4], Aulbach and Hilger [5], Erbe and Hilger [12], Erbe and Peterson [13], Bohner and Peterson [6], and Kaymacalan et al [17]. We also refer to [2],[7], [8], [9], [14, 15], [19], [18] for additional discussion of oscillation and nonoscillation criteria for the continuous and discrete cases.

Definition *Assume* $x : \mathbb{T} \to \mathbb{R}$ *and fix* $t \in \mathbb{T}$; *then we define* $x^{\Delta}(t)$ *to be the number (provided it exists) with the property that given any* $\epsilon > 0$, *there is a neighborhood* U *of* t *such that*

$$|[x(\sigma(t)) - x(s)] - x^{\Delta}(t)[\sigma(t) - s]| \leq \epsilon|\sigma(t) - s|,$$

for all $s \in U$. *We call* $x^{\Delta}(t)$ *the* delta derivative *of* $x(t)$ *at* t.

It can be shown that if $x : \mathbb{T} \to \mathbb{R}$ is continuous at $t \in \mathbb{T}$ and t is right-scattered, then

$$x^{\Delta}(t) = \frac{x(\sigma(t)) - x(t)}{\sigma(t) - t}.$$

Note that if $\mathbb{T} = \mathbb{N}_0$, then

$$x^{\Delta}(t) = \Delta x(t) := x(t + 1) - x(t).$$

If t is right-dense, then

$$x^{\Delta}(t) = \lim_{s \to t} \frac{x(t) - x(s)}{t - s}.$$

In particular, if $\mathbb{T}$ is the real interval $[a, \infty)$, then $x^{\Delta}(t) = x'(t)$.
We assume throughout that the coefficient matrices in the equation $LX = 0$ satisfy $P(t) > 0$ (*positive definite*) for $t \in \mathbb{T}$ and $Q(t) = Q^*(t)$ ($Q(t)$ is Hermitian) for $t \in \mathbb{T}$. If a matrix function $X(t)$ is differentiable, then one may easily establish the formula

$$X^{\sigma}(t) = X(t) + \mu(t)X^{\Delta}(t) \tag{4}$$

for $t \in \mathbb{T}$.

Definition *Let $\mathbb{D}$ denote the set of all $n \times n$ matrix functions X defined on $\mathbb{T}$ such that $X(t)$ is delta differentiable on $\mathbb{T}$ and $(P(t)X^{\Delta}(t))^{\Delta}$ is right-dense continuous on $\mathbb{T}$. We say $X(t)$ is a* solution *of $LX = 0$ on $\mathbb{T}$ provided $X \in \mathbb{D}$ and $LX(t) = 0$ for $t \in \mathbb{T}$. Moreover, if X is a solution of $LX(t) = 0$ satisfying $X^*(t)P(t)X^{\Delta}(t) \equiv [P(t)X^{\Delta}(t)]^*X(t)$, then X is said to be* prepared.

2 Preliminaries

In what follows we shall denote by $\mathcal{M}$ the set of all $n \times n$ Hermitian matrices, and we let $\mathcal{C}$ denote the set of all $d \times d$ Hermitian matrix-valued functions $F(t)$ defined for $t \in \mathbb{T}$ with the property that the integral

$$\lim_{t \to \infty, t \in \mathbb{T}} \int_{t_0}^{t} F(s)\Delta s = \int_{t_0}^{\infty} F(s)\Delta s \quad \text{exists (finite).}$$

The assumption that $F \in \mathcal{C}$ implies that

$$\int_{t}^{\infty} F(s)\Delta s \in \mathcal{M}$$

for all $t \in \mathbb{T}$, $t \geq t_0$. Moreover, if $F(t) \in \mathcal{M}$ and $F(t) \geq 0$ (positive semidefinite), then $F \in \mathcal{C}$ iff $\int_{t_0}^{\infty} tr F(s)\Delta s < \infty$ iff $\int_{t_0}^{\infty} \lambda_1(F(s))\Delta s < \infty$ iff $\int_{t_0}^{\infty} |F(s)|\Delta s < \infty$. Here, and in what follows, we denote by

$$\lambda_1(A) \geq \lambda_2(A) \geq \cdots \geq \lambda_d(A)$$

the real eigenvalues of the matrix $A \in \mathcal{M}$ and we will let the matrix norm $|\cdot|$ be the matrix norm induced by the Euclidean vector norm on $\mathbb{R}^d$. We recall that the trace of A is given by $tr(A) = \sum_{i=1}^{d} \lambda_i(A)$.

We shall assume throughout that the set of right-scattered points, denoted by

$$\hat{\mathbb{T}} := \{t \in \mathbb{T} : \mu(t) > 0\}$$

satisfies

$$\sup \hat{\mathbb{T}} = \infty.$$

That is, there exists a sequence $\{t_k\} \subset \mathbb{T}$ with

$$\lim_{k \to \infty} t_k = \infty$$

and $\mu(t_k) > 0$, $k = 1, 2, 3, \cdots$. We do not assume that $\mu(t)$ is bounded. Our main concern is the oscillatory behavior of equation (1). We recall that equation (1) is *nonoscillatory* on $[a, \infty)$ in case for any prepared solution $Y(t) \not\equiv 0$ there is a $t_1 > a$ such that

$$Y^*(\sigma(t))P(t)Y(t) > 0$$

for $t \in [t_1, \infty)$. (We refer to Erbe and Peterson [14, 15] for an additional discussion of these ideas and some oscillation criteria for the situation when the graininess function $\mu(t)$ is bounded.)

We will denote by $\mathbb{G}_\mu$ the set of all nonnegative locally integrable functions $g : \mathbb{T} \to [0, \infty)$ with $0 \le g(t) \le 1$, $t \in \mathbb{T}$ and such that

$$\int_a^\infty g(s)\mu(s)\Delta s := \lim_{t \to \infty, t \in \mathbb{T}} \int_a^t g(s)\mu(s)\Delta s = \infty.$$

We use the notation

$$G_\mu(t) = \int_a^t g(s)\mu(s)\Delta s$$

and

$$G_\mu(t; \tau) := \int_\tau^t g(s)\mu(s)\Delta s.$$

When we write

$$\frac{1}{G_\mu(t; \tau)} = G(t; \tau)^{-1},$$

we always mean that $t > \tau$ is sufficiently large so that $G_\mu(t; \tau) > 0$. We note that for any fixed $t_0 \in \mathbb{T}$

$$\lim_{t \to \infty} \frac{G_\mu(t; t_0)}{G_\mu(t)} = 1.$$

In addition, if $H(t)$ is a locally integrable $d \times d$ matrix-valued function with $\lim_{t \to \infty} H(t) = C$ (component-wise) where C is a $d \times d$ matrix, some of whose components could be $\pm\infty$ and if $g \in \mathbb{G}_\mu$, then (cf. [15])

$$\lim_{t \to \infty} \frac{\int_{t_0}^t \mu(s)g(s)H(s)\Delta s}{G_\mu(t; t_0)} = C.$$

3   Main results

We introduce the following condition on the trace of $P(t)$ which will be needed below.

(R): There exists a positive constant $M > 0$ and $g \in \mathbb{G}_\mu$ such that for all large t,

$$g(t)\mu(t)tr(P(t)) \leq M.$$

We shall see that condition (R) will allow us to relax the assumptions on the boundedness of the graininess function. (See the examples below.) The main result follows and gives some equivalent conditions for nonoscillation of (1).

Theorem 1 *Assume that (R) holds and that equation (1) is nonoscillatory. Then the following are equivalent:*

(i) $\lim_{t\to\infty} \frac{\int_{t_0}^t \mu(s)g(s)\int_{t_0}^s Q(\eta)\Delta\eta\Delta s}{G_\mu(t;t_0)} = C$, *where C is a constant Hermitian matrix which may depend on $g \in \mathbb{G}_\mu$.*

(ii) $\liminf_{t\to\infty} \frac{\int_{t_0}^t \mu(s)g(s)\int_{t_0}^s tr(Q(\eta))\Delta\eta\Delta s}{G_\mu(t;t_0)} > -\infty,$

(iii) *For any prepared solution Y of (1) with $Y^*(\sigma(t))P(t)Y(t) > 0$ for $t \geq t_0$ the function $Z(t) := P(t)Y^\Delta(t)Y^{-1}(t)$ satisfies*

$$F(t) := Z(t)[P(t) + \mu(t)Z(t)]^{-1}Z(t) \in \mathcal{C} \qquad (5)$$

Proof: $(i) \Rightarrow (ii)$: This is clear because of the linearity of the trace functional.

$(ii) \Rightarrow (iii)$: We assume that (ii) holds and let $Z(t)$ be defined as in (iii). It follows that $Z(t)$ is a Hermitian solution of the Riccati equation

$$Z^\Delta(t) + Q(t) + F(t) = 0, \quad t \geq t_0. \qquad (6)$$

It is easy to see that $F(t) \geq 0$ and that $P(t) + \mu(t)Z(t) > 0$ for $t \geq t_0$. We suppose for the sake of contradiction that $F \notin \mathcal{C}$ and consequently we have

$$\lim_{t\to\infty} \int_{t_0}^t tr(F(s))\Delta s = +\infty. \qquad (7)$$

An integration of the Riccati equation (6) gives

$$-Z(t) = -Z(t_0) + \int_{t_0}^t Q(s)\Delta s + \int_{t_0}^t F(s)\Delta s, \quad t \geq t_0. \qquad (8)$$

Applying the trace functional to both sides of (8), multiplying by $\mu(t)g(t)$, $g \in \mathbb{G}_\mu$, integrating, and dividing by $G_\mu(t;t_0)$ gives, for all large t,

$$\frac{\int_{t_0}^t \mu(s)g(s)tr(-Z(s))\Delta s}{G_\mu(t;t_0)} = -trZ(t_0) + \frac{\int_{t_0}^t \mu(s)g(s)\int_{t_0}^s tr(Q(\eta))\Delta\eta\Delta s}{G_\mu(t;t_0)}$$

$$+ \frac{\int_{t_0}^t \mu(s)g(s)\int_{t_0}^s tr(F(\eta))\Delta\eta\Delta s}{G_\mu(t;t_0)}. \qquad (9)$$

Now since $g \in \mathbb{G}_\mu$, it follows from (7) that the last term on the right side of (9) has limit $+\infty$ as $t \to \infty$. Hence, it follows that

$$\lim_{t \to \infty} \int_{t_0}^t \mu(s)g(s)\,\mathrm{tr}(-Z(s))\Delta s = +\infty \tag{10}$$

(since the second term on the right hand side of (9) is bounded below as $t \to \infty$). Therefore if we denote

$$B(t) := \int_{t_0}^t \mu(s)g(s)(-Z(s))\Delta s \tag{11}$$

we have $\mathrm{tr}\,(B(t)) \to \infty$ as $t \to \infty$. Therefore, for all large t we have, from (9), (10), and (11),

$$\mathrm{tr}\,B(t) \geq \frac{1}{2}\int_{t_0}^t \mu(s)g(s)\int_{t_0}^s \mathrm{tr}\,F(\eta)\Delta\eta\Delta s := \frac{1}{2}S(t). \tag{12}$$

Because of (7) it follows that $\lim_{t \to \infty} S(t) = +\infty$. Also, for all large t, we have by condition (R) and $P(t) > -\mu(t)Z(t)$ that

$$\mu(t)g(t)\mu(t)\mathrm{tr}(-Z(t) \leq g(t)\mu(t)\mathrm{tr}(P(t) \leq M \leq \int_{t_o}^t \mu(s)g(s)\mathrm{tr}(-Z(s))\Delta s. \tag{13}$$

By properties of traces we have

$$\mathrm{tr}(F(t)) = \mathrm{tr}\left((Z(t))^2(P(t) + \mu(t)Z(t))^{-1}\right) \geq$$
$$\geq \mathrm{tr}(Z(t))^2\lambda_d\left(P(t) + \mu(t)Z(t))^{-1}\right) = \frac{\mathrm{tr}(Z(t))^2}{\lambda_1\,(P(t) + \mu(t)Z(t))}. \tag{14}$$

Now

$$\mathrm{tr}(Z(t))^2 = \sum_{i=1}^d \lambda_i(Z(t)^2) \geq \frac{1}{d}\left(\sum_{i=1}^d \lambda_i(Z(t))\right)^2 = \frac{1}{d}\,(\mathrm{tr}(Z(t)))^2. \tag{15}$$

We define $J := \{t \geq t_0 : \mathrm{tr}(-Z(t)) \geq 0\}$ and it follows from (10) that $J \neq \phi$. Therefore, from (13), (14), (15), and the Cauchy-Schwartz inequality, we obtain

$$\left(\int_{t_0}^{t} \mu(s)g(s)(tr(-Z(s)))\Delta s \right)^2 \leq \left(\int_{J} \mu(s)g(s)(tr(-Z(s)))\Delta s \right)^2$$

$$\leq \left(\int_{J} \mu(s)g(s)\sqrt{\lambda_1(P(s)+\mu(s)Z(s))} \frac{tr(-Z(s)}{\sqrt{\lambda_1(P(s)+\mu(s)Z(s))}}\Delta s \right)^2$$

$$\leq \left(\int_{J} (\mu(s))^2(g(s))^2\lambda_1(P(s)+\mu(s)Z(s))\Delta s \right) \times$$

$$\times \left(\int_{J} \frac{(tr(Z(s)))^2}{\lambda_1(P(s)+\mu(s)Z(s))}\Delta s \right)$$

$$\leq \left(\int_{J} (\mu(s))^2(g(s))^2 tr(P(s)+\mu(s)Z(s))\Delta s \right) \left(d \int_{J} tr(F(s))\Delta s \right) \quad (16)$$

$$\leq d \left(\int_{J} (\mu(s))^2(g(s))^2 tr(P(s))\Delta s \right) \left(\int_{J} tr(F(s))\Delta s \right)$$

$$\leq dM \left(\int_{J} \mu(s)g(s)\Delta s \right) \left(\int_{J} tr(F(s))\Delta s \right)$$

$$\leq dM \left(\int_{t_0}^{t} \mu(s)g(s)\Delta s \right) \left(\int_{t_0}^{t} tr F(s))\Delta s \right),$$

where we have also used the fact that (R) holds and that $tr(Z(t)) \leq 0$ on J. We have therefore for all large t,

$$\frac{1}{4}S(t)S(\sigma(t)) \leq tr B(t)tr B(\sigma(t))$$

$$= \left(\int_{t_0}^{t} \mu(s)g(s)tr(-Z(s))\Delta s \right) \times$$

$$\times \left(\int_{t_0}^{t} \mu(s)g(s)tr(-Z(s))\Delta s + \int_{t}^{\sigma(t)} \mu(s)g(s)tr(-Z(s))\Delta s \right)$$

$$\leq \left(\int_{t_0}^{t} \mu(s)g(s)tr(-Z(s))\Delta s \right)^2 +$$

$$+ \left(\int_{t_0}^{t} \mu(s)g(s)tr(-Z(s))\Delta s \right) (\mu(t)g(t)tr(-Z(t))) \mu(t)$$

$$\leq \left(\int_{t_0}^{t} \mu(s)g(s)tr(-Z(s))\Delta s \right)^2 + M \left(\int_{t_0}^{t} \mu(s)g(s)tr(-Z(s))\Delta s \right)$$

$$\leq 2 \left(\int_{t_0}^{t} \mu(s)g(s)tr(-Z(s))\Delta s \right)^2 .$$

Consequently, we have from (16)

$$S(t)S(\sigma(t)) \; \leq \; 8\left(\int_{t_0}^{t} \mu(s)g(s)tr(-Z(s))\Delta s\right)^2$$

$$\leq \; 8dMG_\mu(t;t_0)\left(\int_{t_0}^{t} tr(F(s)\Delta s\right), \tag{17}$$

which gives

$$\frac{\mu(t)g(t)}{G_\mu(t;t_0)} \; \leq \; \frac{8dM\mu(t)g(t)\int_{t_0}^{t} trF(s)\Delta s}{S(t)S(\sigma(t))}$$

$$= \; \frac{8dM\,(S(t))^\Delta}{S(t)S(\sigma(t))} = 8dM\left(-\frac{1}{S(t)}\right)^\Delta. \tag{18}$$

Since $S(t) \to \infty$ we see from (18) and an integration that

$$\int_{t_0}^{\infty} \frac{\mu(s)g(s)}{G_\mu(s;t_0)}\Delta s < \infty, \tag{19}$$

which is a contradiction to the fact that $G_\mu(t;t_0) \to \infty$ (using Theorem 2.6 of [12]). Therefore it follows that (iii) holds.

We next want to prove that $(iii) \Rightarrow (i)$. We define

$$b(t) := \frac{\int_{t_0}^{t} \mu(s)g(s)|Z(s)|\Delta s}{G_\mu(t;t_0)}, \tag{20}$$

where $|Z(t)|$ denotes the norm of $Z(t)$; i.e., $|A| = (\lambda_1(A^*A))^{\frac{1}{2}}$, for any Hermitian matrix A. We shall show that $\lim_{t\to\infty} b(t) = 0$ from which it will follow that

$$\lim_{t\to\infty} \frac{\int_{t_0}^{t} \mu(s)g(s)tr(-Z(s))\Delta s}{G_\mu(t;t_0)} = 0. \tag{21}$$

If $\int_{t_0}^{\infty} \mu(s)g(s)|Z(s)|\Delta s < \infty$, then clearly we have $\lim_{t\to\infty} b(t) = 0$. So we may suppose that $\int_{t_0}^{\infty} \mu(s)g(s)|Z(s)|\Delta s = \infty$.

Notice first that since $\lambda_1(A + B) \leq \lambda_1(A) + \lambda_1(B)$ for Hermitian matrices A and B we have

$$|Z(t)|^2 \leq (tr(Z(t)))^2 \leq (trF(t))\,(\lambda_1(P(t) + \mu(t)Z(t))$$

$$\leq (trF(t))\,(\lambda_1(P(t)) + \lambda_1(\mu(t)Z(t))). \tag{22}$$

We have by the Cauchy-Schwartz inequality

$$\left(\int_{t_0}^{t} \mu(s)g(s)|Z(s)|\Delta s \right)^2$$

$$= \left(\int_{t_0}^{t} \sqrt{\mu(s)g(s)}\sqrt{\mu(s)g(s)}|Z(s)|\Delta s \right)^2$$

$$\leq \left(\int_{t_0}^{t} \mu(s)g(s)\Delta s \right)\left(\int_{t_0}^{t} \mu(s)g(s)|Z(s)|^2\Delta s \right)$$

$$\leq G_\mu(t;t_0)\left(\int_{t_0}^{t} \mu(s)g(s)tr F(s)(\lambda_1(P(s)+\lambda_1(\mu(s)Z(s)))\Delta s \right) \qquad (23)$$

$$\leq G_\mu(t;t_0)\left(\int_{t_0}^{t} \mu(s)g(s)tr F(s)\lambda_1(P(s))\Delta s + \right.$$

$$\left. + \int_{t_0}^{t} \mu(s)g(s)tr F(s)\lambda_1(\mu(s)Z(s))\Delta s \right)$$

$$\leq G_\mu(t;t_0)\left(M\int_{t_0}^{t} tr F(s)\Delta s + \int_{t_0}^{t} \mu(s)g(s)tr F(s)\lambda_1(\mu(s)Z(s))\Delta s \right).$$

Therefore, from (20) and (23) we obtain

$$b(t) = \frac{\int_{t_0}^{t} \mu(s)g(s)|Z(s)|\Delta s}{G_\mu(t;t_0)}$$

$$\leq \frac{M\int_{t_0}^{t} tr F(s)\Delta s}{\int_{t_0}^{t} \mu(s)g(s)|Z(s)|\Delta s} + \frac{\int_{t_0}^{t} \mu(s)g(s)tr F(s)\lambda_1(\mu(s)Z(s))\Delta s)}{\int_{t_0}^{t} \mu(s)g(s)|Z(s)|\Delta s} \qquad (24)$$

Now the first term on the right side of (24) $\to 0$ as $t \to \infty$ because $F \in C$ and $\int_{t_0}^{t} \mu(s)g(s)|Z(s)|\Delta s \to \infty$. We claim next that

$$\frac{\int_{t_0}^{t} \mu(s)g(s)tr F(s)\lambda_1(\mu(s)Z(s))\Delta s)}{\int_{t_0}^{t} \mu(s)g(s)|Z(s)|\Delta s} \to 0 \qquad (25)$$

as $t \to \infty$. Notice that if t_k is any sequence $\in \mathbb{T}$ with $t_k \to \infty$ and $\mu(t_k) > 0$ then

$$\sum_{k=1}^{\infty} \mu(t_k)tr F(t_k) = \sum_{k=1}^{\infty} \int_{t_k}^{\sigma(t_k)} tr F(s)\Delta s \leq \int_{t_0}^{\infty} tr F(s)\Delta s < \infty. \qquad (26)$$

and so we have $\lim \mu(t_k)F(t_k) = 0$. Therefore it follows that

$$\lim \frac{\int_{t_0}^{t_k} (\mu(s))^2 g(s)|Z(s)|\Delta s}{\int_{t_0}^{t_k} \mu(s)g(s)|Z(s)|\Delta s} = \lim \frac{(\mu(t_k))^2 g(t_k)|Z(t_k)|tr F(t_k)}{\mu(t_k)g(t_k)|Z(t_k)|}$$

$$= \lim \mu(t_k)tr F(t_k) = 0, \qquad (27)$$

(by L'Hopital's Rule). Hence, it follows that (21) holds. Consequently, from (9) we see that condition (i) holds. This completes the proof. $\qquad \square$

4   Examples

In this section we give two examples.

1: Let $d = 2$, and $\mathbb{T} = \bigcup_{k=0}^{\infty}\{[a_k, b_k]\}$, where $0 = a_0 < b_0 < a_1 < b_1 < \cdots < a_k < b_k$ for all $k \geq 0$ and $a_k \to \infty$. Therefore, $\mu(t) = 0$, if $a_k \leq t < b_k$ and $\mu(b_k) = a_{k+1} - b_k$. Let $Q(t) = diag(q_1(t), q_2(t))$. Suppose $P(t) \equiv I$, the identity matrix, and we define $q_i(t)$ such that $q_1(t) + q_2(t) \geq 0$ and is such that $\int_0^t q_i(s)\,\Delta s$ is arbitrary but satisfies for $i = 1$ or $i = 2$,

$$\frac{\sum_{j=0}^{k}(a_{j+1} - bj)\int_0^{b_j} q_i(\eta)\Delta\eta}{\sum_{j=0}^{k}(a_{j+1} - b_j)} \to \infty$$

as $k \to \infty$. Here we choose $g(t) = 0$, $t \in [a_k, b_k)$, $g(b_k) = \frac{1}{\mu(b_k)}$. It then follows that the limit in (i) of the theorem does not exist although the $\liminf$ in (ii) is nonnegative. Therefore, $Lx = 0$ is oscillatory.

2. As a second example, let $d = 2$, and consider the time scale $\mathbb{T} = \{2^k\}$ $k = 0, 1, \ldots$. Note that $\mu(t) = \sigma(t) - t = 2^{k+1} - 2^k = 2^k$, $t = 2^k := t_k$. We let $Q(t) := (q_{ij}(t))$ and suppose that for some (i_o, j_o) with $1 \leq i_o, j_o \leq 2$ we have $q_{i_o j_o}(t) := \frac{\gamma}{t}$ where $\gamma > 0$. Assume further that $trQ(t) = q_{11}(t) + q_{22}(t) \geq 0$ and that $P(t)$ satisfies $1 \leq trP(t) \leq t$ for $t \in \mathbb{T}$. Let us define $g(t) := \frac{1}{t^2}$. We note that (R) holds since $\mu(t)g(t)trP(t) \leq 1$. We also have:

$$\int_1^{t_k} q_{i_o j_o}(s)\Delta s = \int_1^{2^k} q_{i_o j_o}(s)\Delta s = \sum_{j=0}^{k-1} q_{i_o j_o}(2^j)\mu(2^j) = \gamma k.$$

Further,

$$G_{\mu}(t_k; t_0) = \int_{t_0}^{t_k} \mu(s)g(s)\,\Delta s = \int_1^{2^k} \mu(s)g(s)\,\Delta s = \sum_{j=0}^{k-1}(\mu(2^j))^2 g(2^j) = k$$

and

$$\int_{t_0}^{t_k} \mu(s)g(s)\int_1^s q_{i_o j_o}(\eta)\Delta\eta\Delta s = \int_1^{2^k} \mu(s)g(s)\int_1^s q_{i_o j_o}(\eta)\Delta\eta\Delta s$$

$$= \sum_{j=0}^{k-1}(\mu(2^j))^2 g(2^j)\int_0^{2^j} q_{i_o j_o}(s)\Delta s = \sum_{j=0}^{k-1}\int_0^{2^j} q_{i_o j_o}(s)\Delta s$$

$$= \sum_{j=0}^{k-1} \gamma j = \gamma(\frac{k(k-1)}{2}).$$

Therefore, it follows that $Lx = 0$ is oscillatory since the quotient appearing in the right hand side of (i) does not converge to a matrix $C \in \mathcal{M}$ whereas the $\liminf$ in (ii) is ≥ 0 .

References

[1] R. Agarwal and M. Bohner, Basic calculus on time scales and some of its applications, *Results Math.*, 35, (1999), 3–22.

[2] C. Ahlbrandt and A. Peterson, *Discrete Hamiltonian Systems: Difference Equations, Continued Fractions, and Riccati Equations*, Kluwer Academic Publishers, Boston, 1996.

[3] R. Agarwal and M. Bohner, Quadratic functionals for second order matrix equations on time scales, *Nonlinear Anal.*, 33 (1998) 675–692.

[4] R. Agarwal, M. Bohner, and P. Wong, Sturm-Liouville eigenvalue problems on time scales, *Appl. Math. Comput.*, 99 (1999), 153–166.

[5] B. Aulbach and S. Hilger, Linear dynamic processes with inhomogeneous time scale, in *Nonlinear Dynamics and Quantum Dynamical Systems*, Akademie Verlag, Berlin, 1990.

[6] M. Bohner and A. Peterson, *Dynamic Equations on Time Scales : An Introduction with Applications*, Birkhauser, Boston, 2001.

[7] G. J. Butler, L. H. Erbe, and A. B. Mingarelli, Riccati techniques and variational principles in oscillation theory for linear systems, *Trans. Amer. Math. Soc.* 303 (1987), 263–282.

[8] R. Byers, B. J. Harris, and M. K. Kwong, Weighted means and oscillation conditions for second order matrix differential equations, *J. Differential Equations*, 61 (1986), 164 177.

[9] S. Chen and L. Erbe, Oscillation and nonoscillation for systems of self-adjoint second order difference equations, *SIAM J. Math. Anal.* 20 (1989), 939 949.

[10] L. Erbe, Oscillation criteria for second order linear equations on a time scale, *Canadian Applied Math Quarterly*, to appear.

[11] L. Erbe, Oscillation results for second order linear equations on a time scale, *J. Difference Eqns and Applications*, 8 (11), 2002, 1061–1071.

[12] L. Erbe and S. Hilger, Sturmian Theory on Measure Chains, *Differential Equations and Dynamical Systems*, 1 (1993), 223–246.

[13] L. H. Erbe and A. Peterson, Green's functions and comparison theorems for differential equations on measure chains, *Dynam. Contin. Discrete Impuls. Systems*, 6, (1999), 121–137.

[14] L. H. Erbe and A. Peterson, Oscillation criteria for second-order matrix dynamic equadtions on a time scale, *Journal of Computational and Applied Mathematics*, 141 (2002), 169–186.

[15] L. H. Erbe and A. Peterson, Averaging Techniques for Self-Adjoint Matrix Equations on a Measure Chain, *J. Math.Anal. Appl.* , 271 (2002), 31–58.

[16] S. Hilger, Analysis on measure chains – a unified approach to continuous and discrete calculus, *Results in Mathematics*, 18 (1990), 18–56.

[17] B. Kaymakcalan, V. Laksmikantham, and S. Sivasundaram, *Dynamical Systems on Measure Chains*, Kluwer Academic Publishers, Boston, 1996.

[18] W. Kelley and A. Peterson, *Difference Equations: An Introduction with Applications*, Academic Press, Second Edition, 2001.

[19] A. Peterson and J. Ridenhour, Oscillation of Second Order Linear Matrix Difference Equations, *J. Differential Equations*, 89 (1991), 69–21.

Continuous Dependence in Time Scale Dynamics

BARNABAS M. GARAY

Mathematics Institute, Technical University
H–1521 Budapest, Hungary
E-mail: garay@math.bme.hu

STEFAN HILGER

Didaktik der Physik und Mathematik
Katholische Universität Eichstätt
D-85071 Eichstätt, Germany
E-mail: stefan.hilger@ku-eichstaett.de

and

PETER E. KLOEDEN

FB Mathematik, Johann Wolfgang Goethe Universität
D-60054 Frankfurt am Main, Germany
E-mail: kloeden@math.uni-frankfurt.de

Abstract The aim of this paper is to extend the well-known principle *"uniqueness $\Rightarrow$ continuous dependence over compact time intervals"* from ordinary differential equations to dynamical equations on time scales of the form

$$x^\Delta = f(t, x), \qquad (t, x) \in \mathbb{T} \times \mathbb{R}^n$$

where x^Δ is the dynamical derivative on the time scale $\mathbb{T}$ (i.e., an arbitrary nonempty closed subset of $\mathbb{R}$) and $f : \mathbb{T} \times \mathbb{R}^n \to \mathbb{R}^n$ is a continuous function. Continuous dependence is understood with respect to the initial values $(t_0, x_0) \in \mathbb{T} \times \mathbb{R}^n$, the function f and the time scale $\mathbb{T}$. In addition to the abstract continuity result, the possibility of Gronwall–type estimates for dynamical equations on different time scales is also discussed.

Keywords Time Scale dynamics, Continuous dependence, Gronwall lemma

AMS Subject Classification 34A12, 39A12

0-415-31675-8/04/$0.00+$1.50

1   Dynamical equations on time scales

A time scale $\mathbb{T}$ is simply a nonempty closed subset of $\mathbb{R}$ representing the time instants that are of interest in some situations [1, 5, 7]. For example, in population dynamics (e.g., [2]) a time scale of the form $\cup_{n\in\mathbb{Z}}[n, n+1-\delta]$ for some $\delta \in (0, 1)$ arises in modelling an insect population, while in numerical dynamics (e.g., [6, 8, 9]) one compares the dynamics of a discrete time dynamical system on a time scale $h\mathbb{Z}$ generated by a one-step numerical scheme (such as a Runge–Kutta scheme) with a constant time step $h > 0$ with the dynamics of a continuous time dynamical system on the time scale $\mathbb{R}$ that is governed by an ordinary differential equation to which the numerical scheme is applied.

To avoid unnecessary complications we will assume that sup $\mathbb{T} = \infty$ and inf $\mathbb{T} = -\infty$, although we will mostly consider a nonempty finite time interval $\mathbb{T} \cap [a, b]$ in $\mathbb{T}$. Following [1, 5], we say that a time instant $t_0 \in \mathbb{T}$ is *right-dense* if $\sigma_{\mathbb{T}}(t_0) = t_0$, where $\sigma_{\mathbb{T}}(t_0) := \inf\{t \in \mathbb{T} \,|\, t > t_0\}$; otherwise we say that t_0 is *right-scattered* and call the open interval $(t_0, \sigma_{\mathbb{T}}(t_0))$ the *time-gap* at t_0. Thus the complement of $\mathbb{T}$ is the countable union of all such time-gaps. Similarly, a time instant $t_1 \in \mathbb{T}$ is called *left-scattered* if $t_1 = \sigma_{\mathbb{T}}(t_0) > t_0$ for some $t_0 \in \mathbb{T}$, otherwise t_1 is *left-dense*.

Consider a function $x : \mathbb{T} \to \mathbb{R}^n$ and fix a point $t_0 \in \mathbb{T}$. If t_0 is right-scattered and x is continuous at t_0, then the *dynamical derivative* of the function x at t_0 is defined as the difference quotient

$$x^\Delta(t_0) = \frac{x(\sigma_{\mathbb{T}}(t_0)) - x(t_0)}{\sigma_{\mathbb{T}}(t_0) - t_0},$$

and if t_0 is right-dense, then the dynamical derivative of the function x at t_0 is defined as

$$x^\Delta(t_0) = \lim_{t\to t_0, t\in\mathbb{T}} \frac{x(t) - x(t_0)}{t - t_0},$$

provided that the limit exists in $\mathbb{R}^n$.

Given $(t_0, x_0) \in \mathbb{T} \times \mathbb{R}^n$, the continuous function $x : \mathbb{T} \to \mathbb{R}^n$ is said to be a (local right) *solution of the initial value problem*

$$x^\Delta = f(t, x), \qquad x(t_0) = x_0, \ t_0 \leq t, \ t \in \mathbb{T} \tag{1}$$

if, for some $T \in \mathbb{T} \cap (\sigma_{\mathbb{T}}(t_0), \infty)$, x is differentiable on $\mathbb{T} \cap [t_0, T)$, its dynamical derivative x^Δ satisfies the *dynamical equation* $x^\Delta(t) = f(t, x(t))$ for each $t \in \mathbb{T} \cap [t_0, T)$, and $x(t_0) = x_0$. The initial value problem (1) can be rewritten as a time scale integral equation over $\mathbb{T}$, namely

$$x(t) = x_0 + \int_{t_0}^{t} f(s, x(s)) \, \Delta_{\mathbb{T}}(s), \qquad t_0 \leq t, \ t_0, t \in \mathbb{T} \tag{2}$$

where $\Delta_{\mathbb{T}}(s)$ stays for the so-called dynamical integration. (Note if $\mathbb{T} = \mathbb{R}$, dynamical equations simplify to ordinary differential equations.)

For arbitrary $t_0 \leq t$ with $t_0, t \in \mathbb{T}$, we note that

$$\int_{t_0}^{t} f(s, x(s)) \, \Delta_{\mathbb{T}}(s) = \int_{t_0}^{t} \bar{f}(s, \bar{x}(s)) \, ds, \tag{3}$$

where ds stands for standard Riemann integration on the $\mathbb{R}$-interval $[t_0, t]$ and the functions $\bar{x} : \mathbb{R} \to \mathbb{R}^n$ and $\bar{f} : \mathbb{R} \times \mathbb{R}^n \to \mathbb{R}^n$ defined by

$$\bar{x}(s) := \begin{cases} x(s) & \text{if } s \in \mathbb{T} \\ x(s_0) & \text{if } s \in (s_0, \sigma_{\mathbb{T}}(s_0)) \text{ for some } s_0 \in \mathbb{T} \end{cases}$$

and

$$\bar{f}(s, x) := \begin{cases} f(s, x) & \text{if } (s, x) \in \mathbb{T} \times \mathbb{R}^n \\ f(s_0, x) & \text{if } (s, x) \in (s_0, \sigma_{\mathbb{T}}(s_0)) \times \mathbb{R}^n \text{ for some } s_0 \in \mathbb{T} \end{cases}$$

are called the *piecewise constant interpolants* of x and f, respectively. Since the collection of all left-scattered points is a countable subset of $\mathbb{T}$, the existence of the Riemann integral on the right-hand side of (3) follows from the simple observation that the piecewise constant interpolants $\bar{x}$ and $\bar{f}$ are continuous on the sets $\{t \in \mathbb{R} \,|\, t \text{ is a left-dense point of } \mathbb{T}\}$ and $\{(t, x) \in \mathbb{R} \times \mathbb{R}^n \,|\, t \text{ is a left-dense point of } \mathbb{T}\}$, respectively.

The simplest continuous extension of a continuous function $x : \mathbb{T} \to \mathbb{R}^n$ to the whole line $\mathbb{R}$ is defined by

$$\hat{x}(s) = \begin{cases} x(s) & \text{if } s \in \mathbb{T} \\ x(s_0) + \dfrac{s - s_0}{\sigma_{\mathbb{T}}(s_0) - s_0} \left(x(\sigma_{\mathbb{T}}(s_0)) - x(s_0) \right) & \text{if } s \in (s_0, \sigma_{\mathbb{T}}(s_0)) \\ & \text{for some } s_0 \in \mathbb{T}. \end{cases}$$

We call the function $\hat{x} : \mathbb{R} \to \mathbb{R}^n$ the *linear interpolant of* x. Using property (3) again, we see that the integral equation (2) on the time scale $\mathbb{T}$ can be reformulated as what we call a *hybrid integral equation* on $\mathbb{R}$, namely

$$\hat{x}(t) = x_0 + \int_{t_0}^{t} \bar{f}(s, \bar{x}(s)) \, ds, \qquad t_0 \leq t, \ t_0 \in \mathbb{T}, \ t \in \mathbb{R}. \tag{4}$$

We note (cf. [1, 5, 7]) that the classical results for ordinary differential equations on the existence, uniqueness and C^k dependence of solutions can be transferred to the setting of dynamical equations on time scales. There are some extra difficulties concerning left sided solutions, but we can ensure the existence and uniqueness of the solutions to the left of t_0 by imposing an additional condition such as the invertibility of the mapping from $\mathbb{R}^n$ into $\mathbb{R}^n$ defined by $x \mapsto x + (\sigma(t) - t)f(t, x)$ for all $t \leq t_0$.

2   Metrics for dynamical equations on different time scales

In order to state our "uniqueness $\Rightarrow$ continuous dependence over compact time intervals" result, we need some further preparation.

First, for the convenience of the reader, we recall the concept of the Hausdorff metric. Let $(\mathcal{M}, d)$ be a compact metric space and let $\mathcal{K}(\mathcal{M})$ denote the collection of all nonempty compact subsets of $\mathcal{M}$. The distance between a point $m \in \mathcal{M}$ and a set $A \in \mathcal{K}(\mathcal{M})$ is defined as $d(m, A) = \inf\{d(m, a) \,|\, a \in A\}$. Then, for $A, B \in \mathcal{K}(\mathcal{M})$, we define a metric on $\mathcal{K}(\mathcal{M})$ by

$$d_H(A, B) = \max\{\max\{d(m, A) \,|\, m \in B\},\ \max\{d(m, B) \,|\, m \in A\}\},$$

which is called the *Hausdorff metric*. Moreover, $(\mathcal{K}(\mathcal{M}), d_H)$ is a compact metric space.

Let us fix a compact interval $[a, b] \subset \mathbb{R}$. We define

$$\mathcal{T}_{[a,b]} = \{\mathbb{T} \subset \mathbb{R} \mid \mathbb{T} \text{ is closed, and } \mathbb{T} \cap [a, b] \neq \emptyset\}$$

and denote the closed ball in $\mathbb{R}^n$ of radius $R > 0$ and centered on the origin by $B_R = \{x \in \mathbb{R}^n : |x| \leq R\}$. Then for positive parameters M and L we denote

$$C((\mathbb{T} \cap [a, b]) \times B_M, B_L) =$$
$$\{f : (\mathbb{T} \cap [a, b]) \times B_M \to B_L \mid f \text{ is continuous }\},$$

$$C_L(\mathbb{T} \cap [a, b], B_M) =$$
$$\{x : (\mathbb{T} \cap [a, b]) \to B_M \mid x \text{ is continuous and}$$
$$|x(s) - x(t| \leq L \,|s - t| \text{ whenever } s, t \in \mathbb{T} \cap [a, b]\},$$

$$C(\mathcal{T}_{[a,b]} \times B_M, B_L) = \cup\{C((\mathbb{T} \cap [a, b]) \times B_M, B_L) \mid \mathbb{T} \in \mathcal{T}_{[a,b]}\},$$

$$C_L(\mathcal{T}_{[a,b]}, B_M) = \cup\{C_L(\mathbb{T} \cap [a, b], B_M) \mid \mathbb{T} \in \mathcal{T}_{[a,b]}\}.$$

We equip the set $C_L(\mathcal{T}_{[a,b]}, B_M)$ with the Hausdorff metric on the graphs of the respective functions. More precisely, given $x,\ y \in C_L(\mathcal{T}_{[a,b]}, B_M)$ with domains $\mathbb{T}_x, \mathbb{T}_y \in \mathcal{T}_{[a,b]}$, we find that both

$$\text{graph}(x) = \{(t, x(t)) \in \mathbb{R} \times \mathbb{R}^n \mid t \in \mathbb{T}_x \cap [a, b]\}$$

and

$$\text{graph}(y) = \{(t, y(t)) \in \mathbb{R} \times \mathbb{R}^n \mid t \in \mathbb{T}_y \cap [a, b]\}$$

are nonempty compact subsets of $[a, b] \times B_M \subset \mathbb{R} \times \mathbb{R}^n$. Then

$$\rho_H(x, y) = d_H(\text{graph}(x), \text{graph}(y)),$$

where d_H stays for the Hausdorff metric on the space $\mathcal{K}([a,b] \times B_M)$ of nonempty compact subsets of $[a,b] \times B_M$, is a metric on $C_L(\mathcal{T}_{[a,b]}, B_M)$. (For simplicity, we suppose that d_H is induced by the product metric $d((s,q),(t,p)) = |s-t| + |q-p|$ on $[a,b] \times B_M$.)

Similarly, we equip the set $C(\mathcal{T}_{[a,b]} \times B_M, B_L)$ with the Hausdorff metric on the graphs of the respective functions: given $f, g \in (\mathcal{T}_{[a,b]} \times B_M, B_L)$ with arbitrary domains $\mathbb{T}_f \times B_M$ and $\mathbb{T}_g \times B_M$, we set

$$\rho_H(f,g) = d_H(\text{graph}(f), \text{graph}(g))$$

where d_H now stands for the Hausdorff metric on the space $\mathcal{K}(([a,b] \times B_M) \times B_L)$ of nonempty compact subsets of $([a,b] \times B_M) \times B_L$, again induced, for simplicity, by the product metric on $([a,b] \times B_M) \times B_L$. (The use of the same symbol d_H on different spaces should cause no confusion.)

Returning to $C_L(\mathcal{T}_{[a,b]}, B_M)$ we observe that $\{\text{graph}(x) \mid x \in C_L(\mathcal{T}_{[a,b]}, B_M)\}$ is a closed subset of $(\mathcal{K}([a,b] \times B_M), d_H)$. (The proof is essentially the same as the one of property (6) below.) On the other hand, $\{\text{graph}(f) \mid f \in C(\mathcal{T}_{[a,b]} \times B_M, B_L)\}$ is not a closed subset of $(\mathcal{K}(([a,b] \times B_M) \times B_L), d_H)$, since, in general, the limits of graphs under the Hausdorff metric need not be graphs.

3   Uniqueness $\Rightarrow$ continuous dependence

We are now in a position to present the main result of this paper.

For $k = 0, 1, \ldots$, let $f_k : \mathbb{T}_k \times \mathbb{R}^n \to \mathbb{R}^n$ be a bounded continuous function and let $\varphi_k : \mathbb{T}_k \cap [t_k, \infty) \to \mathbb{R}^n$ be a solution to the initial value problem

$$x^\Delta = f_k(t,x), \quad x(t_k) = x_k, \quad t_k \leq t \in \mathbb{T} \tag{5}$$

where $\mathbb{T}_k$ is a time scale and $(t_k, x_k) \in \mathbb{T}_k \times \mathbb{R}^n$. We assume that there are positive constants K and L such that $|x_k| \leq K$ and $|f_k(t,x)| \leq L$ whenever $(t,x) \in \mathbb{T}_k \times \mathbb{R}^n$, for $k = 0, 1, \ldots$. In addition, we assume for $k = 0$ that any right local solution to (5) is a restriction of φ_0. Finally, for some compact interval $[a,b]$ we assume that $t_k \in [a,b]$ and $\mathbb{T}_k \in \mathcal{T}_{[a,b]}$, for each $k = 0, 1, \ldots$ and observe that

$$f_k|_{(\mathbb{T}_k \cap [t_k,b]) \times B_M} \in C(\mathcal{T}_{[a,b]} \times B_M, B_L), \qquad k = 0, 1, \ldots$$

and

$$\varphi_k|_{\mathbb{T}_k \cap [t_k,b]} \in C_L(\mathcal{T}_{[a,b]}, B_M), \quad k = 0, 1, \ldots,$$

where $M = K + (b-a)L$.

Theorem 1 *Under the above assumptions,*

$$\rho_H(\varphi_k|_{\mathbb{T}_k \cap [t_k,b]}, \varphi_0|_{\mathbb{T}_0 \cap [t_0,b]}) \to 0,$$

whenever $(t_k, x_k) \to (t_0, x_0)$ *and* $\rho_H(f_k|_{(\mathbb{T}_k \cap [t_k,b]) \times B_M}, f_0|_{(\mathbb{T}_0 \cap [t_0,b]) \times B_M}) \to 0$ *as* $k \to \infty$.

Proof. The standard proof (see, e.g., in [4], Chapter 1) of the corresponding result for ordinary differential equations can be repeated here, except that the application of Arzela Theorem is replaced by a compactness argument involving the Hausdorff metric. As in the case of ordinary differential equations, the compactness part of the proof is followed by applying a convergence theorem for the corresponding family of integral equations.

By passing to a subsequence, we may assume that

$$\lim_{k \to \infty} d_H(\text{graph}(\varphi_k), C^*) = 0 \quad \text{for some } C^* \in \mathcal{K}([a,b] \times B_M).$$

We claim that

$$C^* = \text{graph}(q^*) \quad \text{for some } q^* \in C_L(\mathcal{T}_{[a,b]}, B_M). \tag{6}$$

In fact, the definition of d_H implies that

$$C^* = \left\{ (t^*, w^*) \in [a,b] \times B_M \,\middle|\, \exists \tau_k \in \mathbb{T}_k \cap [t_k, b] \right. \tag{7}$$
$$\left. \text{with } (\tau_k, \varphi_k(\tau_k)) \to (t^*, w^*) \text{ as } k \to \infty \right\}.$$

Using compactness again, it follows that $\{(t^*, w^*) \mid w \in B_M\} \cap C^* \neq \emptyset$ if and only if $\tau_k \to t^*$ as $k \to \infty$ for some sequence $\{\tau_k\}_{k=1}^{\infty}$ with $\tau_k \in \mathbb{T}_k \cap [a,b]$. In other words, we find that $d_H(\mathbb{T}_k \cap [a,b], \mathbb{T}^*) \to 0$ for some nonempty compact subset $\mathbb{T}^*$ of $[a,b]$, where d_H here denotes the Hausdorff metric on the space $\mathcal{K}([a,b])$ of nonempty compact subsets of $[a,b]$. By letting $k \to \infty$ in the inequality $|\varphi_k(\tau_k) - \varphi(\tilde{\tau}_k)| \leq L|\tau_k - \tilde{\tau}_k|$, we conclude that $|w^* - \tilde{w}^*| \leq L|t^* - t^*| = 0$ whenever $(t^*, w^*), (t^*, \tilde{w}^*) \in C^*$. Hence $w^* = \tilde{w}^*$ and thus $C^* = \text{graph}(q^*)$ for some function $q^* : \mathbb{T}^* \to B_M$.

A similar limiting argument implies that function q^* satisfies inequality $|q^*(s) - q^*(t)| \leq L|s - t|$ for each $s, t \in \mathbb{T}^*$. This completes the proof of claim (6) and, by the above construction, it means that $\rho_H(\varphi_k, q^*) \to 0$ as $k \to \infty$.

In view of the uniqueness assumption on φ_0, it remains to prove that $q^*|_{\mathbb{T}_0 \cap [t_0, b]} = \varphi_0|_{\mathbb{T}_0 \cap [t_0, b]}$. Although our integrals are understood in the sense of Riemann, our strategy is to apply Lebesgue's Dominated Convergence Theorem for the family of the hybrid integral equations

$$\hat{\varphi}_k(t) = x_k + \int_{t_k}^{t} \bar{f}_k(s, \bar{\varphi}_k(s)) \, ds,$$
$$t_k \leq t, \ t_k \in \mathbb{T}, \ t \in \mathbb{R} \cap [t_k, b], \ k = 1, 2, \ldots. \tag{8}$$

We begin by observing that, as a further consequence of (7), $\varphi_k(\tau_k) \to q^*(t^*)$ whenever $\tau_k \to t^*$ as $k \to \infty$ for some sequence $\{\tau_k\}_{k=1}^{\infty}$ with $\tau_k \in \mathbb{T}_k \cap [a,b]$. A similar argument yields that

$$\text{graph}(f_0|_{(\mathbb{T}_0 \cap [t_0, b]) \times B_M}) = \tag{9}$$
$$\left\{ (t^*, y, w^*) \in ([a,b] \times B_M) \times B_L \,\middle|\, \exists (\tau_k, y_k) \right.$$
$$\in (\mathbb{T}_k \cap [t_k, b]) \times B_M \text{ with } (\tau_k, y_k, f_k(\tau_k, y_k))$$
$$\left. \to (t^*, y, w^*) \text{ as } k \to \infty \right\}.$$

As a by-product, we obtain that $d_H(\mathbb{T}_k \cap [t_k, b], \mathbb{T}_0 \cap [t_0, b]) \to 0$ as $k \to \infty$, where d_H here is the Hausdorff metric on the space $\mathcal{K}([a, b])$. The uniqueness of the limit implies that $\mathbb{T}^* \cap [t_0, b] = \mathbb{T}_0 \cap [t_0, b]$.

As a further consequence of (9), the function family $\{f_k|_{(\mathbb{T}_k \cap [t_k, b]) \times B_M}\}_{k=1}^{\infty}$ is uniformly equicontinuous. If not, then there would be two sequences $\{(s_k, x_k)\}_{k=1}^{\infty}$, $\{(\tau_k, y_k)\}_{k=1}^{\infty}$ and a constant $\eta_0 > 0$ with s_k, $\tau_k \in \mathbb{T}_k \cap [t_k, b]$ and x_k, $y_k \in B_M$, such that $|s_k - \tau_k| \to 0$, and $|x_k - y_k| \to 0$, but with $|f_k(s_k, x_k) - f_k(\tau_k, y_k)| > \eta_0$ for each $k = 1, 2, \ldots$. Using compactness, there is no loss of generality in assuming that s_k, $\tau_k \to s^*$ and x_k, $y_k \to y$ for some $s^* \in \mathbb{T}_0 \cap [t_0, b]$ and $y \in B_M$. It follows that $0 = |f_0(s^*, y) - f_0(s^*, y)| \geq \eta_0$, which is a contradiction. The uniform continuity of the family of functions $\{\varphi_k|_{\mathbb{T}_k \cap [t_k, b]}\}_{k=1}^{\infty}$ follows directly from the uniform Lipschitz property.

Next we discuss the convergence properties of the respective piecewise constant interpolants. Consider a time instant $s \in S := (t_0, b) \setminus \{s \in \mathbb{T}_0 \mid s$ is a left-scattered point of $\mathbb{T}_0\}$. There are two cases according to whether s belongs to a time-gap of $\mathbb{T}_0$ or s is a left-dense point of $\mathbb{T}_0$. If $s \in (s_0, \sigma_{\mathbb{T}_0}(s_0))$ for some $s_0 \in [t_0, b)$, then $s \in (s_k, \sigma_{\mathbb{T}_k}(s_k))$ for k sufficiently large, where $s_k \in \mathbb{T}_k \cap [t_k, b)$ and $\lim_{k \to \infty} s_k = s_0$. Consequently, $\bar{\varphi}_k(s) = \varphi_k(s_k) \to q^*(s_0) = \bar{q}^*(s)$. If s is a left-dense point of $\mathbb{T}_0 \cap (t_0, b)$, then there exists a sequence $\{s_k\}$ (taking k large enough) such that $s_k \in \mathbb{T}_k \cap [t_k, s)$ and $\lim_{k \to \infty} s_k = s$. Since $t_k \leq s_k < s$, the uniform equicontinuity of the family of functions $\{\varphi_k|_{\mathbb{T}_k \cap [t_k, b]}\}_{k=1}^{\infty}$ implies that $|\varphi_k(s_k) - \varphi_k(s)| \to 0$ as $k \to \infty$. By the construction, however, $\varphi_k(s_k) \to q^*(s)$. Consequently, $\bar{\varphi}_k(s) \to \bar{q}^*(s)$.

Summing the previous considerations, we conclude that

$$\bar{\varphi}_k(s) \to \bar{q}^*(s) \qquad \text{for almost every } s \in [t_0, b] \,,$$

where the exceptional set is countable because the set of all left-scattered points of $\mathbb{T}_0$ is countable. With the same exceptional set, a similar argument implies that

$$\bar{f}_k(s, \bar{\varphi}_k(s)) \to \bar{f}_0(s, q^*(s)) \qquad \text{for almost every } s \in [t_0, b] \,.$$

On the other hand, the linear interpolant functions on the left-hand side of the integral equation (8) converge to $\hat{q}^*|_{[t_0, b]}$, the restriction of the linear interpolant of q^* to the interval $[t_0, b]$. This latter convergence is uniform and can be proven by using a simplified version of the method applied for the piecewise constant interpolants.

By letting $k \to \infty$ in (4), we arrive at the conclusion that

$$\hat{q}^*(t) = x_0 + \int_{t_0}^{t} \bar{f}_0(s, \bar{q}^*(s)) \, ds \,, \qquad t_0 \leq t \,, \ t_0 \in \mathbb{T} \,, \ t \in \mathbb{R} \cap [t_0, b] \,.$$

In other words, $q^*|_{\mathbb{T}_0 \cap [t_0, b]}$ solves the initial value problem $x^{\Delta} = f_0(t, x)$, $x(t_0) = x_0$ on $\mathbb{T}_0 \cap [t_0, b]$.

By the uniqueness assumption, we obtain that $q^*|_{\mathbb{T}_0 \cap [t_0, b]} = \varphi_0|_{\mathbb{T}_0 \cap [t_0, b]}$, which is what remain to be proven. $\qquad \square$

4   Towards a Gronwall Lemma on pairs of time scales

The time-gaps of a time scale can be "filled up" by a great variety of inter-polants, not only by the piecewise constant or piecewise linear functions con-sidered above. Under natural conditions on the time scale, Garay and Hilger [3] used cubic interpolants to embed solutions of dynamical equations into solutions of suitably chosen ordinary differential equations. The embeddings in [3] were understood for the vector fields and for the induced dynamical systems simultaneously. In what follows we point out that the main embed-ding result in [3] leads to an estimate between the solutions of two dynamical equations on different time scales, which is of particular interest in numerical dynamics where one compares corresponding dynamical equations on the time scales $h\mathbb{Z}$ and $\mathbb{R}$. In a number of special cases, this estimate is strong enough to imply continuous dependence with respect to the initial values, the right-hand sides, *and* the time scales. However, the conditions that are natural for embeddability are too restrictive when large time-gaps are present. Thus the problem of finding a genuine counterpart of the classical Gronwall lemma for pairs of general time scales still remains open.

Suppose, for $i = 1, 2$, that a time scale $\mathbb{T}_i$ has the property that the set

$$\Omega_i = \sup\{\,\sigma_{\mathbb{T}_i}(t) - t \mid t \in \mathbb{T}_i \text{ is right-scattered}\}$$

is finite and that a function $f_i : \mathbb{T}_i \times \mathbb{R}^n \to \mathbb{R}^n$ is continuous with

$$|f_i(t, x) - f_i(t, y)| \le L|x - y| \quad \text{whenever} \quad (t, x), (t, y) \in \mathbb{T}_i \times \mathbb{R}^n$$

for some $L \ge 0$. In addition assume that

$$\mathrm{id}_X + (\sigma(t) - t)f(t, \cdot) : \mathbb{R}^n \to \mathbb{R}^n \quad \text{is a self-homeomorphism of } \mathbb{R}^n \quad (10)$$

for each $t \in \mathbb{T}_i$. Then, for $i = 1$ and 2, by the fundamental existence and uniqueness theorem in [1] for dynamical equations on time scales there exists a continuous function $\varphi_i : \mathbb{T}_i \times \mathbb{T}_i \times \mathbb{R}^n \to \mathbb{R}^n$ with the properties that, given arbitrary $(t_0^i, x_0^i) \in \mathbb{T}_i \times \mathbb{R}^n$, the function $\varphi_i(\cdot, t_0^i, x_0^i) : \mathbb{T}_i \to \mathbb{R}^n$ is the unique (noncontinuable) solution to the initial value problem $x^\Delta = f_i(t, x)$, $x(t_0^i) = x_0^i$.

Now let $\Omega = \max\{\Omega_1, \Omega_2\}$ and assume that

$$\Omega L < 1/6.$$

(As it is explained in [1], the role of condition (10) is to ensure backward existence and backward uniqueness. In view of the the Hadamard–Levi global inverse function theorem, condition (10) is implied by inequality $\Omega L < 1$. The role of inequality $\Omega L < 1/6$ is to ensure embeddability. Inequality $\Omega L < 1/6$ is not optimal – it can be weakened e.g. to $31\Omega L + 4\Omega^2 L^2 < 27$ but cannot be replaced by $\Omega L < 2 + \eta$ with $\eta > 0$, see [3].)

Applying Theorem 3 of [3], for $i = 1$ and 2, there exists a continuous function $F_i : \mathbb{R} \times \mathbb{R}^n \to \mathbb{R}^n$ such that $F_i|_{\mathbb{T}_i \times \mathbb{R}^n} = f_i$ and

$$|F_i(t,x) - F_i(t,y)| \le 216L\,|x - y| \quad \text{for} \ (t,x),\ (t,y) \in \mathbb{R} \times \mathbb{R}^n.$$

Moreover, $\Phi_i|_{\mathbb{T}_i \times \mathbb{T}_i \times \mathbb{R}^n} = \varphi_i$ where $\Phi_i : \mathbb{R} \times \mathbb{R} \times \mathbb{R}^n \to \mathbb{R}^n$ is the (nonautonomous) solution mapping of the ordinary differential equation $\dot{x} = F_i(t,x)$.

Although F_i and Φ_i are not uniquely defined, any estimate for the difference $|\Phi_1(t, t_0^1, x_0^1) - \Phi_2(s, t_0^2, x_0^2)|$ can be interpreted as a result on comparing solutions of the dynamical equations $x^\Delta = f_1(t,x)$ and $x^\Delta = f_2(t,x)$. (Of course, the Gronwall inequality is valid for the pair of ordinary differential equations $\dot{x} = F_i(t,x)$, $i = 1$ and 2.)

References

[1] M. Bohner and A. Peterson, *Dynamic Equations on Time Scales*, Birkhäuser, Basel, 2000.

[2] F.B. Christiansen and T.M. Fenchel, *Theories of Populations in Biological Communities*, Volume 20 of *Lecture Notes in Ecological Studies*, Berlin, 1977.

[3] B.M. Garay and S. Hilger, Embeddability of time scales in ode dynamics, *Nonlinear Anal.*, **47** (2001), 1357–1371.

[4] P. Hartman, *Ordinary Differential Equations*, Birkhäuser, Basel, 1973.

[5] S. Hilger, Analysis on measure chains – a unified approach to continuous and discrete calculus, *Results in Mathematics*, **18** (1990), 18–56.

[6] S. Hilger and P.E. Kloeden, Comparative time-grainyness and asymptotic stability in dynamical systems, *Autom. Remote Control* **55**, No.9, Pt. 1, (1994), 1293–1298.

[7] S. Keller, *Asymptotisches Verhalten invarianter Faserbündel bei Diskretisierung und Mittelwertbildung im Rahmen der Analysis auf Zeitskalen*, Augsburger Mathematisch–naturwissenschaftliche Schriften, Wißner-Verlag, Augsburg, 2000.

[8] P.E. Kloeden and J. Lorenz, Stable attracting sets in dynamical systems and their one-step discretizations, *SIAM J. Numer. Anal.*, **23** (1986), 986–995.

[9] A.M. Stuart and A.R. Humphries, *Numerical Analysis and Dynamical Systems*, Cambridge University Press, Cambridge, 1996.

On the Riemann Integration
on Time Scales

GUSEIN SH. GUSEINOV

Department of Mathematics, Atılım University
06836 Incek, Ankara, Turkey
E-mail: guseinov@sci.ege.edu.tr

and

BILLÛR KAYMAKÇALAN[1]

Department of Mathematics and Computer Science
Georgia Southern University
Statesboro, GA 30460, USA
E-mail: billur@gasou.edu

Abstract In this paper we introduce and investigate the concepts of Riemann's delta and nabla integrals on time scales. Main theorems of the integral calculus on time scales are proved.

Keywords Time scales, Delta and nabla derivatives, Delta and nabla integrals

AMS Subject Classification 39A10

1   Introduction

For an acquaintance with the theory of calculus on time scales we refer to the original work by Hilger [10], to the paper by Aulbach and Hilger [5], and to the recently appeared works [1, 2, 3, 6, 7, 8, 11, 13].

In the above cited literature the concept of integral on time scales is introduced and called the Cauchy integral (we remark that such an integral, defined in [12, p.255] by means of an antiderivative of a function is named as the Newton integral).

In this paper a treatment of the Riemann integral on time scales is given. The concepts of the Riemann and Riemann-Stieltjes integrals on time scales were investigated earlier by Sailer in [15]. Our treatment is slightly different.

---

[1]On leave from the Middle East Technical University, Ankara-Turkey during the academic year 2001–2002.

Notice that in [15] only the Darboux definition of the integral was considered. Here we give also the Riemann definition of the integral on time scales and we prove the equivalence of the Darboux and Riemann definitions of the integral.

Our first version of the Fundamental Theorem of Calculus (Theorem 4.1) proven in the present paper for all functions g such that g^Δ is integrable (by using a mean value theorem for the Δ-derivative) is proved in [15] only for functions g such that g^Δ is continuous.

Moreover, in order to meet the requirements in applications we distinguish the concepts of delta and nabla integrals.

The concept of Lebesque integral on time scales is introduced in [3].

2   The Riemann delta and nabla integrals

Let $\mathbb{T}$ be a time scale, $a < b$ be points in $\mathbb{T}$, and $[a, b]$ be the closed interval in $\mathbb{T}$. A *partition* of $[a, b]$ is any finite ordered subset $\mathcal{P} = \{t_i\}_{i=0}^n \subset [a, b]$, where

$$a = t_0 < t_1 < \ldots < t_n = b.$$

Let σ and ρ be respectively the forward and backward jump operators in $\mathbb{T}$. Further, let f be a real-valued bounded function defined on $[a, b]$. Let us set

$$M = \sup\{f(t) : t \in [a, \rho(b)]\}, \qquad m = \inf\{f(t) : t \in [a, \rho(b)]\},$$

$$M_i = \sup\{f(t) : t \in [t_{i-1}, \rho(t_i)]\}, \qquad m_i = \inf\{f(t) : t \in [t_{i-1}, \rho(t_i)]\}.$$

The *upper Darboux Δ-sum* $\mathcal{U}(f, \mathcal{P})$ of f with respect to $\mathcal{P}$ is the sum

$$\mathcal{U}(f, \mathcal{P}) = \sum_{i=1}^n M_i(t_i - t_{i-1})$$

and the *lower Darboux Δ-sum* $\mathcal{L}(f, \mathcal{P})$ is

$$\mathcal{L}(f, \mathcal{P}) = \sum_{i=1}^n m_i(t_i - t_{i-1}).$$

Note that

$$\mathcal{U}(f, \mathcal{P}) \leq \sum_{i=1}^n M(t_i - t_{i-1}) = M(b - a);$$

likewise $\mathcal{L}(f, \mathcal{P}) \geq m(b - a)$ and so

$$m(b - a) \leq \mathcal{L}(f, \mathcal{P}) \leq \mathcal{U}(f, \mathcal{P}) \leq M(b - a) \tag{2.1}$$

holds. The *upper Darboux Δ-integral $\mathcal{U}(f)$* of f from a to b is defined by

$$\mathcal{U}(f) = \inf\{\mathcal{U}(f, \mathcal{P}) : \mathcal{P} \text{ is a partition of } [a, b]\}$$

and the *lower Darboux* Δ*-integral* is

$$\mathcal{L}(f) = \sup\{\mathcal{L}(f, \mathcal{P}) : \mathcal{P} \text{ is a partition of } [a, b]\}.$$

In view of (2.1), $\mathcal{U}(f)$ and $\mathcal{L}(f)$ are finite real numbers.

We will prove in Theorem 2.4 that $\mathcal{L}(f) \leq \mathcal{U}(f)$. This is not obvious from (2.1).

Definition 2.1 *We say that* f *is* Δ-*integrable (delta integrable)* *from* a *to* b *provided* $\mathcal{L}(f) = \mathcal{U}(f)$. *In this case, we write* $\int_a^b f(t) \Delta t$ *for this common value. We call this integral the* Darboux Δ-*integral.*

Riemann's definition of the integral is a little different (Definition 2.8), but we will show in Theorem 2.9 that the definitions are equivalent. For this reason, we will call the integral defined above the *Riemann* Δ-*integral*.

Lemma 2.2 *Let* f *be a bounded function on* $[a, b]$. *If* $\mathcal{P}$ *and* $\mathcal{Q}$ *are partitions of* $[a, b]$ *and* $\mathcal{P} \subset \mathcal{Q}$, *then*

$$\mathcal{L}(f, \mathcal{P}) \leq \mathcal{L}(f, \mathcal{Q}) \leq \mathcal{U}(f, \mathcal{Q}) \leq \mathcal{U}(f, \mathcal{P}). \tag{2.2}$$

Lemma 2.3 *If* f *is a bounded function on* $[a, b]$, *and if* $\mathcal{P}$ *and* $\mathcal{Q}$ *are partitions of* $[a, b]$, *then* $\mathcal{L}(f, \mathcal{P}) \leq \mathcal{U}(f, \mathcal{Q})$.

Theorem 2.4 *If* f *is a bounded function on* $[a, b]$, *then* $\mathcal{L}(f) \leq \mathcal{U}(f)$.

Proof. Fix a partition $\mathcal{P}$ of $[a, b]$. Lemma 2.3 shows that $\mathcal{L}(f, \mathcal{P})$ is a lower bound for the set

$$\{\mathcal{U}(f, \mathcal{Q}) : \mathcal{Q} \text{ is a partition of } [a, b]\}.$$

Therefore $\mathcal{L}(f, \mathcal{P})$ must be less than or equal to the greatest lower bound (infimum) of this set. That is,

$$\mathcal{L}(f, \mathcal{P}) \leq \mathcal{U}(f). \tag{2.3}$$

Now (2.3) shows that $\mathcal{U}(f)$ is an upper bound for the set

$$\{\mathcal{L}(f, \mathcal{P}) : \mathcal{P} \text{ is a partition of } [a, b]\}$$

and so $\mathcal{U}(f) \geq \mathcal{L}(f)$. $\qquad\qquad\square$

The next theorem gives a "Cauchy criterion" for integrability.

Theorem 2.5 *A bounded function* f *on* $[a, b]$ *is* Δ-*integrable if and only if for each* $\epsilon > 0$ *there exists a partition* $\mathcal{P}$ *of* $[a, b]$ *such that*

$$\mathcal{U}(f, \mathcal{P}) - \mathcal{L}(f, \mathcal{P}) < \epsilon. \tag{2.4}$$

Lemma 2.6 *For each $\delta > 0$ there exists a partition $\mathcal{P} = \{a = t_0 < t_1 < \ldots < t_n = b\}$ of $[a,b]$ such that for each $i \in \{1,2,\ldots,n\}$ either $t_i - t_{i-1} \leq \delta$ or $t_i - t_{i-1} > \delta$ and $\rho(t_i) = t_{i-1}$.*

Proof. We make up the desired points $t_0 < t_1 < \ldots < t_n$ by induction setting $t_0 = a$,

$$A_i = \{t \in [a,b] : t_{i-1} < t \leq t_{i-1} + \delta\}, \qquad t_i = \begin{cases} \sup A_i, & \text{if } A_i \neq \emptyset, \\ \sigma(t_{i-1}), & \text{if } A_i = \emptyset. \end{cases}$$

Then $a = t_0 < t_1 < t_2 < \ldots$ and $t_{i+1} - t_{i-1} \geq \delta$ for all $i \geq 1$. Therefore, for some positive integer n we will have $t_n = b$. The lemma is thus proven. $\square$

For each $\delta > 0$ we denote by $\mathcal{P}_\delta([a,b])$ or simply by $\mathcal{P}_\delta$ the set of all partitions $\mathcal{P}$ of $[a,b]$ possessing the property indicated in Lemma 2.6.

Note that if $\mathbb{T} = \mathbb{R}$, then $\mathcal{P}_\delta([a,b])$ consists of all partitions of $[a,b]$, the mesh (norm) of which is less than or equal to δ. In case $\mathbb{T} = \mathbb{Z}$, $\mathcal{P}_\delta([a,b])$ contains for all $\delta < 1$ only one partition that consists of all points of $[a,b]$.

Here is another "Cauchy criterion" for integrability.

Theorem 2.7 *A bounded function f on $[a,b]$ is Δ-integrable if and only if for each $\epsilon > 0$ there exists a $\delta > 0$ such that*

$$\mathcal{P} \in \mathcal{P}_\delta \quad \text{implies} \quad \mathcal{U}(f,\mathcal{P}) - \mathcal{L}(f,\mathcal{P}) < \epsilon \tag{2.5}$$

for all partitions $\mathcal{P}$ of $[a,b]$.

We now give Riemann's definition of integrability.

Definition 2.8 *Let f be a bounded function on $[a,b]$, and let $\mathcal{P} = \{a = t_0 < t_1 < \ldots < t_n = b\}$ be a partition of $[a,b]$. A Riemann Δ-sum of f associated with the partition $\mathcal{P}$ is a sum of the form*

$$S = \sum_{i=1}^{n} f(\xi_i)(t_i - t_{i-1}), \tag{2.6}$$

where $\xi_i \in [t_{i-1}, \rho(t_i)]$ for $i = 1,2,\ldots,n$. The function f is Riemann Δ-integrable from a to b if there exists a number I with the following property. For each $\epsilon > 0$ there exists $\delta > 0$ such that

$$|S - I| < \epsilon$$

for every Riemann Δ-sum S of f associated with a partition $\mathcal{P} \in \mathcal{P}_\delta$. The number I is the Riemann Δ-integral of f from a to b.

Theorem 2.9 *A bounded function f on $[a,b]$ is Riemann Δ-integrable if and only if it is (Darboux) Δ-integrable, in which case the values of the integrals agree.*

Above we have defined the Δ-integral when $a < b$. By definition we put

$$\int_b^a f(t)\Delta t = -\int_a^b f(t)\Delta t \quad \text{and} \quad \int_c^c f(t)\Delta t = 0.$$

It is easy to see that

$$\int_t^{\sigma(t)} f(s)\Delta s = [\sigma(t) - t]f(t). \tag{2.7}$$

Indeed (2.7) is obvious if $\sigma(t) = t$, since both sides of (2.7) are equal to zero in this case. Let now $\sigma(t) > t$. Then a single partition of $[t, \sigma(t)]$ is $\mathcal{P} = \{t = s_0 < s_1 = \sigma(t)\}$ and since $\rho(s_1) = s_0 = t$, we have

$$\mathcal{U}(f, \mathcal{P}) = f(t)[\sigma(t) - t] = \mathcal{L}(f, \mathcal{P}).$$

Hence (2.7) follows as well. Note that the Riemann Δ-sum of f associated with $\mathcal{P}$ is also equal to $[\sigma(t) - t]f(t)$.

Clearly above given Definitions 2.1 and 2.8 of the integral, in the case $\mathbb{T} = \mathbb{R}$ coincide with the usual Darboux's and Riemann's definitions of the integral, respectively (see, for example, [9, 14]). Note that if $\mathbb{T} = \mathbb{R}$, then $\mathcal{P}_\delta([a, b])$ consists of all the partitions of $[a, b]$ the norm (mesh) of which is less than or equal to δ.

Let now $\mathbb{T} = \mathbb{Z}$, $a \in \mathbb{Z}$, and $b = a + p$ for some positive integer p. Consider the partition $\mathcal{P}^*$ of $[a, b]$ defined by

$$\mathcal{P}^* = \{a = t_0 < t_1 < \ldots < t_p - b\},$$

where $t_0 = a$, $t_1 = a + 1$, $t_2 = a + 2$, $\ldots$, $t_p = a + p = b$. $\mathcal{P}^*$ contains all the points of $[a, b]$ and $\rho(t_i) = t_{i-1}$ for each $i \in \{1, 2, \ldots, p\}$. Then

$$\mathcal{U}(f, \mathcal{P}^*) = \sum_{i=1}^p M_i(t_i - t_{i-1}) = \sum_{i=1}^p f(t_{i-1}) = \sum_{k=a}^{b-1} f(k),$$

$$\mathcal{L}(f, \mathcal{P}^*) = \sum_{i=1}^p m_i(t_i - t_{i-1}) = \sum_{i=1}^p f(t_{i-1}) = \sum_{k=a}^{b-1} f(k).$$

So

$$\mathcal{U}(f, \mathcal{P}^*) = \mathcal{L}(f, \mathcal{P}^*) = \sum_{k=a}^{b-1} f(k)$$

and therefore, the function f is Darboux Δ-integrable and

$$\int_a^b f(t)\Delta t = \sum_{k=a}^{b-1} f(k). \tag{2.8}$$

For each positive $\delta < 1$, $\mathcal{P}_\delta([a, b])$ consists of the single partition $\mathcal{P}^*$ and the Riemann Δ-sum associated with the partition $\mathcal{P}^*$ is

$$S = \sum_{i=1}^{p} f(\xi_i)(t_i - t_{i-1}) = \sum_{i=1}^{p} f(t_{i-1}) = \sum_{k=a}^{b-1} f(k).$$

Therefore the function f is Riemann Δ-integrable as well and its Riemann Δ-integral is equal to (2.8).

The concept of the ∇-integral, (nabla integral) on time scales is defined similarly by considering $[\sigma(t_{i-1}), t_i]$ instead of $[t_{i-1}, \rho(t_i)]$.

3   Properties of the Riemann integrals

In this section we establish some basic properties of the Riemann delta integral and we show that many familiar functions, including monotonic and continuous functions, are Riemann integrable.

Theorem 3.1 *Every monotonic function on $[a, b]$ is Δ-integrable.*

Proof. We assume f is nondecreasing on $[a, b]$. Since $f(a) \leq f(t) \leq f(b)$ for all $t \in [a, b]$, f is clearly bounded on $[a, b]$. In order to apply Theorem 2.5, let $\epsilon > 0$ and let $\delta = \frac{\epsilon}{f(b)-f(a)+1}$. For the partition

$$\mathcal{P} = \{a = t_0 < t_1 < \ldots < t_n = b\} \in \mathcal{P}_\delta$$

we have

$$\mathcal{U}(f, \mathcal{P}) - \mathcal{L}(f, \mathcal{P}) = \sum_{i=1}^{n} M_i(t_i - t_{i-1}) - \sum_{i=1}^{n} m_i(t_i - t_{i-1})$$

$$= \sum_{i=1}^{n} f(\rho(t_i))(t_i - t_{i-1}) - \sum_{i=1}^{n} f(t_{i-1})(t_i - t_{i-1})$$

$$= \sum_{i=1}^{n} [f(\rho(t_i)) - f(t_{i-1})](t_i - t_{i-1})$$

$$= \sum_{t_i - t_{i-1} \leq \delta} [f(\rho(t_i)) - f(t_{i-1})](t_i - t_{i-1}) +$$

$$+ \sum_{t_i - t_{i-1} > \delta} [f(\rho(t_i)) - f(t_{i-1})](t_i - t_{i-1}). \tag{3.1}$$

The second sum on the right-hand side of (3.1) is equal to zero, since from the condition $\mathcal{P} \in \mathcal{P}_\delta$ and $t_i - t_{i-1} > \delta$, it follows that $\rho(t_i) = t_{i-1}$. The first sum is less than or equal to

$$\delta \sum_{i=1}^{n} [f(\rho(t_i)) - f(t_{i-1})] \leq \delta \sum_{i=1}^{n} [f(t_i) - f(t_{i-1})] = \delta[f(b) - f(a)] < \epsilon.$$

So $\mathcal{U}(f, \mathcal{P}) - \mathcal{L}(f, \mathcal{P}) < \epsilon$ and Theorem 2.5 shows that f is Δ-integrable. $\square$

Theorem 3.2 *Every continuous (in the time scale topology) function f on $[a, b]$ is Δ-integrable.*

Proof. Again, in order to apply Theorem 2.5, consider $\epsilon > 0$. Since f is continuous, it will be uniformly continuous on the compact subset $[a, b]$ of $\mathbb{T}$. Therefore there exists $\delta > 0$ such that

$$t, \tau \in [a, b] \quad \text{and} \quad |t - \tau| \leq \delta \quad \text{imply} \quad |f(t) - f(\tau)| < \frac{\epsilon}{b - a}. \tag{3.2}$$

Consider any partition $\mathcal{P} = \{a = t_0 < t_1 < \ldots < t_n = b\} \in \mathcal{P}_\delta$. Since f assumes its maximum and minimum on each compact interval $[t_{i-1}, \rho(t_i)]$, it follows from (3.2) that

$$
\begin{aligned}
\mathcal{U}(f, \mathcal{P}) \quad - \quad \mathcal{L}(f, \mathcal{P}) &= \sum_{i=1}^{n} (M_i - m_i)(t_i - t_{i-1}) \\
&= \sum_{t_i - t_{i-1} \leq \delta} (M_i - m_i)(t_i - t_{i-1}) + \sum_{t_i - t_{i-1} > \delta} (M_i - m_i)(t_i - t_{i-1}) \\
&= \sum_{t_i - t_{i-1} \leq \delta} (M_i - m_i)(t_i - t_{i-1}) < \frac{\epsilon}{b - a} \sum_{t_i - t_{i-1} \leq \delta} (t_i - t_{i-1}) \\
&\leq \frac{\epsilon}{b - a} \sum_{i=1}^{n} (t_i - t_{i-1}) = \frac{\epsilon}{b - a} \cdot (b - a) = \epsilon.
\end{aligned}
$$

Here we used the fact that if $t_i - t_{i-1} > \delta$, then $\rho(t_i) = t_{i-1}$ and hence $M_i - m_i = 0$. Thus $\mathcal{U}(f, \mathcal{P}) - \mathcal{L}(f, \mathcal{P}) < \epsilon$ and again Theorem 2.5 yields that f is Δ-integrable. $\qquad\square$

Theorem 3.3 *Let f and g be Δ-integrable functions on $[a, b]$ and c be a real number. Then*
    (i) cf is Δ-integrable and $\int_a^b cf(t)\Delta t = c \int_a^b f(t)\Delta t$;
    (ii) $f + g$ is Δ-integrable and $\int_a^b [f(t) + g(t)]\Delta t = \int_a^b f(t)\Delta t + \int_a^b g(t)\Delta t$.

Theorem 3.4 *If f and g are Δ-integrable on $[a, b]$ and if $f(t) \leq g(t)$ for $t \in [a, b]$, then $\int_a^b f(t)\Delta t \leq \int_a^b g(t)\Delta t$.*

Theorem 3.5 *If f is Δ-integrable on $[a, b]$, then $|f|$ is Δ-integrable on $[a, b]$ and*

$$\left| \int_a^b f(t)\Delta t \right| \leq \int_a^b |f(t)|\Delta t.$$

Theorem 3.6 *Let f be a function on $[a, b]$ and let $c \in \mathbb{T}$ with $a < c < b$. If f is Δ-integrable from a to c and from c to b, then f is Δ-integrable from a to b and*

$$\int_a^b f(t)\Delta t = \int_a^c f(t)\Delta t + \int_c^b f(t)\Delta t.$$

Theorem 3.7 *If f and g are Δ-integrable on $[a, b]$, then so is their product.*

Here we will not dwell with the proofs of Theorems 3.3–3.7.

4   Fundamental Theorem of Calculus

There are two versions of the Fundamental Theorem of Calculus. Each says, roughly speaking, that differentiation and integration are inverse operations. In fact, our first version (Theorem 4.1) says that "the Δ-integral of the Δ-derivative of a function is given by the function" and our second version (Theorem 4.3) says that "the Δ-derivative of the Δ-integral of a continuous function is the function."

In this section we assume that $[a, b] \subset \mathbb{T}^\kappa$.

Theorem 4.1 *(Fundamental Theorem of Calculus I) If g is a Δ-differentiable function on $[a, b]$, and if g^Δ is Δ-integrable on $[a, b]$, then*

$$\int_a^b g^\Delta(t)\Delta t = g(b) - g(a). \tag{4.1}$$

Proof. Let $\epsilon > 0$. By Theorem 2.5, there exists a partition $\mathcal{P} = \{a = t_0 < t_1 < \ldots < t_n = b\}$ of $[a, b]$ such that

$$\mathcal{U}(g^\Delta, \mathcal{P}) - \mathcal{L}(g^\Delta, \mathcal{P}) < \epsilon. \tag{4.2}$$

Now we make use of the following mean value result on time scales (see [4, 8]): Let g be a continuous function on $[\alpha, \beta] \subset \mathbb{T}$ that is Δ-differentiable on $[\alpha, \beta)$. Then there exist $\xi, \tau \in [\alpha, \beta)$ such that

$$g^\Delta(\tau) \le \frac{g(\beta) - g(\alpha)}{\beta - \alpha} \le g^\Delta(\xi).$$

Applying this mean value result to each interval $[t_{i-1}, t_i]$, for each $i = 1, 2, \ldots, n$ we obtain $\xi_i, \tau_i \in [t_{i-1}, t_i)$ for which

$$(t_i - t_{i-1})g^\Delta(\tau_i) \le g(t_i) - g(t_{i-1}) \le (t_i - t_{i-1})g^\Delta(\xi_i).$$

Hence we have

$$\sum_{i=1}^n (t_i - t_{i-1})g^\Delta(\tau_i) \le \sum_{i=1}^n [g(t_i) - g(t_{i-1})] \le \sum_{i=1}^n (t_i - t_{i-1})g^\Delta(\xi_i)$$

or

$$\sum_{i=1}^n (t_i - t_{i-1})g^\Delta(\tau_i) \le g(b) - g(a) \le \sum_{i=1}^n (t_i - t_{i-1})g^\Delta(\xi_i).$$

So the estimate

$$\mathcal{L}(g^\Delta, \mathcal{P}) \le g(b) - g(a) \le \mathcal{U}(g^\Delta, \mathcal{P}) \tag{4.3}$$

follows. Since we have

$$\mathcal{L}(g^\Delta, \mathcal{P}) \le \int_a^b g^\Delta(t)\Delta t \le \mathcal{U}(g^\Delta, \mathcal{P}),$$

inequalities (4.2) and (4.3) imply that

$$\left| \int_a^b g^\Delta(t)\Delta t - [g(b) - g(a)] \right| < \epsilon.$$

The fact that ϵ is arbitrary, gives rise to (4.1) being true. □

Theorem 4.2 *(Integration by Parts) If u and v are Δ-differentiable on $[a, b]$, and if u^Δ and v^Δ are integrable on $[a, b]$, then*

$$\int_a^b u^\Delta(t)v(t)\Delta t + \int_a^b u(\sigma(t))v^\Delta(t)\Delta t = u(b)v(b) - u(a)v(a). \qquad (4.4)$$

Proof. Let $g = uv$; then $g^\Delta(t) = u^\Delta(t)v(t) + u(\sigma(t))v^\Delta(t)$ and g^Δ is integrable. Now Theorem 4.1 shows that

$$\int_a^b g^\Delta(t)\Delta t = g(b) - g(a) = u(b)v(b) - u(a)v(a)$$

and so (4.4) holds. □

In what follows we use the convention

$$\int_a^b f(t)\Delta t = -\int_b^a f(t)\Delta t \qquad \text{for} \quad a > b.$$

Theorem 4.3 *(Fundamental Theorem of Calculus II) Let f be a function which is Δ-integrable from a to b. For $t \in [a, b]$, let*

$$F(t) = \int_a^t f(s)\Delta s.$$

Then F is continuous on $[a, b]$. If f is continuous (in the time scale topology) at $t_0 \in [a, b)$, then F is Δ-differentiable at t_0 and

$$F^\Delta(t_0) = f(t_0).$$

References

[1] R.P. Agarwal and M. Bohner, Basic Calculus on Time Scales and Some of its Applications, *Results Math.* **35** (1999) 3-22.

[2] C.D. Ahlbrandt, M. Bohner and J. Ridenhour, Hamiltonian Systems on Time Scales, *J. Math. Anal. Appl.* **99** (2000) 561-578.

[3] F.M. Atıcı and G. Sh. Guseinov, On Green's Functions and Positive Solutions for Boundary Value Problems on Time Scales, *J. Comput. Appl. Math.* (2001) (to appear).

[4] F. M. Atıcı, G.Sh. Guseinov, and B. Kaymakçalan, On Lyapunov Inequality in Stability Theory for Hill's Equation on Time Scales, *J. Inequal. Appl.* **5**(2000) 603-620.

[5] B. Aulbach and S. Hilger, Linear Dynamic Processes with Inhomogeneous Time Scale, In Nonlinear Dynamics and Quantum Dynamical Systems (Gaussig, 1990), vol. **59** of *Math. Res.* (Akademie Verlag, Berlin, 1990) 9-20.

[6] M. Bohner and J. Castillo, Mimetic methods on Measure Chains, *Computers Math. Appl.*(2001) (to appear).

[7] M. Bohner and A. Peterson, *Dynamic Equations on Time Scales: An Introduction with Applications* (Birkhäuser, Boston, 2001).

[8] G. Sh. Guseinov and B. Kaymakçalan, On a Disconjugacy Criterion for Second Order Dynamic Equations on Time Scales, *J. Comput. Appl. Math.* (2001) (to appear).

[9] E. Fischer, *Intermediate Real Analysis* (Springer-Verlag, New York, 1983).

[10] S. Hilger, Analysis on Measure Chains; A Unified Approach to Continuous and Discrete Calculus, *Results Math.* **18** (1990) 18-56.

[11] S. Hilger, Laplace and Fourier transforms on Measure Chains, *Dynam. Systems Appl.* **8** (1999) 471-488.

[12] J. W. Kitchen, *Calculus of One Variable* (Addison-Wesley, California, 1968).

[13] V. Lakshmikantham, S. Sivasundaram, and B. Kaymakçalan, *Dynamic Systems on Measure Chains* (Kluwer Academic Publishers, Dordrecht, 1996).

[14] K. A. Ross, *Elementary Analysis: The Theory of Calculus* (Springer-Verlag, New York, 1990).

[15] S. Sailer, Riemann-Stieltjes-Integrale auf Zeitmengen, (Schriftliche Hausarbeit, Vorgelegt bei: Prof. Dr. B. Aulbach), Universität Augsburg, 1992.

Cauchy Functions and Taylor's Formula for Time Scales $\mathbb{T}$

RAEGAN J. HIGGINS[1] and ALLAN PETERSON[1]

Department of Mathematics and Statistics
University of Nebraska-Lincoln
Lincoln, NE 68588-0323, USA
E-mails: rhiggins@math.unl.edu, apeterso@math.unl.edu

Abstract In this paper we will be concerned with calculating the "Taylor monomials" that appear in the Taylor's formula for a function defined on a time scale. These Taylor monomials are very important for such Taylor series and are intimately related to Cauchy functions for certain dynamic equations. These Cauchy functions arise in variation of constants formulas and are also important when considering certain Green's functions. We will calculate several of these "Taylor monomials" for different time scales. In the last section we will give a short proof of Taylor's Theorem and give an interesting example.

Keywords Taylor monomials, Time scales, Taylor's formula

AMS Subject Classification 39A10, 34B10

1 Introduction

The theory of time scales was introduced by Stefan Hilger in his 1988 PhD thesis [5] (supervised by Bernd Aulbach) in order to unify continuous and discrete analysis. The study of dynamical equations on time scales reveals discrepancies between continuous and discrete analysis and often helps avoid proving results twice, once for differential equations and once for difference equations. The general idea is to prove a result for a dynamic equation where the domain of the unknown function is a so-called time scale, which is an arbitrary closed subset of the reals. If we choose the time scale to be the set of the real numbers, the general result yields a result concerning an ordinary differential equation. If, on the other hand, we choose the time scale to be the set of integers, the general result yields a result for difference equations. However, since there are many other time scales than just the set of real numbers or the set of integers, one has a more general result. Dynamic equations

---

[1]Research was supported by NSF Grant 0072505.

on time scales have a tremendous potential for applications. For example, it can model insect populations that are continuous while in season, die out in winter while their eggs are incubating or dormant, and then hatch in a new season, giving rise to a nonoverlapping population. In Section 2 we will give several preliminary definitions that we will need in this paper. In Section 3 we will define "Taylor monomials" and give several formulas for these Taylor monomials for various time scales. Finally in Section 4 we will give a short proof of Taylor's formula for a function on a time scale and give an interesting example. Argawal and Bohner [1] give a long proof of this Taylor's theorem but they use only basic results.

2   Preliminaries

First, we introduce some definitions. These definitions can be found in M. Bohner and A. Peterson [3] and R. P. Agarwal and M. Bohner [1].

Definition 1 *A* time scale $\mathbb{T}$ *is a nonempty closed subset of the reals.*

Definition 2 *Let $\mathbb{T}$ be a time scale. We define the* forward jump operator $\sigma : \mathbb{T} \to \mathbb{T}$ *by*

$$\sigma(t) := \inf\{s \in \mathbb{T} : s > t\}, \quad for \quad t \in \mathbb{T},$$

while the backward jump operator $\rho : \mathbb{T} \to \mathbb{T}$ *is defined by*

$$\rho(t) := \sup\{s \in \mathbb{T} : s < t\}, \quad for \quad t \in \mathbb{T}.$$

Here we put $\inf \emptyset = \sup \mathbb{T}$ *and* $\sup \emptyset = \inf \mathbb{T}$, *where $\emptyset$ denotes the empty set. For $t \in \mathbb{T}$ we say that t is* left-scattered *if $\rho(t) < t$, while if $\sigma(t) > t$, we say t is* right-scattered. *A point $t \in \mathbb{T}$ is* isolated *if it is both right-scattered and left-scattered at the same time. For $t \in \mathbb{T}$ we say t is* right-dense *if $t < \sup \mathbb{T}$ and $\sigma(t) = t$, while if $t > \inf \mathbb{T}$ and $\rho(t) = t$, we say t is* left-dense. *The* graininess function $\mu : \mathbb{T} \to [0, \infty)$ *is defined by*

$$\mu(t) := \sigma(t) - t.$$

If $\sup \mathbb{T} < \infty$ and $\sup \mathbb{T}$ is left-scattered, *we let $\mathbb{T}^\kappa := \mathbb{T} \backslash \{\sup \mathbb{T}\}$. Otherwise, we let $\mathbb{T}^\kappa := \mathbb{T}$.*

Definition 3 *Assume $f : \mathbb{T} \to \mathbb{R}$ is a function and let $t \in \mathbb{T}^\kappa$. Then we define $f^\Delta(t)$ to be the number (provided it exists) with the property that given any $\varepsilon > 0$, there is a neighborhood U of t such that*

$$|[f(\sigma(t)) - f(s)] - f^\Delta(t)[\sigma(t) - s]| \leq \epsilon|\sigma(t) - s|$$

for all $s \in U$. We call $f^\Delta(t)$ the delta (or Hilger) derivative *of f at t. If $\mathbb{T} = \mathbb{R}$, $f^\Delta = f'$, whereas if $\mathbb{T} = \mathbb{Z}$ (the integers), then*

$$f^\Delta(t) = \Delta f(t) := f(t+1) - f(t),$$

that is, Δ is the usual foward difference operator.

Definition 4 *A function* $f : \mathbb{T} \to \mathbb{R}$ *is called rd-continuous provided it is continuous at right-dense points in* $\mathbb{T}$ *and its left-sided limits exist (finite) at left-dense points in* $\mathbb{T}$.

If $\mathbb{T} = \mathbb{R}$, then $f : \mathbb{R} \to \mathbb{R}$ is rd-continuous if and only if f is continuous. At the other extreme, if $\mathbb{T} = \mathbb{Z}$, then any function defined on $\mathbb{Z}$ is rd-continuous. It is known [3] that if f is rd-continuous, then there is a function F, called an antiderivative of f, such that $F^{\Delta}(t) = f(t)$. In this case, we define

$$\int_a^b f(t)\Delta t = F(t)\big|_a^b.$$

If $\mathbb{T} = \mathbb{R}$, then

$$\int_a^b f(t)\Delta t = \int_a^b f(t)dt,$$

where the integral on the right hand side is the Riemann integral. If every point in $\mathbb{T}$ is isolated and $a < b$ are in $\mathbb{T}$, then we will use the formula (see [3])

$$\int_a^b f(t)\Delta t = \sum_{t=a}^{\rho(b)} f(t)\mu(t).$$

3   Taylor monomials

In this section we first define what we will call the *Taylor monomials* or *generalized polynomials* as defined originally by Agarwal and Bohner [1] and also in the book by Bohner and Peterson [3]. They are called Taylor monomials because they are important, as we will see, in Taylor's formula for a function defined on a time scale. These Taylor monomials are also important because they are intimately related to Cauchy functions for certain dynamic equations which are important in variation of constants formulas. Since Green's functions are often given in terms of Cauchy functions, these Taylor monomials are important in the study of certain boundary value problems. The Taylor monomials $h_k : \mathbb{T} \times \mathbb{T} \to \mathbb{R}$, $k \in \mathbb{N}_0$, are defined recursively as follows: The function h_0 is defined by

$$h_0(t, s) = 1, \quad \text{for all } s, t \in \mathbb{T},$$

and, given h_k for $k \in \mathbb{N}_0$, the function h_{k+1} is defined by

$$h_{k+1}(t, s) = \int_s^t h_k(\tau, s)\,\Delta\tau, \quad \text{for all } s, t \in \mathbb{T}.$$

If we let $h_k^{\Delta}(t, s)$ denote for each fixed $s \in \mathbb{T}$ the derivatives of $h_k(t, s)$ with respect to t, then

$$h_k^{\Delta}(t, s) = h_{k-1}(t, s), \quad \text{for } k \in \mathbb{N}, \quad t \in \mathbb{T}^{\kappa}.$$

The above definition obviously implies

$$h_1(t,s) = t - s, \quad \text{for all } s, t \in \mathbb{T}.$$

However, in general, finding h_k for $k \geq 2$ is very difficult. In this section we will give formulas for several of these important Taylor monomials for various time scales. In many of the applications it suffices to find formulas for $h_k(t,s)$ for just $t \geq s$. The first three examples that we give below are the formulas that are known in the literature and the other examples are new as far as we know.

Example 5 *If $\mathbb{T} = \mathbb{R}$, then*

$$h_k(t,s) = \frac{(t-s)^k}{k!}, \quad \text{for } t, s \in \mathbb{T}, \quad k \in \mathbb{N}_0.$$

Example 6 *Consider the time scale $\mathbb{T} = \mathbb{Z}$. The factorial function (see Kelley and Peterson [6]) $t^{\underline{k}}$ (read as t to the k falling) for $k \in \mathbb{N}_0$ is defined by $t^{\underline{0}} = 1$ and for $k \in \mathbb{N}$,*

$$t^{\underline{k}} = t(t-1)(t-2)\cdots(t-k+1).$$

In this case

$$h_k(t,s) = \frac{(t-s)^{\underline{k}}}{k!}, \quad \text{for } t, s \in \mathbb{T}, \ t \geq s, \quad k \in \mathbb{N}_0.$$

Example 7 (Agarwal and Bohner[1]) *Consider the time scale*

$$\mathbb{T} = \overline{q^{\mathbb{Z}}}, \quad \text{for some } q > 1.$$

This time scale is very important (see, e.g., Bézivin [2], G. Derfel, E. Romanenko, and A. Sharkovsky [4], Trijtzinsky [7], and Zhang [8]). In this case for $k \in \mathbb{N}_0$,

$$h_k(t,s) = \prod_{m=0}^{k-1} \frac{t - q^m s}{\sum_{j=0}^{m} q^j}, \quad \text{for all } s, t \in \mathbb{T}. \tag{1}$$

Example 8 *Consider the time scale with step size $h > 0$,*

$$\mathbb{T} = h\mathbb{Z} = \{0, \pm h, \pm 2h, \pm 3h, ...\}.$$

We claim that for $k \in \mathbb{N}_0$,

$$h_k(t,s) = \frac{\prod_{i=0}^{k-1}(t - ih - s)}{k!}, \quad \text{for all } s, t \in \mathbb{T}, \ t \geq s \tag{2}$$

Evidently, for $k = 0$, the claim (2) holds (by convention $\prod_{i=0}^{-1} = 1$). Now we assume (2) holds with k replaced by some $m \in \mathbb{N}_0$. Then

$$
\begin{aligned}
h_{m+1}^{\Delta}(t,s) &= \frac{h_{m+1}(\sigma(t),s) - h_{m+1}(t,s)}{\mu(t)} \\
&= \frac{\prod_{i=0}^{m}(t+h-ih-s) - \prod_{i=0}^{m}(t-ih-s)}{(m+1)!h} \\
&= \frac{\prod_{i=-1}^{m-1}(t-ih-s) - \prod_{i=0}^{m}(t-ih-s)}{(m+1)!h} \\
&= \frac{\prod_{i=0}^{m-1}(t-ih-s)\left[(t+h-s) - (t-mh-s)\right]}{(m+1)!h} \\
&= \frac{\prod_{i=0}^{m-1}(t-ih-s)\left[t+h-s-t+mh+s\right]}{(m+1)!h} \\
&= \frac{\prod_{i=0}^{m}{}^{1}(t-ih-s)\,h\lfloor 1+m\rfloor}{(m+1)!h} \quad \frac{\prod_{i=0}^{m-1}(t-ih-s)}{m!} = h_m(t,s)
\end{aligned}
$$

and since $h_{m+1}(s,s) = 0$ we get that (2) follows with k replaced by $m+1$. Hence by the principle of mathematical induction, (2) holds for all $k \in \mathbb{N}_0$.

Example 9 *Assume* $\alpha_0 \in \mathbb{R}$ *and* $\alpha_k > 0$, $k \in \mathbb{N}$. *Let* $S := \{t = \sum_{k=0}^{n}\alpha_k,\ n \in \mathbb{N}_0\}$. *If* $\sum_{k=0}^{\infty}\alpha_k = \infty$, *let* $\mathbb{T}$ *be the time scale* $\mathbb{T} = S$, *whereas if* $L = \sum_{k=0}^{\infty}\alpha_k$ *converges, let* $\mathbb{T} = S \cup \{L\}$. *We claim that for this time scale* $\mathbb{T}$,

$$
h_2(t,t_0) = \sum_{j=n_0+1}^{n-1}\sum_{k=n_0+1}^{j}\alpha_{j+1}\alpha_k, \quad \text{for } t \geq t_0, \tag{3}
$$

where $t = \sum_{k=0}^{n}\alpha_k$ *and* $t_0 = \sum_{k=0}^{n_0}\alpha_k$.
To see this let k_2 *be defined for* $t,t_0 \in \mathbb{T}$, $t \geq t_0$ *by the right hand side of equation (3). Then by our convention on sums,* $k_2(t_0,t_0) = 0$ *and for* $t \in \mathbb{T}$, $t \neq \sup\mathbb{T}$,

$$
\begin{aligned}
k_2^{\Delta}(t,t_0) &= \frac{k_2(\sigma(t),t_0) - k_2(t,t_0)}{\mu(t)} \\
&= \frac{\sum_{j=n_0+1}^{n}\sum_{k=n_0+1}^{j}\alpha_{j+1}\alpha_k - \sum_{j=n_0+1}^{n-1}\sum_{k=n_0+1}^{j}\alpha_{j+1}\alpha_k}{\alpha_{n+1}} \\
&= \frac{\sum_{k=n_0+1}^{n}\alpha_{n+1}\alpha_k}{\alpha_{n+1}} = \sum_{k=n_0+1}^{n}\alpha_k = t - t_0,
\end{aligned}
$$

which implies the desired result.

Example 10 *Consider the time scale*

$$
\mathbb{T} = \mathbb{N}_0^{\frac{1}{2}} = \{\sqrt{n} : n \in \mathbb{N}_0\}.
$$

Note that

$$\sigma(t) = \sqrt{n+1}, \quad \mu(t) = \sqrt{n+1} - \sqrt{n},$$

where $t = \sqrt{n}$, $n \in \mathbb{N}_0$. For this time scale we will just find $h_2(t,0)$.
Consider

$$
\begin{aligned}
h_2(t,0) &= \int_0^\tau 1\Delta\tau \\
&= 0\mu(0) + \sqrt{1}\mu\left(\sqrt{1}\right) + \cdots + \mu\sqrt{n-1}\left(\sqrt{n-1}\right) \\
&= \sqrt{0}\left(\sqrt{1} - \sqrt{0}\right) + \sqrt{1}\left(\sqrt{2} - \sqrt{1}\right) + \ldots + \sqrt{n-1}\left(\sqrt{n} - \sqrt{n-1}\right) \\
&= \sum_{k=0}^{n-1} \sqrt{k}\left(\sqrt{k+1} - \sqrt{k}\right) = \sum_{k=0}^{n-1}\left[\left(\sqrt{k(k+1)}\right) - k^{\underline{1}}\right] \\
&= p(t) - \frac{k^{\underline{2}}}{2}\Big|_0^n = p(t) - \frac{n(n-1)}{2} = p(t) - \frac{t^2(t^2-1)}{2},
\end{aligned}
$$

where $p(t) = \sum_{s\in[0,t)} s\sigma(s)$.

Example 11 *We consider the time scale*

$$\mathbb{T} = \mathbb{N}_0^2 = \{n^2 : n \in \mathbb{N}_0\}.$$

Note that

$$\sigma(t) = (n+1)^2, \quad \mu(t) = 2n+1,$$

where $t = n^2$, $n \in \mathbb{N}_0$.
Consider

$$
\begin{aligned}
h_2(t,0) &= \int_0^t \tau\Delta\tau = 0^2\mu\left(0^2\right) + 1^2\mu\left(1^2\right) + \cdots + (n-1)^2\mu\left((n-1)^2\right) \\
&= 1^2(2(1)+1) + 2^2(2(2)+1) + \cdots + (n-1)^2(2(n-1)+1) \\
&= \sum_{k=1}^{n-1} k^2(2k+1) = \sum_{k=1}^{n-1}\left(2k^3 + k^2\right) = \sum_{k=1}^{n-1}\left(2k^{\underline{3}} + 7k^{\underline{2}} + 3k^{\underline{1}}\right) \\
&= \frac{k^{\underline{4}}}{2} + \frac{7k^{\underline{3}}}{3} + \frac{3k^{\underline{2}}}{2}\Big|_1^n = \frac{n^{\underline{4}}}{2} + \frac{7n^{\underline{3}}}{3} + \frac{3n^{\underline{2}}}{2} \\
&= \frac{n(n-1)(n-2)(n-3)}{2} + \frac{7n(n-1)(n-2)}{3} + \frac{3n(n-1)}{2} \\
&= \frac{n(n-1)(3n^2-n-1)}{6} = \frac{\sqrt{t}(\sqrt{t}-1)(3t-\sqrt{t}-1)}{6}.
\end{aligned}
$$

Similarly,

$$h_3(t,0) = \int_0^t h_2(\tau,0)\Delta\tau = \sum_{k=0}^{n-1} h_2\left(k^2,0\right)\mu\left(k^2\right)$$

$$= \sum_{k=0}^{n-1} \frac{k(k-1)\left(3k^2 - k - 1\right)}{6}(2k+1)$$

$$= \sum_{k=0}^{n-1}\left(k^5 - \frac{5k^4}{6} - \frac{2k^3}{3} + \frac{k^2}{3} + \frac{k}{6}\right)$$

$$= \sum_{k=0}^{n-1}\left(k^{\underline{5}} + \frac{55k^{\underline{4}}}{6} + \frac{58k^{\underline{3}}}{3} + \frac{15k^{\underline{2}}}{2}\right)$$

$$- \left.\left(\frac{k^{\underline{6}}}{6} + \frac{11k^{\underline{5}}}{6} + \frac{29k^{\underline{4}}}{6} + \frac{5k^{\underline{3}}}{3}\right)\right|_0^n$$

$$= \frac{n^{\underline{6}}}{6} + \frac{11n^{\underline{5}}}{6} + \frac{29n^{\underline{4}}}{6} + \frac{5n^{\underline{3}}}{3}$$

$$= \frac{n(n-1)(n-2)\left(n^3 - n^2 - n - 5\right)}{6}$$

$$= \frac{\sqrt{t}\left(\sqrt{t}-1\right)\left(\sqrt{t}-2\right)\left(t^{\frac{3}{2}} - t - t^{\frac{1}{2}} - 5\right)}{6}.$$

Now we will examine $h_2(t,s))$ *for* $t \geq s$. *Let* $m = \sqrt{s}$.

$$h_2(t,s)$$

$$= \sum_{k=m}^{n-1}\left(k^2 - m^2\right)(2k+1)$$

$$= \frac{n^{\underline{4}}}{2} + \frac{7n^{\underline{3}}}{3} + \frac{3n^{\underline{2}}}{2} - \frac{m^{\underline{4}}}{2} - \frac{7m^{\underline{3}}}{3} - \frac{3m^{\underline{2}}}{2} - m^2\left(n^{\underline{2}} + n\right) + m^2\left(m^{\underline{2}} + m\right)$$

$$= \frac{(n-m)^{\underline{2}}}{3!} \times$$

$$\times \left\{3(n-m-2)^{\underline{2}} + (12m+14)(n-m-2) + (12m^2 + 24m + 9)\right\}$$

$$= \frac{(\sqrt{t}-\sqrt{s})^{\underline{2}}}{3!} \times$$

$$\times\left\{3(\sqrt{t}-\sqrt{s}-2)^{\underline{2}} + (12\sqrt{s}+14)(\sqrt{t}-\sqrt{s}-2) + (12s+24\sqrt{s}+9)\right\}$$

4   Taylor's Theorem

In Agarwal and Bohner [1] an elementary but lengthy proof of Taylor's Theorem for a function on a time scale is proved. In this section we will give a short proof of this Taylor's Theorem but it will depend on knowing more results.

Theorem 12 *(Taylor's Formula). Assume $f \in C_{rd}^{n+1}(\mathbb{T})$ and $s \in \mathbb{T}$. Then*

$$f(t) = \sum_{k=0}^{n} f^{\Delta^k}(s) h_k(t, s) + \int_s^t h_n(t, \sigma(\tau)) f^{\Delta^{n+1}}(\tau) \Delta\tau. \qquad (4)$$

Proof. Let g be defined by

$$g(t) := f^{\Delta^{n+1}}(t).$$

Then f is the unique solution of the IVP

$$x^{\Delta^{n+1}} = g(t), \quad x^{\Delta^i}(s) = f^{\Delta^i}(s), \quad 0 \le i \le n.$$

Let

$$u(t) := \sum_{k=0}^{n} f^{\Delta^k}(s) h_n(t, s)$$

and

$$v(t) := \int_s^t h_n(t, \sigma(\tau)) g(\tau) \Delta\tau.$$

Then u solves the IVP

$$u^{\Delta^{n+1}} = 0, \quad u^{\Delta^i}(s) = f^{\Delta^i}(s), \quad 0 \le i \le n.$$

It suffices to show v solves the IVP

$$v^{\Delta^{n+1}} = g(t), \quad v^{\Delta^i}(s) = 0, \quad 0 \le i \le n.$$

Clearly $v(s) = 0$. Also

$$\begin{aligned}
v^{\Delta}(t) &= \int_s^t h_n^{\Delta}(t, \sigma(\tau)) g(\tau)\Delta\tau + h_n(\sigma(t), \sigma(t)) g(t) \\
&= \int_s^t h_n^{\Delta}(t, \sigma(\tau)) g(\tau)\Delta\tau.
\end{aligned}$$

Note $v^{\Delta}(s) = 0$ and

$$\begin{aligned}
v^{\Delta^2}(t) &= \int_s^t h_n^{\Delta^2}(t, \sigma(\tau)) g(\tau)\Delta\tau + h_n(\sigma(t), \sigma(t)) g(t) \\
&= \int_s^t h_n^{\Delta^2}(t, \sigma(\tau)) g(\tau)\Delta\tau.
\end{aligned}$$

Note $v^{\Delta^2}(s) = 0$. Proceeding in this manner we obtain by mathematical induction that

$$v^{\Delta^i}(t) = \int_s^t h_n^{\Delta^i}(t, \sigma(\tau)) g(\tau)\Delta\tau$$

for $0 \leq i \leq n$ and $v^{\Delta^i}(s) = 0$, $0 \leq i \leq n$. Finally,

$$
\begin{aligned}
v^{\Delta^{n+1}}(t) &= \int_s^t h_n^{\Delta^{n+1}}(t, \sigma(\tau)) g(\tau) \Delta \tau + h_n(\sigma(t), \sigma(t)) g(t) \\
&= g(t),
\end{aligned}
$$

and this proof is complete. ∎

The following example was motivated by an example shown to the authors by Douglas Anderson. This is an example of a function on the time scale $\mathbb{T} = \mathbb{Z}$, where the Taylor series of a function about $t = 0$ converges to the function for $t \geq 0$ but diverges for $t < 0$.

Example 13 *For* $\mathbb{T} = \mathbb{Z}$, *consider* $f(t) = e_1(t, 0) = 2^t$ *for* $t \in \mathbb{Z}$. *If we expand* f *about* 0, *then Taylor's formula* (4) *for* f *is given by*

$$
f(t) = 2^t = P_n(t) + E_n(t),
$$

where the Taylor polynomial P_n *is given by*

$$
P_n(t) = \sum_{k=0}^n \frac{t^k}{k!}
$$

and the error term E_n *is given by*

$$
E_n(t) = \begin{cases} \sum_{\tau=1}^{t-1}(t - \tau - 1)^{\underline{n}} 2^\tau & \text{if} \quad t > 0 \\ 0 & \text{if} \quad t = 0 \\ -\sum_{\tau=t+1}^0 (t - \tau - 1)^{\underline{n}} 2^\tau & \text{if} \quad t < 0. \end{cases}
$$

Then $P_n(-1) = \frac{1-(-1)^{n+1}}{2}$ *and* $E_n(-1) = \frac{(-1)^{n+1}}{2}$, *so that the Taylor polynomial will not converge to* f *at* -1 *as* $n \to \infty$. *Note, however, that*

$$
f(t) = 2^t = \sum_{k=0}^\infty \frac{t^k}{k!} = P_t(t) = \sum_{k=0}^t \binom{t}{k}
$$

for any nonnegative integer t. *It is not difficult to show that the Taylor's series for* $f(t) = 2^t$ *with respect to the time scale* $\mathbb{T} = \mathbb{Z}$ *diverges for any integer* $t < 0$.

References

[1] R. P. Agarwal and M. Bohner, Basic calculus on time scales and some of its applications, *Results Math.*, 35 (1999), 3–22.

[2] J. P. Bézivin, Sur les équations fonctionnelles aux q-différences, *Aequationes Math.*, 43 (1993), 159–176.

[3] M. Bohner and A. Peterson, *Dynamic Equations on Time Scales: An Introduction with Applications*, Birkhauser, Boston, 2001.

[4] G. Derfel, E. Romanenko, and A. Sharkovsky, Long-time properties of solutions of simplest nonlinear q-difference equations, *J. Differ. Equations Appl.*, 6 (2000), 485–511.

[5] S. Hilger, Ein Maßkettenkalkül mit Anwendung auf Zentrumsmannig-faltigkeiten, PhD thesis, Universität Würzburg, 1988.

[6] W. Kelley and A. Peterson, *Difference Equations: An Introduction With Applications*, Second Edition, Academic Press, 2001.

[7] W. J. Trijtzinsky, Analytic theory of linear q-difference equations, *Acta Math.*, 61 (1993), 1–38.

[8] C. Zhang, Sur la sommabilité des séries entiè res solutions d'équations aux q-différences, *I. C. R. Acad. Sci. Paris Sér. I Math.*, 327 (1998), 349–352.

Embedding a Class of
Time Scale Dynamics into
O.D.E Dynamics with Impulse Effect

JULIO LÓPEZ FENNER [1]

Departamento de Ingeniería Matemática
Universidad de La Frontera, Temuco, Chile
E-mail: jlopez@ufro.cl

Abstract In this note we show how the original goal of unifying continuous and discrete dynamics in a single setting is also accomplished by o.d.e's with impulse effect. This means that at least a class of time scale dynamics can be described in an alternative fashion, which may prove more intuitive to engineers and applied mathematicians.

Keywords Time scales, Dynamic equations, Differential equations with impulse effect

AMS Subject Classification 34A37, 34K45, 93C55

1   Introduction

Dynamical equations are defined on a time scale $\mathbb{T}$ and have the form $x^\Delta = f(t, x)$, where $\mathbb{T}$ is an (arbitrary) closed subset of $\mathbb{R}$. Since the pioneer work of Aulbach and Hilger [1] in the nineties, much work in this field has been done, and, in particular, connections between dynamics induced by dynamical equations and the dynamics described by flows of systems of ordinary differential equations have been sought.

Recently, Garay and Hilger [3] undertook the study of embedding time scale dynamics in o.d.e dynamics, and stated appropriate conditions for it, as well as some counterexamples. Among these, a beautiful counterexample is given by $x^\Delta = -2x$ on $\mathbb{T} = \mathbb{Z}$, which has no counterpart in continuous dynamics, since first order dynamics do not allow oscillations. This behavior, also called clocks by engineers, is modelled with beautiful simplicity in o.d.e's with impulse effect; see below for an example.

Since the literature on time scales and dynamic equations has already grown to a respectable size, we shall only briefly introduce the basic notions

---

[1]Supported by FONDECYT grant 1020938.

in this field, leaving many details to the standard references. On the other side, we shall assume an audience not completely familiar with the theory of o.d.e's with impulse effect, so that this theory shall be depicted with a higher degree of detail. Readers familiar with both theories can skip immediately to section 4.

2   Preliminaries

2.1   Dynamical equations on time scales

Let $\mathbb{T}$ be a closed subset of $\mathbb{R}$, and assume $\sup \mathbb{T} = \infty$, $\inf \mathbb{T} = -\infty$. The forward jump operator is

$$\sigma(t) = \inf\{s \in \mathbb{T} : s > t, \}$$

which is well defined for all $t \in \mathbb{T}$. A point $t_0 \in \mathbb{T}$ is called *right scattered* if $\sigma(t_0) > t_0$. The *graininess* $\mu : \mathbb{T} \to \mathbb{R}_+ \cup \{0\}$ is defined as $\mu(t) = \sigma(t) - t$. A function $f : \mathbb{T} \to \mathbb{R}^d$, $d \in \mathbb{N}$, is said to be *differentiable* at time $t \in \mathbb{T}$ provided

$$f^\Delta(t) := \lim_{s \to t} \frac{f(\sigma(t)) - f(s)}{\sigma(t) - s},$$

where the limit is taken with values of $s \in \mathbb{T}$, $s \neq \sigma(t)$. The following facts concerning differentiability in time scales are taken from Bohner and Lutz [4]:

i) If $f : \mathbb{T} \to \mathbb{R}^d$ is differentiable in $t \in \mathbb{T}$, then it is continuous in t.

ii) If $t \in \mathbb{T}$ is right scattered and f is continuous in t, then $f^\Delta(t) = (f(\sigma(t)) - f(t))/\mu(t)$.

iii) If $f^\Delta(t)$ exists, then $f(\sigma(t)) = f(t) + \mu(t)f^\Delta(t)$.

We observe also that by assumption, all points on the time scale are non degenerate; thus whenever derivatives exist, they are unique (see [2], theorem 2).

A function $x : \mathbb{T} \to \mathbb{R}^d$ is *rd-continuous* if it is continuous in every right-dense point t of $\mathbb{T}$, and the limit from the left $\lim_{s \to t^-} x(t)$ exists. Now discrete and continuous dynamics are both dealt with inside this formalism, which is alternately done by $\mathbb{T} = \mathbb{R}$ or $\mathbb{T} = \mathbb{Z}$. As for a description of both simultaneously, let us look at it from a point of view more close to o.d.e's with impulse effect: take $I \subset \mathbb{R}$ be a closed interval of the form $I = [a, d]$, with $a < d$, $a, d \in \mathbb{R}$. Assume $I \cap \mathbb{T} = [a, b] \cup [c, d]$, with $b < c$, in fact, $c = \sigma(b)$. Then, if $x(t)$ is a (dynamically) differentiable function of $t \in \mathbb{T}$, then $x^\Delta(t) = \frac{dx}{dt}$ for all $t \in (a, b) \cup (c, d)$, whereas $x^\Delta(b) = \frac{x(c) - x(b)}{c - b}$ at $t = b$, and $x^\Delta(c) = \frac{dx}{dt}^+(c)$, the right derivative of $x(t)$ at $t = c$. This means that if $x(t)$ is a solution of some dynamic equation on $\mathbb{T}$, then the dynamics of x is given by an o.d.e on the interior of $I \cap \mathbb{T}$, with "jumps" at $\partial(I \cap \mathbb{T}) \backslash \partial(\mathbb{T})$. If one were to describe the behavior of $x(t)$ in a neighborhood of the points b

and c, then for U an open set in the topology of $\mathbb{T}$ induced by the topology in $\mathbb{R}$, such that $U \cap [a, d] = (\tilde{a}, b] \cup [c, \tilde{d}),\ a < \tilde{a},\ \tilde{d} < d$, then $x(t)$ is (ordinary) differentiable in $U \cap \mathbb{T} \backslash \{c, d\} = (\tilde{a}, b) \cup (c, \tilde{d})$, and $x(c) = x(b) + x^{\Delta}(b) \cdot (c - b)$.

If $x(t)$ fulfills a dynamic equation of the form

$$x^{\Delta}(t) = f(t, x(t)), \qquad t \in \mathbb{T},$$

then locally in U,

$$\frac{dx}{dt}(t) = f(t, x(t)), \quad t \in (a, b) \cup (c, d)$$

$$x(c) = x(b) + x^{\Delta}(b) \cdot (c - b).$$

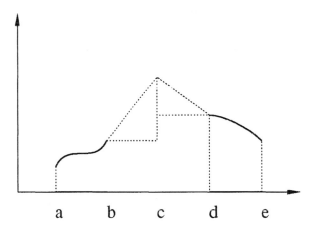

Figure 1: Local view of a solution $x(t)$ of a dynamic equation on a subset $[a, b] \cup \{c\} \cup [d, e]$ of $\mathbb{T}$.

2.2 Differential equations with impulse effect

We consider now a dynamical system described by a differential equation with impulse effect. This approach, as discussed here, is characterized by a family of (fixed and given) time instants $\{t_k\}_{k \in \mathbb{Z}}$ that satisfy $t_k < t_{k+1},\ k \in \mathbb{Z}$, and $\lim_{k \to \infty} t_k = \infty$, $\lim_{k \to -\infty} t_k = -\infty$, and the following set of equations:

$$\dot{x} = f(t, x), \qquad t \neq t_k, \quad k \in \mathbb{Z},$$

$$x(t_k^+) = g(t_k, x(t_k)), \qquad t = t_k,$$

where $x(t_k^+) = \lim_{t \to t_k^+} x(t)$ is the limit from the right at time t_k.

This description models continuous and discrete behavior in a unified setting in the following sense: if we let the system start at time t_k from a state

ξ_k, and let it follow the continuous dynamics, then at any point $t \in (t_k, t_{k+1}]$ the solution of the system corresponds to the solution of the initial value problem $x(t; t_k, \xi_k)$. Once the time t_{k+1} is reached, the state is upgraded and everything starts again, this time from the new state $\xi_{k+1} = g(t_{k+1}, x(t_{k+1}))$. Thus all solutions are left continuous, with $\xi_k = x(t_k^+) = \lim_{t \to t_k^+} x(t)$, for all $k \in \mathbb{Z}$

This approach models pure continuous systems just by putting $g(t_k, \cdot) = I$, the identity transformation, since then the limit of $x(t)$ from the right equals the limit from the left, at time $t = t_k$, which grants continuity (alone) of $x(t)$, but also pure discrete systems as well, since for this it is enough to put $f(t, \cdot) = 0$, $t \neq t_k$. Notice though, that this generalization possesses more structure than the structure of both, the (differentiable) continuous and the discrete case together. This is due to the fact that on one side, instead of having impulses now at fixed times $t_k = k \in \mathbb{Z}$, we allow for a wider class of discrete systems like q systems and others. In principle, one cannot redefine the time in order to get impulses at fixed times $\{k \in \mathbb{Z}\}$, if the sequence $\{t_k\}$ has accumulation points, for example. On the other side, discontinuities on the derivatives but not on the solution itself of such a system are allowed if one takes $g(t_k, \cdot) = I$, $t \neq t_k$. Under a normal Lipschitz condition on the x variable for the function $f(t, x)$, this phenomenon will not appear, so that continuity of the solution plus Lipschitz condition in the x variable of $f(t, x)$ is enough for a description of a pure differentiable system with no impulses. Figure 2 shows a typical graph of a solution of a system with impulse effect.

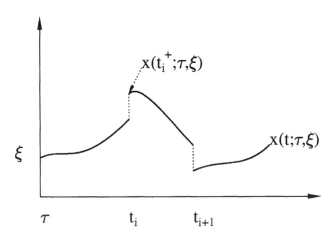

Figure 2: Local view of a solution $x(t)$ of a differential equation with impulse effect.

3   Some examples

Example 3.1 Consider $x^{\Delta}(t) = -2x(t)$, with $t \in \mathbb{T} = \mathbb{Z}$. If a boundary condition ξ is set at time $t = t_0$, then the solution of this dynamic equation is $x(t; t_0, \xi) = (-1)^{(t-t_0)}\xi$. One can extend the domain of definition of the solution to $t \in \mathbb{R}$ by defining $\frac{dx}{dt}(t) = 0$, for $t \in \mathbb{R}\backslash\mathbb{Z}$, thus yielding the solution $x(t; t_0, \xi) = (-1)^{[t-t_0]}\xi$, where $[t]$ is the integer part of $t \in \mathbb{R}$. Thus, this is a special example of the following differential equation with impulse effect:

$$
\begin{aligned}
\dot{x} &= 0, & t \neq t_k, \quad t_k = k \in \mathbb{Z}, \\
x(t_k^+) &= -x(t_k), & t = t_k,
\end{aligned}
$$

This situation describes an embedding of a dynamic equation on $\mathbb{Z}$ into a differential equation with impulse effect on $\mathbb{R}$.

Example 3.2 Consider the differential equation with impulse effect

$$
\begin{aligned}
\dot{x} &= 2a(t \bmod 1) - a, & t \neq t_k, \quad t_k = k \in \mathbb{Z}, \\
x(t_k^+) &= x(t_k), & t = t_k,
\end{aligned}
$$

where $t \bmod 1$ is defined as t if $0 \leq t < 1$, or $t + 1$ if $-1 \leq t < 0$, or $(t - [t]) \bmod 1$ otherwise. This models a periodic system on $\mathbb{R}$, which can be described via a differential equation with impulse effect at times $t \in \mathbb{Z}$ which has no counterpart in the realm of dynamic equations.

Example 3.3 Consider the dynamic equation

$$
x^{\Delta}(t) = f(t, x(t)), \qquad t \in C,
$$

where C is a Cantor set, thus closed in $\mathbb{R}$, thus a time scale. It is not clear whether this equation can be embedded into a context of differential equations with impulse effect.

What these three examples show is that, although both continuous and discrete dynamics can be treated simultaneously by both approaches, namely dynamic equations on time scales and differential equations with impulse effect, there are phenomena which escapes description in both settings simultaneously. Thus, one would expect both kind of results: embedding theorems and no-go theorems.

4   Embedding results for o.d.e.'s with impulse effect and dynamic equations

Theorem 4.1 *Let $\mathbb{T}$ be a non degenerate time scale of the form $\bigcup_{i \in \mathbb{Z}}[a_i, b_i]$, with $a_i \leq b_i < a_{i+1}$, for all $i \in \mathbb{Z}$. Then any solution of the dynamic equation*

$$
x^{\Delta}(t) = f(t, x(t)), \qquad t \in \mathbb{T}
$$

*with $f(t,x)$ rd-continuous in the t variable, and Lipschitz continuous in the x
variable is the restriction to $\mathbb{T}$ of a solution of an o.d.e. on $\mathbb{R}$, with impulses
at times $t_k \in \{a_i : i \in \mathbb{Z}\}$.*

Proof. Assume first that $\lim_{t \to -\infty} a_i = -\infty$, $\lim_{t \to +\infty} a_i = +\infty$, and that
there are no accumulation points. The first assumption is because if it were
not so, and T would be such that either sup $\mathbb{T} = b_{max} \neq \infty$, or inf $\mathbb{T} =
a_{min} \neq -\infty$, or both, then one could just "enlarge" $\mathbb{T}$ by joining the closed
intervals $(-\infty, a_{min}]$, $[b_{max}, +\infty)$. By saying that "there are no accumulation
points", we mean the property that $\forall \epsilon > 0$, $\exists N \in \mathbb{N}$, such that $\forall t \in \mathbb{T}$:

$$[t, t+\epsilon] \cap \mathbb{T} = \begin{cases} \cup_{j=k}^{k+M}[a_j, b_j] \\ \text{or } \emptyset, \end{cases} \quad \text{where } a_k = \max\{a_j : a_j \leq t\}, \text{ and } M \leq N.$$

If this is the case, then let $x(t)$ denote a solution of the dynamic equation. If
$a_i = b_i$ for all $i \in \mathbb{Z}$, the result is immediate, one puts the equation

$$\frac{dx}{dt}(t) = 0, \quad t \neq a_i,$$
$$x(a_{i+1}^+) = x(a_i) + f(a_i, x(a_i))(a_{i+1} - a_i).$$

If $a_i < b_i$ for all $i \in \mathbb{Z}$, then for $t \in (b_j, a_{j+1})$ define $\bar{x}(t)$ to be the unique
solution of the initial value problem

$$\frac{dy}{dt} = f(a_{j+1}, x(a_{j+1})),$$
$$y(b_j) = x(b_j) + [f(b_j, x(b_j)) - f(a_{j+1}, x(a_{j+1}))](a_{j+1} - b_j).$$

Then the function $z(t) = \begin{cases} x(t) & t \in \mathbb{T} \\ \bar{x}(t) & t \in \mathbb{R} \backslash \mathbb{T} \end{cases}$ is solution of the o.d.e. with
impulse effect

$$\frac{dx}{dt} = f(t,x), \qquad\qquad t \in \mathbb{T},$$
$$\frac{dx}{dt} = f(a_{j+1}, x(a_{j+1})), \qquad t \in (b_j, a_{j+1}),$$
$$x(b_j^+) = x(b_j) + $$
$$+ [f(b_j, x(b_j)) - f(a_{j+1}, x(a_{j+1}))](a_{j+1} - b_j), \quad t = b_j.$$

The mixed case is now a combination of the previous cases and will be omitted.
Should $\mathbb{T}$ contain accumulation points, like in the example

$$\mathbb{T} = (-\infty, 0] \cup \{1\} \cup [3/2, 5/3] \cup .. \cup [(4n-1)/(2n), (4n+1)/(2n+1)] \cup .. \cup [2, \infty)$$

then one has to care only about the behavior of the right hand side of the
dynamic equation (respectively o.d.e with impulse effect) at the accumulation
points, for at other points the situation is as before. Let then b denote one
of those accumulation points in $\mathbb{T}$, and assume the sequences $\{a_j\}$, and $\{b_j\}$
satisfy $a_j < b_j < a_{j+1}$, for all j, and $\lim_{j \to \infty} a_j = \lim_{j \to \infty} b_j = b$. Using that
$x(a_{j+1}) = x(b_j) + f(b_j, x(b_j))(a_{j+1} - b_j)$ and the fact that f was assumed to
be rd-continuous, then, since b is left dense, the limit $\lim_{j \to \infty} f(b_j, \xi)$ exists,
and since f was assumed to be Lipschitz in the x variable, then $\lim_{j \to \infty} x(a_j)$

$= \lim_{j \to \infty} x(b_j) = x(b)$. But this means that the same property applies for the induced solution of the constructed o.d.e. with impulse effect. $\square$

The embedding of solutions of o.d.e's with impulse effect into a class of dynamic equations on time scales is more straightforward:

Theorem 4.2 *Let $\{t_k\} \subset \mathbb{R}$ and $x(t)$ be a solution of*

$$\frac{dx}{dt} = f(t, x), \qquad t \neq t_k, \quad k \in \mathbb{Z},$$
$$x(t_k^+) = \tilde{f}(t_k, x(t_k)), \qquad t = t_k.$$

Then there exists a time scale $\mathbb{T}$ and a rd-continuous function $g : \mathbb{T} \times \mathbb{R}^d \to \mathbb{R}$, such that the restriction of $x(t; \tau, \xi)$ to $\mathbb{T}$, is a solution of the equation

$$x^\Delta(t) = g(t, x(t)), \qquad t \in \mathbb{T}.$$

Proof. Assume $t_{i-1} < \tau \leq t_i$. Define $\epsilon = \{\epsilon_j\}_{j=i}^\infty$ such that $t_j + \epsilon_j < t_{j+1}$, for all $j \geq i$, and $\mathbb{T}_\tau - [\tau, t_i] \cup [t_i + \epsilon_i, t_{i+1}] \cup \cdots \cup [t_k + \epsilon_k, t_{k+1}] \cup \cdots$ Recall that $x(t_j^+; \tau, \xi) = \tilde{f}(t_j, x(t_j; \tau, \xi)) + x(t_j; \tau, \xi)$. Define $g : \mathbb{T} \times \mathbb{R}^d \to \mathbb{R}^d$, as

$$g(t, \cdot) = \begin{cases} f(t, \cdot) & \text{if } t \in \text{int}\,\mathbb{T} \cup \{t_j + \epsilon_j\} \\ \frac{1}{\epsilon_j}[x(t^+; \tau, \xi) + \int_t^{t+\epsilon_j} f(s, x(s; \tau, \cdot))ds] & t = t_j \end{cases}$$

(Here, int $\mathbb{T}$ denotes the interior of $\mathbb{T}$.) Then, as in the example above

$$x^\Delta(t_j; \tau, \xi) = \frac{x(t_j + \epsilon_j; \tau, \xi) - x(t_j; \tau, \xi)}{\epsilon_j} = g(t_j, x(t_j; \tau, \xi)),$$

i.e., a solution of the dynamic equation. Moreover, g is rd continuous, since it is continuous on rd points $t \in \mathbb{T}$, and $\lim_{s \to t^-} g(s, \cdot)$ exists for all ld points $t \in \mathbb{T}$. $\square$

Remark 4.3 *This embedding is parameterized by the choices of ϵ and is thus not unique.*

5   Summary and further work

We have shown some embeddability results for two alternate treatments of continuous and discrete dynamics; one provided by the theory of ordinary differential equations with impulse effect, and the recently developed theory of dynamic equations on time scales. It turns out that at least for a class of time scales, both descriptions are equivalent, in the sense that the same dynamics can be modelled by o.d.e's with impulse effect, and the other way round, but certainly not in unique fashion. It is also important to acknowledge that both theories leave some phenomena which cannot or may not be described completely within a single framework. In order to clarify this point, one should address the question of the (rd-) continuity properties of the right hand side of a dynamic equation on a time scale of Cantor-type, or embedabbility of solutions of o.d.e's with impulse effect via restriction of the time domain as families of time scales like those treated here, with an assessment of properties of convergence. This approach may also be used as a different approximation to discretization techniques for ordinary differential equations.

References

[1] Aulbach, B. and Hilger, S., A unified approach to continuous and discrete dynamics, *Qualitative Theory of Differential Equations Szeged 1988 Colloc. Math. Soc. Janos Bolyai*, North Holland, Amsterdam, **53** (1990), 27–56.

[2] Aulbach, B. and Hilger, S., Linear dynamic processes with inhomogeneous time scale. In *Nonlinear Dynamics and Quantum Dynamical Systems*, Akademie Verlag, Berlin, 1990.

[3] Garay, B. and Hilger, S., Embeddability of time scale dynamics in o.d.e dynamics, *Nonlinear Analysis* **47** (2001), 1357–1371.

[4] Bohner, M. and Lutz, D.A., Asymptotic behavior of dynamic equations on time scales, *Journal of Difference Equations and Applications* **7** (2001), 21–50.

An Oscillation Criterion for a Dynamic Sturm-Liouville Equation

ZDENĚK OPLUŠTIL and ZDENĚK POSPÍŠIL[1]

Department of Mathematics, Masaryk University
Janáčkovo nám. 2a, CZ-662 95 Brno, Czech Republic
E-mail: oplustil@math.muni.cz, pospisil@math.muni.cz

Abstract A new oscillation criterion for the Sturm-Liouville homogeneous dynamic equation

$$\left(\frac{1}{q}u^{\Delta}\right)^{\Delta}(t) + p(t)u^{\sigma}(t) = 0$$

with functions $q \in C_{rd}^1(\mathbb{T}, \mathbb{R})$, $p \in C_{rd}(\mathbb{T}, \mathbb{R})$ satisfying the conditions

$$0 < m \le q(t) \le M < \infty \quad \text{for } t \in \mathbb{T}, \quad -\infty < \lim_{t \to \infty} \int^t p(s)\Delta(s) < \infty$$

is established.

Keywords Sturm-Liouville equation, Time scales, Oscillation, Nonoscillation

AMS Subject Classification 39A10

1 Introduction

Dynamic equations (equations on measure chains or time scales) generalize and unify both differential and difference equations [1], [3]. An investigation of a special dynamic equation, the Sturm-Liouville one

$$\left(\frac{1}{q}u^{\Delta}\right)^{\Delta}(t) + p(t)u^{\sigma}(t) = 0, \tag{1}$$

originated in a pioneering paper by Erbe and Hilger [2]. In particular, a sufficient condition for (1) to be oscillatory is established there. The condition requires the assumption $\int^{\infty} p(r)\Delta r = \infty$. In this paper, we present an oscillation criterion for the case when the improper integral $\int^{\infty} p(r)\Delta r$ converges.

What follows is a brief reminding of concepts and formulas concerning the theory of measure chains; the theory is dealt in details in [1] and [3]. The

---

[1]Supported by Grant No. 201/01/0079 of the Grant Agency of Czech Republic.

next section contains the main result of the paper — an oscillatory criterion which unifies some of the recent results [4], [5]. The result was proved using Riccati technique and the proof is sketched in the last section.

Throughout the paper, $\mathbb{T}$ denotes a measure chain with the forward jump operator σ, the growth calibration μ and the graininess μ^*, $\mu^*(t) = \mu(\sigma(t), t)$. As usual, for functions $f : \mathbb{T} \to \mathbb{R}$, we abbreviate $f^\sigma := f \circ \sigma$.

Following [2], we postulate two additional axioms on measure chains:

(i) The growth calibration $\mu(\cdot, s)$, with s fixed, is neither bounded above, nor bounded below.

(ii) The graininess μ^* is bounded on $\mathbb{T}$; i.e., there exists a constant $\bar{\mu} \geq 0$ such that $\mu^*(t) \leq \bar{\mu}$ for each $t \in \mathbb{T}$.

The axiom (i) allows us to consider operators such as $\int^\infty$, $\lim\limits_{t \to \infty}$, $\liminf\limits_{t \to \infty}$ and $\limsup\limits_{t \to \infty}$ on a set of functions $f : \mathbb{T} \to \mathbb{R}$. Moreover, the two additional axioms imply that for each $a \in \mathbb{R}$, $a > \bar{\mu}$ and $t \in \mathbb{R}$ there exists a $T \in \mathbb{T}$ such that $a \leq \mu(T, t) \leq 2a$.

For a function $f : \mathbb{T} \to \mathbb{R}$ we write $f \in C_{rd}(\mathbb{T}, \mathbb{R})$ if it is rd-continuous and $f \in C_{rd}^k(\mathbb{T}, \mathbb{R})$ if it is k-times rd-continuously (delta) differentiable.

Recall some simple formulae concerning the (delta) derivatives and the integrals of functions $f, g : \mathbb{T} \to \mathbb{R}$. In particular, product and quotients rules

$$(fg)^\Delta (t) = f^\Delta(t)g(t) + f^\sigma(t)g^\Delta(t), \tag{2}$$

$$\left(\frac{f}{g}\right)^\Delta (t) = \frac{f^\Delta(t)g(t) - f(t)g^\Delta(t)}{g^\sigma(t)g(t)}, \tag{3}$$

integration by parts

$$\int_a^b f^\Delta(r)g(r)\Delta r = f(b)g(b) - f(a)g(a) - \int_a^b f^\sigma(r)g^\Delta(r)\Delta r \tag{4}$$

and "shift operations"

$$f^\sigma(t) = f(t) + \mu^*(t)f^\Delta(t), \tag{5}$$

$$\int_t^{\sigma(t)} f(r)\Delta r = \mu^*(t)f(t), \qquad \int_a^b f^\sigma(r)\Delta r = \int_{\sigma(a)}^{\sigma(b)} f(r)\Delta r. \tag{6}$$

2   Main result

First, recall some basic results obtained in [2]. The homogenous Sturm-Liouville dynamic equation (1) with $p \in C_{rd}(\mathbb{T}, \mathbb{R})$, $q \in C_{rd}^1(\mathbb{T}, \mathbb{R}^+)$ has a unique solution on the whole time scale $\mathbb{T}$. A solution $u \in C_{rd}^2(\mathbb{T}, \mathbb{R})$ of the equation (1) is called *nonoscillatory (with initial time $t_0 \in \mathbb{T}$)*, if $u^\sigma(t)u(t) > 0$ for each $t \geq t_0$. In particular, a nonoscillatory solution does not change its sign for times greater than the initial time. A solution which is not nonoscillatory, is called *oscillatory*. Either all nontrivial solutions are nonoscillatory or none is. Therefore, we can define: *Equation* (1) is called *nonoscillatory*, if there exists a nonoscillatory solution. Otherwise (1) is called *oscillatory*.

For $t \geq t_0$, we set

$$H(t, t_0) := \frac{\int\limits_{t_0}^{t} p(r) \left(\int\limits_{t_0}^{\sigma(r)} q(s)\Delta s \right)^2 \Delta r}{\int\limits_{t_0}^{t} q(r)\Delta r}, \qquad Q(t, t_0) := \int\limits_{t_0}^{t} q(r)\Delta r \int\limits_{t}^{\infty} p(r)\Delta r.$$

Now, we can state our main result:

Theorem 2.1 *Let the functions $p \in C_{rd}(\mathbb{T}, \mathbb{R})$, $q \in C_{rd}^1(\mathbb{T}, \mathbb{R})$ satisfy the conditions: There exist constants $m, M \in \mathbb{R}$ such that*

$$0 < m \leq q(t) \leq M < \infty \quad \text{for } t \in \mathbb{T}. \tag{7}$$

and

$$-\infty < \lim_{t \to \infty} \int\limits_{\tau}^{t} p(s)\Delta(s) < \infty \quad \text{for each } \tau \in \mathbb{T}. \tag{8}$$

If

$$\limsup_{t \to \infty} [H(t, t_0) + Q(t, t_0)] > 1$$

for each $t_0 \in \mathbb{T}$, then the equation (1) *is oscillatory.*

Corollary 2.2 *Let the function $p \in C_{rd}(\mathbb{T}, \mathbb{R})$ satisfy the condition* (8). *If*

$$\limsup_{t \to \infty} \left[\frac{1}{\mu(t, t_0)} \int\limits_{t_0}^{t} (\mu(\sigma(r), t_0))^2 \, p(r)\Delta r + \mu(t, t_0) \int\limits_{t}^{\infty} p(r)\Delta r \right] > 1$$

for each $t_0 \in \mathbb{T}$, then the equation

$$u^{\Delta\Delta}(t) + p(t)u^\sigma(t) = 0$$

is oscillatory.

The corollary unifies a criterion established in [4] for the case $\mathbb{T} = \mathbb{R}$ and a criterion presented in [5] for the case $\mathbb{T} = \mathbb{Z}$.

3   Sketch of the proof

Throughout this paragraph, we suppose that the equation (1) possesses a nonoscillatory solution u with initial time $t_0 \in \mathbb{T}$ and that the functions q, p satisfy the conditions (7) and (8). The condition (7) allows us to define

$$w(t) := \frac{u^\Delta}{qu}(t) \quad \text{for } t \geq t_0.$$

The formula (5) and the nonoscillatoricity of the function u imply

$$\frac{1}{q(t)} + \mu^*(t)w(t) = \frac{1}{q(t)}\left(1 + \frac{\mu^*(t)u^\Delta(t)}{u(t)}\right) = \frac{1}{q(t)}\frac{u^\sigma(t)}{u(t)}.$$

Therefore

$$1 + \mu^*(t)q(t)w(t) > 0 \quad \text{for each } t \geq t_0. \tag{9}$$

Due to this inequality, we can define

$$R(t) := \frac{qw^2}{1 + \mu^*qw}(t) \quad \text{for } t \geq t_0.$$

The condition (7) together with (9) implies

$$R(t) \geq 0 \quad \text{for each } t \geq t_0, \qquad R(t) = 0 \text{ if and only if } w(t) = 0. \tag{10}$$

Using (3), (1) and (5), we obtain

$$w^\Delta(t) = -p(t) - R(t) \quad \text{for each } t \geq t_0. \tag{11}$$

A simple consequence of (10) is that the improper integral $\int_{t_0}^{\infty} R(r)\Delta r$ either converges or diverges to $+\infty$.

Lemma 3.1

$$\int_{t_0}^{\infty} R(r)\Delta r < \infty.$$

Proof: Let us suppose for contradiction that $\int_{t_0}^{\infty} R(r)\Delta r = \infty$. Integrating (11) from t_0 to t, we get

$$w(t) - w(t_0) = -\int_{t_0}^{t} p(r)\Delta r - \int_{t_0}^{t} R(r)\Delta r. \tag{12}$$

The two equalities together with the conditions (7), (8) imply

$$-\lim_{t \to \infty} w(t) = \int_{t_0}^{\infty} p(r)\Delta r + \int_{t_0}^{\infty} R(r)\Delta r - w(t_0) = \infty.$$

Consequently, there exists a $t_1 \geq t_0$ such that

$$w(t) \leq 0 \quad \text{for } t \geq t_1. \tag{13}$$

This inequality and the additional axiom (ii) allow us to introduce the functions

$$\nu(t) := \sup \{\mu^*(r) : r \geq t\}, \quad \tilde{w}(t) := \sup \{\tilde{w}(r) : r \geq t\} \quad \text{for } t \geq t_1.$$

Then

$$\mu^*(t) \leq \nu(t) \leq \bar{\mu} \quad \text{for } t \geq t_1, \tag{14}$$

$w(t) \leq \tilde{w}(t) < 0$ and both functions are nonincreasing for $t \geq t_1$. Moreover,

$$1 > \mu^*(t)q(t)\,(-w(t)) \geq m\mu^*(t)\,(-w(t))$$

by (7) and (9). Consequently,

$$\mu^*(t) \leq -\frac{1}{mw(t)} \leq -\frac{1}{m\tilde{w}(t)} \leq -\frac{1}{m\tilde{w}(t_1)} \quad \text{for } t \geq t_1.$$

Now, we have

$$\nu(t)\,(-w(t)) = \sup \{\mu^*(r) : r \geq t\}\,(-w(t)) \leq -\frac{(-w(t))}{m\tilde{w}(t_1)} \leq \frac{\tilde{w}(t)}{m\tilde{w}(t_1)} \leq \frac{1}{m},$$

i.e., the function $-\nu w$ is bounded above. The inequality (10) and the identity (12) yields

$$\nu(t)\int_{t_1}^{t} R(r)\Delta r \leq \nu(t)\int_{t_0}^{t} R(r)\Delta r = \nu(t)\left(w(t_0) - \int_{t_0}^{t} p(r)\Delta r\right) - \nu(t)w(t)$$

for $t \geq t_1$. Therefore, there exists a constant $c \in \mathbb{R}$ such that

$$\nu(t)\int_{t_1}^{t} R(r)\Delta r \leq c \quad \text{for } t \geq t_1. \tag{15}$$

The double integration of (11) from t_1 to t gives

$$\int_{t_1}^{t} w(r)\Delta r = w(t_1)\mu(t,t_1) - \int_{t_1}^{t}\left(\int_{t_1}^{r} p(s)\Delta s\right)\Delta r - \int_{t_1}^{t}\left(\int_{t_1}^{r} R(s)\Delta s\right)\Delta r.$$

The division by $\int_{t_1}^{t} q(r)\Delta r > 0$ and a simple manipulation yields

$$\frac{\int_{t_1}^{t} w(r)\Delta r + \frac{1}{2}\int_{t_1}^{t}\left(\int_{t_1}^{r} R(s)\Delta s\right)\Delta r}{\int_{t_1}^{t} q(r)\Delta r} = \frac{A(t) - \frac{1}{2}\int_{t_1}^{t}\left(\int_{t_1}^{r} R(s)\Delta s\right)\Delta r}{\int_{t_1}^{t} q(r)\Delta r}, \tag{16}$$

where

$$A(t) := w(t_1)\mu(t, t_1) - \int\limits_{t_1}^{t} \left(\int\limits_{t_1}^{r} p(s)\Delta s \right) \Delta r.$$

Since $\lim\limits_{t\to\infty} \int\limits_{t_0}^{t} q(r)\Delta r = \infty$ and $\left(\int\limits_{t_0}^{t} q(r)\Delta r \right)^{\Delta} = q(t) > 0$ by (7), we can use the l'Hôspital rule [1, Theorem 1.120] and compute the limit of the right-hand side of (16):

$$\lim\limits_{t\to\infty} \frac{1}{q(t)} \left(w(t_1) - \int\limits_{t_1}^{t} p(r)\Delta r - \frac{1}{2} \int\limits_{t_1}^{t} R(r)\Delta r \right) = -\infty$$

by (7), (8) and our assumption. Therefore, the identity (16) implies that there exists a $t_2 \geq t_1$ such that

$$-\int\limits_{t_1}^{t} w(r)\Delta r > \frac{1}{2} \int\limits_{t_1}^{t} \left(\int\limits_{t_1}^{r} R(s) \right) \Delta r \quad \text{for } t \geq t_2.$$

Since the left hand side is positive by (13), the inequality is equivalent to the following:

$$\left(\int\limits_{t_1}^{t} w(r)\Delta r \right)^2 > \frac{1}{4} \left(\int\limits_{t_1}^{t} \left(\int\limits_{t_1}^{r} R(s)\Delta s \right) \Delta r \right)^2 \quad \text{for } t \geq t_2. \qquad (17)$$

The Cauchy-Schwarz inequality [1, Theorem 6.15] together with (7), (10) and (13) yields

$$\left(\int\limits_{t_1}^{t} w(r)\Delta r \right)^2 \leq \int\limits_{t_1}^{t} \Delta r \int\limits_{t_1}^{t} (w(r))^2 \, \Delta r \leq \frac{\mu(t, t_1)}{m} \int\limits_{t_1}^{t} q(r) \, (w(r))^2 \, \Delta r =$$

$$= \frac{\mu(t, t_1)}{m} \int\limits_{t_1}^{t} R(r) \left(1 + \mu^*(r)q(r)w(r) \right) \Delta r \leq \frac{\mu(t, t_1)}{m} \int\limits_{t_1}^{t} R(r)\Delta r.$$

Denote

$$v(t) := \int\limits_{t_1}^{t} \left(\int\limits_{t_1}^{r} R(s)\Delta s \right) \Delta r.$$

The last inequality and (17) yield

$$\frac{\mu(t, t_1)}{m} v^{\Delta}(t) > \frac{1}{4} (v(t))^2 \quad \text{for } t \geq t_2.$$

Since $(v(t))^2 = v(t)v^\sigma(t) - \mu^*(t)v(t)v^\Delta(t)$ by (5) and there exists a $t_3 \geq t_2$ such that $v(t) > 0$ for $t \geq t_3$ by our assumption, the last inequality can be rearranged:

$$\frac{v^\Delta(t)}{v(t)v^\sigma(t)} > \frac{m}{4\mu(t,t_1) + m\mu^*(t)v(t)} \quad \text{for } t \geq t_3. \tag{18}$$

Without any loss of generality, we can suppose that $\mu(t,t_1) > \bar{\mu}$. Now, let $t \geq t_3$ be arbitrary and $T \in \mathbb{T}$ such that

$$\mu(t,t_1) \leq \mu(T,t) \leq 2\mu(t,t_1). \tag{19}$$

Integrating the left-hand side of (18) from t to T, we obtain the estimation

$$\int_t^T \frac{v^\Delta(r)}{v(r)v^\sigma(r)} \Delta r = -\int_t^T \left(\frac{1}{v(r)}\right)^\Delta \Delta r = -\frac{1}{v(T)} + \frac{1}{v(t)} < \frac{1}{v(t)}.$$

Taking into account (14) and the fact that the functions $\mu(\cdot,t)$, ν, v are nondecreasing, we can estimate the integral of the right hand side of (18):

$$\int_t^T \frac{m\Delta r}{4\mu(r,t_1) + m\mu^*(r)v(r)} \geq \int_t^T \frac{m\Delta r}{4\mu(T,t_1) + m\nu(T)v(T)} =$$

$$= \frac{m\mu(T,t)}{4\mu(T,t_1) + m\nu(T)v(T)}.$$

Therefore, using (19) and (15), we get the inequality

$$\frac{1}{v(t)} > \frac{m\mu(T,t)}{4\mu(T,t_1) + m\nu(T)v(T)} \geq \frac{m\mu(t,t_1)}{4(\mu(T,t) + \mu(t,t_1)) + mc} \geq \frac{m}{12 + m\frac{c}{\mu(t,t_1)}}$$

which is valid for arbitrary $t \geq t_3$. Consequently, letting $t \to \infty$, we obtain the desired contradiction $0 > m/8$, since $v(t) \to \infty$, $\mu(t,t_1) \to \infty$. ∎

This lemma plays a crucial role in the proof of Theorem 2.1. Its counterparts appeared also in the proofs of the corresponding theorems both in continuous and in discrete cases. But the lemma could not be proved by a straightforward translation of arguments used in [4] or [5] into the "measure chain language". (Or, speaking in a more precise way, we did not succeed to do it.) Now, the completion of the proof of the main result consist in a few steps. Each of them can be proved by a modification — with help of the formulas (2)–(6) — of the proof of corresponding proposition for difference equation (for details see [5]).

First,

$$\liminf_{t \to \infty} |w(t)| = 0$$

follows Lemma 3.1. Then, for $t \geq t_0$ and $t_1 \geq t_0$, let us put

$$D(t) := w(t)q(t) \left(2 \int_{t_0}^{\sigma(t)} q(r)\Delta r - \mu^*(t)q(t) \right) - R(t) \left(\int_{t_0}^{\sigma(t)} q(r)\Delta r \right)^2,$$

$$P(t,t_1) := \frac{1}{\int_{t_0}^{t} q(r)\Delta r} \left[w(t_1) \left(\int_{t_0}^{t_1} q(r)\Delta r \right)^2 + \int_{t_0}^{t_1} p(r) \left(\int_{t_0}^{\sigma(t)} q(s)\Delta s \right) \Delta r \right].$$

Subsequently,

$$\lim_{t \to \infty} P(t,t_1) = 0 \quad \text{for each } t_1 \geq t_0,$$

and

$$\int_{t_1}^{t} D(r)\Delta r \leq \int_{t_0}^{t} q(r)\Delta r \quad \text{for each } t \geq t_1 \geq t_0.$$

Moreover

$$w(t) \int_{t_0}^{t} q(r)\Delta r \leq 1 - H(t,t_0) + P(t,t_1) \quad \text{for each } t \geq t_1 \geq t_0$$

and

$$H(t,t_0) + Q(t,t_0) \leq 1 \quad \text{for each } t \geq t_0. \tag{20}$$

Thus, we obtain the statement: *"If the equation* (1) *possesses a nonoscillatory solution with the initial time* $t_0 \in \mathbb{T}$, *then* (20) *holds."* It implies the main theorem immediately.

References

[1] Bohner, M. and Peterson, A. C., *Dynamic Equations on Time Scales. An Introduction with Applications.* Birkhäuser, Boston 2001.

[2] Erbe, L. and Hilger, S., Sturmian theory on measure chains. *Differential Equations Dynam. Systems* **1**, No 3, 223–244, 1993.

[3] Laksmikantham, V., Sivasundaram, S. and Kaymakçalan, B., *Dynamic Systems on Measure Chains.* Kluwer Academic Publishers, Dordrecht-Boston-London, 1996.

[4] Lomtatidze, A., Oscillation and nonoscillation criteria for second order linear differential equation. *Georgian Math. J.* **4**, No 2, 129–138, 1997.

[5] Řehák, P., Oscillation and nonoscillation criteria for second order linear difference equation. *Fasciculi Math.* No 31, 71–89, 2001.

Two Perturbation Results for Semi-Linear Dynamic Equations on Measure Chains

CHRISTIAN PÖTZSCHE[1]

Department of Mathematics, University of Augsburg
D-86135 Augsburg, Germany
E-mail: christian.poetzsche@math.uni-augsburg.de

Abstract In this note we investigate semi-linear parameter dependent dynamic equations on Banach spaces and provide sufficient criteria for them to possess exponentially bounded solutions in forward and backward time. Apart from classical stability theory these results can be applied in the construction of non-autonomous invariant manifolds.

Keywords Time scale, Measure chain, Dynamic equation, Perturbation

AMS Subject Classification 39A11, 39A12, 37C75

1   Introduction and preliminaries

It is well-known that the exponential stability of linear difference or differential equations is robust under sufficiently small perturbations. With regard to this, the present paper has two main goals:

- We can unify the corresponding results for ordinary differential equations (ODEs) and difference equations (OΔEs) within the calculus on measure chains or time scales (cf. [BP01, Hil88, Hil90]).

- Exponential stability is weakened to a certain exponential boundedness of solutions, namely the so-called *quasiboundedness* (see Definition 1).

However, the two results of this paper (Theorems 2 and 4) are originally and basically designed as tools to construct invariant manifolds using a Lyapunov-Perron technique, and an application on general measure chains can be found in [Pöt03], while ODEs and OΔEs are considered in, e.g., [AW96] and [Aul98, APS02], respectively. Indeed the Theorems 2 and 4 carry a certain technical amount in such situations. Related constructions for dynamic equations are

---
[1]Research supported by the "Graduiertenkolleg: Nichtlineare Probleme in Analysis, Geometrie und Physik" (GRK 283) financed by the Deutsche Forschungsgemeinschaft and the State of Bavaria.

provided in [Hil96, Theorem 3.1] and [Kel99, p. 41, Satz 3.2.7, pp. 42–43, Satz 3.2.8]. Both references consider linear regressive equations, while we allow nonlinear perturbations and try to avoid regressivity as far as possible. The general nonlinear case on homogeneous measure chains, i.e., when the graininess μ^* (cf. [Hil90, Section 2.3]) is constant, is treated in [Hil88, pp. 62–63, Satz 10.1, p. 71, Satz 10.3]. Apart from this, our approach has its roots in [Aul87], where finite-dimensional ODEs have been considered; for non-autonomous OΔEs see [Aul95].

Concerning our notation, $\mathbb{R}$ is the real field. Throughout this paper Banach spaces $\mathcal{X}, \mathcal{Y}$ are all real or complex and their norm is denoted by $\|\cdot\|$. $\mathcal{L}(\mathcal{X})$ is the Banach algebra of linear continuous endomorphisms on $\mathcal{X}$ and $I_\mathcal{X}$ the identity mapping on $\mathcal{X}$. We also introduce some notions which are specific to the calculus on measure chains. In all the subsequent considerations we deal with a measure chain $(\mathbb{T}, \preceq, \mu)$ unbounded above and below with bounded graininess μ^*, the forward jump operator $\sigma : \mathbb{T} \to \mathbb{T}$ and with $\tau \in \mathbb{T}$ we write $\mathbb{T}_\tau^+ := \{s \in \mathbb{T} : \tau \preceq s\}$, $\mathbb{T}_\tau^- := \{s \in \mathbb{T} : s \preceq \tau\}$. $\chi_{\mathbb{T}_\tau^+} : \mathbb{T} \to \{0, 1\}$ is the characteristic function of $\mathbb{T}_\tau^+$. $\mathcal{C}_{rd}(I, \mathcal{L}(\mathcal{X}))$ denotes the rd-continuous and $\mathcal{C}_{rd}\mathcal{R}(I, \mathcal{L}(\mathcal{X}))$ the rd-continuous, regressive mappings from a $\mathbb{T}$-interval I into $\mathcal{L}(\mathcal{X})$. Recall that $\mathcal{C}_{rd}^+\mathcal{R}(I, \mathbb{R}) := \{a \in \mathcal{C}_{rd}\mathcal{R}(I, \mathbb{R}) : 1 + \mu^*(t)a(t) > 0 \text{ for } t \in I\}$ forms the so-called *positively regressive group* with respect to the addition $(a \oplus b)(t) := a(t) + b(t) + \mu^*(t)a(t)b(t)$ for $t \in I$ and $a, b \in \mathcal{C}_{rd}^+\mathcal{R}(I, \mathbb{R})$. An element $a \in \mathcal{C}_{rd}^+\mathcal{R}(I, \mathbb{R})$ is denoted as a *growth rate*, if $\sup_{t \in I} \mu^*(t)a(t) < \infty$ holds. Moreover we define the relations $a \lhd b :\Leftrightarrow 0 < \lfloor b - a \rfloor := \inf_{t \in I}(b(t) - a(t))$ and $e_a(t, s) \in \mathbb{R}$ stands for the real exponential function on $\mathbb{T}$ (cf. [Hil90, Section 7]).

Given $A \in \mathcal{C}_{rd}(\mathbb{T}, \mathcal{L}(\mathcal{X}))$, the *transition operator* $\Phi_A(t, \tau) \in \mathcal{L}(\mathcal{X})$, $\tau \preceq t$, of a linear dynamic equation $x^\Delta = A(t)x$ is the solution of the operator-valued initial value problem $X^\Delta = A(t)X$, $X(\tau) = I_\mathcal{X}$ in $\mathcal{L}(\mathcal{X})$ and if A is regressive then $\Phi_A(t, \tau)$ is defined for all $\tau, t \in \mathbb{T}$. The partial derivative of $\Phi_A(t, \tau)$ with respect to the first variable is denoted by $\Delta_1\Phi_A(t, \tau)$.

To bring these preliminaries to an end we introduce the so-called *quasiboundedness* which is a handy notion describing exponential growth of functions. For a further motivation see [AW96, Section 3].

Definition 1. *For a growth rate $c \in \mathcal{C}_{rd}^+\mathcal{R}(\mathbb{T}, \mathbb{R})$, a fixed time $\tau_0 \in \mathbb{T}$, a Banach space $\mathcal{X}$, a $\mathbb{T}$-interval I and a rd-continuous function $\lambda : I \to \mathcal{X}$ we say that λ is*

(a) *c^+-quasibounded if $I = \mathbb{T}_{\tau_0}^+$ and $\|\lambda\|_{\tau,c}^+ := \sup_{t \in \mathbb{T}_\tau^+} \|\lambda(t)\| e_{\ominus c}(t, \tau) < \infty$ for $\tau \in \mathbb{T}_{\tau_0}^+$,*

(b) *c^--quasibounded if $I = \mathbb{T}_{\tau_0}^-$ and $\|\lambda\|_{\tau,c}^- := \sup_{t \in \mathbb{T}_\tau^-} \|\lambda(t)\| e_{\ominus c}(t, \tau) < \infty$ for $\tau \in \mathbb{T}_{\tau_0}^-$,*

(c) *$c^\pm$-quasibounded if $I = \mathbb{T}$ and $\|\lambda\|_{\tau,c}^\pm := \sup_{t \in \mathbb{T}} \|\lambda(t)\| e_{\ominus c}(t, \tau) < \infty$ for $\tau \in \mathbb{T}$.*

With $\mathcal{B}_{\tau,c}^{+}(\mathcal{X}), \mathcal{B}_{\tau,c}^{-}(\mathcal{X})$ and $\mathcal{B}_{c}^{\pm}(\mathcal{X})$ we denote the sets of all c^{+}-, c^{-}- and $c^{\pm}$-quasibounded functions $\lambda : I \to \mathcal{X}$, respectively.

Obviously the three sets $\mathcal{B}_{\tau,c}^{+}(\mathcal{X}), \mathcal{B}_{\tau,c}^{-}(\mathcal{X})$ and $\mathcal{B}_{c}^{\pm}(\mathcal{X})$ are non-empty and using [Hil90, Theorem 4.1(iii)] one can show that they define Banach spaces (cf. [Pöt02, p. 76, Lemma 1.4.3]).

2   Perturbation results

After the above preparations we can tackle the problem of the existence and uniqueness of quasibounded solutions in forward and backward time, respectively. We begin with dynamic equations where the linear part is not necessarily regressive. For difference equations see [Aul98, Lemma 3.3, Lemma 3.2] and [AW96, Lemma 3.2, Lemma 3.4] for Carathéodory differential equations.

Theorem 2. *Assume that $K_1 \geq 1$, $L_1, M \geq 0$, $a \in \mathcal{C}_{rd}^{+}\mathcal{R}(I, \mathbb{R})$ is a growth rate, I denotes some closed $\mathbb{T}$-interval, $\mathcal{X}$ is a Banach space and $\mathcal{P}$ a topological space satisfying the first axiom of countability. Let us consider the parameter dependent dynamic equation*

$$x^{\Delta} = A(t)x + F(t, x, p) + f(t, p), \tag{1}$$

where $A \in \mathcal{C}_{rd}(I, \mathcal{L}(\mathcal{X}))$ and $F : I \times \mathcal{X} \times \mathcal{P} \to \mathcal{X}$, $f : I \times \mathcal{P} \to \mathcal{X}$ are rd-continuous mappings satisfying

$$\|\Phi_A(t, s)\| \leq K_1 e_a(t, s) \quad for \ s, t \in I, \ s \preceq t, \tag{2}$$

$$F(t, 0, p) = 0 \quad for \ t \in I, \ p \in \mathcal{P}, \tag{3}$$

$$\|F(t, x, p) - F(t, \bar{x}, p)\| \leq L_1 \|x - \bar{x}\| \quad for \ t \in I, \ x, \bar{x} \in \mathcal{X}, \ p \in \mathcal{P}. \tag{4}$$

Then for every growth rate $c \in \mathcal{C}_{rd}^{+}\mathcal{R}(I, \mathbb{R})$, $a + K_1 L_1 \lhd c$ and $\tau \in I$ we get the following:

(a) *Supposed I is unbounded above and $f(\cdot, p) \in \mathcal{B}_{\tau,c}^{+}(\mathcal{X})$, $p \in \mathcal{P}$, allows the estimate $\|f(\cdot, p)\|_{\tau,c}^{+} \leq M$ for $p \in \mathcal{P}$, then every solution $\nu(\cdot, p) : I \to \mathcal{X}$, $p \in \mathcal{P}$, of (1) is c^{+}-quasibounded with*

$$\|\nu(\cdot, p)\|_{\tau,c}^{+} \leq K_1 \|\nu(\tau, p)\| + \frac{K_1 M}{\lfloor c - a - K_1 L_1 \rfloor} \quad for \ p \in \mathcal{P}. \tag{5}$$

Moreover, the mapping $\nu : I \times \mathcal{P} \to \mathcal{X}$ is continuous.

(b) *Supposed I is unbounded below and $f(\cdot, p) \in \mathcal{B}_{\tau,c}^{-}(\mathcal{X})$, $p \in \mathcal{P}$, allows the estimate $\|f(\cdot, p)\|_{\tau,c}^{-} \leq M$ for $p \in \mathcal{P}$, then there exists exactly one c^{-}-quasibounded solution $\nu_*(\cdot, p) : I \to \mathcal{X}$, $p \in \mathcal{P}$, of (1), which furthermore satisfies*

$$\|\nu_*(\cdot, p)\|_{\tau,c}^{-} \leq \frac{K_1 M}{\lfloor c - a - K_1 L_1 \rfloor} \quad for \ p \in \mathcal{P}. \tag{6}$$

Moreover, the mapping $\nu_* : I \times \mathcal{P} \to \mathcal{X}$ is continuous.

(c) *Supposed* $I = \mathbb{T}$ *and* $f(\cdot,p) \in \mathcal{B}_c^{\pm}(\mathcal{X})$, $p \in \mathcal{P}$, *allows the estimate* $\|f(\cdot,p)\|_{\tau,c}^{\pm} \leq M$ *for* $p \in \mathcal{P}$, *then there exists exactly one* $c^{\pm}$-*quasi-bounded solution* $\nu_*(\cdot,p) : \mathbb{T} \to \mathcal{X}$, $p \in \mathcal{P}$, *of* (1), *which furthermore satisfies*

$$\|\nu_*(\cdot,p)\|_{\tau,c}^{\pm} \leq \frac{K_1 M}{\lfloor c - a - K_1 L_1 \rfloor} \quad \text{for } p \in \mathcal{P}.$$

Moreover, the mapping $\nu_* : \mathbb{T} \times \mathcal{P} \to \mathcal{X}$ *is continuous.*

Remark 3. *A version of Theorem 2 for parameter independent dynamic equations, where the linear part* $x^{\Delta} = A(t)x$ *is allowed to possess an exponential dichotomy, can be found in [Pöt01, Theorem 3.4] (for assertion (a)) and [Pöt02, p. 111, Satz 2.2.12] (for assertion (c)). Hence we can weaken the assumption* (2) *and the corresponding remark applies to Theorem 4 also.*

Proof. Let $\tau \in \mathbb{T}$ and the growth rate $c \in \mathcal{C}_{rd}^+\mathcal{R}(I,\mathbb{R})$, $a + K_1 L_1 \lhd c$, be given arbitrarily. It is easy to see that the subsequently quoted results from [Hil88, Hil90] are valid in forward time ($t \in \mathbb{T}_{\tau}^+$) without assuming regressivity of the dynamic equation (1).

(a) All solutions of (1) starting at time $\tau \in \mathbb{T}$ exist throughout the $\mathbb{T}$-interval $\mathbb{T}_{\tau}^+$ according to [Hil90, Theorem 5.7]. If $\nu : \mathbb{T}_{\tau}^+ \times \mathcal{P} \to \mathcal{X}$ denotes such an arbitrary solution of (1) then the variation of constants formula (cf. [Hil90, Theorem 6.4(ii)]) yields

$$\nu(t,p) = \Phi_A(t,\tau)\nu(\tau,p) + \int_{\tau}^{t} \Phi_A(t,\sigma(s)) \left[F(s,\nu(s,p),p) + f(s,p) \right] \Delta s$$

for $t \in \mathbb{T}_{\tau}^+$, $p \in \mathcal{P}$ and with (2), (3) as well as (4) we obtain the following estimate

$$\|\nu(t,p)\| \, e_{\ominus a}(t,\tau)$$

$$\leq K_1 \|\nu(\tau,p)\| + K_1 \int_{\tau}^{t} e_a(\tau,\sigma(s)) \, \|F(s,\nu(s,p),p) - F(s,0,p)\| \, \Delta s$$

$$+ K_1 \int_{\tau}^{t} e_a(\tau,\sigma(s)) \, \|f(s,p)\| \, \Delta s$$

$$\leq K_1 \|\nu(\tau,p)\| + K_1 L_1 \int_{\tau}^{t} e_a(\tau,\sigma(s)) \, \|\nu(s,p)\| \, \Delta s$$

$$+ K_1 \int_{\tau}^{t} e_a(\tau,\sigma(s)) e_c(s,\tau) \, \|f(s,p)\| \, e_{\ominus c}(s,\tau) \, \Delta s$$

$$\leq K_1 \|\nu(\tau,p)\| + \int_{\tau}^{t} \frac{K_1 L_1}{1 + \mu^*(s)a(s)} \, \|\nu(s,p)\| \, e_{\ominus a}(s,\tau) \, \Delta s$$

$$+ K_1 M \int_{\tau}^{t} e_a(\tau,\sigma(s)) e_c(s,\tau) \, \Delta s$$

for $t \in \mathbb{T}_\tau^+$ and parameters $p \in \mathcal{P}$. Now Gronwall's Lemma (cf. [Hil88, p. 49, Satz 7.2]) leads to

$$\|\nu(t,p)\| e_{\ominus a}(t,\tau)$$
$$\leq e_{a_0}(t,\tau) \left[K_1 \|\nu(\tau,p)\| + K_1 M \int_\tau^t e_{a+K_1 L_1}(\tau,\sigma(s)) e_c(s,\tau) \Delta s \right]$$
$$\leq e_{a_0}(t,\tau) \left[K_1 \|\nu(\tau,p)\| + \frac{K_1 M}{\lfloor c - a - K_1 L_1 \rfloor} \left(e_{c\ominus(a+K_1 L_1)}(t,\tau) - 1 \right) \right]$$

for $t \in \mathbb{T}_\tau^+$, $p \in \mathcal{P}$, where we have abbreviated $a_0(t) := \frac{K_1 L_1}{1+\mu^*(s)a(s)}$ and used [Pöt01, Lemma 3.1] to evaluate the integral. Finally, multiplying both sides of the above estimate by $e_{a\ominus c}(t,\tau) > 0$ we get

$$\|\nu(t,p)\| e_{\ominus c}(t,\tau)$$
$$\leq K_1 e_{a\oplus a_0\ominus c}(t,\tau) \|\nu(\tau,p)\| + \frac{K_1 M}{\lfloor c - a - K_1 L_1 \rfloor} \left(1 - e_{a_0\oplus a\ominus c}(t,\tau) \right)$$
$$\leq K_1 \|\nu(\tau,p)\| + \frac{K_1 M}{\lfloor c - a - K_1 L_1 \rfloor} \quad \text{for } t \in \mathbb{T}_\tau^+, \ p \in \mathcal{P}$$

with the aid of [Hil90, Theorem 7.4]. This means that $\nu(\cdot,p) : \mathbb{T}_\tau^+ \to \mathcal{X}$, $p \in \mathcal{P}$, is c^+-quasibounded and satisfies (5). The continuity of $\nu : \mathbb{T}_\tau^+ \times \mathcal{P} \to \mathcal{X}$ follows from [Hil88, p. 51, Satz 7.4].

(b) We subdivide the proof of statement (b) into four steps.

Step 1 – **Claim:** *The zero solution of the linear homogeneous equation*

$$x^\Delta = A(t)x \tag{7}$$

is the only solution of (7) in $\mathcal{B}_{\tau,c}^-(\mathcal{X})$.

Because the system (7) evidently possesses an exponential dichotomy on $\mathbb{T}_\tau^-$ with the invariant projector $P(t) \equiv I_\mathcal{X}$ by (2), we can apply [Pöt01, Corollary 2.11(b)] to prove that any c^--quasibounded solution $\nu : \mathbb{T}_\tau^- \to \mathcal{X}$ of (7) vanishes identically.

Step 2 – **Claim:** *There exists exactly one c^--quasibounded solution $\nu_*(\cdot,p) : \mathbb{T}_\tau^- \to \mathcal{X}$, $p \in \mathcal{P}$, of the linear inhomogeneous equation*

$$x^\Delta = A(t)x + f(t,p), \tag{8}$$

which furthermore satisfies

$$\|\nu_*(\cdot,p)\|_{\tau,c}^- \leq \frac{K_1 M}{\lfloor c - a \rfloor} \quad \text{for } p \in \mathcal{P}. \tag{9}$$

Above all the function $\nu_* : \mathbb{T}_\tau^- \times \mathcal{P} \to \mathcal{X}$, $\nu_*(t,p) := \int_{-\infty}^t \Phi_A(t,\sigma(s)) f(s,p) \Delta s$

is well-defined, since the integrand is rd-continuous (in s) and since the estimate

$$\|\nu_*(t,p)\| \, e_{\ominus c}(t,\tau) \leq \int_{-\infty}^{t} \|\Phi_A(t,\sigma(s))\| \, \|f(s,p)\| \, \Delta s e_{\ominus c}(t,\tau)$$

$$\overset{(2)}{\leq} K_1 \int_{-\infty}^{t} e_a(t,\sigma(s)) e_c(s,\tau) \, \Delta s e_{\ominus c}(t,\tau) \, \|f(\cdot,p)\|_{\tau,c}^{+} \leq \frac{K_1 M}{\lfloor c - a \rfloor}$$

for $t \in \mathbb{T}_\tau^-$ holds true, where the improper integral has been evaluated using [Pöt01, Lemma 3.1]. Additionally the inclusion $\nu_*(\cdot,p) \in \mathcal{B}_{\tau,c}^-(\mathcal{X})$, $p \in \mathcal{P}$, yields by passing over to the least upper bound over $t \in \mathbb{T}_\tau^-$. The derivative of ν_* with respect to $t \in \mathbb{T}_\tau^-$ is given by

$$\nu_*^\Delta(t,p) \equiv f(t,p) + \int_{-\infty}^{t} \Delta_1 \Phi_A(t,\sigma(s)) f(s,p) \, \Delta s \equiv A(t)\nu_*(t,p) + f(t,p)$$

on $\mathbb{T}_\tau^-$, and the integral has been differentiated using a result dual to [Pöt01, Lemma 4.2]. Therefore $\nu_*(\cdot,p)$, $p \in \mathcal{P}$, is a c^--quasibounded solution of (8) satisfying the estimate (9). Finally the uniqueness statement immediately results from Step 1, because the difference of two c^--quasibounded solutions of (8) is a c^--quasibounded solution of (7) and consequently identically vanishing. In order to prove the continuity of the mapping $\nu_* : \mathbb{T}_\tau^- \times \mathcal{P} \to \mathcal{X}$, let the pair $(t_0,p_0) \in \mathbb{T}_\tau^- \times \mathcal{P}$ be arbitrarily fixed and consider the alternative representation

$$\nu_*(t,p) = \int_{-\infty}^{\tau} \chi_{\mathbb{T}_t^-}(s)\Phi_A(t,\sigma(s)) f(s,p) \, \Delta s \quad \text{for } t \in \mathbb{T}_\tau^-, \, p \in \mathcal{P}.$$

As $(t,p) \to (t_0,p_0)$ the integrand converges to $\chi_{\mathbb{T}_{t_0}^-}(s)\Phi_A(t_0,\sigma(s)) f(s,p_0)$ for all $s \in \mathbb{T}_\tau^-$ and the inequality

$$\left\| \chi_{\mathbb{T}_t^-}(s)\Phi_A(t,\sigma(s)) f(s,p) \right\| \overset{(2)}{\leq} K_1 e_a(t,\sigma(s)) e_c(s,\tau) \, \|f(\cdot,p_0)\|_{\tau,c}^{-}$$

$$\leq K_1 M e_a(t,\sigma(s)) e_c(s,\tau) \quad \text{for } \sigma(s) \preceq t \tag{10}$$

is valid. Because of $a \lhd c$ we can use [Pöt01, Lemma 3.1] to evaluate the integral from $-\infty$ to τ over the right hand side of the estimate (10), and we may apply Lebesgue's dominated convergence theorem (cf. [Nei01, p. 161, Satz 313]) to get the convergence $\lim_{(t,p)\to(t_0,p_0)} \nu_*(t,p) = \nu_*(t_0,p_0)$, which proves the desired continuity of $\nu_* : \mathbb{T}_\tau^- \times \mathcal{P} \to \mathcal{X}$.

Step 3 – Claim: *There exists exactly one c^--quasibounded solution $\nu_*(\cdot,p) :$ $\mathbb{T}_\tau^- \to \mathcal{X}$, $p \in \mathcal{P}$, of the semi-linear equation (1), which moreover satisfies (6).* In order to set up the framework of Banach's fixed point theorem we define the sets

$$\mathcal{B} := \left\{ \nu : \mathbb{T}_\tau^- \times \mathcal{P} \to \mathcal{X} \,\middle|\, \begin{array}{l} \nu : \mathbb{T}_\tau^- \times \mathcal{P} \to \mathcal{X} \text{ is continuous,} \\ \nu(\cdot,p) \in \mathcal{B}_{\tau,c}^-(\mathcal{X}) \text{ for all } p \in \mathcal{P}, \\ \sup_{p\in\mathcal{P}} \|\nu(\cdot,p)\|_{\tau,c}^- < \infty \end{array} \right\},$$

which are readily seen to be Banach spaces equipped with the norm $\|\nu\|_{\tau,c}^{-,0} :=$ $\sup_{p \in \mathcal{P}} \|\nu(\cdot, p)\|_{\tau,c}^{-}$. As to the construction of an appropriate contraction operator on $\mathcal{B}$ we choose any $\nu \in \mathcal{B}$ and consider the linear inhomogeneous dynamic equation

$$x^{\Delta} = A(t)x + F(t, \nu(t,p), p) + f(t,p). \tag{11}$$

Since (3) and (4) imply the estimate

$$\begin{aligned}
&\|F(t, \nu(t,p), p) + f(t,p)\| \, e_{\ominus c}(t, \tau) \\
&\leq L_1 \|\nu(t,p)\| \, e_{\ominus c}(t, \tau) + \|f(t,p)\| \, e_{\ominus c}(t, \tau) \\
&\leq L_1 \|\nu(\cdot, p)\|_{\tau,c}^{-} + \|f(\cdot, p)\|_{\tau,c}^{-} \leq L_1 \|\nu\|_{\tau,c}^{-,0} + M \quad \text{for } t \in \mathbb{T}_{\tau}^{-}, \, p \in \mathcal{P},
\end{aligned} \tag{12}$$

we may apply Step 2 of the present proof to equation (11). Hence there exists a continuous mapping $\nu_* : \mathbb{T}_{\tau}^{-} \times \mathcal{P} \to \mathcal{X}$ such that the function $\nu_*(\cdot, p)$ is the unique c^{-}-quasibounded solution of (11) to the parameter value $p \in \mathcal{P}$, and for arbitrary $p \in \mathcal{P}$ we get the estimate

$$\|\nu_*(\cdot, p)\|_{\tau,c}^{-} \overset{(9)}{\leq} \frac{K_1}{\lfloor c - a \rfloor} \left(L_1 \|\nu\|_{\tau,c}^{-,0} + M \right) \quad \text{for } p \in \mathcal{P},$$

which shows the inclusion $\nu_* \in \mathcal{B}$. Now we are defining the designated contraction mapping $\mathcal{T} : \mathcal{B} \to \mathcal{B}$, $\nu \mapsto \nu_*$. Then (12) implies the estimate

$$\|(\mathcal{T}\nu)(\cdot, p)\|_{\tau,c}^{-} \leq \frac{K_1}{\lfloor c - a \rfloor} \left[L_1 \|\nu(\cdot, p)\|_{\tau,c}^{-} + \|f(\cdot, p)\|_{\tau,c}^{-} \right] \tag{13}$$

for $p \in \mathcal{P}$, $\nu \in \mathcal{B}$. Now we claim that the mapping $\mathcal{T} : \mathcal{B} \to \mathcal{B}$ actually is a contraction. In order to verify this, let $\nu, \bar{\nu} \in \mathcal{B}$ be arbitrary. Then the difference $(\mathcal{T}\nu - \mathcal{T}\bar{\nu})(\cdot, p)$, $p \in \mathcal{P}$, is a c^{-}-quasibounded solution of the linear inhomogeneous system

$$x^{\Delta} - A(t)x + F(t, \nu(t,p), p) - F(t, \bar{\nu}(t,p), p) \tag{14}$$

for every parameter $p \in \mathcal{P}$ and similar to equation (11) also (14) satisfies all the assumptions of Step 2. Consequently, (4) implies

$$\|(\mathcal{T}\nu - \mathcal{T}\bar{\nu})(\cdot, p)\|_{\tau,c}^{-} \leq \frac{K_1 L_1}{\lfloor c - a \rfloor} \|(\nu - \bar{\nu})(\cdot, p)\|_{\tau,c}^{-} \leq \frac{K_1 L_1}{\lfloor c - a \rfloor} \|\nu - \bar{\nu}\|_{\tau,c}^{0,-}$$

for $p \in \mathcal{P}$, and we therefore get $\|\mathcal{T}\nu - \mathcal{T}\bar{\nu}\|_{\tau,c}^{0,-} \leq \frac{K_1 L_1}{\lfloor c - a \rfloor} \|\nu - \bar{\nu}\|_{\tau,c}^{0,-}$. According to the assumptions we have $a + K_1 L_1 \lhd c$, which is sufficient for $0 \leq \frac{K_1 L_1}{\lfloor c - a \rfloor} < 1$. Thus $\mathcal{T}$ is a contraction and Banach's fixed point theorem implies a unique fixed point $\bar{\nu}_* \in \mathcal{B}$. Applying this fixed point argument to the dynamic equation (1), it is immediately seen that $\bar{\nu}_*$ is a c^{-}-quasibounded solution of (1). Because of $\bar{\nu}_* \in \mathcal{B}$ the mapping $\bar{\nu}_* : \mathbb{T}_{\tau}^{-} \times \mathcal{P} \to \mathcal{X}$ is continuous.

In order to conclude the proof of Step 3 we only have to verify the estimate (6). Since $\bar{\nu}_*$ is a fixed point of $\mathcal{T}$ together with (13) yields $\|(\mathcal{T}\nu)(\cdot,p)\|_{\tau,c}^- \leq \frac{K_1 L_1}{\lceil c-a \rfloor}\|\nu(\cdot,p)\|_{\tau,c}^- + \frac{K_1 L_1}{\lceil c-a \rfloor}\|f(\cdot,p)\|_{\tau,c}^-$ for $p \in \mathcal{P}$ and from this we ultimately get (6).

Step 4: For an arbitrary $\mathbb{T}$-interval I which is unbounded below, Step 3 guarantees the unique existence of the c^--quasibounded solution $\nu_\tau(\cdot,p)$: $\mathbb{T}_\tau^- \to \mathcal{X}$, $p \in \mathcal{P}$, of (1) for every $\tau \in I$ and ν_τ is continuous. Defining $\nu_* : I \times \mathcal{P} \to \mathcal{X}$ as $\nu_*(t,p) := \nu_t(t,p)$ it is easy to see that ν_* is continuous and has all the properties claimed in Theorem 2(b). Just note that for all $p \in \mathcal{P}$ and $\tau_1, \tau_2, t \in \mathbb{T}$, $\tau_1 \preceq \tau_2 \preceq t$ we have $\nu_{\tau_1}(t,p) = \nu_{\tau_2}(t,p)$.

(c) The proof of statement (c) follows along the lines of the Steps 1 to 3 above. One only has to replace the $\mathbb{T}$-interval $\mathbb{T}_\tau^-$ by $\mathbb{T}$ and the c^--quasiboundedness by $c^\pm$-quasibounded functions. $\square$

Now we study dynamic equations where the unperturbed system is assumed to possess quasibounded solutions in backward time. Hence we have to make the hypothesis of its regressivity. Additionally a smallness condition on the Lipschitz constant for the nonlinear perturbation is involved. The next result is similar to [Aul98, Lemma 3.5, Lemma 3.4] and [AW96, Lemma 3.6, Lemma 3.7].

Theorem 4. *Assume that $K_2 \geq 1$, $L_2, M \geq 0$, $b \in \mathcal{C}_{rd}^+\mathcal{R}(I,\mathbb{R})$ is a growth rate, I denotes some closed $\mathbb{T}$-interval, $\mathcal{Y}$ is a Banach space and $\mathcal{P}$ a topological space satisfying the first axiom of countability. Let us consider the parameter dependent dynamic equation*

$$y^\Delta = B(t)y + G(t,y,p) + g(t,p), \qquad (15)$$

where $B \in \mathcal{C}_{rd}\mathcal{R}(I, \mathcal{L}(\mathcal{Y}))$ and $G : I \times \mathcal{Y} \times \mathcal{P} \to \mathcal{Y}$, $g : I \times \mathcal{P} \to \mathcal{Y}$ are rd-continuous mappings satisfying

$$\|\Phi_B(t,s)\| \leq K_2 e_b(t,s) \quad \text{for } s,t \in I, \, t \preceq s, \qquad (16)$$
$$G(t,0,p) = 0 \quad \text{for } t \in I, \, p \in \mathcal{P},$$
$$\|G(t,y,p) - G(t,\bar{y},p)\| \leq L_2 \|y - \bar{y}\| \quad \text{for } t \in I, \, y, \bar{y} \in \mathcal{Y}, \, p \in \mathcal{P}. \qquad (17)$$

Then for every growth rate $d \in \mathcal{C}_{rd}^+\mathcal{R}(I,\mathbb{R})$, $d \lhd b - K_2 L_2$ and $\tau \in I$ we get the following:

(a) Suppose I is unbounded below,

$$\mu^*(t)[K_2 L_2 - b(t)] < 1 \quad \text{for } t \in I \qquad (18)$$

and $g(\cdot,p) \in \mathcal{B}_{\tau,d}^-(\mathcal{Y})$, $p \in \mathcal{P}$, allows the estimate $\|g(\cdot,p)\|_{\tau,d}^- \leq M$ for $p \in \mathcal{P}$, then every solution $v(\cdot,p) : I \to \mathcal{Y}$, $p \in \mathcal{P}$, of (15) is d^--quasibounded with

$$\|v(\cdot,p)\|_{\tau,d}^- \leq K_2 \|v(\tau,p)\| + \frac{K_2 M}{\lfloor b - d + K_2 L_2 \rfloor} \quad \text{for } p \in \mathcal{P}.$$

Moreover, the mapping $v : I \times \mathcal{P} \to \mathcal{Y}$ is continuous.

(b) Supposed I is unbounded above and $g(\cdot, p) \in \mathcal{B}^+_{\tau, d}(\mathcal{Y})$, $p \in \mathcal{P}$, allows the estimate $\|g(\cdot, p)\|^+_{\tau, d} \leq M$ for $p \in \mathcal{P}$, then there exists exactly one d^+-quasibounded solution $v_*(\cdot, p) : I \to \mathcal{Y}$, $p \in \mathcal{P}$, of (15), which furthermore satisfies

$$\|v_*(\cdot, p)\|^+_{\tau, d} \leq \frac{K_2 M}{\lfloor b - d + K_2 L_2 \rfloor} \quad \text{for } p \in \mathcal{P}.$$

Moreover, the mapping $v_* : I \times \mathcal{P} \to \mathcal{Y}$ is continuous.

(c) Supposed $I = \mathbb{T}$ and $g(\cdot, p) \subset \mathcal{B}^\pm_d(\mathcal{Y})$, $p \in \mathcal{P}$, allows the estimate $\|g(\cdot, p)\|^\pm_{\tau, d} \leq M$ for $p \in \mathcal{P}$, then there exists exactly one $d^\pm$-quasibounded solution $v_*(\cdot, p) : \mathbb{T} \to \mathcal{Y}$, $p \in \mathcal{P}$, of (15), which furthermore satisfies

$$\|v_*(\cdot, p)\|^\pm_{\tau, d} \leq \frac{K_2 M}{\lfloor b - d + K_2 L_2 \rfloor} \quad \text{for } p \in \mathcal{P}.$$

Moreover, the mapping $v_* : \mathbb{T} \times \mathcal{P} \to \mathcal{Y}$ is continuous.

Proof. Since the argumentation is dual to Theorem 2 we only give a very rough sketch of the proof. To begin, we prove that the right-hand side of (15) is regressive under the assumption (18). To this end let $t \in I$ be arbitrary. According to the assumptions, the coefficient operator B is regressive and due to the estimates (17) and

$$\left\| [I_\mathcal{Y} + \mu^*(t) B(t)]^{-1} \right\| = \|\Phi_B(t, \sigma(t))\| \overset{(16)}{\leq} K_2 e_b(t, \sigma(t)) = \frac{K_2}{1 + \mu^*(t) b(t)}$$

for $t \in I$, we obtain that $I_\mathcal{Y} + \mu^*(t) B(t) + \mu^*(t) G(t, \cdot, p) + g(t, p) : \mathcal{Y} \to \mathcal{Y}$ is a bijective mapping (cf. (18) and [Aul98, Corollary 6.2]). Therefore equation (15) is regressive and consequently solutions exist and are unique in backward time $(t \in \mathbb{T}^-_\tau)$.

(a) This part of the proof resembles the proof of Theorem 2(a). Hereby growth rates occurring in the Gronwall estimates are positively regressive because of the assumption (18).

(b) One also proceeds in four steps. Above all the zero solution is the unique d^+-quasibounded solution of $y^\Delta = B(t) y$, which follows from [Pöt01, Corollary 2.11(a)], since the above equation trivially possesses an exponential dichotomy on $\mathbb{T}^+_\tau$ with the invariant projector $P(t) \equiv 0$. We define $v_* : \mathbb{T}^+_\tau \times \mathcal{P} \to \mathcal{Y}$, $v_*(t, p) := -\int_t^\infty \Phi_B(t, \sigma(s)) g(s, p) \Delta s$ as the unique d^+-quasibounded solution of $y^\Delta = A(t) x + g(t, p)$. The well-definedness and the solution property can be verified using [Pöt01, Lemma 3.2] and [Pöt01, Lemma 4.2], respectively. Finally a fixed point argument similar to Step 3 gives us the general assertion.

(c) One can verify the statement (c) along the lines of (b). Here one has to replace the $\mathbb{T}$-interval $\mathbb{T}^+_\tau$ by $\mathbb{T}$ and the d^+-quasiboundedness by $d^\pm$-quasibounded functions. $\qquad \square$

References

[Aul87] B. AULBACH, *Hierarchies of invariant manifolds*, Journal of the Nigerian Mathematical Society, 6 (1987), pp. 71–89.

[Aul95] ——, *Hierarchies of invariant fiber bundles*, Southeast Asian Bulletin of Mathematics, 19 (1995), pp. 91–98.

[Aul98] ——, *The fundamental existence theorem on invariant fiber bundles*, Journal of Difference Equations and Applications, 3 (1998), pp. 501–537.

[AW96] B. AULBACH AND T. WANNER, *Integral manifolds for Carathéodory type differential equations in Banach spaces*, in Six Lectures on Dynamical Systems, B. Aulbach and F. Colonius, eds., World Scientific, Singapore, 1996, pp. 45–119.

[APS02] B. AULBACH, C. PÖTZSCHE AND S. SIEGMUND, *A smoothness theorem for invariant fiber bundles*, Journal of Dynamics and Differential Equations, 14(3) (2002), 519–547.

[BP01] M. BOHNER AND A. PETERSON, *Dynamic Equations on Time Scales — An Introduction with Applications*, Birkhäuser, Boston, 2001.

[Hil88] S. HILGER, *Ein Maßkettenkalkül mit Anwendung auf Zentrumsmannigfaltigkeiten* (in german), Ph.D. Thesis, Universität Würzburg, 1988.

[Hil90] ——, *Analysis on measure chains — A unified approach to continuous and discrete calculus*, Results in Mathematics, 18 (1990), pp. 18–56.

[Hil96] ——, *Generalized theorem of Hartman-Grobman on measure chains*, Journal of the Australian Mathematical Society, Series A, 60 (1996), pp. 157–191.

[Kel99] S. KELLER, *Asymptotisches Verhalten invarianter Faserbündel bei Diskretisierung und Mittelwertbildung im Rahmen der Analysis auf Zeitskalen* (in german), Ph.D. Thesis, Universität Augsburg, 1999.

[Nei01] L. NEIDHART, *Integration im Rahmen des Maßkettenkalküls* (in german), Diploma Thesis, Universität Augsburg, 2001.

[Pöt01] C. PÖTZSCHE, *Exponential dichotomies for linear dynamic equations*, Nonlinear Analysis, Theory, Methods & Applications, 47 (2001), pp. 873–884.

[Pöt02] ——, *Langsame Faserbündel dynamischer Gleichungen auf Maßketten* (in german), Ph.D. Thesis, Universität Augsburg, 2002.

[Pöt03] ——, *Pseudo-stable and pseudo-unstable fiber bundles for dynamic equations on measure chains*, Journal of Difference Equations and Applications, to appear (2003).

Miscellaneous on Difference Equations

Conjugate Singular and Nonsingular Discrete
Boundary Value Problems 337
R. P. Agarwal and D. O'Regan

Asymptotic Solutions of a Discrete Schrödinger Equation
Arising from a Dirac Equation with Random Mass 349
B. Aulbach, S. Elaydi and K. Ziegler

Existence of Bounded Solutions of Discrete Delayed Equations 359
J. Baštinec, J. Diblík and B. Zhang

Difference ϕ-Laplacian Periodic Boundary Value Problems:
Existence and Localization of Solutions 367
A. Cabada and V. Otero–Espinar

Asymptotic Behavior of Solutions of $x_{n+1} = p + \frac{x_{n-1}}{x_n}$ 375
E. Camouzis and R. DeVault

Limit Behavior for Quasilinear Difference Equations.....................383
M. Cecchi, Z. Došlá and M. Marini

Difference Equations in the Qualitative Theory of
Delay Differential Equations 391
J. Čermák

Properties of a Class of Numbers Related to the Fibonacci,
Lucas and Pell Numbers 399
F. M. Dannan

Oscillation Theory of a Class of Higher Order Sturm-Liouville
Difference Equations 407
O. Došlý

A Transformation for the Riccati Difference Operator 417
J. Elyseeva

On the Dynamics of $y_{n+1} = \frac{p+y_{n-2}}{qy_{n-1}+y_{n-2}}$ 425
E. A. Grove, G. Ladas and L. C. McGrath

On the Difference Equation $y_{n+1} = \frac{y_{n-(2k+1)}+p}{y_{n-(2k+1)}+qy_{n-2l}}$ 433
E. A. Grove, G. Ladas, L. C. McGrath and H. A. El-Metwally

Almost Periodic Solutions in a Difference Equation 453
Y. Hamaya

Discrete Quadratic Functionals with Jointly Varying Endpoints
via Separable Endpoints 461
R. Hilscher and V. Zeidan

On Finite Difference Potentials 471
A. Hommel

Moment Equations for Stochastic Difference Equations 479
K. Janglajew

Convergence of Solutions in a Nonhyperbolic Case
with Positive Equilibrium 485
C. M. Kent

Strongly Decaying Solutions of Nonlinear Forced Discrete Systems 493
M. Marini, S. Matucci and P. Řehák

Multidimensional Volterra Difference Equations 501
R. Medina and M. Gil'

Constructing Operator-Difference Schemes for Problems
with Matching Boundaries 507
R. V. N. Melnik

On Difference Matrix Equations 515
E. Pereira and J. Vitória

On Some Difference Equations in the Context of q-Fourier Analysis 523
A. Ruffing and M. Simon

Nonoscillation and Oscillation Properties of Fourth Order
Nonlinear Difference Equations 531
E. Schmeidel

A Computational Procedure to Generate Difference Equations
from Differential Equations 539
P. G. Vaidya and S. Angadi

Difference Equations for Multiple Charlier and Meixner Polynomials ... 549
W. Van Assche

Conjugate Singular and Nonsingular Discrete Boundary Value Problems

RAVI P. AGARWAL

Department of Mathematical Sciences
Florida Institute of Technology
Melbourne, Florida 32901–6975, U.S.A.
E-mail: agarwal@fit.edu

and

DONAL O'REGAN

Department of Mathematics
National University of Ireland, Galway, Ireland
E-mail: donal.oregan@nuigalway.ie

Abstract In this paper we shall survey some of our recent existence criteria (single and multiple solutions) for discrete conjugate (two and multi-point) boundary value problems of singular and nonsingular type.

Keywords Conjugate boundary value problems

AMS Subject Classification 39A10, 39A12

1 Introduction

The last fifty years have witnessed several monographs and hundreds of research articles on theory, constructive methods and applications of boundary value problems for ordinary differential equations. In this vast field of research, the conjugate (Hermite) and the right focal (Abel) types of problems have received the most attention. This is largely due to the fact that these types of problems are crucial, in the sense that the methods employed in their study often are easily extendable to other types of problems. The monographs [1–3] present an up-to-date theory of nonsingular conjugate and focal boundary value problems. However, recent studies of real-world phenomena lead to more complex problems which are mostly singular. Unfortunately, for such problems the known existence theory cannot be extended. In fact, to handle such problems we need to develop a new theory for the nonsingular problems which can be extended to singular problems. In this paper we shall present recently established results for the discrete conjugate boundary value problems.

2   Nonsingular positone two-point problems

Consider the nth $(n \geq 2)$ order discrete conjugate problem

$$\begin{cases} (-1)^{n-p}\Delta^n y(k) = f(k, y(k)), & k \in I_0 \\ \Delta^i y(0) = 0, & 0 \leq i \leq p-1 \\ \quad \text{(i.e., } y(0) = \cdots = y(T+n) = 0), & 1 \leq p \leq n-1 \\ \Delta^i y(T+n-i) = 0, & 0 \leq i \leq n-p-1, \\ \quad \text{(i.e., } y(T+p+1) = \cdots = y(T+n) = 0), \end{cases} \quad (2.1)$$

where $T \in \{1, 2, \cdots\}$, $I_0 = \{0, 1, \cdots, T\}$ and $y : I_n = \{0, 1, \cdots, T+n\} \to \mathbf{R}$. We let $C(I_n)$ denote the class of maps w continuous on I_n (discrete topology) with norm $|w|_0 = \max_{i \in I_n} |w(i)|$. By a solution to (2.1) we mean a $w \in C(I_n)$ such that w satisfies the difference equation in (2.1) for $i \in I_0$ and w satisfies the conjugate boundary data.

In what follows, the Green's function $K(k, j)$ for the problem

$$\begin{aligned} \Delta^n y(k) &= 0, & k \in I_0 \\ \Delta^i y(0) &= 0, & 0 \leq i \leq p-1 \\ \Delta^i y(T+n-i) &= 0, & 0 \leq i \leq n-p-1 \end{aligned} \quad (2.2)$$

plays a fundamental role. In [4] it is shown that $K(k, j)$ can be expressed explicitly as

$$\begin{aligned} K(k, j) = & \sum_{l=0}^{p-1} \left[\sum_{i=0}^{p-l-1} \binom{n-p+i-1}{i} \frac{k^{(l+i)}}{(T+n-l)^{(n-p+i)}} \right] \times \\ & \times \frac{(-j-1)^{(n-l-1)}}{l!\,(n-l-1)!} \, (T+n-k)^{(n-p)} \end{aligned}$$

if $j \in \{0, \cdots, k-1\}$, whereas

$$\begin{aligned} K(k, j) = & -\sum_{l=0}^{n-p-1} \left[\sum_{i=0}^{n-p-l-1} \binom{p+i-1}{i} \frac{(T+p+l+i-k)^{(l+i)}}{(T+p+1+l+i)^{(p+i)}} \right] \times \\ & \times \frac{(-1)^l \, (T+p-j)^{(n-l+1)} \, k^{(p)}}{l!\,(n-l-1)!} \end{aligned}$$

if $j \in \{k, k+1, \cdots, T\}$.

It is well known [3] that

$$(-1)^{n-p} K(k, j) \geq 0 \quad \text{for} \quad (k, j) \in I_n \times I_0. \quad (2.3)$$

For $j \in I_0$ let $v(j) \in I_n$ be defined by

$$\max_{k \in I_n} (-1)^{n-p} K(k, j) = \max_{k \in I_n} |K(k, j)| = |K(v(j), j)|. \quad (2.4)$$

We shall also need the following result which is established in [9].

Theorem 2.1. For $k \in J_p = \{p, \cdots, T + p\}$ and $j \in I_0$,

$$(-1)^{n-p} K(k, j) \geq \theta |K(v(j), j)|, \qquad (2.5)$$

here $0 < \theta < 1$ is a constant given by

$$\theta = \min\{b(p), \ b(p+1)\} \qquad (2.6)$$

with

$$b(x) = \frac{\min\{g(x, p), \ g(x, T+p)\}}{\max\{g(x, [\theta(x)]), \ g(x, [\theta(x)]+1), \ g(x, p), \ g(x, T+p)\}}$$

($[\,.\,]$ denotes the greatest integer function) and with

$$g(x, k) = k^{(x-1)} (T + n - k)^{(n-x)} \quad \text{and} \quad \theta(x) = \frac{(x-1)T + (x-2)n + x}{n-1}.$$

Numerous applications of the inequality (2.5) are available in [7].

The following existence principle is a direct consequence of the Leray–Schauder nonlinear alternative.

Theorem 2.2. Suppose $f : I_0 \times \mathbf{R} \to \mathbf{R}$ is continuous (i.e., continuous as a map from the topological space $I_0 \times \mathbf{R}$ into the topological space $\mathbf{R}$ (of course the topology on I_0 will be the discrete topology)). In addition assume there is a constant $M > 0$, independent of λ, with

$$|y|_0 - \sup_{j \in I_n} |y(j)| \neq M$$

for any solution $y \in C(I_n)$ to

$$
\begin{aligned}
&(-1)^{n-p} \Delta^n y(k) = \lambda f(k, y(k)), \quad k \in I_0 \\
&\Delta^i y(0) = 0, \quad 0 \leq i \leq p-1 \\
&\Delta^i y(T+n-i) = 0, \quad 0 \leq i \leq n-p-1
\end{aligned}
\qquad (2.7)_\lambda
$$

for each $\lambda \in (0, 1)$. Then (2.1) has at least one solution $y \in C(I_n)$ with $|y|_0 \leq M$.

This result and an application of Krasnoselskii's fixed point theorem in a cone leads to the following existence principle for twin solutions.

Theorem 2.3. Suppose the following hold:

(1) $f : I_0 \times [0, \infty) \to [0, \infty)$ is continuous (i.e., continuous as a map from the topological space $I_0 \times [0, \infty)$ into the topological space $[0, \infty)$) with $f(i, u) > 0$ for $(i, u) \in I_0 \times (0, \infty)$,

(2) $f(i, u) \leq q(i)\, w(u)$ on $I_0 \times [0, \infty)$ with $q : I_0 \to (0, \infty)$ and $w \geq 0$ continuous and nondecreasing on $[0, \infty)$,

(3) there exists $r > 0$ with

$$\frac{r}{w(r)\, \max_{k \in I_n}\, \sum_{j=0}^{T}(-1)^{n-p}\, K(k, j)\, q(j)} > 1,$$

(4) there exists $\tau : A_p = \{p, p+1, \cdots, T\} \to (0, \infty)$ with $f(i, u) \geq \tau(i)\, w(u)$ on $A_p \times (0, \infty)$,

(5) there exists $R > r$ with

$$\frac{x}{w(x)} \leq \theta \sum_{j=p}^{T}(-1)^{n-p}\, K(\sigma, j)\, \tau(j) \quad \text{for} \quad x \in [\theta\, R, R],$$

here $\sigma \in I_n$ is such that

$$\sum_{j=p}^{T}(-1)^{n-p}\, K(\sigma, j)\, \tau(j) = \max_{k \in I_n}\, \sum_{j=p}^{T}(-1)^{n-p}\, K(k, j)\, \tau(j).$$

Then (2.1) has two solutions $y_1, y_2 \in C(I_n)$ with $y_1 \geq 0$ on I_n, $y_2 > 0$ on J_p and $0 \leq |y_1|_0 < r < |y_2|_0 \leq R$.

Example 2.1. The boundary value problem

$$\Delta^2\, y(i) + \frac{1}{(T+2)^2}\, e^{y(i)} = 0 \quad i \in I_0 \tag{2.8}$$
$$y(0) = y(T+2) = 0$$

has solutions $y_1, y_2 \in C(I_2)$ with $y_1 > 0$, $y_2 > 0$ on $\{1, 2, \cdots, T+1\}$ and $0 < |y_1|_0 < 1 < |y_2|_0$.

To see this we apply Theorem 2.3 with $n = 2$, $p = 1$, $q = \tau = 1$, and

$$w(x) = \frac{1}{(T+2)^2}\, e^x.$$

Notice (1), (2) and (4) are clearly satisfied. It is also easy to see that

$$\max_{k \in I_2}\, \sum_{j=0}^{T}(-1)\, K(k, j)\, q(j) = \max_{k \in I_2}\, \frac{k(T+2-k)}{2} \leq \frac{(T+2)^2}{8}.$$

As a result (3) holds with $r = 1$ since

$$\frac{r}{w(r)\, \max_{k \in I_2}\, \sum_{j=0}^{T}(-1)\, K(k, j)\, q(j)} \geq \frac{8\, r}{e^r}.$$

Finally, since $\frac{x}{e^x} \to 0$ as $x \to \infty$ it is easy to choose $R > r = 1$ so that (5) holds. Theorem 2.3 now guarantees the result (note $-\Delta^2 y_1(i) \geq 1$ on I_0 so $y_1(i) > 0$ on $\{1, 2, \cdots, T+1\}$).

Theorem 2.3 guarantees existence of two solutions y_1, y_2 to (2.1) with $y_1 \geq 0$ on I_n, $y_2 > 0$ on J_p. However, in some situations y_1 may be identically zero. In the next result we put an extra assumption which guarantees that $y_1 > 0$ on J_p for (2.1) even if y identically zero is also a solution.

Theorem 2.4. Suppose in addition to (1) – (5) the following hold

(6) there exists L, $0 < L < r$ with

$$\frac{x}{w(x)} \leq \theta \sum_{j=p}^{T} (-1)^{n-p} K(\sigma, j)\, \tau(j) \quad \text{for} \quad x \in [\theta\, L, L].$$

Then (2.1) has two solutions y_1, $y_2 \in C(I_n)$ with $y_1 > 0$ on J_p, $y_2 > 0$ on J_p and $L \leq |y_1|_0 < r < |y_2|_0 \leq R$.

3 Nonsingular and singular positone two-point problems

Here we shall consider the nth $(n \geq 2)$ order discrete conjugate boundary value problem

$$\begin{aligned}
(-1)^{n-p}\, \Delta^n\, y(k-p) &= f(k, y(k)), \quad i \in J_p \\
\Delta^i\, y(0) &= 0, \quad 0 \leq i \leq p-1 \\
\Delta^i\, y(T+n-i) &= 0, \quad 0 \leq i \leq n-p-1.
\end{aligned} \tag{3.1}$$

By a solution to (3.1) we mean a $w \in C(I_n)$ such that w satisfies the difference equation in (3.1) for $k \in J_p$ and w satisfies the conjugate boundary data.

Since the problem

$$\begin{aligned}
\Delta^n\, y(k-p) &= \phi(k), \quad k \in J_p \\
\Delta^i\, y(0) &= 0, \quad 0 \leq i \leq p-1 \\
\Delta^i\, y(T+n-i) &= 0, \quad 0 \leq i \leq n-p-1
\end{aligned} \tag{3.2}$$

is the same as

$$\begin{aligned}
\Delta^n\, y(k) &= \phi(k+p), \quad k \in I_0 \\
\Delta^i\, y(0) &= 0, \quad 0 \leq i \leq p-1 \\
\Delta^i\, y(T+n-i) &= 0, \quad 0 \leq i \leq n-p-1
\end{aligned}$$

the Green's function $G(k, j)$ of (3.2) satisfies the relation

$$G(k, j) = K(k, j - p).$$

We shall also need the following result from [9].

Theorem 3.1. Suppose $y : I_n \to \mathbf{R}$ is such that

$$(-1)^{n-p} \Delta^n y(k) \geq 0, \quad k \in I_0$$
$$\Delta^i y(0) = 0, \quad 0 \leq i \leq p - 1$$
$$\Delta^i y(T + n - i) = 0, \quad 0 \leq i \leq n - p - 1.$$

Then

$$y(k) \geq \theta \max_{j \in I_n} |y(j)| = \theta \, \|y\| \quad \text{for} \quad k \in J_p,$$

where the constant $0 < \theta < 1$ is the same as in (2.6).

Now suppose $y : I_n \to \mathbf{R}$ satisfies

$$(-1)^{n-p} \Delta^n y(k - p) \geq 0, \quad k \in J_p$$
$$\Delta^i y(0) = 0, \quad 0 \leq i \leq p - 1$$
$$\Delta^i y(T + n - i) = 0, \quad 0 \leq i \leq n - p - 1.$$

Of course $(-1)^{n-p} \Delta^n y(k - p) \geq 0$ for $k \in J_p$ is exactly the same as $(-1)^{n-p} \Delta^n y(k) \geq 0$ for $k \in I_0$ and so

$$y(k) \geq \theta \, \|y\| = \theta \max_{j \in I_n} |y(j)| \quad \text{for} \quad k \in J_p. \tag{3.3}$$

The following existence principle for the discrete conjugate boundary value problems is also a direct consequence of the Leray–Schauder nonlinear alternative.

Theorem 3.2. Suppose $f : J_p \times \mathbf{R} \to \mathbf{R}$ is continuous. Assume there is a constant $M > |a|$, independent of λ, with

$$\|y\| = \max_{j \in I_n} |y(j)| \neq M$$

for any solution $y \in C(I_n)$ to

$$(-1)^{n-p} \Delta^n y(k - p) = \lambda \, f(k, y(k)), \quad k \in J_p$$
$$y(0) = a$$
$$\Delta^i y(0) = 0, \quad 1 \leq i \leq p - 1 \tag{3.4}_\lambda$$
$$y(T + n) = a$$
$$\Delta^i y(T + n - i) = 0, \quad 1 \leq i \leq n - p - 1$$

for each $\lambda \in (0, 1)$. Then $(3.4)_1$ has a solution.

We are now in the position to state the following existence principles for (3.1).

Theorem 3.3. (Nonsingular Problems). Suppose the following conditions are satisfied:

(a) $f : J_p \times [0, \infty) \to [0, \infty)$ is continuous,

(b) there exists a continuous, nondecreasing function $\psi : [0, \infty) \to [0, \infty)$ with $\psi > 0$ on $(0, \infty)$ and a function $q : J_p \to [0, \infty)$ with $f(k, u) \leq q(k)\, \psi(u)$ for all $u \geq 0$ and $k \in J_p$,

(c) $\displaystyle \sup_{c \in (0, \infty)} \left(\frac{c}{\psi(c)} \right) > Q,$ here

$$Q = \max_{k \in I_n} \sum_{j=p}^{T+p} q(j)\, (-1)^{n-p}\, G(k, j).$$

Then (3.1) has a nonnegative solution.

Theorem 3.4. (Singular Problems). Suppose the following conditions are satisfied:

(d) $f : J_p \times (0, \infty) \to (0, \infty)$ is continuous (note $f(k, y)$ may be singular at $y = 0$),

(e) $f(k, u) \leq g(u) + h(u)$ on $J_p \times (0, \infty)$ with $g > 0$ continuous and non-increasing on $(0, \infty)$, $h \geq 0$ continuous on $[0, \infty)$ and h/g nondecreasing on $(0, \infty)$,

(f) for each constant $H > 0$ there exists a continuous function $\psi_H : J_p \to (0, \infty)$ with $f(k, u) \geq \psi_H(k)$ on $J_p \times (0, H]$,

(g) there exists a constant $K_\theta > 0$ with $g(\theta\, u) \leq K_\theta\, g(u)$ for all $u \geq 0$,

(h) $\displaystyle \sup_{c \in (0, \infty)} \left(\frac{c}{g(c) + h(c)} \right) > K_\theta\, Q,$ here

$$Q = \max_{k \in J_p} \sum_{j=p}^{T+p} (-1)^{n-p}\, G(k, j).$$

Then (3.1) has a solution $y \in C(I_n)$ with $y(i) > 0$ for $i \in J_p$.

Example 3.1. Consider the boundary value problem

$$(-1)^{n-p}\, \Delta^n\, y(k-p) = \mu \left([y(k)]^{-\alpha} + A\, [y(k)]^\beta + B \right) \quad \text{for} \quad k \in J_p$$
$$\Delta^i\, y(0) = 0, \quad 0 \leq i \leq p - 1$$
$$\Delta^i\, y(T + n - i) = 0, \quad 0 \leq i \leq n - p - 1$$

$$\tag{3.5}$$

with $\alpha > 0,\ \beta \geq 0,\ A \geq 0,\ B \geq 0$ and $\mu > 0$. If

$$\mu < \frac{\theta^\alpha}{Q} \sup_{c \in (0,\infty)} \left(\frac{c^{\alpha+1}}{1 + A\ c^{\alpha+\beta} + B\ c^\alpha} \right) \tag{3.6}$$

(here θ is as in (2.6) and Q is as in the statement of Theorem 3.4) then (3.5) has a solution $y \in C(I_n)$ with $y(i) > 0$ for $i \in J_p$.

Remark 3.1. If $\beta < 1$ then (3.6) is true for all $\mu > 0$.

The result follows immediately from Theorem 3.4 with $g(u) = \mu\ u^{-\alpha}$ and $h(u) = \mu\ [A\ u^\beta + B]$. Clearly (d), (e), (f) (with $\psi_H = \mu\ H^{-\alpha}$), and (g) (with $K_\theta = \theta^{-\alpha}$) are satisfied. Also (3.6) guarantees that (h) is true.

4   Semipositone two-point problems

Here we shall discuss the discrete conjugate boundary value problem

$$\begin{aligned} (-1)^{n-p}\ \Delta^n\ y(k-p) &= \mu\ f(k, y(k)), \quad k \in J_p \\ \Delta^i\ y(0) &= 0, \quad 0 \leq i \leq p-1 \\ \Delta^i\ y(T+n-i) &= 0, \quad 0 \leq i \leq n-p-1 \end{aligned} \tag{4.1}$$

where $\mu > 0$. We shall assume that the nonlinearity f may take *negative* values.

In the proof of the main result of this section besides the inequality (3.3) we also need the following lemma.

Lemma 4.1.   The boundary value problem

$$\begin{aligned} (-1)^{n-p}\ \Delta^n\ y(k-p) &= 1, \quad k \in J_p \\ \Delta^i\ y(0) &= 0, \quad 0 \leq i \leq p-1 \\ \Delta^i\ y(T+n-i) &= 0, \quad 0 \leq i \leq n-p-1 \end{aligned}$$

has a solution w with

$$w(k) \leq \frac{1}{n!}\ (T+p)^{(p)}\ (T+n-p)^{(n-p)} \quad \text{for} \quad k \in J_p$$

and

$$w(k) \leq \frac{1}{n!}\ (T+n)^{(p)}\ (T+n)^{(n-p)} \quad \text{for} \quad k \in I_n,$$

here $w(0) = \cdots = w(p-1) = w(T+p+1) = \cdots w(T+n) = 0$ with

$$w(k) = \sum_{j=p}^{T+p} (-1)^{n-p}\ G(k, j) = \frac{1}{n!}\ k^{(p)}\ (T+n-k)^{(n-p)}$$

for $k \in I_n$.

We are now in a position to state our main result for (4.1).

Theorem 4.1. Suppose the following conditions are satisfied:

(A) $f : J_p \times [0, \infty) \to \mathbf{R}$ is continuous and there exists a constant $M > 0$ with $f(i, u) + M \geq 0$ for $(i, u) \in J_p \times [0, \infty)$,

(B) $f(i, u) + M \leq \psi(u)$ on $J_p \times [0, \infty)$ with $\psi : [0, \infty) \to [0, \infty)$ continuous and nondecreasing and $\psi(u) > 0$ for $u > 0$,

(C) there exists $r \geq \dfrac{\mu \, M \, (T + p)^{(p)} \, (T + n - p)^{(n-p)}}{n! \, \theta}$ with

$$\frac{r}{\psi(r)} \geq \mu \sup_{k \in I_n} \sum_{j=p}^{T+p} (-1)^{n-p} \, G(k, j),$$

(D) there exists a continuous, nondecreasing function $g : (0, \infty) \to (0, \infty)$ with $f(i, u) + M \geq g(u)$ for $(i, u) \in J_p \times (0, \infty)$,

(E) there exists $R > r$ with

$$\frac{R}{g\,(\epsilon \, R \, \theta)} \leq \mu \sum_{j=p}^{T+p} (-1)^{n-p} \, G(\sigma, j),$$

here $\epsilon > 0$ is any constant (choose and fix it) so that

$$1 - \frac{\mu \, M \, (T + p)^{(p)} \, (T + n - p)^{(n-p)}}{n! \, R \, \theta} \geq \epsilon$$

and $\sigma \in I_n$ is such that

$$\sum_{j=p}^{T+p} (-1)^{n-p} \, G(\sigma, j) = \max_{i \in I_n} \sum_{j=p}^{T+p} (-1)^{n-p} \, G(i, j).$$

Then (4.1) has a solution $y \in C(I_n)$ with $y(i) > 0$ for $i \in J_p$.

5 Nonsingular multi-point problems

Let $Z[c, d] = \{c, c + 1, \cdots, d\}$ for given integers c, d with $d > c$. Here we shall discuss the Hermite discrete problem

$$\begin{aligned}
\Delta^n \, y(m) &= f(m, y(m)), \quad m \in I_N \equiv Z[0, N] \\
\Delta^j \, y(a_i) &= 0, \quad j = 0, \cdots, n_i - 1, \quad i = 1, \cdots, k,
\end{aligned} \tag{5.1}$$

where $N \geq 2$, $n \geq 2$, $2 \leq k \leq n$ are integers, $n_i \geq 1$ for $i = 1, \cdots, k$ with $\sum_{i=1}^{k} n_i = n$, and $0 = a_1 < a_1 + n_1 < a_2 < a_2 + n_2 < \cdots \cdots < a_k \leq a_k + n_k - 1 = N + n$. By a solution to (5.1) we mean a $y \in C(I_{N+n})$

such that y satisfies the difference equation and the boundary data in (5.1) (note in particular $y(m) = 0$ for $m \in Z[a_i, a_i + n_i - 1]$, $i = 1, \cdots, k$). Here $C(I_{N+n})$ denotes the class of maps u continuous on I_{N+n} (discrete topology) with norm $|u|_0 = \max_{m \in I_{N+n}} |u(m)|$.

Let g be the Green's function for

$$\Delta^n y(m) = 0, \quad m \in I_N$$
$$\Delta^j y(a_i) = 0, \quad j = 0, \cdots, n_i - 1, \quad i = 1, \cdots, k. \tag{5.2}$$

Let $\alpha_i = \sum_{j=i+1}^{k} n_j$. We shall need the following inequalities:

$$(-1)^{\alpha_i} g(m, j) \geq 0 \quad \text{for} \quad (m, j) \in Z[a_i, a_{i+1}] \times Z[0, N], \tag{5.3}$$

$$(-1)^{\alpha_i} g(m, j) > 0 \quad \text{for} \quad (m, j) \in Z[a_i + n_i, a_{i+1} - 1] \times Z[0, N], \tag{5.4}$$

$i = 1, \cdots, k - 1$, and

$$(-1)^{\alpha_i} g(m, j) \geq Q_i \max_{m \in I_{N+n}} |g(m, j)| \quad \text{for} \quad (m, j) \in Z[a_i + n_i, a_{i+1} - 1] \times Z[0, N], \tag{5.5}$$

$i = 1, \cdots, k - 1$, here $I_{N+n} = Z[0, N + n]$ and $0 < Q_i \leq 1$ is a constant defined by

$$Q_i = \min \left\{ \frac{\min\{p(a_i + n_i), \ p(a_{i+1} - 1)\}}{\max_{m \in I_{N+n}} p(m)}, \ \frac{\min\{q(a_i + n_i), \ q(a_{i+1} - 1)\}}{\max_{m \in I_{N+n}} q(m)} \right\},$$

where

$$p(m) = \left| \prod_{j=1}^{k-1} (m - a_j)^{(n_j)} \right| (N + n - m)^{(n_k - 1)}$$

and

$$q(m) = m^{(n_1 - 1)} \left| \prod_{j=2}^{k} (m - a_j)^{(n_j)} \right|.$$

Inequalities (5.3) and (5.4) are folklore in the literature, see [3], whereas (5.5) has been established and used in [10].

Now we are in the position to state our existence criteria for (5.1).

Theorem 5.1. Assume $f : I_N \times \mathbf{R} \to \mathbf{R}$ is continuous (i.e., continuous as a map from the topological space $I_N \times \mathbf{R}$ into the topological space $\mathbf{R}$ (of course the topology on I_N is the discrete topology)). Suppose there is a constant $M > 0$ with

$$|y|_0 = \max_{m \in I_{N+n}} |y(m)| \neq M$$

for any solution $y \in C(I_{N+n})$ to

$$\Delta^n y(m) = \lambda f(m, y(m)), \quad m \in I_N$$
$$\Delta^j y(a_i) = 0, \quad j = 0, \cdots, n_i - 1, \quad i = 1, \cdots, k \tag{5.6}_\lambda$$

for each $\lambda \in (0,1)$. Then (5.1) has a solution $y \in C(I_{N+n})$ with $|y|_0 \leq M$ (note $y(m) = 0$ for $m \in Z[a_i, a_i + n_i - 1]$, $i = 1, \cdots, k$).

Theorem 5.2. Suppose the following conditions are satisfied:

(i) $f : I_N \times \mathbf{R} \to [0, \infty)$ is continuous with $f(i, u) > 0$ for $(i, u) \in I_N \times (\mathbf{R} \setminus \{0\})$,

(ii) $|f(m, u)| \leq q(m) \, w(|u|)$ on $I_N \times \mathbf{R}$ with $q : I_N \to (0, \infty)$ and $w \geq 0$ continuous and nondecreasing on $[0, \infty)$,

(iii) there exists $r > 0$ with

$$\frac{r}{w(r) \max_{m \in I_{N+n}} \sum_{j=0}^{N} |g(m,j)| \, q(j)} > 1.$$

Then (5.1) has a solution $y_1 \in C(I_{N+n})$ with $(-1)^{\alpha_i} y_1(m) \geq 0$ for $m \in Z[a_i, a_{i+1}]$, $i = 1, \cdots, k-1$ (note $y_1(m) = 0$ for $m \in Z[a_i, a_i + n_i - 1]$, $i = 1, \cdots, k$) and $|y_1|_0 < r$.

Theorem 5.3. Suppose (i) – (iii) hold. In addition assume there exists $i_0 \in \{1, \cdots, k-1\}$ with the following satisfied:

(iv) there exists $\tau : Z[a_{i_0} + n_{i_0}, a_{i_0+1} - 1] \to (0, \infty)$ with $f(i, u) \geq \tau(i) \, w(|u|)$ on $Z[a_{i_0} + n_{i_0}, a_{i_0+1} - 1] \times \mathbf{R}$,

(v) there exists $R > r$ with

$$\frac{R}{w(Q_{i_0} R)} \leq \sum_{j=a_{i_0}+n_{i_0}}^{a_{i_0+1}-1} (-1)^{\alpha_{i_0}} g(\sigma_{i_0}, j) \, \tau(j),$$

here Q_{i_0} is as in (5.5) and $\sigma_{i_0} \in Z[a_{i_0}, a_{i_0+1}]$ is such that

$$\sum_{j=a_{i_0}+n_{i_0}}^{a_{i_0+1}-1} (-1)^{\alpha_{i_0}} g(\sigma_{i_0}, j) \, \tau(j) = \max_{m \in Z[a_{i_0}, a_{i_0+1}]} \sum_{j=a_{i_0}+n_{i_0}}^{a_{i_0+1}-1} (-1)^{\alpha_{i_0}} g(m, j) \, \tau(j).$$

Then (5.1) has a solution $y_2 \in C(I_{N+n})$ with $(-1)^{\alpha_i} y_2(m) \geq 0$ for $m \in Z[a_i, a_{i+1}]$, $i = 1, \cdots, k-1$, with $(-1)^{\alpha_i} y_2(m) > 0$ for $m \in Z[a_i + n_i, a_{i+1} - 1]$, $i = 1, \cdots, k-1$, and $r < |y_2|_0 \leq R$ (note $y_2(m) = 0$ for $m \in Z[a_i, a_i + n_i - 1]$, $i = 1, \cdots, k$).

Remark 5.1. If in (v) we have $R < r$ then (5.1) has a solution $y \in C(I_{N+n})$ with $R \leq |y|_0 < r$.

Theorem 5.4. Suppose (i) – (v) hold. Then (5.1) has two solutions $y_1, y_2 \in C(I_{N+n})$ with $(-1)^{\alpha_i} y_1(m) \geq 0$, $(-1)^{\alpha_i} y_2(m) \geq 0$ for $m \in Z[a_i, a_{i+1}]$, $i = 1, \cdots, k-1$, with $(-1)^{\alpha_i} y_2(m) > 0$ for $m \in Z[a_i + n_i, a_{i+1} - 1]$, $i = 1, \cdots, k-1$, and with $0 \leq |y_1|_0 < r < |y_2|_0 \leq R$ (note $y_1(m) = y_2(m) = 0$ for $m \in Z[a_i, a_i + n_i - 1]$, $i = 1, \cdots, k$).

Theorem 5.5. Suppose (i) – (v) hold. In addition assume

(vi)   there exists L, $0 < L < r$ with

$$\frac{L}{w\left(Q_{i_0}\, L\right)} \le \sum_{j=a_{i_0}+n_{i_0}}^{a_{i_0+1}-1} (-1)^{\alpha_{i_0}}\, g(\sigma_{i_0}, j)\, \tau(j)$$

is satisfied (here i_0 and σ_{i_0} are as in the statement of Theorem 5.3). Then (5.1) has two solutions y_1, $y_2 \in C(I_{N+n})$ with $(-1)^{\alpha_i}\, y_1(m) \ge 0$, $(-1)^{\alpha_i}\, y_2(m) \ge 0$ for $m \in Z[a_i, a_{i+1}]$, $i = 1, \cdots, k-1$, with $(-1)^{\alpha_i}\, y_1(m) > 0$, $(-1)^{\alpha_i}\, y_2(m) > 0$ for $m \in Z[a_i + n_i, a_{i+1} - 1]$, $i = 1, \cdots, k-1$, and with $L \le |y_1|_0 < r < |y_2|_0 \le R$ (note $y_1(m) = y_2(m) = 0$ for $m \in Z[a_i, a_i + n_i - 1]$, $i = 1, \cdots, k$).

The ideas needed to prove the above results can be easily extended to discuss several other continuous and discrete problems discussed in [2,5,6].

References

[1] R.P. Agarwal, *Boundary Value Problems for Higher Order Differential Equations*, World Scientific, Singapore, 1986.

[2] R.P. Agarwal, *Focal Boundary Value Problems for Differential and Difference Equations*, Kluwer, Dordrecht, 1998.

[3] R.P. Agarwal, *Difference Equations and Inequalities*, 2nd edition, Marcel Dekker, New York, 2000.

[4] R.P. Agarwal, M. Bohner and P.J.Y. Wong, Eigenvalues and eigenfunctions of discrete conjugate boundary value problems, *Computers Math. Applic.* **38**(3–4)(1999), 185–199.

[5] R.P. Agarwal and D. O'Regan, Discrete focal boundary-value problems, *Proc. Edinburgh Math. Soc.* **43**(2000), 155–165.

[6] R.P. Agarwal and D. O'Regan, Multiple nonnegative solutions to singular and nonsingular discrete problems, *Aequationes Math.* **61**(2001), 97–112.

[7] R.P. Agarwal, D. O'Regan and P.J.Y. Wong, *Positive Solutions of Differential, Difference and Integral equations*, Kluwer, Dordrecht, 1999.

[8] R.P. Agarwal and P.J.Y. Wong, *Advanced Topics in Difference Equations*, Kluwer, Dordrecht, 1997.

[9] R.P. Agarwal and P.J.Y. Wong, Extension of continuous and discrete inequalities due to Eloe and Henderson, *Nonlinear Analysis* **34**(1998), 479–487.

[10] R.P. Agarwal and P.J.Y. Wong, Eigenvalue theorems for discrete multipoint conjugate boundary value problems, *J. Computational Applied Mathematics* **113**(2000), 227–240.

Asymptotic Solutions of a Discrete Schrödinger Equation Arising from a Dirac Equation with Random Mass

BERND AULBACH

Department of Mathematics, University of Augsburg
D-86135 Augsburg, Germany
E-mail: aulbach@math.uni-augsburg.de

SABER ELAYDI

Department of Mathematics, Trinity University
San Antonio, Texas 78212, USA
E-mail: selaydi@trinity.edu

and

KLAUS ZIEGLER

Department of Physics, University of Augsburg
D-86135 Augsburg, Germany
E-mail: ziegler@physik.uni-augsburg.de

Abstract For a Dirac particle in one dimension with random mass, the time evolution for the average wavefunction is considered. Using the super-symmetric representation of the average Green's function, we derive a fourth order linear difference equation for the low-energy asymptotics of the average wavefunction. This equation is of Poincaré type, though highly critical and therefore not amenable to standard methods. In this paper we show that, nevertheless, asymptotic expansions of its solutions can be obtained.

Keywords Dirac particle in one dimension with random mass, Green's function, Poincaré type difference equation, Asymptotic expansion

AMS Subject Classification 34L40, 39A10, 39A11, 60H10

1   Introduction

A Dirac particle is considered which can move only in one dimension. It is characterized by a two-component spinor wavefunction. The mass of the particle varies randomly in space. Such a random mass term can be the result of an interaction of the particle with an external random field that affects the kinetic properties. We are interested in the average wavefunction, averaged with respect to the random mass. Therefore, we have to evaluate the average Green's function which describes the evolution of the average wavefunction. A supersymmetric representation of the average Green's function is employed to perform the calculation. It allows us to translate formally the original one-dimensional continuous Dirac Equation with random mass into a discrete deterministic Schrödinger Equation. This equation is a fourth order linear difference equation of Poincaré type, i.e., an equation of the form

$$x_{n+2} + p_1(n)\, x_{n+1} + p_2(n)\, x_n + p_3(n)\, x_{n-1} + p_4(n)\, x_{n-2} = 0$$

where the coefficients $p_1(n), \ldots, p_4(n)$ converge to constants $p_1^*, \ldots, p_4^*$, respectively, as $n \to \infty$. For this kind of equation there is a well-developed classical theory (see, e.g., Elaydi [7, Chapter 8]) providing information on the asymptotic behavior of the solutions. The basic results of the classical theory, however, assume that the four complex roots of the corresponding characteristic equation

$$\lambda^4 + p_1^*\lambda^3 + p_2^*\lambda^2 + p_3^*\lambda + p_4^* = 0$$

have pairwise distinct moduli. This assumption, however, turns out to be drastically violated by the Schrödinger equation arising in this paper. In fact, here the characteristic equation has 1 as a fourfold root.

The way we treat the Schrödinger equation in this paper is to first exploit the particular structure of this equation in order to reduce its order from four to two. The resulting second order equation is of Poincaré type, too, and the roots of the corresponding characteristic equation are $+1$ and -1. They still have the same modulus, however, for this kind of problems there is a general asymptotic theory (even for higher order equations) based on the work of Birkhoff [3, 4], Birkhoff and Trjitzinsky [5] and Adams [1]. Since in our setting the reduced equation is of second order, we can use a simplified version of the general theory which is due to Wong and Li [9, 10], and since in our setting the characteristic roots are simple, the generally very involved expressions in Wong and Li's work become manageable.

2   Derivation of the model equation

We consider the Dirac equation in one dimension for $0 \leq x \leq L$ and periodic boundary conditions

$$H\Psi_E(x) = E\Psi_E(x), \tag{1}$$

where the Dirac operator

$$H = -i\sigma_x \frac{\partial}{\partial x} + m(x)\sigma_y$$

depends on Pauli matrices σ_j and contains a random term $m(x)$. The latter is Gaussian distributed, independently for different sites x with

$$\langle m(x) \rangle = m_0, \quad \langle (m(x) - m_0)^2 \rangle = 2g.$$

$m(x)\sigma_y$ can be considered as a random mass. This physical problem has some interesting properties in terms of the solution $\Psi_E(E)$ [8]. For instance, the solution is localized for $E \neq 0$ and/or $m_0 \neq 0$; i.e., the average wavefunction $\langle \Psi_E(x) \rangle$ decays exponentially on a length scale $\xi(E, m_0)$, and $\xi(E, m_0)$ diverges at $E = m_0 = 0$. The Fourier transform of equation (1) in terms of time t with

$$\Psi(x, t) = \int_{-\infty}^{\infty} \Psi_E(x) e^{iEt} dE \quad \text{and} \quad H\Psi(x, t) = -i\frac{\partial}{\partial t}\Psi(x, t)$$

can be considered as an initial value problem. Starting with the initial function $\Psi(x, 0)$, the time evolution is given as

$$\Psi(x, t) = e^{iHt}\Psi(x, 0).$$

Assuming that the initial function $\Psi(x, 0)$ is given as a non-random function, the time-dependent solution $\Psi(x, t)$ depends on the randomness of H only through the Green's function e^{iHt}. In order to evaluate the average wavefunction

$$\langle \Psi(r, t) \rangle = \langle e^{iHt} \rangle \Psi(r, 0)$$

we need the average Green's function. The latter can be associated with the Greens's function of an effective non-random, translational-invariant Hamiltonian $\mathcal{H}$, since the distribution of the random mass is translational invariant. Then the low-energy asymptotics of the average wavefunction can be determined from the spectral properties of $\mathcal{H}$. The central idea of this approach is that the Dirac equation (1) is a one-dimensional equation which can be represented as a zero-dimensional quantum problem, using the concept of second quantization.

2.1   Effective non-random Hamiltonian

For the following discussion it is convenient to use the Fourier transform of the Green's function, given by the resolvent $G(z) = (z - H)^{-1}$ such that

$$e^{iHt} = \lim_{\epsilon \downarrow 0} \int_{-\infty}^{\infty} G(E - i\epsilon) e^{iEt} dE \quad (t > 0). \tag{2}$$

$\langle G(z) \rangle$ can be formally expressed by the Greens's function of a non-linear supersymmetric theory [6, 11]. Using the notation of Balents and Fisher [2] with f_s ($f_s^\dagger$) annihilation (creation) operator of a fermion and b_s ($b_s^\dagger$) annihilation (creation) operator of a boson with spin $s = \uparrow, \downarrow$, the diagonal elements of the average Green's function, for instance, read

$$\langle G \rangle(\epsilon; x, \uparrow; x, \uparrow) = tr(f_\uparrow f_\uparrow^\dagger e^{-LH}),$$

where tr is the trace with respect to bosonic as well as fermionic states. The new Hamiltonian $\mathcal{H}$ is given as [2]

$$\mathcal{H} = \epsilon(f^\dagger \cdot f + b^\dagger \cdot b) + m_0 A - gA^2 \qquad (3)$$

with Fermi (Bose) operators f (b), $f^\dagger \cdot f = f_\downarrow^\dagger f_\downarrow + f_\uparrow^\dagger f_\uparrow$, and

$$A = f_\uparrow^\dagger f_\downarrow^\dagger - f_\uparrow f_\downarrow + b_\uparrow^\dagger b_\downarrow^\dagger - b_\uparrow b_\downarrow.$$

The supersymmetric Hamiltonian is translational invariant on the interval $[0, L]$. It can be diagonalized with an appropriate unitary transformation. In contrast, the Dirac operator H is not translational invariant, and to diagonalize it we would need a unitary transformation for *each* realization of the random $m(x)$.

2.2   Diagonalization of the supersymmetric Hamiltonian

With the new operators $\Psi = (f, b)$, $\bar{\Psi} = (f^\dagger, B^\dagger \sigma_z)$ and $B^\dagger = (b_\uparrow^\dagger, b_\downarrow)$ we can introduce the current

$$\mathbf{J}_{ab} = \frac{1}{2} \bar{\Psi}_{a\alpha} \sigma_{\alpha\beta} \Psi_{b\beta}$$

to write the Hamiltonian as

$$\mathcal{H} = 2\epsilon J^z + 2m_0 J^x - 4g(J^x)^2$$

with the x and z components of $\mathbf{J}_{ab}$, J^x and J^z. The eigenfunctions of J^z, $J^z |n\rangle_0 = n|n\rangle_0$, form a basis set $\{|n\rangle_0\}$ ($n \geq 0$) with

$$J^x |n\rangle_0 = \frac{1}{2}[(n+1)|n+1\rangle_0 - (n-1)|n-1\rangle_0] \qquad (n \geq 1)$$

and $J^x |0\rangle_0 = \frac{1}{2}|1\rangle_0$. Then the eigenfunction $|0\rangle_R$ of $\mathcal{H}$ with eigenvalue $E = 0$,

$$\mathcal{H}|0\rangle_R = E|0\rangle_R = 0, \qquad (4)$$

can be expanded in terms of the eigenfunctions of J^z as

$$|0\rangle_R = \sum_{n \geq 0} \Phi_n |n\rangle_0.$$

Equation (4) provides for the coefficients with $n \geq 0$ the recursion equation

$$n\big[-(n+1)\Phi_{n+2}+2n\Phi_n-(n-1)\Phi_{n-2}-2M(\Phi_{n+1}-\Phi_{n-1})+2\omega\Phi_n\big] = 0 \quad (5)$$

with the initial conditions $\Phi_{-1} = \Phi_{-2} = 0$. Here we have used $M = m_0/2g$ and $\omega = \epsilon/g$. Equation (5) is called a *discrete Schrödinger equation*. It is the main object of the mathematical study in this paper.

3 Analysis of the Schrödinger equation

Writing the fourth order equation (5) in the form

$$\Phi_{n+2} + \frac{2M}{n+1}\Phi_{n+1} - \frac{2(n-\omega)}{n+1}\Phi_n - \frac{2M}{n+1}\Phi_{n-1} + \frac{n-1}{n+1}\Phi_{n-2} = 0 \quad (6)$$

we see that this equation is of Poincaré type (i.e. the coefficients converge to constants as $n \to \infty$), that the corresponding limiting equation is $\Phi_{n+2} - 2\Phi_n + \Phi_{n-2} = 0$, and that the characteristic equation $\lambda^4 - 2\lambda^2 + 1 = 0$ has 1 as a four-fold root. This is a highly degenerate situation which looks, at first glance, hopeless to be accessible. On the other hand, the particular structure of equation (6) allows us to rewrite this equation in the form

$$(\Phi_{n+2} - \Phi_n) + \frac{2M}{n+1}(\Phi_{n+1} - \Phi_{n-1}) + \frac{1-n}{n+1}(\Phi_n - \Phi_{n-2}) = \frac{-2\omega}{n+1}\Phi_n \quad (7)$$

which, after setting $x_n := \Phi_n - \Phi_{n-2}$, appears as

$$x_{n+2} + \frac{2M}{n+1}x_{n+1} + \frac{1-n}{n+1}x_n = \frac{-2\omega}{n+1}\Phi_n. \quad (8)$$

This equation may be viewed (if Φ_n is considered to be known) as an inhomogeneous linear equation whose homogeneous part

$$x_{n+2} + \frac{2M}{n+1}x_{n+1} + \frac{1-n}{n+1}x_n = 0 \quad (9)$$

is of Poincaré type, too. Moreover, the corresponding characteristic equation $\lambda^2 - 1 = 0$ has two simple roots, namely $\lambda_1 = 1$ and $\lambda_2 = -1$, and this means that we have reduced the given problem to a much simpler one. But still, standard results from the Asymptotic Theory are not applicable since the two characteristic roots have the same modulus. However, employing a simplified version of the Birkhoff-Trjitzinky Theory due to Wong and Li [9, 10] we can tackle equation (9) and consequently (8), (7) and (6).

In order to do so we briefly describe the result which turns out to be useful for the study of (9). For more details we refer to Elaydi [7, Section 8.6].

3.1   The Birkhoff-Adams Theorem

Suppose we are given a second order linear difference equation

$$x_{n+2} + p_1(n)\, x_{n+1} + p_2(n)\, x_n = 0 \qquad (10)$$

whose coefficient functions $p_1(n)$ and $p_2(n)$ have asymptotic expansions

$$p_1(n) \sim \sum_{j=0}^{\infty} \frac{a_j}{n^j} \quad \text{and} \quad p_2(n) \sim \sum_{j=0}^{\infty} \frac{b_j}{n^j} \quad \text{as} \quad n \to \infty$$

with real coefficients $a_0, a_1, \dots, b_0, b_1, \dots$. The limiting equation of (10) is $x_{n+1} + a_0 x_{n+1} + b_0 x_n = 0$, and the corresponding characteristic roots are

$$\lambda_{1/2} = -\frac{a_0}{2} \pm \sqrt{\frac{a_0^2}{4} - b_0^2}\ .$$

If $\lambda_1 \neq \lambda_2$, then equation (10) has two linearly independent solutions $x_{1,n}$ and $x_{2,n}$ of the form

$$x_{i,n} \sim \lambda_i^n n^{\alpha_i} \sum_{r=0}^{\infty} \frac{c_i(r)}{n^r} , \quad i = 1, 2, \quad \text{as} \quad n \to \infty$$

where the α_i and $c_i(r)$ are given as follows: For the α_i we have the explicit formula

$$\alpha_i = \frac{a_1 \lambda_i + b_1}{a_0 \lambda_i + 2 b_0} , \quad i = 1, 2,$$

and for the $c_i(r)$ we have the following recursion: $c_i(0) = 1$ and

$$\sum_{j=0}^{s-1} \left[\lambda_i^2 2^{s-j} \binom{\alpha_i - j}{s - j} + \lambda_i \sum_{r=j}^{s} \binom{\alpha_i - j}{r - j} a_{s-r} + b_{s-j} \right] = 0 , \quad i = 1, 2.$$

This in particular implies

$$c_i(1) = \frac{-2\lambda_i^2 \alpha_i(\alpha_i - 1) - \lambda_i\big(a_2 + \lambda_i a_1 + \alpha_i(\alpha_i - 1)a_0/2\big) - b_2}{2\lambda_i^2(\alpha_i - 1) + \lambda_i\big(a_1 + (\lambda_i - 1)a_0\big) + b_1} , \quad i = 1, 2.$$

The case $\lambda_1 = \lambda_2$, by the way, is more involved (see Elaydi [7, Theorem 8.36]), but it is not needed in this paper.

3.2   The Schrödinger equation with $\omega = 0$

In order to apply the result described in the previous subsection to the homogeneous equation (9) (which represents the case $\omega = 0$ in equation (6)) we have to put the coefficients $\frac{2M}{n+1}$ and $\frac{1-n}{n+1}$ appearing in this equation in the required form. Using the relation $\frac{1}{n+1} = \frac{1}{n} - \frac{1}{n^2} + O(\frac{1}{n^2})$ as $n \to \infty$ we get

$$\frac{2M}{n+1} = \frac{2M}{n} - \frac{2M}{n^2} + O\left(\frac{1}{n^2}\right) \quad \text{and} \quad \frac{1-n}{n+1} = -1 + \frac{2}{n} - \frac{2}{n^2} + O\left(\frac{1}{n^2}\right)$$

as $n \to \infty$, and this implies

$$a_0 = 0 , \quad a_1 = 2M , \quad a_2 = -2M ; \quad b_0 = -1 , \quad b_1 = 2 , \quad b_2 = -2 .$$

Recalling that $\lambda_1 = 1$ and $\lambda_2 = -1$ we therefore get

$$\begin{aligned}
\alpha_1 &= -M - 1 , & c_1(1) &= M^2 + 3M + 1 , \\
\alpha_2 &= M - 1 , & c_2(1) &= M^2 - M + 1 .
\end{aligned}$$

Thus, by the Birkhoff-Adams Theorem, equation (9) has a fundamental set of solutions of the form

$$x_{1,n} \sim n^{-M-1}\left[1 + \frac{M^2 + 3M + 1}{n} + O(n^{-2})\right] , \tag{11}$$

$$x_{2,n} \sim (-1)^n n^{M-1}\left[1 + \frac{M^2 - M + 1}{n} + O(n^{-2})\right] . \tag{12}$$

Going back to the original equation (7) with $\omega = 0$ we get the two relations

$$\Phi_{n+2} - \Phi_n \sim x_{1,n} \quad \text{as} \quad n \to \infty , \tag{13}$$

$$\Phi_{n+2} - \Phi_n \sim x_{2,n} \quad \text{as} \quad n \to \infty . \tag{14}$$

Using the two solutions $y_{1,n} = 1$ and $y_{2,n} = (-1)^n$ of the difference equation $\Phi_{n+2} - \Phi_n = 0$, the variation of constants formula stipulates that a particular solution of (13) or (14) can be written in the form

$$\Phi_n = u_{1,n} y_{1,n} + u_{2,n} y_{2,n}$$

where

$$u_{1,n} = \sum_{r=0}^{n-1} \frac{-g(r) y_{2,r+1}}{W(r+1)} , \quad u_{2,n} = \sum_{r=0}^{n-1} \frac{g(r) y_{1,r+1}}{W(r+1)}$$

with $g(n)$ being $x_{1,n}$ or $x_{2,n}$, respectively, and $W(r+1)$ being the Casoratian

$$W(r+1) = y_{1,r+1} y_{2,r+2} - y_{1,r+2} y_{2,r+1} = (-1)^{r+2} - (-1)^{r+1} = 2(-1)^r .$$

For the asymptotic relation (13) we thus get

$$u_{1,n} = \frac{1}{2}\sum_{r=0}^{n-1} x_{1,r} \quad \text{and} \quad u_{2,n} = \frac{1}{2}\sum_{r=0}^{n-1}(-1)^r x_{1,r} ,$$

and this yields for (13) the particular solution

$$\Phi_n^* \sim \frac{1}{2}\sum_{r=0}^{n-1}\left[1 + (-1)^{n+r}\right] x_{1,r}$$

where $x_{1,r}$ is given in (11). Using the particular form of $y_{1,n}$ and $y_{2,n}$, for arbitrary real constants γ_1, γ_2 we then get the following two solutions of (13):

$$\Phi_{1,n} \sim \Phi_n^* + \gamma_1 \quad \text{and} \quad \Phi_{2,n} \sim \Phi_n^* + (-1)^n \gamma_2 .$$

With the same arguments as before, we get two solutions of (14), namely

$$\Phi_{3,n} \sim \Phi_n^{**} + \gamma_3 \quad \text{and} \quad \Phi_{4,n} \sim \Phi_n^{**} + (-1)^n \gamma_4,$$

where γ_3, γ_4 are arbitrary real constants and Φ_n^{**} is the particular solution

$$\Phi_n^{**} \sim \frac{1}{2} \sum_{r=0}^{n-1} \left[1 + (-1)^{n+r} \right] x_{2,r}$$

of (14) with $x_{2,r}$ given in (12). Thus, altogether we have obtained the four linearly independent solutions $\Phi_{1,n}, \ldots, \Phi_{4,n}$ of the special Schrödinger equation (6) with $\omega = 0$.

3.3   The full Schrödinger equation

We now deal with the full equation (6). Setting again $x_n := \Phi_n - \Phi_{n-2}$ we obtain the equation

$$x_{n+2} + \frac{2M}{n+1} x_{n+1} + \frac{1-n}{n+1} x_n = \frac{-2\omega}{n+1} \Phi_n \qquad (15)$$

which we look at as an inhomogeneous equation for the x_n. The corresponding homogeneous part

$$x_{n+2} + \frac{2M}{n+1} x_{n+1} + \frac{1-n}{n+1} x_n = 0 \qquad (16)$$

has been investigated in the previous section. Using the two linearly independent solutions $x_{1,n}$ and $x_{2,n}$ of (16) (see (11) and (12)) and going back to equation (6) by resubstituting $x_n = \Phi_n - \Phi_{n-2}$, we get the two relations $\Phi_{n+2} - \Phi_n \sim x_{1,n} - \frac{2\omega}{n+1}\Phi_n$ and $\Phi_{n+2} - \Phi_n \sim x_{2,n} - \frac{2\omega}{n+1}\Phi_n$, or equivalently,

$$\Phi_{n+2} + \left(\frac{2\omega}{n+1} - 1 \right) \Phi_n \sim x_{1,n}, \qquad (17)$$

$$\Phi_{n+2} + \left(\frac{2\omega}{n+1} - 1 \right) \Phi_n \sim x_{2,n}. \qquad (18)$$

In order to find the solutions of these relations we first solve the associated homogeneous equation

$$\Phi_{n+2} + \left(\frac{2\omega}{n+1} - 1 \right) \Phi_n = 0. \qquad (19)$$

This equation is of Poincaré type, and the corresponding characteristic equation $\lambda^2 - 1 = 0$ has the two roots $\lambda_1 = 1$ and $\lambda_2 = -1$. Using the relation

$$\frac{2\omega}{n+1} - 1 = -1 + \frac{2\omega}{n} - \frac{2\omega}{n^2} + O\left(\frac{1}{n^2} \right) \quad \text{as} \quad n \to \infty$$

and employing the notation associated with the Birkhoff-Adams Theorem we get

$$a_0 = a_1 = a_2 = 0 \; ; \quad b_0 = -1 \, , \quad b_1 = 2\omega \, , \quad b_2 = -2\omega \, ,$$

and this implies

$$\alpha_1 = \alpha_2 = -\omega \quad \text{and} \quad c_1(1) = c_2(1) = \omega^2 .$$

Applying the Birkhoff-Adams Theorem to equation (19) we therefore get two linearly independent solutions

$$z_{1,n} \;\sim\; n^{-\omega} \left[1 + \frac{\omega^2}{n} + O(n^{-2}) \right] , \tag{20}$$

$$z_{2,n} \;\sim\; (-1)^n \, n^{-\omega} \left[1 + \frac{\omega^2}{n} + O(n^{-2}) \right] . \tag{21}$$

For the inhomogeneous relations (17) and (18) we can then find particular solutions in the form

$$\Psi_n = v_{1,n} \, z_{1,n} + v_{2,n} \, z_{2,n} \tag{22}$$

where

$$v_{1,n} = \sum_{r=0}^{n-1} \frac{-g(r) \, z_{2,r+1}}{V(r+1)} \, , \quad v_{2,n} = \sum_{r=0}^{n-1} \frac{g(r) \, z_{1,r+1}}{V(r+1)} \tag{23}$$

with $g(n)$ being $x_{1,n}$ or $x_{2,n}$, respectively, and $V(r+1)$ being the Casoratian

$$V(r+1) = z_{1,r+1} \, z_{2,r+2} - z_{1,r+2} \, z_{2,r+1} \, .$$

Choosing $g(r) = x_{1,n}$, this allows to compute a particular solution Ψ_n^* for the asymptotic relation (17), and with this we get two linearly independent solutions of (17) of the form

$$\Psi_{1,n} \;\sim\; \Psi_n^* + \gamma_1 \, z_{1,n} \quad \text{and} \quad \Psi_{2,n} \;\sim\; \Psi_n^* + \gamma_2 \, z_{2,n}$$

with real parameters γ_1 and γ_2. Accordingly, we get two linearly independent solutions of (18) in the form

$$\Psi_{3,n} \;\sim\; \Psi_n^{**} + \gamma_3 \, z_{1,n} \quad \text{and} \quad \Psi_{4,n} \;\sim\; \Psi_n^{**} + \gamma_4 \, z_{2,n}$$

where γ_3 and γ_4 are real parameters and Ψ_n^{**} is a particular solution of (18) which can be obtained from (22) and (23) by choosing $g(r) = x_{2,n}$. In summary, we have obtained the four linearly independent solutions $\Psi_{1,n}, \ldots, \Psi_{4,n}$ of the Schrödinger equation (6).

4   Conclusion

The purpose of this paper is to demonstrate how a physical problem of current interest, the motion of a relativistic quantum particle with random mass, can be reduced to a problem on difference equations, and how the resulting difference equation, a highly critical equation of Poincaré type, can be solved. While the derivation of the difference equation is worked out in detail, the limited size of this paper allows to present the procedure of solving this difference equations only in its principal steps. More details will follow in a forthcoming paper.

References

[1] Adams, C. R., On the irregular cases of linear ordinary difference equations, *Trans. Amer. Math. Soc.* **30** (1928), 507-541.

[2] Balents, L. and Fisher, M. P. A., Delocalization transition via supersymmetry in one dimension, *Phys. Rev. B* **56** (1997), 12970-12991.

[3] Birkhoff, G. D., General theory of linear difference equations, *Trans. Amre. Math. Soc.* **12** (1911), 243-284.

[4] Birkhoff, G. D., Formal theory of irregular linear difference equations, *Acta Math.* **54** (1930), 205-246.

[5] Birkhoff, G. D. and Trjitzinsky, W. J., Analytic theory of singular difference equations, *Acta Math.* **60** (1932), 1-89.

[6] Efetov, K. B., Supersymmetry and theory of disordered metals, *Adv. Phys.* **32** (1983), 53-127.

[7] Elaydi, S. N., *An Introduction to Difference Equations*, 2nd Ed., Springer-Verlag, New York 1999.

[8] Takeda, K., Tsurumaru, T., Ichinose, I. and Kimura, M., Localized and extended states in one-dimensional disordered system: Random mass Dirac fermions, *Nucl. Phys. B* **556** (1999), 545-562.

[9] Wong, R. and Li, H., Asymptotic expansions for second order linear difference equations, *J. Comput. Appl. Math.* **41** (1992), 65-94.

[10] Wong, R. and Li, H., Asymptotic expansions for second order linear difference equations II, *Studies Appl. Math.* **87** (1992), 289-324.

[11] Ziegler, K., Disordered system with n orbitals per site: Langrange formulation without replica trick, and scaling law for the density of states, *Z. Phys. B* **48** (1982), 293-304.

Existence of Bounded Solutions
of Discrete Delayed Equations

JAROMÍR BAŠTINEC, JOSEF DIBLÍK

Department of Mathematics
Faculty of Electrical Engineering and Computer Science
Brno University of Technology, Technická 8
616 00 Brno, Czech Republic
E-mails: bastinec@feec.vutbr.cz, diblik@feec.vutbr.cz

and

BINGGEN ZHANG

Department of Mathematics
Ocean University of Qingdao
Qingdao, 266071, China
E-mail: bgzhang@public.qd.sd.cn

Abstract A powerful tool for investigation of various asymptotic, boundary-value and qualitative problems in the theory of ordinary differential equations as well as in the theory of delayed differential equations is the retraction method. The development of this method is discussed in the present contribution in the case of one scalar delayed discrete equation of the form $\Delta u(k + n) = f(k, u(k), u(k + 1), \ldots, u(k + n))$. Conditions that guarantee the existence of at least one solution having its graph in a prescribed set are formulated. The proof is based on the idea of a retract principle. In the construction of a retract mapping the property of continuous dependence of solutions on their initial data is used.

Keywords Retraction method, Discrete delayed equation, Bounded solutions

AMS Subject Classification 39A10, 39A11

1   Introduction

A powerful tool for the investigation of various problems for ordinary differential equations as well as for delayed differential equations is the retraction method (Ważewski's method). For example, with the aid of this principle, some asymptotic problems were considered, e.g., in the papers [3, 4, 14]. For sources we refer to [9, 11] and [15]. In this paper we shall give a construction in which the idea of the retraction principle is developed. Obtained results can be useful for the investigation of asymptotic behavior of solutions of discrete equations of the types indicated. Let us note that the questions concerning asymptotic behavior of solutions of discrete equations were considered in many recent papers, e.g. in [5]–[7], [12, 16].

Let us consider the scalar discrete equation

$$\Delta u(k+n) = f(k, u(k), u(k+1), \ldots, u(k+n)), \tag{1}$$

where $f(k, u_0, u_1, \ldots, u_n)$ is defined on $N(a) \times \mathbb{R}^{n+1}$ with values in $\mathbb{R}$ where $N(a) = \{a, a+1, \ldots\}$, $a \in \mathbb{N}$, $\mathbb{N} = \{0, 1, \ldots\}$ and $n \in \mathbb{N}$. Together with discrete equation (1) we consider an initial problem. It is posed as follows: for a given $s \in \mathbb{N}$ we are seeking the solution of (1) satisfying $n + 1$ initial conditions

$$u(a + s + m) = u^{s+m} \in \mathbb{R}, \ m = 0, 1, \ldots, n \tag{2}$$

with prescribed constants u^{s+m}. Let us recall that the solution of the initial problem (1), (2) is defined as an infinite sequence of numbers

$$\{u(a+s) = u^s, u(a+s+1) = u^{s+1}, \ldots, u(a+s+n) = u^{s+n},$$
$$u(a+s+n+1), u(a+s+n+2), \ldots\}$$

such that for any $k \in N(a+s)$ the equality (1) holds. The existence and uniqueness of the solution of the initial problem (1), (2) is obvious for every $k \in N(a+s)$. We shall suppose that for all $(k, u_0, u_1, \ldots, u_n), (k, v_0, v_1, \ldots, v_n) \in N(a) \times \mathbb{R}^{n+1}$ the Lipschitz condition

$$|f(k, u_0, u_1, \ldots, u_n) - f(k, v_0, v_1, \ldots, v_n)| \leq \lambda(k) \sum_{i=0}^{n} |u_i - v_i| \tag{3}$$

holds with a nonnegative function $\lambda(k)$ defined on $N(a)$. Then the initial problem (1), (2) depends continuously on the initial data (see, e.g., [1]).

2   Formulation of problem

Let us define sets $\omega \subset N(a) \times \mathbb{R}$ and $\omega(k)$ as

$$\omega := \{(k, u) : k \in N(a), \ b(k) < u < c(k)\} \tag{4}$$

with closure $\overline{\omega} := \{(k, u) : k \in N(a), \ b(k) \leq u \leq c(k)\}$ and

$$\omega(k) := \{(u) : b(k) < u < c(k)\}$$

with closure $\overline{\omega}(k) := \{(u) : b(k) \leq u \leq c(k)\}$, where $b(k)$, $c(k)$, $b(k) < c(k)$ are real functions defined on $N(a)$. Obviously $(k, \omega(k)) \subset \omega$ for every $k \in N(a)$

and $\omega = \cup_{k \in N(a)}(k, \omega(k))$. Our aim is to establish a set of sufficient conditions with respect to the right-hand side of equation (1) in order to guarantee the existence of at least one solution $u = u(k)$ defined on $N(a)$ such that $(k, u(k)) \subset (k, \omega(k)) \subset \omega$ for each $k \in N(a)$. More exactly, we formulate the following.

Problem 1 *Suppose that the right-hand side of equation (1) satisfies the Lipschitz condition (3) and that a set ω is defined by (4) with the aid of real functions $b(k)$, $c(k)$ satisfying the inequalities $b(k) < c(k)$ on $N(a)$. Find a set of sufficient conditions which guarantee that there exists an initial problem of (1) with initial data satisfying relations $u_0^* \in \omega(a)$, $u_1^* \in \omega(a+1)$, ..., $u_n^* \in \omega(a+n)$ such that the corresponding solution $u = u^*(k)$ of equation (1) satisfies the inequalities $b(k) < u^*(k) < c(k)$ for every $k \in N(a)$ (i.e., $(k, u^*(k)) \in (k, \omega(k)) \subset \omega$ for every $k \in N(a)$).*

3   Preliminaries

In the proof of Theorem 1 we will test the character of every *"boundary"* point of the set ω. The boundary of this set is defined as

$$\partial\omega := \{(k, u) : k \in N(a), (u - b(k))(u - c(k)) = 0\} = B_1 \cup B_2$$

with

$$B_1 = \{(k, u) : k \in N(a), u = b(k)\} \subset N(a) \times \mathbb{R},$$
$$B_2 = \{(k, u) : k \in N(a), u = c(k)\} \subset N(a) \times \mathbb{R},$$

where functions b, c were defined in Section 2. Let us, moreover, define two auxiliary functions $U_1(k, u) \equiv u - b(k)$, $U_2(k, u) \equiv u - c(k)$ on $N(a) \times \mathbb{R}$ and $\partial\omega(k) := \{b(k), c(k)\}$. The next notion to be defined is the notion of the *full difference of an indicated function with respect to the discrete equation (1) and the sets B_1 and ω (or B_2 and ω)*. For the delay discrete equation (1), we will use the Razumikhin approach known from the theory of delayed functional differential equations (see, e.g., [8, 10, 13]) to compute the full difference taking simultaneously into account the *history* of the solution (i.e. we suppose that if a point of the graph of a solution reaches the boundary $\partial\omega$ than the preceeding $n + 1$ points of this graph lie in ω).

Definition 1 The full difference

$$\Delta U_1(k + n, u)|_{(k+m,u) \in (k+m, \omega(k+m)), m=0,1,\ldots,n-1, (k+n,u) \in B_1}$$

of the function $U_1(k+n, u)$ for a $k \in N(a)$ with respect to the discrete equation (1) and the sets B_1 and ω is defined as

$$\Delta U_1(k + n, u)|_{(k+m,u) \in (k+m, \omega(k+m)), m=0,1,\ldots,n-1, (k+n,u) \in B_1} =$$
$$f(k, u_0, u_1, \ldots, u_{n-1}, b(k+n)) - b(k+n+1) + b(k+n)$$

where $u_0 \in \omega(k), u_1 \in \omega(k+1), \ldots, u_{n-1} \in \omega(k+n-1)$ is assumed.

Definition 2 The full difference

$$\Delta U_2(k + n, u)|_{(k+m,u)\in(k+m,\omega(k+m)),m=0,1,\ldots,n-1,\,(k+n,u)\in B_2}$$

of the function $U_2(k+n, u)$ for a $k \in N(a)$ with respect to the discrete equation (1) and the sets B_2 and ω is defined as

$$\Delta U_2(k + n, u)|_{(k+m,u)\in(k+m,\omega(k+m)),m=0,1,\ldots,n-1,\,(k+n,u)\in B_2} =$$
$$f(k, u_0, u_1, \ldots, u_{n-1}, c(k + n)) - c(k + n + 1) + c(k + n)$$

where $u_0 \in \omega(k), u_1 \in \omega(k + 1), \ldots, u_{n-1} \in \omega(k + n - 1)$ is assumed.

In the following text we will abbreviate corresponding notation and put

$$\Delta U_1^*(k+n, u) \equiv \Delta U_1(k + n, u)|_{(k+m,u)\in(k+m,\omega(k+m)),m=0,1,\ldots,n-1,\,(k+n,u)\in B_1},$$

$$\Delta U_2^*(k+n, u) \equiv \Delta U_2(k + n, u)|_{(k+m,u)\in(k+m,\omega(k+m)),m=0,1,\ldots,n-1,\,(k+n,u)\in B_2}.$$

Definition 3 A point $(k+n, u) \in B_1$ with $k \in N(a)$ is called *the point of the type of strict egress* for the set ω with respect to the discrete equation (1) if

$$\Delta U_1^*(k + n, u) < 0$$

for every $u_0 \in \omega(k), u_1 \in \omega(k + 1), \ldots, u_{n-1} \in \omega(k + n - 1)$.

Definition 4 A point $(k+n, u) \in B_2$ with $k \in N(a)$ is called *the point of the type of strict egress* for the set ω with respect to the discrete equation (1) if

$$\Delta U_2^*(k + n, u) > 0$$

for every $u_0 \in \omega(k), u_1 \in \omega(k + 1), \ldots, u_{n-1} \in \omega(k + n - 1)$.

The following lemma is obvious.

Lemma 1 *A point $(k + n, u) \in B_1 \cup B_2$ with $k \in N(a)$ is the point of the type of strict egress for the set ω with respect to the discrete equation (1) if and only if*

$$f(k, u_0, u_1, \ldots, u_{n-1}, b(k + n)) - b(k + n + 1) + b(k + n) < 0 \qquad (5)$$

for every $u_0 \in \omega(k), u_1 \in \omega(k + 1), \ldots, u_{n-1} \in \omega(k + n - 1)$ in the case when $(k + n, u) \in B_1$, and

$$f(k, u_0, u_1, \ldots, u_{n-1}, c(k + n)) - c(k + n + 1) + c(k + n) > 0 \qquad (6)$$

for every $u_0 \in \omega(k), u_1 \in \omega(k + 1), \ldots, u_{n-1} \in \omega(k + n - 1)$ in the case when $(k + n, u) \in B_2$.

In the proof of Theorem 1 below, the notions *retraction* and *retract* are used too. Let us recall these notions (see e.g. [11]):

Definition 5 If $A \subset B$ are any two sets of a topological space and $\pi : B \to A$ is a continuous mapping from B onto A such that $\pi(p) = p$ for every $p \in A$, then π is said to be a *retraction* of B onto A. When there exists a retraction of B onto A, A is called a *retract* of B.

4   Results

The following theorem represents the main result of this contribution.

Theorem 1 *Let us suppose that $f(k, u_0, u_1, \ldots, u_n)$ is defined on $N(a) \times \mathbb{R}^{n+1}$ with values in $\mathbb{R}$ and satisfies the Lipschitz condition (3). If, moreover, the inequalities (5), (6) hold for every $k \in N(a)$ and every*

$$u_0 \in \omega(k), u_1 \in \omega(k+1), \ldots, u_{n-1} \in \omega(k+n-1),$$

then there exists an initial problem

$$u^*(a+m) = u_m^* \in \omega(a+m), \ m = 0, 1, \ldots, n \tag{7}$$

such that the corresponding solution $u = u^*(k)$ of equation (1) satisfies for every $k \in N(a)$ the inequalities

$$b(k) < u^*(k) < c(k). \tag{8}$$

Proof. Let us suppose that the initial data (7) generating the solution $u = u^*(k)$, $k \in N(a)$ with $u(a+m) = u^*(a+m)$, $m = 0, 1, \ldots, n$ and satisfying the inequalities (8) for every $k \in N(a)$ do not exist. This means, in other words, that for every fixed $u_m^{00} \in \mathbb{R}$, $m = 0, 1, \ldots, n$ such that

$$b(a+m) < u_m^{00} < c(a+m)$$

there exists an integer $k^{00} \in N(a+n+1)$ such that for the corresponding solution $u = u^{00}(k)$ with $u^{00}(a+m) = u_m^{00}$, $m = 0, 1, \ldots, n$ we have $(k^{00}, u^{00}(k^{00})) \notin \omega$ and, moreover, $(a+l, u^{00}(a+l)) \in \omega$, $l = 0, 1, \ldots, k^{00} - a - 1$. Since, in view of inequalities (5), (6) (see Lemma 1) each point $(k+n, u) \in B_1 \cup B_2$, $k \in N(a)$ is the point of the type of strict egress for the set ω with respect to the discrete equation (1), we can, moreover, conclude the following: For every $u_m^0 \in \mathbb{R}$, $m = 0, 1, \ldots, n$ such that

$$b(a+m) < u_m^0 < c(a+m), \ m = 0, 1, \ldots, n-1, \tag{9}$$

$$b(a+n) \leq u_n^0 \leq c(a+n) \tag{10}$$

there exists an integer $k^0 \in N(a+n+1)$ such that for corresponding solution $u = u^0(k)$, $u^0(a+m) = u_m^0$, $m = 0, 1, \ldots, n$ we have $(k^0, u^0(k^0)) \notin \overline{\omega}$ and

$$(a+l, u^0(a+l)) \in \overline{\omega}, \quad l = 0, 1, \ldots, k^0 - a - 1.$$

Obviously, if $u_n^0 = b(a+n)$ or if $u_n^0 = c(a+n)$, the value $k^0 = a+n+1$.

In the next part of the proof we suppose that values $u_m^0 \in \mathbb{R}$, $m = 0, 1, \ldots, n-1$ satisfying (9) are fixed and the value satisfying (10) $u_n^0 \in \mathbb{R}$ varies. In this situation we prove that there exists a retraction of the set $\overline{\omega}(a+n) \equiv [b(a+n), c(a+n)]$ onto the two-point set $\partial\omega(a+n) \equiv \{b(a+n), c(a+n)\}$. (See Definition 5 if $B \equiv [b(a+n), c(a+n)]$ and $A \equiv \{b(a+n), c(a+n)\}$.) In other words, we will prove that there exists a retraction of a closed interval onto its boundary. This is impossible and the contradiction obtained will indicate the wrong supposition at the beginning of the proof. Let us construct this retraction. Define mappings P_1, P_2 and P_3:

$$P_1 : (a+n, u_0^0, u_1^0, \ldots, u_n^0) \rightarrow (k^0, u^0(k^0)),$$

where the value k^0 was defined above;

$$P_2 : (k^0, u^0(k^0)) \rightarrow \begin{cases} (k^0, c(k^0)) & \text{if} \quad u^0(k^0) > c(k^0), \\ (k^0, b(k^0)) & \text{if} \quad u^0(k^0) < b(k^0); \end{cases}$$

and for $(k^0, \tilde{u})$ with $\tilde{u} \in \partial\omega(k^0)$:

$$P_3 : (k^0, \tilde{u}) \rightarrow \begin{cases} (a+n, c(a+n)) & \text{if} \quad \tilde{u} = c(k^0), \\ (a+n, b(a+n)) & \text{if} \quad \tilde{u} = b(k^0). \end{cases}$$

Let us verify that the composite mapping

$$P : (a+n, u_0^0, u_1^0, \ldots, u_n^0) \rightarrow \partial\omega(a+n)$$

where $P = P_3 \circ P_2 \circ P_1$ is continuous with respect to the last coordinate of point $(a+n, u_0^0, u_1^0, \ldots, u_n^0)$, i.e., with respect to u_n^0. In view of construction of mapping P only two result points are possible:

$$\begin{aligned} \textit{either} \quad & P(a+n, u_0^0, u_1^0, \ldots, u_n^0) = (a+n, c(a+n)) \\ \textit{or} \quad & P(a+n, u_0^0, u_1^0, \ldots, u_n^0) = (a+n, b(a+n)). \end{aligned}$$

Let the first possibility hold, i.e., $P(a+n, u_0^0, u_1^0, \ldots, u_n^0) = (a+n, c(a+n))$. Then $P_1(a+n, u_0^0, u_1^0, \ldots, u_n^0) = (k^0, u^0(k^0))$, $(k^0, u^0(k^0)) \notin \overline{\omega}$ and as it follows from definition of mappings P_2, P_3: $u^0(k^0) > c(k^0)$. We remark that in view of our construction there exists a dependence: $k^0 = k^0(u_n^0)$ and $u^0(k^0) = u^0[k^0(u_n^0)]$. The continuity of the mapping P is now a consequence of property of *continuous dependence of initial problem on initial data*. For perturbations $\Delta u_n^0 = u_n^0 + \Delta$ of u_n^0 with sufficiently small Δ we get in view of properties of mappings P_2 and P_3: $P(a+n, u_0^0, u_1^0, \ldots, \Delta u_n^0) = (a+n, c(a+n))$, i.e., the mapping P in this case is continuous. By analogy we proceed in the case when $P(a + n, u_0^0, u_1^0, \ldots, u_n^0) = (a + n, b(a + n))$.

At the end, the continuity of mapping P has been proven if u_n^0 varies within interval $[b(a + n), c(a + n)]$. (Obviously, $P(a + n, u_0^0, \ldots, u_{n-1}^0, c(a + n)) = (a + n, c(a + n))$ and $P(a + n, u_0^0, \ldots, u_{n-1}^0, b(a + n)) = (a + n, b(a + n))$.) So, the desired retraction is realized by the mapping P since (at this moment we take into account that points $u_0^0, u_1^0, \ldots, u_{n-1}^0$ were fixed; in this situation the mapping P depends on u_n^0 only)

$$[b(a + n), c(a + n)] \xrightarrow{P} \{b(a + n), c(a + n)\}$$

is continuous and

$$\{b(a + n)\} \xrightarrow{P} \{b(a + n)\}; \quad \{c(a + n)\} \xrightarrow{P} \{c(a + n)\}.$$

This (as was noted above) is impossible (the boundary of n-dimensional ball is not its retract; see, e.g., [2]). Therefore there exists an initial problem (7) such that the corresponding solution $u = u^*(k)$ satisfies the inequalities (8) for every $k \in N(a)$. This completes the proof.

Theorem 1 can be generalized in the following way. As easy follows from its proof, the assumptions with respect to the function $f(k, u_0, u_1, \ldots, u_n)$ were used only for values

$$(k, u_0, u_1, \ldots, u_n) \in \Omega^* \equiv N(a) \times \omega(k) \times \omega(k+1) \times \cdots \times \omega(k+n-1) \times \overline{\omega}(k+n)$$

although they were supposed to be valid on $N(a) \times \mathbb{R}^{n+1}$. Therefore we reformulate this theorem. Its proof is, in view of this fact, omitted.

Theorem 2 *Let us suppose that the real function $f(k, u_0, u_1, \ldots, u_n)$ is defined for all $(k, u_0, u_1, \ldots, u_n) \in \Omega^*$, and for all $(k, u_0, u_1, \ldots, u_n)$, $(k, v_0, v_1, \ldots, v_n) \in \Omega^*$ we have Lipschitz condition (3). If, moreover, each point $(k, u) \in B_1 \cup B_2$ is the point of the type of strict egress for the set ω with respect to the discrete equation (1), then there exists an initial problem (7) with $u_0^* \in \omega(a), u_1^* \in \omega(a + 1), \ldots, u_n^* \in \omega(a + n)$ such that the corresponding solution $u = u^*(k)$ of discrete equation (1) satisfies the inequalities (8) for every $k \in N(a)$.*

Example 1 Let us consider discrete delayed equation

$$\Delta u(k + 1) = f(k, u(k), u(k + 1)) \equiv \frac{k}{u(k)} \left[\frac{u(k + 1)}{k + 1} \right]^p, \quad p > 2. \tag{11}$$

Let us put $b(k) = \delta k$ with $\delta \in (0, 1)$, $\delta = $ const, $c(k) = \varepsilon k$ with $\varepsilon > 1$, $\varepsilon = $ const and $a = 1$. It is a trivial matter to verify that inequalities (5), (6) hold for $n = 1$ and for every $k \in N(1)$. We get

$$f(k, u_0, b(k + 1)) - b(k + 2) + b(k + 1) = k\delta^p/u_0 - \delta < \delta(\delta^{p-2} - 1) < 0,$$
$$f(k, u_0, c(k + 1)) - c(k + 2) + c(k + 1) = k\varepsilon^p/u_0 - \varepsilon > \varepsilon(\varepsilon^{p-2} - 1) > 0$$

for any $k \in N(1)$. All assumptions of Theorem 2 are satisfied for $n = a = 1$; hence, a solution $u = u^*(k)$ of equation (11) exists such that $\delta k < u^*(k) < \varepsilon k$ for every $k \in N(1)$. Indeed, it is easy to verify that this equation has a solution $u = u(k) = k$.

Acknowledgment. This work was supported by the Grant 201/01/0079 of Czech Grant Agency and by the Project ME423/2001 of The Ministry of Education, Youth and Sports of The Czech Republic.

References

[1] Ravi P. Agarwal, *Differential Equations and Inequalities, Theory, Methods, and Applications*, Marcel Dekker, Inc., 2nd ed., 2000.

[2] K. Borsuk, *Theory of Retracts*, PWN, Warsaw, 1967.

[3] J. Diblík, *Asymptotic representation of solutions of equation $\dot{y}(t) = \beta(t)[y(t) - y(t - \tau(t))]$*, J. Math. Anal. Appl. **217** (1998), 200–215.

[4] J. Diblík, M. Růžičková. *Existence of positive solutions of n-dimensional system of nonlinear differential equations, entering into a singular point*, Arch. Mat. **36** (2000), 435–446.

[5] W. Golda, J. Werbowski, *Oscillation of linear functional equations of the second order*, Funkc. Ekvac. **37** (1994), 221–227.

[6] I. Györi, M. Pituk, *Asymptotic formulae for the solutions of a linear delay difference equation*, J. Math. Anal. Appl. **195** (1995), 376–392.

[7] I. Györi, M. Pituk, *Comparison theorems and asymptotic equilibrium for delay differential and difference equations*, Dyn. Systems and Appl. **5** (1996), 277–302.

[8] J. Hale, S.M.V. Lunel, *Introduction to Functional Differential Equations*, Springer-Verlag, 1993.

[9] Ph. Hartman, *Ordinary differential equations*, Second Edition, Birkhäuser, 1982.

[10] V. Kolmanovskij, A. Myshkis, *Introduction to the Theory and Applications of Functional Differential Equations*, Kluwer Acad. Publ., 1999.

[11] V. Lakshmikantham, S. Leela, *Differential and Integral Inequalities, Vol. I - Ordinary Differential Equations*, Academic Press, New York, London, 1969.

[12] M. Migda, J. Migda, *Asymptotic behaviour of solutions of difference equations of second order*, Demonstr. Math. **XXXII** (1999), 767–773.

[13] B.S. Razumikhin, *Stability of Hereditary Systems*, Nauka, Moscow, 1988. (In Russian)

[14] B. Vrdoljak, *On behaviour of solutions of system of linear differential equations*, Math. Communications **2** (1997), 47–57.

[15] T. Ważewski, *Sur un principe topologique de l'examen de l'allure asymptotique des intégrales des équations différentielles ordinaires*, Ann. Soc. Polon. Math. **20** (1947), 279–313.

[16] S. Zhang, *Stability of infinite delay difference systems*, Nonl. Anal. T.M.A. **22** (1994), 1121–1129.

Difference ϕ-Laplacian Periodic Boundary Value Problems: Existence and Localization of Solutions [1]

ALBERTO CABADA and VICTORIA OTERO-ESPINAR

Departamento de Análise Matemática
Facultade de Matemáticas
Universidade de Santiago de Compostela
15782, Santiago de Compostela, Galicia, Spain
E-mails: cabada@usc.es, vivioe@usc.es

Abstract This paper is devoted to the study of nonlinear difference ϕ – Laplacian problems with periodic boundary value conditions. Using known results that assure the existence of solutions for the considered problems, we obtain sufficient conditions in the nonlinear part of the equations that allow us to define a lower and an upper solution for the treated problems and deduce existence of solutions.

Keywords ϕ – Laplacian problems, lower and upper solutions, comparison results, discontinuous functional dependence.

AMS Subject Classification 39A10, 39A11

1   Introduction

In this paper we study existence results for the following two difference problems

$$-\Delta[\phi(\Delta u(k))] + q(k+1)u(k+1) \quad = \quad f(k, u(k+1)), \qquad k \in I, \quad (1)$$
$$u(0) = u(N), \qquad \Delta u(0) = \Delta u(N), \quad (2)$$

and

$$-\Delta[\phi(\Delta u(k))] + q(k+1)u(k+1) \quad = \quad g(k, u(k+1), u), \qquad k \in I, \quad (3)$$
$$u(0) = u(N), \qquad \Delta u(0) = \Delta u(N). \quad (4)$$

where $N \in \mathbb{N}$ is a fixed positive integer, $I = \{0, 1, \dots, N-1\}$, $f : I \times \mathbb{R} \to \mathbb{R}$ and $g : I \times \mathbb{R} \times \mathbb{R}^{N+2} \to \mathbb{R}$.

We will assume that

---

[1] Supported by M. C. Y. T. – F. E. D. E. R., Spain, project BFM2001-3884-C02-01.

(H_1) $\phi : \mathbb{R} \to \mathbb{R}$ is continuous, strictly increasing in $\mathbb{R}$ and $\phi(\mathbb{R}) = \mathbb{R}$.

(H_2) $q(k+1) > 0$, for all $k \in I$.

We define
$$Q = \min \{ q(k+1) : k \in I \} > 0 \tag{5}$$

Throughout this paper, for each $p \geq 0$ given, if $x = (x_0, \ldots, x_p)$ and $y = (y_0, \ldots, y_p) \in \mathbb{R}^{p+1}$ are such that $x_k \leq y_k$ $(x_k \geq y_k)$ for all $k \in \{0, \ldots, p\}$, we shall denote $x \leq y$ $(x \geq y)$ on $\{0, \ldots, p\}$ and

$$[x, y] = \left\{ z = (z_0, \ldots, z_p) \in \mathbb{R}^{p+1} : x_k \leq z_k \leq y_k, \, k \in \{0, \ldots, p\} \right\}.$$

Furthermore, we shall denote by $J = \{0, \ldots, N+1\}$.

We say that $u \equiv (u(0), \ldots, u(N+1)) \in \Omega \subset \mathbb{R}^{N+2}$ is the maximal solution of problem (1) – (2) (respectively problem (3) – (4)) in Ω, if every solution $v \in \Omega$ of such problem satisfies that $v \leq u$ in J. If the reversed inequalities hold, we say that u is the minimal solution in Ω. We refer to these solutions as extremal solutions.

Now, we introduce the concept of lower and upper solutions for problem (1) – (2) as follows:

Definition 1.1 *A real valued function α on J is a lower solution for problem (1) – (2) if it satisfies*

$$-\Delta[\phi(\Delta\alpha(k))] + q(k+1)\alpha(k+1) \;\leq\; f(k, \alpha(k+1)); \quad k \in I$$
$$\alpha(0) = \alpha(N), \qquad \Delta\alpha(0) \geq \Delta\alpha(N).$$

The concept of an upper solution β is the same assuming the reversed inequalities.

Lower and upper solution for problem (3) – (4) are defined in the same way with obvious notations.

These concepts are classical for ordinary differential equations (see [11] and references therein) but are recent for discrete problems. One can see the paper of Eloe [13], in which existence results for the periodic problem

$$\Delta^2 u_k = f(k, u_k, u_{k+1}); \quad k \in \{1, \ldots, N\}, \quad u(0) = u(N+1), \; u(1) = u(N+2),$$

are obtained.

Second order periodic problems were studied by Atici and Cabada in [2] and by the authors in [5]. The Neumann problem was considered in [6]. Zhuang, Chen and Cheng in [15] study Dirichlet conditions. Higher order problems can be found in the paper of Agarwal and Wong [1], and in the authors' papers [5, 7].

First order problems with nonlinear boundary conditions that include the initial, final and the periodic ones as a particular case are treated in [9].

Problem $-\Delta[\phi(\Delta u(k))] = f(k, u(k+1))$ has been approached in [3]. The ϕ – Laplacian operator arises in the theory of radial solutions for the p –

Laplacian equation $(\phi(x) =| x |^{p-2} x)$ on an annular domain (see [12], and references therein) and has been exhaustively studied recently for differential equations (see, for instance, [4, 10]). There is proved an existence result of extremal solutions when $\alpha \leq \beta$ and allowing nonlinear boundary conditions that include the Dirichlet, Neumann and periodic conditions as a particular cases.

Now we enunciate Theorem 2.2 in [3] adapted to problem (1) – (2).

Theorem 1.2 *Let α and β be a lower and an upper solution respectively for problem (1) – (2), such that $\alpha \leq \beta$ in J. Assume that function f is continuous and that hypothesis (H_1) holds. Then problem (1) – (2) has extremal solutions in the sector $[\alpha, \beta]$.*

Remark 1.3 *Contrary to Theorem 1.2, if $\alpha \geq \beta$ in J, it is possible that problem (1) – (2) has no extremal solutions in $[\beta, \alpha]$. This case has been studied in [8] in which conditions on function f are given to assure the existence of extremal solutions. In any case, for a general continuous function the existence result does not hold, as we can see in Example 3.1 in [2] where an example of nonexistence of solutions in $[\beta, \alpha]$ is introduced.*

It is important to note that even in the case of existence of solutions in $[\beta, \alpha]$, it is not assured the existence of extremal solutions. To see this, consider for instance the following problem:

$$-\Delta^2 u(k) + 2u(k+1) = 4\, u(k+1), \ k \in \{0, 1, 2, 3\},$$

$$u(0) = u(4), \quad \Delta u(0) = \Delta u(4).$$

It is clear that any negative constant sequence β is an upper solution and any positive constant sequence α is a lower solution of this problem. On the other hand, the set of solutions of this problem is given by the expression $\{A, B, -A, -B, A, B\}$, with $A, B \in \mathbb{R}$. Clearly there is no extremal solutions in $[\beta, \alpha]$.

The following comparison result is Proposition 3.4 in [3]:

Proposition 1.4 *Suppose that $h(k, \cdot)$ is a strictly increasing function in $\mathbb{R}$ for all $k \in J$. Suppose $u,\ v \in \mathbb{R}^{N+2}$ are such that the following inequality holds*

$$-\Delta\phi(\Delta u(k)) + h(k, u(k+1)) \leq -\Delta\phi(\Delta v(k)) + h(k, v(k+1)), \text{for all } k \in I$$
$$u(0) - u(N) \leq v(0) - v(N) \quad and \quad u(1) - u(N+1) \geq v(1) - v(N+1).$$

Then $u \leq v$ in J.

This paper is devoted to obtain sufficient conditions in function f that allow us to assure the existence of a pair of lower and upper solutions for problem (1) – (2) and thus, derive existence of solutions for that problem. In

section 3, using the generalized iterative methods of Heikkilä and Lakshmikantham [14] for discontinuous functions, we extend Theorem 1.2 to problem (3)
– (4). After this, we present some conditions in function g that permit us to
assure the existence of a pair of lower and upper solutions for this problem.

2   Existence results

In this section we derive existence and localization results for problem (1) –
(2) by using the preliminary results exposed in section 1. We start assuming
that the nonlinear part is bounded.

Theorem 2.1 *Suppose that conditions (H_1) and (H_2) hold. Assume also that
$f(k, x)$ is a continuous function in $x \in \mathbb{R}$ for every $k \in I$ fixed and there are
two real constants M_1 and M_2 such that $M_1 \le f(k, x) \le M_2$ for all $x \in \mathbb{R}$
and $k \in I$. Then problem (1) – (2) has extremal solutions in $\mathbb{R}^{N+2}$ and all of
them belong to the sector $[-\mid M_1 \mid /Q, \mid M_2 \mid /Q]$, where Q is defined in (5).*

Proof: For $i = 1, 2$, consider the following problems

$$(P_i) \begin{cases} -\Delta[\phi(\Delta u(k))] + q(k+1)u(k+1) = M_i, \ k \in I, \\ \\ u(0) = u(N), \quad \Delta u(0) = \Delta u(N) \end{cases}$$

Since $q(k + 1) > 0$ for all $k \in I$, we know that for $i = 1, 2$ the real numbers
$\alpha_i = - \mid M_i \mid /Q$ and $\beta_i = \mid M_i \mid /Q$ are respectively a lower and an upper
solution of problems (P_i) such that $\alpha_i \le \beta_i$.

From Theorem 1.2 we know that problem (P_i) has extremal solutions in
$[\alpha_i, \beta_i]$. On the other hand, due to the fact that $h(k, x) \equiv q(k+1) x$ is strictly
increasing in x for every $k \in I$, we conclude by Proposition 1.4 that problem
(P_i) has a unique solution u_i, $i = 1, 2$.

Now, from the definition of M_1 and M_2 we have that u_1 is a lower solution
and u_2 is an upper solution of problem (1) – (2). Now, Proposition 1.4
implies that $u_1 \le u_2$ in J. From Theorem 1.2 again, we know that there
exist extremal solutions of problem (1) – (2) in $[u_1, u_2]$. Using Proposition
1.4 again, we obtain that every solution u of problem (1) – (2) is such that
$u_1(k) \le u(k) \le u_2(k)$ for all $k \in J$. In particular, all the solutions of problem
(1) – (2) are lying between the extremal solutions in the sector $[u_1, u_2]$. And
the result holds.   □

Now, we deduce the existence of solutions assuming some monotonicity
properties in f.

Theorem 2.2 *Assume that conditions (H_1) and (H_2) are satisfied and that
$f(k, x)$ is a continuous function and non-increasing in $x \in \mathbb{R}$ for any $k \in I$
fixed. Then problem (1) – (2) has a unique solution in $\mathbb{R}^{N+2}$ and it belongs
to the interval $[-\bar{M}/Q, \bar{M}/Q]$. Where $\bar{M} = \max\{|f(k, 0)|, k \in I\}$ and Q is
given in (5).*

Proof: Let α be a negative constant sequence such that $\alpha \leq -\bar{M}/Q$. Thus

$$-\Delta(\phi(\Delta\alpha(k)) + q(k+1)\alpha(k+1) \leq Q\alpha(k+1) \leq -\bar{M} \leq f(k,0) \leq f(k,\alpha(k+1)).$$

Clearly the reversed inequalities hold for any positive constant $\beta \geq \bar{M}/Q$. In consequence α and β are a pair of lower and upper solutions respectively for problem (1) – (2). From Theorem 1.2 we know that there exist extremal solutions in the sector formed by α and β. Now, from the non-increasing property of function f and Proposition 1.4, we conclude that there is only one solution in that sector.

Finally, since every possible solution of problem (1) – (2) is lying between some sector $[\alpha, \beta]$ we deduce that there is only one solution in $\mathbb{R}^{N+2}$ of the considered problem. $\qquad\square$

3 Discontinuous and functional dependence

In this section we extend Theorem 1.2 to the discontinuous functional problem (3) – (4). The fundamental tool in this case will be the generalized iterative techniques developed by Heikkilä and Lakshmikantham in the monograph [14].

Theorem 3.1 *Let $g : I \times \mathbb{R} \times \mathbb{R}^{N+2} \to \mathbb{R}$ such that $g(k, x, y)$ is continuous in $x \in \mathbb{R}$ for every $y \in \mathbb{R}^{N+2}$ and $k \in I$ fixed, and nondecreasing in y for every $x \in \mathbb{R}$ and $k \in I$ fixed.*

Let α and β be a lower and an upper solution, respectively, for problem (3) – (4) such that $\alpha \leq \beta$ in J. Assume that hypotheses (H_1) and (H_2) are satisfied. Then problem (3) – (4) has extremal solutions in $[\alpha, \beta]$.

Proof: For each $v \in [\alpha, \beta]$ fixed, consider the following problem

$$(P_v) \begin{cases} -\Delta[\phi(\Delta u(k))] + q(k+1)u(k+1) = g(k, u(k+1), v), & k \in I, \\ u(0) = u(N), \quad \Delta u(0) = \Delta u(N). \end{cases}$$

Since $g(k, x, y)$ is non-decreasing in y for every x and k fixed, we have that α and β are a lower and an upper solution of problem (P_v). From Theorem 1.2, problem (P_v) has extremal solutions in $[\alpha, \beta]$.

Define Gv as the minimal solution in $[\alpha, \beta]$ of problem (P_v).

Now, we verify that mapping $G : [\alpha, \beta] \to [\alpha, \beta]$ is non-decreasing in $[\alpha, \beta]$. Indeed, if $v_1, v_2 \in [\alpha, \beta]$ are such that $v_1 \leq v_2$, then Gv_2 is an upper solution of problem (P_{v_1}). In consequence Theorem 1.2 gives the existence of extremal solutions of (P_{v_1}) in $[\alpha, Gv_2]$. Since Gv_1 is the minimal solution in $[\alpha, \beta]$ of the mentioned problem, it must be $Gv_1 \leq Gv_2$.

Clearly G maps monotone sequences of $[\alpha, \beta]$ into convergent ones. Thus, by Theorem 1.2.2 of [14], G has a least fixed point u_* in $[\alpha, \beta]$, which satisfies

$$u_* = \min\{y \in [\alpha, \beta] : Gy \leq y\}. \tag{6}$$

Obviously, u_* is a solution of problem (3) – (4). Let us see that u_* is the minimal solution of that problem in $[\alpha, \beta]$. If y is a solution of (3) – (4) in $[\alpha, \beta]$ then y is a solution of (P_y) and then, by the definition of G, we have that $Gy \leq y$. Now, by (6), $y \geq u_*$.

The existence of the maximal solution of (3) – (4) in $[\alpha, \beta]$ can be proven by similar arguments, redefining the mapping G in the obvious way. $\square$

Theorem 3.2 *Let $g : I \times \mathbb{R} \times \mathbb{R}^{N+2} \to \mathbb{R}$ such that $g(k, x, y)$ is continuous and non-increasing in $x \in \mathbb{R}$ for every $y \in \mathbb{R}^{N+2}$ and $k \in I$ fixed, and non-decreasing in y for every $x \in \mathbb{R}$ and $k \in I$ fixed. Assume that hypotheses (H_1) and (H_2) are satisfied together with (Q defined in (5))*

(H3) There exists $R > 0$ such that for all constant sequence $y \in \mathbb{R}^{N+2}$ satisfying $0 < R \leq \|y\|$, the following inequality holds

$$M_y = \max_{k \in J} |\, g(k, 0, y)\,| \leq Q\,\|y\|.$$

Then problem (3) – (4) has extremal solutions in any sector $[\alpha, \beta]$. Where α and β are constant sequences such that the sector $[-M_{-R}/Q, M_R/Q] \subset [\alpha, \beta]$.

Proof: For each $v \in \mathbb{R}^{n+2}$ fixed, consider problem (P_v) defined in the proof of the previous result. From Theorem 2.2, we know that problem (P_v) admits a unique solution that we denote by $G\,v$ and that belongs to the sector $[-M_v/Q, M_v/Q]$.

Now, let α and β constant sequences in $\mathbb{R}^{N+2}$ such that, $\alpha \leq -R < 0 < R \leq \beta$ in J. From condition (H_3) we know that $M_\beta \leq Q\,\beta$ and $M_\alpha \leq -Q\,\alpha$.

Let $G\alpha$ and $G\beta$ be the unique solutions of (P_α) and (P_β), respectively. Such solutions, by Theorem 2.2, belong to the sectors $[-M_\alpha/Q, M_\alpha/Q]$ and $[-M_\beta/Q, M_\beta/Q]$, respectively. Thus

$$G\beta(k) \leq M_\beta/Q \leq \beta \quad \text{and} \quad G\alpha(k) \geq -M_\alpha/Q \geq \alpha, \ k \in J$$

Let's see that $G\alpha \leq G\beta$ in J.

Assume on the contrary, from the boundary conditions of such solutions, there exists $k_0 \in I$ such that

$$(G\alpha - G\beta)(k_0 + 1) = \max_{k \in J} (G\alpha - G\beta)(k) > 0.$$

As consequence $\Delta(G\alpha - G\beta)(k_0) \geq 0 \geq \Delta(G\alpha - G\beta)(k_0 + 1)$. Thus, condition (H_1) shows that $-\Delta[\phi(\Delta G\alpha(k_0))] + \Delta[\phi(\Delta G\beta(k_0))] \geq 0$

However

$$-\Delta[\phi(\Delta G\alpha(k_0))] + q(k_0+1)\,G\alpha(k_0+1) + \Delta[\phi(\Delta G\beta(k_0))] - q(k_0+1)\,G\beta(k_0+1) =$$

$$= g(k_0, G\alpha(k_0 + 1), \alpha) - g(k_0, G\beta(k_0 + 1), \beta) \leq 0.$$

As consequence $-\Delta[\phi(\Delta G\alpha(k_0))] + \Delta[\phi(\Delta G\beta(k_0))] < 0$, which contradicts the previous inequality.

It is easy to verify that for every $\alpha \leq x \leq \beta$ in J, functions $G\alpha$ is a lower solution for (P_x) and $G\beta$ is an upper solution for (P_x). Thus, the unique solution Gx of (P_x) belongs to the sector $[G\alpha, G\beta]$.

Following the proof of the previous result one can verify that mapping $G : [\alpha, \beta] \rightarrow [G\alpha, G\beta] \subset [\alpha, \beta]$ has extremal fixed points u_* and u^* in $[\alpha, \beta]$ that are the extremal solutions of problem $(3) - (4)$ in that interval, and they are lying between $-M_\alpha/Q$ and M_β/Q.

When $-R < \alpha \leq -M_{-R}/Q$ and $M_R/Q \leq \beta < R$, since $[\alpha, \beta] \subset [-R, R]$ and we know that there exist extremal solutions in $[-R, R]$ and such solutions belong to the sector $[-M_{-R}/Q, M_R/Q] \subset [\alpha, \beta]$, we conclude the assertion of the theorem for this new situation.

The cases $[-R, R] \not\subset [\alpha, \beta]$ and $[\alpha, \beta] \not\subset [-R, R]$ hold analogously. $\qquad\square$

Remark 3.3 *Note that, despite in the previous theorem the existence of extremal solutions is guaranteed in any compact set of $\mathbb{R}^{N+2}$, we cannot assure the existence of extremal solutions in $\mathbb{R}^{N+2}$. To see this we can consider the following problem for any ϕ satisfying (H_1) and $N \in \mathbb{N}$:*

$$-\Delta[\phi(\Delta u(k))] + u(k+1) = \min_{k \in J} u(k), \quad k \in I, u(0) = u(N), \Delta u(0) = \Delta u(N).$$

It is clear that $g(k, x, y) \equiv \min_{k \in J} y(k)$ satisfies the hypothesis of Theorem 3.2.

Clearly every constant function is a solution of this problem, that is there is no extremal solutions in $\mathbb{R}^{N+2}$.

References

[1] Agarwal, R. P. and Wong, P., Upper and lower solutions method for higher-order discrete boundary value problems, *Mathematical Inequalities and Applications* **1** (1998), 551–557.

[2] Atici, F. M. and Cabada, A., Existence and uniqueness results for discrete second order periodic boundary value problems, *Computers and Mathematics with Applications*, to appear.

[3] Cabada, A., Extremal Solutions for the difference ϕ – Laplacian Problem with Nonlinear Functional Boundary Conditions, *Computers and Mathematics with Applications* **42** (2001), 593–601.

[4] Cabada, A., Habets, P. and Pouso, R. L., Optimal existence conditions for ϕ – Laplacian equations with upper and lower solutions in the reversed order, *Journal of Differential Equations* **166** (2000), 385–401.

[5] Cabada, A. and Otero–Espinar, V., Optimal existence Results for n-th Order Periodic Boundary Value Difference Problems, *Journal of Mathematical Analysis and Applications* **247** (2000), 67–86 .

[6] Cabada, A. and Otero–Espinar, V., Fixed sign solutions of second order difference equations with Neumann boundary conditions, *Computers and Mathematics with Applications*, to appear.

[7] Cabada, A. and Otero–Espinar, V., Comparison results for n-th Order Periodic Difference Equations, *Nonlinear Analysis* **47** (2001), 2395–2406.

[8] Cabada, A. and Otero–Espinar, V., Existence and comparison results for difference ϕ – Laplacian boundary value problems with lower and upper solutions in the reversed order, *Journal of Mathematical Analysis and Applications* **267** (2002), 501–521.

[9] Cabada, A., Otero–Espinar, V. and Pouso, R. L., Existence and approximation of solutions for discontinuous first order difference problems with nonlinear functional boundary conditions in the presence of lower and upper solutions, *Computers and Mathematics with Applications* **39** (2000), 21–33.

[10] Cabada, A. and Pouso, R. L., Extremal solutions of strongly nonlinear discontinuous second order equations with nonlinear boundary conditions, *Nonlinear Analysis* **42** (2000), 1377–1396.

[11] Cherpion, M., De Coster, C. and Habets, P., Monotone iterative methods for boundary value problems, *Differential and Integral Equations* **12** (1999), 309–338.

[12] Dang, H. and Oppenheimer, S. F., Existence and uniqueness results for some nonlinear boundary value problems, *Journal of Mathematical Analysis and Applications* **198** (1996), 35–48.

[13] Eloe, P. W., A boundary value problem for a system of difference equations, *Nonlinear Analysis* **7** (1983), 813–820.

[14] Heikkilä, S. and Lakshmikantham, V., *Monotone Iterative Techniques for Discontinuous Nonlinear Differential Equations*, Marcel Dekker Inc., New York, 1994.

[15] Zhuang, W., Chen, Y. and Cheng, S. S., Monotone methods for a discrete boundary problem, *Computers and Mathematics with Applications* **32** (1996), 41–49.

Asymptotic Behavior
of Solutions of $x_{n+1} = p + \dfrac{x_{n-1}}{x_n}$

ELIAS CAMOUZIS

The American College of Greece, Deree College
6 Gravias Street, Aghia Paraskevi, 15342 Athens, Greece
E-mail: e_camouzis@yahoo.com

and

RICHARD DEVAULT

Northwestern State University
Natchitoches, LA 71497, USA
E-mail: richarddevault@yahoo.com

Abstract In this paper we investigate the existence, asymptotic behavior and periodic nature of solutions of the equation

$$x_{n+1} = p + \frac{x_{n-1}}{x_n}, \ n = 0, 1, \dots ,$$

where the parameter p and the initial conditions are nonzero real numbers.

Keywords Difference equation, Asymptotic behavior, Periodic solutions

AMS Subject Classification 39A10, 37B55

1 Introduction and preliminaries

Our goal in this paper is to investigate the existence, asymptotic behavior and periodic nature of solutions of the equation

$$x_{n+1} = p + \frac{x_{n-1}}{x_n}, \ \ n = 0, 1, \dots , \tag{1}$$

where the parameter p is a nonzero real number and the initial conditions are arbitrary nonzero real numbers.

Eq.(1) is investigated in [1] under the assumption that the parameter p is a nonnegative real number and the initial conditions x_{-1} and x_0 are arbitrary positive real numbers. The case $p = -1$, which is the equation

$$x_{n+1} = -1 + \frac{x_{n-1}}{x_n}, \ \ n = 0, 1, \dots , \tag{2}$$

0-415-31675-8/04/$0.00+$1.50

was considered in [2]. In this case it was shown that Eq.(2) has a unique three cycle, and a region of existence of solutions of Eq.(2) was also given. In [2] several open problems about the existence and behavior of solutions of Eq.(1) were posed. In this paper we address open problem 3 (a) and 3 (b) of [2], which we restate.

Open Problem 3 *(a) Find the good set G of Eq.(1), that is, the set of all initial conditions $(x_{-1}, x_0) \in R^2$ through which Eq.(1) is well defined for all $n \geq 0$.*
(b) Find the basin of attraction of the equilibrium $\bar{x} = 1 + p$ of Eq.(1) when $p < -2$ or when $p > 1$.

A **positive semicycle** of a solution $\{x_n\}$ of Eq.(1) consists of a "string" of terms $\{x_l, x_{l+1}, \ldots, x_m\}$, all greater than or equal to $\bar{x}$, with $l \geq -1$ and $m \leq \infty$ and such that either $l = -1$, or $l > -1$ and $x_{l-1} < \bar{x}$ and either $m = \infty$, or $m < \infty$ and $x_{m+1} < \bar{x}$.

A **negative semicycle** of a solution $\{x_n\}$ of Eq.(1) consists of a "string" of terms $\{x_l, x_{l+1}, \ldots, x_m\}$, all less than $\bar{x}$, with $l \geq -1$ and $m \leq \infty$ and such that either $l = -1$, or $l > -1$ and $x_{l-1} \geq \bar{x}$ and either $m = \infty$, or $m < \infty$ and $x_{m+1} \geq \bar{x}$.

2   Periodic solutions of equation (1)

In [1] it was seen that Eq.(1) has infinitely many period two solutions. Concerning periodic solutions of higher period we have the following theorems which we state without a proof.

Theorem 2.1 *Equation (1) possesses the "unique" three-cycle*

$$\ldots, \alpha, \beta, \gamma, \alpha, \beta, \gamma, \ldots$$

where α, β, and γ are the three real and distinct roots of the cubic equation

$$x^3 + (p^2 - p + 1)x^2 + (-1 - p^3)x + p - p^2 - 1 = 0 \tag{3}$$

ordered as follows: α is any one of the three roots of Eq.(3),

$$\beta = \frac{-p\alpha(\alpha - p)}{1 - \alpha(\alpha - p)}, \text{ and } \gamma = \frac{-p\beta(\beta - p)}{1 - \beta(\beta - p)}.$$

Theorem 2.2 *Suppose that $p < -1$ or $p > 1$. Then Eq.(1) possesses the "unique" four-cycle*

$$\ldots, \alpha, \beta, \gamma, \delta, \alpha, \beta, \gamma, \delta, \ldots$$

where α, β, γ, and δ are the four real and distinct roots of the quartic equation

$$(p + 1)x^4 + (p^3 + p^2 + p + 1)x^3 + (1 - p - p^3 - p^4)x^2$$

$$+ (1 - 2p - 2p^3 - p^4)x + 1 + p^2 = 0.$$

3   Existence and asymptotic behavior of solutions of equation (1)

In this section we shall exhibit a region of existence for solutions of Eq.(1) in the case where $p < -2$. In this region every solution of Eq.(1) is well defined for all $n \geq 0$ and converges to the equilibrium solution $\bar{x} = 1 + p$.

Set

$$f(x) = p + \frac{x}{1+p}, \quad g(u,v) = p + \frac{u^2}{pu+v}, \quad h(x) = f(f(x)). \tag{4}$$

The following lemma, which we state without proof, will be useful in the sequel.

Lemma 3.1 *For the functions given in (4) the following are true :*

$$f(x) = g(x,x) \ \ for \ x \neq 0, \ \ and \ \ h(1+p) = f(1+p) = 1+p, \tag{5}$$

$$f'(x) < 0 \ \ and \ \ 0 < h'(x) < 1 \ \ for \ all \ real \ numbers \ x, \tag{6}$$

$$x < h(x) < 1+p < f(x) \ \ for \ \ x < 1+p, \tag{7}$$

$$f(x) < 1+p < h(x) < x \ \ for \ \ x > 1+p. \tag{8}$$

Furthermore, if

$$-\frac{p(1+p)}{2} \leq v \leq u < 1+p \tag{9}$$

or

$$1+p < u \leq v \leq \frac{p}{2} \tag{10}$$

then

$$g_u(u,v) < 0 \ \ and \ \ g_v(u,v) < 0. \tag{11}$$

We define the sequences $\{U_n\}_{n=1}^{\infty}$ and $\{L_n\}_{n=1}^{\infty}$ as follows:

$$L_1 = -\frac{p(1+p)}{2} < 1+p < \frac{p}{2} = f(L_1) = U_1 \tag{12}$$

and for $n = 1, 2, \dots$,

$$L_{n+1} = h(L_n) = f(U_n) \ \ and \ \ U_{n+1} = h(U_n) = f(L_{n+1}). \tag{13}$$

In view of (7),(8),(12),(13), we have

$$L_2 = f(U_1) < 1+p \ \ and \ \ U_2 = f(L_2) > 1+p.$$

Using induction we have

$$L_{n+1} < 1+p \ \ and \ \ U_{n+1} > 1+p \ \ for \ \ n = 1, 2, \dots . \tag{14}$$

Lemma 3.2 *The sequences* $\{L_n\}_{n=1}^{\infty}$, $\{U_n\}_{n=1}^{\infty}$, *have finite limits. In fact*

$$\lim_{n\to\infty} L_n = \lim_{n\to\infty} U_n = 1 + p. \tag{15}$$

Proof. For each $n = 1, 2, \ldots$, and in view of (7),(8),(13),(14), we have

$$L_n < L_{n+1} = h(L_n) < 1 + p \ \text{ and } \ 1 + p < U_{n+1} = h(U_n) < U_n. \tag{16}$$

Let $L = \lim_{n\to\infty} L_n$ and $U = \lim_{n\to\infty} U_n$. By taking limits in (16), as $n \to \infty$, we have

$$L = h(L) \leq 1 + p \ \text{ and } \ U = h(U) \geq 1 + p. \tag{17}$$

In view of (7),(8), (17) implies

$$L = 1 + p \ \text{ and } \ U = 1 + p.$$

The proof of Lemma 3.2 is complete.

Lemma 3.3 *Suppose that* $\{x_n\}$ *is a nontrivial solution of Eq.(1), for which there exists* $N > 0$, *and a positive integer* i, *such that*

$$x_{N-1}, x_N \in [L_i, 1 + p] \tag{18}$$

or

$$x_{N-1}, x_N \in [1 + p, U_i]. \tag{19}$$

Then $x_n \in [L_i, U_i]$ *for all* $n \geq N - 1$, *and all the semicycles of the solution, starting with the semicycle which contains the terms* x_{N-1}, x_N, *have two or three terms. Moreover, the extreme in a semicycle occurs in the first term if it is of length two and in the second term if it is of length three.*

Proof. We consider the following four cases:
Case 1:

$$1 + p \leq x_{N-1} \leq x_N \leq U_i < -1. \tag{20}$$

In view of (20), Eq.(1) implies

$$1 + p \leq x_{N+1} = p + \frac{x_{N-1}}{x_N} \leq p + \frac{1+p}{x_N} = \frac{(1+p-x_N)(1+x_N)}{x_N} + x_N \leq$$

$$x_N \leq U_i. \tag{21}$$

In view of (4),(5),(6),(10),(11),(13),(16),(21), Eq.(1) implies

$$p + 1 \geq x_{N+2} = p + \frac{x_N}{x_{N+1}} \geq p + \frac{x_N}{1+p} = f(x_N) \geq f(U_i) = L_{i+1} > L_i$$

and

$$1 + p \geq x_{N+3} = p + \frac{x_{N+1}}{x_{N+2}} = p + \frac{x_{N+1}}{p + \frac{x_N}{x_{N+1}}} = g(x_{N+1}, x_N) \geq$$

$$g(U_i, U_i) = f(U_i) = L_{i+1} > L_i.$$

Case 2:
$$1 + p \le x_N < x_{N-1} \le U_i. \tag{22}$$

In view of (4),(5),(6),(10),(11),(13),(16),(22), Eq.(1) implies

$$p + 1 > x_{N+1} = p + \frac{x_{N-1}}{x_N} \ge p + \frac{x_{N-1}}{1+p} = f(x_{N-1}) \ge f(U_i) = L_{i+1} > L_i$$

and

$$1 + p > x_{N+2} = p + \frac{x_N}{x_{N+1}} = p + \frac{x_N}{p + \frac{x_{N-1}}{x_N}} = g(x_N, x_{N-1}) \ge$$

$$g(U_i, U_i) = f(U_i) = L_i.$$

Case 3:
$$L_i \le x_N \le x_{N-1} \le 1 + p < -1. \tag{23}$$

In view of (23), Eq.(1) implies

$$1 + p \ge x_{N+1} = p + \frac{x_{N-1}}{x_N} \ge p + \frac{1+p}{x_N} = \frac{(1 + p - x_N)(x_N + 1)}{x_N} + x_N$$

$$\ge x_N \ge L_i. \tag{24}$$

In view of (4),(5),(6),(9),(11),(13),(24), Eq.(1) implies

$$1 + p \le x_{N+2} = p + \frac{x_N}{x_{N+1}} \le p + \frac{x_N}{1+p} = f(x_N) \le f(L_1) = U_i$$

and

$$1 + p \le x_{N+3} - p + \frac{x_{N+1}}{x_{N+2}} = p + \frac{x_{N+1}}{p + \frac{x_N}{x_{N+1}}} = g(x_{N+1}, x_N) \le$$

$$g(L_i, L_i) = f(L_i) = U_i.$$

Case 4:
$$L_i \le x_{N-1} < x_N \le 1 + p. \tag{25}$$

In view of (4),(5),(6),(9),(11),(13),(25), Eq.(1) implies

$$1 + p < x_{N+1} = p + \frac{x_{N-1}}{x_N} \le p + \frac{x_{N-1}}{1+p} = f(x_{N-1}) \le f(L_i) = U_i$$

and

$$1 + p < x_{N+2} = p + \frac{x_N}{x_{N+1}} = p + \frac{x_N}{p + \frac{x_{N-1}}{x_N}} = g(x_N, x_{N-1}) \le$$

$$g(L_i, L_i) = f(L_i) = U_i.$$

In Cases 1,3, it follows easily that $x_{N-2} < 1 + p$, $x_{N-2} \geq 1 + p$, respectively. In Cases 2,4, if $x_{N-2} \geq 1 + p$, $x_{N-2} < 1 + p$, it would follow easily that $x_{N-3} < 1 + p$, $x_{N-3} \geq 1 + p$, respectively. The proof of Lemma 3.3 follows by induction.

For each solution $\{x_n\}$ of Eq.(1), which satisfies (18) or (19), and for each $i, j = 1, 2, \ldots$, we define:

$$m_i = \text{the mimimum value of } \{x_n\} \text{ on the } i\text{-th negative semicycle} \qquad (26)$$

and

$$M_j = \text{the maximum value of } \{x_n\} \text{ on the } j\text{-th positive semicycle.} \qquad (27)$$

It is easy to see that

$$\liminf_{i \to \infty} m_i = \liminf_{n \to \infty} x_n \text{ and } \limsup_{j \to \infty} M_j = \limsup_{n \to \infty} x_n. \qquad (28)$$

Remark 3.4 *We consider as the first positive semicycle the one following the first negative semicycle.*

Lemma 3.5 *Suppose that $\{x_n\}$ is a nontrivial solution of Eq.(1). If for some $N > 0$, and a positive integer i, (18) or (19) holds, then*

$$1 + p \leq M_k \leq U_k \text{ and } L_{k+1} \leq m_{k+1} \leq 1 + p \qquad (29)$$

for each $k = i, i + 1, \ldots$.

Proof. We consider the following four cases:
Case 1: In this case we assume (18) holds, and that the i-th positive semicycle has two terms. Furthermore in view of Lemma 3.3, we have

$$M_i = x_{N_i} \text{ and } m_i = x_{N_i - 2} \qquad (30)$$

where $N - 1 \leq N_i - 2 \leq N$. In view of (4),(6),(13),(18),(26),(30), Eq.(1) implies

$$1 + p \leq M_i = p + \frac{m_i}{x_{N_i - 1}} \leq p + \frac{m_i}{1 + p} = f(m_i) \leq f(L_i) = U_i.$$

Case 2: In this case we assume (18) holds, and that the i-th positive semicycle has three terms. Furthermore in view of Lemma 3.3, we have

$$M_i = x_{N_i + 1} \text{ and } m_i = x_{N_i - 2} \qquad (31)$$

where $N - 1 \leq N_i - 2 \leq N$. In view of (4),(5),(9),(11),(13),(18),(31), Eq.(1) implies

$$1 + p \leq M_i = p + \frac{x_{N_i - 1}}{x_{N_i}} = p + \frac{x_{N_i - 1}}{p + \frac{x_{N_i - 2}}{x_{N_i - 1}}} = g(x_{N_i - 1}, x_{N_i - 2}) \leq$$

$$g(L_i, L_i) = f(L_i) = U_i.$$

Case 3: In this case we assume (19) holds, and that the (i+1)-th negative semicycle has two terms. Furthermore in view of Lemma 3.3, we have

$$m_{i+1} = x_{N_i} \text{ and } M_i = x_{N_i-2} \tag{32}$$

where $N - 1 \le N_i - 2 \le N$. In view of (4),(6),(13),(19),(27),(32), Eq.(1) implies

$$1 + p \ge m_{i+1} = p + \frac{M_i}{x_{N_i-1}} \ge p + \frac{M_i}{1+p} = f(M_i) \ge f(U_i) = L_{i+1}.$$

Case 4: In this case we assume (19) holds, and that the (i+1)-th negative semicycle has three terms. Furthermore in view of Lemma 3.3, we have

$$m_{i+1} = x_{N_i+1} \text{ and } M_i = x_{N_i-2} \tag{33}$$

where $N - 1 \le N_i - 2 \le N$. In view of (4),(5),(10),(11),(13),(19),(33), Eq.(1) implies

$$1 + p \ge m_{i+1} = p + \frac{x_{N_i-1}}{x_{N_i}} = p + \frac{x_{N_i-1}}{p + \frac{x_{N_i-2}}{x_{N_i-1}}} = g(x_{N_i-1}, x_{N_i-2}) \ge$$

$$g(U_i, U_i) = f(U_i) = L_{i+1}.$$

The proof of Lemma 3.5 follows by induction.

Theorem 3.6 *Let $\{x_n\}$ be a solution of Eq.(1). Suppose that one of the following statements is true.*

   (i) *There exists $N > 0$ such that $x_{N-1}, x_N \in [L_1, 1+p]$.*

   (ii) *There exists $N > 0$ such that $x_{N-1}, x_N \in [1+p, U_1]$.*

Then

$$\lim_{n \to \infty} x_n = 1 + p.$$

Proof. If either i) or ii) is true, from Lemma 3.5, we have that (29) holds. By taking limits in (29) as $k \to \infty$, and in view of (15), it follows

$$1 + p = \lim_{k \to \infty} L_{k+1} \le \liminf_{k \to \infty} m_{k+1} \le \limsup_{k \to \infty} M_k \le \lim_{k \to \infty} U_k = 1 + p. \tag{34}$$

In view of (28), (34) implies

$$\lim_{n \to \infty} x_n = 1 + p.$$

The proof of Theorem 3.6 is complete.

4   An invariant interval for solutions of eq. (1)

An **invariant interval** for Eq.(1) is an interval I with the property that if two consecutive terms of the solution fall in I then all the subsequent terms of the solution also belong to I.

In this section we find an invariant interval for solutions of Eq.(1), assuming that $p < -3$. We also prove that every solution that falls within that interval converges to the equilibrium solution $\bar{x} = 1 + p$. It is important to note that $[L_1, U_1]$ is not an invariant interval.

Let L, U with $L < U$ be the two distinct solutions of the equation

$$x^2 - (p-1)x - (p-1) = 0. \tag{35}$$

Then it follows that

$$L_1 < L < 1 + p < U. \tag{36}$$

Theorem 4.1 *Suppose that $\{x_n\}$ is a solution of Eq.(1), for which there exists $N > 0$ such that x_{N-1} and x_N are both elements of the interval $[L, U]$. Then*

$$x_n \in [L, U] \text{ for all } n \geq N - 1 \tag{37}$$

and

$$\lim_{n \to \infty} x_n = 1 + p. \tag{38}$$

Proof. If x_{N-1} and x_N are both elements of the interval $[L, U]$ then in view of (35) and (36), Eq.(1) implies

$$L = p + \frac{U}{L} \leq x_{N+1} = p + \frac{x_{N-1}}{x_N} \leq p + \frac{L}{U} = U. \tag{39}$$

The proof of (37) follows by induction.

Next, if the solution $\{x_n\}$ of Eq.(1), is increasing, in view of (37), (38) follows. Otherwise we may assume that $x_N < x_{N-1}$. Then in view of (37), Eq.(1) implies $x_{N+1} \in [L, 1 + p)$. Furthermore, either $x_N \in [L, 1 + p)$ or $x_N \in [1 + p, U]$ and $x_{N+2} \in [L, 1 + p)$, which in view of (36) implies that (18) is satisfied, and so the proof of Theorem 4.1 follows from Theorem 3.6 .

References

[1] A. M. Amleh, D. A. Georgiou, E. A. Grove, and G. Ladas, "On the Recursive Sequence $x_{n+1} = \alpha + \frac{x_{n-1}}{x_n}$", *Journal of Mathematical Analysis and Applications* 233 (1999), 790–798.

[2] E. Camouzis, R. DeVault, and G. Ladas, "On the Recursive Sequence $x_{n+1} = p + \frac{x_n}{x_{n-1}}$", *Journal of Difference Equations and Applications*, Vol. 7, (2001), 477–482 .

Limit Behavior for Quasilinear Difference Equations

MARIELLA CECCHI

Department of Electronics and Telecommunication
University of Florence
Via S. Marta 3, 50139 Florence, Italy
E-mail: cecchi@det.unifi.it

ZUZANA DOŠLÁ [1]

Department of Mathematics, Masaryk University
Janáčkovo nám. 2a, 66295 Brno, Czech Republic
E-mail: dosla@math.muni.cz

and

MAURO MARINI

Department of Electronics and Telecommunication
University of Florence
Via S. Marta 3, 50139 Florence, Italy
E-mail: marini@ing.unifi.it

Abstract The nonlinear difference equation

$$\Delta(a_n \Phi_p(\Delta x_n)) = b_n f(x_{n+1}), \quad \Phi_p(u) = |u|^{p-2}u \quad p > 1,$$

where $\{a_n\}$, $\{b_n\}$ are positive real sequences for $n \geq 1$, $f : \mathbb{R} \to \mathbb{R}$ is continuous with $uf(u) > 0$ for $u \neq 0$, is considered. Solutions approaching zero are classified according to the limit behavior of their quasidifference, and necessary or sufficient conditions for their existence are given.

Keywords Decaying solution, Limit behavior, Quasilinear difference equations.

AMS Subject Classification 39A10

---

[1]Supported by the Czech Grant Agency, grant 201/01/0079.

1   Introduction

Consider the second order nonlinear difference equation

$$\Delta(a_n \Phi_p(\Delta x_n)) = b_n f(x_{n+1}), \tag{1}$$

where $\{a_n\}$ and $\{b_n\}$ are positive real sequences for $n \geq 1$, $f : \mathbb{R} \to \mathbb{R}$ is a continuous function such that $uf(u) > 0$ for $u \neq 0$, and $\Phi_p(u) = |u|^{p-2}u$ with $p > 1$. Recall that the left-hand-side in (1) is the one-dimensional discrete analogue of the p-Laplacian $\Delta_p(u) = div(|\nabla u|^{p-2}\nabla u)$, that models a variety of physical problems, such as, for instance, flow through porous media, nonlinear elasticity and reaction-diffusion problems. We refer to [6] for an extensive bibliography. Equation (1) and its special cases have been deeply investigated; see, e.g., [1, 2, 3, 4, 5, 7, 8].

It is easy to prove that any nontrivial solution $\{x_n\}$ of (1) is eventually increasing or decreasing and belongs to one of the two classes:

$$\mathbb{M}^+ = \{\{x_n\} \text{ solution of } (1) : \exists n_x \geq 1 : \ x_n \Delta x_n > 0 \text{ for } n \geq n_x\},$$
$$\mathbb{M}^- = \{\{x_n\} \text{ solution of } (1) : \ x_n \Delta x_n < 0 \text{ for } n \geq 1\},$$

(see [2, Lemma 1]; see also [7], [8] for special cases). In [2] the qualitative behavior of solutions in the class $\mathbb{M}^-$ is investigated and two problems suggested in [7, p.297] are resolved. More precisely the following two subclasses

$$\mathbb{M}_B^- = \{\{x_n\} \in \mathbb{M}^- : \ \lim_{n \to \infty} x_n = \ell_x \neq 0\},$$
$$\mathbb{D} = \{\{x_n\} \in \mathbb{M}^- : \ \lim_{n \to \infty} x_n = 0\},$$

are considered: solutions in $\mathbb{M}_B^-$ and $\mathbb{D}$ are called *asymptotically constant solutions* and *decaying solutions*, respectively. Both classes $\mathbb{M}_B^-, \mathbb{D}$ are characterized in terms of the convergence or divergence of two series

$$Y_1 = \lim_{m \to \infty} \sum_{n=1}^{m} \Phi_{p^*}\left(\sum_{k=1}^{n} \frac{b_k}{a_n}\right), \quad Y_2 = \lim_{m \to \infty} \sum_{n=1}^{m} \Phi_{p^*}\left(\sum_{k=n}^{m} \frac{b_k}{a_n}\right),$$

where $p^* = \frac{p}{p-1}$.

In particular the following result is proved in [2]:

Theorem A. Consider equation (1). Then it holds that $\mathbb{M}^- \neq \emptyset$. In addition:
(a) If $Y_2 = \infty$, then $\mathbb{M}^- = \mathbb{D} \neq \emptyset$ and $\mathbb{M}_B^- = \emptyset$.
(b) If $Y_2 < \infty$, then $\mathbb{M}_B^- \neq \emptyset$.
(c) If $Y_1 < \infty$, $Y_2 < \infty$, then $\mathbb{D} \neq \emptyset$ and $\mathbb{M}_B^- \neq \emptyset$.

We note that no monotonicity assumptions on f are needed for the validity of Theorem A, as some physical applications suggest (see, e.g., [6]).

In the discretization of certain elliptic problems with free boundaries, a crucial role is played by the limit behavior of decaying solutions. This fact suggests considering a more complete classification of decaying solutions. In view of (1), the quasidifference

$$x_n^{[1]} = a_n \Phi_p(\Delta x_n) \tag{2}$$

of any solution $\{x_n\} \in \mathbb{M}^-$ is either positive decreasing or negative increasing. Hence $\lim_{n \to \infty} x_n^{[1]}$ must be finite and the class $\mathbb{D}$ can be divided, *a priori*, into the following two subsets:

$$\mathbb{D}_R = \{\{x_n\} \in \mathbb{M}^- : \lim x_n = 0, \lim x_n^{[1]} \neq 0\},$$
$$\mathbb{D}_S = \{\{x_n\} \in \mathbb{M}^- : \lim x_n = 0, \lim x_n^{[1]} = 0\}.$$

Solutions in $\mathbb{D}_R$ and $\mathbb{D}_S$ are called *regularly decaying solutions* and *strongly decaying solutions*, respectively. Clearly, $\mathbb{M}^- = \mathbb{M}_B^- \cup \mathbb{D}_R \cup \mathbb{D}_S$. Necessary and sufficient conditions fully characterizing the existence of regularly/strongly decaying solutions are given and the role of the nonlinearity of f is also investigated.

2 Necessary conditions

Let $q > 1$ and, jointly with Y_1, Y_2, consider the series

$$Y_q = \lim_{m \to \infty} \sum_{n=1}^{m} b_n \Phi_q \left(\sum_{k=n}^{m} \Phi_{p^*} \left(\frac{1}{a_{k+1}} \right) \right),$$

$$Y_3 = \lim_{m \to \infty} \sum_{n=1}^{m} \Phi_{p^*} \left(\frac{1}{a_n} \right), \qquad Y_4 = \lim_{m \to \infty} \sum_{n-1}^{m} b_n.$$

Relations between Y_1, Y_2, Y_3, Y_4, Y_q are described by the following:

Lemma 2.1.
(a) *If $Y_1 < \infty$, then $Y_3 < \infty$.*
(b) *If $Y_2 < \infty$, then $Y_4 < \infty$.*
(c) *If $Y_q < \infty$, then $Y_3 < \infty$.*
(d) *If $Y_q = \infty$, then $Y_3 = \infty$ or $Y_4 = \infty$.*
(e) *$Y_3 < \infty$ and $Y_4 < \infty$ if and only if $Y_2 < \infty$ and $Y_q < \infty$.*

Proof. Claims a), b) – see [2, Lemma 2]. Claim c) follows from the inequality

$$\sum_{n=1}^{m} b_n \Phi_q \left(\sum_{k=n}^{m} \Phi_{p^*} \left(\frac{1}{a_{k+1}} \right) \right) \geq b_1 \Phi_q \left(\sum_{k=1}^{m} \Phi_{p^*} \left(\frac{1}{a_{k+1}} \right) \right)$$

and claim d) from the inequality

$$\sum_{n=1}^{m} b_n \Phi_q \left(\sum_{k=n}^{m} \Phi_{p^*} \left(\frac{1}{a_{k+1}} \right) \right) \leq \left(\sum_{n=1}^{m} b_n \right) \Phi_q \left(\sum_{n=1}^{m} \Phi_{p^*} \left(\frac{1}{a_{n+1}} \right) \right).$$

Claim e) can be proved in a similar way.   □

The next two results give conditions ensuring that the classes $\mathbb{D}_R$ and $\mathbb{D}_S$ are empty.

Theorem 2.2. *The class $\mathbb{D}_R$ is empty if any of the following conditions is satisfied:*

a) $Y_3 = \infty$;

b) *there exists $q > 1$ such that*

$$\liminf_{u \to 0} \frac{f(u)}{\Phi_q(u)} > 0 \tag{3}$$

and $Y_q = \infty$.

Proof. Claim a) follows from [2, Lemma 3] (see also [8]).

Claim b). Assume $Y_3 < \infty$; otherwise the assertion follows from claim a). Without loss of generality, let $\{x_n\}$ be a positive solution of (1) in the class $\mathbb{D}_R$. Then there exists $\ell_x > 0$ such that $x_n^{[1]} < -\ell_x$ and so

$$x_n \geq \Phi_{p*}(\ell_x) \sum_{k=n}^{\infty} \Phi_{p*}\left(\frac{1}{a_k}\right). \tag{4}$$

In view of (3), there exists a positive constant h such that for any $u \in (0, x_0]$ we have $f(u) \geq h\Phi_q(u)$. From this fact and (4) we obtain

$$f(x_{n+1}) \geq h\Phi_q\left(\Phi_{p*}(\ell_x) \sum_{k=n}^{\infty} \Phi_{p*}\left(\frac{1}{a_{k+1}}\right)\right)$$

or, from (1),

$$\Delta(a_n \Phi_p(\Delta x_n)) \geq h_1 b_n \Phi_q\left(\sum_{k=n}^{\infty} \Phi_{p*}\left(\frac{1}{a_{k+1}}\right)\right),$$

where $h_1 = h\Phi_{qp*}(\ell_x)$. By summing from n to ∞ we obtain a contradiction to the boundedness of $x_n^{[1]}$.   □

Theorem 2.3. *The class $\mathbb{D}_S$ is empty if any of following conditions is satisfied:*

a) $Y_2 < \infty$ *and*

$$\limsup_{u \to 0} \frac{f(u)}{\Phi_p(u)} < \infty; \tag{5}$$

b) *there exists $q \geq p$ such that*

$$\limsup_{u \to 0} \frac{f(u)}{\Phi_q(u)} < \infty \tag{6}$$

and $Y_q < \infty$.

Proof. Claim a). Let $\{x_n\} \in \mathbb{D}_S$ and, without loss of generality, assume $x_n > 0$. In view of (5), there exists a positive constant H_1 such that for any $u \in (0, x_0]$ it holds that $f(u) \le H_1 \Phi_p(u)$. Taking into account this fact and by the summation of (1) twice from n to ∞ we obtain

$$x_n = \sum_{k=n}^{\infty} \Phi_{p^*} \left(\frac{1}{a_k} \left(\sum_{i=k}^{\infty} b_i f(x_{i+1}) \right) \right) \le H_1 \Phi_{p^*}(\Phi_p(x_{n+1})) \sum_{k=n}^{\infty} \Phi_{p^*} \left(\frac{1}{a_k} \left(\sum_{i=k}^{\infty} b_i \right) \right).$$

Thus

$$1 < \frac{x_n}{x_{n+1}} \le H_1 \sum_{k=n}^{\infty} \Phi_{p^*} \left(\sum_{i=k}^{\infty} \frac{b_i}{a_k} \right)$$

and we obtain a contradiction as $n \to \infty$.

Claim b). Let $\{x_n\} \in \mathbb{D}_S$ and again, without loss of generality, assume $x_n > 0$. From (2), upon summing from n to ∞, we have

$$-x_n = \sum_{k=n}^{\infty} \Phi_{p^*} \left(\frac{1}{a_k} \right) \Phi_{p^*}(x_k^{[1]}) \ge \Phi_{p^*}(x_n^{[1]}) \sum_{k=n}^{\infty} \Phi_{p^*} \left(\frac{1}{a_k} \right)$$

or

$$\Phi_q(x_{n+1}) \le \Phi_q(\Phi_{p^*}(-x_{n+1}^{[1]})) \Phi_q \left(\sum_{k=n}^{\infty} \Phi_{p^*} \left(\frac{1}{a_{k+1}} \right) \right). \tag{7}$$

Since the sequence $\{-x_i^{[1]}\}$ is positive decreasing, from (7) we obtain

$$\Phi_q(x_{n+1}) \le \Phi_q(\Phi_{p^*}(-x_n^{[1]})) \Phi_q \left(\sum_{k=n}^{\infty} \Phi_{p^*} \left(\frac{1}{a_{k+1}} \right) \right). \tag{8}$$

In view of (6), there exists a positive constant H_2 such that for any $u \in (0, x_0]$ it holds that $f(u) \le H_2 \Phi_q(u)$. From this fact, (1) and (8), we obtain

$$\Delta(a_n \Phi_p(\Delta x_n)) \le H_2 b_n \Phi_q(x_{n+1}) \le H_2 b_n \Phi_q(\Phi_{p^*}(-x_n^{[1]})) \Phi_q \left(\sum_{k=n}^{\infty} \Phi_{p^*} \left(\frac{1}{a_{k+1}} \right) \right)$$

and, by summation from n to ∞,

$$-x_n^{[1]} \le H_2 \sum_{i=n}^{\infty} b_i \Phi_q(\Phi_{p^*}(-x_i^{[1]})) \Phi_q \left(\sum_{k=i}^{\infty} \Phi_{p^*} \left(\frac{1}{a_{k+1}} \right) \right).$$

As already claimed, the sequence $\{-x_i^{[1]}\}$ is positive decreasing; hence

$$\frac{-x_n^{[1]}}{\Phi_q(\Phi_{p^*}(-x_n^{[1]}))} \le H_2 \sum_{i=n}^{\infty} b_i \Phi_q \left(\sum_{k=i}^{\infty} \Phi_{p^*} \left(\frac{1}{a_{k+1}} \right) \right).$$

In view of the fact that $Y_q < \infty$ and

$$\frac{u}{\Phi_q(\Phi_{p^*}(u))} = \frac{1}{|u|^{(q-p)/p-1}},$$

we get a contradiction to the assumption that $\{x_n\} \in \mathbb{D}_S$. $\qquad \square$

3   Main results

In this section we present sufficient conditions for existence of solutions in the classes $\mathbb{D}_R$ and $\mathbb{D}_S$. The following holds:

Theorem 3.1. *The class* $\mathbb{D}_R$ *is not empty if any of following conditions is satisfied:*
  a) $Y_2 < \infty$ *and* $Y_q < \infty$;
  b) there exists $q > 1$ *such that (6) holds and* $Y_q < \infty$.

Proof. Claim a) follows from Theorem 3 and its proof in [2].

Claim b). In view of (6), there exists a positive constant M such that for any $u \in (0, 1]$ it holds that $f(u) \leq M\Phi_q(u)$. Since $Y_q < \infty$, from Lemma 2.1c we have $Y_3 < \infty$. Choose n_0 so large that

$$\sum_{n=n_0}^{\infty} b_n \Phi_q \left(\sum_{k=n+1}^{\infty} \Phi_{p^*} \left(\frac{1}{a_k} \right) \right) < \frac{1}{2M}. \tag{9}$$

Denote by $\ell_{n_0}^{\infty}$ the Banach space of all bounded sequences defined for all integer $n \geq n_0$ and endowed with the topology of the supremum norm. Let Ω be the nonempty subset of $\ell_{n_0}^{\infty}$ given by

$$\Omega = \left\{ \{u_n\} \in \ell_{n_0}^{\infty} : \sum_{j=n}^{\infty} \Phi_{p^*} \left(\frac{1}{2a_j} \right) \leq u_n \leq \sum_{j=n}^{\infty} \Phi_{p^*} \left(\frac{1}{a_j} \right) \right\}.$$

Clearly Ω is bounded, closed, and convex. Now consider the operator

$$T : \Omega \longrightarrow \ell_{n_0}^{\infty}$$

which assigns to any $U = \{u_n\} \in \Omega$ the sequence $W = T(U) = \{w_n\}$ given by

$$w_n = \sum_{j=n}^{\infty} \Phi_{p^*} \left(\frac{1}{a_j} \left(\frac{1}{2} + \sum_{i=j}^{\infty} b_i f(u_{i+1}) \right) \right). \tag{10}$$

Obviously, $w_n \geq \sum_{j=n}^{\infty} 1/\Phi_{p^*}(2a_j)$. In view of (9) and the fact that $f(u) \leq M\Phi_q(u)$, it holds that

$$\sum_{j=n_0}^{\infty} b_j f(u_{j+1}) \leq \sum_{j=n_0}^{\infty} M b_j \Phi_q(u_{j+1}) \leq M \sum_{j=n_0}^{\infty} b_j \Phi_q \left(\sum_{i=j+1}^{\infty} \Phi_{p^*} \left(\frac{1}{a_i} \right) \right) \leq \frac{1}{2}.$$

From the previous inequality, (10), and the monotonicity of Φ_{p^*}, we obtain

$$w_n \leq \sum_{j=n}^{\infty} \Phi_{p^*} \left(\frac{1}{a_j} \right),$$

i.e., $T(\Omega) \subseteq \Omega$. In order to complete the proof it is sufficient to prove the relative compactness of $T(\Omega)$, the continuity of T in Ω, and to apply the Schauder fixed point theorem. As regards the compactness of $T(\Omega)$, by a result in [5, Theorem 3.3], it is sufficient to show that for any $\varepsilon > 0$ there exists $N \geq n_0$ such that $|w_k - w_\ell| < \varepsilon$ whenever $k, \ell \geq N$ for any $W = \{w_n\} \in T(\Omega)$. Without loss of generality assume $k < \ell$. We have

$$|w_k - w_\ell| = \sum_{j=k}^{\ell-1} \Phi_{p^*} \left(\frac{1}{a_j} \left(\frac{1}{2} + \sum_{i=j}^{\infty} b_i f(u_{i+1}) \right) \right) \leq \sum_{j=k}^{\ell-1} \Phi_{p^*} \left(\frac{1}{a_j} \right).$$

Choosing N large, we obtain the assertion. The continuity of T in $T(\Omega)$ can be proved using the discrete analog of the Lebesgue dominated convergence theorem (see, e.g., [2, Lemma A]) and an argument similar to that given in [2, Theorem 2]. The details are left to the reader. $\qquad\Box$

Theorem 3.2. *The class* $\mathbb{D}_S$ *is not empty if there exists* $q > 1$ *such that (3) holds and* $Y_2 = Y_q = \infty$.

Proof. From Theorem A we have $\mathbb{M}_B = \emptyset$ and $\mathbb{D} \neq \emptyset$. By Theorem 2.2 it holds that $\mathbb{D}_R = \emptyset$ and the conclusion follows. $\qquad\Box$

Applying Theorems 2.2, 2.3, 3.1 and 3.2 to the equation

$$\Delta(a_n \Phi_p(\Delta x_n)) = b_n \Phi_q(x_{n+1}) \tag{11}$$

and using Lemma 2.1 we get the following result.

Corollary 3.3. *For equation (11) the following holds:*
(a) If $Y_2 - \infty$ *and* $Y_q = \infty$ *then* $\mathbb{M}_B^- = \emptyset$, $\mathbb{D}_R = \emptyset$ *and* $\mathbb{D}_S \neq \emptyset$.
(b) If $Y_2 = \infty$ *and* $Y_q < \infty$ *then* $\mathbb{M}_B^- = \emptyset$ *and* $\mathbb{D}_R \neq \emptyset$.
(c) If $Y_2 < \infty$ *and* $Y_q = \infty$ *then* $\mathbb{M}_B^- \neq \emptyset$ *and* $\mathbb{D}_R = \emptyset$.
(d) If $Y_2 < \infty$ *and* $Y_q < \infty$ *then* $\mathbb{M}_B^- \neq \emptyset$ *and* $\mathbb{D}_R \neq \emptyset$.
In addition, for $p \leq q$ *it holds that* $\mathbb{D}_S = \emptyset$ *in claims b), c) and d).*

Concluding remarks.
(1) The conclusions of Theorems 2.2 and 3.1 remain valid under the following slightly modified assumptions: in Theorem 2.2b, the assumptions (3) and $Y_q = \infty$ can be replaced by assuming f is nondecreasing and

$$Y_f(\lambda) \equiv \lim_{m \to \infty} \sum_{n=1}^{m} b_n f \left(\lambda \sum_{k=n}^{m} \Phi_{p^*} \left(\frac{1}{a_{k+1}} \right) \right) = \infty$$

for every positive constant λ. Similarly in Theorem 3.1b, the assumptions (6) and $Y_q < \infty$ can be replaced by assuming f is nondecreasing and $Y_f(\lambda_0) < \infty$ for some $\lambda_0 > 0$. The details are left to the reader.

(2) In the linear case, an important role in the qualitative theory is played by the so-called recessive solutions, i.e., solutions $y = \{y_k\}$ such that

$$\lim_{k \to \infty} \frac{y_k}{x_k} = 0$$

for any solution $x = \{x_k\}$ which is linearly independent of y. Our results can be used to extend such a property to the equation (11) with $p = q$ (so-called half-linear equation). This will be given elsewhere.

Acknowledgment. The authors thank the referee for his useful comments.

References

[1] R.P. Agarwal, *Difference Equations and Inequalities*, 2nd Edition, Pure Appl. Math. 228, Marcel Dekker, New York, 2000.

[2] M. Cecchi, Z. Došlá, M. Marini, Positive decreasing solutions of quasilinear difference equations, *Computers Math. Appl.* **42** (2001), 1401–1410.

[3] M. Cecchi, Z. Došlá, M. Marini, Unbounded solutions of quasilinear difference equations, *Computers Math. Appl.*, to appear.

[4] S. Cheng, H.J. Li, W.T. Patula, Bounded and zero convergent solutions of second order difference equations, *J. Math. Anal. Appl.* **141** (1989), 463–483.

[5] S.S. Cheng, W.T. Patula, An existence theorem for a nonlinear difference equation, *Nonlinear Anal.* **20** (1993), 193–203.

[6] J. I. Díaz, *Nonlinear partial differential equations and free boundaries*, Vol.I: Elliptic equations, Pitman Advanced Publ., **106**, (1985).

[7] E. Thandapani, M. M. S. Manuel, J. R. Graef, P. Spikes, Monotone properties of certain classes of solutions of second order nonlinear difference equations, *Computers Math. Appl.* **36** (1998) No.10–12, 291–297.

[8] E. Thandapani, K. Ravi, Bounded and monotone properties of solutions of second-order quasilinear difference equations, *Computers Math. Appl.* **38** (1999), 113–121.

Difference Equations in the Qualitative Theory of Delay Differential Equations

JAN ČERMÁK [1]

Institute of Mathematics, Faculty of Mechanical Engineering
Brno University of Technology
Technická 2, 616 69 Brno, Czech Republic
E-mail: cermakh@um.fme.vutbr.cz

Abstract In this contribution we present a brief survey of some connections between difference equations and delay differential equations. As a new result, we derive the asymptotic bound of all solutions of a certain linear delay differential equation by means of a solution of an appropriate linear difference equation.

Keywords Delay differential and difference equation, Asymptotic behavior of solutions.

AMS Subject Classification 34K25, 39A99

1   Introduction

Difference equations appear in the qualitative theory of delay differential equations in many contexts. The close connection between both types of equations can be observed especially in the oscillation and asymptotic theory.

Many oscillation results concerning delay difference equations seem to be a discrete analogue of the corresponding continuous case (the survey of the relevant results can be found in [4]). The role of difference equations in the asymptotic theory of delay differential equations consists in several aspects. We mention at least a few examples of these connections.

The estimates of solutions of certain delay differential equations can be derived in terms of solutions of some auxiliary difference (or functional) equations (see, e.g., [3], [6] or [8]). The methods utilized in the problem of the asymptotic equilibrium for delay differential equations have been successfully employed in [5] to obtain the sufficient conditions for the existence of asymptotic equilibrium of certain delay difference equations. Some asymptotic problems concerning vector delay differential equations can be reduced to the discussion of the positive solutions of the auxiliary vector difference equations (see [1] and [2]).

---

[1]The research was supported by the grant # 201/01/0079 of the Czech Grant Agency.

The main goal of this paper is to demonstrate another application of difference equations to the asymptotic theory of differential equations with an unbounded lag. Using the simple asymptotic property of an auxiliary difference equation we derive new asymptotic estimates for all solutions of the equation

$$\dot{x}(t) = -c(t)[x(t) - \gamma x(\lambda t)] + f(t), \qquad t \geq 0, \tag{1}$$

where $\gamma \neq 0$, $0 < \lambda < 1$ are real constants and $c > 0$, f are continuous functions fulfilling some additional requirements.

2   Preliminaries

By a solution of (1) we understand a real valued continuous function x defined in some interval $[t_0, \infty)$, where $t_0 \geq 0$ and satisfying (1) for all $t \geq \lambda^{-1}t_0$. Similarly we introduce the notion of a solution for other delay equations occurring in this paper.

To explain the main idea we look in an intuitive manner at the following special case of (1). Consider the equation

$$\dot{x}(t) = -c[x(t) - x(\lambda t)] + f(t), \tag{2}$$

where $c > 0$, $0 < \lambda < 1$ are constants and f is a continuously differentiable function fulfilling $f(t) = O(t^\beta)$, $\dot{f}(t) = O(t^{\beta-1})$ as $t \to \infty$ for a real parameter β. Setting

$$s = \frac{\log t}{\log \lambda^{-1}}, \qquad w(s) = t^{-\delta}x(t) \tag{3}$$

we obtain the differential-difference equation

$$w'(s) = \log \lambda [c\lambda^{-s} + \delta]w(s) + c\lambda^{\delta-s} \log \lambda^{-1} w(s-1) + f(\lambda^{-s})\lambda^{\delta s - s} \log \lambda^{-1}. \tag{4}$$

From equation (4) we show that there is a $\delta > \beta$ such that w and w' are bounded as $s \to \infty$. Equation (4) is then considered as a difference equation

$$w(s) - \lambda^\delta w(s - 1) = h(s), \tag{5}$$

where $h(s) = O(\lambda^{\mu s})$ as $s \to \infty$, $\mu = \min(1, \delta - \beta) > 0$. Then the asymptotic behavior of (5) determines via substitution (3) the asymptotic behavior of (2).

3   The asymptotic behavior of solutions

We start with the simple statement describing the asymptotics of solutions of the auxiliary linear difference equation (for the similar situation see [7]).

Lemma 3.1 *Let w be a solution of the difference equation*

$$w(s) = g(s)w(s - 1) + h(s), \qquad s \geq s_0, \tag{6}$$

where $g, h \in C([s_0, \infty))$, $|g(s)| \le r$ *for all $s \ge s_0$ and $h(s) = O(\exp\{-\rho s\})$ as $s \to \infty$ for a suitable real $\rho > 0$. Then*

$$
\begin{array}{llll}
w(s) = O(r^s) & as\ s \to \infty & for\ \rho > \log r^{-1}, & (7i) \\
w(s) = O(sr^s) & as\ s \to \infty & for\ \rho = \log r^{-1}, & (7ii) \\
w(s) = O(\exp\{-\rho s\}) & as\ s \to \infty & for\ \rho < \log r^{-1}. & (7iii)
\end{array}
$$

Proof. Denote $s_j = s_0 + j$, $I_j = [s_{j-1}, s_j]$, $M_j = \sup\{|w(s)|,\ s \in I_j\}$ and let $s \in I_{j+1}$, $j = 1, 2, \ldots$. Then

$$|w(s)| \le rM_j + K_1 \exp\{-\rho s_j\},$$

$K_1 > 0$ is a suitable real, i.e.,

$$M_{j+1} \le rM_j + K_1 \exp\{-\rho s_j\}.$$

Repeating this procedure we obtain

$$M_{j+1} \le M_1 r^j + K_2 r^j \sum_{k=1}^{j} \frac{r^k}{\exp\{\rho k\}}, \qquad K_2 - K_1 \exp\{-\rho s_0\}, \qquad (8)$$

$j = 1, 2, \ldots$. Now we distinguish three cases with the respect to the sign of $\rho + \log r$.

(i) If $\rho + \log r > 0$, then

$$M_{j+1} \le M_1 r^j + K_3 r^j, \qquad K_3 = \frac{K_2}{r \exp\{\rho\} - 1},$$

i.e., M_j / r^j is bounded as $j \to \infty$. This implies the property (7i).

(ii) If $\rho + \log r = 0$, then

$$M_{j+1} \le M_1 r^j + K_2 j r^j,$$

i.e., $M_j / (j r^j)$ is bounded as $j \to \infty$ and (7ii) holds.

(iii) If $\rho + \log r < 0$, then rewrite (8) as

$$M_{j+1} \le M_1 r^j + K_4(\exp\{-\rho j\} - r^j), \qquad K_4 = \frac{K_2}{1 - r \exp\{\rho\}},$$

i.e.,

$$\frac{M_{j+1}}{\exp\{-\rho j\}} \le M_1 r^j \exp\{\rho j\} + K_4(1 - r^j \exp\{\rho j\}).$$

Letting $j \to \infty$ we can easily verify the boundedness of $M_j / \exp\{-\rho j\}$ as $j \to \infty$ and this proves (7iii). $\square$

The following auxiliary asymptotic estimate concerns the delay differential equations having more general form than (1).

Lemma 3.2 *Let x be a solution of the equation*

$$\dot{x}(t) = -a(t)x(t) + b(t)x(\lambda t) + q(t), \qquad t \in I = [0, \infty), \tag{9}$$

where $0 < \lambda < 1$, $a, b, q \in C(I)$, $a(t) \geq K/t^{\omega}$, $0 < |b(t)| \leq La(t)$ for all $t \in I$ and suitable reals $0 \leq \omega < 1$, $K > 0$, $L > 0$. If $q(t) = O(t^{\beta})$ as $t \to \infty$ for a real β, then

$$x(t) = O(t^{\delta}) \quad as \; t \to \infty, \qquad \delta > \max\left(\beta + \omega, \frac{\log L}{\log \lambda^{-1}}\right). \tag{10}$$

Proof. Choose $t_0 > 0$ such that $\delta + a(t)t > 0$ for all $t \geq t_0$ and consider a solution x of (9) defined on $[t_0, \infty)$. The substitution (3) converts x into a function w satisfying the differential-difference equation

$$w'(s) = \log \lambda [a(\lambda^{-s})\lambda^{-s} + \delta]w(s) + b(\lambda^{-s})\lambda^{\delta-s}\log\lambda^{-1}w(s-1)$$
$$+ q(\lambda^{-s})\lambda^{\delta s-s}\log\lambda^{-1}. \tag{11}$$

Multiply (11) by $\lambda^{-\delta s}\exp\{\int_{s_0}^{\lambda^{-s}} a(u)\}\,du$, $s_0 = \log t_0/\log\lambda^{-1}$ to obtain

$$\frac{d}{ds}\left[\lambda^{-\delta s}\exp\{\int_{s_0}^{\lambda^{-s}} a(u)\,du\}w(s)\right]$$
$$= b(\lambda^{-s})\lambda^{\delta-s-\delta s}\exp\{\int_{s_0}^{\lambda^{-s}} a(u)\,du\}\log\lambda^{-1}w(s-1)$$
$$+ q(\lambda^{-s})\lambda^{-s}\log\lambda^{-1}\exp\{\int_{s_0}^{\lambda^{-s}} a(u)\,du\}.$$

Put $s_j = s_0 + j$, $I_j = [s_{j-1}, s_j]$, $M_j = \sup\{|w(s)|, \; s \in I_j\}$ and let $s^* \in I_{j+1}$, $j = 1, 2, \ldots$. Then integrating the last equality over $[s_j, s^*]$ we have

$$w(s^*) = \lambda^{\delta(s^*-s_j)}\exp\{-\int_{\lambda^{-s_j}}^{\lambda^{-s^*}} a(u)\,du\}w(s_j)$$
$$+ \lambda^{\delta s^*}\exp\{-\int_{s_0}^{\lambda^{-s^*}} a(u)\,du\}$$
$$\times \int_{s_j}^{s^*} b(\lambda^{-s})\lambda^{\delta-s-\delta s}\exp\{\int_{s_0}^{\lambda^{-s}} a(u)\,du\}\log\lambda^{-1}w(s-1)\,ds$$
$$+ \lambda^{\delta s^*}\exp\{-\int_{s_0}^{\lambda^{-s^*}} a(u)\,du\}$$
$$\times \int_{s_j}^{s^*} q(\lambda^{-s})\lambda^{-s}\log\lambda^{-1}\exp\{\int_{s_0}^{\lambda^{-s}} a(u)\,du\}\,ds. \tag{12}$$

Now rewrite the assumptions $|b(t)| \leq La(t)$ and $q(t) = O(t^\beta)$ as $t \to \infty$ as

$$|b(\lambda^{-s})| < \lambda^{-\delta} a(\lambda^{-s}), \quad |q(\lambda^{-s})| \leq K_1 \lambda^{-\beta s}, \tag{13}$$

where $K_1 > 0$ is a constant. Using (12) and (13) we have

$$
\begin{aligned}
|w(s^*)| \leq\ & M_j \lambda^{\delta(s^*-s_j)} \exp\{-\int_{\lambda^{-s_j}}^{\lambda^{-s^*}} a(u)\,du\} \\
& + M_j \lambda^{\delta s^*} \exp\{-\int_{s_0}^{\lambda^{-s^*}} a(u)\,du\} \\
& \times \int_{s_j}^{s^*} a(\lambda^{-s}) \lambda^{-s-\delta s} \log\lambda^{-1} \exp\{\int_{s_0}^{\lambda^{-s}} a(u)\,du\}\,ds \\
& + K_1 \lambda^{\delta \varepsilon^*} \exp\{-\int_{s_0}^{\lambda^{-s^*}} a(u)\,du\} \\
& \times \int_{s_j}^{s^*} \lambda^{-s-\beta s} \log\lambda^{-1} \exp\{\int_{s_0}^{\lambda^{-s}} a(u)\,du\}\,ds\,.
\end{aligned}
$$

Since

$$
\int_{s_j}^{s^*} \lambda^{-s-\beta s} \log\lambda^{-1} \exp\{\int_{s_0}^{\lambda^{-s}} a(u)\,du\}\,ds
$$

$$
\leq \lambda^{(\delta-\beta)s_j} \int_{s_j}^{s^*} \lambda^{(-1-\delta)s} \log\lambda^{-1} \exp\{\int_{s_0}^{\lambda^{-s}} a(u)\,du\}\,ds
$$

and

$$K \leq a(\lambda^{-s}) \lambda^{-\omega s}, \tag{14}$$

then

$$
\begin{aligned}
|w(s^*)| \leq\ & M_j \lambda^{\delta(s^*-s_j)} \exp\{-\int_{\lambda^{-s_j}}^{\lambda^{-s^*}} a(u)\,du\} \\
& + M_j \lambda^{\delta s^*} \exp\{-\int_{s_0}^{\lambda^{-s^*}} a(u)\,du\} \\
& \times \int_{s_j}^{s^*} a(\lambda^{-s}) \lambda^{(-1-\delta)s} \log\lambda^{-1} \exp\{\int_{s_0}^{\lambda^{-s}} a(u)\,du\}\,ds \\
& + K_2 \lambda^{\delta s^*} \exp\{-\int_{s_0}^{\lambda^{-s^*}} a(u)\,du\} \lambda^{(\delta-\beta)s_j} \\
& \times \int_{s_j}^{s^*} a(\lambda^{-s}) \lambda^{-(1+\delta+\omega)s} \log\lambda^{-1} \exp\{\int_{s_0}^{\lambda^{-s}} a(u)\,du\}\,ds,
\end{aligned}
$$

where $K_2 = K_1/K$. From here we get

$$|w(s^*)| \leq M_j \lambda^{\delta(s^*-s_j)} \exp\{-\int_{\lambda^{-s_j}}^{\lambda^{-s^*}} a(u)\,du\}$$

$$+ (M_j + K_2\lambda^{-\omega}\lambda^{(\delta-\beta-\omega)s_j})\lambda^{\delta s^*} \exp\{-\int_{s_0}^{\lambda^{-s^*}} a(u)\,du\}$$

$$\times \int_{s_j}^{s^*} a(\lambda^{-s})\lambda^{(-1-\delta)s} \log\lambda^{-1} \exp\{\int_{s_0}^{\lambda^{-s}} a(u)\,du\}\,ds\,. \qquad (15)$$

Integrating by parts and using (14) we can estimate the last line of (15) as

$$\lambda^{-\delta s} \exp\{\int_{s_0}^{\lambda^{-s}} a(u)\,du\}\Big|_{s_j}^{s^*} (1 + K_3 \exp\{-\theta s_j\})\,,$$

where $\theta = 1 - \omega > 0$ and $K_3 > 0$ is a real constant. Using this

$$|w(s^*)| \leq M_j \lambda^{\delta(s^*-s_j)} \exp\{-\int_{\lambda^{-s_j}}^{\lambda^{-s^*}} a(u)\,du\}$$

$$+ (M_j + K_2\lambda^{-\omega}\lambda^{(\delta-\beta-\omega)s_j})\lambda^{\delta s^*} \exp\{-\int_{s_0}^{\lambda^{-s^*}} a(u)\,du\}$$

$$\times \lambda^{-\delta s} \exp\{\int_{s_0}^{\lambda^{-s}} a(u)\,du\}\Big|_{s_j}^{s^*} (1 + K_3 \exp\{-\theta s_j\})$$

$$\leq M_j(1 + K_3\exp\{-\theta s_j\}) + K_2\lambda^{-\omega}\lambda^{(\delta-\beta-\omega)s_j}(1 + K_3\exp\{-\theta s_j\})$$

$$\leq M_j^*(1 + K_4\exp\{-\kappa s_j\})\,,$$

where $M_j^* = \max(M_j, K_2\lambda^{-\omega})$, $\kappa = \min(\theta, (\delta - \beta - \omega)\log\lambda^{-1}) > 0$ and $K_4 > 0$ is a real. The choice of $s^* \in M_{j+1}$ was arbitrary; hence

$$M_{j+1}^* \leq M_j^*(1 + K_4\exp\{-\kappa s_j\}) \leq M_1^* \prod_{k=1}^{j}(1 + K_4\exp\{-\kappa s_k\})\,.$$

Now the convergence of the corresponding infinite product implies that M_j^* are bounded as $j \to \infty$ and this proves (10). $\square$

Using Lemma 3.1, Lemma 3.2 and the ideas suggested in Section 2 we prove the following result.

Theorem 3.3 *Let x be a solution of the equation*

$$\dot{x}(t) = -c(t)[x(t) - \gamma x(\lambda t)] + f(t)\,, \qquad t \in I = [0, \infty)\,, \qquad (1)$$

where $\gamma \neq 0$, $0 < \lambda < 1$, $c, f \in C^1(I)$, $c(t) \geq K/t^\omega$, $\dot{c}(t) \leq Mc(t)/t$ for all $t \in I$ and suitable constants $K > 0$, $M \geq 0$, $0 \leq \omega < 1$. Further, let

$$\alpha = \frac{\log|\gamma|}{\log\lambda^{-1}}\,.$$

If $f(t) = O(t^\beta)$ and $\dot{f}(t) = O(t^{\beta-1})$ as $t \to \infty$ for a real β, then

$$
\begin{array}{llll}
x(t) = O(t^\alpha) & \text{as } t \to \infty & \text{for } \beta + \omega < \alpha, & (16\text{i}) \\
x(t) = O(t^\alpha \log t) & \text{as } t \to \infty & \text{for } \beta + \omega = \alpha, & (16\text{ii}) \\
x(t) = O(t^{\beta+\omega}) & \text{as } t \to \infty & \text{for } \beta + \omega > \alpha. & (16\text{iii})
\end{array}
$$

Proof. Let $\delta > \max(\beta + \omega, \alpha)$. Then, by Lemma 3.2, $x(t) = O(t^\delta)$ as $t \to \infty$. Differentiating (1) we obtain

$$
\ddot{x}(t) = -\Big[c(t) - \frac{\dot{c}(t)}{c(t)}\Big]\dot{x}(t) + \gamma\lambda c(t)\dot{x}(\lambda t) + \dot{f}(t) - \frac{\dot{c}(t)}{c(t)}f(t). \qquad (17)
$$

Equation (17) is equation (9) with $a(t) = c(t) - \dot{c}(t)/c(t)$, $b(t) = \gamma\lambda c(t)$, $q(t) = \dot{f}(t) - f(t)\dot{c}(t)/c(t)$ and the unknown function $\dot{x}$. We verify the validity of the assumptions of Lemma 3.2 for equation (17). It holds

$$
c(t) - \frac{\dot{c}(t)}{c(t)} \geq c(t) - \frac{M}{t} \geq \frac{K}{t^\omega} - \frac{M}{t} \geq \frac{K^*}{t^\omega}
$$

for a real constant $K^* > 0$ and all t large enough. Further, since

$$
\frac{\dot{c}(t)}{c(t)} \leq \frac{M}{t} \leq M^* c(t)
$$

for a real constant $0 \leq M^* < 1 - \lambda^{1-\omega}$ and all t large enough, then

$$
|\gamma|\lambda c(t) \leq \frac{|\gamma|\lambda}{1 - M^*}\Big(c(t) - \frac{\dot{c}(t)}{c(t)}\Big).
$$

Finally,

$$
\dot{f}(t) - f(t)\frac{\dot{c}(t)}{c(t)} = O(t^{\beta-1}) \qquad \text{as } t \to \infty.
$$

Then the repeated application of Lemma 3.2 to equation (17) yields

$$
\dot{x}(t) = O(t^{\delta-\zeta}) \qquad \text{as } t \to \infty, \qquad \omega < \zeta \leq 1.
$$

Following the proof of Lemma 3.2, we introduce the change of variables (3) in (1) to obtain the equation

$$
w(s) - \gamma\lambda^\delta w(s-1) = \frac{w'(s) + \delta\log\lambda^{-1}w(s) - f(\lambda^{-s})\lambda^{(\delta-1)s}\log\lambda^{-1}}{-c(\lambda^{-s})\lambda^{-s}\log\lambda^{-1}}. \qquad (18)
$$

The relations $x(t) = O(t^\delta)$ and $\dot{x}(t) = O(t^{\delta-\zeta})$ as $t \to \infty$ imply the properties $w(s) = O(1)$ and $w'(s) = O(\lambda^{(\zeta-1)s})$ as $s \to \infty$. Hence, equation (18) can be viewed as the difference equation (6), where $g(s) \equiv \gamma\lambda^\delta$ and $h(s) = O(\exp\{-\rho s\})$ as $s \to \infty$, $\rho = \min((\zeta - \omega)\log\lambda^{-1}, (\delta - \beta - \omega)\log\lambda^{-1}) > 0$.

Since $\log r^{-1} = (\delta - \alpha) \log \lambda^{-1}$, by Lemma 3.1 we distinguish three cases with the respect to the relation between $\beta + \omega$ and α.

Case $\beta + \omega < \alpha$: First let $\alpha < \beta + \zeta$. Then $\rho = (\delta - \beta - \omega) \log \lambda^{-1}$ for a suitable δ, $\alpha < \delta < \beta + \zeta$. Hence, by (7i), $w(s) = O(\lambda^{(\delta - \alpha)s})$ as $s \to \infty$.
If $\alpha \geq \beta + \zeta$, then $\rho = (\zeta - \omega) \log \lambda^{-1}$ for a δ, $\alpha < \delta < \alpha + \zeta - \omega$. Asymptotic relation (16i) now follows from (7i).

Case $\beta + \omega = \alpha$: Choose δ, $\alpha < \delta < \alpha + \zeta - \omega$. Then $\rho = (\delta - \beta - \omega) \log \lambda^{-1}$ and $w(s) = O(\lambda^{(\delta - \alpha)s} s)$ as $s \to \infty$. This proves (16ii).

Case $\beta + \omega > \alpha$: Choose δ, $\beta + \omega < \delta < \beta + \zeta$. Then $\rho = (\delta - \beta - \omega) \log \lambda^{-1}$ and $w(s) = O(\lambda^{(\delta - \beta - \omega)s})$ as $s \to \infty$. The proof is completed. $\square$

Remark The conclusions of the previous statement generalize some known results derived for particular cases of equation (1). Note., e.g., that equation (1) with constant coefficients has been studied in [7]. If we assume c to be constant in our investigations, then $\omega = 0$ and the asymptotic formulae (16i)–(16iii) coincide with those derived in [7]. Similarly, the asymptotic properties of homogeneous equation (1) with $\gamma = 1$ have been described in [8]. It can be shown that our results generalize and extend some parts of this paper.

References

[1] Čermák, J., The asymptotic bounds of solutions of linear delay systems, *J. Math. Anal. Appl.* **225** (1998), 373–388.

[2] Čermák, J., The Abel equation in the asymptotic theory of functional differential equations, *Aequationes Math.*, to appear.

[3] Diblík, J., Asymptotic behaviour of solutions of linear differential equations with delay, *Ann. Polon. Math.* **58** (1993), 131–137.

[4] Győri, I. and Ladas, G., *Oscillation Theory of Delay Differential Equations*, Clarendon Press, 1991.

[5] Győri, I. and Pituk, M., Comparison theorems and asymptotic equilibrium for delay differential and difference equations, *Dynam. Systems Appl.* **5** (1996), 277–302.

[6] Heard, M. L., A change of variables for functional differential equations, *J. Differential Equations* **18** (1975), 1–10.

[7] Lim, E. B., Asymptotic bounds of solutions of the functional differential equation $x'(t) = a x(\lambda t) + b x(t) + f(t)$, $0 < \lambda < 1$, *SIAM J. Math. Anal.* **9** (1978), 915–920.

[8] Makay, G. and Terjéki, J., On the asymptotic behavior of pantograph equations, *E.J.Qualitative Theory of Diff. Equ.* **2** (1998), 1–12.

Properties of a Class of Numbers Related to the Fibonacci, Lucas and Pell Numbers

FOZI. M. DANNAN

Department of Mathematics, Faculty of Science
Qatar University, Doha - QATAR
E-mail: fmdannan@qu.edu.qa

Abstract In this article we investigate a class of numbers that contains the Fibonacci, Lucas and Pell numbers. We obtain some properties that are satisfied by these numbers.

Keywords Fibonacci numbers, Lucas numbers, Pell numbers

AMS Subject Classification 11B39

1 Introduction

The numbers we are going to consider satisfy the recurrence relation

$$U(n+2, u_1, u_2, a, b) = aU(n+1, u_1, u_2, a, b) + bU(n, u_1, u_2, a, b)$$
$$U(1, u_1, u_2, a, b) = u_1, \quad aU(2, u_1, u_2, a, b) = u_2, \quad ab \neq 0 \tag{1}$$

for integers $n \geq 2$ and real u_1, u_2, a and b. We observe that

(i) The Fibonacci numbers $F(n)$ defined by $F(n+2) = F(n+1) + F(n)$ and $F(1) = F(2) = 1$ satisfy

$$U(n, 1, 1, 1, 1) = F(n). \tag{2}$$

(ii) The Lucas numbers $L(n)$ defined by $L(n+2) = L(n+1) + L(n)$ and $L(1) = 1$, $L(2) = 3$ satisfy

$$U(n, 1, 3, 1, 1) = L(n). \tag{3}$$

(iii) The Pell numbers $P(n)$ defined by $P(n+2) = 2P(n+1) + P(n)$ and $P(1) = 1$, $P(2) = 2$ satisfy

$$U(n, 1, 2, 2, 1) = P(n). \tag{4}$$

(iv) The numbers $R(a, n)$ defined by $R(a, n + 2) = aR(a, n + 1) + R(a, n)$, for integers $a \geq 1$, $R(a, 0) = 0$, $R(a, 1) = 1$ satisfy

$$U(n, 1, a, a, 1) = R(a, n). \tag{5}$$

This class of numbers had been introduced by Entringer and Slater [3] while investigating the problem of information dissemination through telegraphs. In 1986, N. H. Bong [1] obtained several interesting properties for $R(a, n)$. In spite of the articles and books that had been published about the Fibonacci, Lucas and Pell numbers and their applications [2, 4–8], these numbers are still an interesting subject for investigation.

In this article we introduce new properties of the function $U(n, u_1, u_2, a, b)$ that will be denoted by $U(n)$ when the appearance of the arguments is not necessary. This type of function had been studied extensively by Horadam [5].

2 Preliminary results

In this section we give some properties of the function $U(n, u_1, u_2, a, b)$. We assume, as a first case, that the equation

$$\lambda^2 - a\lambda - b = 0 \tag{6}$$

has two different roots

$$\lambda_1 = \frac{a + \varepsilon\sqrt{\Delta}}{2}, \quad \lambda_2 = \frac{a - \varepsilon\sqrt{\Delta}}{2} \tag{7}$$

where $\Delta = |a^2 + 4b|$, $\varepsilon = 1$ when $a^2 + 4b > 0$ and $\varepsilon = \sqrt{-1}$ when $a^2 + 4b < 0$.

Lemma 2.1. *Assume that $\lambda_1 \neq \lambda_2$, then for $\alpha, \beta, \gamma, \delta$ and $k \in \mathcal{R}$, the following relations hold true:*

(i) $U(n, \alpha u_1, \alpha u_2, a, b) = \alpha U(n, u_1, u_2, a, b)$,

(ii) $U(n, u_1, ku_2, ka, k^2 b) = k^{n-1} U(n, u_1, u_2, a, b)$,

(iii) $U(n, \alpha \nu_1 + \beta \nu_2, \gamma w_1 + \delta w_2, a, b) =$
$\qquad = U(n, \alpha \nu_1, \gamma w_1, a, b) + U(n, \beta \nu_2, \delta w_2, a, b)$.

Proof. The solution of (1) can be written in the form

$$U(n) = c_1 \lambda_1^n + c_2 \lambda_2^n \tag{8}$$

where

$$u_1 = U(1), \quad u_2 = U(2) \tag{9}$$

and the constants $c_1 = c_1(u_1, u_2, a, b)$ and $c_2 = c_2(u_1, u_2, a, b)$ can be obtained from

$$c_1 \lambda_1 + c_2 \lambda_2 = u_1,$$
$$c_1 \lambda_1^2 + c_2 \lambda_2^2 = u_2. \tag{10}$$

By solving (10) and substituting in (8) we get

$$U(n) = \frac{u_2 - \lambda_2 u_1}{\lambda_1 - \lambda_2} \lambda_1^{n-1} - \frac{u_2 - \lambda_1 u_1}{\lambda_1 - \lambda_2} \lambda_2^{n-1}. \tag{11}$$

Using (7) it follows from (9) that

$$U(n, u_1, u_2, a, b) = \frac{1}{\varepsilon \sqrt{\Delta}} [b u_1 (\lambda_1^{n-2} - \lambda_2^{n-2}) + u_2 (\lambda_1^{n-1} - \lambda_2^{n-1})]. \tag{12}$$

If we notice that $U(n, u_1, u_2, a, b)$ is a linear and homogeneous function in u_1 and u_2, the relations (i) and (iii) follow directly.

Since $\lambda = \lambda_1(a, b)$ and $\lambda_2 = \lambda_2(a, b)$ satisfy

$$\lambda_i(ka, k^2 b) = k\lambda_i(a, b), \quad i = 1, 2, \tag{13}$$

therefore

$$U(n, u_1, ku_2, ka, k^2 b)$$
$$= \frac{1}{\varepsilon k \sqrt{\Delta}} [k^2 u_1(k)^{n-2} (\lambda_1^{n-2} - \lambda_2^{n-2}) + k u_2(k)^{n-1} (\lambda_1^{n-1} - \lambda_2^{n-1})],$$

and (ii) follows directly .

Lemma 2.2. *Assume that* $\lambda_1 \neq \lambda_2$. *Then*

$$U(n, u_1, u_2, a, b) = b u_1 U(n - 2, 1, a, a, b) + u_2 U(n - 1, 1, a, a, b). \tag{14}$$

Proof. Let

$$Q(p) = \lambda_1^p - \lambda_2^p. \tag{15}$$

Then from (6) it follows that $Q(p)$ satisfies

$$Q(p + 1) = aQ(p + 1) + bQ(p) \tag{16}$$

and

$$Q(1) = \varepsilon\sqrt{\Delta}, \quad Q(2) = a\varepsilon\sqrt{\Delta}. \tag{17}$$

Therefore (16) and (17) have the solution

$$Q(p) = U(p, \varepsilon\sqrt{\Delta}, a\varepsilon\sqrt{\Delta}, a, b). \tag{18}$$

Using (18) it follows from (12) that

$$U(n, u_1, u_2, a, b) = \frac{1}{\varepsilon\sqrt{\Delta}} [b u_1 Q(n - 2) + u_2 Q(n - 1)]. \tag{19}$$

Applying Lemma 2.1(i), we get (14).

Now,

$$
\begin{aligned}
D(n+m+1)D(n-m) &- D(n+m)D(n-m+1) = \\
&= \lambda_1^{2n+1}[c_1 + c_2(n+m+1)][c_1 + c_2(n-m)] - \\
&- \lambda_1^{2n+1}[c_1 + c_2(n+m)][c_1 + c_2(n-m+1)] = \\
&= -2mc_2^2\lambda_1^{2n+1}.
\end{aligned}
\tag{33}
$$

If we notice that $\lambda_1 = \dfrac{a}{2}$, we obtain from (32)

$$
c_1 = \frac{2p_1\lambda_1 - p_2}{\lambda_1^2}, \qquad c_1 = \frac{p_2 - p_1\lambda_1}{\lambda_1^2},
\tag{34}
$$

hence

$$
c_2^2\lambda_1^{2n+1} = (p_2 - p_1\lambda_1)^2\lambda_1^{2n-3} = \frac{1}{4}(2p_2 - ap_1)^2(\frac{a}{2})^{2n-3}.
\tag{35}
$$

The required relation follows directly from (33) and (35). The relation (30) can be proved similarly.

Theorem 3.4. *Assume that $\lambda_1 = \lambda_2$. Then the following relations hold true:*

$$
\begin{aligned}
D(n+k+1)D(n-k) &- D(n+1)D(n) = \\
&= -\frac{k(k+1)}{2}(\frac{a}{2})^{2n}(2p_2 - ap_1)^2, \quad (n > k \geq 1)
\end{aligned}
\tag{36}
$$

and

$$
\begin{aligned}
D(n+k)D(n-k) &- D(n+1)D(n-1) = \\
&- -\frac{(k^2-1)}{4}(\frac{a}{2})^{2n-4}(2p_2 - ap_1)^2, \quad (n > k \geq 2).
\end{aligned}
\tag{37}
$$

Proof. Putting $m = 1, 2, \ldots, k$ in (29) and adding the resulting relations we obtain

$$
\begin{aligned}
D(n+k+1)D(n-k) &- D(n+1)D(n) = \\
&= -\frac{1}{2}(\frac{a}{2})^{2n}(2p_2 - ap_1)^2 \sum_{i=1}^{k} i = \\
&= -\frac{k(k+1)}{4}(\frac{a}{2})^{2n}(2p_2 - ap_1)^2.
\end{aligned}
$$

Similarly from (30) we obtain

$$
\begin{aligned}
D(n+k)D(n-k) &- D(n+1)D(n-1) = \\
&= -\frac{1}{4}(\frac{a}{2})^{2n-4}(2p_2 - ap_1)^2 \sum_{i=1}^{k}(2i-1) = \\
&= -\frac{(k^2-1)}{4}(\frac{a}{2})^{2n-4}(2p_2 - ap_1)^2.
\end{aligned}
$$

The proof is complete.

Theorem 3.2. *Assume that* $\lambda_1 \neq \lambda_2$ *and* $n > m \geq 2$. *Then the following relations hold true:*

$$U(n+m)U(n-m) - U(n+m-1)U(n-m+1) =$$
$$= (-b)^{n-m-1}(bu_1^2 + au_1u_2 - u_2^2)U(2m-1,1,a,a,b) \qquad (25)$$

and

$$U(n+k)U(n-k) - U(n+1)U(n-1) =$$
$$= (bu_1^2 + au_1u_2 - u_2^2)\sum_{i=2}^{k}(-b)^{n-i-1}V(2i-1), \quad (n > k \geq 2). \qquad (26)$$

In the case when Equation (6) has equal roots, we notice that $b = -\frac{a^2}{4}$ and $\lambda_1 \quad \lambda_2 \quad \frac{a}{2}$. If we denote the solution of (1) by $D(n, p_1, p_2, a)$, then we have

$$D(n \mid 2, p_1, p_2, a) = aD(n \mid 1, p_1, p_2, a) - \frac{a^2}{4}D(n, p_1, p_2, a) \qquad (27)$$

and

$$D(1, p_1, p_2, a) = p_1, \quad D(2, p_1, p_2, a) = p_2. \qquad (28)$$

In what follows we write $D(n)$ instead of $D(n, p_1, p_2, a)$ when the appearance of the arguments is not necessary.

Theorem 3.3. *Assume that Equation (6) has equal roots. Then the following relations hold true:*

$$D(n+m+1)D(n-m) - D(n+m)D(n-m+1) =$$
$$= -\frac{m}{2}(\frac{a}{2})^{2n-3}(2p_2 - ap_1)^2, \quad (n > m \geq 1) \qquad (29)$$

and

$$D(n+m)D(n-m) - D(n+m-1)D(n-m+1) =$$
$$= -\frac{2m-1}{4}(\frac{a}{2})^{2n-4}(2p_2 - ap_1)^2, \quad (n > m \geq 2). \qquad (30)$$

Proof. The solution of (27) and (28) can be written as follows:

$$D(n) = \lambda_1^n(c_1 + c_2n), \quad D(1) = p_1, \quad D(2) = p_2 \qquad (31)$$

where c_1 and c_2 satisfy

$$(c_1 + c_2)\lambda_1 = p_1, \quad (c_1 + 2c_2)\lambda_1^2 = p_2. \qquad (32)$$

Therefore,

$$c_1 = \frac{2p_1\lambda_1 - p_2}{\lambda_1^2}, \quad c_1 = \frac{p_2 - p_1\lambda_1}{\lambda_1^2}. \qquad (33)$$

Now,

$$
\begin{aligned}
D(n+m+1)&D(n-m) - D(n+m)D(n-m+1) = \\
&= \lambda_1^{2n+1}[c_1 + c_2(n+m+1)][c_1 + c_2(n-m)] - \\
&\quad - \lambda_1^{2n+1}[c_1 + c_2(n+m)][c_1 + c_2(n-m+1)] = \\
&= -2mc_2^2\lambda_1^{2n+1}.
\end{aligned} \tag{34}
$$

If we notice that $\lambda_1 = \dfrac{a}{2}$, we obtain from (33) that

$$
c_2^2\lambda_1^{2n+1} = (p_2 - p_1\lambda_1)^2\lambda_1^{2n-3} = \frac{1}{4}(2p_2 - ap_1)^2(\frac{a}{2})^{2n-3}. \tag{35}
$$

The required relation follows directly from (34) and (35). The relation (30) can be proved similarly.

Theorem 3.4. *Assume that $\lambda_1 = \lambda_2$. Then the following relations hold true:*

$$
\begin{aligned}
D(n+k+1)&D(n-k) - D(n+1)D(n) = \\
&= -\frac{k(k+1)}{2}(\frac{a}{2})^{2n-3}(2p_2 - ap_1)^2, \quad (n > k \geq 1)
\end{aligned} \tag{36}
$$

and

$$
\begin{aligned}
D(n+k)&D(n-k) - D(n+1)D(n-1) = \\
&= -\frac{(k^2-1)}{4}(\frac{a}{2})^{2n-4}(2p_2 - ap_1)^2, \quad (n > k \geq 2).
\end{aligned} \tag{37}
$$

Proof. Putting $m = 1, 2, \ldots, k$ in (29) and adding the resulting relations we obtain

$$
\begin{aligned}
D(n+k+1)&D(n-k) - D(n+1)D(n) = \\
&= -\frac{1}{2}(\frac{a}{2})^{2n-3}(2p_2 - ap_1)^2 \sum_{i=1}^{k} i = \\
&= -\frac{k(k+1)}{4}(\frac{a}{2})^{2n-3}(2p_2 - ap_1)^2.
\end{aligned}
$$

Similarly from (30) we obtain

$$
\begin{aligned}
D(n+k)&D(n-k) - D(n+1)D(n-1) = \\
&= -\frac{1}{4}(\frac{a}{2})^{2n-4}(2p_2 - ap_1)^2 \sum_{i=1}^{k}(2i-1) = \\
&= -\frac{(k^2-1)}{4}(\frac{a}{2})^{2n-4}(2p_2 - ap_1)^2.
\end{aligned}
$$

The proof is complete.

4   Applications

1. *Fibonacci numbers.* Since $F(n) = U(n, 1, 1, 1, 1)$, we get from Theorems 3.1 and 3.2 the following properties:

$$F(n + m + 1)F(n - m) - F(n + m)F(n - m + 1) =$$
$$= (-1)^{n-m-1}F(2m), \quad n > m \geq 1, \tag{38}$$

$$F(n + m)F(n - m) - F(n + m - 1)F(n - m + 1) =$$
$$= (-1)^{n-m-1}F(2m - 1), \quad n > m \geq 2, \tag{39}$$

$$F(n + k + 1)F(n - k) - F(n + 1)F(n) =$$
$$= \sum_{i=1}^{k}(-1)^{n-i-1}F(2i), \quad n > k \geq 1, \tag{40}$$

$$F(n + k)F(n - k) - F(n + 1)F(n - 1) =$$
$$= \sum_{i=2}^{k}(-1)^{n-i-1}F(2i - 1), \quad n > k \geq 2. \tag{41}$$

2. *Lucas numbers.* Since $L(n) = U(n, 1, 3, 1, 1)$, we get from Theorems 3.1 and 3.2 the following properties:

$$L(n + m + 1)L(n - m) - L(n + m)L(n - m + 1) =$$
$$= (-1)^{n-m}5F(2m), \quad n > m \geq 1 \tag{42}$$

and

$$L(n + m)L(n - m) - L(n + m - 1)L(n - m + 1) =$$
$$= (-1)^{n-m}5F(2m - 1), \quad n > m \geq 2. \tag{43}$$

Similar properties to (40) and (41) can be written for $L(n)$.

3. *Pell numbers.* Since $P(n) = U(n, 1, 2, 2, 1)$, we get from Theorems 3.1 and 3.2 the following properties:

$$P(n + m + 1)P(n - m) - P(n + m)P(n - m + 1) =$$
$$= (-1)^{n-m-1}P(2m), \quad n > m \geq 1 \tag{44}$$

and

$$P(n+m)P(n-m) - P(n+m-1)P(n-m+1) =$$
$$= (-1)^{n-m-1}P(2m-1), \quad n > m \geq 2. \tag{45}$$

Two more properties can be obtained from Theorems 3.1 and 3.2.

4. *The class* $R(n, a)$. Since $R(n, a) = U(n, 1, a, a, 1)$, we obtain from Theorems 3.1 and 3.2 the following properties:

$$R(n+m+1, a)R(n-m, a) - R(n+m, a)R(n-m+1, a) =$$
$$= (-1)^{n-m-1}R(2m, a), \quad n > m \geq 1 \tag{46}$$

and

$$R(n+m, a)R(n-m, a) - R(n+m-1, a)R(n-m+1, a) =$$
$$= (-1)^{n-m-1}R(2m-1, a), \quad n > m \geq 2. \tag{47}$$

Two more properties similar to (40) and (41) can be obtained from Theorems 3.1 and 3.2.

References

[1] Bong, N. H., On the class of numbers related to both the Fibonacci and Pell numbers and their applications, *Fibonacci Numbers and Their Applications*, Math. Appl. 28, Reidel, Dordrecht, 1986, 9–37.

[2] Dubner, H. and Keller, W., New Fibonacci and Lucas primes, *Math. Comp.* 68, 225 (1999), 417–427.

[3] Entringer, R. C. and Slater, P. J., Gossips and telegraphs, *J. Franklin Institute* **307**, n. 6 (June) 1979, 353–359.

[4] Hoggatt, V. E. Jr., *The Fibonacci and Lucas Numbers*, Boston, MA, Houghton Mifflin, 1969.

[5] Horadam, A. F., Special properties of the sequence $w(n, a, b; p, q)$, *Fibonacci Quarterly* 5, 1967, 424–434.

[6] Philipou, A. N., Bergum, G. E. and Horadam, A. F. (Eds.), *Fibonacci Numbers and Their Applications*, D. Reidel Publishing Co., 1986.

[7] Sloane, N. J. A. and Plouffe, S., *Encyclopedia of Integer Sequences*, Academic Press, 1995.

[8] Vorob'ev, N. N., *Fibonacci numbers*, New York, Blaisdell Publishing Co., 1961.

Oscillation Theory
of a Class of Higher Order
Sturm-Liouville Difference Equations

ONDŘEJ DOŠLÝ [1]

Department of Mathematics, Masaryk University,
Janáčkovo nám, 2a, CZ-662 95 Brno, Czech Republic
E-mail: dosly@math.muni.cz

Abstract In this paper we study oscillatory properties of the higher order Sturm-Liouville difference equation

$$(-1)^n \Delta^n \left(k^{(\alpha)} \Delta^n y_k \right) = q_k y_{k+n}, \quad \alpha \in \mathbb{R}, \quad k^{(\alpha)} := \frac{\Gamma(k+1)}{\Gamma(k+1-\alpha)}. \qquad (*)$$

We present a brief survey of the recent results concerning oscillation of $(*)$, and then we use them to discuss the problem of the oscillation constant in the Euler-type difference equation.

Keywords Sturm-Liouville difference operator, Oscillation criteria, Nonoscillation criteria, Oscillation constant, Factorization of disconjugate operators

AMS Subject Classification 39A10, 39A12

1 Introduction

In this contribution we deal with oscillatory properties of the $2n$-order Sturm-Liouville difference equation

$$L(y) := \sum_{\nu=0}^{n} (-1)^\nu \Delta^\nu \left(r_k^{[\nu]} \Delta^\nu y_{k+n-\nu} \right) = 0, \qquad (1)$$

where $r_k^{[\nu]}$, $\nu = 0, \dots, n$, are real-valued sequences with $r_k^{[n]} > 0$. Comparing oscillation theory of (1) with oscillation theory of its continuous counterpart

$$\sum_{\nu=0}^{n} (-1)^n \left(r_\nu(t) y^{(\nu)} \right)^{(\nu)} = 0, \qquad (2)$$

[1]Supported by the Grant 201/01/0079 of the Czech Grant Agency.

the basic facts of the higher-order discrete oscillation theory have been established only relatively recently, whereas the basic results concerning oscillation of (2) can be found already in [7]. The reason is that the fundamental paper of Bohner [3], where the so-called Roundabout theorem for linear Hamiltonian difference systems

$$\Delta x_k = A_k x_{k+1} + B_k u_k \quad \Delta u_k = C_k x_{k+1} - A_k^T u_k \tag{3}$$

is proved, comes only from 1996.

In our paper we recall first some results of that paper, in particular, the relationship between (1) and (3) (Section 2). In Section 3 we present general oscillation and nonoscillation criteria for (1) whose proofs are based on the so-called variational principle. Section 4 contains the specification of these general results to a special class of equations (1). In the last section we discuss the problem of the oscillation constant in Euler-type difference equations.

2   Preliminaries

Let y be a solution of (1) and

$$x_k = \begin{pmatrix} y_{k+n-1} \\ \Delta y_{k+n-2} \\ \vdots \\ \Delta^{n-1} y_k \end{pmatrix}, u_k = \begin{pmatrix} \sum_{\nu=1}^n (-1)^{\nu-1} \Delta^{\nu-1} (r_k^{[\nu]} \Delta^\nu y_{k+n-\nu}) \\ \vdots \\ -\Delta(r_k^{[n]} \Delta^n y_k) + r_k^{[n-1]} \Delta^{n-1} y_{k+1} \\ r_k^{[n]} \Delta^n y_k \end{pmatrix}.$$

Then (x, u) is a solution of (3) with

$$B_k = \text{diag}\{0, \ldots, 0, \tfrac{1}{r_k^{[n]}}\}, \quad C_k = \text{diag}\{r_k^{[0]}, r_k^{[1]}, \ldots, r_k^{[n-1]}\},$$

$$A = A_{ij} = \begin{cases} 1 & \text{if } j = i+1, \ i = 1, \ldots, n-1, \\ 0 & \text{elsewhere;} \end{cases}$$

see [2]. A $2n \times n$ matrix solution of (3) is said to be the *conjoined basis* of (3) if $X^T U \equiv U^T X$ and $\text{rank}(X, U) = n$. We say that an interval $(m, m+1]$ contains a *focal point* of (X, U) if

$$\text{Ker } X_{m+1} \subseteq \text{Ker } X_m \quad \text{and} \quad D_m := X_m X_{m+1}^\dagger (I - A_m)^{-1} B_m \geq 0$$

fails to hold. Here Ker, † and $\geq$ stand for the kernel, the Moore-Penrose generalized inverse and nonnegative definiteness of the matrix indicated. System (3) is said to be *nonoscillatory* if there exists $N \in \mathbb{N}$ such that the solution (X, U) given by the initial condition $X_N = 0$, $U_N = I$ has no focal point in (N, ∞) and it is said to be *oscillatory* in the opposite case. Oscillatory properties of (1) are defined via the corresponding properties of the associated linear Hamiltonian system.

One of the main results of [3] is the statement which relates nonoscillation of (3) to positivity of a certain discrete quadratic functional and this result specified to (1) reads as follows.

Proposition 2.1 *Equation* (1) *is nonoscillatory iff there exists $N \in \mathbb{N}$ such that*

$$\mathcal{F}(y; N, \infty) := \sum_{k=N}^{\infty} \left[\sum_{\nu=0}^{n} r_k^{[\nu]} (\Delta^\nu y_{k+n-\nu})^2 \right] > 0 \qquad (4)$$

for every nontrivial $y \in \mathcal{D}_n(N)$, where

$$\mathcal{D}_n(N) : \quad = \quad \{y = \{y_k\}_N^\infty : \ y_N = y_{N+1} = \cdots = y_{N+n-1} = 0, \ \exists m \geq N :$$
$$y_k = 0, \ k \geq m\}.$$

Suppose that (1) is nonoscillatory. Then there exist solutions $y^{[1]}, \ldots, y^{[n]}$ of this equation such that for any other system $\tilde{y}^{[1]}, \ldots, \tilde{y}^{[n]}$ which together with $y^{[1]}, \ldots, y^{[n]}$ forms the basis of the solution space of (1) and

$$\lim_{k \to \infty} \frac{C(y^{[1]}, \ldots, y^{[n]})}{C(\tilde{y}^{[1]}, \ldots, \tilde{y}^{[n]})} = 0,$$

where $C(\cdot)$ denotes the Casoratian of the sequences in brackets, see [2]. The system $y^{[1]}, \ldots, y^{[n]}$ ($\tilde{y}^{[1]}, \ldots, \tilde{y}^{[n]}$) is said to be the *recessive* (*dominant*) system of solutions of (1).

In our paper we will need also another definition of nonoscillation of higher order linear difference equations, introduced by Hartman [8]. Consider the n-th order linear difference equation

$$\mathcal{L}(y) := x_{k+n} + a_k^{[n-1]} x_{k+n-1} + \ldots a_k^{[1]} x_{k+1} + a_k^{[0]} x_k = 0. \qquad (5)$$

An integer $k + m$ is said to be the *generalized zero point of multiplicity* m of a sequence x_k if $x_k \neq 0$, $x_{k+1} = \cdots = x_{k+m-1} = 0$ and $(-1)^{m-1} x_{k+m} x_k \leq 0$. Equation (5) is said to be *H-nonoscillatory* if there exists $M \in \mathbb{N}$ such that every nontrivial solution of (5) has at most $n - 1$ generalized zeros (counting multiplicity) on $[M, \infty)$. H-nonoscillatory operators $\mathcal{L}$ admit Polya's factorization for large k

$$\mathcal{L}(y) = \frac{1}{\alpha_k^{[n]}} \Delta \left\{ \frac{1}{\alpha_k^{[n-1]}} \Delta \left[\cdots \frac{1}{a_k^{[1]}} \left(\Delta \left(\frac{y_k}{\alpha_k^{[0]}} \right) \right) \cdots \right] \right\}.$$

The sequences $\alpha^{[j]}$ can be computed explicitly via Casoratians of solutions of (1); see [8]. Moreover, if $n = 2m$ is even and the operator L is (formally) self-adjoint, the sequences $\alpha^{[j]}$ can be taken in such a way that $\alpha^{[n-j]} = \alpha^{[j]}$, $j = 0, \ldots, m - 1$.

3   General (non)oscillation criteria

In this section we present oscillation and nonoscillation criteria for the equation

$$L(y) = q_k y_{k+n}, \tag{6}$$

where the difference operator L is given in (1). Equation (1) is supposed to be nonoscillatory, and (6) is viewed as a perturbation of equation (1). The quadratic functional corresponding to (6) has the form

$$\tilde{\mathcal{F}}(y; N, \infty) := \mathcal{F}(y; N, \infty) - \sum_{k=N}^{\infty} q_k y_{k+n}^2. \tag{7}$$

According to Proposition 2.1, if the sequence q is "sufficiently positive" ("not too positive") compared with "the measure of nonoscillation of L", then (6) is oscillatory (nonoscillatory). In the next two general theorems we specify the meaning of the above vague expressions.

Theorem 3.1 ([4]). *Let $y^{[1]}, \ldots, y^{[n]}$, $\tilde{y}^{[1]}, \ldots \tilde{y}^{[n]}$ be the recessive and dominant systems of solutions of (1) and denote by (X, U), $(\tilde{X}, \tilde{U})$ the matrix solutions of the associated LHS generated by these solutions (i.e., columns of (X, U) resp. $(\tilde{X}, \tilde{U})$ are formed by differences and quasidifferences of $y^{[i]}$, $\tilde{y}^{[i]}$, $i = 1, \ldots, n$). Equation (6) is oscillatory provided there exists $c = (c_1, \ldots, c_n)^T \in \mathbb{R}^n$ such that one of the following conditions holds.*

(i) Leighton-Wintner type criterion:

$$\sum_{}^{\infty} q_k \left(c_1 y_k^{[1]} + \cdots + c_n y_k^{[n]} \right)^2 = \infty, \tag{8}$$

(ii) Nehari-type criterion with recessive system:

$$\limsup_{k \to \infty} \frac{\sum_{j=k}^{\infty} q_j \left(c_1 y_j^{[1]} + \cdots + c_n y_j^{[n]} \right)^2}{c^T \left(\sum^k X_{j+1}^{-1} (I - A)^{-1} B_j X_j^{T-1} \right)^{-1} c} > 1 \tag{9}$$

(iii) Nehari-type criterion with dominant system:

$$\limsup_{k \to \infty} \frac{\sum^k q_j \left(c_1 \tilde{y}_j^{[1]} + \cdots + c_n \tilde{y}_j^{[n]} \right)^2}{c^T \left(\sum_{j=k}^{\infty} \tilde{X}_{j+1}^{-1} (I - A)^{-1} B_j \tilde{X}_j^{T-1} \right)^{-1} c} > 1. \tag{10}$$

Under the stronger assumption that (1) is H-nonoscillatory, the statement of theorem remains valid without the assumption of nonnegativity of q, but $\limsup$ in (9) and (10) must be replaced by $\liminf$.

   The next statement is the nonoscillatory counterpart of Theorem 3.1. This is a new statement, so we present also its proof.

Theorem 3.2 *Suppose that* (1) *is H-nonoscillatory and*

$$L(y) = \frac{\Delta}{a_{k+n}^{[0]}}\left(\frac{\Delta}{a_{k+n-1}^{[1]}}\left(\cdots\frac{\Delta}{a_{k+1}^{[n-1]}}\left(\frac{\Delta}{a_k^{[n]}}\left(\frac{\Delta}{a_k^{[n-1]}}\left(\cdots\frac{\Delta}{a_k^{[1]}}\left(\frac{y_k}{a_k^{[0]}}\right)\cdots\right)\right)\right)\cdots\right)\right), \quad (11)$$

where $a_k^{[j]} > 0$, $j = 0, \ldots, n$, *is its Polya's factorization. If there exist sequences* $b^{[j]}$, $j = 1, \ldots, n-1$, *such that the second order difference equations*

$$\Delta\left(\frac{b_k^{[j]}\Delta z_k}{(a_{k+n-j}^{[j]})^2}\right) + b_k^{[j-1]}z_{k+1} = 0, \quad j = 1, \ldots, n, \quad (12)$$

with $b_k^{[n]} = a_k^{[n]}$, $b_k^{[0]} = q_k(a_{k+n}^{[0]})^2$, *are nonoscillatory, then* (6) *is nonoscillatory.*

Proof. First of all observe that using summation by parts (applied n times and then again n times) and the factorization formula (11), for any $N \in \mathbb{N}$ and $y \in \mathcal{D}_n(N)$ we have

$$\mathcal{F}(y, N, \infty) = \sum_{k=N}^{\infty}\left(\sum_{\nu=0}^{n} r_k^{[\nu]}(\Delta y_{k+n-\nu})^2\right) - \sum_{k=N}^{\infty} y_{k+n}L(y)$$

$$= \sum_{k=N}^{\infty}\frac{1}{a_k^{[n]}}\left\{\Delta\left[\frac{1}{a_k^{[n-1]}}\cdots\Delta\left(\frac{1}{a_k^{[1]}}\Delta\left(\frac{y_k}{a_k^{[0]}}\right)\right)\cdots\right]\right\}^2.$$

Since second order difference equations (12) are nonoscillatory, according to Proposition 2.1 (with $n = 1$) there exist $N^{[j]} \in \mathbb{N}$, $j = 1, \ldots, n$, such that for $z \in \mathcal{D}_1(N^{[j]})$

$$\sum_{k=N}^{\infty}\frac{b_k^{[j]}(\Delta z_k)^2}{(a_{k+n-j}^{[j]})^2} > \sum_{k=N}^{\infty} b_k^{[j-1]}z_{k+1}^2.$$

Let $N = \max\{N^{[1]}, \ldots, N^{[n-1]}\}$ and $0 \not\equiv y \in \mathcal{D}_n(N)$ be arbitrary. Denote $z^{[0]} = \frac{y}{a^{[0]}}$, $z^{[j]} = \frac{1}{a^{[j]}}\Delta z^{[j-1]}$, $j = 1, \ldots, n-1$. Then $z^{[j]} \in \mathcal{D}_{n-j}(N)$ and using the previous computation

$$\tilde{\mathcal{F}}(y; N, \infty) = \mathcal{F}(y; N, \infty) - \sum_{k=N}^{\infty} q_k y_{k+n}^2$$

$$= \sum_{k=N}^{\infty}\frac{1}{a_k^{[n]}}\left\{\Delta\left[\frac{1}{a_k^{[n-1]}}\cdots\Delta\left(\frac{1}{a_k^{[1]}}\Delta\left(\frac{y_k}{a_k^{[0]}}\right)\right)\cdots\right]\right\}^2 - \sum_{k=N}^{\infty} q_k y_{k+n}^2$$

$$= \sum_{k=N}^{\infty}\frac{(\Delta z_k^{[n-1]})^2}{a_k^{[n]}} - \sum_{k=N}^{\infty} q_k y_{k+n}^2 > \sum_{k=N}^{\infty} b_k^{[n-1]}(z_{k+1}^{[n-1]})^2 - \sum_{k=N}^{\infty} q_k y_{k+n}^2$$

$$= \sum_{k=N}^{\infty}\frac{b_k^{[n-1]}(\Delta z_{k+1}^{[n-2]})^2}{(a_{k+1}^{[n-1]})^2} - \sum_{k=N}^{\infty} q_k y_{k+n}^2 > \cdots$$

$$\ldots > \sum_{k=N}^{\infty} \frac{b_k^{[1]}(\Delta z_{k+n-1}^{[0]})^2}{(a_{k+n-1}^{[1]})^2} - \sum_{k=N}^{\infty} q_k (a_{k+n}^{[0]})^2 z_{k+n}^{[0]} > 0.$$

Consequently, again by Proposition 2.1, equation (6) is nonoscillatory. $\qquad\square$

4   Two-terms Sturm-Liouville equations

In this section we specify the previous general results to the two-terms difference equation of the form

$$(-1)^n \Delta^n \left(k^{(\alpha)} \Delta^n y_k \right) = q_k y_{n+k}, \tag{13}$$

where $k^{(\alpha)} := \frac{\Gamma(k+1)}{\Gamma(k-\alpha+1)}$, Γ being the usual Euler gamma function. The below given oscillation and nonoscillation criteria are proved in [6]. Here we present these criteria in a slightly modified form in order to introduce better the problem discussed in the next section.

Theorem 4.1 *Suppose that $q_k \geq 0$ for large k.*

(i) $\alpha \notin \{1, 3, \ldots, 2n-1\}$, $m \in \{1, \ldots, n\}$, $\alpha < 2n-1$ and

$$M := \lim_{k \to \infty} k^{2n+1-\alpha-2m} \left(\sum_{j=k}^{\infty} q_j j^{(2m-2)} \right). \tag{14}$$

Equation (13) is oscillatory if

$$M > \mu_{n,\alpha,m} := \frac{[(2n-m-\alpha)^{(m)}(n-m)!]^2}{2n-2m+1-\alpha},$$

and nonoscillatory if

$$M < \nu_{n,\alpha,m} := \frac{\prod_{j=0}^{n-1}(2n-\alpha-1-j)^2}{4^n(2n-\alpha-1-2m)}.$$

(ii) Let $\alpha \in \{1, 3, \ldots, 2n-1\}$, $m = \frac{2n-1-\alpha}{2}$ and

$$M := \lim_{k \to \infty} \lg k \left(\sum_{j=k}^{\infty} q_j j^{(2m)} \right).$$

Equation (13) is oscillatory if

$$M > \rho_{n,m} := [m!(n-m-1)!]^2$$

and nonoscillatory if

$$M < \frac{\rho_{n,m}}{4}.$$

Note that a criterion similar to that given in part (i) of the previous theorem can be formulated also in the case $\alpha > 2n - 1$. Since $\nu_{n,\alpha,m} < \mu_{n,\alpha,m}$ (and, of course, $\rho_{n,m}/4 < \rho_{n,m}$) can be verified by a simple calculation, we see the gap between oscillation and nonoscillation criteria in the sense that based on Theorem 4.1 we cannot decide about the oscillation nature of (13) if the limit M is between nonoscillation and oscillation constant. In the last section we show that in some particular cases we are able to remove this gap and to prove that the "correct" oscillation constant is the constant appearing in the nonoscillation criterion.

Note that the assumption of nonnegativity of the sequence q in the previous theorem is not actually necessary. If this assumption is not satisfied, the statement remains valid if the sequence q_k is replaced by its nonnegative part $q_k^+ := \max\{0, q_k\}$ in the "nonoscillatory" part of the theorem. Also, the limit in the definition of the constant M can be replaced by $\liminf$ in the oscillatory part and by $\limsup$ in the nonoscillatory part. We preferred here the modified formulation of Theorem 4.1 since in the form presented here it better illustrates the gap between the constants in oscillation and nonoscillation criteria.

5   The exact value of the oscillation constant

In the recent paper [5] dealing with the Sturm-Liouville *differential* equation

$$(-1)^n \left(t^\alpha y^{(n)} \right)^{(n)} = q(t)y \tag{15}$$

we were able to remove the gap between oscillation and nonoscillation constants (the constants $\mu_{n,\alpha,m}, \nu_{n,\alpha,m}, \ldots$ appearing in the continuous counterparts of the criteria given in Theorem 4.1 have the same values as in the discrete case). We succeeded in proving that (15) with $\alpha \not\in \{1, 3, \ldots, 2n-1\}$ is oscillatory if $m \in \{1, \ldots, n\}$, $\alpha < 2n - 1$ and

$$\lim_{t \to \infty} t^{2n-1-\alpha-2m} \int_t^\infty q(s)s^{2m}\, ds > \nu_{n,\alpha,m},$$

or if $m \in \{1, \ldots, n\}$, $\alpha > 2m - 1$ and

$$\lim_{t \to \infty} t^{2m-1-\alpha} \int_t^\infty q(s)s^{2n-2m}\, ds > \zeta_{n,\alpha,m}$$

(the oscillation criterion similarly can be improved also in the case $\alpha \in \{1, 3, \ldots, 2n-1\}$). In the continuous case we used the fact that for $\alpha \not\in \{1, 3, \ldots, 2n-1\}$ we know at least two different solutions of the Euler equation

$$(-1)^n \left(t^\alpha y^{(n)} \right)^{(n)} = \frac{\gamma_{n,\alpha}}{t^{2n-\alpha}} y, \quad \gamma_{n,\alpha} := 4^{-n} \prod_{j=0}^{n-1} (2n - \alpha - 1 - 2j)^2, \tag{16}$$

namely $y = t^{\frac{2n-1-\alpha}{2}}$, $\tilde{y} = t^{\frac{2n-1-\alpha}{2}} \lg t$. Equation (15) is then rewritten into the form

$$(-1)^n \left(t^\alpha y^{(n)}\right)^{(n)} - \frac{\gamma_{n,\alpha}}{t^{2n-\alpha}} y = \left(q(t) - \frac{\gamma_{n,\alpha}}{t^{2n-\alpha}}\right) y \tag{17}$$

and the continuous analogue of the part (i) of Proposition 2.1 with the operator $L(y) = (-1)^n \left(t^\alpha y^{(n)}\right)^{(n)} - \frac{\gamma_{n,\alpha}}{t^{2n-\alpha}} y$ and $\left(q(t) - \frac{\gamma_{n,\alpha}}{t^{2n-\alpha}}\right)$ instead of q is applied to (17).

The discrete analogue of (16) with $\alpha = 0$ is the equation

$$(-1)^n \Delta^{2n} y_k + \frac{\gamma}{(k+2n-1)^{(2n)}} y_k = 0, \tag{18}$$

see [1]. However, this equation is not in self-adjoint form (self-adjoint form requires the index $k+n$ by y in the second term on the left-hand side of (18)), and for this reason the above mentioned continuous method does not extend directly to difference equations. A subject of the present investigation is to find how to overcome this difficulty. Till now we have been able to treat only the case $\alpha = 0$ in (13). We present here this partial result with the outline of the proof. We hope to prove the statement in the full generality in the future and to publish the complete proof in a subsequent paper.

Theorem 5.1 *Suppose that $\alpha = 0$ in (13) and M is given by (14) with $\alpha = 0$. Equation (13) is oscillatory provided*

$$\lim_{k \to \infty} k^{2n-1} \left(\sum_{j=k}^{\infty} q_j\right) > \nu_{n,0,1} = \frac{[(2n-1)!!]^2}{4^n(2n-1)}. \tag{19}$$

Proof. The proof is divided into three steps. The first step consists of proving that equation (13) with $\alpha = 0$ and

$$q_k = \gamma_{n,0} k^{-2n} + O(k^{-(2n+1)})$$

is nonoscillatory. This is proved essentially using Theorem 3.2 with suitably chosen sequences $b^{[j]}$, $j = 1, \ldots, n-1$. The next step is the proof of the claim that (13) is oscillatory provided $q_k - \gamma_{n,0} k^{-2n} \geq 0$ for large k and

$$\sum^{\infty} \left(q_k - \frac{\gamma_{n,0}}{k^{2n}}\right) k^{2n-1} = \infty. \tag{20}$$

Here, in contrast to the continuous case, we cannot use the fact that $y_k = k^{\frac{2n-1}{2}}$ is a solution of (13) with $q_k = \gamma_{n,0} k^{-2n}$. We have to use directly Proposition 2.1; we construct a sequence $y \in \mathcal{D}_n(N)$ for which the quadratic functional corresponding to (13) is negative. In this construction we use the result of the first step, and one has to overcome certain technical difficulties which do not appear in the continuous case.

Finally, in the last step it is proved that (19) implies (20). This part of the proof requires, compared with the computations given in [5], only minor technical modifications. $\square$

References

[1] R. P. AGARWAL, *Difference Equations and Inequalities: Theory, Methods and Applications*, Pure and Applied Mathematics, M. Dekker, New York, 1992.

[2] C. D. AHLBRANDT, A. C. PETERSON, *Discrete Hamiltonian Systems: Difference Equations, Continued Fractions, and Riccati Equations*, Kluwer Academic Publishers, Boston, 1996.

[3] M. BOHNER, *Linear Hamiltonian difference systems: disconjugacy and Jacobi-type conditions*, J. Math. Anal. Appl. **199** (1996), 804–826.

[4] O. DOŠLÝ, *Oscillation criteria for higher order Sturm-Liouville difference equations*, J. Differ. Equations Appl. **4** (1998), no. 5, 425–450.

[5] O. DOŠLÝ, *Constants in the oscillation theory of higher order Sturm-Liouville differential equations*, Electron J. Differ. Equ. **2002** (2002), No. 34, 12 p.

[6] O. DOŠLÝ, R. HILSCHER, *A class of Sturm-Liouville difference equations: (Non)oscillation constants and property BD*, to appear in Comput. Math. Appl.

[7] I. M. GLAZMAN, *Direct Methods of Qualitative Analysis of Singular Differential Operators*, Davey, Jerusalem 1965.

[8] P. HARTMAN, *Difference equations: disconjugacy, principal solutions, Green's function, complete monotonicity*, Trans. Amer. Math. Soc. **246** (1978), 1–30.

[9] R. HILSCHER, *Discrete spectra criteria for certain class of singular differential and difference operators*, Comput. Appl. Math. **42** (2001), 465–471.

A Transformation for the Riccati Difference Operator

JULIA ELYSEEVA [1]

Department of Mathematics
Moscow State University of Technology
101472 Vadkovskii per.3a, Moscow, Russia
E-mail: elyseeva@mtu-net.ru

Abstract We examine transformations for symplectic difference systems and Riccati difference operators connected with permutations of rows of a conjoined basis. The concept of an integration path for a conjoined basis is developed to formulate the definition of a focal point and disconjugacy in terms of solutions of the transformed Riccati equation.

Keywords Symplectic systems, Riccati difference equation, Focal points

AMS Subject Classification 39A10

1 Introduction

We consider transformations for the symplectic difference system

$$Y_{i+1} = W_i \, Y_i \,, \quad W_i^T J_{2n} \, W_i = J_{2n}, \quad i = 0, 1, \dots, N, \tag{1.1}$$

and the Riccati difference operator

$$R_W[Q] = W_{21}^i - Q_{i+1} \, W_{11}^i + W_{22}^i \, Q_i - Q_{i+1} \, W_{12}^i \, Q_i,$$

where W_i, Y_i, J_{2n} are real partitioned matrices with $n \times n$ blocks

$$W_i = \left(W_{kp}^i \right), \; k,p = 1,2, \; Y_i = \begin{bmatrix} X_i \\ U_i \end{bmatrix}, \; J_{2n} = \begin{bmatrix} 0_n & I_n \\ -I_n & 0_n \end{bmatrix},$$

and I_n, 0_n are the identity and zero matrices. It is well known that the Riccati matrix difference equation

$$R_W[Q] = 0_n, \; i = 0, \dots, N \tag{1.2}$$

---

[1]Supported by the Federal Goal Program "Integration" under grant number 43.

has a symmetric solution $Q_i^T = Q_i$ iff there exists a conjoined basis of system (1.1)

$$Y_i^T J_{2n} Y_i = 0_n, \ \mathrm{rank} Y_i = n$$

such that the condition

$$\det X_i \neq 0, \ i = 0, \dots, N+1 \tag{1.3}$$

holds. If (1.3) does not hold, we have to consider generalized solutions of equation (1.2). According to the definition (see Bohner and Došlý [2, Definition 3]), a conjoined basis of (1.1) is said to have a focal point in $(i, i+1]$ if the conditions

$$\mathrm{Ker} X_{i+1} \subseteq \mathrm{Ker} X_i, \tag{1.4}$$

$$X_i X_{i+1}^\dagger W_{12}^i \geq 0 \tag{1.5}$$

do not hold (here, $\dagger$ denotes the Moore-Penrose inverse of the matrix A; $\mathrm{Ker} A$ denotes the kernel of A; for a symmetric matrix A we write $A \geq 0$ if A is positive semidefinite). If a conjoined basis without focal points in $(0, N+1]$ is considered, then conditions (1.4), (1.5) are equivalent to the existence of a symmetric solution of the "implicit Riccati equation" (see Bohner and Došlý [2])

$$R_W[Q] X_i = 0_n, \ i = 0, \dots, N \tag{1.6}$$

such that the condition

$$\left(W_{22}^{i\,T} - W_{12}^{i\,T} Q_{i+1} \right) W_{12}^i \geq 0, \ i = 0, \dots N \tag{1.7}$$

holds. If the matrix X_i is nonsingular, condition (1.7) may be rewritten as

$$\left(W_{11}^i + W_{12}^i Q_i \right)^{-1} W_{12}^i \geq 0 \tag{1.8}$$

for the solution of the Riccati equation (1.2). System (1.1) is disconjugate on $(0, N+1]$ if the solution with initial conditions $X_0 = 0_n$, $U_n = I_n$ (the principal solution at 0) has no focal points in $(0, N+1]$. By Bohner and Došlý [2], we have to know normalized conjoined bases to get the solution of (1.6). We offer to consider the solutions of a transformed Riccati equation

$$R_{\widetilde{W}}[Q_j] = 0_n, \tag{1.9}$$

where the matrices $\widetilde{W}_i$ are defined by the formula $\widetilde{W}_i = \left(\mathfrak{N}_{j(i+1)} \right)^T W_i \, \mathfrak{N}_{j(i)}$, and $\mathfrak{N}_{j(i)}$ are the symplectic orthogonal matrices uniquely determined by the values of a function $j = j(i)$, $i = 0, \dots, N+1$ (it is called an integration path for a conjoined basis Y_i (see Eliseeva [4])). The transformations with matrices $\mathfrak{N}_{j(i)}$ are connected with permutations of rows of a conjoined basis in such a manner that condition (1.3) holds for the transformed basis for any i. The idea of this approach was offered in Taufer [6] for linear differential systems and was developed in Eliseeva [5] for Hamiltonian differential systems. In this work we introduce a special integration path $j = j(i)$, $i = 0, \ldots, N+1$ connected with rankX_i to formulate the definition of a focal point in terms of solutions of (1.9). The main result is the following theorem.

Theorem 1.1. *Let* rankX_i = rank$F_j X_i$ = rankF_j, *where* Y_i *is a conjoined basis of system* (1.1), F_j, G_j *are diagonal* $n \times n$ *matrices*, $F_j + G_j = I_n$, *and the diagonal of* G_j *is composed of the zeros and ones that constitute the binary representation for the number* $j = j(i)$, $i = 0, \ldots, N+1$. *Then condition* (1.4) *is equivalent to the condition*

$$F_{j(i)} \left(\widetilde{W}_{11}^i + \widetilde{W}_{12}^i Q_{j(i)} \right)^{-1} G_{j(i+1)} = 0_n, \tag{1.10}$$

and if (1.4) *(or* (1.10)*) holds, then* (1.5) *is equivalent to*

$$F_{j(i)} \left(\widetilde{W}_{11}^i + \widetilde{W}_{12}^i Q_{j(i)} \right)^{-1} \widetilde{W}_{12}^i F_{j(i)} \geq 0. \tag{1.11}$$

The matrix $Q_{j(i)}^i = Q_{j(i)} = \left(G_{j(i)} X_i + F_{j(i)} U_i \right) \left(F_{j(i)} X_i - G_{j(i)} U_i \right)^{-1}$ *is the solution of* (1.9) *for* $j = j(i)$, $i = 0, \ldots, N+1$.

Remark 1.2. *Note that if* (1.3) *holds for* $i = 0, \ldots, N+1$, *then* $F_j \equiv I_n$, $G_j \equiv 0_n, j = j(i)$, $i = 0, \ldots, N+1$; *hence, the integration path defined by the matrix* G_j *is trivial* $(j(i) \equiv 0)$, *and the solution along this path* Q_0^i *coincides with the classical solution of* (1.2). *In this case, condition* (1.10) *is trivial, and* (1.11) *passes into* (1.8).

Corollary 1.3. *Let* Y_i *be the principal solution at* 0. *Then system* (1.1) *is disconjugate on* $(0, N+1]$ *iff conditions* (1.10), (1.11) *hold for the solution of the equation* (1.9) *with the initial conditions* $Q_{j(0)}^0 = 0_n$, $j(0) = 2^n - 1$; *the path* $j = j(i)$, $i = 0, \ldots, N+1$ *is defined in Theorem 1.1.*

2   Preliminaries

We introduce the transformations

$$Y_i = \mathfrak{N}_j Y_j^i, \quad \mathfrak{N}_j = \begin{bmatrix} F_j & G_j \\ -G_j & F_j \end{bmatrix}, \quad Y_j^i = \begin{bmatrix} X_j^i \\ U_j^i \end{bmatrix}, \quad j = j(i). \tag{2.1}$$

When writing $Y_{j(i)}$, $X_{j(i)}$, $U_{j(i)}$, $Q_{j(i)}$, we always mean $Y_{j(i)}^i$, $X_{j(i)}^i$, $U_{j(i)}^i$, $Q_{j(i)}^i$. For the matrices $Y_{j(i)}$ we have the transformed system

$$Y_{j(i+1)} = \widetilde{W}_i Y_{j(i)}, \quad \widetilde{W}_i = \mathfrak{N}_{j(i+1)}^T W_i \mathfrak{N}_{j(i)}. \tag{2.2}$$

Consider the definition of the matrices $\mathfrak{N}_j$. We say that a matrix $\mathfrak{N} \in \Omega_{\mathfrak{N}}$ if it may be written in form (2.1) with $n \times n$ diagonal blocks F, G which obey the conditions:

$$F^2 + G^2 = I_n, \; FG = 0_n, \qquad (2.3)$$

$$F \geqslant 0, \; G \geqslant 0. \qquad (2.4)$$

Condition (2.3) defines a group of symplectic orthogonal matrices. It is easy to verify that there exist only 2^n symplectic orthogonal matrices $\mathfrak{N}_j$ defined by (2.3), (2.4). The number $j = j(i)$ of any $\mathfrak{N}_j$ takes the values from the set $\{0, 1, \ldots, 2^n - 1\}$, and the diagonal of G_j is composed of the zeros and ones that constitute the binary representations of $j = j(i)$. In this case, we have $\mathfrak{N}_0 = I_{2n}, \; \mathfrak{N}_{2^n - 1} = J_{2n}, \; \mathfrak{N}_j = J_{2n} \mathfrak{N}_{2^n - 1 - j}^T$.

The treatment of the set $\Omega_{\mathfrak{N}}$ is justified by the following theorem.

Theorem 2.1. *For any conjoined basis of system* (1.1) *there exists a function* $j = j(i), \; i = 0, \ldots, N + 1$ *such that the matrix* $X_{j(i)}$ *in* (2.1) *is nonsingular:*

$$\det \left(X_{j(i)} \right) = \det \left(F_j X_i - G_j U_i \right) \neq 0, \; i = 0, \ldots, N + 1. \qquad (2.5)$$

Hence, there exists a symmetric solution $Q_j = U_j X_j^{-1}$ *of the Riccati equation* (1.9) *associated with transformed system* (2.2).

Proof. See Eliseeva [5], where the relations between Plucker's (Grassmann's) coordinates of the Lagrangean plane in the symplectic space (see Zelikin [7]) are used. $\qquad \square$

Definition 2.2. *A function* $j = j(i), \; i = 0, \ldots, N + 1$ *is called an integration path for a conjoined basis of system* (1.1) *if condition* (2.5) *holds. In this case, one can consider* $Q_j^i = Q_j = U_j X_j^{-1}$ *as the solution along the path* $j = j(i), \; i = 0, \ldots, N + 1$.

It does not follow from this definition that an integration path is uniquely defined. But if we know an integration path $j(i)$ for a conjoined basis and Q_j, we possess all information on the existence of other integration paths for given basis and rank X_i. To show this, consider the properties of transformations (2.1). Let $l = l(i), \; i = 0, \ldots, N + 1$ be another function and matrices $\mathfrak{N}_l$ belonging to $\Omega_{\mathfrak{N}}$ with blocks F, G. Then we have

$$Y_l = \mathfrak{N}_p Y_j, \quad \mathfrak{N}_p = \mathfrak{N}_l^T \mathfrak{N}_j = \begin{bmatrix} F_p & G_p \\ -G_p & F_p \end{bmatrix}, \qquad (2.6)$$

where $Y_i = \mathfrak{N}_j Y_j = \mathfrak{N}_l Y_l$. Certainly, if $l = 0$, (2.6) passes into (2.1) (we have $Y_0^i \equiv Y_i$). The diagonal matrices F_p, G_p obey conditions (2.3), but generally speaking conditions (2.4) do not hold for F_p, G_p. We can formulate the following proposition.

Proposition 2.3. *Let* $\mathfrak{N}_j, \mathfrak{N}_l$ *belong to* $\Omega_{\mathfrak{N}}$. *The matrix* $\mathfrak{N}_p \in \Omega_{\mathfrak{N}}$ *iff* $G_j \geqslant G_l \; (F_l \geqslant F_j)$. *The matrix* $\mathfrak{N}_p^T \in \Omega_{\mathfrak{N}}$ *iff* $G_l \geqslant G_j \; (F_j \geqslant F_l)$.

Proof. It follows from (2.6) that

$$F_p = F_l F_j + G_l G_j, \qquad G_p = F_l G_j - F_j G_l, \qquad (2.7)$$

or

$$F_l = F_p F_j + G_p G_j, \qquad G_l = F_p G_j - F_j G_p, \qquad (2.8)$$

where we use $\mathfrak{N}_l = \mathfrak{N}_j \mathfrak{N}_p^T$. By (2.7) and the definition of $\Omega_\mathfrak{N}$, we have that $F_p \geqslant 0$, and $G_p \geqslant 0$ iff $F_j G_l = 0_n$. But $G_j - G_l = F_l - F_j = G_j F_l - G_l F_j \geqslant 0$ iff $F_j G_l = 0_n$. In this case, $G_j - G_l = F_l - F_j = G_p \geqslant 0$, or

$$F_l = F_j + G_p, \qquad G_l = G_j - G_p. \qquad (2.9)$$

We have proved the first claim of Proposition 2.3. The second claim can be proved in the same manner. Using (2.9), we also have that the condition $\mathfrak{N}_p \in \Omega_\mathfrak{N}$ is equivalent to the condition $G_j \geqslant G_p$ ($F_p \geqslant F_j$). $\qquad \square$

Let $j - j(i)$, $i - 0, \ldots, N+1$ be an integration path for a conjoined basis Y_i. Consider the formula which connects all principal minors of the matrix $Q_j = U_j X_j^{-1}$ with the corresponding minors of order n of Y_i. We have

$$M\left[G_p Q G_p\right] = (-1)^{\operatorname{rank}(G_l F_j)} \frac{\det\left(F_l X_i - G_l U_i\right)}{\det X_j}, \qquad (2.10)$$

where $M\left[G_p Q G_p\right]$ denotes the principal minor of $Q_j = U_j X_j^{-1}$ located in rows and columns defined by the positions of nonzero elements in the diagonal of matrix G_p. The minor $\det\left(F_l X_i - G_l U_i\right)$ of order n accurate to a sign coincides with the minor of Y_i. The location of this minor is uniquely determined by F_l, G_l and formulae (2.8). It follows from (2.10) that another integration path $l = l(i)$, $i = 0, \ldots, N+1$ exists if and only if the principal minor $M\left[G_p Q G_p\right]$ with G_p defined by (2.7) is nonzero for any $i = 0, \ldots, N+1$ or

$$\operatorname{rank}\left(G_p Q_j G_p\right) = \operatorname{rank} G_p > 0. \qquad (2.11)$$

If condition (2.11) holds, one can consider the formula that connects the solutions $Q_{l(i)}^i = Q_{l(i)}$, $Q_{j(i)}^i = Q_{j(i)}$ for the different integration paths $Q_l = (I_n - F_p Q_j G_p)\left(F_p Q_j F_p - (G_p Q_j G_p)^\dagger\right)(I_n - G_p Q_j F_p)$. For example, if $j(i) \equiv 0$, $l(i) \equiv 2^n - 1$, we have $Q_{2^n-1}^i = -\left(Q_0^i\right)^{-1} \equiv -Q_i^{-1}$; moreover, $Q_{2^n-1-j}^i = -\left(G_p Q_j^i G_p\right)^{-1}$, $G_p = G_j - F_j$ (it follows from the last formula and (2.11) that Q_j is invertible).

Now we introduce a special integration path for a conjoined basis.

Lemma 2.4. *Let the matrices* $\mathfrak{N}_j \in \Omega_\mathfrak{N}$, $j = j(i)$, $i = 0, \ldots, N+1$, *and*

$$\operatorname{rank} X_i = \operatorname{rank} F_j X_i = \operatorname{rank} F_j, \qquad (2.12)$$

where X_i *is the upper block of a conjoined basis* Y_i. *Then* $j = j(i)$, $i = 0, \ldots, N+1$ *is the integration path for* Y_i, *and the solution along this path* Q_j *satisfies the condition*

$$G_j Q_j G_j = 0_n. \qquad (2.13)$$

Proof. Let condition (2.12) hold. We have to prove that (2.5) holds. Assume without loss of generality that

$$F_j = \text{diag}\left[\underbrace{11\ldots1}_{k}00\ldots0\right], \quad G_j = \text{diag}\left[\underbrace{00\ldots0}_{k}11\ldots1\right].$$

In the general case one can consider the matrix LY_i, where the block-diagonal orthogonal symplectic matrix L carries out the same permutations of rows of X_i, U_i. As is well known (see Zelikin [7]), if we multiply Y_i by a nonsingular $n \times n$ matrix M, Plucker's coordinates of Y_i will multiply by $\det M$. Hence, we can consider the new conjoined basis $\widetilde{Y}_i = Y_i M_1$, $\det M_1 \neq 0$ such that $\widetilde{X}_i = \begin{bmatrix} X_{11} & 0 \\ X_{21} & 0 \end{bmatrix}$, $\det X_{11} \neq 0$, where X_{11} is a $k \times k$ matrix. If $\widetilde{U}_i$ is separated into blocks $\widetilde{U}_i = (U_{lp})$, $l, p = 1, 2$ in the same manner, then we have to prove that $\det \begin{pmatrix} X_{11} & 0 \\ -U_{21} & -U_{22} \end{pmatrix} \neq 0$ or $\det U_{22} \neq 0$. If rank $U_{22} = q < n - k$, it is possible to introduce the block-diagonal matrix $M_2 = \text{diag}\,[I_k, P]$, $\det P \neq 0$ such that $\widehat{U}_{22} = U_{22}P$ has the same structure as $\widetilde{X}_i$ with $n - k - q$ zero columns. Then the new conjoined basis $\widehat{Y}_i = Y_i M_1 M_2$, $\det(M_1 M_2) \neq 0$ has $n - k - q$ zero columns as well because, by the condition $\widehat{X}_i^T \widehat{U}_i = \widehat{U}_i^T \widehat{X}_i$, we have for the blocks of $\widehat{X}_i = \widetilde{X}_i$, $\widehat{U}_i = \left(\widehat{U}_{lp}\right)$, $l, p = 1, 2$ that $X_{11}^T \widehat{U}_{12} + X_{21}^T \widehat{U}_{22} = 0$ or $\widehat{U}_{12} = -\left(X_{11}^T\right)^{-1} X_{21}^T \widehat{U}_{22}$, and we arrive at a contradiction with the definition of Y_i. Then the path defined by (2.12) does exist.

To prove (2.13), consider the factorization

$$\begin{aligned} X_i &= (F_j X_j + G_j U_j) = (F_j + G_j Q_j) X_j = \\ &= (I_n + G_j Q_j F_j)(F_j + G_j Q_j G_j) X_j, \end{aligned} \quad (2.14)$$

where $j = j(i)$ is an integration path. Since the matrices $I_n + G_j Q_j F_j$, X_j are nonsingular, we have that

$$\text{rank}\,X_i = \text{rank}\,(F_j + G_j Q_j G_j) = \text{rank}\,F_j + \text{rank}\,(G_j Q_j G_j),$$

and $\text{rank}\,X_i = \text{rank}\,F_j$ iff $G_j Q_j G_j = 0_n$.   □

Remark 2.5. *It was possible to prove the existence of the integration path $j = j(i)$, $i = 0, \ldots, N + 1$ defined by (2.12) using only the definition of a conjoined basis Y_i. But it is important that if we have a solution Q_j along an integration path, we can construct another path defined by (2.12). So, if in (2.14) rank $(G_j Q_j G_j) > 0$, one can find the principal minor $M\,[G_p Q G_p]$ of Q_j such that $G_j \geqslant G_p \geqslant 0$, rank $(G_j Q_j G_j) = \text{rank}\,(G_p Q_j G_p) = \text{rank}\,G_p > 0$. Hence, we can introduce the new integration path $l = l(i)$ determined by (2.9) with $\text{rank}\,X_i = \text{rank}\,(F_l X_i) = \text{rank}\,F_l$, and $G_l Q_l G_l = 0_n$.*

Definition 2.6. *The function $j = j(i)$, $i = 0, \ldots, N + 1$ defined by condition (2.12) is called a special path for a conjoined basis Y_i.*

3   Proof of the main theorem

Lemma 3.1. *Let* $j = j(i)$, $i = 0, \ldots, N+1$ *be a special integration path for a conjoined basis* Y_i. *Then we have*

$$\operatorname{Im} X_i = \operatorname{Im}(F_j + G_j Q_j F_j), \quad \operatorname{Ker} X_i = \operatorname{Ker}(F_j X_j).$$

Proof. Note that factorization (2.14) for the case (2.13) may be rewritten as

$$X_i = (F_j + G_j Q_j F_j)(F_j X_j) = AB. \tag{3.1}$$

Then (3.1) is the analog of the skeleton factorization (Gantmacher [3]) such that

$$\operatorname{rank} X_i = \operatorname{rank} A = \operatorname{rank} B = \operatorname{rank} F_j, \quad A = AF_j, \quad B = F_j B,$$

$$BB^{\dagger} = F_j, \quad A^{\dagger} A = F_j, \quad X_i^{\dagger} = B^{\dagger} A^{\dagger}$$

so that $X_i B^{\dagger} = A$, $A^{\dagger} X_i = B$. Hence, by the last conditions, we have that $\operatorname{Im} X_i = \operatorname{Im} A$, $\operatorname{Ker} X_i = \operatorname{Ker} B$.   □

Using (3.1), one can get the following proposition.

Proposition 3.2. *Let* $j = j(i)$, $i = 0, \ldots, N+1$ *be a special integration path for a conjoined basis* Y_i. *Then for any* $n \times n$ *matrix* C *we have*

$$\operatorname{Ker} X_{i+1} \subseteq \operatorname{Ker} C \qquad \Leftrightarrow \qquad C X_{j(i+1)}^{-1} = C X_{j(i+1)}^{-1} F_{j(i+1)}. \tag{3.2}$$

Proof. One direction of (3.2) is trivial because it follows from the right-hand side of (3.2) that $C = C X_{j(i+1)}^{-1} F_{j(i+1)} X_{j(i+1)}$, and, by Lemma 3.1, $\operatorname{Ker} X_{i+1} = \operatorname{Ker}(F_{j(i+1)} X_{j(i+1)})$. If $\operatorname{Ker} X_{i+1} \subseteq \operatorname{Ker} C$, then, by Bohner [1], we have $C = C(F_j X_j)^{\dagger} F_j X_j$, $j = j(i+1)$ or $C X_j^{-1} = C(F_j X_j)^{\dagger} F_j$, $j - j(i+1)$. Then for $j - j(i+1)$ it follows that

$$C X_j^{-1} = C(F_j X_j)^{\dagger} = C X_j^{-1} F_j. \tag{3.3}$$

□

Now we can prove the first claim of Theorem 1.1. It is sufficient to put $C = F_{j(i)} X_{j(i)}$ in equivalence (3.2). Hence, we have

$$F_{j(i)} X_{j(i)} X_{j(i+1)}^{-1} = F_{j(i)} X_{j(i)} X_{j(i+1)}^{-1} F_{j(i+1)},$$

where $X_{j(i)} X_{j(i+1)}^{-1} = \left(\widetilde{W}_{11}^i + \widetilde{W}_{12}^i Q_{j(i)} \right)^{-1}$, by the first equation of the transformed symplectic systems (2.2).

Consider the transformation of Y_i

$$\widetilde{Y}_i = A_j Y_i = \begin{bmatrix} F_j X_j \\ -G_j X_j \end{bmatrix}$$

with the symplectic matrix

$$A_j = \begin{bmatrix} I_n - G_j Q_j F_j & G_j Q_j G_j \\ -F_j Q_j F_j & I_n + F_j Q_j G_j \end{bmatrix} = \mathfrak{N}_j \begin{bmatrix} I_n & 0_n \\ -Q_j & I_n \end{bmatrix} \mathfrak{N}_j^T,$$

where Q_j is the solution along an integration path for Y_i. If $j = j(i)$ is a special path, then by (2.13), A_j is the lower triangular block matrix, and by Bohner and Došlý [2, Lemma 7], $\widetilde{Y}_i$ does not have focal points iff Y_i does not have ones. It is easy to verify that the matrix of the new symplectic system for $\widetilde{Y}_i$ may be written as $\mathfrak{N}_{j(i+1)} \begin{bmatrix} B & \widetilde{W}_{12}^i \\ 0_n & (B^T)^{-1} \end{bmatrix} \mathfrak{N}_{j(i)}^T$, where

$B = \left(\widetilde{W}_{11}^i + \widetilde{W}_{12}^i Q_{j(i)} \right)$. Condition (1.4) for $\widetilde{Y}_i$ is equivalent to (1.10) because the special integration paths for $\widetilde{Y}_i, Y_i$ coincide. Then, if (1.10) holds, using (3.3), we have the left-hand side of (1.5) for $\widetilde{Y}_i$ in form

$$F_{j(i)} X_{j(i)} \left(F_{j(i+1)} X_{j(i+1)} \right)^\dagger \left(-F_{j(i+1)} B G_{j(i)} + F_{j(i+1)} \widetilde{W}_{12}^i F_{j(i)} \right) =$$
$$= F_{j(i)} X_{j(i)} X_{j(i+1)}^{-1} F_{j(i+1)} \left(-B G_{j(i)} + \widetilde{W}_{12}^i F_{j(i)} \right) = F_{j(i)} B^{-1} \widetilde{W}_{12}^i F_{j(i)}.$$

We have proved the second part of Theorem 1.1. □

References

[1] Bohner, M., Riccati matrix difference equations and linear Hamiltonian difference system, *Dynamics of Continuous, Discrete and Impulsive Systems* **2** (1996), 147–159.

[2] Bohner, M. and Došlý, O., Disconjugacy and transformations for symplectic systems, *Rocky Mountain Journal of Mathematics* **27** (1997), 707–743.

[3] Gantmacher, F., *The Theory of Matrices*, v.1, Chelsea Publishing Company, New York, 1959.

[4] Eliseeva, Y., An algorithm for solving the matrix difference Riccati equation, *Computational Mathematics and Mathematical Physics* **39** (1999), 187–194.

[5] Eliseeva, Y., On an algorithm for solving the symplectic matrix Riccati equation, *Moscow University Computational Mathematics and Cybernetics* **2** (1990), 14–19.

[6] Taufer, J., On Factorization Method, *Aplikace Matematiky* **11** (1966), 427–451.

[7] Zelikin, M., *Uniform spaces and the Riccati equation in calculus of variations*, Factorial, Moscow, 1998.

On the Dynamics of $y_{n+1} = \dfrac{p+y_{n-2}}{q\,y_{n-1}+y_{n-2}}$

E. A. GROVE, G. LADAS and L. C. McGRATH

Department of Mathematics, University of Rhode Island
Kingston, Rhode Island, 02881 USA
E-mail: grove@math.uri.edu, gladas@math.uri.edu

Abstract We investigate the global character of solutions of the equation in the title, with positive parameters and positive initial conditions.

Keywords Asymptotic stability, Attracting intervals, Boundedness, Difference equations, Global asymptotic stability, Invariant intervals, Semicycles, Permanence, Persistence

AMS Subject Classification 39A10, 39A11

1 Introduction and preliminaries

In this paper we investigate the global character of solutions of the third order rational difference equation

$$y_{n+1} = \frac{p+y_{n-2}}{qy_{n-1}+y_{n-2}} \quad , \quad n = 0,1,\dots \tag{1}$$

with positive parameters and positive initial conditions.

The study of rational difference equations is quite challenging and rewarding and is still in its infancy. See [1] and [2] for some basic results in this area and for open problems and conjectures.

We believe that results about rational difference equations are of paramount importance in their own right, and furthermore, we believe that these results offer prototypes towards the development of the basic theory of the global behavior of solutions of nonlinear difference equations of order greater than one. The techniques and results which we develop in this paper to understand the dynamics of Eq.(1) are also useful in analyzing the equations in the mathematical models of various biological systems and other applications.

We now present some definitions and known results which will be useful in our investigation of Eq.(1).

Let I be an interval of real numbers, and let $f : I^3 \to I$ be a continuous function. Consider the difference equation

$$y_{n+1} = f(y_n, y_{n-1}, y_{n-2}) \quad , \quad n = 0,1,\dots \tag{2}$$

with initial conditions $y_{-2}, y_{-1}, y_0 \in I$. We say that $\bar{y}$ is an *equilibrium point* of Eq.(2) if

$$f(\bar{y}, \bar{y}, \bar{y}) = \bar{y}.$$

The following two theorems are found in [1].

Theorem A *Let $[a, b]$ be an interval of real numbers, and assume that*

$$f : [a, b] \times [a, b] \times [a, b] \to [a, b]$$

is a continuous function satisfying the following properties:

(a) $f(x, y, z)$ *is non-increasing in each of its arguments.*

(b) *If $(m, M) \in [a, b] \times [a, b]$ is a solution of the system*

$$M = f(m, m, m) \quad and \quad m = f(M, M, M),$$

  then $m = M$.

Then Eq.(2) has a unique equilibrium point $\bar{y} \in [a, b]$, and every solution of Eq.(2) converges to $\bar{y}$.

Theorem B *Let $[a, b]$ be an interval of real numbers, and assume that*

$$f : [a, b] \times [a, b] \times [a, b] \to [a, b]$$

is a continuous function satisfying the following properties:

(a) $f(x, y, z)$ *is non-decreasing in $x \in [a, b]$ for each $(y, z) \in [a, b] \times [a, b]$.*

(b) $f(x, y, z)$ *is non-increasing in $y \in [a, b]$ for each $(x, z) \in [a, b] \times [a, b]$.*

(c) $f(x, y, z)$ *is non-decreasing in $z \in [a, b]$ for each $(x, y) \in [a, b] \times [a, b]$.*

(d) *If $(m, M) \in [a, b] \times [a, b]$ is a solution of the system*

$$M = f(M, m, M) \quad and \quad m = f(m, M, m),$$

  then $m = M$.

Then Eq.(2) has a unique equilibrium point $\bar{y} \in [a, b]$, and every solution of Eq.(2) converges to $\bar{y}$.

2   Local stability, identities, and invariant and attracting intervals

In this section we develop some preliminary results which will be useful in establishing the global character of solutions in certain regions of the parameters.

2.1   Local stability analysis

Eq.(1) has the unique positive equilibrium point

$$\bar{y} = \frac{1 + \sqrt{1 + 4p(1+q)}}{2(1+q)}.$$

The linearized equation associated with Eq.(1) about $\bar{y}$ is

$$z_{n+1} + \frac{q}{1+q}z_{n-1} + \frac{p - q\bar{y}}{(\bar{y}+p)(1+q)}z_{n-2} = 0, \quad n \geq 0. \tag{3}$$

The following result provides a sufficient condition for the equilibrium $\bar{y}$ of Eq.(1) to be locally asymptotically stable. The proof is straightforward and will be omitted.

Theorem 2.1 *Suppose $q \leq 4p+1$. Then the equilibrium $\bar{y}$ of Eq.(1) is locally asymptotically stable.*

2.2   Identities and invariant intervals

The following identities play an important role in understanding the character of solutions of Eq.(1).

Lemma 2.2 *Let $\{y_n\}_{n=-2}^{\infty}$ be a solution of Eq.(1). Then the following statements are true:*

$$y_{n+1} - 1 \;=\; q\left(\frac{\frac{p}{q} - y_{n-1}}{qy_{n-1} + y_{n-2}}\right), \quad n \geq 0 \tag{4}$$

$$y_{n+1} - \frac{p}{q} \;=\; \frac{pq(1 - y_{n-1}) + (q - p)y_{n-2}}{q(qy_{n-1} + y_{n-2})}, \quad n \geq 0 \tag{5}$$

$$y_n - y_{n+4} \;=\; \frac{qy_{n-1}(y_n - \frac{p}{q}) + y_{n+1}(y_{n-1} + qy_n)(y_n - 1)}{q(p + y_{n-1}) + y_{n+1}(qy_n + y_{n-1})}, \quad n \geq 0. \tag{6}$$

The proofs of the following three lemmas follow directly from the identities (4)–(6). These results show that, in particular, the interval with endpoints 1 amd $\frac{p}{q}$ is invariant.

Lemma 2.3 *Assume that $p > q$ and let $\{y_n\}_{n=-2}^{\infty}$ be a solution of Eq.(1). Then the following statements are true:*

(i) *If $y_N \leq \dfrac{p}{q}$ for some $N \geq 0$, then $y_{N+2} \geq 1$.*

(ii) If $y_N \geq 1$ for some $N \geq 0$, then $y_{N+2} \leq \frac{p}{q}$.

(iii) If $y_N \in \left[1, \frac{p}{q}\right]$ for some $N \geq 0$, then $y_{N+2k} \in \left[1, \frac{p}{q}\right]$ for all $k \geq 0$.

(iv) If $y_N < 1$ for some $N \geq 0$, then $y_{N+2} > 1$.

(v) If $y_N > \frac{p}{q}$ for some $N \geq 0$, then $y_{N+2} < 1$.

(vi) If $y_N, y_{N+1} \in \left[1, \frac{p}{q}\right]$ for some $N \geq 0$, then $y_n \in \left[1, \frac{p}{q}\right]$ for all $n \geq N$.

Lemma 2.4 *Assume that $p = q$ and let $\{y_n\}_{n=-2}^{\infty}$ be a solution of Eq.(1). Then the following statements are true:*

(i) If $y_N < 1$ for some $N \geq 0$, then $y_{N+2} > 1$.

(ii) If $y_N = 1$ for some $N \geq 0$, then $y_{N+2} = 1$.

(iii) If $y_N > 1$ for some $N \geq 0$, then $y_{N+2} < 1$.

Lemma 2.5 *Assume that $p < q$ and let $\{y_n\}_{n=-2}^{\infty}$ be a solution of Eq.(1). Then the following statements are true:*

(i) If $y_N \leq 1$ for some $N \geq 0$, then $y_{N+2} \geq \frac{p}{q}$.

(ii) If $y_N \geq \frac{p}{q}$ for some $N \geq 0$, then $y_{N+2} \leq 1$.

(iii) If $y_N \in \left[\frac{p}{q}, 1\right]$ for some $N \geq 0$, then $y_{N+2k} \in \left[\frac{p}{q}, 1\right]$ for all $k \geq 0$.

(iv) If $y_N < \frac{p}{q}$ for some $N \geq 0$, then $y_{N+2} > 1$.

(v) If $y_N > 1$ for some $N \geq 0$, then $y_{N+2} < 1$.

(vi) If $y_N, y_{N+1} \in \left[\frac{p}{q}, 1\right]$ for some $N \geq 0$, then $y_n \in \left[\frac{p}{q}, 1\right]$ for all $n \geq N$.

2.3   Attracting Intervals

In this section we show that, when $p \neq q$, the interval with endpoints 1 and $\frac{p}{q}$ attracts all solutions of Eq.(1)

Lemma 2.6 *Suppose $p > q$. Then every solution of Eq.(1) eventually enters and remains in the interval $\left[1, \frac{p}{q}\right]$.*

Proof: Let $M \geq 0$. By Lemma 2.3, it suffices to show that there exists $k \geq 0$ such that $y_{M+2k} \in \left[1, \frac{p}{q}\right]$. So for the sake of contradiction, suppose no such k exists. It follows by Lemma 2.3 that without loss of generality, we may assume that for all $k \geq 0$,

$$y_{M+4k} < 1 \quad \text{and} \quad y_{M+4k+2} > \frac{p}{q}.$$

Thus by Lemma 2.2, we see that

$$y_M < y_{M+4} < y_{M+8} < \cdots < y_{M+4k} < \cdots < 1$$

and

$$y_{M+2} > y_{M+6} > \cdots > y_{M+4k+2} > \cdots > \frac{p}{q}.$$

Hence there exists $L_0 \in (0, 1]$ and $L_2 \in \left[\frac{p}{q}, \infty\right)$ such that

$$\lim_{k \to \infty} y_{M+4k} = L_0 \quad \text{and} \quad \lim_{k \to \infty} y_{M+4k+2} = L_2.$$

For $k \geq 0$, we see that

$$y_{M+4k+2} = \frac{p + y_{M+4k-1}}{qy_{M+4k} + y_{M+4k-1}}$$

from which it follows that

$$y_{M+4k-1} = \frac{p - qy_{M+4k}y_{M+4k+2}}{y_{M+4k+2} - 1}$$

and so

$$\lim_{k \to \infty} y_{M+4k-1} = L_3$$

where $L_3 = \frac{p - qL_0L_2}{L_2 - 1} \in [0, \infty)$. Similarily

$$y_{M+4k+1} = \frac{p + y_{M+4k-2}}{qy_{M+4k-1} + y_{M+4k-2}}$$

and so

$$\lim_{k \to \infty} y_{M+4k+1} = L_1$$

where $L_1 = \frac{p+L_2}{qL_3+L_2} \in (0, \infty)$. It follows that the solution $\{z_n\}_{n=-2}^{\infty}$ of Eq.(1) with initial conditions $z_{-2} = L_0, z_{-1} = L_1, z_0 = L_2$ is periodic with period four, and that

$$L_3 = z_1 = \frac{p + L_0}{qL_1 + L_0} > 0.$$

Thus $\{z_n\}_{n=-2}^{\infty}$ is indeed a positive solution of Eq.(1), and, as $z_0 = L_2 \geq \frac{p}{q} > 1$, it follows by Eq.(6) that

$$0 = L_2 - L_2 = z_0 - z_4 > 0$$

which is impossible, and so the proof is complete. $\square$

The proof of the following lemma is similar to that of Lemma 2.6 and will be omitted.

Lemma 2.7 *Suppose $p < q$. Then every solution of Eq.(1) eventually enters and remains in the interval $\left[\frac{p}{q}, 1\right]$.*

3   Global stability when $p > q$

Theorem 3.1 *Suppose $p > q$. Then the equilibrium $\bar{y}$ of Eq.(1) is globally asymptotically stable.*

Proof: We know by Theorem 2.1 that $\bar{y}$ is a locally aymptotically stable equilibrium point of Eq.(1), and so it suffices to show that $\bar{y}$ is a global attractor of Eq.(1). Let $f : (0, \infty)^3 \longrightarrow (0, \infty)$ be given by

$$f(u, v, w) = \frac{p + w}{qv + w}.$$

Since $q < p$, it follows that $f\left[\left[1, \frac{p}{q}\right]^3\right] \subset \left[1, \frac{p}{q}\right]$. Note that for $(u, v, w) \in \left[1, \frac{p}{q}\right]^3$ we have

$$\frac{\partial f}{\partial u}(u, v, w) = 0, \quad \frac{\partial f}{\partial v}(u, v, w) = \frac{-q(p + w)}{(qv + w)^2} < 0,$$

$$\frac{\partial f}{\partial w}(u, v, w) = \frac{q(v - \frac{p}{q})}{(qv + w)^2} \leq 0,$$

and so the result follows by Theorem A and Lemma 2.6.   □

4   Global stability when $p = q$

Theorem 4.1 *Suppose $p = q$. Then the equilibrium $\bar{y} = 1$ of Eq.(1) is globally asymptotically stable.*

Proof: We know that by Theorem 2.1 that $\bar{y}$ is a locally asymptotically stable equilibrium point of Eq.(1), and so it suffices to show that $\bar{y}$ is a global attractor of Eq.(1). So let $\{y_n\}_{n=-2}^{\infty}$ be a solution of Eq.(1). It suffices to show that $\lim_{n \to \infty} y_n = 1$. To this end, it follows by Identities (4) and (6) that

$$\lim_{n \to \infty} y_{4n} = \lim_{n \to \infty} y_{4n+1} = \lim_{n \to \infty} y_{4n+2} = \lim_{n \to \infty} y_{4n+3} = 1.$$ □

5   Global stability when $p < q \leq 2p$

Theorem 5.1 *Suppose $p < q \leq 2p$. Then the equilibrium $\bar{y}$ of Eq.(1) is globally asymptotically stable.*

Proof: We know by Theorem 2.1 that $\bar{y}$ is a locally asymptotically stable equilibrium point of Eq.(1), and so it suffices to show that $\bar{y}$ is a global attractor of Eq.(1). Let $f : (0, \infty)^3 \longrightarrow (0, \infty)$ be given by

$$f(u, v, w) = \frac{p+w}{qv+w}.$$

Since $p < q$, it follows that $f\left[\left[\frac{p}{q}, 1\right]^3\right] \subset \left[\frac{p}{q}, 1\right]$. Note that for $(u, v, w) \in \left[1, \frac{p}{q}\right]^3$ we have

$$\frac{\partial f}{\partial u}(u, v, w) = 0, \quad \frac{\partial f}{\partial v}(u, v, w) = \frac{-q(p+w)}{(qv+w)^2} < 0,$$

$$\frac{\partial f}{\partial w}(u, v, w) = \frac{q(v - \frac{p}{q})}{(qv+w)^2} \geq 0,$$

and so the result follows from Lemma 2.7 and Theorem B.   □

References

[1] M. R. S. Kulenovic and G. Ladas, *Dynamics of Second Order Rational Difference Equations, With Open Problems and Conjectures*, Chapman and Hall/CRC, Boca Raton, 2001.

[2] V. L. Kocic and G. Ladas, *Global Behavior of Nonlinear Difference Equations of Higher Order with Applications*, Kluwer Academic Publishers, Dordrecht, 1993.

On the Difference Equation
$$y_{n+1} = \frac{y_{n-(2k+1)}+p}{y_{n-(2k+1)}+qy_{n-2l}}$$

E. A. GROVE, G. LADAS [1], L. C. McGRATH

Department of Mathematics, University of Rhode Island
Kingston, Rhode Island, 02881-0816 USA
E-mail: grove@math.uri.edu, gladas@math.uri.edu

and

H. A. EL-METWALLY

Mathematics Department, Faculty of Science
Mansoura University, Mansoura, Egypt

Abstract We investigate the global stability, the periodic nature, and the boundedness character of the solutions of the equation in the title, where p and q are positive real numbers, k and l are non-negative integers, and the initial conditions are positive real numbers.

Keywords Boundedness, Global convergence, Periodic solutions, Rational recursive sequences

AMS Subject Classification 39A10, 39A11

1 Introduction and preliminaries

Consider the non-linear, rational difference equation

$$y_{n+1} = \frac{y_{n-(2k+1)}+p}{y_{n-(2k+1)}+qy_{n-2l}} \quad , \quad n = 0, 1, \ldots \tag{1}$$

where the parameters p and q are positive real numbers, k and l are non-negative integers, and the initial conditions $y_0, y_{-1}, \ldots, y_{-\max\{2k+1,2l\}}$ are arbitrary positive real numbers.

Our goal in this paper is to investigate the global stability, the periodic nature, and the boundedness character of the solutions of Eq.(1). The case $k = l = 0$ was treated in [14]. For similar investigations about second order rational difference equations see [1], [2], [8]-[11], and [14].

---

[1]Corresponding author. Email:gladas@math.uri.edu

We now present some definitions and known results which will be useful in our investigation of Eq.(1).

Let I be an interval of real numbers, let $j \geq 1$ be a positive integer, and let $f : I^{j+1} \to I$ be a continuously differentiable function. Consider the difference equation

$$y_{n+1} = f(y_n, y_{n-1}, \ldots, y_{n-j}) \quad , \quad n = 0, 1, \ldots \quad (2)$$

with initial conditions $y_0, y_{-1}, \ldots, y_{-j} \in I$.

We say that $\bar{y}$ is an *equilibrium point* of Eq.(2) if $f(\bar{y}, \bar{y}, \ldots, \bar{y}) = \bar{y}$. That is, the constant sequence $\{y_n\}_{n=-j}^{\infty}$ (called an *equilibrium solution*) with $y_n = \bar{y}$ for all $n \geq -j$ is a solution of Eq.(2). The *linearized equation* of Eq.(2) about the equilibrium point $\bar{y}$ is the linear difference equation

$$z_{n+1} = p_0 z_n + p_1 z_{n-1} + \cdots + p_j z_{n-j} = 0 \quad , \quad n = 0, 1, \ldots \quad (3)$$

where $p_i = \frac{\partial f}{\partial x_i}(\bar{y}, \bar{y}, \ldots, \bar{y})$ for $i = 0, 1, \ldots, j$. The *characteristic equation* of Eq.(3) is the equation

$$\lambda^{j+1} - p_0 \lambda^j - \cdots - p_{j-1}\lambda - p_j = 0. \quad (4)$$

The following well-known theorem, called the *Linearized Stability Theorem*, is very useful in determining the local stability character of the equilibrium solution $\bar{y}$ of Eq.(2

Theorem A (The Linearized Stability Theorem) *The following statements are true.*

1. *If all roots of Eq.(4) have absolute value less than one, then the equilibrium point $\bar{y}$ of Eq.(2) is locally asymptotically stable.*

2. *If at least one of the roots of Eq.(4) has absolute value greater than one, then the equilibrium point $\bar{y}$ of Eq.(2) is unstable.*

If all roots of Eq.(4) have absolute value less than one, then the equilibrium point $\bar{y}$ of Eq.(2) is called a *sink*. If at least one root of Eq.(4) has absolute value less than one, and at least one root of Eq.(4) has absolute value greater than one, then the equilibrium point $\bar{y}$ of Eq.(2) is called a *saddle point equilibrium*. In particular, a saddle point equilibrium is unstable.

The next theorem is due to C.W. Clark (see [3]) and gives a readily verifiable sufficient condition that the equilibrium point $\bar{y}$ of Eq.(2) be a sink.

Theorem B (Clark's Theorem) *If $|p_0| + |p_1| + \cdots + |p_j| < 1$ then the equilibrium point $\bar{y}$ of Eq.(2) is a sink.*

Definition 1 We say that Eq.(2) is *permanent* if there exist real numbers m and M with $0 < m \leq M$ such that for every solution $\{y_n\}_{n=-j}^{\infty}$ of Eq.(2), there exists an integer $N \geq -j$ (which depends upon the initial conditions $y_0, y_{-1}, \ldots, y_{-j}$) such that $m \leq y_n \leq M$ for all $n \geq N$.

The importance of permanence for biological systems was thoroughly reviewed by Hutson and Schmitt. See [12].

The following theorems, which are minor modifications of Theorem 5.2 in [11], will be useful in establishing the global stability of the positive equilibrium.

Theorem C *Let a and b be real numbers with $a < b$, and assume that $f \in C\left[[a,b]^2, [a,b]\right]$ satisfies the following conditions:*

1. *$f(u,v)$ is non-increasing in $u \in [a,b]$ for each fixed $v \in [a,b]$, and $f(u,v)$ is non-increasing in $v \in [a,b]$ for each fixed $u \in [a,b]$.*

2. *If $m, M \in [a,b]$ is a solution of $m = f(M,M)$ and $M = f(m,m)$ then $m = M$.*

Then Eq.(2) has a unique equilibrium point $\bar{y} \in [a,b]$, and every solution of Eq.(2) converges to $\bar{y}$.

Theorem D *Let a and b be real numbers with $a < b$, and assume that $f \in C\left[[a,b]^2, [a,b]\right]$ satisfies the following conditions:*

1. *$f(u,v)$ is non-decreasing in $u \in [a,b]$ for each fixed $v \in [a,b]$, and $f(u,v)$ is non-increasing in $v \in [a,b]$ for each fixed $u \in [a,b]$.*

2. *If $m, M \in [a,b]$ is a solution of $m = f(m,M)$ and $M = f(M,m)$ then $m = M$.*

Then Eq.(2) has a unique equilibrium point $\bar{y} \in [a,b]$, and every solution of Eq.(2) converges to $\bar{y}$.

2 Local stability of equation (1)

Eq.(1) has a unique positive equilibrium point $\bar{y}$, and $\bar{y}$ satsifies the *equilibrium equation*

$$(1+q)\bar{y}^2 - \bar{y} - p = 0. \tag{5}$$

The linearized equation of Eq.(1) about $\bar{y}$ is

$$z_{n+1} + \frac{p-q\bar{y}}{(1+q)(\bar{y}+p)}z_{n-(2k+1)} + \frac{q}{1+q}z_{n-2l} = 0 \quad , \quad n = 0, 1, \dots . \tag{6}$$

Let $j = \max\{2k+1, 2l\}$. Then the characteristic equation of Eq.(6) is

$$\lambda^{j+1} + \frac{p-q\bar{y}}{(1+q)(\bar{y}+p)}\lambda^{j-(2k+1)} + \frac{q}{1+q}\lambda^{j-2l} = 0. \tag{7}$$

Let $g : (0, \infty) \to (0, \infty)$ be given by $g(x) = (1 + q)x^2 - x - p$. Then for $x \in (0, \infty)$, we have

$$x < \bar{y} \quad \text{if and only if} \quad g(x) < 0.$$
$$x = \bar{y} \quad \text{if and only if} \quad g(x) = 0.$$
$$x > \bar{y} \quad \text{if and only if} \quad g(x) > 0.$$

It follows easily by the above that

$$\frac{p - q\bar{y}}{(1 + q)(\bar{y} + p)} < 0 \quad \text{if and only if} \quad p < q.$$

$$\frac{p - q\bar{y}}{(1 + q)(\bar{y} + p)} = 0 \quad \text{if and only if} \quad p = q.$$

$$\frac{p - q\bar{y}}{(1 + q)(\bar{y} + p)} > 0 \quad \text{if and only if} \quad p > q.$$

Moreover, suppose $q \neq 1$. Then it follows directly by a computation that

$$g\left(\frac{2p}{q - 1}\right) = -\frac{p(q - (4p + 1))(1 + q)}{(q - 1)^2}.$$

We are now ready for the main result of this section.

Theorem 2.1 *The following statements are true.*

1. *Suppose $q < 4p + 1$. Then the equilibrium point $\bar{y}$ of Eq.(6) is a sink.*

2. *Suppose $q > 4p + 1$. Then the equilibrium point $\bar{y}$ of Eq.(6) is unstable, and in fact is a saddle point equilibrium.*

Proof: Suppose $0 < q \leq p$. Then $\dfrac{p - q\bar{y}}{(1 + q)(\bar{y} + p)} \geq 0$, and so

$$\left|\frac{p - q\bar{y}}{(1 + q)(\bar{y} + p)}\right| + \left|\frac{q}{1 + q}\right| = \frac{p - q\bar{y}}{(1 + q)(\bar{y} + p)} + \frac{q}{1 + q} = \frac{p + pq}{(1 + q)(\bar{y} + p)}$$

$$= \frac{p(1 + q)}{(1 + q)(\bar{y} + p)} = \frac{p}{\bar{y} + p} < 1.$$

It follows by Theorem B that $\bar{y}$ is a sink.

Next suppose that $p < q < 4p + 1$. Then $\dfrac{p - q\bar{y}}{(1 + q)(\bar{y} + p)} < 0$, and so

$$\left|\frac{p - q\bar{y}}{(1 + q)(\bar{y} + p)}\right| + \left|\frac{q}{1 + q}\right| = \frac{q\bar{y} - p}{(1 + q)(\bar{y} + p)} + \frac{q}{1 + q} = \frac{q\bar{y} - p + q\bar{y} + pq}{\bar{y} - p + q\bar{y} + pq} < 1$$

if and only if $q\bar{y} - p < \bar{y} - p$ if and only if $(q - 1)\bar{y} < 2p$. Thus if $q \leq 1$, it follows by Theorem B that $\bar{y}$ is a sink.

So suppose $1 < q$. Then $(q-1)\bar{y} < 2p$ if and only if $\bar{y} < \frac{2p}{q-1}$ which is true as

$$g\left(\frac{2p}{q-1}\right) = -\frac{p(q-(4p+1))(1+q)}{(q-1)^2} > 0.$$

Thus it follows by Theorem B that $\bar{y}$ is a sink in this case also.

Finally, suppose that $q < 4p+1$. We shall show that $\bar{y}$ is a saddle point equilibrium. First note that $\frac{2p}{q-1} < \bar{y}$ because

$$g\left(\frac{2p}{q-1}\right) = -\frac{p(q-(4p+1))(1+q)}{(q-1)^2} < 0.$$

Suppose $k < l$. Then the characteristic equation of Eq.(6) is

$$h(\lambda) = \lambda^{2l+1} + \frac{p-q\bar{y}}{(1+q)(\bar{y}+p)}\lambda^{2l-(2k+1)} + \frac{q}{1+q} = 0.$$

Note that

$$h(-1) = -1 - \left(\frac{p-q\bar{y}}{(1+q)(\bar{y}+p)}\right) + \frac{q}{1+q} = \frac{-2p+(q-1)\bar{y}}{(1+q)(\bar{y}+p)} > 0.$$

It follows that there exists a real root $\lambda_1 < -1$ of the characteristic equation of Eq.(6). Now $\frac{q}{1+q} < 1$, and so there also exists a root λ_2 of the characteristic equation of Eq.(6) with modulus less than one, and so $\bar{y}$ is a saddle point equilibrium of Eq.(1).

Suppose that $l \leq k$. Then the characteristic equation of Eq.(6) is

$$h(\lambda) = \lambda^{2k+2} + \frac{q}{1+q}\lambda^{(2k+1)-2l} + \frac{p-q\bar{y}}{(1+q)(y+p)} = 0.$$

Note that

$$h(-1) = 1 - \frac{q}{1+q} + \frac{p-q\bar{y}}{(1+q)(\bar{y}+p)} = \frac{2p-(q-1)\bar{y}}{(1+q)(\bar{y}+p)} < 0.$$

It follows that there exists a real root $\lambda_1 < -1$ of the characteristic equation of Eq.(6). Note also that as $p < q$, we have $0 < \frac{p-q\bar{y}}{(1+q)(\bar{y}+p)} < 1$, and so there also exists a root λ_2 of the characteristic equation of Eq.(6) with modulus less than one, and so $\bar{y}$ is a saddle point equilibrium of Eq.(1). $\square$

3   The prime period two-cycle of equation (1)

In this section we study the prime period two-cycle of Eq.(1).

Lemma 3.1 *Let p and q be positive real numbers. Then the following statements are true:*

1. Equation (1) has a positive prime period two solution if and only if $4p + 1 < q$.

2. Suppose $4p + 1 < q$. Then Eq.(1) has the (essentially) unique prime period two solution $\phi, \psi, \phi, \psi, \ldots$ where ϕ and ψ are the roots of the quadratic equation

$$t^2 - t + \frac{p}{q-1} = 0. \tag{8}$$

Proof: Suppose $2k + 1 < 2l$. The case where $2l < 2k + 1$ is similar and will be omitted.

(a) Suppose that Eq.(1) has the positive prime period two solution ϕ, ψ, ϕ, ψ, $\ldots$. We shall show that $4p + 1 < q$, and that ϕ and ψ are the roots of Eq.(8). Now

$$\psi = y_1 = \frac{y_{-(2k+1)} + p}{y_{-(2k+1)} + qy_{-2l}} = \frac{\psi + p}{\psi + q\phi}$$

and

$$\phi = y_2 = \frac{y_{-2k} + p}{y_{-2k} + qy_{-2l+1}} = \frac{\phi + p}{\phi + q\psi},$$

and so we see that $\psi^2 + q\phi\psi = \psi + p$ and $\phi^2 + q\phi\psi = \phi + p$. Hence $(\phi + \psi)(\phi - \psi) = \phi^2 - \psi^2 = \phi - \psi$, and so as $\phi \neq \psi$, we see that $\phi + \psi = 1$. Thus $\psi = 1 - \phi$, and so $\phi^2 + q\phi(1 - \phi) = \phi + p$. That is, $(q - 1)\phi^2 + (1 - q)\phi + p = 0$. We similarly have

$$(q - 1)\psi^2 + (1 - q)\psi + p = 0.$$

It follows that $q > 1$, and that ϕ and ψ are roots of Eq.(8). In particular, $0 < 1 - \frac{4p}{q-1}$ and so $4p + 1 < q$.

(b) Next assume that $4p + 1 < q$, and set $\phi = \left(1 - \sqrt{1 - \frac{4p}{q-1}}\right)/2$ and $\psi = \left(1 + \sqrt{1 - \frac{4p}{q-1}}\right)/2$. It suffices to show that the solution $\{y_n\}_{n=-2l}^{\infty}$ of Eq.(1) with initial conditions

$$y_{-2l} = \phi, y_{-2l+1} = \psi, \ldots, y_0 = \phi$$

is a prime period two-cycle. Note that $0 < \phi < \psi$. Now ϕ and ψ are the roots of Eq.(8), and so

$$\psi^2 = \psi - \frac{p}{q-1} \quad \text{and} \quad \phi\psi = \frac{p}{q-1}.$$

Thus

$$y_1 = \frac{y_{-(2k+1)} + p}{y_{-(2k+1)1} + qy_{-2l}} = \frac{\psi + p}{\psi + q\phi} = \frac{\psi^2 + p\psi}{\psi^2 + q\phi\psi}$$

$$= \frac{\psi - \frac{p}{q-1} + p\psi}{\psi - \frac{p}{q-1} + \frac{pq}{q-1}} = \frac{(1+p)\psi - \frac{p}{q-1}}{\psi + \frac{pq-p}{q-1}} = \frac{(1+p)\psi - \frac{p}{q-1}}{\psi + p}$$

and so we see that $y_1 = \psi$ if and only if

$$\frac{(1+p)\psi - \frac{p}{q-1}}{\psi + p} = \psi$$

if and only if $(1+p)\psi - \frac{p}{q-1} = \psi^2 + p\psi = \psi - \frac{p}{q-1} + p\psi$ which is true. The proof that $y_2 = \phi$ is similar and will be omitted. $\quad\square$

The following lemma will be useful in establishing the local asymptotic stability of the two-cycle $\phi, \psi, \phi, \psi, \ldots$ of Eq.(8).

Lemma 3.2 *Let p and q be positive real numbers, and suppose $4p+1 < q$. Let $\phi, \psi, \phi, \psi, \ldots$ be the (essentially) unique prime period two solution of Eq.(1). Then the following statements are true:*

1. $\phi + \psi = 1$.

2. $\phi\psi = \dfrac{p}{q-1}$.

3. $\phi^2 = \phi - \dfrac{p}{q-1}$.

4. $\psi^2 = \psi - \dfrac{p}{q-1}$.

5. $\phi^2 + \psi^2 = 1 - \dfrac{2p}{q-1}$.

6. $(\phi + p)(\psi + p) = \dfrac{pq}{q-1} + p^2$.

7. $(\phi + q\psi)(\psi + q\phi) = pq - p + q$.

8. $(q\phi - p)(q\psi - p) = \dfrac{pq}{q-1} + p^2$.

9. $(\phi + q\psi)^2 = (q^2 - 1)\psi + 1 - pq + p$.

10. $(\psi + q\phi)^2 = (q^2 - 1)\phi + 1 - pq + p$.

Proof: The proof of Lemma 3.2 follows from the fact that ϕ and ψ are the two roots of Eq.(8) $\quad\square$

In [14], it was shown that the two-cycle $\phi, \psi, \phi, \psi, \ldots$ of Eq.(1) is a sink when $k = 0$ and $l = 0$. We are now ready to consider the nature of the stability of the two-cycle of Eq.(1) in the case of $k = 0$ and $l = 1$. The general case $k \geq 0$ and $l \geq 0$ is still open.

For the remainder of this section, assume that $k = 0$ and $l = 1$.

For $n \geq 0$, consider the transformation

$$u_n = y_{n-2}, \quad v_n = y_{n-1}, \quad w_n = y_n.$$

Then
$$u_{n+1} = v_n , \quad v_{n+1} = w_n , \quad w_{n+1} = \frac{v_n + p}{v_n + qu_n} .$$

Let $T : (0,\infty)^3 \to (0,\infty)$ be given by

$$T(u, v, w) = \left(v, w, \frac{v + p}{v + qu} \right) .$$

Then $T^2(u, v, w) = \left(w, \frac{v+p}{v+qu}, \frac{w+p}{w+qv} \right)$ and so

$$JT^2|_{(\phi,\psi,\phi)} = \begin{bmatrix} 0 & 0 & 1 \\ -\frac{q(\psi+p)}{(\psi+q\phi)^2} & \frac{q\phi-p}{(\psi+q\phi)^2} & 0 \\ 0 & -\frac{q(\phi+p)}{(\phi+q\psi)^2} & \frac{q\psi-p}{(\phi+q\psi)^2} \end{bmatrix} .$$

Lemma 3.3 Let $k = 0$ and $l = 1$. Then the characteristic equation of the two-cycle $(\phi, \psi, \phi, \ldots)$ is

$$0 = \lambda^3 - \frac{2p+1}{pq - p + q}\lambda^2 + \frac{p}{(pq - p + q)(q - 1)}\lambda - \frac{pq^2}{(pq - p + q)(q - 1)} . \quad (9)$$

Proof: The characteristic equation of the two-cycle $(\phi, \psi, \phi, \ldots)$ is

$$\begin{aligned}
0 &= \det\left(\lambda I - JT^2|_{(\phi,\psi,\phi)} \right) \\
&= \lambda\left(\lambda - \frac{q\phi - p}{(\psi + q\phi)^2} \right)\left(\lambda - \frac{q\psi - p}{(\phi + q\psi)^2} \right) - \frac{q^2(\phi + p)(\psi + p)}{(\phi + q\psi)^2(\psi + q\phi)^2} \\
&= \lambda\left(\lambda^2 - \left[\frac{q\phi - p}{(\psi + q\phi)^2} + \frac{q\psi - p}{(\phi + q\psi)^2} \right]\lambda + \frac{(q\phi - p)(q\psi - p)}{(\psi + q\phi)^2(\phi + q\psi)^2} \right) \\
&\quad - \frac{q^2(\phi + p)(\psi + p)}{(\phi + q\psi)^2(\psi + q\phi)^2} \\
&= \lambda^3 - \left(\frac{q\phi - p}{(\psi + q\phi)^2} + \frac{q\psi - p}{(\phi + q\psi)^2} \right)\lambda^2 + \left(\frac{(q\phi - p)(q\psi - p)}{(\psi + q\phi)^2(\phi + q\psi)^2} \right)\lambda \\
&\quad - \frac{q^2(\phi + p)(\psi + p)}{(\phi + q\psi)^2(\psi + q\phi)^2} \\
&= \lambda^3 - \left(\frac{(q\phi - p)(\phi + q\psi)^2 + (q\psi - p)(\psi + q\phi)^2}{(\psi + q\phi)^2(\phi + q\psi)^2} \right)\lambda^2 \\
&\quad + \frac{(q\phi - p)(q\psi - p)}{(\psi + q\phi)^2(\phi + q\psi)^2}\lambda - \frac{q^2(\phi + p)(\psi + p)}{(\phi + q\psi)^2(\psi + q\phi)^2}
\end{aligned}$$

$$= \lambda^3 + \frac{(q\phi - p)(q\psi - p)}{[(\phi + q\psi)(\psi + q\phi)]^2}\lambda - \frac{q^2(\phi + p)(\psi + p)}{[(\phi + q\psi)(\psi + q\phi)]^2}$$

$$-\left(\frac{(q\phi-p)[(q^2-1)\psi + 1-pq+p] + (q\psi-p)[(q^2-1)\phi+1-pq+p]}{[(\phi + q\psi)(\psi + q\phi)]^2}\right)\lambda^2$$

$$= \lambda^3 + \frac{\frac{pq}{q-1} + p^2}{(pq - p + q)^2}\lambda - \frac{q^2\left(\frac{pq}{q-1} + p^2\right)}{(pq - p + q)^2}$$

$$-\frac{q(q^2-1)(\phi\psi+\psi\phi)+(q-pq^2+pq-p(q^2-1))(\phi+\psi) + 2p(pq-p-1)}{(pq-p+q)^2}\lambda^2$$

$$-\lambda^3 - \frac{\frac{2q(q^2-1)p}{q-1} - p(q^2-1) + q - 2p - pq^2 + 2p^2q + pq - 2p^2}{(pq - p + q)^2}\lambda^2$$

$$\mid \frac{pq + p^2q - p^2}{(pq - p + q)^2(q - 1)}\lambda - \frac{pq^3 + p^2q^3 - p^2q^2}{(pq - p + q)^2(q - 1)}$$

$$= \lambda^3 - \frac{\frac{2pq^3 - 2pq}{q-1} - pq^2 + p + q - 2p - pq^2 + 2p^2q + pq - 2p^2}{(pq - p + q)^2}\lambda^2$$

$$+\frac{p(q + pq - p)}{(pq - p + q)^2(q - 1)}\lambda - \frac{pq^2(q + pq - p)}{(pq - p + q)^2(q - 1)},$$

and so

$$0 = \lambda^3 - \frac{\frac{2pq^3 - 2pq}{q-1} - 2pq^2 - p + q + 2p^2q + pq - 2p^2}{(pq - p + q)^2}\lambda^2$$

$$+\frac{p}{(pq - p + q)(q - 1)}\lambda - \frac{pq^2}{(pq - p + q)(q - 1)}$$

$$= \lambda^3 + \frac{(2p + 1)(1 - q)}{(pq - p + q)(q - 1)}\lambda^2 + \frac{p}{(pq - p + q)(q - 1)}\lambda$$

$$-\frac{pq^2}{(pq - p + q)(q - 1)}$$

$$= \lambda^3 - \frac{2p + 1}{pq - p + q}\lambda^2 + \frac{p}{(pq - p + q)(q - 1)}\lambda - \frac{pq^2}{(pq - p + q)(q - 1)}.$$

This completes the proof of Lemma 3.3.   □

We are now ready for the main result of this section.

Theorem 3.4 *Let $k = 0$ and $l = 1$. Assume that p and q are positive real numbers such that $4p+1 < q$. Then the (essentially) unique prime period two solution $\phi, \psi, \phi, \psi, \ldots$ of Eq.(1) is a sink.*

Proof: Recall that every root of $\lambda^3 + a_2\lambda^2 + a_1\lambda + a_0 = 0$ has modulus less than one if and only if

$$
\begin{array}{ll}
\text{(I)} & |a_2 + a_0| < 1 + a_1 \\
\text{(II)} & |a_2 - 3a_0| < 3 - a_1 \\
\text{(III)} & a_0^2 + a_1 - a_0 a_2 < 1.
\end{array}
$$

In our case, $a_2 = -\frac{2p+1}{pq-p+q}$, $a_1 = \frac{p}{(pq-p+q)(q-1)}$ and $a_0 = -\frac{pq^2}{(pq-p+q)(q-1)}$. We first consider Condition (I).

$$
\left| \left(-\frac{2p+1}{pq-p+q} \right) + \left(-\frac{pq^2}{(pq-p+q)(q-1)} \right) \right| < 1 + \frac{p}{(pq-p+q)(q-1)}
$$

if and only if $0 < q^2 - (4p+2)q + (4p+1)$. Set $f(x) = x^2 - (4p+2)x + (4p+1)$. Note that

$$
\begin{array}{rclcl}
f(4p+1) & = & 16p^2 + 8p + 1 - 16p^2 - 4p - 8p - 2 + 4p + 1 & = & 0 \\
f'(4p+1) & = & 8p + 2 - 4p - 2 & > & 0.
\end{array}
$$

So as $q > 4p+1$, it follows that Condition (I) is satisfied.
   We next consider Condition (II).

$$
\left| \left(-\frac{2p+1}{pq-p+q} \right) - 3 \left(-\frac{pq^2}{(pq-p+q)(q-1)} \right) \right| < 3 - \frac{p}{(pq-p+q)(q-1)}
$$

if and only if $|(2p+1)(q-1) - 3pq^2| < 3(pq-p+q)(q-1) - p$ if and only if

$$
-3pq^2 + 6pq - 2p - 3q^2 + 3q < 2pq - 2p + q - 1 - 3pq^2 < 3pq^2 - 6pq + 2p + 3q^2 - 3q.
$$

Recall that $q = 4p+1+\delta$ for some $\delta > 0$. Consider the left hand inequality:

$$
-3pq^2 + 6pq - 2p - 3q^2 + 3q < 2pq - 2p + q - 1 - 3pq^2
$$

if and only if $4pq + 2q < 3q^2 + 1$ if and only if

$$
4p(4p+1+\delta) + 2(4p+1+\delta) < 3(4p+1+\delta)^2 + 1
$$

if and only if $0 < 32p^2 + (12 + 20\delta)p + 2 + 4\delta + 3\delta^2$ which is true. Consider the right hand inequality:

$$
2pq - 2p + q - 1 - 3pq^2 < 3pq^2 - 6pq + 2p + 3q^2 - 3q
$$

if and only if $8pq + 4q < 4p + 1 + 6pq^2 + 3q^2$ if and only if

$$
8p(4p+1+\delta) + 4(4p+1+\delta) < 4p + 1 + 6p(4p+1+\delta)^2 + 3(4p+1+\delta)^2
$$

if and only if $0 < 96p^3 + 64p^2 + 10p + 48\delta p^2 + 6\delta^2 p + 28\delta p + 3\delta^2 + 2\delta$ which is true. Thus Condition (II) is satisfied.

Finally, we consider Condition (III).

$$\left(-\frac{pq^2}{(pq-p+q)(q-1)}\right)^2 + \frac{p}{(pq-p+q)(q-1)}$$

$$-\left(-\frac{2p+1}{(pq-p+q)}\right)\left(-\frac{pq^2}{(pq-p+q)(q-1)}\right) < 1$$

if and only if $p^2q^4 + p(pq-p+q)(q-1) - (2p+1)pq^2(q-1) < (pq-p-q)^2(q-1)^2$
if and only if $0 < -pq + p^2q^2 - 5pq^3 + q^2(q-1)^2 - 2p^2q + 2pq^4 + 4pq^2$ if and only if

$$
\begin{aligned}
0 \quad < \quad & -p(4p+1+\delta) + p^2(4p+1+\delta)^2 - 5p(4p+1+\delta)^3 + \\
& (4p+1+\delta)^2(4p+1+\delta-1)^2 - 2p^2(4p+1+\delta) + \\
& 2p(4p+1+\delta)^4 + 4p(4p+1+\delta)^2 \\
= \quad & 23p^2 + 8\delta p + 464p^4 + 105\delta^2 p^2 + 288p^3 + \\
& 408\delta p^3 + 104\delta p^2 + 21\delta^2 p + 19\delta^3 p + \delta^2 + \delta^4 + 2\delta^3 \\
& + 512p^5 + 2\delta^4 p + 192\delta^2 p^3 + 512\delta p^4
\end{aligned}
$$

which is true. Therefore Condition (III) is satisfied. Hence the period-two cycle of Eq.(1) is locally asymptotically stable.   □

Conjecture: We offer the following two conjectures. Let k and l be non-negative integers such that $4p + 1 < q$.

1. The (essentially) unique period two-cycle of Eq.(1) is a sink.

2. Every positive solution of Eq.(1) converges to a (not necessarily prime) period two solution of Eq.(1).

4   Permanence

In this section we show that Eq.(1) is permanent. We also establish some lemmas which will be useful in the sequel. The proofs of the following four lemmas are straightforward and will be omitted.

Lemma 4.1 *Let p and q be positive real numbers, and suppose that $\{y_n\}_{n=-2}^{\infty}$ is a positive solution of Eq.(1). Let $n \geq 0$. Then the following identities hold:*

$$y_{n+1} - 1 = \frac{q\left(\frac{p}{q} - y_{n-2l}\right)}{y_{n-(2k+1)} + qy_{n-2l}}.$$

$$y_{n+1} - \frac{p}{q} = \frac{\left(1 - \frac{p}{q}\right)y_{n-(2k+1)} + p(1 - y_{n-2l})}{y_{n-(2k+1)} + qy_{n-2l}}. \tag{10}$$

$$y_n - y_{n+(4l+2)} = \tag{11}$$

$$\frac{y_{n+4l-2k}\left(y_{n+2l-(2k+1)} + qy_n\right)(y_n - 1) + qy_{n+2l-(2k+1)}\left(y_n - \frac{p}{q}\right)}{y_{n+4l-2k}\left(y_{n+2l-(2k+1)} + qy_n\right) + pq + qy_{n-2l-(2k+1)}}.$$

Lemma 4.2 *Let p and q be positive real numbers with $p > q$, and suppose that $\{y_n\}_{n=-\max\{2k+1,2l\}}^{\infty}$ is a positive solution of Eq.(1). Let $N \geq 0$ be a non-negative integer. Then the following statements are true:*

1. If $y_N < \frac{p}{q}$,   then $y_{N+(2l+1)} > 1$.

2. If $y_N = \frac{p}{q}$,   then $y_{N+(2l+1)} = 1$.

3. If $y_N > \frac{p}{q}$,   then $y_{N+(2l+1)} < 1$.

4. If $y_N \geq 1$,   then $y_{N+(2l+1)} < \frac{p}{q}$.

5. If $y_N < 1$,   then $y_{N+(2l+1)} > 1$.

6. If $y_N \geq \frac{p}{q}$,   then $y_{N+(4l+2)} < y_N$.

7. If $y_N \leq 1$,   then $y_N < y_{N+(4l+2)}$.

8. If $y_N \in \left[1, \frac{p}{q}\right]$,   then $y_{N+(2l+1)j} \in \left[1, \frac{p}{q}\right]$ for all $j \geq 0$.

9. If $y_N \in \left[1, \frac{p}{q}\right]$,   then $y_{N+(2l+1)j} \in \left(1, \frac{p}{q}\right)$ for all $j \geq 2$.

10. $\bar{y} \in \left(1, \frac{p}{q}\right)$.

Lemma 4.3 *Let p and q be positive real numbers with $p = q$, and suppose that $\{y_n\}_{n=-\max\{2k+1,2l\}}^{\infty}$ is a positive solution of Eq.(1). Let $N \geq 0$ be a non-negative integer. Then the following statements are true:*

1. If $y_N < 1$,   then $y_{N+(2l+1)} > 1$.

2. If $y_N = 1$,   then $y_{N+(2l+1)} = 1$.

3. If $y_N > 1$,   then $y_{N+(2l+1)} < 1$.

4. If $y_N > 1$,   then $1 < y_{N+(4l+2)} < y_N$.

5. If $y_N < 1$,   then $y_N < y_{N+(4l+2)} < 1$.

6. $\bar{y} = 1$.

Lemma 4.4 *Let p and q be positive real numbers with $p < q$, and suppose that $\{y_n\}_{n=-\max\{2k+1,2l\}}^{\infty}$ is a positive solution of Eq.(1). Let $N \geq 0$ be a non-negative integer. Then the following statements are true:*

1. If $y_N < \dfrac{p}{q}$,   then $y_{N+(2l+1)} > 1$.

2. If $y_N = \dfrac{p}{q}$,   then $y_{N+(2l+1)} = 1$.

3. If $y_N > \dfrac{p}{q}$,   then $y_{N+(2l+1)} < 1$.

4. If $y_N \leq 1$,   then $y_{N+(2l+1)} > \dfrac{p}{q}$.

5. If $y_N > 1$,   then $y_{N+(2l+1)} < 1$.

6. If $y_N \leq \dfrac{p}{q}$,   then $y_N < y_{N+(4l+2)}$.

7. If $y_N \geq 1$,   then $y_N > y_{N+(4l+2)}$.

8. If $y_N \in \left[\dfrac{p}{q}, 1\right]$,   then $y_{N+(2l+1)j} \in \left[1, \dfrac{p}{q}\right]$ *for all* $j \geq 0$.

9. If $y_N \in \left[\dfrac{p}{q}, 1\right]$,   then $y_{N+(2l+1)j} \in \left(1, \dfrac{p}{q}\right)$ *for all* $j \geq 2$.

10. $\bar{y} \in \left(\dfrac{p}{q}, 1\right)$.

Lemma 4.5 *Let p and q be positive real numbers with $p > q$, and suppose that $\{y_n\}_{n=-\max\{2k+1,2l\}}^{\infty}$ is a positive solution of Eq.(1). Then there exists $N \geq 0$ such that*

$$y_n \in \left(1, \frac{p}{q}\right) \qquad \text{for all} \qquad n \geq N.$$

Proof: We first claim that

$$1 \leq \liminf_{n\to\infty} y_n \leq \limsup_{n\to\infty} y_n \leq \frac{p}{q}.$$

With this in mind, let $M \geq 0$. It suffices to show that

$$1 \leq \liminf_{j\to\infty} y_{M+(2l+1)j} \leq \limsup_{j\to\infty} y_{M+(2l+1)j} \leq \frac{p}{q}.$$

Case 1. Suppose there exists $j_0 \geq 0$ such that $y_{M+(2l+1)j_0} \in \left[1, \frac{p}{q}\right]$. Then it follows by Lemma 4.2 that

$$1 \leq \liminf_{j\to\infty} y_{M+(2l+1)j} \leq \limsup_{j\to\infty} y_{M+(2l+1)j} \leq \frac{p}{q}.$$

Case 2. Suppose $y_{M+(2l+1)j} \notin \left[1, \frac{p}{q}\right]$ for all $j \geq 0$. Then in particular, $y_M \notin \left[1, \frac{p}{q}\right]$. If $y_M < 1$, then it follows by Lemma 4.2 that $y_{M+(2l+1)} > \frac{p}{q}$, and so without loss of generality, we may assume that $y_M > \frac{p}{q}$. It follows by Lemma 4.2 that there exist positive real numbers $L_0 \in \left[\frac{p}{q}, \infty\right)$ and $L_{2l+1} \in (0, 1]$ such that $\{y_{M+(4l+2)j}\}_{n=0}^{\infty}$ is a strictly monotonically decreasing sequence which converges to L_0, and $\{y_{M+(2l+1)+(4l+2)j}\}_{j=0}^{\infty}$ is a strictly monotonically increasing sequence which converges to L_{2l+1}. In order to prove the claim, it suffices to show that

$$L_0 = \frac{p}{q} \quad \text{and} \quad L_{2l+1} = 1.$$

For the sake of contradiction, suppose this is not true. It suffices to consider the following two cases:

Case a.   Suppose $L_0 > \frac{p}{q}$ and $L_{2l+1} = 1$. Note that if $j \geq 0$,

$$y_{M+(4l+2)j+(4l+2)} = \frac{y_{M+(4l+2)j+(4l-2k)} + p}{y_{M+(4l+2)j+(4l-2k)} + q y_{M+(4l+2)j+(2l+1)}}$$

and so

$$y_{M+(4l+2)j+(4l-2k)} = \frac{p - q y_{M+(4l+2)j+(4l+2)} y_{M+(4l+2)j+(2l+1)}}{y_{M+(4l+2)j+(4l+2)} - 1}.$$

Thus

$$\lim_{j \to \infty} y_{M+(4l+2)j+(4l-2k)} = \frac{p - q L_0 L_{2l+1}}{L_0 - 1} = \frac{p - q L_0}{L_0 - 1} < 0$$

which is a contradiction.

Case b.   Suppose $L_0 \geq \frac{p}{q}$ and $L_{2l+1} < 1$. As in Case a,

$$\lim_{j \to \infty} y_{M+(4l+2)j+(4l-2k)} = \frac{p - q L_0 L_{2l+1}}{L_0 - 1} = L_{4l-2k}.$$

For $j \geq 0$,

$$y_{M+(4l+2)j+(2l+1)} = \frac{y_{M+(4l+2)j+(2l-2k-1)} + p}{y_{M+(4l+2)j+(2l-2k-1)} + q y_{M+(4l+2)j}}$$

from which it follows that

$$y_{M+(4l+2)j+(2l-2k-1)} = \frac{p - q y_{M+(4l+2)j+(2l+1)} y_{M+(4l+2)j}}{y_{M+(4l+2)j+(2l+1)} - 1}.$$

Thus we see that

$$\lim_{j \to \infty} y_{M+(4l+2)j+(2l-2k-1)} = \frac{p - q L_{2l+1} L_0}{L_{2l+1} - 1} = L_{2l-2k-1}.$$

Now

$$\frac{p - qL_0L_{2l+1}}{L_0 - 1} = L_{4l-2k} \geq 0 \quad \text{and} \quad \frac{p - qL_{2l+1}L_0}{L_{2l+1} - 1} = L_{2l-2k} \geq 0.$$

So as $L_0 \geq \frac{p}{q} > 1$ and $L_{2l+1} < 1$, it follows that $L_{4l-2k} = L_1 = 0$. Thus there exists $M_1 \geq 0$ such that if $j \geq M_1$, then $y_{M+(4l+2)j+(4l-2k)} < 1$. Hence by Lemma 4.2 we see that $\{y_{M+(4l+2)j+(4l+2)}\}_{j=M_1}^{\infty}$ is a positive, strictly increasing sequence which converges to 0. This is impossible. Thus it is true that

$$1 \leq \liminf_{n\to\infty} y_n \leq \limsup_{n\to\infty} y_n \leq \frac{p}{q}.$$

Let $P \geq 0$. It follows by Lemma 4.2 that the proof will be complete if we show there exists $j_0 \geq 0$ such that

$$y_{P+(2l+1)j_0} \in \left[1, \frac{p}{q}\right].$$

So for the sake of contradiction, suppose $y_{P+(2l+1)j} \notin \left[1, \frac{p}{q}\right]$ for all $j \geq 0$. Then exactly as above, it follows that without loss of generality we may assume that

$$\lim_{j\to\infty} y_{P+(4l+2)j} = L_0 = \frac{p}{q}$$

and

$$\lim_{j\to\infty} y_{P+(2l+1)+(4l+2)j} = L_{2l+1} = 1.$$

As before,

$$\lim_{j\to\infty} y_{P+(4l+2)j+(4l-2k)} = \frac{p - qL_0L_{2l+1}}{L_0 - 1} = 0$$

which again is impossible. $\qquad\square$

The proof of the following lemma is similar to that of Lemma 4.5 and will be omitted.

Lemma 4.6 *Let p and q be positive real numbers with $p < q$, and suppose that $\{y_n\}_{n=-\max\{2k+1,2l\}}^{\infty}$ is a positive solution of Eq.(1). Then there exists $N \geq 0$ such that*

$$y_n \in \left(\frac{p}{q}, 1\right) \qquad \text{for all} \qquad n \geq N.$$

Theorem 4.7 *Let p and q be positive real numbers. Then Eq.(1) is permanent.*

Proof: The proof follows by Lemmas 4.5, 5.2 (which we prove in the next section), and 4.6. $\qquad\square$

5   Global attractivity of equation (1)

Recall that $\bar{y}$ is the unique equilibrium point of Eq.(1). The main result of this section is that the equilibrium point $\bar{y}$ is a global attractor of Eq.(1) if and only if $0 < q \le 4p + 1$.

Lemma 5.1 *Let p and q be positive real numbers with $p > q$. Then the equilibrium point $\bar{y}$ is a global attractor of Eq.(1).*

Proof: Let $\{y_n\}_{n=-\max\{2k+1,2l\}}^{\infty}$ be a positive solution of Eq.(1). It suffices to show

$$\lim_{n \to \infty} y_n = \bar{y}.$$

We know by Lemma 4.5 that there exists $N \ge 0$ such that $1 < y_n < \frac{p}{q}$ for all $n \ge N$. Let $f : \left[1, \frac{p}{q}\right] \times \left[1, \frac{p}{q}\right] \to (0, \infty)$ be given by

$$f(u, v) = \frac{u + p}{u + qv}.$$

Note that

$$\frac{\partial f}{\partial u}(u, v) = \frac{q\left(v - \frac{p}{q}\right)}{(u + qv)^2}$$

and so we see that f is decreasing in each argument. It follows immediately that

$$1 \le f(u, v) \le \frac{p}{q} \qquad \text{for all} \qquad (u, v) \in \left[1, \frac{p}{q}\right] \times \left[1, \frac{p}{q}\right].$$

Suppose $1 \le m \le M \le \frac{p}{q}$ is a solution of the system

$$m = f(M, M) \qquad \text{and} \qquad M = f(m, m).$$

We shall show that $m = M$. Indeed,

$$m = f(M, M) = \frac{M + p}{M + qM} \qquad \text{and} \qquad M = f(m, m) = \frac{m + p}{m + qm}$$

from which it follows that

$$mM(1 + q) = M + p \qquad \text{and} \qquad Mm(1 + q) = m + p.$$

Thus $m = M$. The result follows by Theorem C.   □

Lemma 5.2 *Let $p = q > 0$. Then the equilibrium point $\bar{y}$ is a global attractor of Eq.(1)*

Proof: Let $\{y_n\}_{n=-\max\{2k+1,2l\}}^{\infty}$ be a positive solution of Eq.(1). It suffices to show

$$\lim_{n \to \infty} y_n = \bar{y}.$$

It follows by Lemma 4.3 that the the subsequences

$$\{y_{(4l+2)j}\}_{j=0}^{\infty}, \{y_{1+(4l+2)j}\}_{j=0}^{\infty}, \cdots, \{y_{(4l+1)+(4l+2)j}\}_{j=0}^{\infty}$$

all converge to the same value $\bar{y} = 1$, and hence that $\lim\limits_{n\to\infty} y_n = \bar{y}$.   □

Lemma 5.3 *Let p and q be positive real numbers with $p < q \leq 4p + 1$. Then the equilibrium point $\bar{y}$ is a global attractor of Eq.(1).*

Proof: Let $\{y_n\}_{n=-\max\{2k+1,2l\}}^{\infty}$ be a positive solution of Eq.(1). It suffices to show

$$\lim_{n\to\infty} y_n = \bar{y}.$$

We know by Lemma 4.6 that there exists $N \geq 0$ such that $\frac{p}{q} < y_n < 1$ for all $n \geq N$. Let $f : \left[\frac{p}{q}, 1\right]^2 \to (0, \infty)$ be given by

$$f(u, v) = \frac{u + p}{u + qv} \qquad \text{for} \qquad u, v \in \left[\frac{p}{q}, 1\right].$$

Note that for $u, v \in \left[\frac{p}{q}, 1\right]$,

$$\frac{\partial f}{\partial u} = \frac{q\left(v - \frac{p}{q}\right)}{(u + qv)^2},$$

and so we see that $f(u, v)$ is increasing in u for each fixed $v \in \left[\frac{p}{q}, 1\right]$, and $f(u, v)$ is decreasing in v for each fixed $u \in \left[\frac{p}{q}, 1\right]$.

We shall show that $\frac{p}{q} \leq f(u, v) \leq 1$ for all $u, v \in \left[\frac{p}{q}, 1\right]$. Observe that for $u, v \in \left[\frac{p}{q}, 1\right]$,

$$f(u, v) \leq f\left(1, \frac{p}{q}\right) = \frac{1 + p}{1 + q\left(\frac{p}{q}\right)} = \frac{1 + p}{1 + p} = 1$$

and

$$f(u, v) \geq f\left(\frac{p}{q}, 1\right) = \frac{\frac{p}{q} + p}{\frac{p}{q} + q \cdot 1} = \frac{p + pq}{p + q^2} = \frac{p}{q}\left(\frac{q + q^2}{p + q^2}\right) > \frac{p}{q}.$$

We claim that if $\frac{p}{q} \leq m \leq M \leq 1$ is a solution of the system

$$m = f(m, M) \qquad \text{and} \qquad M = f(M, m)$$

then $m = M$. For the sake of contradiction, suppose that $m < M$. Now

$$m = \frac{m + p}{m + qM} \qquad \text{and} \qquad M = \frac{M + p}{M + qm}$$

from which it follows that

$$m^2 + qmM = m + p \quad \text{and} \quad M^2 + qMm = M + p.$$

By subtracting the first equation from the second, we see that

$$M^2 - m^2 = M - m.$$

So as $m < M$, we have $m + M = 1$, and hence

$$m^2 + qm(1 - m) = m + p.$$

Thus m is a root of the equation

$$(1 - q)t^2 - (1 - q)t - p = 0. \tag{12}$$

In particular, $q \neq 1$ since $p > 0$.
Now either

$$m = \frac{(1 - q) - \sqrt{(1 - q)^2 + 4(1 - q)p}}{2(1 - q)} \quad \text{or}$$

$$m = \frac{(1 - q) + \sqrt{(1 - q)^2 + 4(1 - q)p}}{2(1 - q)}.$$

Case (a): Suppose $0 < q < 1$. This is impossible, because

$$\frac{(1 - q) - \sqrt{(1 - q)^2 + 4(1 - q)p}}{2(1 - q)} < 0 \quad \text{and}$$

$$\frac{(1 - q) + \sqrt{(1 - q)^2 + 4(1 - q)p}}{2(1 - q)} > 1.$$

Case (b): Suppose $q > 1$. Recall that $p < q \leq 4p + 1$, and so

$$D = (1 - q)^2 + 4(1 - q)p = (1 - q)(1 - q + 4p) \leq 0.$$

If $q < 4p + 1$, then $D < 0$, which is impossible, because then Eq.(12) would have no real roots. Thus we must have $q = 4p + 1$. But this means that $m = \frac{1}{2}$, and so as $m + M = 1$, we see that $M = \frac{1}{2}$ also. This is also a contradiction, since we assumed $m < M$.
Thus we see that $m = M$. The proof follows by Theorem D. $\qquad \square$

Theorem 5.4 *Let p and q be positive real numbers. Then the equilibrium point $\bar{y}$ of Eq.(1) is a global attractor of Eq.(1) if and only if $0 < p < q \leq 4p+1$.*

Proof: The proof follows by Lemmas 3.1, 5.1, 5.2, and 5.3. $\qquad \square$

References

[1] C.H. Gibbons, M.R.S. Kulenovic, and G. Ladas, On the Recursive Sequence

$$x_{n+1} = \frac{\alpha + \beta x_{n-1}}{\gamma + x_n} \,,$$

Math. Sci. Res. Hot-Line 4 (2) (2000), 1-11.

[2] C.H. Gibbons, M.R.S. Kulenovic, and G. Ladas, On the Recursive Sequence

$$y_{n+1} = \frac{p + qy_n + ry_{n-1}}{1 + y_n} \,,$$

Proceedings of the Fifth International Conference on Difference Equations and Applications, Temuco, Chile Jan. 3–7, 2000, Gordon and Breach Science Publishers.

[3] C.W. Clark, A delayed recruitment model of population dynamics with an application to baleen whale populations, *J. Math. Bipl.*, 3:381–391, 1976.

[4] E.A. Grove, G. Ladas, and L.C. McGrath, On the dynamics of

$$y_{n+1} = \frac{p + y_{n-2}}{qy_{n-1} + y_{n-2}} \,.$$

This volume, 425–431.

[5] E.A. Grove, G. Ladas, L.C. McGrath, and C.T. Teixeira, Existence and behavior of solutions of a rational system, *Journal of Communications on Applied Nonlinear Analysis* 8(2001), no. 1, 1–25.

[6] H. El-Metwally, E.A. Grove, and G. Ladas, A global convergence result with applications to periodic solutions, *J. Math. Anal. Appl.*, (2000), 161–170.

[7] H. El-Metwally, E.A. Grove, G. Ladas, and H. Voulov, On the global attractivity and periodic character of some difference equations, *J. Difference Eqns. Appl.*, 7 (2000).

[8] M.R.S. Kulenovic and G. Ladas, On period two solutions of

$$x_{n+1} = \frac{\alpha + \beta x_n + \gamma x_{n-1}}{A + B x_n + C x_{n-1}} \,,$$

J. Difference. Eqns. Appl. 6 (2000), 641–646.

[9] M.R.S. Kulenovic and G. Ladas, *Dynamics of Second Order Rational Difference Equations, with Open Problems and Conjectures*, Chapman and Hall/CRC, Boca Raton, Florida, USA 2001.

[10] M.R.S. Kulenovic, G. Ladas and N.R. Prokup, On the recursive sequence

$$x_{n+1} = \frac{\alpha x_n + \beta x_{n-1}}{1 + x_n} ,$$

J. Difference Eqns. Appl. 6 (2000), 563–576.

[11] M.R.S. Kulenovic, G. Ladas, and W.S. Sizer, On the recursive sequence

$$x_{n+1} = \frac{\alpha x_n + \beta x_{n-1}}{\gamma x_n + \delta x_{n-1}} ,$$

Math. Sci. Res. Hot-Line 2 (5) (1998), 1–16.

[12] V. Hutson and K. Schmidt, Persistence and the dynamics of biological systems, *Math. Biosciences,* 111:1–71, 1992.

[13] V.L. Kocic and G. Ladas, *Global Behavior of Nonlinear Difference Equations of Higher Order with Applications,* Kluwer Academic Publishers, Dordrecht, 1993.

[14] W. Kosmala, M.R.S. Kulenovic, G. Ladas, and C.T. Teixera, On the recursive sequence

$$y_{n+1} = \frac{p + y_{n-1}}{q y_n + y_{n-1}} ,$$

J. Math. Anal. Appl. 251(2000), 571–586.

Almost Periodic Solutions in a Difference Equation

YOSHIHIRO HAMAYA

Department of Information Science
Okayama University of Science
1-1 Ridai-cho, Okayama 700-0005, Japan
E-mail: hamaya@mis.ous.ac.jp, hamaya@icity.or.jp

Abstract In order to obtain the existence of an almost periodic solution for an almost periodic functional difference equation $x(n+1) = f(n, x_n), n \in Z^+$ and where x_n is defined by $x_n(s) = x(n+s)$ for $s \in Z^-$, on a fading memory space B, we consider a certain stability property, which is referred to as BS-stable under disturbances from $\Omega(f)$ with respect to K, this stability implies ρ-stable under disturbances from $\Omega(f)$ with respect to compact set K.

Keywords Almost periodic solutions, Totally stable, Stable under disturbances from hull

AMS Subject Classification 39A10, 39A11

1 Introduction

For ordinary differential equations and functional differential equations, the existence of almost periodic solutions of almost periodic systems has been studied by many authors. One of the most popular methods is to assume certain stability properties [3,8,9,11]. In this paper, in order to obtain an existence theorem for an almost periodic solution to a functional difference equation with infinite delay, we will employ to change Hamaya's results [4,5] and Murakami and Yoshizawa's result [10] for the integrodifferential equations and the functional differential equations, respectively, into theorems for the functional difference equation. This paper is based on Hamaya's result [6].

Let R^m denote Euclidean m-space, Z is the set of integers, Z^+ is the set of nonnegative integers and $|\cdot|$ will denote the Euclidean norm in R^m. For any interval $I \subset Z := (-\infty, \infty)$, we denote by $BS(I)$ the set of all bounded functions mapping I into R^m and set $|\phi|_I = \sup\{|\phi(s)| : s \in I\}$.

Now, for any function $x : (-\infty, a) \to R^m$ and $n < a$, define a function $x_n : Z^- = (-\infty, 0] \to R^m$ by $x_n(s) = x(n+s)$ for $s \in Z^-$. Let B be a real linear space of functions mapping Z^- into R^m with a complete seminorm $|\cdot|_B$. We assume the following conditions on the space B.

(A1) There exist positive constant J, L and M with the property that if $x :$ $(-\infty, a) \to R^m$ is defined on $[\sigma, a)$ with $x_\sigma \in B$ for some $\sigma < a, \sigma, a \in Z$, then for all $n \in [\sigma, a)$,

(i) $x_n \in B$,

(ii) $J|x(n)| \le |x_n|_B \le L \sup_{\sigma \le s \le n} |x(s)| + M|x_\sigma|_B$,

(A2) If $\{\phi^k\}$ is a sequence in $B \cap BS$ converging to a function ϕ uniformly on any compact interval in Z^- and $\sup_k |\phi^k|_{BS} < \infty$, then $\phi \in B$ and $|\phi^k - \phi|_B \to 0$ as $k \to \infty$.

We hold that the space B contains BS and that there is a constant $l > 0$ such that

$$|\phi|_B \le l|\phi|_{BS} \qquad \text{for all} \quad \phi \in BS. \tag{1}$$

The space B is called a fading memory space for Z if it satisfies the following fading memory condition together with (A1) and (A2):

(A3) If $x : Z \to R^m$ is a function such that $x_0 \in B$, and $x(n) \equiv 0$ on Z^+, then $|x_n|_B \to 0$ as $n \to \infty$.

We introduce an almost periodic function $f(n, x) : Z \times D \to R^m$, where D is an open set in B.

Definition 1 $f(n, x)$ *is said to be almost periodic in n uniformly for $x \in D$, if for any $\epsilon > 0$ and any compact set K in D, there exists a positive integer $L^*(\epsilon, K)$ such that any interval of length $L^*(\epsilon, K)$ contains an integer τ for which*

$$|f(n + \tau, x) - f(n, x)| \le \epsilon \tag{2}$$

for all $n \in Z$ and all $x \in K$. Such a number τ in (2) is called an ϵ-translation number of $f(n, x)$.

In order to formulate a property of almost periodic functions, which is equivalent to the above definition, we discuss the concept of the normality of almost periodic functions. Namely, let $f(n, x)$ be almost periodic in n uniformly for $x \in D$. Then, for any sequence $\{h'_k\} \subset Z$, there exists a subsequence $\{h_k\}$ of $\{h'_k\}$ and function $g(n, x)$ such that

$$f(n + h_k, x) \to g(n, x) \tag{3}$$

uniformly on $Z \times K$ as $k \to \infty$, where K is any compact set in D. There are many properties of the discrete almost periodic functions [1,2], which are corresponding properties of the continuous almost periodic functions $f(t, x) \in C(R \times D, R^m)$ [3,11]. We shall denote by $T(f)$ the function space consisting of all translates of f, that is, $f_\tau \in T(f)$, where

$$f_\tau(n, x) = f(n + \tau, x), \qquad \tau \in Z \tag{4}$$

Let $H(f)$ denote the closure of $T(f)$ in the sense of (3). $H(f)$ is called the hull of f. In particular, we denote by $\Omega(f)$ the set of all limit functions $g \in H(f)$ such that for some sequence $\{n_k\}, n_k \to \infty$ as $k \to \infty$ and $f(n + n_k, x) \to g(n, x)$ uniformly on $Z \times S$ for any compact subset S in R^m. By (3), if $f : Z \times D \to R^m$ is almost periodic in n uniformly for $x \in D$, so is a function in $\Omega(f)$. The following concept of asymptotic almost periodicity was introduced by Frechet in the case of continuous function (cf. [3,11]).

Definition 2 $u(n)$ *is said to be asymptotically almost periodic if it is a sum of an almost periodic function $p(n)$ and a function $q(n)$ defined on $I^* = [a, \infty) \subset Z^+$ which tends to zero as $n \to \infty$, that is,*

$$u(n) = p(n) + q(n).$$

A function $u(n)$ is asymptotically almost periodic if and only if for any sequence $\{n_k\}$ such that $n_k \to \infty$ as $k \to \infty$ there exists a subsequence $\{n_{k_j}\}$ for which $u(n + n_{k_j})$ converges uniformly on $a \leq n < \infty$.

2   Existence of almost periodic solutions

We consider the almost periodic solution of a functional difference equation

$$x(n + 1) = f(n, x_n), \qquad n \in Z^+, \tag{5}$$

where $f : Z^+ \times B \to R^m$. We impose the following assumptions:

(H1) $\sup\{|f(n, \phi)| : n \in Z^+, |\phi|_B \leq H\} =: L_0(H) < \infty$ for each $H > 0$.

(H2) $f(n, \phi)$ is uniformly continuous at second variable $\phi \in S$ for any compact set S in B and almost periodic in n uniformly for $x \in D$.

(H3) Eq.(5) has a bounded solution $u(n)$ defined on Z^+ which passes through $(0, u_0)$, that is $\sup_{n \geq 0} |u(n)| < \infty$ and $u_0 \in BS$.

From (H3) and axiom (A1) it follows that $\sup_{n \geq 0} |u_n|_B < \infty$; hence $\sup_{n \geq 0} |u(n + 1)| < \infty$ by (H1). Thus the set

$$\Gamma(u) := \text{the closure of } \{u_n : n \in Z^+\}$$

is compact in B and consider any compact set K in R^m such that the interior of $K \supset \Gamma(u)$. (cf. [8,9]).

Now we introduce BS-stability properties and ρ-stability properties with respect to the compact set K and the metric ρ.

Definition 3 *The bounded solution $u(n)$ of Eq.(5) is said to be BS-totally stable (in short, BS-TS) if for any $\epsilon > 0$ there exists a $\delta(\epsilon) > 0$ such that if $n_0 \geq 0, |x_{n_0} - u_{n_0}|_{BS} < \delta(\epsilon)$ and $h \in BS([n_0, \infty))$ which satisfies $|h|_{[n_0, \infty)} < \delta(\epsilon)$, then $|x(n) - u(n)| < \epsilon$ for all $n \geq n_0$, where $x(n)$ is a solution of*

$$x(n + 1) = f(n, x_n) + h(n) \qquad n \geq 0 \tag{6}$$

through (n_0, ϕ) such that $x_{n_0}(s) = \phi(s)$ for all $s \leq 0$.

Let K be the compact set in R^m such that $u(n) \in K$ for all $n \in Z$, where $u(n) = \phi^0(n)$ for $n < 0$. For any $\theta, \psi \in BS$, we set

$$\rho(\theta, \psi) = \sum_{j=1}^{\infty} \rho_j(\theta, \psi)/[2^j(1 + \rho_j(\theta, \psi))] \quad \text{where}$$

$$\rho_j(\theta, \psi) = \sup_{-j \le s \le 0} |\theta(s) - \psi(s)|.$$

Clearly, $\rho(\theta^k, \theta) \to 0$ as $k \to \infty$ if and only if $\theta^k(s) \to \theta(s)$ uniformly on any compact subset of $(-\infty, 0]$ as $k \to \infty$.

Definition 4 *The bounded solution $u(n)$ of Eq.(5) is said to be (K, ρ)-totally stable (in short, (K, ρ)-TS) if for any $\epsilon > 0$ there exists a $\delta(\epsilon) > 0$ such that if $n_0 \ge 0$, $\rho(x_{n_0}, u_{n_0}) < \delta(\epsilon)$ and $h \in BS([n_0, \infty))$ which satisfies $|h|_{|[n_0, \infty)} < \delta(\epsilon)$, then $\rho(x_n, u_n) < \epsilon$ for all $n \ge n_0$, where $x(n)$ is a solution of (6) through (n_0, ϕ) such that $x_{n_0}(s) = \phi(s) \in K$ for all $s \le 0$.*

If the above term $\rho(x_n, u_n)$ is replaced by $|x(n) - u(n)|$, then we have another concept of (K, ρ)-total stability; which will be referred to as the $((K, \rho), R^m)$-total stability (in short, $((K, \rho), R^m)$-TS).

Next, we shall consider the week stability concept rather than the total stability. For the compact set K, $(P, Q) \in \Omega(f)$, we define $\pi(P, Q)$ by

$$\pi(P, Q) = \sum_{j=1}^{\infty} \pi_j(P, Q)/[2^j(1 + \pi_j(P, Q))],$$

where $\pi_j(P, Q) = \sup\{|P(n, x(s)) - Q(n, x(s))| : n \in Z, s \in [-j, 0], x(s) \in K\}$.

Definition 5 *The bounded solution $u(n)$ of Eq.(5) is said to be BS-stable under disturbances from $\Omega(f)$ with respect to K (in short, BS-s.d.$\Omega(f)$) if for any $\epsilon > 0$ there exists an $\eta(\epsilon) > 0$ such that $|x(n) - u(n)| < \epsilon$ for all $n \ge n_0$, whenever $g \in \Omega(f)$, $\pi(f, g) \le \eta(\epsilon)$ and $|x_{n_0} - u_{n_0}|_{BS} < \eta(\epsilon)$ for some $n_0 \ge 0$, where $x(n)$ is a solution through (n_0, ϕ) of*

$$x(n+1) = g(n, x_n), \qquad n \ge 0 \tag{7}$$

such that $x_{n_0}(s) = \phi(s) \in K$ for all $s \le 0$.

Definition 6 *The bounded solution $u(n)$ of Eq.(5) is said to be (K, ρ)-stable under disturbances from $\Omega(f)$ (in short, (K, ρ)-s.d.$\Omega(f)$) if for any $\epsilon > 0$ there exists an $\eta(\epsilon) > 0$ such that $\rho(x_n, u_n) < \epsilon$ for all $n \ge n_0$, whenever $g \in \Omega(f)$, $\pi(f, g) \le \eta(\epsilon)$ and $\rho(x_{n_0}, u_{n_0}) < \eta(\epsilon)$ and for some $n_0 \ge 0$, where $x(n)$ is a solution of (7) through (n_0, ϕ) such that $x_{n_0}(s) = \phi(s) \in K$ for all $s \le 0$.*

If the above term $\rho(x_n, u_n)$ is replaced by $|x(n) - u(n)|$, then we have another concept of (K, ρ)-stable under disturbances from $\Omega(f)$; which will be referred to as the $((K, \rho), R^m)$-stable under disturbances from $\Omega(f)$ (in short, $((K, \rho), R^m)$-s.d.$\Omega(f)$).

Therefore the (K, ρ)-s.d.$\Omega(f)$ implies the BS-s.d.$\Omega(f)$, because we have $\rho(\phi, \psi) \leq |\phi - \psi|_{BS}$ for $\phi, \psi \in BS$. In Theorem 3, we discuss the opposite implications.

Now, we have the main results, and see [6] for the proofs of theorems.

Theorem 1 *Under the assumptions (H1),(H2) and (H3), if the bounded solution $u(n)$ of Eq.(5) is (K, ρ)-TS, then it is (K, ρ)-s.d.$\Omega(f)$.*

Theorem 2 *Under the assumptions (H1),(H2) and (H3), if the bounded solution $u(n)$ of Eq.(5) is (K, ρ)-s.d.$\Omega(f)$, then it is an asymptotically almost periodic solution of Eq.(5). Consequently, Eq.(5) has an almost periodic solution.*

Corollary 1 *Under the assumptions (H1),(H2) and (H3), if the bounded solution $u(n)$ of Eq.(5) is (K, ρ)-TS, then it is an asymptotically almost periodic solution of Eq.(5) and Eq.(5) has an almost periodic solution.*

Proof. It is easy from Theorem 1 and 2 to prove this corollary. Indeed, the (K, ρ)-TS of $u(n)$ yields the (K, ρ)-s.d.$\Omega(f)$ of $u(n)$ by Theorem 1, and then $u(n)$ is an asymptotic almost periodic solution of Eq.(5) by Theorem 2. Therefore Eq.(5) has an almost periodic solution.

Theorem 3 *Let B be a fading memory space, and assume conditions (H1), (H2) and (H3). Then the solution $u(n)$ of Eq.(5) is BS-s.d.$\Omega(f)$ implies the solution $u(n)$ of Eq.(5) is (K, ρ)-s.d.$\Omega(f)$.*

Outline of proof. First, we have the following claim.

Claim 1. Under the above assumption, the solution $u(n)$ of Eq.(5) is $((K, \rho), R^m)$-s.d.$\Omega(f)$ implies (K, ρ)-s.d.$\Omega(f)$.

Now, in order to complete the proof of Theorem 3, we shall accomplish it by contradiction. By claim 1, we assume that the solution $u(n)$ of Eq.(5) is BS-s.d.$\Omega(f)$ but not $((K, \rho), R^m)$-s.d.$\Omega(f)$ here, $K \subset \{x \in R^m : |x| \leq \alpha\}$ for some $\alpha > 0$. Since the solution $u(n)$ of Eq.(5) is not $((K, \rho), R^m)$-s.d.$\Omega(f)$, there exists an $\epsilon \in (0, 1)$, sequence $\{\tau_m\} \subset Z^+, \{n_m\}(n_m > \tau_m), \{\phi^m\} \subset BS$ with $\phi^m(s) = x_{\tau_m}(s) \in K$ for all $s \leq 0$, $\{g_m\}$ with $g_m \in \Omega(f)$, and solutions $\{x(n)\}$ through (τ_m, ϕ^m) of

$$x(n + 1) = g_m(n, x_n) \tag{8}$$

such that

$$\rho(\phi^m, u_{n_m}) < 1/m \quad \text{and} \quad \pi(f_m, g_m) < 1/m \tag{9}$$

and that

$$|x(n_m) - u(n_m)| = \epsilon \quad \text{and} \quad |x(n) - u(n)| < \epsilon \quad \text{on} \quad [\tau_m, n_m) \tag{10}$$

for $m \in N$ (N denotes the set of all positive integers), where $x(n_m)$ is a solution through (τ_m, ϕ^m) of (8). For each $m \in N$ and $T \in Z^+$, we define $\phi^{m,T} \in BS$ by

$$\phi^{m,T}(\theta) = \begin{cases} \phi^m(\theta) & \text{if} \quad -T \le \theta \le 0, \\ \phi^m(-T) + u(\tau_m + \theta) - u(\tau_m - T) & \text{if} \quad \theta < -T. \end{cases}$$

Notice that $|\phi^{m,T} - u_{\tau_m}|_{BS} = |\phi^m - u_{\tau_m}|_{[-T,0]}$.

Next, we have the following claims.

Claim 2. $\sup\{|\phi^{m,T} - \phi^m|_B : m \in N\} \to 0$ as $T \to \infty$.

Claim 3. The set $\{\phi^{m,T}, \phi^m : m \in N, T \in Z^+\}$ is relatively compact in B.

Now, for any $m \in N$, set the solution $x^m(n) = x(n + \tau_m)$ of (8) if $n \le n_m - \tau_m$ and $x^m(n) = x^m(n_m - \tau_m)$ if $n > n_m - \tau_m$. Moreover, set $x^{m,T}(n) = \phi^{m,T}(n)$ if $n \in Z^-$ and $x^{m,T}(n) = x^m(n)$ if $n \in Z^+$. Since $x_0^m = \phi^m$ and $|x^m(n)| < 1 + |u|_{[0,\infty)} =: d < \infty$ for $n \in Z^+$, we have

$$|x_n^m|_B \le Ld + M|\phi^m|_B \le Ld + Ml|\phi^m|_{BS} \le Ld + Ml\alpha$$

by (1) and axiom (A1); hence, if $0 \le n < n_m - \tau_m$, then

$$\begin{aligned} |x^m(n+1)| &\le |f(n + \tau_m, x_n^m)| + |g_m(n + \tau_m, x_n^m) - f(n + \tau_m, x_n^m)| \\ &\le L_0(Ld + Ml\alpha) + 1/m \le L_1 \quad \text{(independent of } m \in N) \end{aligned}$$

by (9) and (H1). Consequently,

$$|x^m(s_1) - x^m(s_2)| \le 2L_1, \qquad s_1, s_2 \in Z^+, m \in N. \tag{11}$$

Set

$$W = \text{the closure of } \{x_n^{m,T}, x_n^m : m \in N, n \in Z^+, T \in Z^+\}.$$

Combining (11) with Claim 3, we see by (cf. [8,9,11]) that the set W is compact in B; hence $f(n, \phi)$ is uniformly continuous at the second variable on W by (H2). Define a function $q_{m,T}$ on Z^+ by $q_{m,T}(n) = f(n + \tau_m, x_n^m) - f(n + \tau_m, x_n^{m,T})$ if $0 \le n \le n_m - \tau_m$, and $q_{m,T}(n) = q_{m,T}(n_m - \tau_m)$ if $n > n_m - \tau_m$. Since $|x_n^{m,T} - x_n^m|_B \le M|\phi^{m,T} - \phi^m|_B$ ($n \in Z^+, m \in N$) by axiom (A1), it follows from Claim 2 that $\sup\{|x_n^{m,T} - x_n^m|_B : n \in Z^+, m \in N\} \to 0$ as $T \to \infty$; hence one can choose $T = T(\epsilon) \in N$ in such a way that

$$\sup\{|\hat{q}_{m,T}(n)| : m \in N, n \in Z^+\} < \delta(\epsilon/2)/2,$$

where $\delta(\cdot)$ is the one for BS-s.d.$\Omega(f)$ of the solution $u(n)$ of Eq.(5). Moreover, for this T, select an $m \in N$ such that $m > 2^T(1 + \delta(\epsilon/2))/\delta(\epsilon/2)$. Then $2^{-T}|\phi^m - u_{\tau_m}|_T/[1 + |\phi^m - u_{\tau_m}|_T] \le \rho(\phi^m, u_{\tau_m}) < 2^{-T}\delta(\epsilon/2)/[1 + \delta(\epsilon/2)]$ by (9), which implies that

$$|\phi^m - u_{\tau_m}|_T < \delta(\epsilon/2) \quad \text{or} \quad |\phi^{m,T} - u_{\tau_m}|_{BS} < \delta(\epsilon/2).$$

The function $x^{m,T}$ satisfies $x_0^{m,T} = \phi^{m,T}$ and

$$
\begin{aligned}
x^{m,T}(n+1) &= x^m(n+1) = \\
&= f_m(n+\tau_m, x_n^m) + (g_m(n+\tau_m, x_n^m) - f_m(n+\tau_m, x_n^m)) = \\
&= f_m(n+\tau_m, x_n^{m,T}) + q_{m,T}(n) + (g_m(n+\tau_m, x_n^m) - f_m(n+\tau_m, x_n^m))
\end{aligned}
$$

for $n \in [0, n_m - \tau_m)$. Since $u^m(n) = u(n+\tau_m)$ is a BS-s.d.$\Omega(f)$ solution of $x_{n+1} = f_m(n+\tau_m, x_n^m)$ with the same $\delta(\cdot)$ as the one for $u(n)$, from the fact that $\pi(f_m, g_m) < 1/m$ and hence $\sup_{n \geq 0} |q_{m,T}(n) + (g_m(n+\tau_m, x_n^m) - f_m(n+\tau_m, x_n^m))| < \delta(\epsilon/2)/2 + 1/m < \delta(\epsilon/2)$ it follows that $|x^{m,T}(n) - u(n+\tau_m)| < \epsilon/2$ on $[0, n_m - \tau_m)$. In particular, we have $|x^{m,T}(n_m - \tau_m) - u(n_m)| < \epsilon$ or $|x(n_m) - u(n_m)| < \epsilon$, which contradicts (10). This completes the proof.

Theorem 3 is true for functional differential equations with infinite delay [7]. By Theorem 3 and Theorem 2, we have the following corollary.

Corollary 2 *Let B be a fading memory space. Under the assumptions (H1), (H2) and (H3), if the bounded solution $u(n)$ of Eq.(5) is BS-s.d.$\Omega(f)$, then the Eq.(5) has an almost periodic solution.*

[10] has shown that if the bounded solution $u(t)$ of functional differential equation with infinite delay

$$
\dot{x} = f(t, x_t), \qquad t \geq 0 \tag{12}
$$

is BC-TS then it is (K, ρ)-TS, and it also is well known that if the bounded solution $u(t)$ of Eq.(12) is BC-TS, then it is BC-s.d.$\Omega(f)$ [8]. We can improve these to our Equation (5). Then, we have the following corollary.

Corollary 3 *Let B be a fading memory space. Under the assumptions (H1), (H2) and (H3), if the bounded solution $u(n)$ of Eq.(5) is BS-TS, then Eq.(5) has an almost periodic solution.*

Proof. The BS-TS of $u(n)$ implies the BS-s.d.$\Omega(f)$ of $u(n)$ by (cf. [Corollary 1,10 in 8] and Theorem 1), then $u(n)$ is an asymptotically almost periodic solution of Eq.(5) by Corollary 2. Therefore Eq.(5) has an almost periodic solution.

References

[1] R. P. Agarwal, *Difference Equations and Inequalities*, 2nd Edition, Marcel Dekker, 2000.

[2] C. Corduneanu, *Almost periodic discrete processes*, Libertas Mathematica, 2 (1982) 159–169.

[3] A. M. Fink, *Almost Periodic Differential Equations*, Lecture Notes in Mathematics 377, Springer-Verlag, 1974.

[4] Y. Hamaya, *Periodic solutions of nonlinear integrodifferential equations*, Tohoku Math. J., 41 (1989), 105–116.

[5] Y. Hamaya, *Stability property for an integrodifferential equation*, Differential and Integral Equations, 6 (1993), 1313–1324.

[6] Y. Hamaya, *Existence of an almost periodic solution in a difference equation with infinite delay*, Journal of Difference Equations and Applications 9 (2003), 227–237.

[7] Y. Hamaya, *Relationships between BC-$s.d.\Omega(f)$ and (K,ρ)-$s.d.\Omega(f)$ in an abstract functional differential equation with infinite delay*, International Journal of Differential Equations and Applications 4 (2002), 303–321.

[8] Y. Hino, S Murakami and T. Naito, *Functional Differential Equations with Infinite Delay*, Lecture Notes in Mathematics 1473, Springer-Verlag, 1991.

[9] J. Kato, A. A Martynyuk and A. A Shestakov, *Stability of Motion of Nonautonomous Systems (Method of Limiting Equations)*, Gordon and Breach Publishers, 1996.

[10] S. Murakami and T. Yoshizawa, *Relationships between BC-stabilities and ρ-stabilities in functional differential equations with infinite delay*, Tohoku Math. J., 44 (1992), 45–57.

[11] T. Yoshizawa, *Stability Theory and the Existence of Periodic Solutions and Almost Periodic Solutions*, Applied Mathematical Sciences 14, Springer-Verlag, 1975.

Discrete Quadratic Functionals with Jointly Varying Endpoints via Separable Endpoints

ROMAN HILSCHER [1]

Department of Mathematics, Michigan State University
East Lansing, MI 48824-1027, U.S.A.
E-mail: hilscher@math.msu.edu

and

VERA ZEIDAN [2]

Department of Mathematics, Michigan State University
East Lansing, MI 48824-1027, U.S.A.
E-mail: zeidan@math.msu.edu

Abstract A characterization of the positivity of a discrete quadratic functional with mixed state endpoints is presented. The functional is transformed in two different ways into problems with separable endpoints to which previous results are applied. The outcome are conditions that include augmented conjoined bases and implicit and explicit Riccati equations with corresponding augmented equality or inequality boundary conditions. The results of this paper extend and complete known ones.

Keywords Discrete quadratic functional, Linear Hamiltonian difference system, Conjoined basis, Riccati difference equation, Focal point

AMS Subject Classification 39A12, 49N10

1 Introduction

In this paper we study the positivity of the discrete quadratic functional

$$\mathcal{I}(\eta, q) := \begin{pmatrix} \eta_0 \\ \eta_{N+1} \end{pmatrix}^T \Gamma \begin{pmatrix} \eta_0 \\ \eta_{N+1} \end{pmatrix} + \sum_{k=0}^{N} \{ \eta_{k+1}^T C_k \, \eta_{k+1} + q_k^T B_k \, q_k \}$$

[1] Research partially supported by the Czech Grant Agency under grant 201/01/0079.
[2] Research supported by the National Science Foundation under grant DMS – 0072598.

over $\Delta\eta_k = A_k\eta_{k+1} + B_k q_k$, $k \in [0, N]$, and the joint boundary conditions

$$\mathcal{M}\begin{pmatrix} \eta_0 \\ \eta_{N+1} \end{pmatrix} = 0. \tag{1}$$

The quadratic functional $\mathcal{I}$ arises as the second variation in the discrete calculus of variations and optimal control theory (see, e.g., [4]), and hence the characterization of the positivity of $\mathcal{I}$ yields sufficiency criteria for nonlinear discrete problems. When the state endpoints vary jointly as in (1), the positivity of $\mathcal{I}$ is characterized in [1, Theorem 3] in terms of these notions: (i) disconjugacy, (ii) a special conjoined basis, i.e., the principal solution of the associated linear Hamiltonian difference system, with certain augmented boundary conditions at $N+1$, and (iii) an augmented implicit Riccati equation with augmented boundary conditions at $N + 1$. This result required a *controllability* assumption. The positivity of $\mathcal{I}$ without controllability was later characterized in [3, Theorem 2.3] via the disconjugacy and via the principal solution of the Hamiltonian system with an augmented boundary conditions of the *Riccati type*. When the boundary conditions are separated, i.e., when

$$\Gamma = \begin{pmatrix} \Gamma_0 & 0 \\ 0 & \Gamma_1 \end{pmatrix} \quad \text{and} \quad \mathcal{M} = \begin{pmatrix} \mathcal{M}_0 & 0 \\ 0 & \mathcal{M}_1 \end{pmatrix}, \tag{2}$$

improved results were obtained in [2, Theorem 3], [3, Theorem 3.2], and recently in [5], where a *natural* conjoined basis is used. However, when specialized to (2) the known work on the joint boundary conditions (1) does not directly yield the above mentioned results for the separable endpoints case.

The goal of this paper is to provide a complete characterization of the positivity of $\mathcal{I}$ for the general case (1) in terms of the disconjugacy (no generalized zeros), certain conjoined bases with augmented boundary conditions, and augmented implicit and explicit Riccati equations with endpoints constraints. The functional $\mathcal{I}$ is transformed into augmented quadratic functionals $\mathcal{I}^*$ and $\mathcal{I}^{\#}$ with separable endpoints. The original joint boundary conditions are augmented and moved either to $N + 1$ (for $\mathcal{I}^*$ in Section 3) or to 0 (for $\mathcal{I}^{\#}$ in Section 4). Then we apply to these augmented problems our results on separated boundary conditions from [5] (see Propositions 2.2–2.4) to obtain the characterization of the positivity of $\mathcal{I}$. The most significant contributions of this work reside in Theorems 3.2 and 4.1. When the boundary conditions of $\mathcal{I}$ are moved to $N + 1$, we show in Theorem 3.2 that the controllability assumption used in [1, Theorem 3] can be removed by restricting the augmented implicit Riccati equation onto a subspace. This theorem also completes [3, Theorem 2.3] in a sense that the boundary conditions for the augmented conjoined basis $(\hat{X}^*, \hat{U}^*)$ and the augmented implicit Riccati equation are derived. In Theorem 4.1, the original boundary conditions are moved to 0. We show then in Corollary 4.2 that this is the correct approach in order to extend directly the results for separable endpoints (2) to joint endpoints (1). Next, under certain normality assumptions, the positivity of $\mathcal{I}$ is characterized in Theorems 3.3, 4.3 in terms of augmented conjoined bases and augmented *explicit* Riccati equations with initial equality and final inequality endpoint constraints. Finally, without any assumption, Theorems 3.4, 4.4 provide the characterization of the positivity of $\mathcal{I}$ via augmented conjoined bases and augmented *explicit* Riccati equations with both boundary conditions in the form of strict inequalities.

2 Preliminaries

Given $n, N \in \mathbb{N}$ with $N \geq 2$, we denote by $J := [0, N]$ and $J^* := [0, N+1]$ the intervals of integers between the indicated endpoints. We assume that A_k, B_k, C_k, $k \in J$, are $n \times n$-matrices, Γ and $\mathcal{M}$ are $2n \times 2n$-matrices such that $B_k, C_k, \Gamma, \mathcal{M}$ are symmetric, and $\tilde{A}_k := (I - A_k)^{-1}$ exists. Without loss of generality, $\mathcal{M}$ is a projection and $\Gamma = (I - \mathcal{M})\Gamma(I - \mathcal{M})$. All quantities are supposed to be real valued. The forward difference operator is denoted by Δ, i.e., $\Delta y_k = y_{k+1} - y_k$. We will use the notation Ker, Im, ^{T}, †, ≥ 0, and > 0 to denote the kernel, image, transpose, Moore-Penrose inverse, nonnegative definiteness, and positive definiteness of a given matrix, respectively.

The sequences $\{\eta_k\}_{k=0}^{N+1}$ and $\{q_k\}_{k=0}^{N}$ of n-vectors form an *admissible pair* (η, q) if they satisfy the equation of motion for $\mathcal{I}$, i.e., $\Delta\eta_k = A_k\eta_{k+1} + B_kq_k$, $k \in J$. The quadratic functional $\mathcal{I}$ is *positive definite* ($\mathcal{I} > 0$) if $\mathcal{I}(\eta, q) > 0$ for all admissible (η, q) satisfying (1) and $\eta \not\equiv 0$. The corresponding *linear Hamiltonian difference system* is

$$\Delta\eta_k = A_k\eta_{k+1} + B_kq_k, \quad \Delta q_k = C_k\eta_{k+1} - A_k^Tq_k. \tag{H}$$

As usual, the vector solutions of (H) will be denoted by small letters and the $n \times n$-matrix solutions by capital ones. Let (X, U), $(\tilde{X}, \tilde{U})$ be solutions of (H). Then $X_k^T\tilde{U}_k - U_k^T\tilde{X}_k \equiv W$, where W is a constant $n \times n$ matrix. If $W = I$ then these solutions are called *normalized*. A solution (X, U) is said to be a *conjoined basis* if X^TU is symmetric and rank $\binom{X}{U} = n$. The conjoined basis $(\hat{X}, \hat{U})$ of (H) given by the initial conditions $\hat{X}_0 = 0$, $\hat{U}_0 = I$, is called the *principal solution* of (H). Following [1], a solution (X, U) of (H) is said to have *no focal points in* $(0, N+1]$, provided Ker $X_{k+1} \subseteq$ Ker X_k and $D_k := X_kX_{k+1}^\dagger\tilde{A}_kB_k \geq 0$ holds for all $k \in J$.

Remark 2.1 (i) *If $\mathcal{A}$ is a (constant) nonsingular $2n \times 2n$-matrix then (X, U) is a conjoined basis of* (H) *iff $(\tilde{X}, \tilde{U}) := (X\mathcal{A}, U\mathcal{A})$ is a conjoined basis of* (H). *Moreover,* Ker $X_{k+1} \subseteq$ Ker X_k *iff* Ker $\tilde{X}_{k+1} \subseteq$ Ker $\tilde{X}_k$. *In this case, by* $\tilde{X} = \tilde{X}\tilde{X}_{k+1}^\dagger\tilde{X}_{k+1}$ *and* $\tilde{X}\tilde{X}^\dagger = XX^\dagger$, *we have* $\tilde{D}_k = D_k$.

(ii) *In view of (i), (X, U) has no focal points in* $(0, N+1]$ *iff $(\tilde{X}, \tilde{U})$ defined as in (i) has no focal points in* $(0, N+1]$.

A solution (η, q) of (H) has a *generalized zero* in the interval $(m, m+1]$, provided $\eta_m \neq 0$, $\eta_{m+1} \in$ Im $\tilde{A}_mB_m$, and $\eta_m^TB_m^\dagger(I - A_m)\eta_{m+1} \leq 0$. When the right endpoint is fixed, i.e., $\mathcal{M}_1 = I$ in (2), the generalized zero concept is used to define conjugate intervals to 0. Let $m \in J$. An interval $(m, m+1]$ is said to be *conjugate to* 0 if there exists a solution (η, q) of (H) having a generalized zero in $(m, m+1]$ and, for some $\gamma_0 \in \mathbb{R}^n$, satisfying the initial boundary and transversality conditions $\mathcal{M}_0 \eta_0 = 0$ and $q_0 = \Gamma_0 \eta_0 + \mathcal{M}_0 \gamma_0$. With (H) it is associated the *Riccati matrix difference equation*

$$R[W]_k \equiv \Delta W_k - C_k + A_k^T W_k + (W_{k+1} - C_k)\tilde{A}_k(A_k + B_k W_k) = 0. \quad \text{(R)}$$

We will also discuss certain implicit Riccati equations.

In this paper, the following normality will be used. A pair (A, B) is called $\mathcal{M}$-*normal on* J^* if the system $-\Delta q_k = A_k^T q_k$, $B_k q_k = 0$, $k \in J$, $\binom{-q_0}{q_{N+1}} = \mathcal{M}\gamma$, possesses only the zero solution $q_k \equiv 0$ on J^*. In the case of separated boundary conditions, $\mathcal{M}$-normality will be denoted by $(\mathcal{M}_0 : \mathcal{M}_1)$-normality.

Next, similarly as in [1, Remark 3(ii)], we define the transition matrices $\Phi_{k,m}$, $\Psi_{k,m}$ and controllability matrices G_k, $\tilde{G}_k$ as follows: set $G_0 := 0$, $\tilde{G}_{N+1} := 0$, $\Phi_{0,0} := I$, $\Psi_{N+1,N} := I$, $\Phi_{k,m} := \tilde{A}_{k-1}\tilde{A}_{k-2}\ldots\tilde{A}_m$ for $k, m \in J$, $k > m$, $\Psi_{k,m} := (I - A_k)(I - A_{k+1})\ldots(I - A_m)$ for $k, m \in J$, $k \leq m$, $G_k := \begin{pmatrix} \Phi_{k,0}B_0 & \Phi_{k,1}B_1 & \ldots & \Phi_{k,k-1}B_{k-1} \end{pmatrix}$, and finally $\tilde{G}_k := \begin{pmatrix} B_k & \Psi_{k,k}B_{k+1} & \ldots & \Psi_{k,N-1}B_N \end{pmatrix}$. For the separated boundary conditions (2), a pair (η, q) with $\mathcal{M}_0\eta_0 = 0$ is admissible iff $\eta_k = (\Phi_{k,0}(I - \mathcal{M}_0) \quad G_k\mathcal{P}_k)\binom{\alpha}{q}$ for all $k \in J$, where $q := (q_0^T \ldots q_N^T)^T$, $\alpha := \eta_0$, and $\mathcal{P}_k : \mathbb{R}^{(N+1)n} \to \mathbb{R}^{kn}$ is the restriction operator onto the first k entries of q. Similarly, a pair (η, q) with $\mathcal{M}_1\eta_{N+1} = 0$ is admissible iff $\eta_k = (-\tilde{G}_k\tilde{\mathcal{P}}_k \quad \Psi_{k,N}(I - \mathcal{M}_1))\binom{q}{\beta}$ for all $k \in J$, where $\beta := \eta_{N+1}$, and $\tilde{\mathcal{P}}_k : \mathbb{R}^{(N+1)n} \to \mathbb{R}^{(N-k+1)n}$ is the restriction operator onto the last $N - k + 1$ entries of q, i.e., cutting the first k entries. Note that $\mathcal{P}_{N+1}$ and $\tilde{\mathcal{P}}_0$ are the identity matrices.

The following results on separated boundary conditions will be needed.

Proposition 2.2 *[5, Theorem 1] The following are equivalent.*

(i) $\mathcal{I} > 0$ *over* $\mathcal{M}_0\eta_0 = 0$, $\mathcal{M}_1\eta_{N+1} = 0$, *and* $\eta \not\equiv 0$.

(ii) *There is no interval* $(m, m+1] \subseteq (0, N+1]$ *conjugate to* 0 *and any solution* (η, q) *of* (H) *with* $\mathcal{M}_0\eta_0 = 0$, $q_0 = \Gamma_0\eta_0 + \mathcal{M}_0\gamma_0$, $\mathcal{M}_1\eta_{N+1} = 0$, *and* $\eta_{N+1} \neq 0$ *satisfies* $\eta_{N+1}^T(\Gamma_1\eta_{N+1} + q_{N+1}) > 0$.

(iii) *The conjoined basis* $(\bar{X}, \bar{U})$ *of* (H) *given by the initial conditions*

$$\bar{X}_0 = I - \mathcal{M}_0, \quad \bar{U}_0 = \Gamma_0 + \mathcal{M}_0$$

has no focal points in $(0, N+1]$ *and satisfies*

$$\bar{X}_{N+1}^T(\Gamma_1\bar{X}_{N+1} + \bar{U}_{N+1}) \geq 0 \quad \text{on } \operatorname{Ker}\mathcal{M}_1\bar{X}_{N+1}, \quad \text{(3)}$$

$$\operatorname{Ker}(I - \mathcal{M}_1)(\Gamma_1\bar{X}_{N+1} + \bar{U}_{N+1}) \cap \operatorname{Ker}\mathcal{M}_1\bar{X}_{N+1} \subseteq \operatorname{Ker}\bar{X}_{N+1}. \quad \text{(4)}$$

(iv) *The implicit Riccati matrix equation*

$$R[\bar{W}]_k(\Phi_{k,0}(I - \mathcal{M}_0) \quad G_k\mathcal{P}_k) = 0 \quad \text{on } \operatorname{Ker}\mathcal{M}_1(\Phi_{N+1,0}(I - \mathcal{M}_0) \quad G_{N+1}), \quad \text{(5)}$$

$k \in J$, *or equivalently,*

$$R[\bar{W}]_k(-\tilde{G}_k\tilde{\mathcal{P}}_k \ \Psi_{k,N}(I-\mathcal{M}_1)) = 0 \ on \ \mathrm{Ker}\,\mathcal{M}_0(-\tilde{G}_0 \ \Psi_{0,N}(I-\mathcal{M}_1)), \quad (6)$$

$k \in J$, *has a symmetric solution* $\bar{W}_k$ *on* J^* *satisfying* $\bar{W}_0 = \Gamma_0$,

$$\Gamma_1 + \bar{W}_{N+1} > 0 \quad on \ \mathrm{Ker}\,\mathcal{M}_1 \cap \mathrm{Im}\,\bar{X}_{N+1}, \quad (7)$$

and $\bar{\mathcal{D}}_k := B_k - B_k\tilde{A}_k^T(\bar{W}_{k+1} - C_k)\tilde{A}_k B_k \geq 0$ *for all* $k \in J$.

Proposition 2.3 *[5, Theorem 2] Assume that* (A, B) *is* $(\mathcal{M}_0 : I)$-*normal on* J^*. *Then the following are equivalent.*

(i) $\mathcal{I} > 0$ *over* $\mathcal{M}_0\eta_0 = 0$, $\mathcal{M}_1\eta_{N+1} = 0$, *and* $\eta \not\equiv 0$.

(ii) *There exists a conjoined basis* (X, U) *of* (H) *with no focal points in* $(0, N+1]$, X_k *invertible for all* $k \in J^*$, *and satisfying* $(I - \mathcal{M}_0)(\Gamma_0 X_0 - U_0) = 0$ *and* $X_{N+1}^T(\Gamma_1 X_{N+1} + U_{N+1}) > 0$ *on* $\mathrm{Ker}\,\mathcal{M}_1 X_{N+1}$.

(iii) *There exists a symmetric solution* W_k *on* J^* *of the explicit Riccati matrix equation* (R) *with* $I + B_k W_k$ *nonsingular and* $(I + B_k W_k)^{-1}B_k \geq 0$ *for all* $k \in J$, *and satisfying* $(I - \mathcal{M}_0)W_0 - \Gamma_0 = 0$ *and* $\Gamma_1 + W_{N+1} > 0$ *on* $\mathrm{Ker}\,\mathcal{M}_1$.

Proposition 2.4 *[5, Theorem 3] The following are equivalent.*

(i) $\mathcal{I} > 0$ *over* $\mathcal{M}_0\eta_0 = 0$, $\mathcal{M}_1\eta_{N+1} = 0$, *and* $\eta \not\equiv 0$.

(ii) *There exists a conjoined basis* (X, U) *of* (H) *with no focal points in* $(0, N+1]$, X_k *invertible for all* $k \in J^*$, *and satisfying* $X_0^T(\Gamma_0 X_0 - U_0) > 0$ *on* $\mathrm{Ker}\,\mathcal{M}_0 X_0$ *and* $X_{N+1}^T(\Gamma_1 X_{N+1} + U_{N+1}) > 0$ *on* $\mathrm{Ker}\,\mathcal{M}_1 X_{N+1}$.

(iii) *There exists a symmetric solution* W_k *on* J^* *of the explicit Riccati matrix equation* (R) *with* $I + B_k W_k$ *invertible and* $(I + B_k W_k)^{-1}B_k \geq 0$ *for all* $k \in J$, *and satisfying* $\Gamma_0 - W_0 > 0$ *on* $\mathrm{Ker}\,\mathcal{M}_0$ *and* $\Gamma_1 + W_{N+1} > 0$ *on* $\mathrm{Ker}\,\mathcal{M}_1$.

3 Augmented problem: right endpoint

We consider an augmented problem with $2n \times 2n$ coefficients, for which the boundary conditions are separated. The original cost and boundary conditions are now moved to the right endpoint $N + 1$. Define the matrices $\Gamma_0^* := 0$, $\Gamma_1^* := \Gamma$, $\mathcal{M}_0^* = \frac{1}{2}\begin{pmatrix} I & -I \\ -I & I \end{pmatrix}$, $\mathcal{M}_1^* := \mathcal{M}$, $A_k^* := \begin{pmatrix} 0 & 0 \\ 0 & A_k \end{pmatrix}$, $B_k^* := \begin{pmatrix} 0 & 0 \\ 0 & B_k \end{pmatrix}$, $C_k^* := \begin{pmatrix} 0 & 0 \\ 0 & C_k \end{pmatrix}$, $\tilde{A}_k^* := \begin{pmatrix} I & 0 \\ 0 & \tilde{A}_k \end{pmatrix}$, $\eta_k^* := \begin{pmatrix} \alpha \\ \eta_k \end{pmatrix}$, $q_k^* := \begin{pmatrix} \beta \\ q_k \end{pmatrix}$. Note that $\mathrm{Ker}\,\mathcal{M}_0^* = \mathrm{Im}\begin{pmatrix} I \\ I \end{pmatrix}$. Consider the discrete quadratic functional

$$\mathcal{I}^*(\eta^*, q^*) := (\eta_{N+1}^*)^T \Gamma_1^* \eta_{N+1}^* + \sum_{k=0}^{N} \left\{ (\eta_{k+1}^*)^T C_k^* \eta_{k+1}^* + (q_k^*)^T B_k^* q_k^* \right\}$$

over $\Delta\eta_k^* = A_k^*\eta_{k+1}^* + B_k^* q_k^*$, $k \in J$, and the separated boundary conditions

$$\mathcal{M}_0^* \eta_0^* = 0, \quad \mathcal{M}_1^* \eta_{N+1}^* = 0. \quad (8)$$

It is an easy computation to show that $\mathcal{I} > 0$ over the joint boundary conditions (1) iff $\mathcal{I}^* > 0$ over the separated boundary conditions (8). Thus, by applying to $\mathcal{I}^*$ the results in Propositions 2.2-2.4, we obtain the characterization of the positivity of the original functional $\mathcal{I}$. With $\mathcal{I}^*$ we consider the corresponding augmented linear Hamiltonian difference system

$$\Delta \eta_k^* = A_k^* \eta_{k+1}^* + B_k^* q_k^*, \quad \Delta q_k^* = C_k^* \eta_{k+1}^* - (A_k^*)^T q_k^*, \tag{H*}$$

and the augmented Riccati difference equation

$$R^*[W^*]_k \equiv \Delta W_k^* - C_k^* + (A_k^*)^T W_k^* + (W_{k+1}^* - C_k^*) \tilde{A}_k^* (A_k^* + B_k^* W_k^*) = 0. \tag{R*}$$

Remark 3.1 (i) *A pair* (η^*, q^*) *is admissible for* $\mathcal{I}^*$ *iff it has the form* $\eta_k^* = \left(\begin{smallmatrix} \alpha \\ \eta_k \end{smallmatrix} \right)$, $q_k^* = \left(\begin{smallmatrix} \beta_k \\ q_k \end{smallmatrix} \right)$, *where* (η, q) *is admissible for* $\mathcal{I}$. *Moreover,* (η^*, q^*) *satisfies* $\mathcal{M}_0^* \eta_0^* = 0$ *iff* $\alpha = \eta_0$, *and* $\mathcal{M}_1^* \eta_{N+1}^* = 0$ *iff* $\mathcal{M} \left(\begin{smallmatrix} \eta_0 \\ \eta_{N+1} \end{smallmatrix} \right) = 0$.

(ii) *It is readily seen that any solution* (η^*, q^*) *of* (H*) *is of the form* $\eta_k^* := \left(\begin{smallmatrix} \alpha \\ \eta_k \end{smallmatrix} \right)$, $q_k^* := \left(\begin{smallmatrix} \beta \\ q_k \end{smallmatrix} \right)$, *where* (η, q) *solves* (H), *and any solution* (X^*, U^*) *of* (H*) *is of the form* $X_k^* = \left(\begin{smallmatrix} K & L \\ X_k & \tilde{X}_k \end{smallmatrix} \right)$, $U_k^* = \left(\begin{smallmatrix} M & N \\ U_k & \tilde{U}_k \end{smallmatrix} \right)$, *where* (X, U) *and* $(\tilde{X}, \tilde{U})$ *are solutions of* (H). *The (constant) matrices* K, L, M, N *are determined by the initial conditions* (X_0^*, U_0^*).

In our first result we complete [3, Theorem 2.3], where the conditions (i)-(iii) in Theorem 3.2 below are shown to be equivalent, but the augmented boundary conditions (10)–(11) in (iii) are replaced by the Riccati boundary conditions (13) from (iv). We derive both the augmented boundary conditions for (X^*, U^*) in (iii) and also the corresponding augmented implicit Riccati equation and its boundary conditions in (iv). This approach also provides an alternative proof of [3, Theorem 2.3]. Our Theorem 3.2 also generalizes [1, Theorem 3], where the positivity of $\mathcal{I}$ is characterized in terms of an augmented implicit Riccati equation and its boundary conditions under a controllability (or $\operatorname{Ker} \mathcal{M} \subseteq \operatorname{Im} X_{N+1}^*$) assumption. We remove this assumption by restricting the Riccati equation onto a subspace.

Theorem 3.2 (Characterization of $\mathcal{I} > 0$) *The following are equivalent.*
(i) $\mathcal{I} > 0$ *over admissible* (η, q) *with* $\mathcal{M} \left(\begin{smallmatrix} \eta_0 \\ \eta_{N+1} \end{smallmatrix} \right) = 0$ *and* $\eta \not\equiv 0$.
(ii) *No solution* (η, q) *of* (H) *with* $\eta_0 = 0$ *has any generalized zero in* $(0, N+1]$, *and every solution* (η, q) *of* (H) *with* $\mathcal{M} \left(\begin{smallmatrix} \eta_0 \\ \eta_{N+1} \end{smallmatrix} \right) = 0$ *and* $\left(\begin{smallmatrix} \eta_0 \\ \eta_{N+1} \end{smallmatrix} \right) \neq 0$ *satisfies*

$$\left(\begin{matrix} \eta_0 \\ \eta_{N+1} \end{matrix} \right)^T \left\{ \Gamma \left(\begin{matrix} \eta_0 \\ \eta_{N+1} \end{matrix} \right) + \left(\begin{matrix} -q_0 \\ q_{N+1} \end{matrix} \right) \right\} > 0.$$

(iii) *The principal solution* $(\hat{X}, \hat{U})$ *of* (H), *i.e.,* $(\hat{X}_0, \hat{U}_0) = (0, I)$, *has no focal points in* $(0, N+1]$ *and the* $2n \times 2n$-*matrices*

$$\hat{X}_k^* := \left(\begin{matrix} 0 & I \\ \hat{X}_k & \tilde{X}_k \end{matrix} \right), \quad \hat{U}_k^* := \left(\begin{matrix} -I & -I \\ \hat{U}_k & \tilde{U}_k \end{matrix} \right), \tag{9}$$

where $(\tilde{X}, \tilde{U})$ *solves* (H) *with* $(\tilde{X}_0, \tilde{U}_0) = (I, I)$, *satisfy*

$$(\hat{X}^*_{N+1})^T (\Gamma \hat{X}^*_{N+1} + \hat{U}^*_{N+1}) \geq 0 \quad on \ \operatorname{Ker} \mathcal{M} \hat{X}^*_{N+1}, \tag{10}$$

$$\operatorname{Ker}(I - \mathcal{M})(\Gamma \hat{X}^*_{N+1} + \hat{U}^*_{N+1}) \cap \operatorname{Ker} \mathcal{M} \hat{X}^*_{N+1} \subseteq \operatorname{Ker} \hat{X}^*_{N+1}. \tag{11}$$

(iv) *The augmented implicit Riccati matrix equation*

$$R^* [\hat{W}^*]_k \begin{pmatrix} I & 0 \\ \Phi_{k,0} & G_k \mathcal{P}_k \end{pmatrix} = 0 \quad on \ \operatorname{Ker} \mathcal{M} \begin{pmatrix} I & 0 \\ \Phi_{N+1,0} & G_{N+1} \end{pmatrix}, \tag{12}$$

$k \in J$, *has a symmetric solution* $\hat{W}^*_k = \left(\begin{smallmatrix} * & * \\ * & \hat{W}_k \end{smallmatrix} \right)$ *on* J^* *satisfying* $\hat{\mathcal{D}}_k := B_k - B_k \tilde{A}^T_k (\hat{W}_{k+1} - C_k) \tilde{A}_k B_k \geq 0$ *for all* $k \in J$, $\hat{W}^*_0 = 0$, *and*

$$\Gamma + \hat{W}^*_{N+1} > 0 \quad on \ \operatorname{Ker} \mathcal{M} \cap \operatorname{Im} \hat{X}^*_{N+1}. \tag{13}$$

Proof. Apply Proposition 2.2 to the functional $\mathcal{I}^*$.

"(ii)" Condition (ii) of Proposition 2.2 means that $(0, N+1]$ contains no interval conjugate to 0 for (H*), and every solution (η^*, q^*) of (H*) with $\mathcal{M}^*_0 \eta^*_0 = 0$, $q^*_0 = \mathcal{M}^*_0 \gamma^*$, $\mathcal{M}^*_1 \eta^*_{N+1} = 0$, and $\eta^*_{N+1} \neq 0$ satisfies $(\eta^*_{N+1})^T (\Gamma^*_1 \eta^*_{N+1} + q^*_{N+1}) > 0$. The result then follows from Remark 3.1.

"(iii)" Condition (iii) of Proposition 2.2 states that the conjoined basis $(\bar{X}^*, \bar{U}^*)$ of (II*) given by the initial conditions $\bar{X}^*_0 = I - \mathcal{M}^*_0 = \frac{1}{2} \left(\begin{smallmatrix} I & I \\ I & I \end{smallmatrix} \right)$, $\bar{U}^*_0 = \mathcal{M}^*_0 = \frac{1}{2} \left(\begin{smallmatrix} I & -I \\ -I & I \end{smallmatrix} \right)$ has no focal points in $(0, N+1]$. Define $(\hat{X}^*, \hat{U}^*) := (\bar{X}^* \mathcal{A}, \bar{U}^* \mathcal{A})$, where $\mathcal{A} = \left(\begin{smallmatrix} -I & 0 \\ I & 2I \end{smallmatrix} \right)$. Then it follows that $(\hat{X}^*, \hat{U}^*)$ has the form (9) and, by Remark 2.1(ii), $(\hat{X}^*, \hat{U}^*)$ has no focal points in $(0, N + 1]$. A direct proof shows that the latter condition is equivalent to $(\hat{X}, \hat{U})$ having no focal points. Next, the inequality $(\bar{X}^*_{N+1})^T (\Gamma^*_1 \bar{X}^*_{N+1} + \bar{U}^*_{N+1}) \geq 0$ on $\operatorname{Ker} \mathcal{M}^*_1 \bar{X}^*_{N+1}$ from (3) is equivalent to (10). Finally, condition $(I - \mathcal{M}^*_1)(\Gamma^*_1 \bar{X}^*_{N+1} + \bar{U}^*_{N+1})\beta^* = 0$ and $\mathcal{M}^*_1 \bar{X}^*_{N+1} \beta^* = 0$ imply $\bar{X}^*_{N+1} \beta^* = 0$ from (4) is equivalent to (11).

"(iv)" First, Riccati equation (12) is (5), where each matrix has the superscript $*$, i.e., $R^*[\hat{W}^*]$ as in (R*), $\Phi^*_{k,m} = \left(\begin{smallmatrix} I & 0 \\ 0 & \Phi_{k,m} \end{smallmatrix} \right)$, $G^*_k \mathcal{P}^*_k q^* = \left(\begin{smallmatrix} 0 \\ G_k \mathcal{P}_k q \end{smallmatrix} \right)$, $\left(\begin{smallmatrix} \alpha^* \\ q^* \end{smallmatrix} \right) \in \operatorname{Ker} \mathcal{M}^*_1 (\Phi^*_{N+1,0}(I - \mathcal{M}^*_0) \ G^*_{N+1})$ iff $\left(\begin{smallmatrix} \alpha \\ q \end{smallmatrix} \right) \in \operatorname{Ker} \mathcal{M} \left(\begin{smallmatrix} I & 0 \\ \Phi_{N+1,0} & G_{N+1} \end{smallmatrix} \right)$, where $\alpha^* = \left(\begin{smallmatrix} \alpha_0 \\ \alpha_1 \end{smallmatrix} \right)$, $\alpha = \frac{1}{2}(\alpha_0 + \alpha_1)$, and $q^*_k = \left(\begin{smallmatrix} \beta_k \\ q_k \end{smallmatrix} \right)$, $k \in J$. Next, $\hat{\mathcal{D}}^* \geq 0$ and $\hat{\mathcal{D}}^* = \left(\begin{smallmatrix} 0 & 0 \\ 0 & \hat{\mathcal{D}} \end{smallmatrix} \right)$ yield $\hat{\mathcal{D}}_k \geq 0$. Finally, $\bar{W}^*_0 = \Gamma^*_0 = 0$ and (13) is just (7). ∎

The $(\mathcal{M}^*_0 : I)$-normality of (A^*, B^*) needed in Proposition 2.3 translates as $(I : I)$-normality of (A, B). The next two results are obtained by applying Propositions 2.3 and 2.4 to the functional $\mathcal{I}^*$.

Theorem 3.3 (Characterization of $\mathcal{I} > 0$) *Assume that* (A, B) *is* $(I : I)$-*normal on* J^*. *Then the following are equivalent.*

(i) $\mathcal{I} > 0$ *over admissible* (η, q) *with* $\mathcal{M} \left(\begin{smallmatrix} \eta_0 \\ \eta_{N+1} \end{smallmatrix} \right) = 0$ *and* $\eta \not\equiv 0$.

(ii) *There exists a conjoined basis* (X^*, U^*) *of* (H*) *with no focal points in* $(0, N + 1]$, X^*_k *invertible for all* $k \in J^*$, *and satisfying* $\left(\begin{smallmatrix} I & I \\ I & I \end{smallmatrix} \right) U^*_0 = 0$ *and* $(X^*_{N+1})^T (\Gamma X^*_{N+1} + U^*_{N+1}) > 0$ *on* $\operatorname{Ker} \mathcal{M} X^*_{N+1}$.

(iii) *There exists a symmetric solution* $W_k^* = (\begin{smallmatrix} * & * \\ * & W_k \end{smallmatrix})$ *on* J^* *of the augmented explicit Riccati matrix equation* (R*) *with* $I + B_k W_k$ *invertible and* $(I + B_k W_k)^{-1} B_k \geq 0$ *for all* $k \in J$, *and satisfying* $W_0^* = (\begin{smallmatrix} W_0 & -W_0 \\ -W_0 & W_0 \end{smallmatrix})$ *and* $\Gamma + W_{N+1}^* > 0$ *on* $\mathrm{Ker}\,\mathcal{M}$.

Theorem 3.4 (Characterization of $\mathcal{I} > 0$) *The following are equivalent.*

(i) $\mathcal{I} > 0$ *over admissible* (η, q) *with* $\mathcal{M}(\begin{smallmatrix} \eta_0 \\ \eta_{N+1} \end{smallmatrix}) = 0$ *and* $\eta \not\equiv 0$.

(ii) *There exists a conjoined basis* (X^*, U^*) *of* (H*) *with no focal points in* $(0, N+1]$, X_k^* *invertible for all* $k \in J^*$, *and satisfying* $-(X_0^*)^T U_0^* > 0$ *on* $\mathrm{Ker}\,(\begin{smallmatrix} I & -I \\ -I & I \end{smallmatrix}) X_0^*$ *and* $(X_{N+1}^*)^T (\Gamma X_{N+1}^* + U_{N+1}^*) > 0$ *on* $\mathrm{Ker}\,\mathcal{M} X_{N+1}^*$.

(iii) *There exists a symmetric solution* $W_k^* = (\begin{smallmatrix} * & * \\ * & W_k \end{smallmatrix})$ *on* J^* *of the augmented explicit Riccati matrix equation* (R*) *with* $I + B_k W_k$ *invertible and* $(I + B_k W_k)^{-1} B_k \geq 0$ *for all* $k \in J$, *and satisfying* $-W_0^* > 0$ *on* $\mathrm{Im}\,(\begin{smallmatrix} I \\ I \end{smallmatrix})$ *and* $\Gamma + W_{N+1}^* > 0$ *on* $\mathrm{Ker}\,\mathcal{M}$.

4   Augmented problem: left endpoint

The original cost and boundary conditions are now moved to the left endpoint 0. Define the matrices $\Gamma_0^\# := \Gamma$, $\Gamma_1^\# := 0$, $\mathcal{M}_0^\# = \mathcal{M}$, $\mathcal{M}_1^\# := \frac{1}{2}(\begin{smallmatrix} I & I \\ -I & I \end{smallmatrix})$, $A_k^\# := (\begin{smallmatrix} A_k & 0 \\ 0 & 0 \end{smallmatrix})$, $B_k^\# := (\begin{smallmatrix} B_k & 0 \\ 0 & 0 \end{smallmatrix})$, $C_k^\# := (\begin{smallmatrix} C_k & 0 \\ 0 & 0 \end{smallmatrix})$, $\tilde{A}_k^\# := (\begin{smallmatrix} \tilde{A}_k & 0 \\ 0 & I \end{smallmatrix})$, $\eta_k^\# := (\begin{smallmatrix} \eta_k \\ \alpha \end{smallmatrix})$, $q_k^\# := (\begin{smallmatrix} q_k \\ \beta \end{smallmatrix})$. Consider the discrete quadratic functional

$$\mathcal{I}^\#(\eta^\#, q^\#) := (\eta_0^\#)^T \Gamma_0^\# \eta_0^\# + \sum_{k=0}^{N} \left\{ (\eta_{k+1}^\#)^T C_k^\# \eta_{k+1}^\# + (q_k^\#)^T B_k^\# q_k^\# \right\}$$

over $\Delta \eta_k^\# = A_k^\# \eta_{k+1}^\# + B_k^\# q_k^\#$, $k \in J$, and the separated boundary conditions

$$\mathcal{M}_0^\# \eta_0^\# = 0, \quad \mathcal{M}_1^\# \eta_{N+1}^\# = 0. \tag{14}$$

Then, as in the previous section, $\mathcal{I} > 0$ over the joint boundary conditions (1) iff $\mathcal{I}^\# > 0$ over the separated boundary conditions (14). Thus, by applying to $\mathcal{I}^\#$ the results in Propositions 2.2–2.4 we obtain the characterization of the positivity of $\mathcal{I}$. With $\mathcal{I}^\#$ we consider the corresponding augmented linear Hamiltonian difference system

$$\Delta \eta_k^\# = A_k^\# \eta_{k+1}^\# + B_k^\# q_k^\#, \quad \Delta q_k^\# = C_k^\# \eta_{k+1}^\# - (A_k^\#)^T q_k^\#, \tag{H\#}$$

and the augmented Riccati difference equation

$$R^\#[W^\#]_k \equiv \Delta W_k^\# - C_k^\# + (A_k^\#)^T W_k^\# + (W_{k+1}^\# - C_k^\#)\tilde{A}_k^\#(A_k^\# + B_k^\# W_k^\#) = 0. \tag{R\#}$$

The conclusions as in Remark 3.1 now hold for the solutions of (H#).

Theorem 4.1 (Characterization of $\mathcal{I} > 0$) *The following are equivalent.*

(i) $\mathcal{I} > 0$ *over admissible* (η, q) *with* $\mathcal{M}(\begin{smallmatrix} \eta_0 \\ \eta_{N+1} \end{smallmatrix}) = 0$ *and* $\eta \not\equiv 0$.

(ii) *No solution* (η, q) *of* (H) *with* $\mathcal{M}\left(\begin{smallmatrix} \eta_0 \\ 0 \end{smallmatrix}\right) = 0$, $\left(\begin{smallmatrix} q_0 \\ \beta \end{smallmatrix}\right) = \Gamma\left(\begin{smallmatrix} \eta_0 \\ 0 \end{smallmatrix}\right) + \mathcal{M}\gamma$, *for some* $\beta \in \mathbb{R}^n$, $\gamma \in \mathbb{R}^{2n}$, *has any generalized zero in* $(0, N+1]$, *and every solution* (η, q) *of* (H) *with* $\mathcal{M}\left(\begin{smallmatrix} \eta_0 \\ \eta_{N+1} \end{smallmatrix}\right) = 0$, $\left(\begin{smallmatrix} q_0 \\ \beta \end{smallmatrix}\right) = \Gamma\left(\begin{smallmatrix} \eta_0 \\ \eta_{N+1} \end{smallmatrix}\right) + \mathcal{M}\gamma$ *for some* $\beta \in \mathbb{R}^n$, $\gamma \in \mathbb{R}^{2n}$, *and* $\eta_{N+1} \neq 0$ *satisfies* $\eta_{N+1}^T(\beta + q_{N+1}) > 0$.

(iii) *The conjoined basis* $(\bar{X}^\#, \bar{U}^\#)$ *of* (H$^\#$) *given by the initial conditions*

$$\bar{X}_0^\# = I - \mathcal{M}, \quad \bar{U}_0^\# = \Gamma + \mathcal{M}$$

has no focal points in $(0, N+1]$ *and satisfies*

$$(\bar{X}_{N+1}^\#)^T \bar{U}_{N+1}^\# \geq 0 \quad \text{on } \mathrm{Ker}\left(\begin{smallmatrix} I & -I \\ -I & I \end{smallmatrix}\right) \bar{X}_{N+1}^\#,$$

$$\mathrm{Ker}\left(\begin{smallmatrix} I & I \\ I & I \end{smallmatrix}\right) \bar{U}_{N+1}^\# \cap \mathrm{Ker}\left(\begin{smallmatrix} I & -I \\ -I & I \end{smallmatrix}\right) \bar{X}_{N+1}^\# \subseteq \mathrm{Ker}\, \bar{X}_{N+1}^\#.$$

(iv) *The augmented implicit Riccati matrix equation*

$$R^\#[\bar{W}^\#]_k \begin{pmatrix} -\tilde{G}_k \tilde{P}_k & \Psi_{k,N} \\ 0 & I \end{pmatrix} = 0 \quad \text{on } \mathrm{Ker}\,\mathcal{M}\begin{pmatrix} -\tilde{G}_0 & \Psi_{0,N} \\ 0 & I \end{pmatrix}, \tag{15}$$

$k \subset J$, *has a symmetric solution* $\bar{W}_k^\# = \left(\begin{smallmatrix} \bar{W}_k & * \\ * & * \end{smallmatrix}\right)$ *on* J^* *satisfying* $\bar{\mathcal{D}}_k := B_k - B_k \tilde{A}_k^T(\bar{W}_{k+1} - C_k)\tilde{A}_k B_k \geq 0$ *for all* $k \in J$, $\bar{W}_0^\# = \Gamma$, *and*

$$\bar{W}_{N+1}^\# > 0 \quad \text{on } \mathrm{Im}\left(\begin{smallmatrix} I \\ I \end{smallmatrix}\right) \cap \mathrm{Im}\, \bar{X}_{N+1}^\#.$$

Proof. Apply Proposition 2.2 to $\mathcal{I}^*$. Note that (15) is derived from (6). ∎

Next we show that Theorem 4.1 is a direct extension of the separable endpoints case. Its proof is based on technical computations needed to rephrase each of the conditions of Theorem 4.1.

Corollary 4.2 *If the boundary conditions are separated, i.e., if* $\mathcal{M}$ *and* Γ *are given by* (2), *then Theorem 4.1 reduces to Proposition 2.2.*

The notion of $(\mathcal{M}_0^\# : I)$-normality of $(A^\#, B^\#)$ translates as follows:

$$-\Delta q_k = A_k^T q_k, \quad B_k q_k = 0, \quad k \in J, \quad \left(\begin{smallmatrix} q_0 \\ \beta \end{smallmatrix}\right) = \mathcal{M}\gamma, \tag{16}$$

possesses only the zero solution $q_k \equiv 0$ on J^* and $\beta = 0$. This type of normality is, however, stronger that the usual $\mathcal{M}$-normality of (A, B). Applying Propositions 2.3 and 2.4 to $\mathcal{I}^*$ we obtain the last two results of this paper.

Theorem 4.3 (Characterization of $\mathcal{I} > 0$) *Assume that* (A, B) *is normal on* J^* *as in* (16). *Then the following are equivalent.*

(i) $\mathcal{I} > 0$ *over admissible* (η, q) *with* $\mathcal{M}\left(\begin{smallmatrix} \eta_0 \\ \eta_{N+1} \end{smallmatrix}\right) = 0$ *and* $\eta \not\equiv 0$.

(ii) *There exists a conjoined basis* $(X^\#, U^\#)$ *of* (H$^\#$) *with no focal points in* $(0, N+1]$, $X_k^\#$ *invertible for all* $k \in J^*$, *and satisfying* $(I - \mathcal{M})(\Gamma X_0^\# - U_0^\#) = 0$ *and* $(X_{N+1}^\#)^T U_{N+1}^\# > 0$ *on* $\mathrm{Ker}\left(\begin{smallmatrix} I & -I \\ -I & I \end{smallmatrix}\right) X_{N+1}^\#$.

(iii) *There exists a symmetric solution* $W_k^\# = \left(\begin{smallmatrix} W_k & * \\ * & * \end{smallmatrix}\right)$ *on* J^* *of the augmented explicit Riccati matrix equation* (R$^\#$) *with* $I + B_k W_k$ *invertible and* $(I + B_k W_k)^{-1} B_k \geq 0$ *for all* $k \in J$, *and satisfying* $(I - \mathcal{M})W_0^\# - \Gamma = 0$ *and* $W_{N+1}^\# > 0$ *on* $\mathrm{Im}\left(\begin{smallmatrix} I \\ I \end{smallmatrix}\right)$.

Theorem 4.4 (Characterization of $\mathcal{I} > 0$) *The following are equivalent.*

(i) $\mathcal{I} > 0$ *over admissible* (η, q) *with* $\mathcal{M}\left(\begin{smallmatrix} \eta_0 \\ \eta_{N+1} \end{smallmatrix}\right) = 0$ *and* $\eta \not\equiv 0$.

(ii) *There exists a conjoined basis* $(X^\#, U^\#)$ *of* $(\mathrm{H}^\#)$ *with no focal points in* $(0, N+1]$, $X_k^\#$ *invertible for all* $k \in J^*$, *and satisfying* $(X_0^\#)^T (\Gamma X_0^\# - U_0^\#) > 0$ *on* $\operatorname{Ker} \mathcal{M} X_0^\#$ *and* $(X_{N+1}^\#)^T U_{N+1}^\# > 0$ *on* $\operatorname{Ker}\left(\begin{smallmatrix} I & -I \\ -I & I \end{smallmatrix}\right) X_{N+1}^\#$.

(iii) *There exists a symmetric solution* $W_k^\# = \left(\begin{smallmatrix} W_k & \star \\ \star & \star \end{smallmatrix}\right)$ *on* J^* *of the augmented explicit Riccati matrix equation* $(\mathrm{R}^\#)$ *with* $I + B_k W_k$ *invertible and* $(I + B_k W_k)^{-1} B_k \geq 0$ *for all* $k \in J$, *and satisfying* $\Gamma - W_0^\# > 0$ *on* $\operatorname{Ker} \mathcal{M}$ *and* $W_{N+1}^\# > 0$ *on* $\operatorname{Im}\left(\begin{smallmatrix} I \\ I \end{smallmatrix}\right)$.

References

[1] M. Bohner, Linear Hamiltonian difference systems: disconjugacy and Jacobi-type conditions, *Journal of Mathematical Analysis and Applications* **199** (1996), 804–826.

[2] M. Bohner, Riccati matrix difference equations and linear Hamiltonian difference systems, *Dynamics of Continuous, Discrete and Impulsive Systems* **2** (1996), no. 2, 147–159.

[3] M. Bohner, O. Došlý, W. Kratz, Discrete Reid roundabout theorems, in: *Discrete and Continuous Hamiltonian Systems* (Eds., Agarwal, R. P., and Bohner, M.), *Dynamic Systems and Applications* **8** (1999), no. 3-4, 345–352.

[4] R. Hilscher, V. Zeidan, Second order sufficiency criteria for a discrete optimal control problem, *Journal of Difference Equations and Applications* **8** (2002), no. 6, 573–602.

[5] R. Hilscher, V. Zeidan, A remark on discrete quadratic functionals with separable endpoints, *Rocky Mountain Journal of Mathematics*, to appear.

On Finite Difference Potentials

ANGELA HOMMEL

Institute of Mathematics and Physics,
Bauhaus–University of Weimar
D-99421 Weimar, Germany
E-mail: angela.hommel@bauing.uni-weimar.de

Abstract In this article the method of difference potentials is presented for the discrete Laplace equation in the three-dimensional case. At first important properties of the discrete fundamental solution are described. The theory is based on a discrete analogue of the integral representation for functions in C^2 and the possibility to split the difference potential on the boundary into a discrete single- and double-layer potential. In addition, a uniqueness theorem for the interior Neumann problem is formulated and the solvability of the system of linear equations on the boundary is studied.

Keywords Finite difference operator, Fundamental solution, Difference potential

AMS Subject Classification 39A70, 39A12

1 Introduction

The classical potential theory is a well-known method to solve boundary value problems. The differential equation inside the domain can be solved by the help of a special integral equation on the boundary, such that both equations are equivalent with respect to solvability. From the analytical point of view we prefer the study of the integral equation. In addition, problems in exterior domains can be solved more efficiently and the reduction of the dimension is an advantage for numerical calculations. In general the integral equation on the boundary can not immediately be solved. Often a quadrature formula is used to calculate an approximate solution. By this way the equivalence to the original problem is lost and a loss of potential theoretical information can appear.

The approximation of the boundary value problem by a finite difference boundary value problem on a uniform lattice with the mesh width h can be considered as an idea to overcome these difficulties. Ryabenkij already proved in [5], that it is possible to describe a discrete theory, where with respect to solvability the equivalence between the difference equation and the equation on the boundary is preserved. The fact that in the discrete case the boundary equation is a system of linear equations is an important advantage. While in [5] the considered system on the boundary was overdetermined, it could be demonstrated in [3] how efficient the discrete method can be used, if the difference potential on the boundary is split into a discrete single- and double-layer potential.

In order to construct a discrete potential theory, the existence of a discrete fundamental solution or of a Green's function is necessary. An integral representation for the discrete fundamental solution can be obtained by the help of the discrete Fourier transform (see [8]). In the literature the existence of a discrete fundamental solution, given as integral representation, was, for instance, investigated in [9] and in more general cases in [1]. For special properties of the discrete fundamental solutions, which are most important for numerical calculations, we refer to [7, 10]. Besides the existence of the discrete fundamental solution, we are first of all interested in convergence results.

2   Discrete fundamental solution

Let $\mathbb{R}^3$ be the three-dimensional Euclidean space. An equidistant lattice of the mesh width $h > 0$ is defined by $\mathbb{R}_h^3 = \{mh = (m_1 h, m_2 h, m_3 h) : m_i \in \mathbf{Z}, \; i = 1, 2, 3\}$. We consider a bounded and simply connected domain $G \subset \mathbb{R}^3$ with a piecewise smooth boundary Γ. To describe the method of difference potentials the sets $M = \{m = (m_1, m_2, m_3) : \; m_i \in \mathbf{Z}, \; i = 1, 2, 3, \; (m_1 h, m_2 h, m_3 h) \in (G \cap \mathbb{R}_h^3)\}$ and $K = \{(0, 0, 0), (-1, 0, 0), (1, 0, 0), (0, -1, 0), (0, 1, 0), (0, 0, -1), (0, 0, 1)\}$ are introduced. For all points $m \in M$ the seven-point star $N_m = \{m + k : \; k \in K\}$ is defined. The union $\bigcup_{m \in M} N_m$ is denoted by N. At all points $r = (r_1, r_2, r_3) \in N$ the set $K_r = \{k \in K : \; r + k \notin M\}$ is analyzed. Similar to the bounded domain G, the discrete domain $G_h = \{(m_1 h, m_2 h, m_3 h) : \; m = (m_1, m_2, m_3) \in M\}$ with the double-layer boundary $\gamma_h = \{rh : \; r \in N \text{ and } K_r \neq \emptyset\}$ will be studied. In more detail, all points rh with $k = (0, 0, 0) \in K_r$ are mesh points of the outer boundary layer γ_h^-. The mesh points $rh \in \gamma_h \setminus \gamma_h^-$ we call points of the inner boundary layer γ_h^+. Outer edges of the domain do not play any role in this theory.

In the following we consider the difference equation

$$-\Delta_h u_h(mh) = \sum_{k \in K} a_k u_h(mh - kh) = f_h(mh) \qquad \forall \, mh \in G_h$$

with the coefficients $a_k = \begin{cases} -1/h^2 & \text{for } k \in K, \; k \neq (0, 0, 0) \\ 6/h^2 & \text{for } k = (0, 0, 0). \end{cases}$

Each solution $E_h(mh)$ of the equation

$$-\Delta_h E_h(mh) = \begin{cases} 1/h^3 & \text{if } mh = (0, 0, 0) \\ 0 & \text{if } mh \neq (0, 0, 0) \end{cases}$$

is called a discrete fundamental solution if this solution does not grow more quickly than a power of $|mh|$ at infinity. Such a solution can be calculated with the help of the discrete Fourier transform

$$(F_h u_h)(\xi) = \begin{cases} \frac{h^3}{\sqrt{2\pi}^3} \sum_{mh \in \mathbb{R}^3_h} u_h(mh) e^{ih<m,\xi>} & \xi \in Q_h \\ 0 & \xi \in \mathbb{R}^3 \setminus Q_h \end{cases}$$

and the inverse Fourier transform $F_h^{-1} = R_h F$, where $Q_h = \{\xi \in \mathbb{R}^3 : -\frac{\pi}{h} < \xi_j < \frac{\pi}{h}, \ j = 1,2,3\}$, $< m,\xi > = m_1\xi_1 + m_2\xi_2 + m_3\xi_3$ and $R_h u$ denotes the restriction of the function $u(x)$ to the lattice $\mathbb{R}^3_h$. The symbol F characterizes the classical Fourier transform. As an integral representation we obtain

$$E_h(mh) = \frac{1}{(2\pi)^3} \int_{Q_h} \frac{e^{-ih<m,\xi>}}{d^2} d\xi \quad \text{with} \quad d^2 = \frac{4}{h^2}\left(\sin^2 \frac{h\xi_1}{2} + \sin^2 \frac{h\xi_2}{2} + \sin^2 \frac{h\xi_3}{2}\right).$$

This integral exists as an improper integral. In the following some important properties of the discrete fundamental solution are presented. For the idea of the proofs we refer to [3], where the case $n = 2$ is considered.

Lemma 2.1 *At each mesh point* $mh \in \mathbb{R}^3_h$ *the discrete fundamental solution can be estimated in the form* $|E_h(mh)| \leq C (|mh| + h)^{-1}$.

In order to get convergence results we estimate the difference between the discrete fundamental solution $E_h(mh)$ and the continuous fundamental solution $E(x) = \frac{1}{(2\pi)^3} \int_{\mathbb{R}^3} \frac{e^{-i<x,\xi>}}{|\xi|^2} d\xi - \frac{1}{4\pi|x|}$.

Theorem 2.2 *Let* $E^*(x) = \begin{cases} E(x) & x \neq (0,0,0) \\ 0 & x = (0,0,0) \end{cases}$.

At each mesh point $mh \in \mathbb{R}^3_h$ *the approximation error of the discrete fundamental solution can be estimated by*

$$|E_h(mh) - E_h^*(mh)| \leq \begin{cases} C\,h\,|mh|^{-2} & mh \neq (0,0,0) \\ C\,h^{-1} & mh = (0,0,0). \end{cases}$$

We present now convergence results in the spaces l_p and L_p. Let $Q(G_h)$ be the smallest cube parallel to the axis with the center at $(0,0,0)$, which covers the domain G_h. The length of this cube is denoted by $L = 2lh$.

Theorem 2.3 *For each mesh width* $h \leq e^{-1}$ *it can be proved*

$$\|E_h - E^*\|_{l_p(G_h)} \leq \begin{cases} C(L)\,h & 1 \leq p < \frac{3}{2} \\ C\,h\,|\ln h|^{2/3} & p = \frac{3}{2} \\ C\,h^{-1+3/p} & \frac{3}{2} < p < 3. \end{cases}$$

For the idea of the proof we refer to theorem 1.2 in [3]. To get a convergence result in the space L_p, the restriction operator of the inverse discrete Fourier transform is neglected. For $2 \leq p < 3$ we obtain

$$\begin{aligned} E_h &\longrightarrow E && \text{in } L_p(G) && \text{for } h \longrightarrow 0 && \text{and} \\ E_h - E &\longrightarrow 0 && \text{in } L_p(\mathbb{R}^3) && \text{for } h \longrightarrow 0. \end{aligned}$$

3   Difference potentials

In analogy to the integral representation for functions in C^2 it can be proved

$$\sum_{rh\in\gamma_h} \Big(\sum_{k\in K_r} E_h(lh-(r+k)h)\,a_k\,h^3 \Big)\,u_h(rh)$$

$$- \sum_{m\in M} E_h(lh-mh)\,\Delta_h\,u_h(mh)\,h^3 = \begin{cases} u_h(lh) & l\in N \\ 0 & l\notin N. \end{cases}$$

For numerical calculations it is advantageous to split the difference potential on the boundary into a discrete single- and double-layer potential.

Theorem 3.1 *We denote the boundary values on γ_h^- by $u_R(rh)=u_h(rh)$ and the normal derivatives by $u_A(rh)=h^{-1}\sum_{k\in K\backslash K_r}(u_h(rh)-u_h((r+k)h))$.*
The difference potential on the boundary can be written in the form

$$\sum_{rh\in\gamma_h} \Big(\sum_{k\in K_r} E_h(lh-(r+k)h)a_k h^3 \Big)u_h(rh) = (P_h^E u_A(lh) - (P_h^D u_R)(lh)$$

with the single-layer potential $(P_h^E u_A)(lh) = \sum_{rh\in\gamma_h^-} u_A(rh)\,E_h(lh-rh)\,h^2$

and the discrete double-layer potential

$$(P_h^D u_R)(lh) = \sum_{rh\in\gamma_h^-}\sum_{k\in K\backslash K_r} \frac{E_h(lh-rh)-E_h(lh-(r+k)h)}{h}u_R(rh)h^2$$
$$-\kappa(lh)u_R(lh).$$

$\kappa(lh)$ is the characteristic function of γ_h^-.

Proof: We split the boundary γ_h into the parts γ_h^- and γ_h^+. For all $rh\in\gamma_h^-$ we get

$$(P_h^\alpha u_h)(lh) := \sum_{rh\in\gamma_h^-} \Big(\sum_{k\in K_r} E_h(lh-(r+k)h)\,a_k\,h^3 \Big)\,u_h(rh)$$

$$= \sum_{rh\in\gamma_h^-}(-\Delta_h E_h(lh-rh)\,h^3)u_h(rh) - \sum_{rh\in\gamma_h^-}\Big(\sum_{k\in K\backslash K_r} E_h(lh-(r+k)h)a_k h^3 \Big)u_h(rh)$$

$$= \kappa(lh)u_h(lh) + \sum_{rh\in\gamma_h^-}\Big(\sum_{k\in K\backslash K_r} E_h(lh-(r+k)h)\,h \Big)\,u_h(rh).$$

Using the substitution $r=s+k$ with $sh\in\gamma_h^+$ and $k\in K_s$ it follows in the other case

$$(P_h^\beta u_h)(lh) := \sum_{sh\in\gamma_h^+}\Big(\sum_{k\in K_s} E_h(lh-(s+k)h)\,a_k\,h^3 \Big)\,u_h(sh)$$

$$= -\sum_{sh\in\gamma_h^+}\left(\sum_{k\in K_s} E_h(lh-(s+k)h)\,h\right)u_h(sh)$$

$$= -\sum_{rh\in\gamma_h^-}\left(\sum_{-k\in K\backslash K_r} u_h((r+(-k))h))\right)E_h(lh-rh)h.$$

Based on the relations

$$-(P_h^D u_R)(lh) \;=\; (P_h^\alpha u_h)(lh) \;-\; \sum_{rh\in\gamma_h^-}\sum_{k\in K\backslash K_r} E_h(lh-rh)\,u_h(rh)\,h \quad\text{and}$$

$$(P_h^E u_A)(lh) \;=\; (P_h^\beta u_h)(lh) \;+\; \sum_{rh\in\gamma_h^-}\sum_{k\in K\backslash K_r} E_h(lh-rh)\,u_h(rh)\,h$$

Theorem 3.1 is completely proved ∎

The potential $(P_h^E u_A)(lh)$ as well as the potential $(P_h^D u_R)(lh)$ are *discrete harmonic functions in the domain* G_h, because they fulfill at each mesh point $lh \in G_h$ the difference equation $-\Delta_h(P_h^E u_A)(lh) = 0$ and $-\Delta_h(P_h^D u_R)(lh) = 0$.

4   Uniqueness and the system of equations

In the following we analyze the discrete Neumann problem in interior domains with respect to uniqueness. For the discrete Dirichlet problem and the problems in exterior domains we refer to [4]. In order to prove the uniqueness theorem, the first discrete Green's formula is used. The proof of this formula is presented in [3] in the case of $n = 2$. In the three-dimensional situation we split the outer boundary layer into 6 parts $\gamma_{hi}^- = \{rh \in \gamma_h^- : r+k_i \in M\}$, $i = 1,\ldots,6$ with $k_1 = (-1,0,0)$, $k_2 = (0,-1,0)$, $k_3 = (0,0,-1)$, $k_4 = (1,0,0)$, $k_5 = (0,1,0)$ and $k_6 = (0,0,1)$. Different parts of γ_h^- can overlap.

Theorem 4.1 (First Discrete Green's Formula) *For two arbitrary mesh functions* w_h *and* u_h *it can be proved*

$$\sum_{m\in M} w_h(mh)\,\Delta_h u_h(mh)\,h^3$$

$$= -\sum_{m\in M}\sum_{i=1}^{3}\left(\frac{w_h(mh)-w_h((m+k_i)h)}{h}\right)\left(\frac{u_h(mh)-u_h((m+k_i)h)}{h}\right)h^3$$

$$+\sum_{rh\in\gamma_h^-}\sum_{k\in K\backslash K_r} w_h(rh)\left(\frac{u_h(rh)-u_h((r+k)h)}{h}\right)h^2$$

$$-\sum_{i=1}^{3}\sum_{rh\in\gamma_{hi}^-}\left(\frac{w_h(rh)-w_h((r+k_i)h)}{h}\right)\left(\frac{u_h(rh)-u_h((r+k_i)h)}{h}\right)h^3.$$

The difference quotients of w_h and u_h in the first summand on the right-hand side approximate the gradient in the first Green's formula of the classical theory. The second term can be considered as an approximation of the integral on the boundary in the Green's formula because each summand is a product of the function value of w_h and the discrete normal derivative of u_h. Only the last term is a special issue of the discrete theory. It tends to zero for $h \to 0$, if w_h und u_h are restrictions to the lattice of sufficiently smooth functions.

Theorem 4.2 *If the necessary condition* $\sum\limits_{rh \in \gamma_h^-} \psi_h(rh)\, h^2 = 0$ *is fulfilled, then the solution of the interior Neumann problem* (N_i)

$$-\Delta_h\, u_h(mh) \;=\; 0 \qquad \forall\, mh \in G_h$$

$$h^{-1} \sum_{k \in K \backslash K_r} (u_h(rh) - u_h((r+k)h)) \;=\; \psi_h(rh) \qquad \forall\, rh \in \gamma_h^-$$

is unique up to a constant.

Proof: The necessary condition follows from Theorem 4.1, if $w_h = 1$. Let us assume, that two solutions of the problem (N_i) exist. In this case the difference $u_h^* = u_h^1 - u_h^2$ is a solution of the homogeneous problem (N_i). From the first Green's formula results for $u_h^* = w_h = u_h$

$$0 \;=\; -\sum_{m \in M} \sum_{i=1}^{3} \left(\frac{u_h^*(mh) - u_h^*((m+k_i)h)}{h} \right)^2 h^3$$
$$-\sum_{i=1}^{3} \sum_{rh \in \gamma_{hi}^-} \left(\frac{u_h^*(rh) - u_h^*((r+k_i)h)}{h} \right)^2 h^3,$$

where all summands on the right-hand side have the same sign. Therefore it follows $u_h^*(mh) = u_h^*((m+k_i)h)$ as well as $u_h^*(rh) = u_h^*((r+k_i)h)$ for $m \in M$, $i = 1, 2, 3$ and $rh \in \gamma_{hi}^-$. Using these relations and the boundary condition, it can be proved $u_h^*(nh) = u_h^1(nh) - u_h^2(nh) = C$ at all mesh points $nh \in G_h \cup \gamma_h^-$ with an arbitrary constant $C < \infty$ ∎

In order to calculate the solution of the interior Neumann problem we study a system of linear equations on the boundary, which can immediately be used for numerical calculations.

Theorem 4.3 *If the necessary condition is fulfilled and the system*

$$\psi_h(lh) = \sum_{k \in K \backslash K_l} \sum_{rh \in \gamma_h^-} (E_h(lh - rh) - E_h((l+k)h - rh))\, v_h(rh)\, h \qquad \forall\, lh \in \gamma_h^-$$

can be solved, then the potential $(P_h^E v_h)(mh) = \sum\limits_{rh \in \gamma_h^-} v_h(rh)\, E_h(mh - rh)\, h^2$

is for all $mh \in (G_h \cup \gamma_h^-)$ *a solution of the interior Neumann problem.*

Proof: If the system of equations is solvable, then at each mesh point $lh \in \gamma_h^-$ the discrete normal derivative of the single-layer potential has the property $h^{-1} \sum\limits_{k \in K \setminus K_l} \left((P_h^E v_h)(lh) - (P_h^E v_h)((l+k)h) \right) = \psi_h(lh)$. The difference equation $-\Delta_h(P_h^E v_h)(mh) = 0$ is fulfilled at all mesh points $mh \in G_h$ ∎

Theorem 4.4 *The condition* $\sum\limits_{rh \in \gamma_h^-} \psi_h(rh)\, h^2 = 0$ *is necessary and sufficient for the solvability of the system of linear equations in theorem 4.3.*

Proof: The necessary condition results from

$$\sum_{lh \in \gamma_h^-} \psi_h(lh)\, h^2 = \sum_{lh \in \gamma_h^-} \left(\sum_{k \in K \setminus K_l} \sum_{rh \in \gamma_h^-} \left(E_h(lh-rh) - E_h((l+k)h-rh) \right) v_h(rh)\, h \right) h^2$$

$$= \sum_{rh \in \gamma_h^-} v_h(rh)\, h^2 \left(\sum_{lh \in \gamma_h^-} \sum_{k \in K \setminus K_l} \left(E_h(rh - lh) - E_h(rh - (l+k)h) \right) h \right)$$

and the formula

$$\sum_{lh \in \gamma_h^-} \sum_{k \in K \setminus K_l} (E_h(rh - lh) - E_h(rh - (l+k)h))h = 0 \quad \forall\, rh \in \gamma_h^-.$$

We show now that the condition is also sufficient. In relation to the last formula it can be proved that the system

$$0 = \sum_{rh \in \gamma_h^-} \sum_{k \subset K \setminus K_r} \left(E_h(lh - rh) - E_h(lh - (r+k)h) \right) w_h(rh)\, h \qquad \forall\, lh \in \gamma_h^-,$$

which is adjunct to the homogeneous system in Theorem 4.3, has at the points $rh \in \gamma_h^-$ a nontrivial solution $w_h(rh) = 1$. Therefore there exists at least one nontrivial solution $v_h^*(rh)$ of the homogeneous system of equations in Theorem 4.3. It can be proved, that $\sum\limits_{rh \in \gamma_h^-} v_h^*(rh)\, h^2 = C_1 \neq 0$. Moreover, the homogeneous system of Theorem 4.3 has no nontrivial solution which is linearly independent from v_h^*. If v_h^{**} would be such a nontrivial solution, then it could be proved that $\sum\limits_{rh \in \gamma_h^-} v_h^{**}(rh)\, h^2 = C_2 \neq 0$. Furthermore, $v_h^\circ = v_h^* \, C_2 - v_h^{**} \, C_1$ is a solution of the homogeneous system in Theorem 4.3. From $\sum\limits_{rh \in \gamma_h^-} v_h^\circ(rh)\, h^2 = 0$ we conclude $v_h^\circ(rh) = 0$ for all $rh \in \gamma_h^-$ and obtain $v_h^{**}(rh) = C_2/C_1 \, v_h^*(rh)$. Using Fredholm's theorems, the proof is complete ∎

Finally we remark, that in the discrete case it is not easy to get information about the solvability of the system of linear equations for Dirichlet problems if we use an ansatz with the double-layer potential. This fact results from the special structure of the double-layer potential, which includes the term $\kappa(lh)u_R(lh)$. Because this term exists it is not possible to use the relation between Dirichlet and Neumann problems, which is in the continuous case based on the adjunct equation of the homogeneous equation of the Neumann problem. With a similar idea we were able to prove the solvability of the described system of equations in the two–dimensional case, where we are confronted with the same situation.

References

[1] Boor, C., Höllig, K. and Riemenschneider, S., Fundamental solutions for multivariate difference equations, *Amer. J. Math.* **111** (1989), 403–415.

[2] Duffin, R. J., Discrete potential theory, *Duke Math. J.* **20** (1953), 233–251.

[3] Hommel, A., Fundamentallösungen partieller Differenzenoperatoren und die Lösung diskreter Randwertprobleme mit Hilfe von Differenzenpotentialen, Dissertation, Bauhaus-Universität Weimar, 1998.

[4] Hommel, A., The theory of difference potentials in the three-dimensional case, *Proceedings to the conference IKM 2000 in Weimar*

[5] Ryabenkij, V. S., *The Method of Difference Potentials for Some Problems of Continuum Mechanics*, (in Russian), Moscow, Nauka 1987.

[6] Sobolev, S. L., Über die Eindeutigkeit der Lösung von Differenzengleichungen des elliptischen Typs, *Doklady Akad. Nauk SSSR* **87** (1952), No.2, 179–182 (in Russian)

[7] Sobolev, S. L., Über eine Differenzengleichung, *Doklady Akad. Nauk SSSR* **87** (1952), No.3, 341–343 (in Russian)

[8] Stummel, F., Elliptische Differenzenoperatoren unter Dirichletrandbedingungen, *Math. Z.* **97** (1967), 169–211.

[9] Thomée, V., Discrete interior Schauder estimates for elliptic difference operators, *SIAM J. Numer. Anal.* **5** (1968), 626–645.

[10] Van der Pol, B., The finite-difference analogy of the periodic wave equation and the potential equation, Appendix IV in *Prohability and Related Topics in Physical Sciences*, Marc Kac, Interscience Publishers, New York 1957.

Moment Equations for Stochastic Difference Equations

KLARA JANGLAJEW

Institute of Mathematics, University of Bialystok
15-267 Bialystok, Poland
E-mail: jang@math.uwb.edu.pl

Abstract A system of linear difference equations with coefficients dependent on Markov chains is studied. For this system is proposed a method of the construction of a system of difference equations for the first moment of a solution.

Keywords Difference equation, Linear system, Markov chain, First moment

AMS Subject Classification 39A99

1 Introduction

Consider the system of difference equations

$$X_{n+1} = A(n, \zeta_n)X_n \qquad (n = 0, \pm 1, ...) \qquad (1)$$

where $dim X_n = m$ and ζ_n is a Markov chain with a finite number of states $\theta_1, ..., \theta_q$. Let $p_s(n)$ denote the probability of θ_s

$$p_s(n) = P\{\zeta_n = \theta_s\} \qquad (s = 1, ..., q) \qquad (2)$$

We assume that the probabilities $p_s(n)$ $(s = 1, ..., q)$ satisfy a system of linear difference equations [1]

$$p_s(n + 1) = \sum_{j=1}^{q} \pi_{sj}(n)p_j(n) \qquad (s = 1, ..., q) \qquad (3)$$

where coefficients satisfy the conditions

$$\pi_{sj}(n) \geq 0, \qquad \sum_{s=1}^{q} \pi_{sj}(n) = 1 .$$

Let $f(n, X, \zeta)$ denote the density distribution of the system of random values (X_n, ζ_n). Since ζ_n is a finite-valued random variable,

$$f(n, X, \zeta) = \sum_{s=1}^{q} f_s(n, X)\delta(\zeta - \theta_s) \qquad (4)$$

where $\delta(\zeta)$ is the Dirac delta function. The functions $f_s(n, X)$ are called the particular density distributions [5]. For simplicity of notation, we write $A_s(n)$ instead of $A(n, \theta_s)$.

For $\zeta_n = \theta_s$, system (1) assumes the form

$$X_{n+1} = A_s(n)X_n, \det A_s(n) \neq 0 \qquad (5)$$

and for fixed s we have the transformation of the density distribution

$$f(n + 1, X) = f(n, A_s^{-1}(n)X) \mid \det A_s^{-1}(n) \mid .$$

Let us introduce the stochastic operators

$$R_s(n)f(X) = f(A_s^{-1}(n)X) \mid \det A_s^{-1}(n) \mid \qquad (s = 1, ..., q) .$$

Since the Markov chain ζ_n goes into an arbitrary state θ_j from the state θ_s with the probability $\pi_{js}(n)$, it follows that

$$f_j(n + 1, X) = \sum_{s=1}^{q} \pi_{js} R_s(n)f_s(n, X) \quad (j = 1, ..., q) .$$

The last equality may be written in the form

$$f_j(n + 1, X) = \sum_{s=1}^{q} \pi_{js}(n)f_s(n, A_s^{-1}(n)X) \mid \det A_s^{-1}(n) \mid . \qquad (6)$$

Next we form the system of equations for first moments

$$Y_n = E(X_n) \qquad (n = 0, \pm1, \pm2, ...) \qquad (7)$$

where

$$Y_n = \underbrace{\int \int ... \int}_{m} X f(n, X)dx_1...dx_m, \quad X = \begin{pmatrix} x_1 \\ \vdots \\ x_m \end{pmatrix}, \quad f(n, X) = \sum_{j=1}^{q} f_j(n, X) .$$

Multiply system (6) by X and integrate over the m-dimensional space variables $x_1, x_2, ..., x_m$. Next we set

$$Y_{nj} = \underbrace{\int \int ... \int}_{m} X f_j(n, X)dx_1 dx_2...dx_m \qquad (j = 1, ..., q) .$$

Using the change of integration variables $Z = A_s^{-1}(n)X$ to compute the multiple integral, we obtain the system

$$Y_{n+1,j} = \sum_{s=1}^{q} \pi_{js}(n)A_s(n)Y_{ns} \quad (j = 1, ..., q). \tag{8}$$

From the formula

$$Y_n = \sum_{j=1}^{q} Y_{nj}$$

it is possible to find the first moment of a stochastic solution. Analogously, we can get systems of matrix difference equations for matrices of second moments [5].

2   Construction of a system of moment equations

We will use the above result while considering the system of linear difference equations, with piecewise constant coefficients and with a small parameter,

$$X_{n+1} = X_n + \mu A(\zeta_n)X_n \quad (n = 0 \pm 1 \pm 2, ...) \tag{9}$$

where ζ_n is a finite-valued Markov process taking values $\theta_1, ..., \theta_q$ with probabilities $p_s(n)$ $(s = 1, ...q)$ (see (2)) that satisfy the system of difference equations

$$p_s(n+1) = \sum_{j=1}^{q} \pi_{sj}p_j(n) \quad (s = 1, ..., q).$$

We assume that the Markov process is ergodic, i.e., the matrix

$$\Pi = \begin{pmatrix} \pi_{11} & \cdots & \pi_{1q} \\ \cdots & \cdots & \cdots \\ \pi_{q1} & \cdots & \pi_{qq} \end{pmatrix}$$

has a simple eigenvalue equal to 1 and all other eigenvalues lie inside the unit disk.

The system of difference equations (8) takes the form

$$Y_{n+1,j} = \sum_{s=1}^{q} \pi_{js}(I + \mu A_s)Y_{ns} \quad (j = 1, ..., q) \tag{10}$$

where $A_s = A(\theta_s)$, $s = 1, ..., q$. For sufficiently small values of $|\mu| > 0$ it is possible from system (10) to obtain a system of difference equations for the first moments

$$Y_n = E(X_n) = \sum_{j=1}^{q} Y_{nj}.$$

For this let us make in system (10) the change

$$Y_n = Y_{n1} + Y_{n2} + \ldots + Y_{nq}, \quad V_{nj} = Y_{nj} \ (j = 2, ..., q), \tag{11}$$

which has the inverse

$$Y_{n1} = Y_n - V_{n2} - \ldots - V_{nq}, \quad Y_{nj} = V_{nj} \ (j = 2, ..., q). \tag{12}$$

Then system (10) assumes the form

$$Y_{n+1} = Y_n + \mu(A_2 - A_1)V_{n2} + \ldots + \mu(A_q - A_1)V_{nq},$$

$$V_{n+1,j} = \pi_{j1}IY_n + \mu\pi_{j1}A_1Y_n + \sum_{s=2}^{q}(\pi_{js} - \pi_{j1})V_{nj} +$$

$$+\mu\sum_{s=2}^{q}(\pi_{js}A_s - \pi_{j1}A_1)V_{ns} \ (j = 2, ..., q). \tag{13}$$

In order to obtain a system of difference equations for Y_n we look for integral manifolds of solutions of system (13) defined by vector equations

$$V_{nj} = K_jY_n, \quad K_j = K_j(\mu), \ (j = 2, ..., q; \ n = 0, \pm 1, ...). \tag{14}$$

Eliminating vectors $V_{nj} \ (j = 2, ..., q)$ from system (13) gives a system equations for the matrices $K_j \ (j = 2, ..., q)$

$$K_j - \sum_{s=2}^{q}(\pi_{js} - \pi_{j1})K_s = \pi_{j1}I - \mu K_j\sum_{s=2}^{q}(A_s - A_1)K_s +$$

$$+\mu\pi_{j1}A_1 + \mu\sum_{s=2}^{q}(\pi_{js}A_s - \pi_{j1}A_1)K_s. \tag{15}$$

The last system may be solved by a method of successive approximations. Let

$$\alpha = \max_s \|A_s\|, \quad k = \max_s \|K_s\| \ (s = 1, 2, ..., q).$$

Since the Markov process is ergodic, the system of linear algebraic equations

$$K_j - \sum_{s=2}^{q}(\pi_{js} - \pi_{j1})K_s = L_j \ (j = 2, ..., q)$$

can always be solved for the matrices $K_j \ (j = 2, ..., q)$. Moreover there exists $\varrho > 0$ such that

$$k \le \varrho\max_s \|L_j\| \ (s = 2, ..., q).$$

The decomposition (15) has a solution if

$$k \le \varrho(1 + \gamma k + \gamma k^2), \quad \gamma = 2\alpha q|\mu| \tag{16}$$

for $k > 0$. The inequality (16) certainly has a positive solution for

$$|\mu| < [2\alpha q(p + \sqrt{p_2 + p})^2]^{-1}. \tag{17}$$

In this case we obtain the system of moment equations for $Y_n = E(X_n)$

$$Y_{n+1} = (I + \mu \sum_{s=2}^{q}(A_s - A_1)K_s)y_n \quad (n = 0, \pm 1, \pm 2, ...). \tag{18}$$

Summarizing, we have

Theorem. *If in system (9) the matrix $A(\zeta_n)$ depends on a finite-valued homogeneous ergodic Markov process ζ_n, then for sufficiently small values of $\mu > 0$ there is the linear system of difference equations for the first moments $Y_n = E(X_n)$ in the form (18).*

Remark. In [2] an asymptotic method of the construction of moment equations for a linear system with random coefficients has been proposed.
Moment equations in the form of a system of difference equations has been used for computing the mean value of income from sales of bus tickets in [3].

Example. Consider the difference equation

$$x_{n+1} = x_n + \mu a(\zeta_n)x_n \quad (n = 0, \pm 1, ...) \tag{19}$$

where ζ_n is a Markov process which takes two states a_1, a_2 with probabilities $p_1(n)$, $p_2(n)$ defined by the system

$$
\begin{aligned}
p_1(n+1) &= (1 - \alpha)p_1(n) + \beta p_2(n) \\
p_2(n+1) &= \alpha p_1(n) + (1 - \beta)p_2(n), \quad 0 < \alpha, \beta < 1.
\end{aligned}
$$

In this case system (10) has the form

$$
\begin{aligned}
y_{n+1,1} &= (1 - \alpha)(1 + \mu a_1)y_{n1} + \beta(1 + \mu a_2)y_{n2} \\
y_{n+1,2} &= \alpha(1 + \mu a_1)y_{n1} + (1 - \beta)(1 + \mu a_2)y_{n2}. \tag{20}
\end{aligned}
$$

Make the change of variables

$$y_n = y_{n1} + y_{n2}, \quad v_n = -\frac{\alpha}{\alpha + \beta}y_{n1} + \frac{\beta}{\alpha + \beta}y_{n2},$$

which splits system (20) for $\mu = 0$. The system of difference equations for variables y_n, v_n takes the form

$$
\begin{aligned}
y_{n+1} &= (1 + \mu\frac{a_1\beta + a_2\alpha}{\alpha + \beta})y_n + \mu(a_2 - a_1)v_n, \\
v_{n+1} &= \mu(a_2 - a_1)\frac{\alpha\beta(1 - \alpha - \beta)}{(\alpha + \beta)^2}y_n + \\
&\quad + [1 - \alpha - \beta + \mu\frac{1 - \alpha - \beta}{\alpha + \beta}(\alpha a_1 + \beta a_2)]v_n.
\end{aligned}
$$

We are looking for the integral manifold of the last system of the form

$$v_n = ky_n, \quad k = k(\mu).$$

For k we have the equation

$$\mu(a_2 - a_1)\frac{\alpha\beta(1-\alpha-\beta)}{(\alpha+\beta)^2} + [1 - \alpha - \beta + \mu\frac{1-\alpha-\beta}{\alpha+\beta}(\alpha a_1 + \beta a_2)]k =$$

$$= k[1 + \mu\frac{a_1\beta + a_2\alpha}{\alpha+\beta} + \mu(a_2 - a_1)]k.$$

From this we obtain

$$k = \mu(a_2 - a_1)\frac{\alpha\beta(1-\alpha-\beta)}{(\alpha+\beta)^3} + O(\mu^2).$$

For the variable $y_n = E(x_n)$ we obtain the moment equation

$$y_{n+1} = \left[1 + \mu\frac{a_1\beta + a_2\alpha}{\alpha+\beta} + \mu^2(a_2 - a_1)^2\frac{\alpha\beta(1-\alpha-\beta)}{(\alpha+\beta)^3} + O(\mu^3)\right]y_n.$$

Moment equations may be used for the investigation of stability of random solutions (see for instance [4]).

References

[1] S.N. Elaydi, *An Introduction to Difference Equations*, Second Edition, Springer, New York, 1999.

[2] K. Janglajew, O. Lavrenyk, On the asymptotic method for construction of moment equations, in: *Control and Self-Organization in Nonlinear Systems (Bialystok,2000)*, Technical University of Bialystok Press, 2000, 57–62.

[3] K. Janglajew, O. Lavrenyk, Mathematical Model of Passenger Transport, Miscellanea Algebraicae, WZiA Akademiii Swietokrzyskiej, no. 2, 2001, 55–62.

[4] K. Janglajew, K. Valeev, On the stability of solutions of a system of difference equations with random coefficients, *Miscellanea Methodological*, WSP,Kielce, no. 3, 1/1998, 35–41 (in Polish).

[5] K.G. Valeev, O.L. Karelova, W.I. Gorelov, The optimization of linear systems with random coefficients, Moscow, 1996 (in Russian).

Convergence of Solutions in a Nonhyperbolic Case with Positive Equilibrium

C. M. KENT

Department of Mathematics and Applied Mathematics
Virginia Commonwealth University
Oliver Hall, 1001 West Main Street, P.O. Box 842014
Richmond, Virginia 23284-2014, U.S.A.
E-mail: cmkent@mail1.vcu.edu

Abstract We study a family of second-order difference equations of the form

$$x_{n+1} = f(x_n, x_{n-1}), \quad n = 0, 1, \dots,$$

for which there exists a unique positive equilibrium and all positive solutions converge to period-two solutions. We find sufficient conditions for the existence of positive solutions which converge to the unique positive equilibrium and for the existence of positive solutions which converge to prime period-two solutions.

Keywords Difference equations, Period-two solutions

AMS Subject Classification 39A10

1 Introduction and preliminaries

We investigate the behavior of solutions of a family of second-order difference equations of the form

$$x_{n+1} = f(x_n, x_{n-1}), \quad n = 0, 1, \dots, \tag{1}$$

where there exists a nonnegative number a and a function h such that the following is true:

(P1) $f \in C\left[(0, \infty)^2, (a, \infty)\right]$.

(P2) (i) $h \in C\left[(a, \infty), (a, \infty)\right]$;

  (ii) $h \circ h = I_{(a, \infty)}$, the identity function on (a, ∞);

  (iii) h is strictly decreasing on (a, ∞);

0-415-31675-8/04/$0.00+$1.50

(iv) $\lim\limits_{v\to a^+} h(v) = \infty;$

(v) $\lim\limits_{v\to\infty} h(v) = a;$

(vi) For $u \in (0,\infty)$ and $v \in (a,\infty)$, $h(v) \left\{ \begin{array}{lll} > u & \text{iff} & f(v,u) > u, \\ = u & \text{iff} & f(v,u) = u, \\ < u & \text{iff} & f(v,u) < u. \end{array} \right\}$

For convenience, we will say that $f \in \mathcal{G}$ if f satisfies Properties (P1) and (P2). For any equation in this family, we first show that the following is true:

1. there exists a unique positive equilibrium $\bar{x} > a$;

2. all positive solutions converge to (not necessarily prime) period-two solutions.

We then find, as our main result, sufficient conditions that guarantee the existence of nontrivial positive solutions which converge to the unique positive equilibrium $\bar{x}$. Finally, we apply our results to two particular difference equations, both of which possess a nonhyperbolic positive equilibrium. The significance of our results lies in the fact that currently available methods fail in the determination of the stability nature of a nonhyperbolic positive equilibrium. In [2], similar results, involving fewer conditions on f, however, have been discovered and applied to difference equations which possess a nonhyperbolic *zero* equilibrium.

2   A unique positive equilibrium and convergence to period-two solutions

In this section, we establish that if $f \in \mathcal{G}$, then Eq.(1) possesses a unique positive equilibrium and the property that all positive solutions converge to (*not necessarily prime*) period-two solutions. To show that all positive solutions converge to (*not necessarily prime*) period-two solutions, we need to use the fact that there exists a unique positive equilibrium and that every solution is bounded.

Lemma 2.1 *Let $f \in \mathcal{G}$. For $u \in (0,\infty)$ and $v \in (a,\infty)$,*

$$f(v,u) \left\{ \begin{array}{lll} > u & \text{if and only if} & f(f(v,u),v) > v, \\ = u & \text{if and only if} & f(f(v,u),v) = v, \\ < u & \text{if and only if} & f(f(v,u),v) < v. \end{array} \right.$$

Proof The proof follows from Properties (P2) (ii), (iv), and (vi).   □

Lemma 2.2 (Existence of a Unique Positive Equilibrium) *Let $f \in \mathcal{G}$ and let a be as defined in Properties (P1) and (P2). Then Eq.(1) possesses a unique positive equilibrium $\bar{x} > a$.*

Proof The proof follows from Properties (P1) and (P2) (i), (iii), (iv), (v), and (vi). $\square$

Remark 2.3 *Let $f \in \mathcal{G}$, let a be as defined in Properties (P1) and (P2), and let $\bar{x}$ be the unique positive equilibrium of Eq.(1). Then we have the following useful identities from Lemmas 2.1 and 2.2:*

1. $f(v, h(v)) = h(v)$ *for $v \in (a, \infty)$.*

2. $f(h(v), v) = v$ *for $v \in (a, \infty)$.*

3. $h(\bar{x}) = \bar{x}$.

Lemma 2.4 (Boundedness of Solutions) *Let $f \in \mathcal{G}$ and let a be as defined in Properties (P1) and (P2). Let $\{x_n\}_{n=-1}^{\infty}$ be a positive solution of Eq.(1). Then there exist $a < m < M < \infty$ such that*

$$m \leq x_n \leq M, \quad \forall \ n \geq 1.$$

Proof The proof follows from Properties (P1) and (P2) (iii) and (iv), Remark 2.3, and Lemma 2.2. $\square$

Theorem 2.5 (Convergence to Period-Two Solutions) *Let $f \in \mathcal{G}$ and let a be as defined in Properties (P1) and (P2). Suppose that $\{x_n\}_{n=-1}^{\infty}$ is a positive solution of Eq.(1). Then there exists $a < L_o, L_e < \infty$ such that the following statements are true.*

1. *Let $N \geq 0$ be the smallest nonnegative integer such that $x_{N-1} > a$.*

   a. *If $x_{N-1} > x_{N+1}$, then $x_{N+2n-1} \downarrow L_o$ and $x_{N+2n} \downarrow L_e$ as $n \to \infty$.*

   b. *If $x_{N-1} < x_{N+1}$, then $x_{N+2n-1} \uparrow L_o$ and $x_{N+2n} \uparrow L_e$ as $n \to \infty$.*

   c. *If $x_{N-1} = x_{N+1}$, then $x_{N+2n-1} = L_o$ and $x_{N+2n} = L_e \ \forall n \geq 0$.*

2. *$L_o = h(L_e)$, $L_e = h(L_o)$.*

3. *One of the following holds: $L_o < \bar{x} < L_e$, $L_e < \bar{x} < L_o$, or $L_o = L_e = \bar{x}$.*

4. *$(L_o, L_e, L_o, L_e, \ldots)$ is a solution of Eq.(1).*

Proof The Proof follows from Properties (P1) and (P2) (ii), (iii), and (vi), Remark 2.3, and Lemmas 2.1 and 2.4. $\square$

3   Existence of solutions that converge to the unique positive equilibrium

In this section, we present the first of our two main results; namely, that given Eq.(1) with $f \in \mathcal{G}$, we can then impose further conditions upon f that are simple to check and that guarantee the existence of positive solutions which converge to the unique positive equilibrium.

There are four such conditions, which we now define.

Definition 3.1 *A function $f : (0, \infty) \times (0, \infty) \to (a, \infty)$, with $a \in [0, \infty)$, is said to* satisfy Hypothesis (H1) *if f is differentiable.*

Definition 3.2 *A function $f : (0, \infty) \times (0, \infty) \to (a, \infty)$, with $a \in [0, \infty)$, is said to* satisfy Hypothesis (H2) *if for all $u, v \in (0, \infty)$, $f(v, u)$ is decreasing in v and increasing in u.*

Definition 3.3 *A function $f : (0, \infty) \times (0, \infty) \to (a, \infty)$, with $a \in [0, \infty)$, is said to* satisfy Hypothesis (H3) *if for all $u \in (0, \infty)$, $\lim_{v \to a^+} f(v, u) > u$.*

Definition 3.4 *A function $f : (0, \infty) \times (0, \infty) \to (a, \infty)$, with $a \in [0, \infty)$, is said to* satisfy Hypothesis (H4) *if there exists a function $g : (0, \infty) \times (0, \infty) \to (-\infty, \infty)$ such that*

(i) *if $\{x_n\}_{n=-1}^{\infty}$ is a positive solution of Eq.(1), then*

$$x_{n-1} = g(x_{n+1}, x_n), \quad n = 0, 1, \ldots ;$$

(ii) *g is differentiable;*

(iii) *for all $u, v \in (0, \infty)$, $g(v, u)$ is increasing in both v and u;*

(iv) *for all $u, v \in (0, \infty)$, $f(v, g(u, v)) = u$ and $g(f(v, u), v) = u$.*

Remark 3.5 *Observe that if $u \in (0, \infty)$ and $v > f(u, 0)$, then by Hypothesis (H4)(iv), $g(v, u) > g(f(u, 0), u) = 0$. Hence, if $\{x_n\}_{n=-1}^{\infty}$ is a positive solution of Eq.(1), then $g(x_{n+1}, x_n) > 0$ for $n = 0, 1, \ldots$.*

If Hypothesis (H4) holds, we obtain the following result relating f and g to the unique positive equilibrium, $\bar{x}$, and, thus, in addition, to each other.

Lemma 3.6 *Let $f \in \mathcal{G}$ and suppose f satisfies Hypotheses (H2) and (H4). Then, for $u \in (a, \infty)$,*

$$u \begin{cases} > \bar{x} & implies \quad f(u, u) < u \quad and \quad g(u, u) > u, \\ = \bar{x} & implies \quad f(u, u) = u \quad and \quad g(u, u) = u, \\ < \bar{x} & implies \quad f(u, u) > u \quad and \quad g(u, u) < u. \end{cases}$$

Proof The proof follows from Property (P2) (vi), Hypotheses (H2) and (H4) (iii) and (iv), Remark 2.3, and Lemma 2.1. □

We are now ready to state our main result.

Theorem 3.7 (Convergence to the Positive Equilibrium) *Let $\bar{x}$ be the unique positive equilibrium of Eq.(1) and let $u \in (\bar{x}, \infty)$ and $f \in \mathcal{G}$. Suppose f satisfies Hypotheses (H1), (H2), (H3), and (H4). Then there exists $v \in (\bar{x}, \infty)$ such that the positive solution $\{x_n\}_{n=-1}^{\infty}$ of Eq.(1), with $x_{-1} = u$ and $x_0 = v$, is a strictly monotonically decreasing sequence which converges to the unique positive equilibrium, i.e., $x_{-1} > x_0 > x_1 \cdots > \bar{x}$ and $\lim_{n \to \infty} x_n = \bar{x}$.*

Proof The proof of this theorem is a consequence of the following lemmas.

Because of the following lemma, to prove the existence of a positive solution which converges to the unique positive equilibrium, $\bar{x}$, we need only show the existence of strictly monotonically decreasing solutions.

Lemma 3.8 *Let $\bar{x}$ be the unique positive equilibrium of Eq.(1) and let $f \in \mathcal{G}$. Also let $\{x_n\}_{n=-1}^{\infty}$ be a positive solution of Eq.(1), with $x_{-1} > \bar{x}$. Suppose that there exists $N \geq 0$ such that for all $n \geq N$, $x_{n-1} > x_n > \bar{x}$. Then $\lim_{n \to \infty} x_n = \bar{x}$.*

Proof The result follows immediately from the fact that $\bar{x}$ is the only positive equilibrium solution of Eq.(1). □

Remark 3.9 *Let $\{x_n\}_{n=-1}^{\infty}$ be a positive solution of Eq.(1) and let $f \in \mathcal{G}$. Suppose f satisfies (H4). Observe that monotonicity, where $x_{n-1} > x_n$ (for all $n \geq 0$), is equivalent to the condition $x_{n-1} < g(x_n, x_n)$ (for all $n \geq 0$). This equivalence follows from Hypothesis (H4)(i) and (iii) where*

1. *for any $n \geq 0$, $x_{n-1} = g(x_{n+1}, x_n) < g(x_n, x_n)$ if and only if $x_n > x_{n+1}$.*

2. *if $x_0 > x_1 > x_2$, then $x_{-1} = g(x_1, x_0) < g(x_2, x_1) = x_0$.*

Therefore, it must be that monotonicity is lost by a solution when there exists $N \geq 0$ such that $x_{N-1} \geq g(x_N, x_N)$. In fact, we have the following result.

Lemma 3.10 (Non-monotonicity) *Let $\bar{x}$ be the unique positive equilibrium of Eq.(1) and let $f \in \mathcal{G}$. Suppose f satisfies Hypotheses (H2) and (H4). Also let $\{x_n\}_{n=-1}^{\infty}$ be a positive solution of Eq.(1), with $x_{-1} > \bar{x}$. Suppose that there exists $N \geq 0$ such that $x_{N-1} \geq g(x_N, x_N)$. Then, for all $n \geq 0$,*

$$x_{N+2n} < g(x_{N+2n+1}, x_{N+2n+1}) \quad \text{and} \quad x_{N+2n+1} > g(x_{N+2n+2}, x_{N+2n+2}).$$

Proof The proof is similar to the proof of Lemma 3.2 in [2] and will be omitted. □

Given $v \in (a, \infty)$, for the rest of this section we adopt the notation that $\{x_n^v\}_{n=-1}^{\infty}$ is that positive solution $\{x_n\}_{n=-1}^{\infty}$ of Eq.(1) with initial conditions $x_{-1} = u$ and $x_0 = v$. We also assume for the remainder of this section that Eq.(1) is such that $f \in \mathcal{G}$ and f satisfies Hypotheses (H1), (H2), (H3), and (H4).

With Lemma 3.10 in mind, we make the following definitions. For each $N \in \{-1, 0, \ldots\}$, let

$$H_N = \left\{ v \in (a, \infty) \;\middle|\; \begin{array}{l} x_n^v < g(x_{n+1}^v, x_{n+1}^v), \quad n = -1, 0, \ldots, N-1 \\ x_N^v \geq g(x_{N+1}^v, x_{N+1}^v) \end{array} \right\}$$

and

$$K = \{v \in (a, \infty) | x_n^v < g(x_{n+1}^v, x_{n+1}^v) \ \forall \, n = -1, 0, \ldots\}.$$

Remark 3.11 Let $N \in \{0, 1, \ldots\}$. Then the following statements are clearly true.

1. $H_i \cap H_j = \emptyset$ for all $i, j \in \{-1, 0, \ldots\}$ with $i \neq j$.

2. $H_i \cap K = \emptyset$ for all $i \in \{-1, 0, \ldots\}$.

3. $\left(\bigcup_{N=0}^{\infty} H_{2N-1}\right) \cup \left(\bigcup_{N=0}^{\infty} H_{2N}\right) \cup K = (a, \infty)$.

4. Suppose $v \in K$. Then

   (a) $x_{-1} = u > x_0 = v > x_1 > \cdots > \bar{x}$;
   (b) $x_n^v \downarrow \bar{x}$.

Set $I = \bigcup_{N=0}^{\infty} H_{2N-1}$ and $J = \bigcup_{N=0}^{\infty} H_{2N}$.

Lemma 3.12 $I \neq \emptyset$ and $J \neq \emptyset$.

Proof Given $u \in (\bar{x}, \infty)$, we can find $c, d \in (a, \infty)$ such that $u = g(c, c)$ and $d = g(f(d, u), f(d, u))$, by continuity of g, Hypothesis (H4)(iv), and Lemma 3.6. Then we have that $c \in H_{-1}$ and $d \in H_0$. □

Lemma 3.13 I and J are open subsets of (a, ∞).

Proof The proof is similar to the proof of Lemma 3.5 in [2] and will be omitted. □

Thus, we see that since $(0, \infty)$ is connected, we must have $K \neq \emptyset$. The proof of Theorem 3.7 follows from Remark 3.11. □

4 Existence of solutions that converge to prime period-two solutions

In this section, we present the second of our two main results, which turns out to be relatively trivial to prove: If we are given Eq.(1) with $f \in \mathcal{G}$, then we are guaranteed the existence of positive solutions which converge to *prime* period-two solutions. In [2], the situation was different in that it was not trivial to show the existence of solutions converging to prime period-two solutions.

Theorem 4.1 *Consider Eq.(1) where $f \in \mathcal{G}$ and f satisfies Hypotheses (H2) and (H4). Let $\bar{x}$ be the unique positive equilibrium of Eq.(1). Then there exists a positive solution $\{x_n\}_{n=-1}^{\infty}$ of Eq.(1) which converges to a prime period-two solution.*

Proof Set $x_{-1} = x_0 > \bar{x}$. Then, by Lemma 3.6, $x_1 = f(x_{-1}, x_{-1}) < x_{-1}$. This, together with Theorem 2.5, implies that $x_{2n-1} \downarrow L_o$ and $x_{2n} \downarrow L_e$ as $n \to \infty$, with $L_e < \bar{x} < L_o$. $\square$

5 Applications

We now apply our results to two particular difference equations.

Example 5.1 *Equation:* $x_{n+1} = \dfrac{A + x_{n-1}}{1 + x_n}$, $A \in (0, \infty)$.
This equation was investigated previously by M.R.S. Kulenović and G. Ladas in [3]. Note that when we let $A = 1$ and make the change of variables $y_n = x_n - 1$ in the first equation, we obtain

$$y_{n+1} = 1 + \frac{y_{n-1}}{y_n}, \quad n = 0, 1, \ldots.$$

This version of the above equation was investigated previously by A.M. Amleh, D.A. Georgiou, E.A. Grove, and G. Ladas [1].
Observe that $f(v, u) = \dfrac{A + u}{1 + v}$ and $g(v, u) = v(u + 1) - A$.
It is easy to check that $f \in \mathcal{G}$ and f and g satisfy Hypotheses (H1), (H2), (H3), and (H4).

Example 5.2 *Equation:* $x_{n+1} = \dfrac{1}{x_n} + \alpha x_{n-1}$, $\alpha \in (0, 1)$.
Observe that $f(v, u) = \dfrac{1}{v} + \alpha u$ and $g(v, u) = \dfrac{1}{\alpha} \left(v - \dfrac{1}{u} \right)$.
It is easy to check that $f \in \mathcal{G}$ and f and g satisfy Hypotheses (H1), (H2), (H3), and (H4).

Both equations possess a unique *nonhyperbolic* positive equilibrium, thereby making the local stability nature of the equilibrium difficult to determine using currently available methods. Nevertheless, using the results from Sections 4 and 3, we are able to conclude that both equations are characterized by the following.

1. The unique positive equilibrium, $\bar{x} > a$, is (locally) stable. Indeed, given $\epsilon > 0$, Property (P2) allows us to find an M such that $\max\{|M - \bar{x}|, |h(M) - \bar{x}|\} < \epsilon$. Then, if we let $\delta < \min\{|M - \bar{x}|, |h(M) - \bar{x}|\}$, we have that $\|(x_{-1}, x_0) - (\bar{x}, \bar{x})\| < \delta$ implies $\|(x_{n-1}, x_n) - (\bar{x}, \bar{x})\| < \epsilon$ for all $n \geq 0$ by Remark 2.3.

2. While the equilibrium attracts infinitely many positive solutions (Section 3), it does not attract *all* positive solutions (Section 4). Therefore, it is *not* (locally) asymptotically stable.

References

[1] Amleh, A. M., Georgiou D. A., Grove E. A., and Ladas G., On the recursive sequence $x_{n+1} = \alpha + \dfrac{x_{n-1}}{x_n}$, *J. Math. Anal. Appl.* **233** (1999), 790–798.

[2] Kent, C. M., Convergence of solutions in a nonhyperbolic case, in *Proceedings of the Third World Congress of Nonlinear Analysts, Catania/Sicily/Italy, July 19-26, 2000* (Editor-in-Chief Lakshmikantham, V.), Elsevier Science, Amsterdam, **47** (2001), 4651–4665.

[3] Kulenović, M. R. S. and Ladas, G., *Dynamics of Second-Order Rational Difference Equations with Open Problems and Conjectures*, Chapman & Hall/CRC Press, Boca Raton, Forida, 2001.

Strongly Decaying Solutions of Nonlinear Forced Discrete Systems

MAURO MARINI

Department of Electronics and Telecommunications,
University of Florence, I-50139 Florence, Italy
E-mail: marini@ing.unifi.it

SERENA MATUCCI

Department of Electronics and Telecommunications,
University of Florence, I-50139 Florence, Italy
E-mail: matucci@det.unifi.it

and

PAVEL ŘEHÁK [1]

Department of Mathematics, Masaryk University Brno
CZ-60300 Brno, Czech Republic
E-mail: rehak@math.muni.cz

Abstract The nonlinear forced difference system

$$\Delta(r_k \Phi_\alpha(\Delta x_k)) - \sigma \varphi_k f(y_{k+1}) = \sigma \hat{\varphi}_k,$$
$$\Delta(q_k \Phi_\beta(\Delta y_k)) - \psi_k g(x_{k+1}) = \hat{\psi}_k$$

is considered. Necessary and sufficient conditions for existence of positive decreasing solutions with zero limit are given. Some extensions and open problems complete the paper.

Keywords (Singular) Nonlinear difference system, Asymptotic behavior, Positive decreasing solution.

AMS Subject Classification 39A10, 39A11

[1] Supported by the Grants No. 201/01/P041 and No. 201/01/0079 of the Czech Grant Agency and by C.N.R. of Italy.

1   Introduction

The aim of this paper is to examine certain discrete asymptotic boundary value problems on the discrete interval $[m, \infty) := \{m, m+1, ..\}, m \in \mathbb{Z}$, associated to the nonlinear forced difference system

$$\begin{aligned}
\Delta(r_k \Phi_\alpha(\Delta x_k)) - \sigma \varphi_k f(y_{k+1}) &= \sigma \hat{\varphi}_k, \\
\Delta(q_k \Phi_\beta(\Delta y_k)) - \psi_k g(x_{k+1}) &= \hat{\psi}_k
\end{aligned} \tag{1}$$

where $\{\varphi_k\}, \{\psi_k\}, \{r_k\}, \{q_k\}$ are real positive sequences defined for $k \geq m$, the forcing terms $\{\hat{\varphi}_k\}, \{\hat{\psi}_k\}$ are real nonnegative sequences defined for $k \geq m$, $\Phi_\lambda(u) = |u|^{\lambda-1}\text{sgn}\,u$ with $\lambda > 1$, $f, g : (0, \infty) \to (0, \infty)$ are monotone continuous functions and $\sigma \in \{-1, 1\}$.

System (1) occurs in the discretization process of differential systems with p-Laplacian operator. For results concerning nonlinear differential systems and physical motivations, we refer the reader to [6, 10, 13] and to the references therein. Good examples of nontrivial prototypes of nonlinearities f, g are the one-dimensional Laplacians Φ_γ, Φ_δ, respectively, with $\gamma, \delta \neq 1$ (i.e., with possible singularities at zero). Then (1) leads to the Emden-Fowler type system

$$\begin{aligned}
\Delta(r_k \Phi_\alpha(\Delta x_k)) &= \sigma \varphi_k \Phi_\gamma(y_{k+1}), \\
\Delta(q_k \Phi_\beta(\Delta y_k)) &= \psi_k \Phi_\delta(x_{k+1})
\end{aligned} \tag{2}$$

as a special case.

By a *nonoscillatory solution* of (1) we mean a vector sequence $(\{x_k\}, \{y_k\})$ satisfying (1) for $k \geq m$ with both components $\{x_k\}, \{y_k\}$ of eventually fixed sign. Denote with $x_k^{[1]}$ and $y_k^{[1]}$ quasi-differences of x and y respectively: $x_k^{[1]} = r_k \Phi_a(\Delta x_k), y_k^{[1]} = q_k \Phi_\beta(\Delta y_k)$. A nonoscillatory solution $(\{x_k\}, \{y_k\})$ of (1) is said to be a *strongly decaying solution* if the sequences $\{x_k\}, \{y_k\}$ are eventually positive decreasing, $\lim_k x_k = \lim_k y_k = 0$ and

$$\lim_k y_k^{[1]} = 0, \quad \lim_k x_k^{[1]} = \begin{cases} -d_x < 0, & \text{for } \sigma = -1 \\ 0 & \text{for } \sigma = 1. \end{cases} \tag{3}$$

Note that the constant d_x in (3) cannot be zero when $\sigma = -1$, since, in this case, the quasidifference $x_k^{[1]}$ is eventually negative decreasing.

The main purpose of this paper is to give necessary and sufficient conditions for existence of strongly decaying solutions. Both the regular case (i.e., when the nonlinearities f, g are bounded in a neighborhood of zero) and the singular case (i.e., when f or g are unbounded in a right neighborhood of zero) will be considered. Some remarks, comments and suggestions are presented in the last section, jointly with some examples.

Finally the study of system (1) is motivated by the crucial role played by strongly decaying solutions in a variety of physical applications (see, e.g., [5, 6]) and also by some results for scalar difference equations, which attracted a considerable attention in recent years; see, e.g., [2, 3, 4, 7, 9, 11, 14]. Other interesting contributions can be found also in the monograph [1].

2 Main results

Denote with α^* the conjugate number of α, i.e. $1/\alpha + 1/\alpha^* = 1$, and similarly for β^*. We start with a necessary and sufficient criterion for the existence of strongly decaying solutions of (1) with $\sigma = -1$.

Theorem 2.1 *System (1) with $\sigma = -1$ has strongly decaying solutions if and only if the following conditions are satisfied:*

(i) $\sum_{k=m}^{\infty} \hat{\varphi}_k < \infty, \quad \sum_{k=m}^{\infty} \hat{\psi}_k < \infty$

(ii) $\sum_{k=m}^{\infty} \Phi_{\alpha^*} (1/r_k) < \infty$

(iii) there exists a costant $A > 0$ such that
$$\omega_k(A) = \Phi_{\beta^*} \left\{ (1/q_k) \sum_{j=k}^{\infty} \left[\psi_j \, g \left(\sum_{i=j+1}^{\infty} \Phi_{\alpha^*} (A/r_j) \right) + \hat{\psi}_j \right] \right\} < \infty, \quad \text{and}$$
$$\sum_{k=m}^{\infty} \varphi_k f \left(\sum_{j=k+1}^{\infty} \omega_j(A) \right) < \infty.$$

Proof The "if part". We examine the case f and g nondecreasing since the remaining cases are similar. Choose an integer $T \geq m$ such that

$$\sum_{k=T}^{\infty} \left[\varphi_k f \left(\sum_{j=k+1}^{\infty} \omega_j \right) + \hat{\varphi}_k \right] \leq \frac{A}{2}. \tag{4}$$

Denote with ℓ_T^{∞} the Banach space of all bounded sequences defined for every integer $n \geq T$, endowed with the topology of the supremum norm, and consider the set $\Omega \subset \ell_T^{\infty} \times \ell_T^{\infty}$ given by

$$\Omega = \left\{ (u, v) = (\{u_k\}, \{v_k\}) \in \ell_T^{\infty} \times \ell_T^{\infty} : \right.$$
$$\left. \sum_{j=k}^{\infty} \Phi_{\alpha^*} \left(\frac{A/2}{r_j} \right) \leq u_k \leq \sum_{j=k}^{\infty} \Phi_{\alpha^*} \left(\frac{A}{r_j} \right), \quad 0 \leq v_k \leq \sum_{j=k}^{\infty} \omega_j(A) \right\}.$$

Consider the operator $\mathcal{T} : \Omega \to \ell_T^{\infty} \times \ell_T^{\infty}$ defined by $\mathcal{T}(u, v) = (\mathcal{T}_1(v), \mathcal{T}_2(u)) = (\{(\mathcal{T}_1(v))_k\}, \{(\mathcal{T}_2(u))_k\})$, where

$$(\mathcal{T}_1(v))_k = \sum_{j=k}^{\infty} \Phi_{\alpha^*} \left(\frac{A}{r_j} - \frac{1}{r_j} \sum_{i=j}^{\infty} (\varphi_i f(v_{i+1}) + \hat{\varphi}_i) \right),$$

$$(\mathcal{T}_2(u))_k = \sum_{j=k}^{\infty} \Phi_{\beta^*} \left(\frac{1}{q_j} \sum_{i=j}^{\infty} (\psi_i g(u_{i+1}) + \hat{\psi}_i) \right).$$

Ω is a nonempty, closed, bounded and convex subset of $\ell_T^{\infty} \times \ell_T^{\infty}$. In order to show that $\mathcal{T}$ has a fixed point in Ω it is sufficient to verify that the hypotheses of the Schauder fixed point theorem are satisfied. Obviously $T(\Omega) \subset \Omega$. To show that $T(\Omega)$ is relatively compact, from [4, Theorem 3.3], it is sufficient to prove that $T(\Omega)$ is uniformly Cauchy in the topology of $\ell_T^\infty \times \ell_T^\infty$; i.e., for any $\varepsilon > 0$ there exists $N \geq T$ such that for any $k, l \geq N$ it holds

$$|(T_1(v))_k - (T_1(v))_l| < \varepsilon \quad \text{and} \quad |(T_2(u))_k - (T_2(u))_l| < \varepsilon \tag{5}$$

for $(u, v) \in \Omega$. Without loss of generality suppose $k < l$. Then

$$
|(T_1(v))_k - (T_1(v))_l| = \sum_{j=k}^{l-1} \Phi_{\alpha^*} \left(\frac{A}{r_j} - \frac{1}{r_j} \sum_{i=j}^{\infty} (\varphi_i f(v_{i+1}) + \hat{\varphi}_i) \right)
$$
$$
\leq \sum_{j=N}^{\infty} \Phi_{\alpha^*} \left(\frac{A}{r_j} \right) < \varepsilon,
$$

if N is sufficiently large, since (ii) in Theorem 2.1 holds. Similarly the second inequality in (5) can be proved and the relative compactness of $T(\Omega)$ follows.

Now we prove the continuity of T in Ω. Let $\{(u^n, v^n)\}_{n \in \mathbb{N}}$ be a vector sequence in Ω converging to (u, v) in the topology of $\ell_T^\infty \times \ell_T^\infty$. The set Ω is closed and hence $(u, v) \in \Omega$. We have to show that $\|T(u^n, v^n) - T(u, v)\|_{\ell_T^\infty \times \ell_T^\infty} \to 0$ as $n \to \infty$. We have

$$
\|T_2(u^n) - T_2(u)\|_{\ell_T^\infty} = \sup_{k \geq T} |(T_2(u^n))_k - (T_2(u))_k| \leq \sum_{k=T}^{\infty} |(G(u^n))_k - (G(u))_k|,
$$

where

$$
(G(u))_k = \Phi_{\beta^*} \left(\frac{1}{q_k} \sum_{i=k}^{\infty} \left(\psi_i g(u_{i+1}) + \hat{\psi}_i \right) \right).
$$

First we show that G is continuous on

$$
\Omega_1 = \left\{ u = \{u_k\} \in \ell_T^\infty : \sum_{j=k}^{\infty} \Phi_{\alpha^*} \left(\frac{A/2}{r_j} \right) \leq u_k \leq \sum_{j=k}^{\infty} \Phi_{\alpha^*} \left(\frac{A}{r_j} \right) \right\}.
$$

In virtue of (i) and (iii), the series $\sum_{j=k}^{\infty} \psi_j g(u_{j+1}^n) + \hat{\psi}_j$ is totally convergent for $u \in \Omega_1$. The continuity of g implies $\lim_{n \to \infty} g(u_k^n) = g(u_k)$ for $k \geq T+1$. Hence, by the discrete analogue of the Lebesgue dominated convergence theorem, we get $\lim_{n \to \infty} |(G(u^n))_k - (G(u))_k| = 0$ for $k \geq T$, which proves the continuity of G. Further, $|(G(u^n))_k - (G(u))_k| \leq (G(u^n))_k + (G(u))_k \leq 2\omega_k$ for $k \geq T$, which implies the total convergence of the series $\sum_{k=T}^{\infty} |(G(u^n))_k - (G(u))_k|$ since $\sum_{k=T}^{\infty} \omega_j < \infty$. Applying again the discrete analogue of the Lebesgue dominated convergence theorem and using the continuity of G, the continuity of T_2 is obtained and similarly for T_1. Then the hypotheses of the Schauder fixed point theorem are satisfied and T has a fixed point $(x, y) \in \Omega$. It is easy to see that (x, y) is a strongly decaying solution of (1) for $n \geq T$.

The other possible cases can be proved with minor changes, using similar arguments as above.

The "only if part". Let (x, y) be a strongly decaying solution of (1) with $\sigma = -1$. Then from (3) there exist positive constants M_1, M_2 and an integer $T \geq m$ such that, for $k \geq T$

$$\sum_{j=k}^{\infty} \Phi_{\alpha^*}\left(\frac{M_1}{r_j}\right) \leq x_k \leq \sum_{j=k}^{\infty} \Phi_{\alpha^*}\left(\frac{M_2}{r_j}\right). \tag{6}$$

Again, assume f, g nondecreasing since the remaining cases can be treated similarly. Summing twice the second equation in (1) from k to ∞ we have

$$y_k = \sum_{j=k}^{\infty} \Phi_{\beta^*}\left(\frac{1}{q_j}\sum_{i=j}^{\infty}(\psi_i g(x_{i+1}) + \hat{\psi}_i)\right). \tag{7}$$

Now (6) and (7) imply $\sum_{j=k}^{\infty} \omega_j(M_1) \leq y_k < \infty$. Summing the first equation in (1) from k to ∞ we get $\sum_{j=k}^{\infty} (\varphi_i f(y_{i+1}) + \hat{\varphi}_i) < \infty$ and from (6) and (7) we obtain

$$\sum_{j=k}^{\infty}\left[\varphi_j f\left(\sum_{i=j+1}^{\infty} \omega_i(M_1)\right) + \hat{\varphi}_j\right] < \infty$$

that is the assertion.   □

Remark 2.2 *The condition*

$$\sum_{k=m}^{\infty} \Phi_{\beta^*}\left(\frac{1}{q_k}\right) < \infty \tag{8}$$

is not necessary for the validity of Theorem 2.1. If (8) holds, then the assumption (iii) from Theorem 2.1 can be relaxed by simpler (but only sufficient) conditions, namely, that there exist $A > 0, B > 0$ such that

$$\sum_{k=m}^{\infty} \varphi_k f\left(\sum_{j=k+1}^{\infty} \Phi_{\beta^*}\left(\frac{A}{q_j}\right)\right) < \infty, \quad \sum_{k=m}^{\infty} \psi_k g\left(\sum_{j=k+1}^{\infty} \Phi_{\alpha^*}\left(\frac{B}{r_j}\right)\right) < \infty. \tag{9}$$

Remark 2.3 *The proof of Theorem 2.1 gives also an asymptotic estimate for the first component $\{x_k\}$ of strongly decaying solution of (1) with $\sigma = -1$. In fact, x_k is asymptotic to $\sum_{j=k}^{\infty} \Phi_{\alpha^*}(1/r_j)$.*

In the case of system (1) with $\sigma = 1$ the following existence result holds:

Theorem 2.4 *Let* f, g *be nondecreasing. Suppose that at least one of the forcing terms* $\{\hat{\varphi}_k\}, \{\hat{\psi}_k\}$ *is eventually nontrivial. If*

$$\sum_{k=m}^{\infty} \Phi_{\alpha^*} \left(\frac{1}{r_k} \sum_{j=k}^{\infty} (\varphi_j + \hat{\varphi}_j) \right) < \infty, \quad \sum_{k=m}^{\infty} \Phi_{\beta^*} \left(\frac{1}{q_k} \sum_{j=k}^{\infty} (\psi_j + \hat{\psi}_j) \right) < \infty,$$

(10)

then system (1) with $\sigma = 1$ *has a strongly decaying solution, which is decreasing for every integer* $n \geq m$.

Proof   Let $A > 0$ and choose an integer $T \geq m$ such that

$$\sum_{k=T}^{\infty} \Phi_{\alpha^*} \left(\frac{1}{r_k} \sum_{j=k}^{\infty} (\varphi_j f(A) + \hat{\varphi}_j) \right) \leq A, \quad \sum_{k=T}^{\infty} \Phi_{\beta^*} \left(\frac{1}{q_k} \sum_{j=k}^{\infty} (\psi_j g(A) + \hat{\psi}_j) \right) \leq A.$$

Consider the set

$$\Omega = \{(u, v) = (\{u_k\}, \{v_k\}) \in \ell_T^{\infty} \times \ell_T^{\infty} : 0 \leq u_k \leq A, \, 0 \leq v_k \leq A\}$$

and the operator $\mathcal{T} : \Omega \to \ell_T^{\infty} \times \ell_T^{\infty}$ defined by $\mathcal{T}(u, v) = (\mathcal{T}_1(v), \mathcal{T}_2(u)) = (\{(\mathcal{T}_1(v))_k\}, \{(\mathcal{T}_2(u))_k\})$, where

$$(\mathcal{T}_1(v))_k = \sum_{j=k}^{\infty} \Phi_{\alpha^*} \left(\frac{1}{r_j} \sum_{i=j}^{\infty} (\varphi_i f(v_{i+1}) + \hat{\varphi}_i) \right),$$

$$(\mathcal{T}_2(u))_k = \sum_{j=k}^{\infty} \Phi_{\beta^*} \left(\frac{1}{q_j} \sum_{i=j}^{\infty} (\psi_i g(u_{i+1}) + \hat{\psi}_i) \right).$$

Using an argument similar to that in the proof of Theorem 1 we obtain that the operator $\mathcal{T}$ has a fixed point $(x, y) \in \Omega$. Since at least one of the sequences $\{\hat{\varphi}_k\}, \{\hat{\psi}_k\}$ is eventually nontrivial, (x, y) is a strongly decaying solution of (1) with $\sigma = 1$. It remains to be proved that (x, y) can be extended to the left for any $n \geq m$ and x, y are decreasing for every $n \geq m$. For this purpose we rewrite (1) into the system of the form

$$\Delta x_k = \Phi_{\alpha^*} \left((r_{k+1} \Phi_{\alpha}(\Delta x_{k+1}) - \varphi_k f(y_{k+1}) - \hat{\varphi}_k) / r_k \right),$$
$$\Delta y_k = \Phi_{\beta^*} \left((q_{k+1} \Phi_{\beta}(\Delta y_{k+1}) - \psi_k g(x_{k+1}) - \hat{\psi}_k) / q_k \right).$$

Since $\Delta x_T < 0, \Delta y_T < 0, x_T > 0$ and $y_T > 0$, setting $k = T - 1$, we obtain $\Delta x_{T-1} < 0$ and $\Delta y_{T-1} < 0$. Continuing this process step by step, we obtain the assertion. $\qquad\square$

Theorem 2.4 requires the convergence of the series $\sum_{j=m}^{\infty} \varphi_j$ and $\sum_{j=m}^{\infty} \psi_j$. When at least one of these series diverges, we can state the following theorem, whose proof is analogous to the previous one.

Theorem 2.5 *Let f, g be nondecreasing. Assume (i), (ii) from Theorem 2.1 and (8), (9). If at least one of the sequences $\{\hat{\varphi}_k\}, \{\hat{\psi}_k\}$ is eventually nontrivial, then system (1) with $\sigma = 1$ has a strongly decaying solution that is decreasing for every $n \geq m$.*

3   Remarks and open problems

1)   In contrast to Theorem 2.1, the positiveness of forcing terms plays an important role in Theorems 2.4–2.5 for the existence of strongly decaying solutions. On the other hand, there exist systems of type (1) without forcing terms that have such solutions. For instance, the system

$$\Delta((k+1)\Delta x_k) = \frac{1}{(k+1)\sqrt{k}}\sqrt{y_{k+1}},$$

$$\Delta((k-1)\Delta y_k) = \frac{1}{k\sqrt{k+1}}\sqrt{x_{k+1}}$$

has the strongly decaying solution $(1/k, 1/(k-1))$ for $k \geq 2$. Simple computations show that the two inequalities in (10) are fulfilled. We recall that in the correspondent continuous case the existence of strongly decaying solutions can be proved by using the concept of *singular solution* ([13]) that, however, does not have discrete counterpart.

2)   The system

$$\Delta((k+1)\Delta x_k) = \frac{1}{k\sqrt{k+1}}\sqrt[4]{y_{k+1}},$$

$$\Delta(k^2\Delta y_k) = \frac{2k^2 + 4k + 1}{(k+2)^2}x_{k+1}^2$$

has the strongly decaying solution $(1/k, 1/k^2)$ on $\mathbb{N}$. On the other hand, the assumptions of neither Theorem 2.4 nor Theorem 2.5 are fulfilled. This fact suggests we should look for further criteria in which the cases $\sum_{k=m}^{\infty} \Phi_{\alpha^*}(1/r_k) = \infty$ and $\sum_{k=m}^{\infty} \Phi_{\beta^*}(1/q_k) < \infty$ are considered.

3)   Theorems 2.4–2.5 are here stated for nonlinear terms f and g both nondecreasing. Analogous results can be proved if f and g are both nonincreasing, or in the "mixed case" in which f is nondecreasing and g nonincreasing or vice versa.

4)   Using Riccati techniques, in [12] it is shown that some basic properties of the second order half-linear differential and difference equations are essentially the same as those of linear equations. All this suggests we should investigate if it occurs also for half-linear Emden-Fowler systems, i.e., systems of type (2) with $\alpha = \beta = \gamma = \delta$.

5)   As we have already mentioned, there are certain differences between the discrete and the continuous case. Especially in such cases it should be of interest to try to unify, in some sense, these results. It turns out that a good tool to reach this aim could be the theory of time scales [8].

References

[1] R.P. Agarwal, Difference Equations and Inequalities, 2nd edition, *Pure Appl. Math. 228*, Marcel Dekker, New York, 2000.

[2] M. Cecchi, Z. Došlá, M. Marini, Positive decreasing solutions of quasilinear difference equations, *Comput. Math. Appl.* **20** (2001), 1401–1410.

[3] M. Cecchi, Z. Došlá, M. Marini, Unbounded solutions of quasilinear difference equations, *Comput. Math. Appl.* (2002).

[4] S. S. Cheng, W. T. Patula, An existence theorem for a nonlinear difference equation, *Nonlinear Anal., T.M.A.* **20** (1993), 193–203.

[5] P. Clément, J. Fleckinger, E. Mitidieri, F. de Thélin, Existence of positive solutions for a nonvariational quasilinear elliptic system, *J. Differ. Equations* **166** (2000), 455–477.

[6] J. I. Díaz, Nonlinear Partial Differential Equations and Free Boundaries, Volume I: Elliptic equations, *Research Notes in Math. 106*, Pitman Publ. Ltd., London, 1985.

[7] J. R. Graef, E. Thandapani, Oscillation of two-dimensional difference systems, *Comput. Math. Appl.* **36** (1999), 157–165.

[8] S. Hilger, Analysis on measure chains – a unified approach to continuous and discrete calculus, *Result. Math.* **18** (1990), 18–56.

[9] B. Liu, S. S. Cheng, Positive solutions of second order nonlinear difference equations, *J. Math. Anal. Appl.* **204** (1996), 482–493.

[10] M. Marini, S. Matucci, P. Řehák, On decay of coupled nonlinear systems, *Adv. Math. Sci. Appl.* **12** (2002), 521–533.

[11] R. Medina, Nonoscillatory solutions for the one-dimensional p-laplacian, *J. Comput. Appl. Math.* **98** (1998), 27–33.

[12] P. Řehák, Oscillatory properties of second order half-linear difference equations, *Czech. Math. J.* **51** (2001), 303–321.

[13] T. Tanigawa, Positive decreasing solutions of systems of second order quasilinear differential equations, *Funkc. Ekvacioj* **43** (2000), 361–380.

[14] P. J. Y Wong, R. P. Agarwal, Oscillation and monotone solutions of second order quasilinear difference equations, *Funkc. Ekvacioj* **39** (1996), 491–517.

Multidimensional Volterra Difference Equations

RIGOBERTO MEDINA [1]

Departmento de Ciencias Exactas, Universidad de Los Lagos
Casilla 933, Osorno, Chile
E-mail: rmedina@ulagos.cl

and

MICHAEL GIL'

Department of Mathematics, Ben Gurion University of the Negev
P. O. Box 653, Beer-Sheva 84105, Israel
E-mail: gilmi@cs.bgu.ac.il

Abstract Some classes of discrete-time Volterra equations are considered. Conditions for boundedness of solutions are derived. The main tool in this work is the use of recent estimates for the matrix resolvent.

Keywords Volterra difference equations, Boundedness, Norm estimate

AMS Subject Classification 39A10

1 Introduction

Volterra equations with discrete time arise mainly in the process of modelling some real phenomena or by applying a numerical method to a Volterra integral equation. At present much of their general quantitative and especially their qualitative theory remains to be developed. There are, at present, only a few papers dealing with their asymptotic behavior. In this paper, we will investigate the asymptotic behavior of the solutions, and we will develop appropriate methods for estimating the bounds of the solutions.

A variety of methods have been used to investigate stability of Volterra difference equations, such as the direct Lyapunov method (see, e.g., [1–3] and the references therein), comparison theorems (see, e.g., [8]). In [6], topological methods were used to study stability of some nonlinear Volterra difference equations. In this paper, to establish the boundedness of the solutions, we

---

[1] Research supported by Fondecyt under Grant No. 1000023.

will interpret the Volterra equations as operator equations in appropriate spaces. Such an approach for Volterra difference equations has been used by Kolmanovskii et al. [5], Medina [9], and Kwapisz [7].

Let A be an $n \times n$-matrix, and let $\lambda_k (A)$ ($k = 1, 2, ..., n$) denote the eigenvalues of A including their multiplicities. The following quantity plays an essential role in our paper:

$$g(A) = \left(N^2(A) - \sum_{k=1}^{n} |\lambda_k(A)|^2 \right)^{\frac{1}{2}},$$

where $N(A)$ is the Hilbert-Schmidt (Frobenius) norm of A, that is, $N^2(A) = Trace\,(A\,A^*)$. There are a number of properties of $g(A)$ which are useful (see Gil' [4, Section 1.2]). Here, we note that if A is normal, i.e., $AA^* = A^*A$, then $g(A) = 0$. If $A = (a_{ij})$ is a triangular matrix such that $a_{ij} = 0$ for $1 \leq j < i \leq n$, then $g^2(A) = \sum_{1 \leq i < j \leq n} |a_{ij}|^2$.

Let C^n be an n-dimensional complex Euclidean space with the Euclidean norm $|\cdot|_{C^n}$.

Our main tool in this paper is the following estimate for the resolvent of an $n \times n$-matrix A:

$$\left| (A - \lambda I)^{-1} \right|_{C^n} \leq \sum_{k=0}^{n-1} \frac{g^k(A)}{\sqrt{k!}\rho^{k+1}(A, \lambda)} \quad \text{for all regular } \lambda,$$

where $\rho(A, \lambda)$ is the distance between the spectrum of A and a complex λ (see [4, Corollary 1.2.4]).

2   Preliminary facts

For a positive r, put $\omega_r = \{h \in C^n : |h|_{C^n} \leq r\}$. Let $c_0 = c_0(C^n)$ be the Banach space of sequences of vectors from C^n equipped with the norm $|h|_{c_0} = \sup_k |h_k|_{C^n}$, ($h = (h_k)_{k=1}^{\infty} \in c_0$; $h_k \in C^n$, $k = 1, 2, ...$). In addition, $l^2 = l^2(C^n)$ is the Banach space of sequences of vectors from C^n equipped with the norm $|h|_{l^2} = [\sum_{k=1}^{\infty} |h_k|_{C^n}^2]^{1/2}$ ($h = (h_k)_{k=1}^{\infty} \in l^2$, $h_k \in C^n$, $k = 1, 2, ...$). Let a_{jk} ($j, k = 1, 2, ...$) be $n \times n$-matrices, and $f_j \in C^n$ ($j = 1, 2, ...$).

Consider the equation

$$x_j = f_j + \sum_{k=1}^{j-1} a_{jk} [x_k + G_k(x_k)], \quad (j = 1, 2, ...) \tag{1}$$

where the mappings $G_k : C^n \to C^n$ have the property

$$|G_k(h)|_{C^n} \leq q_k \, |h|_{C^n} \quad (h \in \omega_r \; ; k = 1, 2, ... \text{ for some } r \leq \infty). \tag{2}$$

Rewrite equation (1) as

$$x = f + Vx + F(x) \ , \tag{3}$$

where $f = (f_k)$, V and F are defined in l^2 by $[Vx]_j = \sum_{k=1}^{j-1} a_{jk} x_k$ and $[Fx]_j = \sum_{k=1}^{j-1} a_{jk} G_k(x_k)$. Here $[h]_j$ means the j-th coordinate of the element $h \in l^2$. It is assumed that

$$N(V) \equiv \left[\sum_{j=1}^{\infty} \sum_{k=1}^{j-1} |a_{jk}|_{C^n}^2 \right]^{\frac{1}{2}} < \infty, \tag{4}$$

$$|f|_{l^2} \equiv \left[\sum_{k=1}^{\infty} |f_k|_{C^n}^2 \right]^{\frac{1}{2}} < \infty \tag{5}$$

and

$$N_q = \left[\sum_{j=1}^{\infty} \sum_{k=1}^{j-1} |a_{jk}|_{C^n}^2 \, q_k^2 \right]^{\frac{1}{2}} < \infty. \tag{6}$$

Clearly, if $q_k \leq b_0 < \infty$ ($k = 1, 2, ...$), then (4) implies $N_q \leq b_0 N(V)$. Furthermore, denote

$$m(V) = \sum_{k=0}^{\infty} \frac{N^k(V)}{\sqrt{k!}} \ .$$

By the Schwarz inequality it is not hard to show that

$$m(V) \leq \sqrt{2} \exp\left[N^2(V)\right].$$

3 Main results

Now, we are in a position to establish our main results.

Theorem 1 *Let the condition (2) hold with $r = \infty$. Then under the conditions (4–6), and*

$$m(V) \, N_q < 1 \tag{7}$$

a solution $x = (x_k)$ of (1) is in l^2. Moreover, it satisfies the inequality

$$|x|_{l^2} \leq m(V) \, (1 - m(V) \, N_q)^{-1} \, |f|_{l^2} \ . \tag{8}$$

Proof. By Theorem 2.4.1, Gil' [4],

$$\left|(I - V)^{-1}\right|_{l^2} \leq m(V). \tag{9}$$

Moreover, by (2) and the Schwarz inequality

$$|F(x)|_{l^2}^2 = \sum_{j=1}^{\infty} \left| \sum_{k=1}^{j-1} a_{jk} \, G_k(x_k) \right|_{C^n}^2 \le \sum_{j=1}^{\infty} \left[\sum_{k=1}^{j-1} |a_{jk}|_{C^n} \, q_k \, |x_k|_{C^n} \right]^2$$

$$\le \sum_{j=1}^{\infty} \sum_{k=1}^{j-1} |a_{jk}|_{C^n}^2 \, q_k^2 \sum_{k=1}^{j-1} |x_k|_{C^n}^2 .$$

Hence,

$$|F(x)|_{l^2} \le N_q \, |x|_{l^2} . \tag{10}$$

Rewrite (3) as

$$x = (I - V)^{-1} (f + F(x)). \tag{11}$$

By (9) and (10), it follows that $|x|_{l^2} \le m(V) \, [|f|_{l^2} + N_q \, |x|_{l^2}]$. Thus, by (7) we get (8), concluding the proof. $\square$

Put $M_1(V, q) = \max_j \left[\sum_{k=1}^{j-1} |(1 + q_k) \, a_{jk}|_{C^n}^2 \right]^{\frac{1}{2}}$ and

$$c(f) = |f|_{c_0} + |f|_{l^2} \, M_1(V, q) \, m(V) \, (1 - m(V) \, N_q)^{-1} .$$

Theorem 2 *Under the hypotheses of Theorem 1, the estimate*

$$|x|_{c_0} \le c(f) \tag{12}$$

is valid.

Proof. By (1) and (2), $|x|_{c_0} \le |f|_{c_0} + \max_j \sum_{k=1}^{j-1} |a_{jk}|_{C^n} (1 + q_k) \, |x_k|_{C^n}$. The Schwarz inequality yields

$$|x|_{c_0} \le |f|_{c_0} + \max_j \left[\sum_{k=1}^{j-1} |(1 + q_k) \, a_{jk}|_{C^n}^2 \sum_{k=1}^{j-1} |x_k|^2 \right]^{\frac{1}{2}}$$

$$\le |f|_{c_0} + M_1(V, q) \, |x|_{l^2} .$$

From Theorem 1 we infer that

$$|x|_{c_0} \le |f|_{c_0} + m(V) \, (1 - m(V) \, N_q)^{-1} |f|_{l^2} \, M_1(V, q) = c(f)$$

concluding the proof. $\square$

In the next theorem, the condition $r = \infty$ is not assumed.

Theorem 3 *Under the Conditions (2, 4–7) let us assume*

$$c(f) \le r. \tag{13}$$

Then, a solution $x = (x_k)$ of (1) is bounded. Moreover, the inequality (12) is valid.

The proof is a simple application of Urysohn's lemma and Theorem 2.

4   Extensions

The main results established in the previous sections can be extended to many kinds of Volterra difference equations; for example, consider the equation

$$x_j = U_j(x_j) + \sum_{k=1}^{j-1} a_{jk}[x_k + G_k(x_k)], j = 1, 2, \dots \tag{14}$$

where the mappings $G_k : C^n \to C^n$ have the property (2) and the mappings $U_j : C^n \to C^n$ have the property

$$|U_j(x_j)|_{C^n} \leq \sum_{k=1}^{j-1} \beta_{jk} |x_k|_{C^n} \quad j = 1, 2, \dots \tag{15}$$

and $\beta_{jk} \geq 0$ are nonnegative numbers. Rewrite equation (14) as

$$x - U(x) + Vx + F(x), \tag{16}$$

where U, V and F are defined in l^2 by $U(x) = (U_j(x_j))$, $[Vx]_j = \sum_{k=1}^{j-1} a_{jk}x_k$ and $[Fx]_j - \sum_{k=1}^{j-1} a_{jk}G_k(x_k)$. Suppose that

$$N_\beta = \Big[\sum_{j=1}^{\infty}\sum_{k-1}^{j-1}[\beta_{jk}]^2\Big]^{\frac{1}{2}} < \infty. \tag{17}$$

Theorem 4 *Let conditions (2) and (15) hold with $r = \infty$. Then, under the conditions (4), (6) and (17), and if additionally*

$$m(V)(N_\beta + N_q) < 1, \tag{18}$$

a solution $x = (x_k)$ of equation (14) is in l^2. Moreover, it satisfies the inequality

$$|x|_{l^2} \leq m(V)[1 - m(V)(N_\beta + N_q)]^{-1}.$$

Proof. We have

$$|U_j(x_j)|_{C^n}^2 \leq \Big(\sum_{k=1}^{j-1} \beta_{jk}|x_k|_{C^n}\Big)^2 \leq \Big(\sum_{k=1}^{j-1}[\beta_{jk}]^2\Big)\Big(\sum_{k=1}^{j-1}|x_k|_{C^n}^2\Big).$$

This yields

$$|U(x)|_{l^2}^2 = \sum_{j=1}^{\infty}|U_j(x_j)|_{C^n}^2 \leq \sum_{j=1}^{\infty}\Big[\sum_{k=1}^{j-1}[\beta_{jk}]^2\sum_{k=1}^{j-1}|x_k|_{C^n}^2\Big]$$

$$\leq \Big(\sum_{j=1}^{\infty}\sum_{k=1}^{j-1}[\beta_{jk}]^2\Big)\sum_{k=1}^{\infty}|x_k|_{C^n}^2 = N_\beta |x|_{l^2}^2.$$

On the other hand, $x = (I - V)^{-1}(U(x) + F(x))$, thus

$$|x|_{l^2} \leq m(V)[|U|_{l^2} + |F|_{l^2}] \leq m(V)[N_\beta |x|_{l^2} + N_q |x|_{l^2} + 1].$$

By (18) it follows that $|x|_{l^2} \leq m(V)[1 - m(V)(N_\beta + N_q)]^{-1}$. □

5   Conclusion

In this work some possibilities for using the operator method for investigating the asymptotic behavior of Volterra difference equations have been demonstrated. Thus, we have derived some explicit estimates for the solutions in terms of the coefficients of the equation.

References

[1] M.R. Crisci, V.B. Kolmanovskii, E. Russo, A. Vecchio, Stability of continuous and discrete Volterra integro-differential equations by Lyapunov approach, *J. Integral Equations Appl.* 7(4)(1995), 393–411.

[2] S. Elaydi, *An Introduction to Difference Equations*, Springer, New York, 1996.

[3] S. Elaydi, S. Murakami, Asymptotic stability versus exponential stability in linear Volterra difference equations of convolution type, *J. Difference Equations Appl.* 2(1996), 401–410.

[4] M.I. Gil', *Norm Estimates for Operator-Valued Functions and Applications*, Marcel Dekker, New York, 1995.

[5] V.B. Kolmanovskii, A.D. Myshkis, J.-P. Richard, Estimate of solutions for some Volterra difference equations, *Nonlinear Analysis* 40(2000), 345–363.

[6] V.B. Kolmanovskii, A.D. Myshkis, Stability in the first approximation of some Volterra difference equations, *J. Difference Equations Appl.* 3(1998), 563–569.

[7] M. Kwapisz, On l_p solutions of discrete Volterra equations, *Aequationes Mathematicae*, University of Waterloo, 43(1992), 191–197.

[8] R. Medina, Stability results for nonlinear difference equations, *Nonlinear Studies* 6(1)(1999), 73–83.

[9] R. Medina, Solvability of discrete Volterra equations in weighted spaces, *Dynamic Systems Appl.* 5(1996), 407–422.

[10] Q. Sheng, R.P. Agarwal, On nonlinear variation of parameters methods for summary difference equations, *Dynamic Systems Appl.* 2(1993), 227–242.

Constructing Operator-Difference Schemes for Problems with Matching Boundaries

RODERICK V. N. MELNIK

University of Southern Denmark
Mads Clausen Institute, DK-6400, Denmark
E-mail: rmelnik@mci.sdu.dk

Abstract In this paper we construct a system of difference equations resulting from the approximation of a dynamic model based on a strongly coupled system of partial differential equations. The model is an extension of the previously studied case of dynamic electromechanical vibrations of hollow piezoceramic cylinders, but here we account for the acoustic coupling of such cylinders with the surrounding media. The coupling between three media of distinctively different physical natures brings new features in the analysis of the discrete model. We propose an algorithm for the solution of the operator-difference equations resulting from our approximations. Error estimates for our approximations are discussed in the framework of non-smooth solutions.

Keywords Operator-difference schemes, Matching boundaries, Discrete negative norms in space-time.

AMS Subject Classification 35A35, 65M06

1   Introduction

Many important classes of systems of difference equations come from discretization procedures of coupled time-dependent models based on partial differential equations. The systems containing hyperbolic-type equations represent the most challenging and arguably most interesting class of such models. It is often an intrinsic interplay between numerical dissipation and dispersion that create major challenges in investigating discrete approximations for such models. In addition, in a number of applications one has to respond adequately to challenges coming from a strong coupling of the resulting system due to the physical essence of the problem where, for example, the mutual influence of thermal and elastic, or elastic and electric, fields in a solid is very important. The situation becomes even more complicated if we have to derive discrete schemes for problems where such solids are surrounded by media with distinctively different physical properties, e.g., by fluids. This would require dynamic internal boundary conditions at the interface between the solid and the fluid. It is these types of applied problems that have led us to the study of systems of difference equations arising from the full discretization of such coupled time-dependent mathematical models. Such difference approximations can often be defined as operator-difference equations in a sense that they are difference equations with respect to time but they are operator equations with respect to space (e.g., [6]). Operators of these schemes are defined in a certain linear normed space H_h dependent on a vector parameter h which is an analogue of the spatial step.

We organize our further discussion as follows. First, we give a brief background of the application area where the class of difference approximations we are interested in came from. Then, we recall some basic results obtained previously in [6] for a simplified problem and derive a system of difference equations approximating the new fully coupled time-dependent problem considered here. Next, we propose an algorithm for the solution of the difference equations obtained. Finally, we discuss the analysis of the operator-difference system in the framework of the negative norm technique.

2 Application area and coupled systems of PDEs

Our major interest in this paper is the solution procedure for a system of difference equations obtained as a result of the discretization of the following problem.

Consider a piezoelectric solid shell. In order to describe the dynamics of this shell, subject to various loading conditions, one has to incorporate some mechanism of *coupling* between mechanical and electric fields. It is due to this electromechanical coupling that piezoelectrics have become very popular materials in such areas as transducer applications, biomedical engineering, smart materials and structure technology, to name just a few. This coupling can be introduced into the model by constitutive equations. In what follows we consider a solid made of piezoelectric ceramics, meaning that it is a polycrystallic solid with dipoles in each crystal. Those dipoles are oriented stochastically in general, but we can order them (in some small domains) by applying an electric field to the sample. This process is called a preliminary polarization, and we note that if the cylindrical geometry is used (as it is the case here where we consider a hollow piezoceramic cylinder surrounded by media with different physical properties), then most typical types of preliminary polarizations are axial, radial, and circular. From a mathematical point of view, all three types can be obtained from a simple re-ordering of axes. We do not want to go into any further technical detail here except for noting that the radial preliminary polarization provides opportunities for the strongest coupling in the case of this geometry [6]. Putting the above discussion in the modelling context, in order to describe the dynamic behavior of our system one has to solve 22 equations with respect to 3 components of displacements $\mathbf{u} = (u_1, u_2, u_3)^T$, the electric field strength $\mathbf{E}$, electric field induction vectors $\mathbf{D}$, 6 components of stress $\vec{\sigma}_v$ and strain tensors $\vec{\epsilon}_v$, and the electrostatic potential φ. More precisely, in what follows we consider the axisymmetric geometry which leads to the following constitutive equations

$$\vec{\sigma}_v = \mathbf{c}\vec{\epsilon}_v - \mathbf{e}^T \mathbf{E}, \quad \mathbf{D} = \bar{\varepsilon}\mathbf{E} + \mathbf{e}\vec{\epsilon}_v, \tag{1}$$

where the symmetric mechanical stress and strain tensors are presented here in schematic vector forms as $\vec{\sigma} = (\sigma_\theta, \sigma_z, \sigma_r, \sigma_{rz}, \sigma_{r\theta}, \sigma_{\theta z})^T$, $\vec{\epsilon} = (\epsilon_\theta, \epsilon_z, \epsilon_r, 2\epsilon_{rz}, 2\epsilon_{r\theta}, 2\epsilon_{\theta z})^T$, electric field strength and electric induction are given as $\mathbf{E} = (E_\theta, E_z, E_r)^T$, $\mathbf{D}^r = (D_\theta, D_z, D_r)^T$, respectively, $\mathbf{E} = -\nabla\varphi$, and the matrices of elastic, dielectric, and piezoelectric coefficients are given in the forms

$$\mathbf{c} = \begin{pmatrix} c_{33} & c_{13} & c_{13} & 0 & 0 & 0 \\ c_{13} & c_{11} & c_{12} & 0 & 0 & 0 \\ c_{13} & c_{12} & c_{11} & 0 & 0 & 0 \\ 0 & 0 & 0 & c_{66} & 0 & 0 \\ 0 & 0 & 0 & 0 & c_{55} & 0 \\ 0 & 0 & 0 & 0 & 0 & c_{55} \end{pmatrix}, \tag{2}$$

$$\bar{\varepsilon} = \begin{pmatrix} \varepsilon_{33} & 0 & 0 \\ 0 & \varepsilon_{11} & 0 \\ 0 & 0 & \varepsilon_{11} \end{pmatrix}, \quad \mathbf{e} = \begin{pmatrix} e_{33} & e_{31} & e_{31} & 0 & 0 & 0 \\ 0 & 0 & 0 & 0 & 0 & e_{15} \\ 0 & 0 & 0 & e_{15} & 0 & 0 \end{pmatrix}, \tag{3}$$

respectively. These constitutive equations couple two partial differential equations describing the electromechanical dynamics of the hollow cylinder (the equation of motion and the Maxwell equation):

$$\rho\frac{\partial^2 \mathbf{u}}{\partial t^2} = \nabla \cdot \vec{\sigma} + \mathbf{F}, \quad \mathrm{div}\mathbf{D} = \mathbf{G}. \tag{4}$$

In (4) ρ is the density of piezoelectric material, $\mathbf{G}$ and $\mathbf{F}$ are electric and body forces on the piezoelectric.

Discrete schemes for this model have been already studied in [6, 7]. The present paper is not merely a contribution to a further development of the results for systems of difference equations obtained from space-time discretizations of model (1)–(4). Here we consider a much more complicated case where one has to account for the coupling of the solid to the surrounding media, for example fluids, in addition to a strong electromechanical coupling exhibited by the piezoelectric cylinder itself. Let ρ_i, c_i, and p_i, $i = 1, 2$ denote densities, speeds of sound, and acoustic pressures in the internal and external media, surrounding the piezoelectric body. Then, in terms of the differential model this means that (1)–(4) has to be coupled to two pairs (for the interior and exterior of the shell) of three equations which we consider here with respect to components of the fluid particle velocity $\mathbf{v}^{(i)} = (v_r^{(i)}, v_z^{(i)})$, $i = 1, 2$

$$\frac{1}{\rho_i c_i^2} \frac{\partial p_i}{\partial t} + \frac{1}{r} \left(r \frac{\partial v_r^{(i)}}{\partial r} \right) + \frac{\partial v_z^{(i)}}{\partial z} = 0, \tag{5}$$

$$\rho_i \frac{\partial v_r^{(i)}}{\partial t} + \frac{\partial p_i}{\partial r} = 0, \quad \rho_i \frac{\partial v_z^{(i)}}{\partial t} + \frac{\partial p_i}{\partial z} = 0. \tag{6}$$

Equations (5)–(6) are coupled to (1)–(4) by the *matching boundary conditions*. These conditions are dynamic in a sense that they require (a) pressure continuity at the interface, leading to the time derivative of the velocity potentials, and (b) continuity of the radial velocity, leading to time derivatives of the displacements at the interface between the solid and the fluid.

The key challenges can be demonstrated by considering an infinitely long cylinder, in which case the one-dimensional (in the radial direction) solution is sufficient. In the acoustic approximation limit equations (5)–(6) can be transformed into wave equations with respect to acoustic pressures of the surrounding media. If we assume that the particle fluid velocities can be represented as $v_i = -\nabla \varphi_i$, and that $p_i = -\rho_i \frac{\partial \varphi_i}{\partial t}$, $i = 1, 2$, we arrive at the following acoustic approximations for the surrounding media

$$\frac{\partial^2 \varphi_i}{\partial t^2} - c_i^2 \frac{1}{r} \frac{\partial}{\partial r} \left(r \frac{\partial \varphi_i}{\partial r} \right) = 0, \quad i = 1, 2. \tag{7}$$

These equations are coupled to the electromechanical part of the system in the following way. The continuity conditions for pressures and velocities require $\sigma_r = -p_i(t)$, $\frac{\partial u}{\partial t} = \frac{\partial \varphi_i}{\partial r}$. Recall that we have confined ourselves to the infinitely-long-cylinder case, so our geometry is the spatial-temporal region $Q_T = G \times [0, T]$ with $G = [R_0, R_3]$, where $R_0 \leq R_1 < R_2 < R_3$ (in (R_0, R_1) we have a fluid region denoted by index 1, in (R_1, R_2) we have a piezoelectric solid region, and in (R_2, R_3) we have a fluid region denoted by index 2). This means that in the above interfacial boundary conditions $i = 1$ for $r = R_1$, and $i = 2$ for $r = R_2$. Other conditions (electrical boundary conditions, initial conditions, and the limiting conditions for the behavior of the surrounding fluids) will not be discussed here; they are analogous to those considered in [6, 7]. The electromechanical part of the system is described by (1)–(4) which in this specific case will have the form (for region $(R_1, R_2) \times [0, T]$)

$$\rho_0 \frac{\partial^2 u}{\partial t^2} = \frac{1}{r} \frac{\partial}{\partial r} (r \sigma_r) - \frac{\sigma_\theta}{r} + F(r, t), \quad \frac{1}{r} \frac{\partial}{\partial r} (r D_r) = G(r, t). \tag{8}$$

Equations (8) are coupled by the constitutive equations for the radial preliminary polarisation $\sigma_r = c_{33} \epsilon_r + c_{13} \epsilon_\theta - e_{33} E_r$, $\sigma_\theta = c_{13} \epsilon_r + c_{11} \epsilon_\theta - e_{31} E_r$,

$D_r = \varepsilon_{33}E_r + e_{31}\epsilon_\theta + e_{33}\epsilon_r$. This case has not been considered in the literature before in such a general setting as it is proposed here. At the same time, the development of discrete models based on operator-difference schemes represents an important task in theory and applications of difference equations arising from approximations of time-dependent coupled systems of PDEs.

3 Discretization procedures and the solution of the resulting system of difference equations

We aim at developing an efficient procedure for the solution of the problem formulated in Section 2.

For this purpose we make use of the already developed discrete model applicable in region (R_1, R_2) of our problem (see [6]). The resulting system of difference equations can be written in the following form

$$
\begin{cases}
\rho_2 \dfrac{\Delta y^n}{\tau^2} - \dfrac{1}{r}\dfrac{(\bar{r}\bar{\sigma}_r)^{(+1)} - \bar{r}\bar{\sigma}_r}{h_2} - \dfrac{(\bar{\sigma}_\theta)^{(+1)} + \bar{\sigma}_\theta}{2r} + F, \\[2ex]
\dfrac{1}{r}\dfrac{(\bar{r}\bar{D}_r)^{(+1)} - \bar{r}\bar{D}_r}{h_2} = G.
\end{cases}
\tag{9}
$$

This system is a result of the application of the variational approach to system (8) on computational grid $\bar{\omega}_\tau \times \bar{\omega}_2$ in $[R_1, R_2] \times [0, T]$, where, for example, for the locally uniform grids we have $\bar{\omega}_2 = \{r_i^{(2)} = R_1 + ih_2; i = 0, 1, ..., N_2; h_2 = (R_2 - R_1)/N_2\}$, and $\bar{\omega}_\tau = \{t^n = n\tau, n = 0, 1, ..., K\}$, where $K\tau = T$. In a similar way, the grids are introduced in the fluid regions $\bar{\omega}_k = \{r_i^{(k)} = R_{k-1} + ih_k; i = 0, 1, ..., N_k; h_k = (R_k - R_{k-1})/N_k\}$, $k = 1, 3$, where $r_{N_1}^{(1)} = r_0^{(2)}$, and $r_{N_2}^{(1)} = r_0^{(3)}$. In (9) we take approximations of the constitutive equations in the "flux" grid points as $\bar{\epsilon}_r = (y - y^{(-1)})/h_2$, $\epsilon_\theta = (y + y^{(-1)})/2r$, $\Delta y^n = y^{n+1} - 2y^n + y^{n-1}$, super-indices "$\pm$" indicate the right/left shift for one grid point in the spatial direction, $\bar{E}_r = -(\mu - \mu^{(-1)})/h_2$, $\bar{\sigma}_r = c_{33}\bar{\epsilon}_r + c_{13}\bar{\epsilon}_\theta - e_{33}\bar{E}_r$, $\bar{\sigma}_\theta = c_{13}\bar{\epsilon}_r + c_{11}\bar{\epsilon}_\theta - e_{31}\bar{E}_r$, $\bar{D}_r = \varepsilon_{33}\bar{E}_r + e_{31}\bar{\epsilon}_\theta + e_{33}\bar{\epsilon}_r$. Approximations of hyperbolic equations (7) are carried out in a standard manner, while the matching boundary conditions are approximated by using the integro-interpolational approach (e.g., [2, 11]). In particular, to get difference equations for the continuity of pressure at the exterior of the cylinder we integrate (7) in $[R_2, R_2 + h_3/2])$ and then use second-order approximations for all terms in the resulting expression, which, after some additional work omitted here, leads to the following approximation

$$
\frac{h_3}{2}(R_2 + h_3/4)\Delta\omega_0^n/\tau^2 = c_2^2\left((R_2 + h_3/2)\frac{w_1 - w_0}{h_3} - R_2\frac{y_{N_2}^{n+1} - y_{N_2}^{n-1}}{2\tau}\right). \tag{10}
$$

In a similar manner, we get approximations for the second matching boundary conditions by integrating the equation of motion in the appropriate limits, and by using second order approximations for the resulting terms.

The complete set of difference equations is written in terms of discrete functions (v, y, μ, w) giving approximations to $(\varphi_1, u, \varphi, \varphi_2)$, as it is explained in Section 2. It is important to notice that function v is defined on $\bar{\omega}_\tau \times \bar{\omega}_1$ and is coupled to function y defined on $\bar{\omega}_\tau \times \bar{\omega}_2$ by matching conditions at $r_{N_1}^{(1)} = r_0^{(2)}$. Functions y and μ are coupled on $\bar{\omega}_\tau \times \bar{\omega}_2$. In addition, function y is coupled to function w defined on $\bar{\omega}_\tau \times \bar{\omega}_3$ by matching conditions at $r_{N_2}^{(1)} = r_0^{(3)}$. To resolve these difficulties, one can apply the idea similar to that proposed in [2] for the uncoupled case of circular preliminary polarization. In particular, four approximations derived from approximations of the matching interface boundary conditions represent a pair of systems of linear equations $Ax = b$ with $x \in \mathbb{R}^2$, $b \in \mathbb{R}^2$. The first system of this pair should be solved with respect to $(v_{N_1}^{n+1}, y_0^{n+1})^T$ for matrix $A = A_1$ where

$$
A_1 = \begin{pmatrix} \dfrac{h_1}{2\tau^2}(R_1 - \dfrac{h_1}{4}) & -\dfrac{c_1^2 R_1}{2\tau} \\[3mm] \dfrac{R_1}{2\tau} & \rho_2 \dfrac{h_2}{2\tau^2}(R_1 + \dfrac{h_2}{4}) \end{pmatrix}, \tag{11}
$$

and the second system should be solved with respect to $(y_{N_2}^{n+1}, w_0^{n+1})^T$ for matrix $A = A_2$ where

$$
A_2 = \begin{pmatrix} c_2^2 \dfrac{R_2}{2\tau} & \dfrac{h_3}{2\tau^2}(R_2 + \dfrac{h_3}{4}) \\[3mm] \rho_2 \dfrac{h_2}{2\tau^2}(R_2 - \dfrac{h_2}{4}) & \dfrac{R_2}{2\tau} \end{pmatrix}. \tag{12}
$$

The resulting systems have positive determinants and their solution, supplemented by the procedure described in [6], allows us to find all unknown functions on the time-layer $n + 1$, which in its turn allows us to compute the stresses and the acoustic pressure, and proceed to the next time level. Computational experiments with the resulting system of difference equations will be reported in a separate paper [7].

Note that the described procedure puts us into the framework of operator-difference scheme

$$
\begin{cases} D_1 \Delta y^n / \tau^2 + A_1 y + C_1 \mu = \varphi_1, \\[2mm] A_2 \mu + C_2 y = \varphi_2, \\[2mm] y = y_0, \ D_1(y^1 - y^0)/\tau = y_1, \ t = 0, \end{cases} \tag{13}
$$

with operators of the scheme determined in a way similar to that described in [6]. The stability condition for this scheme is a generalization of the CFL stability condition to the case of coupled electromechanical waves (see detail in [6]). A scale of a priori estimates in Sobolev spaces has been obtained for (13) in the above cited paper. We also obtained an estimate in class $V(\bar{Q}_T) = C(\bar{Q}_T) \cap Q_1(\bar{Q}_T)$ of continuous functions with piecewise first derivatives which have square integrable generalized derivatives in the continuity region. This result can be improved further by considering the negative norm technique which becomes popular in the context of post-processing of numerical algorithms.

4   Discrete negative norms and further development

In a number of application areas the problems described above require dealing with non-smooth or generalized data, and when a variational-type approach used for the construction of systems of difference equations approximating the original differential problem, this can lead to norm equivalent functionals that are meaningful for less regular solutions than L^2-type functionals traditionally used in obtaining error estimates. In such situations negative norms become a very useful tool for the analysis because $L^2(\Omega) \subsetneq H^s(\Omega)$ for $s < 0$, where $H^s(\Omega)$ is the space of linear functionals with the finite norm, e.g. for $s = -1$ it is $\|v\|_{-1} = \sup\limits_{\varphi \in W} \dfrac{(v, \varphi)}{\|\varphi\|_1}$ with W being a closure space with respect to norm in $H^1(\Omega)$. The interest to such negative norms have been recently revitalized in the context of effective post-processing of approximations resulting from discretizations of PDEs due to the possibility of devising preconditioners for discretized equations, and of improving error estimates under relaxed regularity assumptions (e.g., [1, 4, 3]). In most situations considered in the literature the negative norm technique has been applied for stationary problems only, with just a few exceptions (e.g. [1, 9]) in which, however, negative norms were considered with respect to the space but not time. For discrete models considered in this paper it seems natural to apply negative norms in both space and time. Such results for difference schemes approximating the wave equations go back to work [8] and have been recently discussed by the authors of [10]. One of the basic results can be formulated as follows. Consider the operator-difference scheme in the form $D\Delta y^j/\tau^2 + B(y^{j+1} - y^{j-1})/2\tau + Ay^j = \varphi$ given the same initial conditions as in (13). Then, as soon as $A^* = A > 0$, $D^* = D > \beta E$, $\beta > 0$, $B \geq 0$, $BD^{-1}A \geq 0$, $D \geq (1 + \epsilon)/4\tau^2 A$, $\epsilon > 0$, by applying the negative norm technique we can obtain the following estimate for the solution of this system of difference equations $\|y(t)\| \leq M \left\{ \|y(0)\|_D + \|Dy_t(0)\|_{A^{-1}} + \|\varphi_1\|_{0,-1} + \|\varphi_2\|_{A^{-1},0} \right\}$, where $y(t) = \xi_{\bar{t}}$, $\xi(0) = 0$, $\|y(t)\|_{A^{-1},0} = \sum_{t'=0}^{T} \tau\|y(t')\|$, $\|y(t)\|_{0,-1}^2 = \sum_{t'=\tau}^{T} \tau\|\xi(t')\|$. This result can be relaxed further since in some special cases the condition of self-adjointness of A and D can be dropped. In conclusion, we note that in a special case where the electromechanical coupling is assumed negligible at the interface boundaries of the cylinder (see [5]) the operator-difference schemes considered here can be reduced to the above form, and therefore the general theory for obtaining error estimates in negative norms in time and space can be applied.

References

[1] Bales, L.A., Semidiscrete and single step fully discrete FE approximations for second order hyperbolic equations with nonsmooth solutions, *Math. Model. Numer. Anal.* **27** (1993), 55–63.

[2] Belova, M.M., Moskalkov, M.N., and Savin, V.G., Numerical solution of the problem of sound emission by a cylindrical piezoelectric vibrator excited by electric pulses, *J. Math. Sciences* **63** (1993), 427–432.

[3] Bertoluzza, S., Canuto, C., and Tabacco, A., Negative norm stabilization of convection-diffusion problems, *Appl. Math. Let.* **13** (2000), 121–127.

[4] Bramble, J.H. and Sun, T., A negative-norm least square method for Reissner-Mindlin plates, *Math. Comp.* **67** (1998), 901–916.

[5] Melnik, R.V.N. and Melnik, K.N., A note on the class of weakly coupled problems of non-stationary piezoelectricity, *Commun. Numer. Meth. Engng.* **14** (1998), 839–847.

[6] Melnik, R.V.N., Convergence of the operator-difference scheme to generalized solutions of a coupled field theory problem, *J. Differ. Equ. Appl.* **4** (1998), 185–212.

[7] Melnik, R.V.N., Numerical analysis of dynamic characteristics of coupled piezoelectric systems in acoustic media, *Math. Comp. Sim.*, to appear.

[8] Moskalkov, M.N., On accuracy of difference schemes for approximations of the wave equation with piecewise coefficients, *Comp. Math. Math Phys.* **14** (1974) 390–401.

[9] Nochetto, R.H., Schmidt, A., and Verdi, C., A posteriori error estimation and adaptivity for degenerate parabolic problems, *Math. Comp.* **69** (1999), 1–24.

[10] Samarskii, A.A. Vabischevich, P.N., and Matus, P., *Difference Schemes with Operator Factors*, Institute of Mathematics, Minsk, 1998.

[11] Samarskii, A.A., *The Theory of Difference Schemes*, Marcel Dekker, N.Y., 2001.

On Difference Matrix Equations

EDGAR PEREIRA [1]

Department of Informatics, Universidade da Beira Interior
6200-Covilhã, Portugal
E-mail: edgar@noe.ubi.pt

and

JOSÉ VITÓRIA [2]

Department of Mathematics, Universidade de Coimbra
3000-Coimbra, Portugal
E-mail: jvitoria@mat.uc.pt

Abstract We study general solutions of difference matrix equations, first in terms of solvents of matrix polynomials and then with block eigenvalues of companion matrices. According to the properties of the matrix polynomial associated with the difference matrix equation we show how the alternative methods can be used to compute explicit solutions of such equations.

Keywords Difference matrix equations, block eigenvalues

AMS Subject Classification 39A10, 15A24, 65Q05

1 Introduction

A homogeneous difference equation of order m with constant $n \times n$ matrix coefficients

$$p_j = x_{j+m} + A_1 x_{j+m-1} + \ldots + A_m x_j = 0, \quad j = 0, 1, \ldots, \qquad (1)$$

is said to be a difference linear matrix equation [4]. This equation is equivalent to a system of difference equations with constant coefficients, which has an important gamma of applications (see [3], [11]). Solutions of equation (1) can be obtained in terms of solvents of (monic) matrix polynomials $P(X)$ having degree m and the same matricial coefficients of (1).

---

[1] Partially supported by ISR (Coimbra).
[2] Partially supported by ISR (Coimbra).

Let us consider a block matrix constructed with the matrices $A_1, \ldots, A_m$, that are the coefficients of $P(X)$,

$$
C = \begin{bmatrix}
0_n & I_n & \cdots & 0_n \\
\vdots & & \ddots & \\
0_n & 0_n & & I_n \\
-A_m & -A_{m-1} & \cdots & -A_1
\end{bmatrix},
$$

of a dimension of $m \times m$ blocks of order n, with 0_n and I_n being the null and identity matrices of order n, respectively. This matrix is called the companion matrix of $P(X)$.

A method dealing with the block eigenvalues of the matrix C using weaker conditions than the one with solvents of $P(X)$ will be presented. In this method, a general solution of (1) is obtained with a complete set of block eigenvalues [9] of the companion matrix C of the matrix polynomial $P(X)$ associated with equation (1). The definition and basic theory of block eigenvalues can be found in [8] and [9].

Another approach is to compute the solutions of (1), using a block bidiagonal form for the companion matrix of $P(X)$, that appears in [4].

2   Solutions in terms of solvents

An important application of the study of solvents of matrix polynomials is to get explicit solutions of differential and difference matrix equations. Next we develop such approach for difference equations. The equivalent development for differential equations can be found in [7].

To each difference matrix equation (1) it is associated a matrix polynomial

$$
P(X) = X^m + A_1 X^{m-1} + \ldots + A_m, \tag{2}
$$

of degree m, where the $A_1, A_2, \ldots, A_m$, of order n, are the same coefficient matrices of (1). An $n \times n$ matrix S is a solvent of $P(X)$ if $P(S) = 0$. Furthermore, if $S_1, S_2, \ldots, S_m$ are m solvents of $P(X)$ and the block Vandermonde matrix

$$
V_S = \begin{bmatrix}
I & I & \cdots & I \\
S_1 & S_2 & \cdots & S_m \\
\vdots & \vdots & & \vdots \\
S_1^{m-1} & S_2^{m-1} & \cdots & S_m^{m-1}
\end{bmatrix},
$$

of order mn ($m \times m$ blocks of order n), is nonsingular, then

$$
CV_S = diag(S_1, S_2, \ldots, S_m)V_S, \tag{3}
$$

where $diag(S_1, S_2, \ldots, S_m)$ is a block diagonal matrix with the $S_1, S_2, \ldots, S_m$ in the diagonal. In this case we say that $S_1, S_2, \ldots, S_m$ are a complete set of solvents of $P(X)$.

Remark 2.1 A necessary and sufficient condition, for a matrix polynomial $P(X)$ to have a solvent, is the existence of a matrix with a form TJT^{-1}, where the matrix T, having eigenvectors of the lambda matrix $P(\lambda)$ as columns, is nonsingular and the matrix J, having eigenvavlues of $P(\lambda)$, is in a Jordan canonical form, see [7] page 520.

Lemma 2.2 ([7]) *Given a difference matrix equation (1) then the general solution is given by*
$$x_j = PC^j z, \quad j = 1, 2, \ldots,$$
where $P = \begin{bmatrix} I_n & 0_n & \ldots & 0_n \end{bmatrix}$ *, C is the companion matrix of the associated $P(X)$ and $z \in \mathbb{C}^{mn}$ is arbitrary.*

From the Lemma 2.2 and (3), a general solution of (1) then can be obtained in terms of solvents, that is:

Theorem 2.3 *Given a difference matrix equation (1) and its associated matrix polynomial $P(X)$, if $S_1, S_2, \ldots, S_m$ are a complete set of solvents of $P(X)$ then a general solution of equation (1) is given by*
$$x_j = S_1^j z_1 + S_2^j z_2 + \ldots + S_m^j z_m, \quad j = 0, 1, 2, \ldots$$
where $z_1, z_2, \ldots, z_m \in \mathbb{C}^n$.

Proof. Using (3) in Lemma 2.2 we have
$$\begin{aligned} x_j = PC^j z &= PV_S diag(S_1, S_2, \ldots, S_m)^j V_S^{-1} z \\ &= PV_S diag(S_1^j, S_2^j, \ldots, S_m^j) V_S^{-1} z. \end{aligned}$$

Since $PV_S = \begin{bmatrix} I_n & I_n & \ldots & I_n \end{bmatrix}$ and by writing $V_S^{-1} z = \begin{bmatrix} z_1 \\ z_2 \\ \vdots \\ z_m \end{bmatrix}$ it follows that

$$x_j = \begin{bmatrix} S_1^j & S_2^j & \ldots & S_m^j \end{bmatrix} \begin{bmatrix} z_1 \\ z_2 \\ \vdots \\ z_m \end{bmatrix} = S_1^j z_1 + S_2^j z_2 + \ldots + S_m^j z_m. \quad \blacksquare$$

3 Solutions in terms of block eigenvalues

Under weaker conditions the block eigenvalues of the block companion matrix C can be used to obtain a general solution of (1). The equivalent procedures for differential equations appear in [9].

Lemma 3.1 *([9]) Let $X_1, X_2, \ldots, X_r$ be block eigenvalues of the companion matrix C and let $V_1, V_2, \ldots, V_r$ be the corresponding block eigenvectors, if the matrix $[V_1, V_2, \ldots, V_r]$ is nonsingular then*

$$C\,[V_1, V_2, \ldots, V_r] = [V_1, V_2, \ldots, V_r]\,diag(X_1, X_2, \ldots, X_r),$$

and we say that these r block eigenvalues are a complete set of block eigenvalues of C.

Now using Lemma 2.2 and Lemma 3.1 we have:

Theorem 3.2 *If $X_1, X_2, \ldots, X_r$ are a complete set of block eigenvalues of the companion matrix C of the matrix polynomial $P(X)$ associated with the difference matrix equation (1) and $V_1, V_2, \ldots, V_r$ are the corresponding block eigenvectors, then the general solution of (1) is given by*

$$x_j = (V_1)_n X_1^j z_1 + (V_2)_n X_2^j z_2 + \ldots + (V_r)_n X_r^j z_r, \quad j = 0, 1, 2, \ldots$$

where for $i = 1, 2, \ldots, r$: $z_i \in \mathbb{C}^{q_i}$ is arbitrary, with q_i being the order of X_i, and $(V_i)_n$ is the $n \times q_i$ submatrix in the top of the block eigenvector V_i.

Proof. It follows from Lemma 2.2 substituting the matrix C by using Lemma 3.1, thus we have

$$x_j = PC^j z = P\,[V_1, V_2, \ldots, V_r]\,diag(X_1, X_2, \ldots, X_r)^j\,[V_1, V_2, \ldots, V_r]^{-1}\,z.$$

Since

$$P\,[V_1, V_2, \ldots, V_r] = \begin{bmatrix} (V_1)_n & (V_2)_n & \cdots & (V_r)_n \end{bmatrix},$$

$$diag(X_1, X_2, \ldots, X_r)^j = diag(X_1^j, X_2^j, \ldots, X_r^j)$$

and by writing

$$[V_1, V_2, \ldots, V_r]^{-1}\,z = \begin{bmatrix} z_1 \\ z_2 \\ \vdots \\ z_r \end{bmatrix},$$

it follows that

$$
\begin{aligned}
x_j &= \begin{bmatrix} (V_1)_n X_1^j & (V_2)_n X_2^j & \cdots & (V_r)_n X_r^j \end{bmatrix} \begin{bmatrix} z_1 \\ z_2 \\ \vdots \\ z_r \end{bmatrix} \\
&= (V_1)_n X_1^j z_1 + (V_2)_n X_2^j z_2 + \ldots + (V_r)_n X_r^j. \quad \blacksquare
\end{aligned}
$$

If the block eigenvalues are considered to be all matrices of order n, then any complete set have m block eigenvalues. In this case we have,

Corollary 3.3 *If the $n \times n$ matrices $X_1, X_2, \ldots, X_m$ are a complete set of block eigenvalues of C, then the general solution of (1) is given by*

$$x_j = (V_1)_n X_1^j z_1 + (V_2)_n X_2^j z_2 + \ldots + (V_m)_n X_m^j z_m, \quad j = 0, 1, 2, \ldots$$

where $z_1, z_2, \ldots, z_m \in \mathbb{C}^n$ are arbitrary and $(V_i)_n, i = 1, 2, \ldots, m$, is the $n \times n$ submatrix in the top of the block eigenvalue V_i.

The goal of this paper is to present a general solution of the difference matrix equation (1) even when it is not possible to achieve it with solvents considering that, in general, the associated matrix polynomial may not have a complete set of solvents. Furthermore, in Corollary 3.3 we consider the block eigenvalues of a fixed order, the same order of the solvents, this simplifies the using of numerical methods to obtain the block eigenvalues of the companion matrix C, see [8]. In the next example, we have a matrix polynomial with no solvents.

Example 3.4 *Consider the homogeneous difference equation*

$$x_{j+2} + A_1 x_{j+1} + \ldots + A_2 x_j = 0, \quad j = 0, 1, \ldots,$$

where A_1 and A_2 are given by

$$A_1 = \begin{bmatrix} -177/35 & -8/35 & 82/35 \\ 10/7 & -16/7 & -32/7 \\ 33/35 & 62/35 & -268/35 \end{bmatrix},$$

$$A_2 = \begin{bmatrix} 244/35 & 66/35 & -274/35 \\ -23/7 & 6/7 & 68/7 \\ -71/35 & -109/35 & 391/35 \end{bmatrix}.$$

The associated matrix polynomial $P(X)$, with $m = 2$ and $n = 3$ is

$$P(X) = X^2 + A_1 X + A_2.$$

If we compute the eigenvectors and eigenvalues of $P(\lambda)$ we note that $P(X)$ has no solvents; see Remark 2.1. Computing a complete set of block eigenvalues of order 3 of the companion matrix C of $P(X)$, we get

$$X_1 = \begin{bmatrix} 5/2 & -5/2 & 1/2 \\ 1/2 & -1/2 & 1/2 \\ 1 & -9 & 4 \end{bmatrix}, X_2 = \begin{bmatrix} 0 & 3/8 & 7/8 \\ 18 & 7/12 & -65/12 \\ -18 & 29/12 & 101/12 \end{bmatrix},$$

and the corresponding block eigenvectors are

$$V_1 = \begin{bmatrix} -1/2 & 5/2 & 0 \\ 1 & 0 & 0 \\ 0 & 1 & 0 \\ 0 & 0 & 1 \\ 5/2 & -5/2 & 1/2 \\ 1/2 & -1/2 & 1/2 \end{bmatrix}, V_2 = \begin{bmatrix} -2 & 7/12 & 7/12 \\ 2 & -1/4 & -1/4 \\ 1 & 0 & 0 \\ 0 & 1 & 0 \\ 0 & 0 & 1 \\ 0 & 3/8 & 7/8 \end{bmatrix}$$

where

$$(V_1)_3 = \begin{bmatrix} -1/2 & 5/2 & 0 \\ 1 & 0 & 0 \\ 0 & 1 & 0 \end{bmatrix} \text{ and } (V_2)_3 = \begin{bmatrix} -2 & 7/12 & 7/12 \\ 2 & -1/4 & -1/4 \\ 1 & 0 & 0 \end{bmatrix}.$$

Thus we have the following general solution

$$x_j = (V_1)_3 X_1^j z_1 + (V_2)_3 X_2^j z_2, \ j = 1, 2, \ldots,$$

with $z_1, z_2 \in \mathbb{C}^3$ being arbitrary vectors.

When it is not possible to achieve a solution with block eigenvalues of order n, it is always possible to use block eigenvalues of any order. The disadvantage here is that the computational efforts are equivalent to solving the spectral problem for C, that is to compute its eigenvalues and eigenvectors. We see this next.

Example 3.5 *Consider the homogeneous difference equation*

$$x_{j+2} + A_1 x_{j+1} + \ldots + A_2 x_j = 0, \ \ j = 0, 1, \ldots,$$

where A_1 and A_2 are given by

$$A_1 = \begin{bmatrix} -282/49 & 318/49 \\ 43/49 & -355/49 \end{bmatrix} \text{ and } A_2 = \begin{bmatrix} 298/49 & -776/49 \\ -51/49 & 584/49 \end{bmatrix}.$$

The associated matrix polynomial, with $m = 2$ and $n = 2$, is

$$P(X) = X^2 + A_1 X + A_2.$$

$P(X)$ has only one solvent, that is

$$S_1 = \begin{bmatrix} 12/7 & -4/7 \\ 1/7 & 16/7 \end{bmatrix}.$$

The block companion matrix C of $P(X)$ has no complete set of block eigenvalues of order 2. We note that in a complete set of block eigenvalues of a fixed order k, any block of the Jordan canonical form of the matrix C has to have an order not greater than k ([9]). Next, we get a complete set of block eigenvalues of the matrix C by solving its spectral problem, that is

$$X_1 = 7 \text{ and } X_2 = \begin{bmatrix} 2 & 1 & 0 \\ 0 & 2 & 1 \\ 0 & 0 & 2 \end{bmatrix}.$$

The corresponding block eigenvectors are

$$V_1 = \begin{bmatrix} -2 \\ 1 \\ -14 \\ 7 \end{bmatrix}, V_2 = \begin{bmatrix} -2 & 1 & 3 \\ 1 & 3 & 4 \\ -4 & 0 & 7 \\ 2 & 7 & 11 \end{bmatrix}.$$

where

$$(V_1)_2 = \begin{bmatrix} -2 \\ 1 \end{bmatrix} \text{ and } (V_2)_2 = \begin{bmatrix} -2 & 1 & 3 \\ 1 & 3 & 4 \end{bmatrix}.$$

Thus, we have the following general solution

$$x_j = (V_1)_2 X_1^j z_1 + (V_2)_2 X_2^j z_2, \quad j = 1, 2, \ldots,$$

with $z_1 \in \mathbb{C}$ and $z_2 \in \mathbb{C}^3$ being arbitrary vectors.

4   Remarks

Most of the methods to compute solvents of matrix polynomials are iterative and generalizations of classical methods to solve complex polynomials: power method [10], Traub's and Bernoulli's methods [2] and Newton's methods [6]. Some of these methods are restricted to the computation of one solvent. The principal limitation that we find is that the examples, presented in these studies, involve matrices of order no greater than 2 or 3. There are also some works with the quadratic equation [1], [5], but the same limitations that were mentioned before still exist.

For the computation of block eigenvalues, we used the block power method, and with a generalization of the Wielandt's deflation, a complete set can be obtained [8]. Here again we have limitations on the size of the matrices involved.

The great advantage of the block eigenvalues over the solvents is that the existence of a complete set of solvents is a very special case. Otherwise a complete set of block eigenvalues of a fixed order is a more common situation, and, in general, a complete set of any order always exists as in Example 3.5.

We thank the referee for the helpful suggestions.

References

[1] G. J. Davis, Numerical solution of a quadratic matrix equation, *SIAM Journal of Scientific Computing*, 2(1981), 164–175.

[2] J. E. Dennis, J. F. Traub and R. P. Weber, Algorithms for solvents of matrix polynomials, *SIAM Journal of Numerical Analysis*, 15(1978), 523–533.

[3] I. Gohberg, P. Lancaster and L. Rodman, *Matrix Polynomials*, Academic Press, New York, 1982.

[4] V. Hernández and F. Incertis, A block bidiagonal form for block companion matrices, *Linear Algebra and Applications*, 75(1986), 241–256.

[5] N. J. Higham and H. M. Kim, Solving a quadratic matrix equation by Newton's method with exact line searchers, to appear in *SIAM Journal of Matrix Analysis and Applications*.

[6] W. Kratz and E. Stickel, Numerical solution of matrix polynomial equations by Newton's method, *IMA Journal of Numerical Analysis*, 7(1987), 355–369.

[7] P. Lancaster, M. Tismenetsky, *The Theory of Matrices*, 2nd edition, Academic Press, New York, 1985.

[8] E. Pereira and J. Vitória, Deflation of block eigenvalues of block partitioned matrices with an application to matrix polynomials of commuting matrices, *Computers and Mathematics with Applications*, 42(2001), 1177–1188.

[9] E. Pereira, Block eigenvalues and solutions of differential matrix equations, to appear in *Mathematical Notes*, Miskolc.

[10] J.S.H. Tsai, L.S. Shieh and T.T.C. Shen, Block power method for computing solvents and spectral factors of matrix polynomials, *Computers and Mathematics with Applications*, 16(1988), 683–699.

[11] T. P. Lima, J. C. N. Clímaco and J. Vitória, On the study of differential and difference equations using lambda-matrices, *Linear Algebra and Applications*, 121(1989), 639–643.

On Some Difference Equations in the Context of q-Fourier Analysis

ANDREAS RUFFING and MORITZ SIMON

Department of Mathematics, Munich University of Technology
Arcisstrasse 21, D-80333 München, Germany
E-mail: ruffing@ma.tum.de

Abstract A discrete Fourier transform on a q-linear grid is presented. Several of its analytic properties are discussed and compared with the continuum situation. We recognize that a special invariant of the related Fourier operator is closely connected to discretizations of the Hermite functions. This result is similar to the continuum scenario and reveals a key role of the q-Fourier transform to the understanding of difference equations. The continuum limit $q \to 1$ in the sense of strong L^2-convergence is investigated for the derived q-Fourier invariant generalization of the Gauss curve. Applications to Schrödinger difference equations are briefly mentioned.

Keywords q-Difference operators, q-Fourier transforms, Orthogonal polynomials

AMS Subject Classification 39A10, 39A13, 33D45, 42A38, 47B39, 47B36

1 Introduction and motivation of the subject

Jean Baptiste Joseph Fourier published in 1822 his "Théorie analytique de la chaleur", in which he asks a fundamental question: To what extent can a periodic function of one variable, whether continuous or discontinuous, be expanded into series of sines of multiples of the variable? Investigating this question, he was creating the development of a new branch in analysis which until today is named for him as "Fourier analysis". Much research within this area is dedicated to the powerful technique of **Fourier transforms** which have a crucial meaning to a broad field of problems across the sciences. For instance, important applications of the Fourier transforms occur in quantum mechanics when it comes to considering the transform as a linear operator, acting on suitable domains in $L^2(\mathbb{R})$, namely $F : D(F) \subseteq L^2(\mathbb{R}) \to L^2(\mathbb{R})$, given by

$$g \mapsto Fg, \qquad \forall x \in \mathbb{R} : \ (Fg)(x) := \frac{1}{\sqrt{2\pi}} \int_{-\infty}^{\infty} (\cos(xt) - i\sin(xt))\, g(t)\, dt \quad (1)$$

There are, in general, close connections between Fourier analytic structures and differential equations. Looking for example at the function $\varphi : \mathbb{R} \to \mathbb{R}$, $x \mapsto \varphi(x) := e^{-\frac{1}{2}x^2}$, one can check that φ is a fixed point of F; hence

$$F\varphi = \varphi \qquad D\varphi = -X\varphi \tag{2}$$

where D denotes the classical derivative and X the multiplication of φ by its argument – both operators satisfying a so-called Heisenberg like commutation relation

$$DX - XD = I \tag{3}$$

on suitable domains in $L^2(\mathbb{R})$, the symbol I denoting the identity map. In the sequel, we will revise several similar situations in discrete q-Fourier analysis, for a general survey see also [3]. We will ask the following question: When replacing the continuously acting Fourier operator F by a suitable discrete Fourier operator F_δ, what are the related **difference equations** behind this transform? We will investigate this problem through several particular Fourier analytic scenarios. The **difference equations** under consideration will be q-difference equations for suitable functions acting on a geometric progression of type $\{+q^n, -q^n \mid n \in \mathbb{Z}\}$ where q is a real parameter, $0 < q < 1$. In the following section, we prepare the involved discrete tools.

2   Basic properties of a q-Fourier transform

We discretize the one-dimensional real axis $\mathbb{R}$ as follows:

$$\mathbb{R}_q := \{+q^n, -q^n \mid n \in \mathbb{Z}\} \qquad 0 < q < 1 \tag{4}$$

In addition, we consider the Hilbert space of all square integrable functions

$$L^2(\mathbb{R}_q) := \{f : \mathbb{R}_q \to \mathbb{C} \mid (1-q) \sum_{n=-\infty}^{+\infty} q^n (f(q^n)\overline{f(q^n)} + f(-q^n)\overline{f(-q^n)}) < \infty\} \tag{5}$$

A Euclidean scalar product in $L^2(\mathbb{R}_q)$ is canonically provided by

$$(f,g) := (1-q) \sum_{n=-\infty}^{+\infty} q^n (f(q^n)\overline{g(q^n)} + f(-q^n)\overline{g(-q^n)}) \tag{6}$$

An orthonormal basis of $L^2(\mathbb{R}_q)$ with respect to (6) is given by

$$e_n^\sigma(\tau q^m) := q^{-\frac{n}{2}}(1-q)^{-\frac{1}{2}}\delta_{mn}\delta_{\sigma\tau}, \tag{7}$$

$m,n \in \mathbb{Z}, \sigma,\tau \in \{+1,-1\}$ and δ_{mn} denoting the Kronecker symbol. We will also use the monolateral difference operator D_q, being densely defined in $L^2(\mathbb{R}_q)$ (z ranging in $\mathbb{R}_q$):

$$f \mapsto D_q f, \quad (D_q f)(z) := \frac{f(qz) - f(z)}{qz - z} \tag{8}$$

The multiplication operator and the right resp. left shift operator will be defined on their respective definition ranges in $L^2(\mathbb{R}_q)$ by

$$(Xf)(z) := zf(z) \qquad (R_qf)(z) := f(qz) \qquad (L_qf)(z) := f(q^{-1}z) \qquad (9)$$

Let us now arrive with the announced q-discrete Fourier transform: A q-deformed cosine resp. sine function is introduced in [5] by

$$\cos_q(z) = \sum_{k=0}^{\infty}(-1)^k\frac{q^{k(k+1)}z^{2k}}{(q;q)_{2k}} \qquad \sin_q(z) = \sum_{k=0}^{\infty}(-1)^k\frac{q^{k(k+1)}z^{2k+1}}{(q;q)_{2k+1}} \qquad (10)$$

Let μ be the number defined by $\mu := \frac{(q;q^2)_\infty}{(q^2;q^2)_\infty}$ where

$$(a;q)_k = \prod_{n=0}^{k-1}(1 - aq^n) \qquad (a;q)_\infty = \lim_{k\to\infty}\prod_{n=0}^{k-1}(1 - aq^n) \qquad (11)$$

If for a special $f \in L^2(\mathbb{R}_q)$ the following expressions exist, we will call them q-cosine transform resp. q-sine transform of f on $\mathbb{R}_q \cap \mathbb{R}^+$; compare [5]:

$$(C_qf)(q^n) := \mu \sum_{k=-\infty}^{\infty} q^k\cos_q(q^{k+n})f(q^k) \qquad (12)$$

$$(S_qf)(q^n) := \mu \sum_{k=-\infty}^{\infty} q^k\sin_q(q^{k+n})f(q^k) \qquad (13)$$

Using the q-difference operator D_q one obtains for any $z \in \mathbb{R}_q$:

$$(1 - q)\,(D_q\cos_q)(z) = -q\,\sin_q(qz) \qquad (1 - q)\,(D_q\sin_q)(z) = \cos_q(z) \qquad (14)$$

3   Comparing discrete and continuous results

Lemma 1

Let $\varphi_q : \mathbb{R}_q \cap \mathbb{R}^+ \to \mathbb{R}$ with $\forall z \in \mathbb{R}_q \cap \mathbb{R}^+ : \varphi_q(z) \neq 0$ fulfill the difference equation

$$\forall z \in \mathbb{R}_q \cap \mathbb{R}^+ : \quad q^{\frac{1}{2}}z\varphi_q(z) + (1 - q)(D_q\varphi_q)(z) = 0 \qquad (15)$$

The function $C_q\varphi_q$ then satisfies the same difference equation, generalizing the continuum result on the discretized real half line: For the function $\varphi : \mathbb{R} \to \mathbb{R}$ fulfilling the differential equation

$$\forall z \in \mathbb{R} : \quad z\varphi(z) + (D\varphi)(z) = 0, \qquad (16)$$

the Fourier transform $F\varphi$ satisfies the same differential equation (16).

Proof

We follow an original idea by T. H. Koornwinder and make first use of the relations (14): Provided, the C_q-cosine transform $f = C_q g$ exists for a function $g : \mathbb{R}_q \cap \mathbb{R}^+ \to \mathbb{R}$, one obtains the following results under S_q transform:

$$q^{k-1}g(q^{k-1}) \xrightarrow{S_q} \frac{f(q^{n+1}) - f(q^n)}{q^n} \qquad \frac{g(q^k) - g(q^{k-1})}{q^k} \xrightarrow{S_q} q^n f(q^n) \quad (17)$$

where $k, n \in \mathbb{Z}$. Let $\alpha \in \mathbb{C}$ and let g fulfill

$$\alpha q^{k-1} g(q^{k-1}) + \frac{g(q^{k-1}) - g(q^k)}{q^{k-1}} = 0 \qquad k \in \mathbb{Z} \quad (18)$$

Due to (17) now follows that $f = C_q g$ fulfills the difference equation

$$\alpha \frac{f(q^{n+1}) - f(q^n)}{q^n} - q\, q^n f(q^n) = 0 \qquad n \in \mathbb{Z} \quad (19)$$

Rewriting (18)(19) in terms of the operator D_q, one obtains the statement after (15) by choosing $\alpha = q^{\frac{1}{2}}$. Concerning the existence of the C_q-transform $f = C_q g$ of a solution to (18), one can easily show that any solutions to (18) with $\alpha = q^{\frac{1}{2}}$ are in $L^2(\mathbb{R}_q)$, hence also $f = C_q g \in L^2(\mathbb{R}_q)$ due to (19). This concludes the proof of Lemma 1.

So far, we have seen that the classical continuum situation of Fourier invariance (16) survives in the sense of Lemma 1 on $\mathbb{R}_q \cap \mathbb{R}^+$. To elaborate more analogies between the continuum situation and the q-discrete situation and to study suitable limit transitions $q \to 1$ of (15), let us rescale $z \mapsto y := \sqrt{2(1-q)}q^{-\frac{1}{4}}z$ and extend the new equation to the whole $\mathbb{R}_q$ to obtain the rescaled difference equation

$$(D_q g)(y) = -2y g(y) \qquad y \in \mathbb{R}_q \quad (20)$$

Lemma 2

Let ψ_q be a positive and nontrivial solution to equation (20), satisfying in addition $\psi_q(q^k) = \psi_q(-q^k)$ where $k \in \mathbb{Z}$. Let ψ be a positive and nontrivial solution to the corresponding differential equation

$$(D\psi)(x) = -2x\,\psi(x) \qquad x \in \mathbb{R} \quad (21)$$

Let the functions $H_n : \mathbb{R} \to \mathbb{R}$, $H_n^q : \mathbb{R}_q \to \mathbb{R}$, $n \in \mathbb{N}_0$ be given by

$$H_n(x) := (-1)^n (\psi(x))^{-1}(D^n \psi)(x) \quad H_n^q(y) := (-1)^n (\psi_q(y))^{-1}(D_q^n \psi_q)(y) \quad (22)$$

where $x \in \mathbb{R}$, $y \in \mathbb{R}_q$. These functions satisfy the recurrence relations

$$H_{n+1}(x) - 2x H_n(x) + 2n H_{n-1}(x) = 0 \qquad n \in \mathbb{N}_0 \qquad x \in \mathbb{R} \quad (23)$$

$$H_{n+1}^q(y) - 2q^n y H_n^q(y) + 2\frac{1-q^n}{1-q} H_{n-1}^q(y) = 0 \quad n \in \mathbb{N}_0 \qquad y \in \mathbb{R}_q \quad (24)$$

thus yielding the recurrence relations for the classical continuous Hermite polynomials and some of its special discrete q-modifications. In addition, the following mutual orthogonality relations hold if $m, n \in \mathbb{N}_0, m \neq n$

$$\int_{-\infty}^{\infty} H_m(x) \, H_n(x) \, \psi(x) \, dx = 0 \tag{25}$$

and in the q-discrete scenario (the parameter $c \in \mathbb{R} \setminus \{0\}$ being arbitrary):

$$\sum_{k=-\infty}^{\infty} \left(H_m^q(cq^k) \, H_n^q(cq^k) + H_m^q(-cq^k) \, H_n^q(-cq^k) \right) \psi_q(cq^k) \, (1-q)q^k = 0 \tag{26}$$

Proof

The proof for (25) is standard. Proving (26) is done by referring to the so-called discrete q-Hermite polynomials of type II which are defined on $\mathbb{R}$ by the requirements $h_{-1}(x) := 0$, $h_0(x) = 1$ and by the recurrence relation

$$xh_n(x) = h_{n+1}(x) + q^{-2n+1}(1 - q^n)h_{n-1}(x) \quad x \in \mathbb{R} \quad n \in \mathbb{N}_0 \tag{27}$$

Using rescaling arguments and the discrete orthogonality relation for the polynomials (27) — see the classification report [4] — one arrives with (26). The recurrence relation (24) is derived by successive use of the q-product rule

$$D_q(fg) = (D_q f)(R_q g) + (D_q g)f \tag{28}$$

which holds for any functions $f, g : \mathbb{R}_q \to \mathbb{C}$. This concludes the proof for Lemma 2.

To investigate the limit transition $q \to 1$ of non-vanishing symmetric solutions to (20), we need some tools which are introduced in the following.

Definition 1

Let $\rho \in (0,1)$ and let $f \in L^2(\mathbb{R})$. Let $S_\rho f : \mathbb{R} \setminus \{0\} \to \mathbb{R}$ be given by

$$(S_\rho f)(x) := f(sign(x)\rho^k) \quad \text{if} \quad |x| \in (\rho^{k+1}, \rho^k] \quad k \in \mathbb{Z}$$

Moreover we define the bounded linear operator $T_\rho : L^2(\mathbb{R}) \to L^2(\mathbb{R})$ and the embedding $J_\rho : L^2(\mathbb{R}_\rho) \to L^2(\mathbb{R})$ by

$$T_\rho f := S_\rho f \quad \text{if} \quad S_\rho f \in L^2(\mathbb{R}) \quad \text{and} \quad \text{else:} \quad T_\rho f := f$$

$$x \mapsto (J_\rho g)(x) := g(\rho^k) \quad \text{if} \quad x \in (\rho^{k+1}, \rho^k] \quad k \in \mathbb{Z}$$

$$x \mapsto (J_\rho g)(x) := g(-\rho^k) \quad \text{if} \quad x \in [-\rho^k, -\rho^{k+1}) \quad k \in \mathbb{Z}$$

These tools now enable us to establish a theorem for the limit transition $q \to 1$ of positive symmetric solutions to (20).

Theorem 1

Let $(q_n)_{n\in\mathbb{N}} \subseteq (0,1)$ be a monotonously increasing sequence of numbers in $(0,1)$ with $\lim_{n\to\infty} q_n = 1$ and let for $n \in \mathbb{N}$ the functions $\psi_n \in L^2(\mathbb{R}_{q_n})$ solve

$$(D_{q_n}\psi_n)(y) + 2y\psi_n(y) = 0 \qquad y \in \mathbb{R}_{q_n} \qquad n \in \mathbb{N} \qquad (29)$$

with additional properties

$$\forall z \in \mathbb{R}_{q_n} : \ \psi_n(z) = \psi_n(-z) \qquad \lim_{z\to 0} \psi_n(z) = 1 \qquad n \in \mathbb{N} \qquad (30)$$

Let moreover $\psi : \mathbb{R} \to \mathbb{R}$, $x \mapsto \psi(x) := e^{-x^2}$. Then for all $n \in \mathbb{N}$ the following L^2-convergence situation holds:

$$||J_{q_n}\psi_n - \psi||_{L^2(\mathbb{R})} < \infty \qquad \lim_{n\to\infty} ||J_{q_n}\psi_n - \psi||_{L^2(\mathbb{R})} = 0 \qquad (31)$$

Proof

During the proof, we will omit the $L^2(\mathbb{R})$-symbol at the norms $||\circ||$. First, the theory of analytic functions guarantees that there exists a unique sequence of holomorphic even functions $u_n : \mathbb{C} \to \mathbb{C}$ with $u_n(0) = 1$, $n \in \mathbb{N}$ and

$$u_n(z) = (1 + 2(1 - q_n) z^2) u_n(q_n z) \qquad z \in \mathbb{C} \qquad (32)$$

The functions $\varphi_n : \mathbb{R} \to \mathbb{R}$, $x \mapsto \varphi_n(x) := (u_n(x))^{-1}$ are in $L^2(\mathbb{R})$ satisfying

$$\varphi_n(q_n x) = (1 + 2(1 - q_n) x^2) \varphi_n(x) \qquad n \in \mathbb{N} \qquad x \in \mathbb{R} \qquad (33)$$

$$\forall x \in \mathbb{R} : \ \lim_{n\to\infty} \varphi_n(x) = \psi(x) = e^{-x^2} \quad and \quad \lim_{n\to\infty} ||\varphi_n - \psi|| = 0 \qquad (34)$$

We define now $\psi_n : \mathbb{R}_{q_n} \to \mathbb{R}$ by

$$y \mapsto \psi_n(y) := \varphi_n(y) \qquad n \in \mathbb{N} \qquad y \in \mathbb{R}_{q_n} \qquad (35)$$

The ψ_n, $n \in \mathbb{N}$ satisfy (29) and we have

$$(T_{q_n}\varphi_n)(x) = (J_{q_n}\psi_n)(x) \qquad x \in \mathbb{R} \setminus \{0\} \qquad (36)$$

$$\forall n \in \mathbb{N}, \ \forall x \in \mathbb{R} \setminus \{0\} : \ (J_{q_n}\psi_n)(x) - \psi(x) = (T_{q_n}\varphi_n)(x) - \psi(x) \qquad (37)$$

$$\forall n \in \mathbb{N} : \ ||J_{q_n}\psi_n - \psi|| \leq ||T_{q_n}\varphi_n - T_{q_n}\psi|| + ||T_{q_n}\psi - \psi|| \qquad (38)$$

It can be shown that the linear operators T_{q_n} are bounded with induced operator norm $||T_{q_n}||$ and hence we have

$$||J_{q_n}\psi_n - \psi|| \leq ||T_{q_n}|| \ ||\varphi_n - \psi|| + ||T_{q_n}\psi - \psi|| \qquad (39)$$

We have L^2-convergence of $T_{q_n}u$ to u for any $u \in L^2(\mathbb{R})$. Therefore, according to the Banach-Steinhaus theorem, the sequence of norms $||T_{q_n}||$ is bounded. Using the second statement in (34) and taking into account that $\lim_{n\to\infty} ||T_{q_n}\psi - \psi|| = 0$, we finally arrive with the result

$$\lim_{n\to\infty} ||J_{q_n}\psi_n - \psi|| = 0 \qquad (40)$$

This completes the proof of the theorem demonstrating that the embeddings of the discrete functions ψ_n, $n \in \mathbb{N}$ converge to the function $x \mapsto \psi(x) = e^{-x^2}$ even in the strong L^2-sense.

Having stated so many analogies between the q-discrete difference scenario and the continuum situation, we now want to work out a crucial difference between both approaches which shall be done in Theorem 2. It will reveal some interesting aspects of the q-discrete Fourier theory in applications to research on special partial differential equations.

Definition 2

*Let $\psi_q : \mathbb{R}_q \to \mathbb{R}$ resp. $\psi : \mathbb{R} \to \mathbb{R}$ be positive symmetric solutions to (20) resp. (21) and let the continuous Hermite polynomials H_n resp. q-discrete Hermite polynomials H_n^q be given by (22). We denote the finite linear complex spans of the first n continuum **Hermite functions** $H_j \psi, j = 1...n$ by V_n, the finite linear complex spans of the first n **discrete q-Hermite functions** $H_j^q \psi_q, j = 1...n$ by V_n^q, of the first n continuum functions $H_j \sqrt{\psi}, j = 1...n$ by W_n and of the first n discrete functions $H_j^q \sqrt{\psi_q}, j = 1...n$ by W_n^q.*

Theorem 2

*The following statements on **formally self-adjoint** equations hold:*

$$(-D + 2X)(D + 2X)h_n = 4nh_n \qquad n \in \mathbb{N}_0 \qquad (41)$$

with $h_n := (-D + 2X)^n \psi \in V_n$ and therefore $(-D + 2X)(D + 2X)V_n \subseteq V_n$. However, for the discrete situation, there is no such polynomial mapping property but a shift situation occurs in the index:

$$(-q^{-1}L_q D_q + 2X)(D_q + 2X)V_n^q \subset V_{n+2}^q \qquad n \in \mathbb{N}_0 \qquad (42)$$

For all functions $f, g \in L^2(\mathbb{R}_q)$ with compact support there is a formal symmetry property with respect to the $L^2(\mathbb{R}_q)$-scalar product:

$$((-q^{-1}L_q D_q + 2X)(D_q + 2X)f, g) = (f, (-q^{-1}L_q D_q + 2X)(D_q + 2X)g)$$

Moreover, we have the following situation in the continuum:

$$\left(-D - \left(\frac{D\sqrt{\psi}}{\sqrt{\psi}}\right)(X)\right)\left(D - \left(\frac{D\sqrt{\psi}}{\sqrt{\psi}}\right)(X)\right)v_n = 2nv_n \qquad n \in \mathbb{N}_0 \qquad (43)$$

where $v_n := \left(-D - \left(\frac{D\sqrt{\psi}}{\sqrt{\psi}}\right)(X)\right)^n \sqrt{\psi} \in W_n$, $(-D + X)(D + X)W_n \subseteq W_n$ and there exists a discrete q-analog (n ranging again in $\mathbb{N}_0$):

$$\left(-D_q - \left(\frac{D_q\sqrt{\psi_q}}{\sqrt{\psi_q}}\right)(X)R_q\right)\left(q^{-1}L_q D_q - q^{-1}L_q\left(\frac{D_q\sqrt{\psi_q}}{\sqrt{\psi_q}}\right)(X)\right)v_n^q = \frac{q^{2n} - 1}{q - 1}v_n^q \tag{44}$$

where $v_n^q := \left(-D_q - \left(\frac{D_q\sqrt{\psi_q}}{\sqrt{\psi_q}}\right)(X)R_q\right)^n \sqrt{\psi_q} \subseteq W_n^q \subseteq L^2(\mathbb{R}_q)$, and hence finally

$$\left(-D_q - \left(\frac{D_q\sqrt{\psi_q}}{\sqrt{\psi_q}}\right)(X)R_q\right)\left(q^{-1}L_q D_q - q^{-1}L_q\left(\frac{D_q\sqrt{\psi_q}}{\sqrt{\psi_q}}\right)(X)\right)W_n^q \subseteq W_n^q \tag{45}$$

Proof

The proof of the different statements consists of straightforward calculations which are elementary, using essentially the q-product rule (28), but quite tedious.

Future perspectives and interpretations

The results in Theorem 2 basically reveal different scenarios of eigensolutions to q-discrete Schrödinger equations. The considered equations are all formally self-adjoint. We see that in the q-discrete scenario, there seem to be much more "integrable Schrödinger potentials" than in the corresponding continuum situation. New scenarios with respect to eigensolutions occur and give hope to understanding q-difference Schrödinger equations as regularizations to their continuous counterparts. This will eventually offer an approach to applications of q-difference equations to concrete current research on molecular dynamics. In particular, various topics of conventional quantum mechanics like the convergence behavior of Rayleigh-Schrödinger perturbation series should be investigated in context of q-difference equations. Much work has to be done on this new area.

References

[1] R. Askey, N. M. Atakishiyev, S. K. Suslov: An analog of the Fourier transformation for a Q-harmonic oscillator, In: *Symmetries in Science VI* (B. Gruber, Ed.), Plenum Press, New York, (1993), 57–63

[2] C. Berg, A. Ruffing: Generalized q-Hermite polynomials, *Communications in Mathematical Physics* **223**, (2001), 29–46

[3] J. Bustoz, S. K. Suslov: Basic analog of Fourier series on a q-quadratic grid, *Methods and Applications of Analysis 5*, (1998), No. 1, 1–38

[4] R. Koekoek, R.F. Swarttouw: The Askey-scheme of hypergeometric orthogonal polynomials and its q-analogue, Report 98–17, Delft University of Technology, Faculty TWI (1998)

[5] T. H. Koornwinder, R. F. Swarttouw: On q-analogues of the Fourier and Hankel transforms, *Trans AMS* 333 1, (1992), 445

[6] R. Lasser, A. Ruffing: Continuous orthogonality measures and difference ladder operators for discrete q-Hermite polynomials II, preprint 2001

[7] A. Ruffing, M. Witt: On the integrability of q-oscillators based on invariants of discrete Fourier transforms, *Letters in Mathematical Physics* **42**, No 2., (1997), 167–181

Nonoscillation and Oscillation Properties of Fourth Order Nonlinear Difference Equations

EWA SCHMEIDEL

Institute of Mathematics, Poznań University of Technology
ul. Piotrowo 3a, 60-695 Poznań, Poland
e-mail: eschmeid@math.put.poznan.pl

Abstract. We study the equation

$$\Delta^2(r_n\Delta^2 y_n) = f(n, y_{n+2}). \qquad (E)$$

In the Introduction we divide the solutions of (E) into two types: F_+- and F_--solutions. Next, we prove an existence theorem for the linear case. Then some properties of F_+-solutions are given. In Section 2 we investigate the nonoscillatory solutions of (E) and the relations between F_+- and F_--solutions and their nonoscillatory behavior. In Section 3 we turn our attention to some oscillatory properties of (E).

Keywords Difference equations, Fourth order difference equations, Oscillatory solutions, Nonoscillatory solutions

AMS Subject classification 39A10

1 Introduction

Like second order difference equations, the oscillatory and asymptotic behavior of solutions of third order difference equations have been discussed in many papers. For example, note the following papers: [11], [12], [13], [14] and [16]. However, the study of difference equations for orders greater than three has received little attention. Fourth order linear difference equations were considered in [3], [4], [15], [17] and [18]. Fourth order nonlinear difference equations were studied in [6], [8], [10] and [19].

We will study fourth order difference equations of the form

$$\Delta^2(r_n\Delta^2 y_n) = f(n, y_{n+2}), \quad n \in N. \qquad (E)$$

We denote by N the set of positive integers, by R the set of real numbers, by R_+ the set of real positive numbers and by R_- the set of negative numbers.

0-415-31675-8/04/$0.00+$1.50

For a function $x : N \to R$, the forward difference operators are defined as follows: $\Delta x_n = x_{n+1} - x_n, n \in N$ and $\Delta^k x_n = \Delta(\Delta^{k-1} x_n)$, for $k = 2, 3, \ldots$. The sequence y is the trivial sequence if there exists $n_0 \in N$, such that $y_n = 0$, for all $n \geq n_0$. By a solution of (E) we mean any nontrivial sequence y satisfying equation (E) for all $n \in N$.

A solution is oscillatory if for every $m \in N$, there exists $n \geq m$, such that $y_n y_{n+1} \leq 0$. A sequence y is termed quickly oscillatory if and only if $y_n = (-1)^n a_n$ where a is a sequence of positive numbers or negative numbers. We assume that the void sum is equal to zero. In this paper we assume that the function $f : N \times R \to R$ satisfies the condition

$$x f(n, x) < 0, \; for \; n \in N, \; x \in R \setminus \{0\}. \qquad (*)$$

Throughout this paper we assume also that $r : N \to R_+$ and $\sum_{n=1}^{\infty} \frac{n}{r_n} = \infty$.

Following Taylor, we begin our study of equation (E) by considering an operator which plays a vital role in our investigation. Set:

$$F(x_n) = x_{n+1}\Delta(r_n \Delta^2 x_n) - (\Delta x_n)(r_n \Delta^2 x_n).$$

For $r_n \equiv 1$, we obtain the operator used by Popenda and Schmeidel in [8].

Lemma 1.1.

$$\Delta F(x_n) = x_{n+2}\Delta^2(r_n\Delta^2 x_n) - r_n(\Delta^2 x_n)^2.$$

Proof. The proof is obvious and will be omitted.

Lemma 1.2. *Let f satisfy Condition $(*)$. Then the operator $F(y_n)$ is non-increasing for every solution y of (E).*

Proof. Lemma 1.2 follows directly from Lemma 1.1.

If $F(y_n) \geq 0$ for all $n \in N$, then a solution y of equation (E) is called an F_+-solution. If $F(y_n) < 0$ for some n, then y is called an F_--solution.

We use the operator F to classify the solutions of equation (E). The operator F divides the set of solutions into two disjoint subset F_+ and F_--solutions.

We first prove an existence theorem.

Theorem 1.1. *Assume that $f(n, y_{n+2}) = b_n y_{n+2}$ in (E), where $b : N \to R$. Then equation (E) has F_+-solutions.*

Proof. Assume that X, Y, Z and W are four lineary independent solutions of equation (E).
Let

$$y_n^m = a_1^m X_n + a_2^m Y_n + a_3^m Z_n + a_4^m W_n \, , \; \text{for each } m \in N.$$

Choose a_i^m, $i = 1, 2, 3, 4$ in such a way that

$$y_m^m = y_{m+1}^m = y_{m+2}^m = 0, \text{ and } (a_1^m)^2 + (a_2^m)^2 + (a_3^m)^2 + (a_4^m)^2 = 1.$$

Let

$$A_m = (a_1^m, \, a_2^m, \, a_3^m, \, a_4^m).$$

Then $\| A_m \| = 1$. Since the unit ball in R^4 is compact, A_m contains a convergent subsequence $A_{m_i} \to A = (a_1, \, a_2, \, a_3, \, a_4)$, as $i \to \infty$, and $(a_1)^2 + (a_2)^2 + (a_3)^2 + (a_4)^2 = 1$.

Let

$$y_n = \lim_{i \to \infty} y_n^{m_i} = a_1 X_n + a_2 Y_n + a_3 Z_n + a_4 W_n.$$

Since

$$(a_1)^2 + (a_2)^2 + (a_3)^2 + (a_4)^2 = 1,$$

then y is a nontrivial solution of (E).

Now we show that y is F_+-solution. Suppose the contrary. Then there exists an integer r for which $F(y_r) < 0$. Since $y_r^{m_i} \to y_r$, we can infer that $F(y_r^{m_i}) \to F(y_r) < 0$, as $i \to \infty$. Hence, there exists M for which $F(y_r^{m_i}) < 0$ and $m_i > r$, for all $i > M$. Since $F(y_{m_i}^{m_i}) = 0$ and $F(y_r^{m_i}) < 0$, we have that for $m_i > r$, $0 = F(y_{m_i}^{m_i}) < F(y_r^{m_i}) < 0$ which leads to a contradiction. Hence y is an F_+-solution of (E).

For $r_n \equiv 1$, we obtain Theorem 1.2. from [18].

Theorem 1.2. *Assume that f satisfies the Condition $(*)$, r is a nondecreasing sequence and let y be an F_+-solution of (E). Then for $j = 0, 1, 2, \ldots$*

$$\sum_{n=1}^{\infty} r_n (\Delta^{2+j} y_n)^2 < \infty, \tag{i}$$

$$\lim_{n \to \infty} r_n \Delta^{2+j} y_n = 0. \tag{ii}$$

Proof. Let y be an F_+-solution of (E). Let $j = 0$. Then, from Lemma 1.1. and equation (E) we obtain

$$\Delta F(y_k) = y_{k+2} f(k, y_{k+2}) - r_k (\Delta^2 y_k)^2.$$

By summation, we obtain

$$F(y_n) = F(y_1) + \sum_{k=1}^{n-1} y_{k+2} f(k, y_{k+2}) - \sum_{k=1}^{n-1} r_k (\Delta^2 y_k)^2.$$

Since $F(y_n) \geq 0$, we have

$$0 \leq F(y_1) + \sum_{k=1}^{n-1} y_{k+2} f(k, y_{k+2}) - \sum_{k=1}^{n-1} r_k (\Delta^2 y_k)^2.$$

The function f satisfies $(*)$ and so, $\sum\limits_{k=1}^{n-1} y_{k+2} f(k, y_{k+2}) \leq 0$.

Hence, $$\sum_{k=1}^{n-1} r_k (\Delta^2 y_k)^2 \leq F(y_1).$$

Therefore $$\sum_{k=1}^{\infty} r_k (\Delta^2 y_k)^2 \leq F(y_1) < \infty.$$

We have established Condition (i) for $j = 0$.

For $j > 0$ we will prove Theorem 1.2. by induction.

Assume that $\sum\limits_{k=1}^{\infty} r_k (\Delta^{2+j-1} y_k) < \infty$. We need to express $(\Delta^{2+j} y_k)^2$ in terms of lower order differences.

Since for arbitrary reals a, b we have $-2ab \leq a^2 + b^2$, then

$$\begin{aligned}
(\Delta^{2+j} y_k)^2 &= (\Delta^{2+j-1} y_{k+1} - \Delta^{2+j-1} y_k)^2 \\
&= (\Delta^{2+j-1} y_{k+1})^2 - 2\Delta^{2+j-1} y_{k+1} \Delta^{2+j-1} y_k + (\Delta^{2+j-1} y_k)^2 \\
&\leq 2(\Delta^{2+j-1} y_{k+1})^2 + 2(\Delta^{2+j-1} y_k)^2.
\end{aligned}$$

So, $r_k (\Delta^{2+j} y_k)^2 \leq 2 r_k [(\Delta^{2+j-1} y_{k+1})^2 + (\Delta^{2+j-1} y_k)^2]$. Therefore

$$\sum_{k=1}^{\infty} r_k (\Delta^{2+j} y_k)^2 \leq 2 \sum_{k=1}^{\infty} [r_k (\Delta^{2+j-1} y_{k+1})^2 + r_k (\Delta^{2+j-1} y_k)^2]$$

$$\leq 2 \sum_{k=1}^{\infty} [r_{k+1} (\Delta^{2+j-1} y_{k+1})^2 + r_k (\Delta^{2+j-1} y_k)^2] \leq 4 \sum_{k=1}^{\infty} r_k (\Delta^{2+j-1} y_k)^2 < \infty.$$

We proved that $\sum\limits_{k=1}^{\infty} r_k (\Delta^{2+j} y_k)^2 < \infty$, and so (i) holds for all $j \geq 0$.

Condition (ii) follows directly from (i).

2   Nonoscillation results

In this part some nonoscillation results are given.

The next Lemma which was proved by Taylor in [18] plays an important role in our investigations.

Lemma 2.1. *If $\Delta^M y_n > 0$ and $\Delta^{M+1} y_n > 0$ for all $n \geq K$, then*

$$\lim_{n \to \infty} \Delta^{M-1} y_n = \lim_{n \to \infty} \Delta^{M-2} y_n = \ldots = \lim_{n \to \infty} y_n = \infty.$$

Remark 2.1. *Assume that eventually,* $\Delta^M y_n < 0$ *and* $\Delta^{M+1} y_n < 0$*. Then*

$$\lim_{n \to \infty} \Delta^{M-1} y_n = \lim_{n \to \infty} \Delta^{M-2} y_n = ... = \lim_{n \to \infty} y_n = -\infty.$$

The next result from [6] focuses on the behavior of nonoscillatory solutions.

Theorem 2.1. *Assume that function f satisfies condition $(*)$. Every nonoscillatory solution y of equation (E) can be one of the types:*

$$
\begin{array}{llllll}
y_n > 0, & \Delta y_n > 0, & \Delta^2 y_n > 0, & \Delta(r_n \Delta^2 y_n) > 0, & \Delta^2(r_n \Delta^2 y_n) < 0, & (A4+) \\
y_n < 0, & \Delta y_n < 0, & \Delta^2 y_n < 0, & \Delta(r_n \Delta^2 y_n) < 0, & \Delta^2(r_n \Delta^2 y_n) > 0, & (A4-) \\
y_n > 0, & \Delta y_n > 0, & \Delta^2 y_n < 0, & \Delta(r_n \Delta^2 y_n) > 0, & \Delta^2(r_n \Delta^2 y_n) < 0, & (A2+) \\
y_n < 0, & \Delta y_n < 0, & \Delta^2 y_n > 0, & \Delta(r_n \Delta^2 y_n) < 0, & \Delta^2(r_n \Delta^2 y_n) > 0. & (A2-)
\end{array}
$$

A solution y of (E) is called an $(A2)$-solution, if y is an $(A2+)$-solution or an $(A2-)$-solution.

A solution y of (E) is called an $(A4)$-solution, if y is an $(A4+)$-solution or an $(A4-)$-solution.

Theorem 2.2. *Assume that the function f satisfies the Condition $(*)$ and that r is a nondecreasing sequence. Then every nonoscillatory solution y of (E) is an F_+-solution if and only if y is an $(A2)$-solution.*

Proof. We prove Theorem for an eventually positive solution. (For negative solution the proof is similar.)

Let y be an eventually positive F_+-solution. Suppose for the sake of contradiction that it is an $(A4)$-solution. Then from $\Delta(r_n \Delta^2 y_n) > 0$ we get $r_n \Delta^2 y_n > r_M \Delta^2 y_M > 0$, for $n > M$. This inequality contradicts Condition (ii) of Theorem 1.2. So, y is an $(A2+)$-solution.

Let y be an $(A2+)$-solution. We will show the positivity of the operator F on the whole sequence. Choose m sufficiently large. Then from the definition of $(A2+)$-solution we have $F(y_n) > 0$ for $n \geq m$. By Lemma 1.2. the operator F is nonincreasing. Hence $F(y_j) \geq F(y_m) > 0$ for all $j < m$. Since m was taken arbitrary then $F(y_n) > 0$ for all $n \in N$. So, y is an F_+-solution.

If we set in Theorem 2.2. $r_n \equiv 1$, we get Theorem 2 from [8].

Theorem 2.2. shows that the operator F divides the set of nonoscillatory solutions of (E) into two disjoint subsets: F_+-solutions which are the same as $(A2)$-solutions and F_--solutions which are the same as $(A4)$-solutions.

Theorem 2.3. *Assume that the function f satisfies condition $(*)$ and that r is a nondecreasing sequence. Then every nonoscillatory solution y of (E) such that*

$$\sum_{n=1}^{\infty} r_n |\Delta^2 y_n| < \infty$$

is an F_+-solution.

Proof. Let y be a nonoscillatory solution of (E) such that $\sum\limits_{n=1}^{\infty} r_n|\Delta^2 y_n| < \infty$. Suppose that y is an F_--solution. Then there exists $m \in N$ such that $F(y_m) < 0$. Hence by Lemma 1.2. $F(y_n) \leq F(y_m) \leq 0$ for $n \geq m$. Since y is nonoscillatory, then from Theorem 2.1. y can be an $(A2)$ or $(A4)$-solution. We will exclude both of these cases. For an $(A2)$-solution there exists $M \in N$ such that $F(y_n) > 0$ for all $n \geq M$. This is a contradiction. Next, let y be $(A4+)$-solution. (For $(A4-)$-solution the proof is analogous.) From Theorem 2.1. we have $\Delta(r_n\Delta^2 y_n) > 0$. So we get $\sum\limits_{n=1}^{\infty} r_n\Delta^2 y_n = \infty$. This contradiction completes the proof.

3   Oscillation results

Let f satisfy the condition

$$f(n, x) \leq b_n x, \text{ where } b : N \to R_-. \tag{$**$}$$

Theorem 3.1. *Assume that Condition $(**)$ holds and that r is nondecreasing sequence. If $\sum\limits_{i=1}^{\infty} | b_i |= \infty$, then all solutions of (E) are oscillatory.*

Proof. Let $\sum\limits_{i=1}^{\infty} | b_i |= \infty$. Suppose that (E) has nonoscillatory solutions. Let y be such a solution. Let $y_n > 0$, eventually. (For $y_n < 0$ the proof is similar.) Then from Theorem 2.1., $\Delta y_n > 0$, eventually. So there exists $M \in N$ such that $y_n > 0$ and $\Delta y_n > 0$, for $n > M$.
From (E) and $(**)$ we obtain

$$\Delta(r_n\Delta^2 y_n) - \Delta(r_M\Delta^2 y_M) = \sum_{i=M}^{n-1} f(i, y_{i+2}) \leq \sum_{i=M}^{n-1} b_i y_{i+2} \leq y_M \sum_{i=M}^{n-i} b_i.$$

But $\sum\limits_{i=M}^{\infty} b_i = -\infty$.
Clearly, as $n \to \infty$, then $\Delta(r_n\Delta^2 y_n) \to -\infty$ so, $\Delta(r_n\Delta^2 y_n) < 0$ eventually. But this is impossible since from Theorem 2.1. $\Delta(r_n\Delta^2 y_n) > 0$ eventually. Therefore (E) cannot have nonoscillatory solution.

Theorem 3.2. *Assume that f satisfies Condition $(*)$. Then equation (E) cannot have a quickly oscillation solution.*

Proof. Assume that eventually $a_n > 0$ and suppose that $y_n = (-1)^n a_n$ is a solution of (E).

Then $\Delta^2(r_n\Delta^2 y_n) = (-1)^n[r_{n+2}(a_{n+4} + 2a_{n+3} + a_{n+2})$
$$+2r_{n+1}(a_{n+3} + 2a_{n+2} + a_{n+1})$$
$$+r_n(a_{n+2} + 2a_{n+1} + a_n)].$$

Thus (E) can be written in the form

$(-1)^n[r_{n+2}(a_{n+4} + 2a_{n+3} + a_{n+2}) + 2r_{n+1}(a_{n+3} + 2a_{n+2} + a_{n+1})$
$$+r_n(a_{n+2} + 2a_{n+1} + a_n)] = f(n, (-1)^{n+2}a_{n+2}).$$

Hence for n even,

$[r_{n+2}(a_{n+4} + 2a_{n+3} + a_{n+2}) + 2r_{n+1}(a_{n+3} + 2a_{n+2} + a_{n+1})$
$$+r_n(a_{n+2} + 2a_{n+1} + a_n)] = f(n, a_{n+2}),$$

where

$[r_{n+2}(a_{n+4} + 2a_{n+3} + a_{n+2}) + 2r_{n+1}(a_{n+3} + 2a_{n+2} + a_{n+1})$
$$+r_n(a_{n+2} + 2a_{n+1} + a_n)] > 0 \text{ and from } (*), f(n, a_{n+2}) < 0.$$

On the other hand, for n odd

$-[r_{n+2}(a_{n+4} + 2a_{n+3} + a_{n+2}) + 2r_{n+1}(a_{n+3} + 2a_{n+2} + a_{n+1})$
$$+r_n(a_{n+2} + 2a_{n+1} + a_n)] = f(n, -a_{n+2}),$$

where

$-[r_{n+2}(a_{n+4} + 2a_{n+3} + a_{n+2}) + 2r_{n+1}(a_{n+3} + 2a_{n+2} + a_{n+1})$
$$+r_n(a_{n+2} + 2a_{n+1} + a_n)] < 0 \text{ and } f(n, -a_{n+2}) > 0.$$

These contradictions prove our Theorem.

References

[1] R.P. Agarwal, *Difference Equations and Inequalities. Theory, Methods and Applications*, Marcel Dekker, Inc. New York 1992.

[2] S.N. Elaydi, *An Introduction to Difference Equations*, Springer-Verlag., New York 1996.

[3] J.R. Graef, E. Thandapani, Oscillatory and asymptotic behavior of fourth order nonlinear delay difference equations, *Fasciculi Mathematici*, 31(2001), 23–36.

[4] J.W. Hooker, W.T. Patula, Growth and oscillation properties of solutions of a fourth order linear difference equation, *J. Austral. Math. Soc.* Ser. B 26(1985), 310–328.

[5] W.G. Kelly, A.C. Peterson, *Difference Equations*, Academic Press, Inc., Boston-San Diego 1991.

[6] B. Liu, J. Yan, Oscillatory and asymptotic behavior of fourth order nonlinear difference equations, *Acta Mathematica Sinica*, 13(1)(1997), 105–115.

[7] J. Popenda, E. Schmeidel, Nonoscillatory solutions of third order differ-
     ence equations, *Portugaliae Mathematica*, 49(2)(1992), 233–239.

[8] J. Popenda, E. Schmeidel, On the solution of fourth order difference
     equations, *Rocky Mountain Journal of Mathematics*, 25(4)(1995), 1485–
     1499.

[9] E. Schmeidel, Oscillation and nonoscillation theorems for fourth order
     difference equations, *Rocky Mountain Journal of Mathematics* (to ap-
     pear).

[10] E. Schmeidel, B. Szmanda, Oscillatory and asymptotic behavior of cer-
     tain difference equation, *Nonlinear Analysis*, 47(2001), 4731–4742.

[11] B. Smith, Quasi-adjoint third order difference equations: oscillatory and
     asymptotic behavior, *Internat. J. Math. — Math. Sci.*, 9(4)(1986), 781–
     784.

[12] B. Smith, Oscillation and nonoscillation theorems for third order quasi-
     adjoint difference equations, *Portugaliae Mathematica*, 45(3)(1988), 229–
     243.

[13] B. Smith, Oscillatory and asymptotic behavior in certain third order dif-
     ference equations, *Rocky Mountain Journal of Mathematics*, 17(3)(1987),
     597–606.

[14] B. Smith, Linear third order difference equations: oscillatory and asymp-
     totic behavior, *Rocky Mountain Journal of Mathematics*, 22(4)(1992),
     1559–1564.

[15] B. Smith, W.E. Taylor, Oscillatory and asymptotic behavior of certain
     fourth order difference equations, *Rocky Mountain Journal of Mathemat-
     ics*, 16(2)(1986), 403–406.

[16] B. Smith, W.E. Taylor, Asymptotic behavior of solutions of a third order
     difference equation, *Portugaliae Mathematica*, 44(2)(1987), 113–117.

[17] B. Smith, W.E. Taylor, Oscillation properties of fourth order linear dif-
     ference equations, *Tamkang Journal of Mathematics*, 18(4)(1987), 89–95.

[18] W.E. Taylor, Oscillation properties of fourth order difference equations,
     Portugaliae Mathematica, 45(1)(1988), 105–144.

[19] B.G. Zhang, S.S. Cheng, On a class of nonlinear difference equations,
     Journal of Difference Equations and Applications, 1(1995), 391–411.

A Computational Procedure to Generate Difference Equations from Differential Equations

P. G. VAIDYA and SAVITA ANGADI

Mathematical Modelling Unit
National Institute of Advanced Studies
Indian Institute of Science Campus
Bangalore 560012, India
E-mail: pgvaidya@nias.iisc.ernet.in, savita@nias.iisc.ernet.in

Abstract In this paper we have developed methods to compute maps from differential equations. We take two examples. First is the case of the harmonic oscillator and the second is the case of Duffing's equation. First we convert these equations to a canonical form. This is slightly nontrivial for the Duffing's equation. Then we show a method to extend these differential equations. In the second case, symbolic algebra needs to be used. Once the extensions are accomplished, various maps are generated. The Poincaré sections are seen as a special case of such generated maps. Other applications are also discussed.

Keywords Differential equations, Duffing's equation, Canonical form, Invariant manifold, Derivative extension

AMS Subject Classification 65L06, 34A34

1 Introduction

The motivation behind this paper stems from the fact that while a large number of systems are modeled with the help of ordinary differential equations, the data used in the analysis of such systems is almost always such that more appropriate models would be those which one based on maps. In this paper we have described a computational procedure to obtain maps from differential equations. This process is at the very heart of the reason why differential equations are so popular. In fact, the most basic of all the numerical methods to solve a differential equation uses a trivial case of such a conversion to maps.

We begin in section three by presenting this result formally. More interesting cases of conversion to maps require the concept of the "Extension" of differential equations. There are two types of such extensions. They are explained next followed by an illustration in the single case of the harmonic oscillator. In that case, we can generate an exact map.

Next, we take the case of Duffing's oscillator. Here a combination of the extensions is used. Combined with a symbolic computation algorithm, maps correct to any order can be calculated. The specific case of a map, correct to the second order in step size, follows next. A brief discussion of applications is included in the end.

2   Canonical Form

Let the differential equation be given by

$$\dot{x} = f(x, t)$$

where $x = [X_0, X_1, X_2, \ldots, X_{n-1}]^T$. In this paper we would assume that the equation is in the form

$$\dot{X}_0 = X_1, \quad \dot{X}_1 = X_2, \quad \ldots \quad \dot{X}_{n-1} = f(X_0, X_1, \ldots, X_{n-1}, t). \tag{1}$$

We would call this form the canonical form. In this form, the rest of the calculations are easier. In most cases, equations in non-canonical form can be reduced to this form.

3   Flow and maps generated by the equation

Figure 1 shows the flow and maps generated by a differential equation. If you start with a point x (in n dimension) at $t = t_0$ then at $t = t_0 + h_1$ you get a point y. This relationship of y to x can be written as a map:

$$y = g(x, t_0, h_1).$$

Multiple composition of such maps gives rise to points z etc.

$$g(x, t_0, h_1 + h_2) = g(g(x, t_0, h_1), t_0 + h_1, h_2).$$

Which simply states that going from x to z gives the same result as first going from x to y and then from y to z. However, in practice a composition might lead to better accuracy than the direct computation.

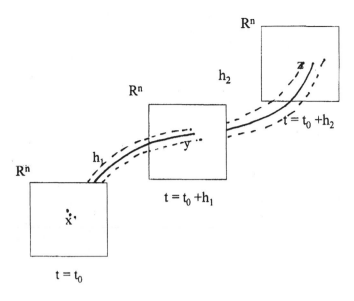

Figure 1. Flows and maps.

4 Euler's method as iterated maps

The most basic of all numerical methods is the Euler's method to compute solutions of differential equations. We can state the procedure as

$$
\begin{bmatrix}
Y_0 \\
Y_1 \\
Y_2 \\
\vdots \\
Y_{n-2} \\
Y_{n-1}
\end{bmatrix}
=
\begin{bmatrix}
X_0 + hX_1 \\
X_1 + hX_2 \\
X_2 + hX_3 \\
\vdots \\
X_{n-2} + hX_{n-1} \\
X_{n-1} + hf(X_0, X_1, X_2 \ldots X_{n-1}, t)
\end{bmatrix}.
$$

This can be seen as a map generated from the equation. The Euler's method consists in repeated compositions of this map. Here the conversion is trivial. For non-trivial conversions the concept of extended differential equations is needed which is discussed next.

5 Extended differential equations

There are two types of extensions. We can illustrate them by taking the simple example of the harmonic oscillator

$$
\dot{X}_0 = X_1, \quad \dot{X}_1 = -\omega^2 X_0.
$$

The first type of extension retains the canonical form but extends the order of the equation. Therefore, the new set of equations could be

$$\dot{X}_0 = X_1, \quad \dot{X}_1 = X_2, \quad \dot{X}_2 = X_3, \quad \dot{X}_3 = -\omega^2 X_2. \tag{2}$$

The apparent order of the equations has increased by two. But, there would be two new invariant manifolds:

$$X_2 + \omega^2 X_0 = 0 \tag{3}$$
$$X_3 + \omega^2 X_1 = 0. \tag{4}$$

The equations of manifolds and the final form of the extended system together represent the system.

In the second type of extension, the canonical form is abandoned for the newer equations. However, the right hand side of the equations continues to be functions of the original set of variables. The information about the invariant manifolds is explicitly carried by the new equations. To continue the same example, we consider

$$\dot{X}_0 = X_1, \quad \dot{X}_1 = -\omega^2 X_0, \quad \dot{X}_2 = -\omega^2 X_1, \quad \dot{X}_3 = -\omega^4 X_0. \tag{5}$$

Later we will choose more interesting examples including Duffing's equation.

Next, we will use the example of the harmonic oscillator to create very accurate maps.

6   Application of the second type of extension to set exact maps for the harmonic oscillator

We would rewrite the second type of extension, indicating in each case the state variable that each of the derivatives represents:

$$\dot{X}_0 = X_1 = X_1, \quad \dot{X}_1 = -\omega^2 X_0 = X_2, \quad \dot{X}_2 = -\omega^2 X_1 = X_3, \tag{6}$$
$$\dot{X}_3 = \omega^4 X_0 = X_4, \quad \dot{X}_4 = \omega^4 X_1 = X_5, \quad \dot{X}_5 = -\omega^6 X_0 = X_6. \tag{7}$$

We now use the Taylor's series to evaluate $Y_0 = X_0(t + h)$, $Y_1 = X_1(t + h)$. These represent the equations for the map x to y, after a time delay of h. From the Taylor's series

$$Y_0 = X_0 + hX_1 + \frac{h^2}{2!}X_2 + \frac{h^3}{3!}X_3 + \cdots \tag{8}$$

$$Y_1 = X_1 + hX_2 + \frac{h^2}{2!}X_3 + \frac{h^3}{3!}X_4 + \cdots \tag{9}$$

After substitution

$$\begin{bmatrix} Y_0 \\ Y_1 \end{bmatrix} = \begin{bmatrix} \cos(\omega h) & \frac{\sin(\omega h)}{\omega} \\ -\omega\sin(\omega h) & \cos(\omega h) \end{bmatrix} \begin{bmatrix} X_0 \\ X_1 \end{bmatrix}.$$

This is an exact map and can be seen as a solution of the equation.

7   Application to Duffing's equation

The canonical (non-autonomous) form of the Duffing's equation is

$$\dot{X}_0 = X_1, \quad \dot{X}_t = -cX_1 - kX_0 - \delta X_0^3 + F\cos(\omega t + \alpha). \tag{10}$$

After a little manipulation we get the first type of extension. This is an autonomous form (still canonical):

$$\dot{X}_0 = X_1, \quad \dot{X}_1 = X_2, \quad \dot{X}_2 = X_3, \quad \dot{X}_3 = f_3(X_0, X_1, X_2, X_3). \tag{11}$$

where

$$f_3 = -cX_3 - kX_2 - 3\delta X_0^2 X_2 - 6\delta X_0 X_1^2 - \omega^2 X_2 \tag{12}$$

$$-\omega^2 cX_1 - \omega^2 kX_0 - \omega^2 \delta X_0^3. \tag{13}$$

Since this is the first type of extension, we would have the following two invariant manifolds which were obtained during the derivation of the above equation:

$$X_2 = cX_1 - kX_0 - \delta(X_0)^3 + F\cos(\omega t + \alpha) \tag{14}$$

$$X_3 = -cX_2 - kX_1 - 3\delta(X_0)^2 X_1 - F\omega\sin(\omega t + \alpha). \tag{15}$$

Thus, although the system has four equations, only the knowledge of two variables is enough to find the four state variables.

Now using the second type of extension, we get

$$\dot{X}_0 = X_1, \quad \dot{X}_1 = X_2, \quad \dot{X}_2 = X_3,$$

$$\dot{X}_3 = f_3(X_0, X_1, X_2, X_3),$$

$$\dot{X}_4 = f_4(X_0, X_1, X_2, X_3). \tag{16}$$

where $f_4(X_0, X_1, X_2, X_3) = \frac{d}{dt}f_3$. Now,

$$\frac{d}{dt}f_3 = \frac{\partial f_3}{\partial X_0}\frac{\partial X_0}{\partial t} + \frac{\partial f_3}{\partial X_1}\frac{\partial X_1}{\partial t} + \frac{\partial f_3}{\partial X_2}\frac{\partial X_2}{\partial t} + \frac{\partial f_3}{\partial X_3}\frac{\partial X_3}{\partial t} \tag{17}$$

Therefore,

$$f_4 = \frac{\partial f_3}{\partial X_0}X_1 + \frac{\partial f_3}{\partial X_1}X_2 + \frac{\partial f_3}{\partial X_2}X_3 + \frac{\partial f_3}{\partial X_3}f_3(X_0, X_1, X_2, X_3). \tag{18}$$

This process can be repeated to any order by symbolical computation.

Symbolic derivative extension scheme:

One possibility of keeping track of all the terms appearing in the symbolic calculation is that a sum of terms can be represented by a matrix. In the matrix each row represents a term of this form:

$$T = \alpha(X_0)^p(X_1)^q(X_2)^r(X_3)^s$$

The coefficient α is in the last column. The first column gives p, the second gives q, the third gives r, the fourth gives r, and the fifth gives s. So that, for example, we could automatically compute

$$\frac{\partial T}{\partial X_1} X_2 = \alpha q (X_0)^p (X_0)^{q-1} (X_2)^{r+1} (X_3)^s$$

We write an algorithm that indicates that α now changes to αq, etc., for example, the right hand side of equation (6).

$$-cX_3 - kX_2 - 3\delta X_0^2 X_2 - 6\delta X_0 X_1^2 - \omega^2 X_2 - \omega^2 c X_1 - \omega^2 k X_0 - \omega^2 \delta X_0^3.$$

This would be represented by

$$\begin{bmatrix} 0 & 0 & 0 & 1 & -c \\ 0 & 0 & 1 & 0 & -k - \omega^2 \\ 2 & 0 & 1 & 0 & -3\delta \\ 1 & 2 & 0 & 0 & -6\delta \\ 0 & 1 & 0 & 0 & -\omega^2 c \\ 1 & 0 & 0 & 0 & -\omega^2 k \\ 3 & 0 & 0 & 0 & -\omega^2 \delta \end{bmatrix}$$

The fourth row represents $-6\delta X_0 X_1^2$. Now

$$\frac{\partial(-6\delta X_0 X_1^2)}{\partial X_1} = (-6\delta)(2)(X_0^1 X_1^1 X_2^0 X_3^0).$$

This operation is seen as reducing the second column by 1 and multiplying the last column by 2. Similarly , the whole expression can be differentiated.

This procedure can then be used to carry out the second type of extension to any desired order: First, using the procedure, derivatives of any order can be computed

$$X_n = q_n((X_0, X_1, X_2, X_3).$$

Then using the invariant manifold equations

$$X_n = q_n[X_0, X_1, X_2(X_0, X_1, t), X_3(X_0, X_1, t)].$$

This can be recast in the form $X_n = \bar{q}_n(X_0, X_1, t)$. Or given X_0, X_1, and t, we can compute X_2, X_3 and follow the algorithm as described above. Full details for the case of the Duffing's equation are given in the longer version of this paper [3].

8   Creating maps of arbitrary accuracy

We now wish to create maps from the Duffing's equation

$$Y_0 = g_0(X_0, X_1) , \quad Y_1 = g_1(X_0, X_1) \tag{19}$$

where the dependence of g on t_0 and h follows from the Taylor's series

$$g_0(X_0, X_1) \;=\; X_0 + hX_1 + \frac{h^2}{2!}X_2 + \frac{h^3}{3!}X_3 + \cdots \tag{20}$$

$$g_1(X_0, X_1) \;=\; X_1 + hX_2 + \frac{h^2}{2!}X_3 + \frac{h^3}{3!}X_4 + \cdots \tag{21}$$

This procedure can be used to any desired order of h. For the sake of brevity we would state the specific result valid up to the second order:

$$g_0(X_0, X_1, t) = \left(h - \frac{1}{2}h^2 c\right)X_1 - \frac{1}{2}h^2 \delta X_0^3 + \left(1 - \frac{1}{2}h^2 k\right)X_0 + \frac{1}{2}h^2 F\cos(\omega t + \alpha)$$

$$g_1(X_0, X_1, t) = \left(h - \frac{1}{2}h^2 c\right)F\cos(\omega t + \alpha) - [cX_1 + kX_0 + \delta X_0^3]$$

$$- \frac{3}{2}h^2 \delta X_0^2 X_1 + \left(1 - \frac{1}{2}h^2 k\right)X_1 - \frac{1}{2}h^2 \omega F\sin(\omega t + \alpha)$$

9   Numerical study

For the Duffing's equation the following parameters were chosen and

$$C = 0.044964 \quad \delta = 1 \quad \omega = 0.44964 \quad F = 1.02 \quad \alpha = 0$$

First, h was chosen to be 0.01397. The error even after a thousand iterations of the map correct only to h^2 is quite small. Figure 2 shows the difference between the predictions from the map vs. the prediction using a Runge Kutta routine. However, if double the step size (h) is used, then the errors increase [3]. Thus as h increases, a greater number of terms is required to reach the same level of accuracy. In theory there are no limitations to the order to which the maps can be calculated.

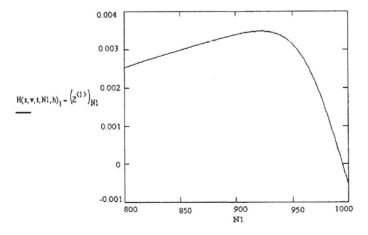

Figure 2. Difference between prediction and calculation.

10   Potential applications

We mentioned earlier that often the equation might involve independent variables, but the data might consist of only part of these variables. When this happens, it is quite common that an estimate of the remaining variables is needed. Thus, for example, in the Duffing's equation, if the data is about position X, information about the velocity V would be needed. We are presenting the results of such a computation below. Consider two samples of X: X_i and X_j, not necessarily consecutive ones.Then, let $h1 = (j - i)h$. From the second order map, we have

$$x_j = (h1 - \frac{h1^2}{2}c)v_i - \frac{h1^2}{2}\delta x_i^3 + (1 - \frac{h1^2}{2}k)x_i + \frac{h1^2}{2}F\cos(\omega t_i + \alpha).$$

Therefore,

$$v_i = \frac{1}{h1 - \frac{h1^2}{2}c}(x_j + \frac{h1^2}{2}\delta x_i^3 - (1 - \frac{h1^2}{2}k)x_i - \frac{h1^2}{2}F\cos(\omega t_i + \alpha)).$$

If we get a map correct to h^3 this expression becomes longer. And the expression, correct to 4th order, is even more lengthy. However, Maple or Mathcad can readily handle it. We have shown below (Figure 4) the results of computations. What is plotted is the difference between the computed value and the "exact" value, as determined by an R-K integration. Here, D2 represents the difference when a map with second order accuracy is used.

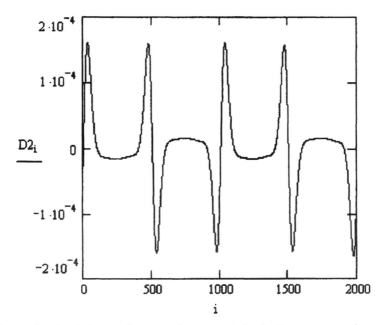

Figure 4. The difference between the computed velocity correct to the second order and the exact value.

When an expression correct to the third order is used (which is easy using computer algebra) errors are of the order of 10^{-6}, and with fourth order accuracy they are of the order of 10^{-7} [3]. Note also that once V is known all other derivatives can be explicitly calculated as discussed before.

There are several other potential applications, with high practical value. We have shown the connection of this procedure to the Euler's method. We can use the higher order maps to create more efficient integration routines. There are many techniques to remove noise and control a dynamical system which is expressed in the form of a map. It is clear that once we convert an equation into maps, those techniques become readily available.

There are certain specific applications. Parameter identification can be carried out quite successfully if the data are sampled at very small intervals [1]. If this method were to be extended to the case when the sampling rate is not very high, maps would be needed. For accurate Fourier analysis, and in chaotic synchronization in flows, high sampling rates are used. Here our recent work is showing further promise using maps, too. In the case of chaotic synchronization the proof of synchronization has so far assumed continuous feedback [2], but in practice often discrete synchronizing pulses are used. The conversion of equations to maps would be helpful.

11 Conclusions

Using the first and second type of extension of differential equations and Taylor series expansion we can construct maps from differential equations to any arbitrary accuracy. This has many potential applications.

12 Acknowledgments

Thanks are due to Professor R. Narasimha of the National Institute of Advanced Studies, Dr. Mythili Ramaswamy of TIFR, Bangalore, for their help in developing the analysis. The work was supported by a grant from the Department of Science and Technology and the Department of Culture.

References

[1] P. G. Vaidya and S. Angadi, Decoding chaotic cryptography without an access to superkey (presented in "Applied Non-linear Dynamics" Conference at Thessaloniki, Greece 27-30/8/2001)

[2] Rong He and P. G. Vaidya, "Analysis and synthesis of synchronous chaotic and periodic systems", *Phys. Rev. A*

[3] P. G. Vaidya and S. Angadi, "A computational procedure to generate difference equations from differential equation and its applications", NIAS Report, (2002) [A longer version of this paper].

Difference Equations for Multiple Charlier and Meixner Polynomials [1]

WALTER VAN ASSCHE

Department of Mathematics
Katholieke Universiteit Leuven
B-3001 Leuven, Belgium
E-mail: walter@wis.kuleuven.ac.be

Abstract We give a third order difference equation for multiple Charlier polynomials and two types of multiple Meixner polynomials. These polynomials are orthogonal with respect to two measures on the integers $\mathbb{N}$.

Keywords Discrete orthogonality, Multiple orthogonal polynomials, Difference equation

AMS Subject Classification 33C45, 39A13, 42C05

1 Discrete orthogonal polynomials

Suppose μ is a discrete positive measure on the real line. We will consider only discrete measures for which the support is (a subset of) $\mathbb{N}$ (the linear lattice). The corresponding monic orthogonal polynomials $\{P_n : n = 0, 1, 2, \ldots\}$ are such that P_n is a monic polynomial of degree n for which

$$\int P_n(x) x^k \, d\mu(x) = 0, \qquad k = 0, 1, 2, \ldots, n - 1.$$

On the linear lattice we have two important operators, namely the *forward difference operator* Δ for which

$$\Delta f(x) = f(x + 1) - f(x),$$

and the *backward difference operator* ∇ for which

$$\nabla f(x) = f(x) - f(x - 1).$$

If we use the Pochhammer symbol $(a)_k = a(a + 1)(a + 2) \cdots (a + k - 1)$, with $(a)_0 = 1$, then the polynomials $(-x)_k$ are much more convenient as building

[1]Research supported by INTAS 00-272 and project G.0184.02 of FWO-Vlaanderen.

blocks than the monomials x^k because of the relation $\Delta(-x)_k = -k(-x)_{k-1}$. The orthogonality is therefore easier to formulate as

$$\int P_n(x)(-x)_k \, d\mu(x) = 0, \qquad k = 0, 1, 2, \ldots, n-1.$$

The classical discrete orthogonal polynomials on the linear lattice consist of the following families [5, 7, 8], where we use δ_k for the Dirac measure with mass 1 at $k \in \mathbb{N}$:

- The Charlier polynomials $C_n(x; a)$ for which

$$\mu = \sum_{k=0}^{\infty} \frac{a^k}{k!} \delta_k,$$

(Poisson distribution), with $a > 0$.

- The Kravchuk polynomials $K_n(x; p, N)$ for which

$$\mu = \sum_{k=0}^{N} \binom{N}{k} p^k (1-p)^{N-k} \delta_k,$$

(binomial distribution), with $0 < p < 1$ and $N \in \mathbb{N}$.

- The Meixner polynomials $M_n(x; \beta, c)$ for which

$$\mu = \sum_{k=0}^{\infty} \frac{(\beta)_k c^k}{k!} \delta_k,$$

(negative binomial distribution), with $\beta > 0$ and $0 < c < 1$.

- The Hahn polynomials $Q_n(x; \alpha, \beta, N)$ for which

$$\mu = \sum_{k=0}^{N} \binom{\alpha+k}{k} \binom{\beta+N-k}{N-k} \delta_k,$$

(hypergeometric distribution), with $\alpha, \beta > -1$ or $\alpha, \beta < -N$, $N \in \mathbb{N}$.

2   Discrete multiple orthogonal polynomials

For multiple orthogonal polynomials one needs $r \geq 2$ measures $\mu_1, \ldots, \mu_r$ on $\mathbb{R}$. The polynomials are indexed by a multi-index $\vec{n} = (n_1, \ldots, n_r) \in \mathbb{N}^r$, with length $|\vec{n}| = n_1 + \ldots + n_r$. A type II multiple orthogonal polynomial is defined as a polynomial $P_{\vec{n}}$ of degree $\leq |\vec{n}|$ so that

$$\int P_{\vec{n}}(x) x^k \, d\mu_1(x) \quad = \quad 0, \quad k = 0, 1, \ldots, n_1 - 1$$

$$\vdots$$

$$\int P_{\vec{n}}(x) x^k \, d\mu_r(x) \quad = \quad 0, \quad k = 0, 1, \ldots, n_r - 1.$$

This gives a linear system of $|\vec{n}|$ homogeneous equations for the $|\vec{n}|+1$ unknown coefficients of $P_{\vec{n}}$. The index $\vec{n}$ is said to be *normal* if $P_{\vec{n}}$ is unique (up to a multiplicative factor) and has exactly degree $|\vec{n}|$. In that case we will always consider *monic* polynomials.

In this paper we will only consider discrete measures on (a subset of) $\mathbb{N}$. The orthogonality conditions are then more conveniently written as

$$\sum_{j=0}^{\infty} P_{\vec{n}}(j)(-j)_k \, w_1(j) \;=\; 0, \qquad k = 0, 1, ..., n_1 - 1,$$

$$\vdots$$

$$\sum_{j=0}^{\infty} P_{\vec{n}}(j)(-j)_k \, w_r(j) \;=\; 0, \qquad k = 0, 1, ..., n_r - 1.$$

Systems of measures $(\mu_1, \ldots, \mu_r)$ for which all the multi-indices are normal are known as perfect systems. A useful sufficient condition was given by Nikishin. Let $w_j : [0, \infty) \to [0, \infty)$ be continuous functions. The system $(w_1, w_2, \ldots, w_r)$ is an *algebraic Chebyshev system* (AT system) on $[0, \infty)$ if

$$w_1(x), x w_1(x), \ldots, x^{n_1 - 1} w_1(x), \;\; \ldots \;\; w_r(x), x w_r(x), \ldots, x^{n_r - 1} w_r(x)$$

is a Chebyshev system on $[0, \infty)$ for each multi-index $\vec{n}$ (with $|\vec{n}| < N$). This means that

$$\sum_{j=1}^{r} Q_{n_j - 1}(x) w_j(x)$$

has at most $|\vec{n}| - 1$ zeros on $[0, \infty)$ for any choice of polynomials Q_{n_j-1} of degree at most $n_j - 1$. In an AT system all the $\vec{n}$ with $|\vec{n}| < N$ are normal. All the weights that we use in this paper are AT systems; hence the corresponding monic multiple orthogonal polynomials are unique.

In [4] we introduced five families of discrete multiple orthogonal polynomials (on a linear lattice). For each of these families we gave a Rodrigues formula, and for $r = 2$ we gave an explicit expression and a four-term recurrence relation, which gives a linear relationship between four multiple orthogonal polynomials with multi-indices $(n + 1, m), (n, m), (n, m - 1), (n - 1, m - 1)$. The multiple orthogonal polynomials in each of these five families also satisfy a difference equation of order $r + 1$. In this paper we will show this for the multiple Charlier and Meixner polynomials for $r = 2$, and we give the difference equation explicitly. The general case will be considered elsewhere. This difference equation gives a linear relationship for the polynomial of a given degree evaluated at x, $x + 1$, $x + 2$ and $x + 3$. This means that these multiple orthogonal polynomials satisfy both a recurrence relation (where the degree changes) and a difference equation (where the variable changes). This makes them interesting from the viewpoint of *bispectral problems*.

2.1   Multiple Charlier polynomials

For multiple Charlier polynomials $C_{\vec{n}}^{\vec{a}}(x)$ we use r Poisson measures with different parameters $a_1, \ldots, a_r$:

$$\mu_j = \sum_{k=0}^{+\infty} \frac{a_j^k}{k!} \delta_k, \qquad a_j > 0, \quad j = 1, ..., r,$$

with $a_i \neq a_j$ whenever $i \neq j$. In [4] we showed that there are r raising operators

$$\frac{a_i}{w_i(x)} \nabla \left(w_i(x) C_{\vec{n}}^{\vec{a}}(x) \right) = -C_{\vec{n}+\vec{e}_i}^{\vec{a}}(x), \qquad i = 1, ..., r, \tag{1}$$

where $w_i(x) = a_i^x / \Gamma(x+1)$. An explicit expression for these polynomials for $r = 2$ was given in [4]:

$$(-a_1)^{-n_1} (-a_2)^{-n_2} C_{n_1,n_2}^{a_1,a_2}(x)$$

$$= \sum_{k=0}^{n_1} \sum_{\ell=0}^{n_2} (-n_1)_k (-n_2)_\ell (-x)_{k+\ell} \frac{\left(-\frac{1}{a_1}\right)^k}{k!} \frac{\left(-\frac{1}{a_2}\right)^\ell}{\ell!}. \tag{2}$$

We will now show that there is also a lowering operator.

Theorem 2.1. *For multiple Charlier polynomials (with $r = 2$) we have*

$$\Delta C_{n_1,n_2}^{a_1,a_2}(x) = n_1 C_{n_1-1,n_2}^{a_1,a_2}(x) + n_2 C_{n_1,n_2-1}^{a_1,a_2}(x). \tag{3}$$

Proof. Since we are working in an AT system, the monic multiple orthogonal polynomials are unique. We will first show that

$$\Delta C_{n_1,n_2}^{a_1,a_2}(x) = (n_1+n_2) C_{n_1,n_2-1}^{a_1,a_2}(x) + n_1(a_1-a_2) C_{n_1-1,n_2-1}^{a_1,a_2}(x). \tag{4}$$

If we use summation by parts, then one sees that $\Delta C_{n_1,n_2}^{a_1,a_2}(x)$ is orthogonal to $(-x)_k$ for $0 \leq k \leq n_1 - 2$ with respect to w_1, and orthogonal to $(-x)_k$ for $0 \leq k \leq n_2 - 2$ with respect to w_2. Both $C_{n_1,n_2-1}^{a_1,a_2}(x)$ and $C_{n_1-1,n_2-1}^{a_1,a_2}(x)$ have the same orthogonality conditions, and they are linearly independent. These $n_1 + n_2 - 2$ orthogonality relations show that

$$\Delta C_{n_1,n_2}^{a_1,a_2}(x) = A C_{n_1,n_2-1}^{a_1,a_2}(x) + B C_{n_1-1,n_2-1}^{a_1,a_2}(x),$$

with coefficients A and B that remain to be determined. Comparing the coefficient of $(-x)_{n_1+n_2-1}$ on both sides of this equation shows that

$$A = n_1 + n_2.$$

Comparing coefficients of $(-x)_{n_1+n_2-2}$, for which we use (2) together with $\Delta(-x)_k = -k(-x)_{k-1}$, gives

$$-(n_1 a_1 + n_2 a_2)(n_1 + n_2 - 1) = -A(n_1 a_1 + (n_2-1)a_2) + B,$$

so that $B = n_1(a_1 - a_2)$. This shows that (4) indeed holds. Interchanging the role of a_1 and a_2 and the role of n_1 and n_2 (observe that $C_{n_1,n_2}^{a_1,a_2}(x) = C_{n_2,n_1}^{a_2,a_1}(x)$ always holds) also gives

$$\Delta C_{n_1,n_2}^{a_1,a_2}(x) = (n_1 + n_2)C_{n_1-1,n_2}^{a_1,a_2}(x) - n_2(a_1 - a_2)C_{n_1-1,n_2-1}^{a_1,a_2}(x). \quad (5)$$

Elimination of $C_{n_1-1,n_2-1}^{a_1,a_2}(x)$ from (4)–(5) then gives the required relation (3). □

A combination of the raising operators (1) and the lowering operator (3) gives the difference equation.

Theorem 2.2. *The multiple Charlier polynomials (with $r = 2$) satisfy the third order difference equation*

$$a_1 a_2 \nabla_1 \nabla_2 \Delta C_{n_1,n_2}^{a_1,a_2}(x) = -\left[a_1 n_2 \nabla_1 + a_2 n_1 \nabla_2\right] C_{n_1,n_2}^{a_1,a_2}(x), \quad (6)$$

where

$$\nabla_i = \frac{\Gamma(x + 1)}{a_i^x} \nabla \frac{a_i^x}{\Gamma(x+1)}.$$

Proof. First of all, observe that $\nabla_1 \nabla_2 = \nabla_2 \nabla_1$, so that ∇_1 and ∇_2 are commuting operators. Apply the product $\nabla_1 \nabla_2$ to (3), then

$$\nabla_1 \nabla_2 \Delta C_{n_1,n_2}^{a_1,a_2}(x) = n_1 \nabla_2 \nabla_1 C_{n_1-1,n_2}^{a_1,a_2}(x) + n_2 \nabla_1 \nabla_2 C_{n_1,n_2-1}^{a_1,a_2}(x).$$

Now use the raising operation (1); then the required result (6) follows. □

2.2 Multiple Meixner polynomials of the first kind

Meixner polynomials have two parameters $\beta > 0$ and $0 < c < 1$. We can define two kinds of multiple Meixner polynomials by keeping one parameter fixed and by changing the other parameter [4]. If we keep β fixed and use the measures

$$\mu_i = \sum_{k=0}^{+\infty} \frac{(\beta)_k c_i^k}{k!} \delta_k, \qquad i = 1, \ldots, r,$$

where $c_i \neq c_j$ whenever $i \neq j$, then we get multiple Meixner polynomials of the first kind $M_{\vec{n}}^{\beta;\vec{c}}(x)$. In [4] we showed the existence of r raising operators

$$\frac{c_i}{(1 - c_i)w_i(x; \beta - 1)} \nabla \left(w_i(x; \beta) M_{\vec{n}}^{\beta;\vec{c}}(x)\right) = -M_{\vec{n}+\vec{e}_i}^{\beta;\vec{c}}(x), \qquad i = 1, \ldots, r, \quad (7)$$

where

$$w_i(x; \beta) = \frac{\Gamma(\beta + x)c_i^x}{\Gamma(x + 1)}.$$

An explicit expression for these polynomials for $r = 2$ was given in [4]:

$$
\left(\frac{c_1 - 1}{c_1}\right)^{n_1} \left(\frac{c_2 - 1}{c_2}\right)^{n_2} M_{n_1,n_2}^{\beta;c_1,c_2}(x)
$$

$$
= (\beta)_{n_1+n_2} \sum_{j=0}^{n_1+n_2} \sum_{k=0}^{j} \frac{(-n_1)_k(-n_2)_{j-k}}{(\beta)_j} \frac{\left(\frac{c_1-1}{c_1}\right)^k}{k!} \frac{\left(\frac{c_2-1}{c_2}\right)^{j-k}}{(j-k)!}(-x)_j. \quad (8)
$$

We will now also give a lowering operator for these polynomials.

Theorem 2.3. *Multiple Meixner polynomials of the first kind (for $r = 2$) have the lowering operator*

$$
\Delta M_{n_1,n_2}^{\beta;c_1,c_2}(x) = n_1 M_{n_1-1,n_2}^{\beta+1;c_1,c_2}(x) + n_2 M_{n_1,n_2-1}^{\beta+1;c_1,c_2}(x), \quad (9)
$$

Proof. We first show that

$$
\Delta M_{n_1,n_2}^{\beta;c_1,c_2}(x) = (n_1 + n_2) M_{n_1,n_2-1}^{\beta+1;c_1,c_2}(x)
$$

$$
+ n_1 \left(\frac{c_1}{1-c_1} - \frac{c_2}{1-c_2}\right)(\beta + n_1 + n_2 - 1) M_{n_1-1,n_2-1}^{\beta+1;c_1,c_2}(x). \quad (10)
$$

Summation by parts shows that $\Delta M_{n_1,n_2}^{\beta;c_1,c_2}(x)$ is orthogonal to $(-x)_k$ for $0 \leq k \leq n_1 - 2$ with respect to $w_1(x; \beta + 1)$, and orthogonal to $(-x)_k$ for $0 \leq k \leq n_2 - 2$ with respect to $w_2(x; \beta + 1)$. Both polynomials $M_{n_1,n_2-1}^{\beta+1;c_1,c_2}(x)$ and $M_{n_1-1,n_2-1}^{\beta+1;c_1,c_2}(x)$ have the same orthogonality properties and are linearly independent. Hence we can write $\Delta M_{n_1,n_2}^{\beta;c_1,c_2}(x)$ as a linear combination of these two polynomials

$$
\Delta M_{n_1,n_2}^{\beta;c_1,c_2}(x) = A M_{n_1,n_2-1}^{\beta+1;c_1,c_2}(c) + B M_{n_1-1,n_2-1}^{\beta+1;c_1,c_2}(x).
$$

The $n_1 + n_2 - 2$ orthogonality conditions and the two parameters A, B then completely determine the $n_1 + n_2$ coefficients of the polynomial $\Delta M_{n_1,n_2}^{\beta;c_1,c_2}(x)$. Comparing coefficients of $(-x)_{n_1+n_2-1}$ on both sides of the equation gives $A = n_1 + n_2$. Comparing coefficients of $(-x)_{n_1+n_2-2}$, which can be done by using (8), gives

$$
-(n_1 + n_2 - 1)(n_1 \frac{c_1}{1-c_1} + n_2 \frac{c_2}{1-c_2})(\beta + n_1 + n_2 - 1)
$$

$$
= -(n_1 + n_2)\left(n_1 \frac{c_1}{1-c_1} + (n_2 - 1)\frac{c_2}{1-c_2}\right)(\beta + n_1 + n_2 - 1) + B,
$$

which gives $B = n_1[c_1/(1-c_1) - c_2/(1-c_2)](\beta + n_1 + n_2 - 1)$, and thus (10). Changing the role of c_1 and c_2 and the role of n_1 and n_2 also gives

$$
\Delta M_{n_1,n_2}^{\beta;c_1,c_2}(x) = (n_1 + n_2) M_{n_1-1,n_2}^{\beta+1;c_1,c_2}(x)
$$

$$
- n_2 \left(\frac{c_1}{1-c_1} - \frac{c_2}{1-c_2}\right)(\beta + n_1 + n_2 - 1) M_{n_1-1,n_2-1}^{\beta+1;c_1,c_2}(x). \quad (11)
$$

Eliminating $M_{n_1-1,n_2-1}^{\beta+1;c_1,c_2}(x)$ from (10)–(11) then gives the required equation (9). $\qquad\square$

A combination of the raising operators and the lowering operator then gives the difference equation.

Theorem 2.4. *Multiple Meixner polynomials of the first kind (for $r = 2$) satisfy the third order difference equation*

$$a_1 a_2 \nabla_1^\beta \nabla_2^{\beta+1} \Delta M_{n_1,n_2}^{\beta;c_1,c_2}(x) = -\left[a_2 n_1 \nabla_2^\beta + a_1 n_2 \nabla_1^\beta \right] M_{n_1,n_2}^{\beta;c_1,c_2}(x), \qquad (12)$$

where $a_i = c_i/(1 - c_i)$ and

$$\nabla_i^\beta = \frac{\Gamma(x+1)}{\Gamma(\beta+x-1)c_i^x} \nabla \frac{c_i^x \Gamma(\beta+x)}{\Gamma(x+1)}.$$

Proof. Observe first of all that

$$\nabla_1^\beta \nabla_2^{\beta+1} = \nabla_2^\beta \nabla_1^{\beta+1}.$$

Applying $\nabla_1^\beta \nabla_2^{\beta+1}$ to (9) then gives

$$\nabla_1^\beta \nabla_2^{\beta+1} \Delta M_{n_1,n_2}^{\beta;c_1,c_2}(x) = n_1 \nabla_2^\beta \nabla_1^{\beta+1} M_{n_1-1,n_2}^{\beta+1;c_1,c_2}(x) + n_2 \nabla_1^\beta \nabla_2^{\beta+1} M_{n_1,n_2-1}^{\beta+1;c_1,c_2}(x).$$

Now use the raising operations (7), then the required difference equation follows. $\qquad\square$

2.3 Multiple Meixner polynomials of the second kind

Another way to get multiple Meixner polynomials is to keep the parameter c fixed and to change the parameter β. Multiple Meixner polynomials of the second kind $M_{\vec{n}}^{\vec{\beta};c}(x)$ are related to the measures

$$\mu_i = \sum_{k=0}^{+\infty} \frac{(\beta_i)_k c^k}{k!} \delta_k, \qquad i = 1, \dots, r,$$

where $\beta_i - \beta_j \notin \mathbb{Z}$ for every $i \neq j$. This condition ensures that these measures form a perfect system, so that all multi-indices are normal. The raising operators are given by [4]

$$\frac{c}{(1-c)w(x;\beta_i-1)} \nabla \left(w(x;\beta_i) M_{\vec{n}}^{\vec{\beta};c}(x) \right) = -M_{\vec{n}+\vec{e}_i}^{\vec{\beta}-\vec{e}_i;c}(x), \qquad i = 1, \dots, r,$$

$$(13)$$

where

$$w(x;\beta) = \frac{\Gamma(\beta+x)c^x}{\Gamma(x+1)}.$$

An explicit expression was given in [4] for $r = 2$:

$$\left(\frac{c-1}{c}\right)^{n_1+n_2} \frac{1}{(\beta_2)_{n_2}(\beta_1)_{n_1}} M_{n_1,n_2}^{\beta_1,\beta_2;c}(x)$$

$$= \sum_{j=0}^{n_1+n_2} \sum_{k=0}^{j} \frac{(-n_1)_k(-n_2)_{j-k}(\beta_1+n_1)_{j-k}}{k!(j-k)!(\beta_2)_{j-k}} \frac{\left(\frac{c-1}{c}\right)^j}{(\beta_1)_j}(-x)_j. \quad (14)$$

The lowering operator for these multiple orthogonal polynomials is given by

Theorem 2.5. *For multiple Meixner polynomials of the second kind, with $r = 2$, one has*

$$(\beta_1 - \beta_2)\Delta M_{n_1,n_2}^{\beta_1,\beta_2;c}(x)$$

$$= n_1(\beta_1 - \beta_2 - n_2)M_{n_1-1,n_2}^{\beta_1+1,\beta_2;c}(x) - n_2(\beta_2 - \beta_1 - n_1)M_{n_1,n_2-1}^{\beta_1,\beta_2+1;c}(x). \quad (15)$$

Proof. If we use summation by parts on the orthogonality relations, then $\Delta M_{n_1,n_2}^{\beta_1,\beta_2;c}(x)$ is seen to be orthogonal to $(-x)_k$ for $0 \leq k \leq n_1 - 2$ with respect to $w(x; \beta_1+1)$ and orthogonal to $(-x)_k$ for $0 \leq k \leq n_2-2$ with respect to $w(x; \beta_2 + 1)$. The polynomials $M_{n_1-1,n_2}^{\beta_1+1,\beta_2;c}(x)$ and $M_{n_1,n_2-1}^{\beta_1,\beta_2+1;c}(x)$ share the same orthogonality conditions. These orthogonality conditions already give n_1+n_2-2 linear conditions for the n_1+n_2 coefficients of $\Delta M_{n_1,n_2}^{\beta_1,\beta_2;c}(x)$; hence we can write

$$\Delta M_{n_1,n_2}^{\beta_1,\beta_2;c}(x) = A M_{n_1-1,n_2}^{\beta_1+1,\beta_2;c}(x) + B M_{n_1,n_2-1}^{\beta_1,\beta_2+1;c}(x), \quad (16)$$

and then we need to determine A and B. Comparing the coefficients of $(-x)_{n_1+n_2-1}$ on both sides of (16) gives

$$A + B = n_1 + n_2,$$

and comparing the coefficients of $(-x)_{n_1+n_2-2}$, for which we use (14), gives

$$(n_1 + n_2 - 1)[n_2(\beta_2 + n_2 - 1) + n_1(\beta_1 + n_1 + n_2 - 1)]$$
$$= A[n_2(\beta_2 + n_2 - 1) + (n_1 - 1)(\beta_1 + n_1 + n_2 - 1)]$$
$$+ B[(n_2 - 1)(\beta_2 + n_2 - 1) + n_1(\beta_1 + n_1 + n_2 - 2)].$$

Solving for A and B gives

$$A = \frac{n_1(\beta_1 - \beta_2 - n_2)}{\beta_1 - \beta_2}, \quad B = \frac{-n_2(\beta_2 - \beta_1 - n_1)}{\beta_1 - \beta_2}.$$

$\square$

A combination of the raising operators and the lowering operator then gives the third order difference equation.

Theorem 2.6. *Multiple Meixner polynomials of the second kind (for $r = 2$) satisfy the third order difference equation*

$$(\beta_1 - \beta_2)\nabla_1\nabla_2\Delta M_{n_1,n_2}^{\beta_1,\beta_2,c}(x)$$

$$= -\frac{1-c}{c}\left[n_1(\beta_1 - \beta_2 - n_2)\nabla_2 - n_2(\beta_2 - \beta_1 - n_1)\nabla_1\right]M_{n_1,n_2}^{\beta_1,\beta_2,c}(x), \quad (17)$$

where

$$\nabla_i = \frac{\Gamma(x+1)}{\Gamma(\beta_i+x)c^x}\nabla\frac{c^x\Gamma(\beta_i+x+1)}{\Gamma(x+1)}.$$

Proof. First observe that $\nabla_1\nabla_2 = \nabla_2\nabla_1$, so that both difference operators are commuting. Apply $\nabla_1\nabla_2$ to (15), then

$$(\beta_1 - \beta_2)\nabla_1\nabla_2\Delta M_{n_1,n_2}^{\beta_1,\beta_2,c}(x)$$

$$= n_1(\beta_1-\beta_2-n_2)\nabla_2\nabla_1 M_{n_1-1,n_2}^{\beta_1+1,\beta_2;c}(x) - n_2(\beta_2-\beta_1-n_1)\nabla_1\nabla_2 M_{n_1,n_2-1}^{\beta_1,\beta_2+1,c}(x).$$

Now apply the raising operations (13); then one gets (17). $\qquad\square$

References

[1] M. Abramowitz, I. A. Stegun, *Handbook of Mathematical Functions*, National Bureau of Standards, Washington, 1964; Dover, New York, 1972.

[2] A. I. Aptekarev, *Multiple orthogonal polynomials*, J. Comput. Appl. Math. 99 (1998), 423–447.

[3] A. I. Aptekarev, A. Branquinho, W. Van Assche, *Multiple orthogonal polynomials for classical weights*, Trans. Amer. Math. Soc. (to appear)

[4] J. Arvesú, J. Coussement, W. Van Assche, *Some discrete multiple orthogonal polynomials*, J. Comput. Appl. Math. 153 (2003), 19–45.

[5] T. S. Chihara, *An Introduction to Orthogonal Polynomials*, Gordon and Breach, New York, 1978.

[6] A. Erdélyi, *Higher Transcendental Functions*, Vol. I, McGraw-Hill Book Company, New York, 1953.

[7] R. Koekoek, R. F. Swarttouw, *The Askey-scheme of hypergeometric orthogonal polynomials and its q-analogue*, Reports of the faculty of Technical Mathematics and Informatics no. 98-17, Delft, 1998 (math. CA/9602214 at arXiv.org).

[8] A. F. Nikiforov, S. K. Suslov, V. B. Uvarov, *Classical Orthogonal Polynomials of a Discrete Variable*, Springer-Verlag, Berlin, 1991.

[9] E. M. Nikishin, V. N. Sorokin, *Rational Approximations and Orthogonality*, Translations of Mathematical Monographs, vol. 92, Amer. Math. Soc., Providence, RI, 1991.

Author Index

A

Adleman, L.   51
Agarwal, R. P.   337
Akin–Bohner, E.   231
Angadi, S.   539
Aulbach, B.   239, 349

B

Baštinec, J.   359
Bohner, M.   231

C

Cabada, A.   367
Camouzis, E.   375
Castillo, S.   253
Cecchi, M.   383
Čermák, J.   391
Cheng, Q.   51
Cheng, S. S.   61
Ciocci, M.-C.   75
Clark, A. S.   95

D

Dannan, F. M.   399
Devaney, R. L.   105
DeVault, R.   375
Diblík, J.   359
Došlá, Z.   383
Došlý, O.   407

E

Elaydi, S.   349
El-Metwally, H. A.   433
Elyseeva, J.   417
Erbe, L.   267

F

Fedorenko, V. V.   123

G

Garay, B. M.   279
Gil', M.   61, 501
Goel, A.   51
Grove, E. A.   425, 433
Guseinov, G. S.   289

H

Hamaya, Y.   453
Higgins, R. J.   299
Hilger, S.   279
Hilscher, R.   461
Hommel, A.   471
Huang, M.-D.   51

I

Iavernaro, F.   35

J

Janglajew, K.   479

K

Kaymakçalan, B.   289
Keller, K.   131
Kent, C. M.   485
Kloeden, P. E.   139, 279
Kozyakin, V.   139, 153
Krause, U.   167

L

Ladas, G.   425, 433
López Fenner, J.   309

M

Marini, M.   383, 493
Matucci, S.   493
Mazzia, F.   35
McCluskey, C. C.   181
McGrath, L. C.   425, 433
Medina, R.   501
Melnik, R. V. N.   507
Muldowney, J. S.   181

N

Neidhart, L.   239
Nishimura, K.   189

O

Oliveira, H.   199
Opluštil, Z.   317
O'Regan, D.   337
Otero-Espinar, V.   367

P

Pereira, E.   515
Peterson, A.   299
Pinto, M.   253
Pospíšil, Z.   317
Pötzsche, C.   325

R

Ramos, J. S.   199
Řehák, P.   493
Ruffing, A.   523

S

Schmeidel, E.   531
Sedaghat, H.   207
Sharkovsky, A. N.   3
Shigoka, T.   189
Siegmund, S.   215
Simon, M.   523

T

Thomas, E. S.   95

Tian, C.-J.   61
Trigiante, D.   35

V

Vaidya, P. G.   539
Van Assche, W.   549
Vanderbauwhede, A.   75
Vitória, J.   515

W

Wasserman, H.   51

Y

Yano, M.   189

Z

Zeidan, V.   461
Zeilberger, D.   25
Zhang, B.   359
Ziegler, K.   349